CONTEMPORARY MATHEMATICS

Titles in this Series

Volume

1 **Markov random fields and their applications,** Ross Kindermann and J. Laurie Snell

2 **Proceedings of the conference on integration, topology, and geometry in linear spaces,** William H. Graves, Editor

3 **The closed graph and P-closed graph properties in general topology,** T. R. Hamlett and L. L. Herrington

4 **Problems of elastic stability and vibrations,** Vadim Komkov, Editor

5 **Rational constructions of modules for simple Lie algebras,** George B. Seligman

6 **Umbral calculus and Hopf algebras,** Robert Morris, Editor

7 **Complex contour integral representation of cardinal spline functions,** Walter Schempp

8 **Ordered fields and real algebraic geometry,** D. W. Dubois and T. Recio, Editors

9 **Papers in algebra, analysis and statistics,** R. Lidl, Editor

10 **Operator algebras and K-theory,** Ronald G. Douglas and Claude Schochet, Editors

11 **Plane ellipticity and related problems,** Robert P. Gilbert, Editor

12 **Symposium on algebraic topology in honor of José Adem,** Samuel Gitler, Editor

13 **Algebraists' homage: Papers in ring theory and related topics,** S. A. Amitsur, D. J. Saltman and G. B. Seligman, Editors

14 **Lectures on Nielsen fixed point theory,** Boju Jiang

15 **Advanced analytic number theory. Part I: Ramification theoretic methods,** Carlos J. Moreno

16 **Complex representations of GL(2, K) for finite fields K,** Ilya Piatetski-Shapiro

17 **Nonlinear partial differential equations,** Joel A. Smoller, Editor

18 **Fixed points and nonexpansive mappings,** Robert C. Sine, Editor

19 **Proceedings of the Northwestern homotopy theory conference,** Haynes R. Miller and Stewart B. Priddy, Editors

20 **Low dimensional topology,** Samuel J. Lomonaco, Jr., Editor

Titles in this Series

Volume

CONTEMPORARY MATHEMATICS

Volume 35

Four-Manifold Theory

Cameron Gordon and
Robion Kirby, Editors

AMERICAN MATHEMATICAL SOCIETY
Providence · Rhode Island

PROCEEDINGS OF THE AMS-IMS-SIAM JOINT SUMMER RESEARCH CONFERENCE IN THE MATHEMATICAL SCIENCES ON FOUR-MANIFOLD THEORY

HELD AT DURHAM, NEW HAMPSHIRE
JULY 4–10, 1982

These proceedings were prepared by the American Mathematical Society with partial support from the National Science Foundation Grant MCS 7924296.

1980 *Mathematics Subject Classification*. Primary 57-06, 53-06, 57MXX.

Library of Congress Cataloging in Publication Data
Main entry under title:

Four-manifold theory.
(Contemporary mathematics; v. 35)
Bibliography: p.
1. Four manifolds (Topology)—Addresses, essays, lectures. I. Gordon, Cameron, 1945– II. Kirby, Robion C., 1938– III. Series: Contemporary mathematics (American Mathematical Society); v. 35.
QA613.2.F68 1984 514′.223 84-24595
ISBN 0-8218-5033-4

TABLE OF CONTENTS

PREFACE

These are the proceedings of the conference on 4-manifolds held at Durham, New Hampshire on 4-10 July 1982 under the auspices of the American Mathematical Society and National Science Foundation. The organizing committee was Sylvain Cappell, Cameron Gordon, and Robion Kirby.

The conference was highlighted by the breakthroughs of Freedman and Donaldson, and Quinn's completion at the conference of the proof of the annulus conjecture (we commend the AMS committee, particularly Julius Shaneson, who had the foresight in spring 1981 to choose the subject, 4-manifolds, in which such remarkable activity was imminent). Freedman and several others spoke on his work and some of their talks are represented by papers in this volume. Donaldson and Taubes gave surveys of their work on gauge theory and 4-manifolds and their papers are here. There were a variety of other lectures, including Quinn's surprise, and a couple of problem sessions which led to the problem list.

We would like to thank the contributors, almost all of whom submitted their papers in very timely fashion, and Carole Kohanski from the AMS who ran the nonmathematical side of things very smoothly, even through 100-degree temperatures. Thanks also to Suzy Crumley for typing all the manuscripts.

Cameron Gordon
Department of Mathematics
University of Texas
Austin, Texas 78712

Robion Kirby
Department of Mathematics
University of California
Berkeley, California 98720

Contemporary Mathematics
Volume 35, 1984

FIBERED KNOTS AND INVOLUTIONS ON HOMOTOPY SPHERES

I. R. Aitchison
J. H. Rubinstein

1. FIBERED KNOTS AND INVOLUTIONS ON HOMOTOPY 4-SPHERES

This paper arises from an attempt to understand and generalize the results of Akbulut and Kirby (AK1). We modify their techniques to investigate the structure of an infinite class of homotopy 4-spheres constructed by Cappell and Shaneson (CS1), two of which are described in Akbulut and Kirby (AK1), and two of which double cover manifolds known to be exotic. All these homotopy spheres are either S^4 or obtained from S^4 by the Gluck construction on a knotted 2-sphere. All have an orientation reversing involution with circle of fixed points, and thus lead to possibly interesting involutions on homotopy $S^2 \times D^2$'s. The double covers of the exotic homotopy projective spaces are shown to be 2-fold covers of S^4, branched over a knotted 2-sphere. All of the above involutions desuspend to Z-homology 3-spheres, and consequently the exotic nature of Cappell and Shaneson's projective spaces is detected by the Fintushel-Stern invariant (FS). We give some more examples of free involutions on homotopy 4-spheres.

The main technique is handle decompositions; we exploit Reidemeister-Singer stabilization of Heegard decompositions of 3-manifolds to show that any 4-manifold with fibered 2-knot splits naturally as the union of two submanifolds built with 0-,1- and 2-handles, and such that the common boundary of these has an induced open book decomposition with binding the equator of the fibered 2-knot.

Cappell and Shaneson's examples involve mapping tori with fiber the punctured 3-torus $T^3 - \mathrm{int}(B^3)$, and thus we analyze the diffeotopy group of T^3. This requires some algebraic results on conjugacy in $SL(3;Z)$, details of which we include in the appendix to preserve continuity of geometric arguments. These results allow us to isotope diffeomorphisms so that they reflect the symmetry natural to T^3.

Finally we consider Cappell and Shaneson's more general modifications of smooth, closed and non-orientable 4-manifolds to obtain exotic homotopy equivalences. In many situations we reduce questions to those concerning the modifications on RP^4 and $RP^2 \times D^2$. In this context we refer the reader to

0271-4132/84 $1.00 + $.25 per page

Akbulut's paper in these proceedings.

We recall the construction in Cappell and Shaneson (CS1): Take $B \in SL(3,\mathbb{Z})$, with $\det(B-1) = \pm 1$. The linear action of B on $\mathbb{R}^3$ induces a diffeomorphism $\varphi_B: T^3 \to T^3$, where $T^3 = S^1 \times S^1 \times S^1$ is the three dimensional torus, the quotient of $\mathbb{R}^3$ under the action of $(2\mathbb{Z})^3$. Isotope φ_B to a diffeomorphism ψ_B, which is the identity on a ball R' T^3, and construct the mapping torus E_{ψ_B} of ψ_B by taking $T^3 \times [-1,1]$ and identifying the ends by ψ_B:

$$E_{\psi_B} = \frac{T^3 \times [-1,1]}{(x,-1) \sim (\psi_B(x),1)}$$

Remove $\mathrm{int}(R' \times S^1) \cong \mathrm{int}(B^3 \times S^1$ and replace by $S^2 \times D^2$ glued in by some diffeomorphism of $S^2 \times S^1$, to obtain a homotopy 4-sphere. Note that the isotopy of φ_B to ψ_B induces an isotopy of φ_B^{-1} to ψ_B^{-1}, which is also the identity on R'.

If B is conjugate to A in $SL(3,\mathbb{Z})$ then the mapping tori of φ_B and φ_A are diffeomorphic. So we can always replace B by a more convenient matrix A in the same conjugacy class as B in $SL(3,\mathbb{Z})$.

If $\det(A-1) = -1$, then $\det(A^{-1}-1) = 1$. Since the mapping torus of $\varphi_{A^{-1}}: T^3 \to T^3$ is diffeomorphic to the mapping torus of φ_A, it suffices to consider only the case $\det(A-1) = 1$. Any such matrix has characteristic polynomial $f_a(x) = x^3 - ax^2 + (a-1)x - 1$, for some $a \in \mathbb{Z}$.

We begin with a variation of Akbulut and Kirby's (AK1) technique, as generalized by Montesinos (Mo), for obtaining handle decompositions for 4-dimensional mapping tori and 4-manifolds with fibered 2-knots. As regards the algebraic structure of conjugacy in $SL(3;Z)$, we present the relevant results as required deferring proofs to the final section of this paper.

2. OPEN BOOK DECOMPOSITIONS OF CLOSED 3- AND 4-MANIFOLDS

Let M^n be a closed orientable n-manifold, $V \subset M^n$ an open ball neighborhood of $m \in M^n$, and $h: M^n \to M^n$ a diffeomorphism which restricts to the identity on V. Construct the mapping torus E_0^{n+1} of h restricted to $M_0^n \cong M^n - \mathrm{int}\,V$; thus $\partial E_0^{n+1} \cong \partial\bar{V} \times S^1 \cong S^{n-1} \times S^1$. Since $\partial(S^{n-1} \times D^2) \cong S^{n-1} \times S^1$, we may obtain a closed n+1-manifold $E_\varphi^{n+1} = E_0^{n+1} \cup_\varphi (S^{n-1} \times D^2)$ by gluing together $S^{n-1} \times D^2$ and E_0^{n+1} by some diffeomorphism $\varphi: S^{n-1} \times S^1 \to S^{n-1} \times S^1$. The image K^{n-1} of $S^{n-1} \times \{0\} \subset E_\varphi^{n+1}$ is thus a knotted (n-1)-sphere in E_φ^{n+1}.

<u>DEFINITION.</u> A closed manifold W^{n+1} is an <u>open book with binding</u> S^{n-1} if it is diffeomorphic to some manifold E_φ^{n+1} described as above. The manifold M_0^n is called the <u>page</u> of the open book decomposition. Equivalently, we say that K^{n-1} is a <u>fibered</u> (n-1)-<u>knot</u>.

It is well-known that every closed 3-manifold admits an open book decomposition with binding S^1.

THEOREM 2.1. *Suppose W^4 is a closed orientable 4-manifold admitting an open book decomposition with binding S^2. Then there exist orientable 4-manifolds M_B, M_D such that $W^4 \cong M_B \cup_g M_D$, where each of M_B and M_D is built with a 0-handle, k 1-handles and k 2-handles, for some $k \in \mathbb{N}$; the gluing map g is some diffeomorphism of $\partial M_B \equiv \partial M_D$, and further there is a natural open book decomposition of ∂M_B with binding S^1 induced by the decomposition of W^4.*

PROOF. By assumption, $W^4 \cong E_0^4 \cup_\varphi S^2 \times D^2$ where E_0^4 is the mapping torus of some diffeomorphism $\rho_0 : M_0^3 \to M_0^3$ which restricts to the identity on $M_0^3 \cong S^2$. We begin by decomposing E_0^4.

Obtain a unique 3-manifold M^3 by closing M_0^3 with a 3-ball V, and extend ρ_0 to a diffeomorphism $\rho : M^3 \to M^3$ by the identity on V. For some $k \in \mathbb{N}$, M^3 admits a genus k Heegard decomposition, i.e., there is a handle presentation

$$M^3 = h^0 \cup \left(\bigcup_{i=1}^{k} h_i^1 \right) \cup \left(\bigcup_{i=1}^{k} h_i^2 \right) \cup h^3$$

with one handle each of index 0 and 3, and k handles each of index 1 and 2. We use subscripts to label a handle, superscripts to indicate the index. $H_B \cong h^0 \cup \left(\bigcup_{i=1}^{k} h_i^1 \right) = \#_k S^1 \times D^2$ is a genus k-handlebody, - the "base" handlebody, and turning the 2- and 3-handles upside down, we obtain the "dual" handlebody H_D, also of genus k. Clearly $\rho(H_B) \cup \rho(H_D)$ gives an alternative Heegard decomposition for M^3 - it is not known whether we may carry out an ambient isotopy of M^3 carrying $\rho(H_B)$ onto H_B and $\rho(H_D)$ onto H_D.

LEMMA 2.2. *Given a diffeomorphism $\rho : M^3 \to M^3$, we may assume ρ is isotopic to a diffeomorphism preserving some Heegard decomposition of M^3.*

PROOF. By the Reidemeister-Singer Theorem -- see for example Singer (S) -- we may assume that for some $s \in \mathbb{N}$, $\rho(H_B)\#_s S^1 \times D^2$ is isotopic to $H_B \#_s S^1 \times D^2$. We carry out this stabilization by adding s complementary 1- and 2-handle pairs to $H_B \cup H_D$, giving a genus-(k+s) Heegard decomposition of M^3. The images under ρ of the new 1-handles h_j^1, $j = k+1, \ldots, k+s$ are added to $\rho(H_B)$ giving $\rho(H_B \#_s S^1 \times D^2) = \rho(H_B) \#_s S^1 \times D^2$. Isotoping $\rho(H_B \#_s S^1 \times D^2)$ onto $H_B \#_s S^1 \times D^2$, the Lemma follows.

Thus we may assume without loss of generality that ρ preserves some genus k Heegard decomposition $M^3 = H_B \cup H_D$. For later applications, we shall be interested in whether certain diffeomorphisms actually preserve a given Heegard decomposition -- there is a practical criterion for this.

A 1-spine C of H_B consists of a bouquet of circles $C_1 \vee \cdots \vee C_k$ disjoint except for a common intersection point, the "base" 0-spine Q. Similarly a 1-spine $\bar{C}$ for H_D consists of a bouquet of circles $\bar{C}_1 \vee \cdots \vee \bar{C}_k$, disjoint except for the point $\bar{Q}$. By a small isotopy, we may assume that $\rho(C) \cap \bar{C} = \emptyset$, and thus there is a neighborhood $\bar{N}$ of $\bar{C}$ disjoint from $\rho(C)$. Isotoping $\bar{N}$ onto H_D, we see that we may always assume $\rho(C) \subset H_B$.

LEMMA 2.3. A diffeomorphism $\rho : M^3 \to M^3$ may be isotoped to preserve any given Heegard decomposition $M^3 = H_B \cup H_D$ iff $\rho(C)$, $\rho(\bar{C})$ can be isotoped simultaneously into H_B and H_D, respectively.

PROOF. Let N, $\bar{N}$ be closed disjoint neighborhoods of C, $\bar{C}$ in H_B, H_D respectively, and $S_k = H_B \cap H_D \cong \#_k S^1 \times S^1$ be the Heegard surface corresponding to the genus k Heegard decomposition of M^3, where we suppose $\rho(N) \cap S_k = \emptyset = \rho(\bar{N}) \cap S_k$. Hence

$$S_k \subset M^3 - \operatorname{int} \rho(N \cup \bar{N}) \cong \#_k S^1 \times S^1 \times [-1,1] .$$

If S_k is incompressible in $M^3 - \operatorname{int} \rho(N \cup \bar{N})$, then it is isotopic to the standard section corresponding to $\#_k S^1 \times S^1 \times \{0\}$, by a result of Waldhausen (Wd). Since S_k separates M^3, we may thus assume that $\rho(H_B)$, $\rho(H_D)$ are isotopic simultaneously to H_B, H_D respectively.

The only alternative is S_k compressible in $M^3 - \operatorname{int} \rho(N \cup \bar{N})$. Compressing S_k gives a separating incompressible surface S_j of strictly lower genus than $\rho(\partial N)$. Hence $\rho(\partial N \cup \partial \bar{N})$ lies on one side of S_j -- and one of H_B or H_D must miss $\rho(C) \cup \rho(\bar{C})$, a contradiction.

Now suppose ρ is a diffeomorphism of $M^3 = H_B \cup H_D$ preserving the Heegard decomposition. We may further assume after an appropriate isotopy that there is an arc joining Q and $\bar{Q}$, intersecting S_k in a single point q, which is left fixed pointwise by ρ; and then that there is a ball neighborhood R of this arc, meeting S_k in a single disc D^2_s, also left pointwise fixed. Choose a ball neighborhood R' of q, with $R' \subset R$, missing Q and Q', and intersecting S_k in a disc in D^2_s (figure 1). We may assume that $R \cap H_B$, $R \cap H_D$ are properly contained in the 0-handles of H_B, H_D respectively. Let $H'_B = \overline{H_B - R'}$, $H'_D = \overline{H_D - R'}$, and $D^2_+ = \partial R' \cap H_B$, $D^2_- = \partial R' \cap H_D$ where $D^2_+ \cap D^2_- = \alpha \cong S^1$. Clearly H'_B, H'_D are each genus k handlebodies.

The mapping torus E^4 of $\rho : M^3 \to M^3$ is obtained by taking the product $M^3 \times [-1,1]$ and identifying $(x,-1)$ with $(\rho(x),1)$. We have thus proved.

LEMMA 2.4. E^4 splits as the union $E^4 = E'_B \cup R' \times S^1 \cup E'_D$ where E'_B, E'_D are the mapping tori corresponding to the restriction of ρ to H'_B and H'_D respectively.

It is clear that we may take $E^4_0 = E^4 - (\operatorname{int} R') \times S^1$. Two closed orientable 4-manifolds W^4, $\tilde{W}^4$ containing fibered 2-knots arise naturally from E^4_0: By a result of Gluck (G), up to isotopy there is only one diffeomorphism

$\sigma: S^2\times S^1 \to S^2\times S^1$, corresponding to the nontrivial element of $\pi_1(SO(3)) \cong \mathbb{Z}_2$, which does not extend to a diffeomorphism of $S^2\times D^2$. The diffeomorphism is given by

$$\sigma(p,\theta) = (\tau_\theta(p),\theta)\ ,\ p\in S^2\ ,\ \theta\in[0,2\pi]\ ,\ S^1 = \frac{[0,2\pi]}{0\sim 2\pi}$$

where τ_θ is rotation of S^2 through an angle θ about the axis through the North and South poles. W^4 results by gluing $S^2\times D^2$ onto E_0^4 by the identity, $\tilde{W}^4$ by gluing in $S^2\times D^2$ with a "twist", corresponding to σ.

REMARKS: Leaving the arc $Q\bar{Q}$ pointwise fixed, we can give the ball R a full twist about this axis (Figure 1). The mapping torus $\tilde{E}^4$ corresponding to this choice of isotopy of ρ differs from E^4 - removing $(\mathrm{int}\,R')\times S^1$ from $\tilde{E}^4$, W^4 is obtained by gluing in $S^2\times D^2$ by σ, since the twist of $\partial(R'\times S^1)\cong S^2\times S^1$ induced by the alternative choice of isotopy may be pushed off into $S^2\times D^2$.

The choice of Heegard decomposition $H_B\cup H_D$ of M^3, and the isotopy of ρ so that $\rho(H_B)=H_B$, determine the splitting. Even if $H_B\cup H_D$ is chosen as a Heegard decomposition of minimal genus (which may be non-unique up to isotopy -- see Birman, Gonzalez-Acuña and Montesinos (BGM)) there are still many different ways of isotoping ρ so that $\rho(H_B)=H_B$. We shall have cause to illustrate this later.

Having chosen a handle-body preserving isotopy of ρ, we glue $S^2\times D^2$ onto E_0^4 as follows: Since $\partial(E_0^4) = \partial(R'\times S^1) = (D_+^2\cup D_-^2)\times S^1$, we split $S^2\times D^2$ as $(D_+^2\cup D_-^2)\times D^2 = D_+^2\times D^2\cup D_-^2\times D^2$, adding $H_+^2\equiv D_+^2\times D^2$ as a 2-handle on E_B' along $D_+^2\times\partial D^2\cong D_+^2\times S^1$ -- with even framing to give W^4, odd framing to give $\tilde{W}^4$. Similarly $H_-^2\equiv D_-^2\times D^2$ is added to E_D' as a 2-handle along $D_-^2\times S^1$ with framing determined by that of H_+^2.

To conclude the proof of Theorem 2.1, we must first describe a handle-decomposition for E_B' and E_D'. The following procedure was introduced by Akbulut and Kirby (AK1), and described explicitly by Montesinos (Mo). Take $h_1^0 = h^0 - \mathrm{int}\,R'$ as 0-handle for H_B'. Then

$$H_B'\times[-1,1] = H_1^0\cup(\bigcup_{i=1}^k H_i^1)$$

where $H_1^0 = h_1^0\times[-1,1]$, $H_i^1 = h_i^1\times[-1,1]$. As a model for $S^3=\partial H_1^0$, we take $\mathbb{R}^3\cup\infty$. $Q\times\{-1\}$ is taken as ∞, $Q\times\{1\}$ the origin of $\mathbb{R}^3$. The 1-handle H_i^1 is attached to small balls B_i, $a(B_i)$, neighborhoods of points b_i, $a(b_i)$ on the unit sphere, where $a:\mathbb{R}^3\to\mathbb{R}^3$ is the antipodal map. After attaching all 1-handles, $\partial(H_B'\times[-1,1])\cong \#_k S^2\times S^1$ is effectively modelled by removing the interiors of the disjoint collection $\{B_i\}_{i=1}^k\cup\{a(B_i)\}_{i=1}^k$, and identifying ∂B_i with $a(\partial B_i)$ by reflection in the plane through $0\in\mathbb{R}^3$, perpendicular to the line segment $b_i\cdot a(b_i)$. (Figure 2a)

This construction arises as follows: Denote the ball of radius r centered at $0 \in \mathbb{R}^3$ by B_r, and $S_r^2 = \partial B_r$. Decompose S^3 as $\tilde{B}_3 \cup S^2 \times [+1,3] \cup B_1$ where $\tilde{B}_r = S^3 - \operatorname{int} B_r$. This gives $S^3 = \partial(h_1^0 \times I) = (h_1^0 \times \partial I) \cup (\partial h_1^0 \times I)$ - the 0-handles of $H_B' \times \{-1\}$, $H_B' \times \{1\}$ are respectively $\tilde{B}_3$ and B_1.

For $i=1,\dots,k$ choose a point $b_i \in S_1^2$ and a 2-disc neighborhood D_i^2 of b_i such that $D_i^2 \cap D_j^2 = \emptyset$ if $i \neq j$ and $D_i^2 \cap a(D_j^2) = \emptyset$ $i,j=1,\dots,k$. Denoting the cone through $D_i^2 \cup a(D_i^2)$ with vertex 0 by C_i, take as attaching tube for $h_i^1 \times \{t\}$ the disjoint 2-discs determined by $C_i \cap S_{2-t}^2$. Thus the 1-handle H_i^1 of $H_B' \times [-1,1]$ has attaching tube $C_i \cap (\bigcup_{r \in [1,3]} S_r^2)$, the disjoint union of balls B_i and $B_i' = a(B_i)$. Consequently $\partial(H_1^0 \cup H_i^1)$ is obtained by removing the interiors of B_i and B_i' and identifying their boundaries by reflection in the hyperplane through 0 perpendicular to $b_i . a(b_i)$. Note that $\partial H_B' \times \{t\}$ is given by the induced identification on $S_{2-t}^2 - \bigcup_{i=1}^k \operatorname{int} C_i$. Smoothing this construction gives the handle structure of $H_B' \times [-1,1]$.

We take $(R \cap H_B') \times \{1\}$ to be a ball neighborhood of $0 \in \mathbb{R}^3$ and $(R \cap H_B') \times \{-1\}$ to be a ball neighborhood of ∞ in $\mathbb{R}^3$. Also $D_+ \times \{1\} \subset \partial(R \cap H_B') \times \{1\}$ can be assumed to be a disk with center on the line $t(\frac{1}{2},\frac{1}{2},\frac{1}{2})$ for $t > 0$, with $D_+ \times \{-1\}$ equal to the intersection of the cone through 0 and $D_+ \times \{1\}$ with $\partial(R \cap H_B') \times \{-1\}$.

To identify $H_B' \times \{-1\}$ with $H_B' \times \{1\}$, begin by adding a 1-handle H_{11}^1 by its ends to $(R \cap H_B') \times \{1\} \cup (R \cap H_B') \times \{-1\}$ (Figure 2b) -- the boundary is modified by removing the interiors of these balls and identifying their boundaries by radial projection from $0 \in \mathbb{R}^3$. To identify $H_i' \times \{-1\}$ with its image in $H_B' \times \{1\}$ we add a 2-handle H_{ii}^2 with attaching sphere $(C_i - (R - \operatorname{int} R')) \times \{-1\} \cup \rho(C_i - (R - \operatorname{int} R')) \times \{1\} \cup \lambda_i \cup \mu_i$, for each $i = 1,\dots,k$, where λ_i, μ_i are arcs running over the 1-handle H_{11}^1. Framings are determined by the annuli $A_i \times \{-1\} \cup \rho(A_i) \times \{1\} \cup \lambda_i \times [-1,1] \cup \mu_i \times [-1,1]$, where A_i is a 2-disc neighborhood of $C_i - (R - \operatorname{int} R')$ (Figure 2c). This completes the construction of E_B'.

The manifold M_B in Theorem 2.1 is obtained from E_B' by adding the 2-handle H_+^2 along $D_+^2 \times S^1$ i.e. with attaching sphere a circle running around H_{11}^1 and along the line $t(\frac{1}{2},\frac{1}{2},\frac{1}{2})$ from $D_+^2 \times \{1\}$ to $D_-^2 \times \{1\}$ in the model, and with zero framing (Figure 2d). Note that without loss of generality the line $t(\frac{1}{2},\frac{1}{2},\frac{1}{2})$ can be assumed to miss the attaching balls B_i, B_i'.

E_D' is constructed in exactly the same way: Let $h_1^3 = h^3 - \operatorname{int} R'$. Then $H_D' = \bigcup_{i=1}^k h_i^2 \cup h_1^3$ -- turning the 2- and 3-handles upside down, we take $\bar{h}_i^1 \equiv h_i^2$, $\bar{h}_1^0 \equiv h_1^3$, and obtain as above

$$E_D' = \bar{H}_1^0 \cup (\bigcup_{i=1}^k \bar{H}_i^1) \cup \bar{H}_{11}^1 \cup (\bigcup_{i=1}^k \bar{H}_{ii}^2)$$

where $\bar{H}_1^0 = \bar{h}_1^0 \times [-1,1]$, $\bar{H}_i^1 = \bar{h}_i^1 \times [-1,1]$, and $\bar{H}_{jj}^k$ identifies $\bar{h}_j^k \times \{-1\}$ with $\rho(\bar{h}_j^k) \times \{1\}$. The manifold M_D is obtained by adding H_-^2 along $D_-^2 \times S^1$, also with zero framing, where the disks $D_-^2 \times \{\pm 1\}$ have centers on the line $t(\frac{1}{2},\frac{1}{2},\frac{1}{2})$ with $t < 0$ in $\mathbb{R}^3$.

It is clear that $\tilde{W}^4 = \tilde{M}_B \cup_\phi \tilde{M}_D$, where $\tilde{M}_B$ and $\tilde{M}_D$ are obtained respectively from E_B' and E_D' by adding the 2-handles H_+^2, H_-^2 with odd framing, and $\phi\colon \partial\tilde{M}_B \to \partial\tilde{M}_D$ is some diffeomorphism.

$\partial M_B \cong \partial M_D$ is obtained as follows: Construct the mapping torus of ρ restricted to $\partial(H_B') \cong \#_k S^1 \times S^1$, and perform 0-framed surgery on the solid torus $D_+^2 \times S^1$. $\partial\tilde{M}_B$ is constructed similarly, but with the $D_+^2 \times S^1$ sewn back with a twist corresponding to the framing of H_+^2. In both cases we have an open book decomposition with connected binding, which we may take as the circle $\alpha = \partial D_+^2$.

The proof of Theorem 2.1 is completed by noting that, in the terminology of handle theory (see Rourke and Sanderson (RS)), the attaching spheres of the 2-handles H_+^2, H_-^2 intersect the belt spheres of the 1-handles H_{11}^1, $\bar{H}_{11}^1$ respectively, once geometrically, thus forming complementary handle pairs in each case, which may be cancelled. In general, if a 2-handle δ passes around a 1-handle d once geometrically, we may slide any other 2-handle δ_i, passing around d, over δ and thus off d, as indicated in Figure 3. Note that the new attaching sphere for δ_i becomes the connect-sum of the old one and a copy of that of δ for each slide performed.

Cancellation of complementary 2- and 3-handle pairs is achieved analogously although by the result of Laudenbach and Poenaru (LP) the attaching spheres of all 3- and 4-handles in a handle decomposition of a closed 4-manifold are uniquely determined up to isotopy, and thus the sliding of 3-handles over each other need not be described explicitly. However, it is of interest to keep track of the geometric intersection of the attaching spheres of 3-handles, after sliding, with the belt spheres of 2-handles which remain uncancelled.

It is mainly in this respect that our splitting technique differs from the following construction of W^4, Montesinos' generalization of that given by Akbulut and Kirby: Using the same model as before, construct

$$M^3 \times [-1,1] = H_1^0 \cup \left(\bigcup_{i=1}^k H_i^1\right) \cup \left(\bigcup_{i=1}^k H_i^2\right) \cup (H_1^3)$$

where $H_i^2 = h_i^2 \times [-1,1]$, $H_1^3 = h_1^3 \times [-1,1]$. The attaching tubes of the 2-handles are obtained by fattening the attaching tubes of h_i^2. By the result of Laudenbach and Poenaru, the attaching sphere of the 3-handle H_1^3, need not be drawn in. Add the handles H_{11}^1, H_{ii}^2 as before, and identify $h_i^2 \times \{-1\}$ with $\rho(h_i^2) \times \{1\}$ by adding a 3-handle H_{ii}^3, $i=1,\ldots,k$. $h_1^3 \times \{-1\}$ is identified with $\rho(h_1^3) \times \{1\}$ by adding a 4-handle H_{11}^4. Again, the attaching spheres of the latter 3- and 4-handles need not be drawn. This gives a complete description of E^4; to construct W^4, remove $\mathrm{int}(R' \times S^1)$ as before -- equivalent to removing a 3- and 4-handle -- and sew in $S^2 \times D^2$ by turning a handle decomposition upside down to give $S^2 \times D^2 = H^2 \cup H^4$. The 2-handle is again added along $D^2_+ \times S^1$ without a twist to give W^4, with a twist to give $\tilde{W}^4$.

This construction takes $h^0_1 - \mathrm{int}\, R'$ as 0-handle for M^3, and R' as 3-handle, which is thus presumed pointwise fixed. So the handle structure in this construction differs from our construction where we have effectively introduced a cancelling 1- and 2-handle pair for R'. These require a 2- and a 3-handle for identification in the mapping torus -- and all four of these handles are then removed so that $S^2 \times D^2$ can be glued in.

3. DIFFEOMORPHISMS OF T^3

In order to investigate Cappell and Shaneson's construction, an explicit description of T^3 is required: The vector space structure on $\mathbb{R}^3$ defines $\mathbb{R}^3$ as a Lie group, T^3 being the homogeneous space arising as the quotient of $\mathbb{R}^3$ under the action of its discrete subgroup $(2\mathbb{Z})^3$. Let $\pi: \mathbb{R}^3 \to T$ denote the quotient map. Taking the standard orthormal basis for $\mathbb{R}^3$, a network L of lines is obtained by taking the image of the coordinate axes under $(2\mathbb{Z})^3$. Two disjoint networks L_B and L_D arise by translating L by the vectors $[-\frac{1}{2}, -\frac{1}{2}, -\frac{1}{2}]^T$ and $[\frac{1}{2}, \frac{1}{2}, \frac{1}{2}]^T$ respectively. Let N_B be a neighborhood of L_B, invariant under the action of $(2\mathbb{Z})^3$, and such that $N_D = \mathbb{R}^3 - \mathrm{int}\, N_B$ is the translate of N_B by the vector $[1, 1, 1]^T$.

Then $\pi(N_B) \cong \pi(N_D)$ is diffeomorphic to $\#_3 S^1 \times D^2$, giving rise to a genus three Heegard decomposition of T^3:

$$T^3 = H_B \cup H_D = (h^0 \cup 3h^1) \cup (3h^2 \cup h^3)$$

where $H_B = \pi(N_B)$, $H_D = \pi(N_D)$ and we have turned the 0- and 1-handles of H_D upside down to give 2- and 3-handles for T^3.

A convenient model for T^3 is provided by a fundamental domain in R^3 for the action of $(2\mathbb{Z})^3$: take the cube $\mathscr{W}$ of edge length two, centered at the origin of $\mathbb{R}^3$, with faces parallel to the coordinate planes. T^3 may be considered as W with opposite faces identified by reflection in the appropriate plane.

A 1-spine C of H_B is provided by $\pi(L_B)$, consisting of a bouquet of circles, $C_1 \vee C_2 \vee C_3$ disjoint except for the common intersection point $Q = \pi(-\frac{1}{2}, -\frac{1}{2}, -\frac{1}{2})$ which we take as 0-spine for H_B. The circle C_i is the image under π of the line through $(-\frac{1}{2}, -\frac{1}{2}, -\frac{1}{2})$ parallel to the x_i-axis, and oriented accordingly (Figure 4a). Similarly, a 1-spine $\bar{C} = \bar{C}_1 \vee \bar{C}_2 \vee \bar{C}_3$ for H_D is given by $\pi(L_D)$, where again the circles $\bar{C}_i$ are disjoint but for their common intersection point at the dual -spine $\bar{Q} = \pi(\frac{1}{2}, \frac{1}{2}, \frac{1}{2})$. $\bar{C}_i$ is the image of the line through $(\frac{1}{2}, \frac{1}{2}, \frac{1}{2})$ parallel to the x_i-axis, but with opposite orientation to that of C_i.

The model W easily provides the attaching spheres for the 1-handles of H_D, viewed as 2-handles in T^3 attached to H_B: denote by h_i^2 the 2-handle of T^3 corresponding to the 1-handle h_i^1 of H_D with core $\bar{C}_i - \bar{Q}$. The attaching sphere of h_i^2 is given by isotoping a small unknotted circle, j_i, linking $\bar{C}_i$ once, onto ∂H_B (Figure 4). Denoting by α_i the class in $\pi_1(H_B)$ represented by the circle C_i, the attaching spheres are given by

$$h_1^2\colon \alpha_2^{-1}\alpha_3^{-1}\alpha_2\alpha_3 \qquad h_2^2\colon \alpha_3^{-1}\alpha_1^{-1}\alpha_3\alpha_1 \qquad h_3^2\colon \alpha_1^{-1}\alpha_2^{-1}\alpha_1\alpha_2 \ .$$

The family of lines in $\mathbb{R}^3$ parallel to the x_i-axis gives a fibering of T^3 by circles -- by abuse of notation we shall refer to an isotopy of T^3, which preserves each such fiber setwise, as an isotopy in direction x_i. We shall also denote the 2-dimensional torus in T^3 covered by the plane $x_i = k \subset \mathbb{R}^3$ by $T^2_{x_i=k}$.

The choice of Heegard decomposition is motivated by the following observation: Parameterizing T^3 as $\{(e^{i\theta\pi}, e^{i\phi\pi}, e^{i\psi\pi})\colon (\theta, \phi, \psi) \in \mathbb{R}^3\}$, the involution g of T^3 is given by

$$g\colon (e^{i\theta\pi}, e^{i\phi\pi}, e^{i\psi\pi}) \equiv e^{i\pi(\theta,\phi,\psi)} \to e^{-i\pi(\theta,\phi,\psi)} \equiv e^{i\pi a(\theta,\phi,\psi)}$$

where $a\colon \mathbb{R}^3 \to \mathbb{R}^3$ is the antipodal map. Hence $g(H_B) = H_D$, $g(H_D) = H_B$ for the chosen Heegard decomposition. Furthermore, the eight fixed points of g

$$\{e^{i\pi(\delta_1,\delta_2,\delta_3)}\colon \delta_i = 0 \text{ or } 1\}$$

$$= \{q = e^{i\pi(0,0,0)},\ \alpha_1 = e^{i\pi(1,0,0)},\ \alpha_2 = e^{i\pi(0,1,0)},\ \alpha_3 = e^{i\pi(0,0,1)},$$
$$\alpha_4 = e^{i\pi(1,1,0)},\ \alpha_5 = e^{i\pi(1,0,1)},\ \alpha_6 = e^{i\pi(0,1,1)},\ \alpha_7 = e^{i\pi(1,1,1)}$$

lie on the Heegard surface $S_H = H_B \cap H_D$ which is also preserved by g (see Figure 5). Note that $g(C_i) = \bar{C}_i$, and g is orientation preserving on S_H. The restriction of g to S_H is an involution, necessarily that shown in Figure 5, i.e. rotation about some axis. Now any diffeomorphism of T^3 to itself is uniquely determined by its action on $\pi_1(T^3) \cong \mathbb{Z}^3$, up to isotopy (see e.g. (Wd)). Thus every such diffeomorphism arises from the linear action on $\mathbb{R}^3$ of a matrix $A \in GL(3,\mathbb{Z})$ -- the corresponding diffeomorphism $\phi_A\colon T^3 \to T^3$ is defined by

$$\phi_A\left(e^{i\pi(\theta,\phi,\psi)}\right) = e^{i\pi A(\theta,\phi,\psi)}$$

and thus satisfies $g \circ \phi_A = \phi_A \circ g$.

DEFINITION. We shall call a diffeomorphism $\phi\colon T^3 \to T^3$ _symmetric_ if

(i) $\phi \circ g = g \circ \phi$

(ii) $\phi(H_B) = H_B$ if ϕ preserves orientation
or $\phi(H_B) = H_D$ if ϕ reverses orientation

(iii) $\phi\left(e^{i\pi(t,t,t)}\right) = e^{-i\pi(t,t,t)}\ \forall t \in [-\frac{1}{2}, \frac{1}{2}]$ i.e. ϕ preserves the arc joining Q and $\bar{Q}$ pointwise if ϕ preserves orientation, or $\phi\left(e^{i\pi(t,t,t)}\right) = e^{-i\pi(t,t,t)}\ \forall t \in [-\frac{1}{2}, \frac{1}{2}]$ if ϕ reverses orientation.

To determine whether a given diffeomorphism is symmetric, we need a canonical form for matrices in $SL(3;Z)$. Proofs of the following theorems can be found in the final section:

THEOREM A1. Let X $SL(3;Z)$; if ± 1 is an eigenvalue of X, then X is conjugate in $GL(3;Z)$ to $\begin{bmatrix} \pm 1 & 0 & 0 \\ a & b & c \\ d & e & f \end{bmatrix}$. If ± 1 is not an eigenvalue, then X is conjugate to $\lambda I + \mu Y$, where $Y = \begin{bmatrix} 0 & 0 & 1 \\ m & 0 & 0 \\ n & p & q \end{bmatrix}$.

COROLLARY A3. Suppose X satisfies $f_a(x) = 0$. Then X is conjugate to a matrix of form

$$A_{a,\lambda,p} = \begin{bmatrix} 0 & 0 & 1 \\ m & \lambda & 0 \\ n & p & a-\lambda \end{bmatrix} \qquad \begin{aligned} n &= \lambda(a-\lambda) - (a-1) \\ mp &= 1 + n\lambda \ , \ 0 \leq \lambda < p \end{aligned}$$

We shall refer to such matrices as "Cappell-Shaneson" matrices (CS matrices).

LEMMA 3.1. For each Cappell — Shaneson matrix $A = \begin{bmatrix} 0 & 0 & 1 \\ m & \lambda & 0 \\ n & p & a-\lambda \end{bmatrix}$, $m > 0$, $\lambda \geq 0$, the diffeomorphism $\phi_A : T^3 \to T^3$ induced by the linear action of A on $\mathbb{R}^3$ is isotopic to a symmetric diffeomorphism $\psi_A : T^3 \to T^3$.

PROOF. It is clear that $\phi_A \circ g(x) = g \circ \phi_A(x)\ \forall x \in T^3$. We isotope ϕ_A by moving the images of C, $\bar{C}$ into the respective handlebodies in such a way as to satisfy (i) at all stages: Hence it suffices to describe the isotopy of $\bar{C}$. Notice that

$$\phi_A(C) = \pi \circ A(L_B) = \pi(W \cap A(L_B)),\ \phi_A(\bar{C}) = \pi \circ A(L_D) = \pi(W \cap A(L_D))$$

providing visual representation for the isotopy in the model W. We proceed in stages, isotopies of T^3 induced by isotopies of $\mathbb{R}^3$ commuting with the action of $(2\mathbb{Z})^3$:

(a) Let $\psi_t : T^3 \to T^3$, $t \in [0,1]$ denote the isotopy induced by isotoping L_B $(\frac{1}{2}-\varepsilon)$ units in direction $[1, 1, 0]^T$, and L_D $(\frac{1}{2}-\varepsilon)$ units in direction $[-1, -1, 0]^T$ (Figure 6a). Since $A(0, 0, \frac{1}{2}) = (\frac{1}{2}, 0, \frac{a-\lambda}{2})$, choose $\varepsilon = \frac{1}{2(m+\lambda)}$ -- thus the image $\pi \circ A \circ \psi_1(\frac{1}{2}, \frac{1}{2}, \frac{1}{2}) = \pi\left(\frac{1}{2}, \frac{1}{2}, \frac{a-\lambda}{2} + \frac{n+p}{2(m+\lambda)}\right)$ of $\bar{Q}$ lies on a fiber through $\bar{Q}$ in direction x_3. The images of $\bar{C}_1$, $\bar{C}_2$ lie on the torus $T^2_{x_1=\frac{1}{2}}$, which is preserved setwise ψ_t, $t \in [0,1]$.

(b) The image of C_3 intersects the torus $T^2_{x_1=\frac{1}{2}}$ at one point: carry out an isotopy of T^3 whose support is a small neighborhood of the tori

$T^2_{x_1=\frac{1}{2}}$, $T^2_{x_1=-\frac{1}{2}}$ leaving the images of $C_3, \bar{C}_3$ fixed except for in a neighborhood of their respective intersections with these tori (Figures 6bc). This isotopy rotates the torus $T^2_{x_1=\frac{1}{2}}$ in direction $-x_3$, through a distance $\left(\frac{(a-\lambda)}{2} + \frac{(n+p)}{2(m+\lambda)} - \frac{1}{2}\right)$. This returns the arc $Q, \bar{Q}$ to its original position, henceforth kept fixed for a suitable choice of the isotopy. We parameterize

$$\bar{C}_1 = [\tfrac{1}{2}+s, \tfrac{1}{2}, \tfrac{1}{2}]^T, \ \bar{C}_2 = [\tfrac{1}{2}, \tfrac{1}{2}+t, \tfrac{1}{2}]^T, \ \bar{C}_3 = [\tfrac{1}{2}, \tfrac{1}{2}, \tfrac{1}{2}+u]^T.$$

Then the images after the isotopies (a) and (b) are

$$\bar{C}_1 \to [\tfrac{1}{2}, \tfrac{1}{2}+ms, \tfrac{1}{2}+ns]^T, \ \bar{C}_2 \to [\tfrac{1}{2}, \tfrac{1}{2}+\lambda t, \tfrac{1}{2}+pt]^T,$$

$\bar{C}_3 \to [\frac{1}{2}+u, \frac{1}{2}, \frac{1}{2}+f(u)]^T$, where f is a function with $f(0)=0$.

(c) Isotope the intersection point of the image of $\bar{C}_3$ with the torus $T^2_{x_1=-\frac{1}{2}}$ in direction $-x_3$ until it lies on the torus $T^2_{x_3=\frac{1}{2}}$, keeping the torus $T^2_{x_1=-\frac{1}{2}}$ fixed setwise. Now isotope the image of $\bar{C}_3$, lying on the torus $T^2_{x_2=\frac{1}{2}}$, into the handlebody H_D, in the essentially unique way forced by requiring that the support of the isotopy misses the tori $T^2_{x_1=\frac{1}{2}}$, $T^2_{x_1=-\frac{1}{2}}$ (Figure 6d).

(d) Leaving the images of $C_3, \bar{C}_3$ pointwise fixed, isotope the images of $\bar{C}_1, \bar{C}_2$ on the torus $T^2_{x_1=\frac{1}{2}}$ into the handlebody H_D -- again, this isotopy is essentially unique (Figure 6e).

In order to construct the homotopy 4-spheres using these diffeomorphisms, a characterization of the isotopy of a neighborhood of the fixed point $q=\pi(0, 0, 0)$ is required: For convenience we shall work in $\mathbb{R}^3$.

<u>MINIMAL STRAIGHTENING</u>

<u>LEMMA 3.2</u>. <u>There is a canonical straightening to the identity for each diffeomorphism</u> $\phi_B: T^3 \to T^3$, B <u>a Cappell-Shaneson matrix, in a neighborhood of the fixed point</u> $\pi(0, 0, 0)$. <u>This is called minimal straightening.</u>

<u>PROOF</u>. Begin with matrices of the form $A = \begin{bmatrix} 0 & 0 & 1 \\ m & \lambda & 0 \\ n & p & a-\lambda \end{bmatrix}$, $p > \lambda \geq 1$.

We describe the isotopy in three steps.

(1) $A[0, 1, 0,]^T = [0, \lambda, p]^T$ lies in the 1st quadrant of the x_2x_3-plane. The image of the x_2-axis divides this plane into two open sets; $A[1, 0, 0]^T$ lies in the same component as the $(-x_3)$-axis, since $\det A=1$ and the image of the vector $[0, 0, 1]^T$ lies in the half space $\mathbb{R}^3_+ = \{x, y, z): x \geq 0\}$. Carry out an isotopy given by

$$A_s = \begin{bmatrix} 0 & 0 & 1 \\ m-ms & \lambda & 0 \\ n-ns-s & p & v \end{bmatrix}, \quad \det A_s = 1 + s(\lambda-1) \neq 0, \ s \in [0,1]$$

leaving the images of $[0, 1, 0]^T$, $[0, 0, 1]^T$ fixed and sending $[1, 0, 0]^T \to [0, 0, -1]^T$.

(2) Now isotope the image of $[0, 1, 0]^T$ back to its original position, and simultaneously straighten the image of $[0, 0, 1]^T$ so that it lies along the $+x_1$-axis. This may be described by

$$A_{1t} = \begin{bmatrix} 0 & 0 & 1 \\ 0 & \lambda-\lambda t+t & 0 \\ -1 & p-pt & v-vt \end{bmatrix}, \quad t \in [0,1] \quad \text{with determinant} \quad 1+(\lambda-1)(1-t) \neq 0 .$$

(3) Finally, leaving the x_2-axis fixed, rotate the images of $[1, 0, 0]^T$ and $[0, 0, 1]^T$ back to their initial positions, described by

$$A_{11r} = \begin{bmatrix} r & 0 & 1-r \\ 0 & 1 & 0 \\ r-1 & 0 & r \end{bmatrix}, \quad \det(A_{11r}) = 2r^2 - 2r + 1 , \; r \in [0,1] .$$

We illustrate the procedure for the matrices $\begin{bmatrix} 0 & 0 & 1 \\ 1 & 1 & 0 \\ 0 & 1 & v \end{bmatrix}$, $v \in \mathbb{Z}$, in Figure 7.

II. Since an arbitrary Cappell-Shaneson matrix is conjugate to one of this form, minimal straightening of such a matrix is defined by conjugating the isotopy at each state.

<u>EXAMPLE</u>. We illustrate in Figure 8 with a representative of the nontrivial class when $a = -5$ i.e.,

$$\begin{bmatrix} 1 & 0 & 1 \\ 1 & -1 & -1 \\ 2 & -3 & -5 \end{bmatrix} = \begin{bmatrix} -1 & 0 & 0 \\ -5 & 1 & -1 \\ 1 & 0 & -1 \end{bmatrix} \begin{bmatrix} 0 & 0 & 1 \\ -5 & 2 & 0 \\ -8 & 3 & -7 \end{bmatrix} \begin{bmatrix} -1 & 0 & 0 \\ -6 & 1 & -1 \\ -1 & 0 & -1 \end{bmatrix}$$

<u>REMARKS</u>

1. It is important to keep track of the image of $[1, 1, 1]^T$ during this isotopy. For the example above, the images are given by

$$[2, 2s-1, 7s-6]^T, \; [2, 2t+1, 1-4t]^T, \; [2-r, 3-2r, 4r-3]^T .$$

2. There are several choices for λ corresponding to a given conjugacy class of matrices. We may remove this ambiguity by requiring that $\lambda \geq 1$ be minimal, and similarly p. In case minimal straightening can be achieved by one linear matrix isotopy, any conjugation has entries linear in the isotopy parameter, and no ambiguity can arise.

3. An isotopy of the inverse of matrices of the type above is determined by taking the inverses of the isotopy giving minimal straightening.

4. CAPPELL AND SHANESON'S HOMOTOPY SPHERES

Extend the straightening of a ball neighborhood of q to a straightening of a ball neighborhood of $Q\bar{Q}$, again g-symmetrically. Denote by Σ_A (respectively $\tilde{\Sigma}_A$) the homotopy 4-sphere constructed by sewing in $S^2 \times D^2$ to the mapping torus of $(T^3 - \text{int}\,V)$, under this final diffeomorphism, with framing 0 (respectively framing +1).

THEOREM 4.1. (i) For each $A \in SL(3,\mathbb{Z})$, $\det(A-1) = 1$, the homotopy 4-sphere Σ_A (resp. $\tilde{\Sigma}_A$) decomposes as the union of two copies of a homology ball M_A (resp. $\tilde{M}_A$). M_A (resp. $\tilde{M}_A$) has a handle-decomposition with a 0-handle, k 1-handles and k 2-handles, where k is at most 2.

(ii) The homology sphere ∂M_A (resp. $\partial\tilde{M}_A$) is a 2-fold branched cover of S^3, branched over a knot.

(iii) The two copies of M_A are glued together by the 2-fold branched covering transformation g_A on ∂M_A.

PROOF. (i) To decompose Σ_A (resp. $\tilde{\Sigma}_A$) as in Theorem 2.1, take the symmetric diffeomorphism $\psi_A \cong \phi_A$ in each case, with minimal straightening. This gives $H_B \cong H_D$ (resp. $\tilde{H}_B \cong \tilde{H}_D$), and hence we take $M_A = H_B$ (resp. $\tilde{M}_A = \tilde{H}_B$). From Theorem 2.1, we may suppose that k is at most three. However, the 2-handle H^2_{33} geometrically cancels the 1-handle H^1_1. The homology type of the pair $M_A, \partial M_A$ (resp. $\tilde{M}_A, \partial\tilde{M}_A$) is determined by a simple argument using the Mayer-Vietoris and relative homology sequences.

(ii) The involution $g: T^3 \to T^3$ induces an orientation reversing diffeomorphism $G: \Sigma_A \to \Sigma_A$ (resp. $\tilde{G}: \tilde{\Sigma}_A \to \tilde{\Sigma}_A$) defined by

$$G(x,t) = (g(x),t) \qquad \forall (x,t) \in \Sigma_A - S^2 \times D^2$$

$$G(\alpha,\beta) = (-\alpha,\beta) \qquad \forall (\alpha,\beta) \in S^2 \times D^2$$

($\tilde{G}$ defined similarly). Hence G (resp. $\tilde{G}$) interchanges the two copies of M_A (resp. $\tilde{M}_A$) leaving ∂M_A (resp. $\partial\tilde{M}_A$) setwise fixed.

The fixed point set of G is $\{(x,t) \in \Sigma_A - S^2 \times D^2 : g(x) = x\}$ which by Smith Theory (see, e.g. Bredon (B)) consists of a circle C_G.

C_G consists of the arcs $\alpha_j \times I \subset T^3 \times I$, $1 \leq j \leq 7$, joined end to end in the mapping torus $(T^3 - V) \times I/(x,t) \sim (\psi_A(x),t)$, i.e. ψ_A acts as a permutation of order seven on the set $\{a_j\}_{j=1}^7$,

The involution $g: S_H \to S_H$ expresses $S_H \cong \#_3 S^1 \times S^1$ as a 2-fold branched cover of S^2, branched over 8 points (Figure 5). Hence the quotient of $(S_H \times_{\psi_A} S^1)$ under $g \times \text{identity}$ is $S^2 \times S^1$. Surgery along the standard generator of $\pi_1(S^2 \times S^1) \cong \mathbb{Z}$ always gives S^3, regardless of the framing. Hence ∂M_A is a 2-fold branched cover of S^3, branched over the image $\bar{C}_G = \rho_G(C_G)$, where $\rho_G: \partial M_A \to S^3$ is the quotient map.

Since C_G lies in $S_H \times_{\psi_A} S_1 - V \times S^1$, $\bar{C}_G$ lies in an unknotted solid torus $T_G \subset S^3$. In Figure 9a, we show the relation of T_G to the surgery description of S^3 obtained above. This enables us to view $\bar{C}_G$ as a knot K in S^3, using "Kirby Calculus" (K2) to slide T_G off the link which gives S^3.

The construction of $\partial\tilde{M}_A$ is exactly the same, except that the relation of T_G to the link description of S^3 is as indicated in Figure 9b. Hence sliding T_G as in Figure 9a, we obtain the knot $\tilde{K} \subset S^3$, differing from K by a complete twist, due to the twist in T_G.

That K is a 7-bridge knot may be seen as follows: G preserves each $(S_H - V) \times t \subset \partial M_A$, $t \in [0]$ and hence $\bar{C}_G$ intersects $D^2 \times \{t\} \subset D^2 \times S^1 \equiv T_G$ in seven points.

It is not clear whether M_A is in fact contractible -- and if so, whether the words describing the 2-handle attaching maps give a representation of the trivial group, trivializable by Andrews-Curtis moves (AC).

However, we observe that

$$A_{\lambda,a,m} = \begin{bmatrix} 0 & -1 & 1 \\ m-\lambda & \lambda-1 & 1 \\ m+n-\lambda-p & p+2\lambda-a-1 & a+1-\lambda \end{bmatrix} = \begin{bmatrix} 1 & 0 & 0 \\ 1 & 1 & 0 \\ 1 & 1 & 1 \end{bmatrix} \begin{bmatrix} 0 & 0 & 1 \\ m & \lambda & 0 \\ n & p & (a-\lambda) \end{bmatrix} \begin{bmatrix} 1 & 0 & 0 \\ -1 & 1 & 0 \\ 0 & -1 & 1 \end{bmatrix}$$

Hence if there is a symmetric isotopy of the diffeomorphism of T^3 induced by $A_{\lambda,a,m}$ - which is probable, although it would be more difficult to describe -- then the homology ball resulting from the splitting as in the Theorem 2.1 would in fact be a Mazur manifold: contractible, with one handle each of index ≤ 2. (Mazur (M)). Writing $A_{\lambda,a,m}$ as a product of elementary matrices will probably suffice.

Simpler symmetric isotopies are possible in specific cases. We illustrate for the rational canonical forms:

<u>Lemma 4.2</u>. <u>For</u> $A_v = \begin{bmatrix} 0 & 0 & 1 \\ 1 & 1 & 0 \\ 0 & 1 & v \end{bmatrix}$, $v \in \mathbb{Z}$, <u>there is a symmetric diffeomorphism</u> ψ_v <u>isotopic to</u> $\phi_v \equiv \phi_{A_v}$, <u>such that, taking</u> $\alpha_i = [C_i] \in \pi_1(H_B) \cong F_3$,

$$\psi_{v*}\alpha_1 = \alpha_2, \quad \psi_{v*}\alpha_2 = \alpha_3\alpha_2, \quad \psi_{v*}\alpha_3 = \alpha_1\alpha_3^v$$

<u>where</u> $\psi_{v*}: \pi_1(H_D) \to \pi_1(H_D)$, <u>and the images of the</u> C_i <u>in</u> H_D <u>are determined analogously</u>.

<u>PROOF</u>. Observe $\phi_v(\frac{1}{2}, \frac{1}{2}, \frac{1}{2}) = (\frac{1}{2}, 1, \frac{v+1}{2})$, $\phi_v(\frac{1}{2}, \frac{1}{2}, 0) = (0, 1, \frac{1}{2})$. Thus isotoping $\phi_v(\frac{1}{2}, \frac{1}{2}, \frac{1}{2})$ along $\phi_v[0, 0, -1]^T$, and $\phi_v(-\frac{1}{2}, -\frac{1}{2}, -\frac{1}{2})$ along $\phi_v[0, 0, 1]^T$, carries the image of the arc $Q\bar{Q}$ into the cube W, linearly. Let $\psi_t: T^3 \to T^3$, $t \in [0,1]$ be the isotopy depicted in Figure 10a. The point

$q = \exp i\pi(0, 0, 0)$ is kept fixed at each stage and

$$\psi_t \exp i\pi(x, y, z) = \exp i\pi(x, y, z + t(\tfrac{1}{2} - \varepsilon)) \quad \forall(x, y, z) \in L_B$$

$$\psi_t \exp i\pi(x', y', z') = \exp i\pi(x', y', z' - t(\tfrac{1}{2} - \varepsilon)) \quad \forall(x', y', z') \in L_D$$

for some small $\varepsilon > 0$.

We parametrize $\bar{C}_1 = (\tfrac{1}{2} + s, \tfrac{1}{2}, \tfrac{1}{2})$, $\bar{C}_2 = (\tfrac{1}{2}, \tfrac{1}{2} + t, \tfrac{1}{2})$, $\bar{C}_3 = (\tfrac{1}{2}, \tfrac{1}{2}, \tfrac{1}{2} + u)$. Then the images $\phi_v \bar{C}_1 = (\tfrac{1}{2}, 1 + s, \frac{v+1}{2})$, $\phi_v \bar{C}_2 = (\tfrac{1}{2}, 1 + t, \frac{v+1}{2} + t)$, $\phi_v \bar{C}_3 = (u + \tfrac{1}{2}, 1, vu + \frac{v+1}{2})$, and $\phi_v \psi_1 \bar{C}_1 = (\varepsilon, 1 + s, \tfrac{1}{2} + \varepsilon v)$, $\phi_v \psi \bar{C}_2 = (\varepsilon, 1 + t, v + \tfrac{1}{2} + t)$, $\phi_v \psi_1 \bar{C}_3 = (\varepsilon + u, 1, \tfrac{1}{2} + v\varepsilon + vu)$.

The images of C and $\bar{C}$ under ϕ_v are depicted in Figure 10b, where we specifically illustrate for $v = 8$ -- it will be clear that the choices of isotopy apply to all v. After the isotopy $\phi_v \circ \psi_t$, $t \in [0,1]$, the images of C and $\bar{C}$ lie as in Figure 10c. The isotopy corresponds to winding $\phi_v(\bar{Q})$ around the torus $T^2_{x_2=1}$ $\frac{v}{4}$ times in direction $A_v[0, 0, -1]^T$, so that the images of $\bar{C}_1$, $\bar{C}_2$ at each stage lie on some torus $T^2_{x_2=k}$, $k \in (0, \tfrac{1}{2}]$. $\varepsilon > 0$ is chosen so that $\phi_v \circ \psi_1(\bar{Q})$ lies almost on the torus $T^2_{x_1=0}$. Note that $\phi_v \circ \psi_1(\bar{Q}) = (\varepsilon, 1, \tfrac{1}{2} + \varepsilon v)$. The isotopy of C is the image under g of that of $\bar{C}$.

Now isotope $\phi_v \psi_1(\bar{Q})$, $\phi_v \psi_1(Q)$ in direction $-x_2$, x_2 respectively, so that they lie on the tori $T^2_{x_2=\frac{1}{2}}$, $T^2_{x_2=-\frac{1}{2}}$ respectively, simultaneously isotoping C_2, $\bar{C}_2$ as indicated in Figure 10d. This enables the image of $Q\bar{Q}$ to be eventually returned pointwise to its original position. However, we first isotope the images of C_3, $\bar{C}_3$ onto the tori $T^2_{x_2=-\frac{1}{2}}$, $T^2_{x_2=\frac{1}{2}}$ respectively, keeping the tori $T^2_{x_1=k}$ setwise fixed at all stages. (Figure 10e)

Isotope the images of Q, $\bar{Q}$ back to their original positions, keeping the images of C_1, $\bar{C}_1$ lying along some fiber in direction x_2 at each stage. The images of C and $\bar{C}$ may now be isotoped into the appropriate handlebodies, as indicated in Figure 10f -- the images of C_3 and $\bar{C}_3$ are kept on the tori $T^2_{x_2=-\frac{1}{2}}$, $T^2_{x_2=\frac{1}{2}}$ at all stages.

It is clear that this procedure may be carried out for arbitrary $v \in \mathbb{Z}$. For $v < 0$, we obtain an isotopy as depicted in Figure 10g. The images of C_1, C_2, C_3 represent the words in $\pi_1(H_B)$ given by

$$[C_1] = \alpha_2$$

$$[C_2] = \alpha_3\alpha_2$$

$$[C_3] = \alpha_1\alpha_3^v .$$

ψ_v is the symmetric diffeomorphism which follows by extending the straightening of a ball neighborhood of $q = \pi(0, 0, 0)$ to a straightening of a ball neighborhood R of $Q\bar{Q}$, again g-symmetrically.

Denote by Σ_v the homotopy 4-sphere constructed from the mapping torus of ϕ_v, with minimal straightening and $S^2 \times D^2$ sewn in with 0-framing. Sewing in $S^2 \times D^2$ with odd framing gives a homotopy 4-sphere $\tilde{\Sigma}_v$.

<u>THEOREM 4.3.</u> <u>For each</u> $v \in \mathbb{Z}$, Σ_v <u>is diffeomorphic to</u> S^4.

<u>PROOF.</u> The symmetry of ψ_v must first be broken: Returning to Figure 10c, instead of isotoping the image of $\bar{c}_3$ in direction $-x_2$, carry out an isotopy in direction $+x_2$, as indicated in Figures 11a,b for $v = 0$, and Figures 11c,d for $v \neq 0$. Simultaneously isotope the image of c_2, keeping it on the torus $T^2_{x_1=-\frac{1}{2}}$, so that it feeds into H_B first in direction $+x_2$, then $+x_3$. The images of C and $\bar{C}$ are then fed into the handlebodies. The image of C in H_B is depicted in Figures 11e,f, that of $\bar{C}$ in H_D is shown in Figures 11g,h. Mapping the latter image of $\bar{C}$ into H_B by g, we obtain a diagram more easily visualized for the construction of Σ_v. (Figures 12a,b)

<u>REMARK.</u> It is clear that the diffeomorphism of T^3 given above is isotopic to ψ_v, leaving R pointwise fixed at all stages: thus the homotopy 4-spheres, constructed by either choice of 1-spine feeding, are diffeomorphic.

Using the model of Theorem 2.1, we construct the manifolds M_B, M_D. The attaching tubes for 1-handles are balls centered at points on the coordinate axes -- and we thus use $+H_i^!$, $-H_i^!$ to indicate the 3-ball of $H_i^!$ lying in $x_i \geq 0$, $x_i \leq 0$ respectively (Figure 14a). Furthermore, we shall maintain the same name for a 2-handle, even after it has been slid over another 2-handle and thus has a new attaching sphere. It is also convenient to note that framings of a 2-dimensional representation of a knot or link are changed if loops of a component are turned over, as depicted in Figure 13. Prospective framing changes about to arise in this way shall be placed in brackets next to the loop crossover point in question. Non zero framings are indicated where necessary -- in general, we leave inessential framings to the reader for evaluation.

The convention we have used for describing framings of 2-handles is to take as reference -- 0-framing -- the annulus obtained from the attaching sphere and a push-off parallel in the plane of representation. Hence a framing annulus for a +1-framed 2-handle twists once clockwise.

Although framings determined by this convention are not invariant under change of attaching sphere representation, they are convenient to use when little rearrangement of a diagram is carried out.

In the diagrams we have used for representing the mapping tori of the diffeomorphisms $\phi_A : T^3 \to T^3$, $A \in SL(3,\mathbb{Z})$, $\det(A-1) = 1$, 2-handles obtained by fattenning those of T^3 are 0-framed by the annuli of the latter used for gluing onto the boundary of the 0- and 1-handles of T^3.

The framings for the 2-handles H^2_{ii} used in identifying handlebodies in T^3 with their images under ϕ_A are determined as follows: In the universal cover $\mathbb{R}^3$ of T^3, take the standard coordinate axes and push off parallel in the direction $[1, 1, 1]^T$, to obtain three infinite strips intersecting transversely in an arc along a ray through the origin. Linearity of the map $A: \mathbb{R}^3 \to \mathbb{R}^3$ ensures that the images of the boundary components of any one of these strips are parallel. Projecting to T^3, the strips become annuli which have parallel boundary components in the model we have used -- and we have taken care in subsequent descriptions of isotopies to maintain this property -- thus determining 0-framing for H^2_{ii}.

STRUCTURE OF M_D. (i) $v = 0$. The diagram is shown in Figure 14a after minimal straightening and the addition of all handles in

$$M_D = \bar{H}^0 \cup \bar{H}^1_1 \cup \bar{H}^1_2 \cup \bar{H}^1_3 \cup \bar{H}^1_{11} \cup \bar{H}^2_{11} \cup \bar{H}^2_{22} \cup \bar{H}^2_{33} \cup \bar{H}^2_- .$$

Framings for all 2-handles are zero. Slide $\bar{H}^2_{11}, \bar{H}^2_{22}, \bar{H}^2_{33}$ off $\bar{H}^1_{11}$ using $\bar{H}^2_-$ -- equivalent to band-connect-summing with 6 pushed off copies of the attaching sphere of $\bar{H}^2_-$ (Figure 14b). The loops of attaching spheres protruding from the ball at ∞ to which $\bar{H}^1_{11}$ is attached pull through to give Figure 14c. Now cancel $\bar{H}^2_-$ and $\bar{H}^1_{11}$ to obtain Figure 14d. In Figure 14e, $\bar{H}^2_{33}$ has been slid over $\bar{H}^2_{11}$ at $+\bar{H}^1_1$, and the loop of $\bar{H}^2_{22}$ protruding from $-\bar{H}^1_2$ rearranged. Pull the loop of $\bar{H}^2_{33}$ at $+\bar{H}^1_2$ around this 1-handle and off -- Figure 14f. Cancelling $\bar{H}^1_1$ with $\bar{H}^2_{11}$, and pulling the loop of $\bar{H}^2_{22}$ through $\bar{H}^1_2$ gives Figure 14g, where the loop of $\bar{H}^2_{33}$ at $+\bar{H}^1_2$ is about to slide through. This gives Figure 14h, where $\bar{H}^1_2$ and $\bar{H}^2_{33}$ cancel to give Figure 14i. Cancelling the last complementary handle pair gives M_D diffeomorphic to B^4.

(ii) for $v \neq 0$, the procedure is much the same -- illustrated in Figure 15a for $v > 0$, Figure 15b for $v < 0$, to verify that geometric linking does not prevent any of the loops from sliding around the appropriate 1-handles. Thus for all cases, we obtain $M_D \cong B^4$.

STRUCTURE OF M_B. Again, a consistent procedure gives M_B diffeomorphic to B^4 in all cases, with minimal straightening. After sliding each of H^2_{11}, H^2_{22} and H^2_{33} off H^1_+, there is a loop of H^2_{22} protruding from $+H^1_2$. Sliding this around H^1_2 and off allows H^2_{11} and H^1_2 to cancel -- after which H^2_{33} cancels H^1_1, and finally H^2_{22} cancels H^1_3. The procedure begins with Figures 16a,b,c for the cases $v < 0$, $v = 0$, $v > 0$ respectively.

Hence Σ_v is diffeomorphic to $B^4 \cup B^4$, which is necessarily S^4 -- two balls in dimension four are glued together in an essentially unique way.

From Table 1 we know there are many homotopy 4-spheres in the construction, with minimal straightening, which correspond to classes of matrices not represented by the rational form (see Appendix). However, we conjecture that all such are actually diffeomorphic to S^4. As evidence for this we prove

THEOREM 4.4. S^4 may be constructed by Cappell and Shaneson's procedure using minimal straightening of the matrix $A = \begin{bmatrix} 1 & 0 & 1 \\ 1 & -1 & -1 \\ 2 & -3 & -5 \end{bmatrix}$ representing the non-trivial ideal class in $C(\mathbb{Z}[\Theta_{-5}])$.

PROOF. The reason for this choice of representative is the following: The images of C_1, C_2, C_3 represent the words $\alpha_1\alpha_2\alpha_3^2$, $\alpha_2^{-1}\alpha_3^{-3}$ and $\alpha_1\alpha_2^{-1}\alpha_3^{-5}$ respectively, with respect to the base point $\phi_A(Q)$ in $\pi_1(T) \cong \mathbb{Z}^3$. Using the abelian group structure, we isotope ϕ_A so that the new images represent the words $\alpha_1\alpha_3^2\alpha_2$, $\alpha_2^{-1}\alpha_3^{-3}$, $\alpha_2^{-1}\alpha_3^{-3}\alpha_1\alpha_3^{-2}$ in $\pi_1(H_B)$. Hence when we construct M_B, H_{11}^2 slides off H_{11}^1, and then off H_1^1, leaving H_1^1 and H_{33}^2 as a complementary pair. Cancelling these, the diagram is easily recognized as B^4 -- hence we may expect M_D to give B^4 also.

For convenience, take $Q = \pi(0, 0, 0)$, and $\bar{Q} = \pi(1, 1, 1)$. As before, we isotope so that $Q\bar{Q}$ is preserved -- the isotopy achieving this, and giving the desired images for C_1, C_2 and C_3 is shown in Figures 17a,b. Thus with minimal straightening, Figure 17b is obtained.

REMARK. As the dual spine $\bar{C}$ has been omitted from the diagrams, it must be verified that Figure 17b actually corresponds to a diffeomorphism that preserves the handlebodies. By fattening up the images of C_1, C_2 and C_3, we may view the image as a genus 3-handlebody. Since diffeomorphisms of such are generated by sliding and twisting handles, we proceed to slide these around, until we obtain an obvious image of H_B under some diffeomorphism. By staying inside H_B, while doing this, the image of the dual spine is forced into H_D, verifying that Figure 17b corresponds to a diffeomorphism of T^3 that preserves the splitting.

The diagram for M_B is given in Figure 17c, after sliding 2-handles off H_{11}^1 and cancelling the latter with H_+^2. The loop of H_{11}^2 protruding from $+H_1^1$ pulls off around H_1^1, leaving H_1^1 and H_{33}^2 complementary handles, which we cancel. Sliding H_{22}^2 off H_2^1, using H_{11}^2, we obtain that complementary handles H_{11}^2 and H_2^1 may be cancelled. That the remaining diagram represents B^4 is clear.

We leave to the reader the determination of the structure of M_D!

THEOREM 4.5. For each $v \in \mathbb{Z}$, Σ_v decomposes as the union of two copies of Mazur manifold $\tilde{M}_v$, glued together by an involution G of $\partial\tilde{M}_v$, representing $\partial\tilde{M}_v$ as a 2-fold branched cover of S^3.

PROOF. Using the symmetric splitting ψ_v of Lemma 4.2, the diagram for $\tilde{M}_v \equiv \tilde{H}_B$ is shown in Figure 18a, where we have added H_+^2 with framing $+1$, and used this to slide other 2-handles off H_{11}^1 -- thus introducing a full $+1$ twist in the 6 strands as shown.

Now slide H^2_{33} off H^1_1 using H^2_{11}, and then slide H^2_{33} off H^1_3 using H^2_{22}, to obtain Figure 18b, as Heegard diagram for $\tilde{M}_v$. Framings may be calculated using the Kirby Calculus (K2).

REMARKS. 1. If G extends over $\operatorname{int}\tilde{M}_v$, then $\tilde{\Sigma}_v$ is diffeomorphic to S^4. On the other hand, non-existence of an extension gives a counter-example to the smooth s-cobordism theorem in dimension 5 -- c.f. Akbulut and Kirby's remarks in Kirby (K1).

2. We show in a subsequent chapter that $\tilde{\Sigma}_2$, $\tilde{\Sigma}_6$ double cover manifolds known to be exotic; hence there is some chance they are not S^4. Certainly our splitting method has not succeeded for them.

Using the non-symmetric diffeomorphism of T^3, as in Figures 11, 12, we again obtain $\tilde{\Sigma}_v$ as the union of two (possibly distinct) Mazur manifolds $\tilde{M}_B$, $\tilde{M}_D$. This enables us to describe $\partial\tilde{M}_B = \partial\tilde{M}_D$ as obtained from S^3 by surgery on a knot: we illustrate for the case $v = 0$.

A slice of $S^2 \times \{0\} \subset S^2 \times D^2$ may be obtained by keeping sight of a meridian of the attaching tube of H^2_+ in the decomposition of Σ_v given in Theorem 4.3 -- we obtain a knot in S^3, after following the procedure indicated in Figure 19 (suppressing the "framing" attached to the meridian).

$\tilde{\Sigma}_v$ arises by removing $S^2 \times D^2$ from S^4, and replacing with a twist; the diagrams for $\tilde{M}_B$, $\tilde{M}_D$ differ from M_B, M_D in that H^2_+, H^2_- are added with framing +1 (after inverting the diagram for M_D). However, if we connect sum $\tilde{M}_B$ with $\mathbb{C}P^2$, we do not change the boundary -- and choosing -1 intersection form gives a new 4-manifold whose diagram is given in Figure 19b after sliding H^2_+ over the -1-framed 2-handle of $\mathbb{C}P^2$ (Figure 19a). We can now slide and cancel exactly as in obtaining a slice of $S^2 \times D^2$ in S^4 -- thus it is clear that $\partial\tilde{M}_B = \partial\tilde{M}_D$ is obtained from S^3 by surgery on the slice of $S^2 \times D^2$ (Figure 19c). Furthermore, this implies that the procedure, carried out on $\partial\tilde{M}_D$, gives exactly the same knot (being the slice of $S^2 \times \{0\}$) and thus cannot furnish further examples of inequivalent knots producing the same 3-manifold, as in Lickorish (L1). However, the ribbons naturally obtained may differ -- see (AK1) in this regard.

Returning to the symmetric splitting, it is not too hard to obtain an explicit picture of $\partial\tilde{M}_v$ as an open book decomposition: we note that

$$\begin{bmatrix} 0 & 0 & 1 \\ 1 & 1 & 0 \\ 0 & 1 & v \end{bmatrix} = \begin{bmatrix} 1 & 0 & 0 \\ 0 & 1 & 0 \\ 1 & 0 & 1 \end{bmatrix}^v \begin{bmatrix} 0 & 1 & 0 \\ 1 & 0 & 1 \\ 1 & 0 & 0 \end{bmatrix} \begin{bmatrix} 0 & 1 & 0 \\ 0 & 0 & 1 \\ 1 & 0 & 0 \end{bmatrix} = A^v BD$$

It is not difficult to see that $\phi_A, \phi_B, \phi_D : T^3 \to T^3$ can be isotoped to symmetric diffeomorphisms -- allowing explicit calculation, by iteration, of the handle structure and position of C_G in the mapping torus of $\psi_v\big|_{\partial H_B}$, from which $\partial\tilde{M}_v$ arises.

CONJECTURE (Gluck). If we remove a neighborhood of a knotted 2-sphere K in S^4, and glue back in by the diffeomorphism of $S^2 \times S^1$ corresponding to the non-trivial element of $\pi_1(SO(3)) \cong \mathbb{Z}_2$, then the resulting manifold X^4 is diffeomorphic to S^4.

The manifolds $\tilde{\Sigma}_v$, $v \in \mathbb{Z}$ arise in this way: P. Melvin remarks in Kirby (K1) that $X^4 \cong S^4 \Leftrightarrow (X^4 \# \mathbb{C}P^2, K \# \mathbb{C}P^1) \cong (\mathbb{C}P^2, \mathbb{C}P^1)$, pairwise.

THEOREM 4.6. *Let X^4 be obtained from S^4 by removing $S^2 \times D^2$ and sewing in by a twist. Then $X^4 \# \mathbb{C}P^2$ is diffeomorphic to $\mathbb{C}P^2$.*

PROOF. Let $N \cong S^2 \times D^2$ be a neighborhood of the knot $K \subset S^4$. Then $S^4 = Y^4 \cup H_0^2 \cup H^4$, where $Y^4 = S^4 - \operatorname{int} N$, and H_0^2 is a 0-framed 2-handle. Thus $X^4 \# \mathbb{C}P^2$ can be viewed as the union of $\mathbb{C}P^2 - \operatorname{int} B^4$ and $Y^4 \cup H_1^2$, where the 2-handle now has framing 1. There exists a 0-framed 2-handle $\bar{H}_0^2$ in $X^4 \# \mathbb{C}P^2 - \operatorname{int} Y^4$ with the same attaching sphere as H_1^2 in ∂Y^4. Adding this 2-handle to Y^4 gives $S^4 - \operatorname{int} B^4 = B^4$.

On the other hand, if Y^4 is replaced by $B^3 \times S^1 = h^0 \cup H^1$,

$$\begin{aligned} B^3 \times S^1 \cup H_1^2 \cup (\mathbb{C}P^2 - \operatorname{int} B^4) &= H^0 \cup H^1 \cup H_1^2 \cup (\mathbb{C}P^2 - \operatorname{int} B^4) \\ &= B^4 \cup \{\mathbb{C}P^2 - \operatorname{int} B^4\} \\ &= \mathbb{C}P^2 . \end{aligned}$$

There is a diffeomorphism between $X^4 \# \mathbb{C}P^2$ and $\mathbb{C}P^2$, defined as follows:

$$Y^4 \cup \bar{H}_0^2 \cong B^4$$

$$\begin{aligned} X^4 \# \mathbb{C}P^2 - \operatorname{int}\{Y^4 \cup \bar{H}_0^2\} &= (\mathbb{C}P^2 - \operatorname{int} B^4) \cup H_1^2 - \operatorname{int} \bar{H}_0^2 \\ &= (\mathbb{C}P^2 - \operatorname{int} B^4) \cup H_1^2 \cup B^3 \times S^1 - \operatorname{int}(B^3 \times S^1 \cup \bar{H}_0^2) \\ &= \mathbb{C}P^2 - \operatorname{int} B_0^4 \end{aligned}$$

where $$B_0^4 = \bar{H}_0^2 \cup S^1 \times B^3 .$$

We thus obtain $X^4 \# \mathbb{C}P^2 - \operatorname{int} B^4 \cong \mathbb{C}P^2 - \operatorname{int} B_0^4$, which gives the diffeomorphism desired.

CONCLUDING REMARKS. Suppose Σ^4 is a homotopy 4-sphere with an open book decomposition with S^2 binding.

1. Is there a Heegard decomposition of the page and an isotopy of the monodromy ρ so that Σ^4 is split into $E_B \cup E_D$, with E_B, E_D homology balls -- or more preferable, contractible?

In the latter case, can one also obtain E_B, E_D with fundamental group presentation trivializable by Andrews-Curtis moves, so that $E_B \cup_{id} E_B \cong E_D \cup_{id} E_D \cong S^4$?

2. Which 3-manifolds M^3 have Heegard decompositions $H_B \cup H_D$ such that the homeotopy group $\mathscr{H}(M^3)$ contains a central element interchanging H_B and H_D? (In analyzing Cappell and Shaneson's constructions we use such an element in $\mathscr{H}(S^1 \times S^1 \times S^1)$). Products of S^1 and a surface certainly do -- and perhaps some other Seifert fibre spaces.

5. INVOLUTIONS ON HOMOTOPY 4-SPHERES

We consider several constructions of closed non-orientable smooth 4-manifolds homotopy equivalent to RP^4, real projective 4-space. Wall (Wa) has shown there are two smooth s-cobordism classes of such manifolds, the first example of a representative for the non-trivial class having been constructed by Cappell and Shaneson (CS2).

CONSTRUCTION 1: The decomposition $S^4 = D^2 \times S^2 \cup S^1 \times B^3$ induces a decomposition $RP^4 = D^2 \tilde{\times} RP^2 \cup S^1 \tilde{\times} B^3$. We remove $S^1 \tilde{\times} B^3$ and glue in its place $T_0^3 \times_{\phi_A} S^1$, where $A: R^3 \to R^3$ has $\det(A) = -1$ and $\det(A^2 - I) = 1$, and the diffeomorphism $\phi_A: T^3 \to T^3$ has been isotoped to the antipodal map in a neighborhood R' of the fixed point q. The resulting 4-manifold represents the non-trivial s-cobordism class of homotopy real projective spaces.

CONSTRUCTION 2: The decomposition $S^4 = B^4 \cup_a B^4$, where a is the antipodal map on S^3, yields the decomposition $RP^4 = B^4 \cup N(RP^3)$, where $N(RP^3)$ is the twisted line bundle over RP^3. Suppose now that N^3 is a Z-homology 3-sphere with free involution τ, and that $M^3 = \partial W^4$ where $W^4 \cup_\tau W^4$ is a homotopy 4-sphere Σ_τ. Then Σ_τ has free involution with quotient Q^4, whose s-cobordism class is determined via

THEOREM (Fintushel-Stern (FS)): *Let T be a free involution on the homotopy 4-sphere Σ_τ which desuspends to an involution τ on a Z-homology 3-sphere M^3. Then there is an almost framing $\mathscr{F}$ for M^3/τ such that*

$\rho(T) = \mu(M^3/\tau, \mathscr{F}) + \frac{1}{2}\alpha(M^3, \tau) \equiv \pm 1 \pmod{16}$ if Σ_τ/T is s-cobordant to $\mathbb{R}P^4$

and $\rho(T) = \mu(M^3/\tau, \mathscr{F}) + \frac{1}{2}\alpha(M^3, \tau) \equiv \pm 9 \pmod{16}$ if Σ_τ/T is s-cobordant to an exotic homotopy projective space. Hence in this case Σ_τ/T is exotic.

Here $\alpha(M^3, t)$ is the Browder-Livesay invariant for the free involution t. For details, see Lopez de Medrano (LM). As stated, this theorem is not as it appears in (FS), but the details may be filled in easily by the reader, bearing in mind that Yoshida (Y) has shown that $\alpha(M^3, \tau) \equiv \mu(M^3) \pmod{16}$. Fintushel and Stern show that the Brieskorn sphere $\Sigma(3,5,19)$ bounds a contractible manifold built with a single 1- and 2-handle, and has a free involution which is part of a circle action. It follows that there is an exotic involution on S^4.

We shall show that Cappell and Shaneson's construction is a special case of this construction:

THEOREM 5.1. The Cappell-Shaneson involutuions on homotopy 4-spheres desuspend to involutions on Z-homology 3-spheres. Hence the exotic nature of the quotient is detected by the Fintushel-Stern invariant ρ.

PROOF. We continue with previously established notation.

(a) Algebraic preliminaries: If A $GL(3;Z)$ has $\det(A) = -1$ and $\det(A^2 - I) = 1$, then A has characteristic polynomial either $h_0(x) = x^3 - x+1$ or $h_1(x) = x^3 - 2x^2 - x+1$. A^2 then has characteristic polynomial $f_2(x)$ or $f_6(x)$ respectively. Conversely, if B has characteristic polynomial $f_2(x)$, then $h_0(-B^2 + B) = 0$, and if B has characteristic polynomial $f_6(x)$, then $h_1(B^2 - 5B + 2I) = 0$. It follows that there is a unique conjugacy class in $GL(3;Z)$ for matrices with characteristic polynomials $h_0(x)$ and $h_1(x)$. Thus

LEMMA 5.2. Any matrix A as above is conjugate in $GL(3;Z)$ to one of

$$B_0 = \begin{bmatrix} 0 & -1 & 0 \\ 0 & 0 & -1 \\ -1 & -1 & 0 \end{bmatrix} \quad \underline{\text{or}} \quad B_1 = \begin{bmatrix} 0 & -1 & 0 \\ 0 & 0 & -1 \\ -1 & -1 & 2 \end{bmatrix}$$

Hence there are only two mapping tori to consider, and we use the techniques of the previous section to describe a canonical isotopy to a symmetric diffeomorphism in each case. Isotopy of the 1-spine is illustrated in Figure 20. To straighten to the antipodal map in a neighborhood of the fixed point, observe that the matrices $D_0 = (-B_0)$ and $D_1 = (-B_1)^{-1}$ have characteristic polynomials $f_0(x)$ and $f_{-1}(x)$ respectively, and thus we may use minimal straightening as defined previously. Carrying this out we obtain the symmetric diffeomorphism illustrated in Figure 21. Let ϕ_{ij} denote the diffeomorphism which differs from this by j complete twists about the fixed point, and denote by M_{ij} the corresponding mapping tori, $M_{ij} = T_0^3 \times_{\phi_{ij}} S^1$. Here $i = 0,1$ and $j \in Z$.

To glue M_{ij} to $RP^2 \tilde{\times} D^2$, we note that the boundary $S^2 \tilde{\times} S^1$ has group of diffeomorphisms, modulo those isotopic to the identity, given by $Z_2 \oplus Z_2$. Let ρ_k denote the diffeomorphism which rotates the S^2 factor k times in going around S^1, and let

$$Q_{ijk} = T_0^3 \times_{\phi_{ij}} S^1 \cup_{\rho_k} RP^2 \tilde{\times} D^2 .$$

Then the twisting in the gluing map and in the mapping torus may be absorbed together and reduced mod 2, leaving us with 4 possibilities Q_{ij}, $i,j = 0,1$, with $Q_{ij} \equiv Q_{ijo}$.

The double cover $\tilde{Q}_{ij}$ of Q_{ij} has corresponding decomposition $\tilde{Q}_{ij} = T_0^3 \times_{\phi_{ij}^2} S^1 \cup_{id} S^2 \times S^1$. The spine-feeding and straightening for ϕ_{ij}^2 are illustrated in Figure 22 in the case $j=0$, obtained by iterating the diffeomorphism ϕ_{ij}. This differs from minimal straightening by a full twist, and we see that $\tilde{Q}_{0j} = \tilde{\Sigma}_2$ and $\tilde{Q}_{ij} = \tilde{\Sigma}_6$. Hence we have proved.

THEOREM 5.3: *Let Q^4 be a homotopy projective space arising from Cappell and Shaneson's construction. Then the double cover $\tilde{Q}^4$ is obtained from S^4 by the Gluck construction on a knot. Moreover, $\tilde{Q}^4 = \tilde{\Sigma}_2$ or $\tilde{\Sigma}_6$ in the notation of Section 4.*

NOTE: The twist corresponding to $j=1$ lifts to two full twists in the boundary of $S^2 \times D^2$ and thus gives a gluing map which extends over $S^2 \times D^2$. Thus $\tilde{Q}_{i0} = \tilde{Q}_{i1}$, $i=0,1$.

Since ϕ_{ij} is symmetric, so is ϕ^2_{ij}, and thus we have a decomposition for $\tilde{Q}_{ij}$ as in the previous section: $\tilde{Q}_{ij} = W_{ij} \cup W_{ij}$. Here W_{ij} is a contractible 4-manifold, and the gluing map is the restriction to $N_{ij} = \partial W_{ij}$ of the involution $G:\tilde{Q}_{ij} \to \tilde{Q}_{ij}$ defined by

$$\begin{aligned} G(x,t) &= (g(x),t) \qquad (x,t) \in T^3_0 \times_{\phi^2_{ij}} S^1 \\ &= (-x,t) \qquad (x,t) \in S^2 \times D^2 . \end{aligned}$$

G has a circle C_G of fixed points in N_{ij}, and the restriction of G to N_{ij} represents N_{ij} as a 2-fold cover of S^3 branched over the image $\bar{C}_G$ of C_G.

Furthermore, the decomposition described is also preserved under the covering transformation $H:\tilde{Q}_{ij} \to \tilde{Q}_{ij}$, which is defined by

$$\begin{aligned} H(x,t) &= (\phi_{ij}(x),-t)) \qquad (x,t) \in T^3_0 \times_{\phi_{ij}} S^1 \\ &= (-x,-t) \qquad (x,t) \in S^2 \times D^2 . \end{aligned}$$

Hence we may equally well describe $\tilde{Q}_{ij}$ as the union of two copies of W_{ij}, glued together by the restriction of H to N_{ij}. Hence $Q_{ij} = W_{ij} \cup N(N_{ij}/H)$ where $N(N_{ij}/H)$ is a twisted line bundle over N_{ij}/H, and the proof of the theorem is complete.

We now observe that the involutions G,H in fact commute, and thus we have a third involution GH. This has fixed point set $S_{GH} = S^2 \times \{0\} \subset S^2 \times D^2$, which is the binding of the open book decomposition of $\tilde{Q}_{ij}$. We thus have further counterexamples to the higher dimensional Smith Conjecture. The quotient of $\tilde{Q}_{ij}$ by GH is $T^3_0 \times_{g\phi_{ij}} S^1 \cup S^2 \times D^2$, and since $g\phi_{ij}$ is a symmetric diffeomorphism isotopic to ϕ_{-B_i} with minimal straightening when $j=0$, and a full twist when $j=1$, we see that $\tilde{Q}_{ij}/GH = S^4$ when $j=0$, and $\tilde{Q}_{ij}/GH = \tilde{\Sigma}_i$ when $j=1$. Consider the case $j=0$. As remarked the commuting involutions G,H and GH all preserve C_G, C_{GH} and S_{GH}, and choosing any one involution, the remaining two pass to the quotient to define the same induced involution: we obtain

(i) $H_0 = G_0 : S^4 = \tilde{Q}_{i0}/GH \to S^4$. The fixed point set is $\hat{C}_G = C_G/GH$, and thus the quotient is not a manifold. However, restricting G_0 to $N'_{i0} = N_{i0}/GH$, we see N'_{i0} as a 2-fold cover of S^3 with branch set $\hat{C}_G$. Hence N_{i0} is a 4-fold cover of S^3 with branch set $C_G \cup C_{GH}$.

(ii) $H_1 = (GH)_1 : \tilde{Q}_{i0}/G \to \tilde{Q}_{i0}/G$, with fixed point set $S_{GH}/G \cong RP^2$. Restricting to N_{i0}/G (which is S^3), we obtain S^3 as a 2-fold cover of S^3 with branch set $\bar{C}_{GH} = C_{GH}/G$. Thus C_{GH} is unknotted in S^3.

(iii) $G_2 = (GH)_2 : Q_{i0} \to Q_{i0}$, with fixed point set $S_{GH}/H \cup C_G/H$. This expresses N_{i0}/H as a 2-fold branched cover of S^3 with branch set the link $C_G/H \cup C_{GH}/H$.

For the case Q_{i1}, the only difference is that $\tilde{Q}_{i1}/GH = \tilde{\Sigma}_{-i}$ rather than S^4.

As the smooth Poincare conjecture remains unresolved, it is possible that $\tilde{Q}_{ij}$ is an example of a non-standard differential structure on S^4. We would then have a counterexample to Gluck's conjecture, the smooth s-cobordism theorem in dimension 4 (by removing two disjoint 4-balls in Q_{ij} to obtain a homotopy $S^3 \times I^1$), and also the smooth 5-dimensional relative s-cobordism theorem -- cf. Kirby (K1). To determine which of these alternatives holds, it would be fruitful to investigate the extension problem: do either of the involutions G or H, restricted to N_{ij}, extend to a diffeomorphism of W_{ik}? Since the double of any Mazur manifold is S^4 -- we offer an alternative proof of this later -- an affirmative answer would give $\tilde{Q}_{ij}$ diffeomorphic to S^4.

We remark that Matumoto and Siebenmann (MS) have shown that the TOP s-bobordism theorem fails in at least one of dimensions 4 or 5.

Identification of N_{ij} would be useful in resolving this question: our investigations offer several alternative descriptions.

(i) a surgery description on a 2-component link in S^3, one of whose components is unknotted and 0-framed.

(ii) an open book decomposition, obtainable by constructing the mapping torus of ϕ_{ij} restricted to $\partial H_B = \partial H_D \subset T^3$, as in the previous chapter.

(iii) a description as 2-fold branched cover of S^3.

(iv) a description as 4-fold branched cover of S^3, branched over $\hat{C}_G \cup \hat{C}_{GH}$, the image of $\bar{C}_G \cup \bar{C}_{GH}$: If we branch over $\hat{C}_G, \hat{C}_{GH}$ separately the other curve lifts to a connected component. Hence $\hat{C}_G, \hat{C}_{GH}$ link each other an odd number of times algebraically. Furthermore, $\bar{C}_{GH}$ and $\hat{C}_{GH}$ are unknotted, and $\hat{C}_G$ is a bridge knot -- in fact, $\hat{C}_G, \hat{C}_{GH}$ have linking number ± 7.

We remark that given such a link, we can always construct a diagram as in Figure 23: Since $H_1(S^3 - \hat{C}_G \cup \hat{C}_{GH}) = \mathbb{Z} \times \mathbb{Z}$, we can take the 4-fold cover corresponding to the epimorphism $\pi_1(S^3 - \hat{C}_G \cup \hat{C}_{GH}) \to \mathbb{Z}_2 \times \mathbb{Z}_2$. This 4-fold cover has covering transformation group Γ generated by two commuting involutions. Two of the involutions in Γ have a circle of fixed points, and the third is free.

In view of the structure obtained for Q_{ik} -- the union of a Mazur manifold and the mapping cylinder of a free involution on its boundary -- it is of interest whether $\mathbb{R}P^4$ can be similarly decomposed. To resolve this we present a new proof of the well known

LEMMA 5.4. *Let* M^4 *be a Mazur manifold. Then* $M^4 \cup_{id} M^4 \cong S^4$.

PROOF. M^4 has a handle decomposition

$$M^4 = H^0 \cup H^1 \cup H^2$$

where H^2 is attached to $S^2 \times S^1 = \partial(H^0 + H^1)$ along a solid torus $C \times D^2$ with C a knotted circle homotopic to the generator of $\pi_1(S^1 \times S^2)$. The complement of $C \times D^2$ in $S^3 = \partial H^2$ is another unknotted solid torus $D^2 \times S^1$. Thus gluing the two copies of M^4 together by the identity, we obtain a homotopy 4-sphere Σ^4 with handle decomposition

$$\Sigma^4 = H^0 \cup H^1 \cup H^2 \cup \bar{H}^2 \cup \bar{H}^3 \cup H^4 .$$

The involution $\sigma: \Sigma^4 \to \Sigma^4$, which interchanges the two copies of M^4, also interchanges the two 2-handles H^2 and $\bar{H}^2$. Hence these 4-balls H^2 and $\bar{H}^2$ are glued together in Σ^4 by identifying the solid tori complementary in their boundaries to the attaching tubes which glue them onto $H \cup H^1$, $H^3 \cup H^4$ respectively -- $H^2 \cup \bar{H}^2$ thus gives $S^2 \times D^2$. Remove $S^2 \times D^2$ from Σ^4, and replace with $S^1 \times B^3$ – the result being $S^1 \times S^3$ as this procedure effectively collapses the boundaries of $H^0 \cup H^1$, $H^3 \cup H^4$ onto each other. As $S^1 \times B^3$ unknots in $S^1 \times S^3$, the complement is again $S^1 \times B^3$. Now remove $S^1 \times B^3$ and replace with the original $S^2 \times D^2$, to obtain

$$\begin{aligned}\Sigma^4 &= ((H^0 \cup H^1 \cup H^3 \cup H^4 \cup S^1 \times B^3) - S^1 \times B^3) \cup H^2 \cup \bar{H}^2 \\ &\cong (S^1 \times S^3 - S^1 \times B^3) \cup H^2 \cup \bar{H}^2 \\ &\cong S^1 \times B^3 \cup S^2 \times D^2 \cong S^4\end{aligned}$$

by Laudenbach and Poenaru.

THEOREM 5.5. *There are infinitely many distinct Mazur manifolds which are characteristic submanifolds for the antipodal map* $a: S^4 \to S^4$.

PROOF. Let C be an arbitrary embedded circle in $S^1 \times S^2$, homotopic to the generator of $\pi_1(S^1 \times S^2)$. $S^1 \times S^2$ is the quotient of $S^1 \times S^2$ under the covering projection $\rho: S^1 \times S^2 \to S^1 \times S^2$ defined by the covering transformation

$$g(x,y) = (a(x),y) \quad \forall (x,y) \in S^1 \times S^2$$

where $a: S^1 \to S^1$ is the antipodal map.

Thus $\rho^{-1}(C)$ has one component $\bar{C}$, homotopic to the generator of $\pi_1(S^1 \times S^2)$. We illustrate in Figure 24 with C given by Mazur's original example (Mazur (M)).

Let M^4 be the Mazur manifold obtained by adding a 2-handle H^2 to $S^1 \times B^3$ with attaching circle $\bar{C}$. M^4 is thus obtained by doing $(1,n)$ surgery on the solid torus $\bar{C} \times D^2 \subset S^1 \times S^2$, where n is the framing of H^2. Take $C' \subset \partial(\bar{C} \times D^2)$ a framing curve for H^2. In order to extend

$$g: S^2 \times S^1 - \bar{C} \times D^2 \to S^2 \times S^1 - \bar{C} \times D^2$$

to a free involution $\bar{g}: \partial M^4 \to \partial M^4$, we require $n \in \mathbb{Z}$ for H^2 to be odd, since $g(C') \cap C' = \emptyset$ iff C' is the lift of a curve in $\partial(C \times D^2)$, which must be a $(2,n)$ curve. The extension to H^2 is then uniquely determined.

Now take two copies of M^4, $M_1^4 = H^0 \cup H^1 \cup H^2$, $M_2^4 = \bar{H}^2 \cup \bar{H}^3 \cup \bar{H}^4$, where we have turned the handle decomposition of M_2^4 upside down. Gluing these together by $\bar{g}$ on the boundary, we obtain S^4 -- since the diffeomorphism $\bar{g}$ may be extended to

$$P: M^4 \to M^4$$

by putting

$$P(x,y) = (a(x),y) \quad \forall (x,y) \in S^1 \times B^3 \cong H^0 \cup H^1$$

$$P(\alpha,\beta) = (\alpha,a(\beta)) \quad \forall (\alpha,\beta) \in D^2 \times D^2 \cong H^2 .$$

There is thus a free involution on

$$S^4 = H^0 \cup H^1 \cup H^2 \cup \bar{H}^2 \cup \bar{H}^3 \cup \bar{H}^4$$

determined by $P\sigma = \sigma P$ where σ interchanges M_1^4 and M_2^4 as in Lemma 5.4. As P preserves the handle decomposition of M^4, $P\sigma$ preserves $H^2 \cup \bar{H}^2 \cong S^2 \times D^2$, and restricts to

$$(x,y) \to (a(x),a(y)) \qquad (x,y) \in S^2 \times S^1 = \partial(S^2 \times D^2) .$$

As in Lemma 5.4, we remove $S^2 \times D^2$ and replace by $B^3 \times S^1$, extending the involution on $\partial(S^2 \times D^2)$ in the obvious manner to obtain a free involution χ on $S^3 \times S^1$.

Thus $$B^3 \times S^1/\chi \cong B^3 \underset{\sim}{\times} S^1 \subset S^3 \times S^1/\chi \cong S^3 \underset{\sim}{\times} S^1 .$$

Again as in Lemma 5.4 we may remove $B^3 \underset{\sim}{\times} S^1$ from $S^3 \underset{\sim}{\times} S^1$ by isotopic unknotting to obtain $B^3 \underset{\sim}{\times} S^1 = (H^0 \cup H^1 \cup \bar{H}^3 \cup \bar{H}^4)/\chi = (H^0 \cup H^1 \cup \bar{H}^3 \cup \bar{H}^4)/P\sigma$.

Hence $$\begin{aligned} S^4/P\sigma &\cong B^3 \underset{\sim}{\times} S^1 \cup \mathbb{R}P^2 \underset{\sim}{\times} D^2 \\ &\cong \mathbb{R}P^4 . \end{aligned}$$

COROLLARY 5.6. *Let M^3 be the homology $\mathbb{R}P^3$ obtained by $(2,n)$ surgery on an arbitrary embedded circle $C \subset S^2 \times S^1$, with C homotopic to the generator of $\pi_1(S^1 \times S^2)$ and n odd. Then M^3 embeds one-sidedly in $\mathbb{R}P^4$.*

This follows immediately, since M^3 is the quotient of M^4 by σP -- hence RP^4 is the union of M^4 and $N(M^3)$.

If we double $\#_k S^1 \times D^2$, we obtain $\#_k S^1 \times S^2$ with orientation reversing involution τ interchanging the handlebodies and restricting to the identity on the boundary of each. Suppose now that $\phi: \#_k S^1 \times D^2 \to \#_k S^1 \times D^2$ is a diffeomorphism which is the identity in a neighborhood of a fixed point $q \in \partial(\#_k S^1 \times D^2)$.

Then there is an induced diffeomorphism, also denoted ϕ, on $\#_k S^1 \times S^2$ by doing ϕ on each handlebody, which restricts to the identity in a neighborhood R of the fixed point q. Forming the mapping torus of ϕ^2 and surgering $R \times S^1$ we obtain a manifold M_{ϕ^2} with involution T, defined as follows:

$$T(x,y) = (\tau\phi(x),-y)), \quad (x,y) \in (\#_k S^1 \times S^2)_0 \times_{\phi^2} S^1$$

$$= (r(x),-y)), \quad (x,y) \in S^2 \times D^2$$

where r is reflection of S^2 in the equator. There is thus a circle of fixed points. If M is a homotopy sphere we obtain examples of homotopy $RP^2 \times D^2$'s.

On the other hand, doing the Gluck construction on the knotted 2-sphere will give a 4-manifold with involution which is now free. This corresponds to replacing ϕ in the above construction of M by the diffeomorphism $\bar{\phi}$ which differs from ϕ by a rotation of π in a neighborhood of the fixed point q. Hence if $M_{\bar{\phi}}$ is a homotopy sphere we obtain some more potentially exotic homotopy RP^4's.

However, we can readily see that the quotient is s-cobordant to RP^4 as follows:

Thicken the construction of the mapping torus by taking the product with I to obtain a homology circle. Let $R \cap \#_k S^1 \times S^1 = D$. By adding a 2-handle $B^2 \times B^3$ along $(S^1 \times D) \times I$ we obtain a homotopy 5-ball with involution

$$T(x,y,z) = (\phi(x),-y,-z)) \quad \text{on the mapping torus}$$

$$= (-x,-y,-z) \quad \text{on the 2-handle}.$$

Flipping the I-factor...$z \to -z$...corresponds on the boundary to interchanging the handlebodies of $\#_k S^2 \times S^1$, i.e. τ. Hence we obtain the required extension over a homotopy 5-ball, which clearly has a unique fixed point $(0,0)$ in the 2-handle, and about which it is the antipodal map.

Removing an open neighborhood of $(0,0)$ and taking the quotient we obtain the desired s-cobordism to RP^4.

Clearly there are many possible choices for the diffeomorphism ϕ, but we have not determined which gives the standard RP^4 as quotient. The construction works because the involution τ and $\bar{\phi}$ commute. If we take some other commuting involution τ', then $M^3 = \#_k S^1 \times D^2 \cup_{\tau'} \#_k (S^1 \times D^2$ gives a 3-manifold with involution, and the above construction goes through. The Cappell and Shaneson examples are exactly of this kind, and it is clear that the techniques of the previous sections enable us to determine whether or not the quotients are s-cobordant to RP^4.

6. THE GENERAL CONSTRUCTION

Let X^4 be a nonorientable 4-manifold, with embedded circle C representing an orientation-reversing element of order two in $\pi_1(X^4)$. Denote by $\tilde{X}^4$ the orientable double cover of X^4, $p:\tilde{X}^4 \to X^4$ the projection, and $\sigma:\tilde{X}^4 \to X^4$ the covering transformation which restricts to the antipodal map on the circle $C = p^{-1}(C)$.

1. There is a tubular neighborhood $\tilde{N}(\tilde{C})$ of $\tilde{C}$ invariant under σ, and on which the restriction of σ is given by

$$\sigma(x,y) = (a(x),\mu(y)) \quad \forall (x,y) \in S^1 \times B^3 \cong \tilde{N}(\tilde{C})$$

where $\mu:B^3 \to B^3$ is orientation reversing, $\mu^2 = 1$, and a is the antipodal map. Thus $S^1 \times B^3/\sigma \cong S^1 \tilde{\times} B^3$. Since the diffeomorphism type of a bundle over S^1 depends only on the isotopy class of the monodromy, we may replace μ by the more convenient orientation-reversing involution a on B^3, which is isotopic to μ. Hence without loss of generality we may assume that $\sigma:\tilde{X}^4 \to \tilde{X}^4$ restricts to

$$\sigma(x,y) = (a(x),a(u)) \quad \forall (x,y) \in S^1 \times B^3 \cong \tilde{N}(\tilde{C})$$

2. There is an unknotted embedded solid torus $T \subset p(\partial\tilde{N}(\tilde{C})) \cong S^2 \underset{\sim}{\times} S^1$, whose core is a circle C' such that $[C'] = [C]^2 \in \pi_1)X^4)$. Thus $p^{-1}(C')$ has two components $\tilde{C}_1'$, $\tilde{C}_2' \subset \partial\tilde{N}(\tilde{C})$, with C_1' bounding a disc D in $\tilde{X}^4 - \operatorname{int}\tilde{N}(\tilde{C})$ which by Norman's argument (No) may be assumed to be locally flatly embedded. A neighborhood of D in $\tilde{X}^4 - \operatorname{int}\tilde{N}(\tilde{C})$ is a 4-ball H_+^2 which may be considered as a 2-handle attached to $\tilde{N}(\tilde{C})$, along the component of $p^{-1}(T)$ containing $\tilde{C}_1'$, with framing some integer m. This gives

$$\tilde{X}^4 - (H_+^2 \cup \tilde{N}(\tilde{C})) \cong \tilde{X}^4 - B^4$$

since adding a 2-handle along $S^1 \times \{x\} \subset S^1 \times S^2 = \partial(S^1 \times B^3)$ always gives B^4, regardless of the framing.

3. Let $X_{ik}^4 = (X^4 - \operatorname{int}N(C)) \cup_{\rho_0} M_{ik} \cong (X^4 - \operatorname{int}N(C) \cup_{\rho_n} M_{ik+n}$.

4. $\tilde{X}_{ik}^4$ is obtained from $\tilde{X}^4$ by removing $\tilde{N}(\tilde{C})$ and gluing in $\tilde{M}_{ik} = (T^3 - R') \times_{\phi_{ik}^2} S^1$ by the identity on the boundary $S^2 \times S^1$. Hence

$$\begin{aligned} \tilde{X}_{ik}^4 &= \tilde{X}^4 - (H_+^2 \cup \tilde{N}(\tilde{C})) \cup_{id} (H_+^2 \cup \tilde{M}_{ik}) \\ &= (\tilde{X}^4 - B^4) \cup_{id} (H_+^2 \cup \tilde{M}_{ik}) . \end{aligned}$$

THEOREM 6.1. *If m is even, $\tilde{X}^4_{ik} \cong \tilde{X}^4 \# \tilde{Q}_{ik}$ for $i = 0,1$ and $k = 0,1$. If m is odd $\tilde{X}^4_{ik} \cong \tilde{X}^4$, where m is the framing of the 2-handle.*

PROOF. 1. We first observe that $\tilde{M}_{ik}$ differs from $\tilde{M}_{i,0}$ by $2k$ twists in the boundary $\partial R' \times S^1 \cong S^2 \times S^1$. Again because $\pi_1(SO(3)) \cong \mathbb{Z}_2$, it is the case that the group of orientation-preserving diffeomorphisms from $S^2 \times S^1$ to itself, modulo isotopy, is $\mathbb{Z}_2 \times \mathbb{Z}_2$ (see (G)). Hence $\tilde{X}^4_{ik} \cong \tilde{X}^4_{i,0}$.

2. From the previous chapter, there are two possibilities for $\tilde{M}_{i,0} \cup S^2 \times D^2$; S^4 or $\tilde{Q}_{i0}$.

Adding $S^2 \times D^2 = H^4 \cup H^2$ to $\tilde{M}_{i0}$ with H^2 attached by odd framing gives S^4 -- whereas even framing gives $\tilde{Q}_{i0}$ by Theorem 5.3. Thus

$$\tilde{X}^4_i = \tilde{X}^4_{ik} = \begin{cases} (\tilde{X}^4 - B^4) \cup_{id} (S^4 - B^4) & \text{if } m \text{ is odd} \\ (\tilde{X}^4 - B^4) \cup_{id} (\tilde{Q}_{i0} - B^4) & \text{if } m \text{ is even} \end{cases}$$

completing the proof.

If C' bounds a locally flat embedded disc D^2, $p^{-1}(D^2)$ consists of disjoint discs $\tilde{D}^2_1, \tilde{D}^2_2$ in $\tilde{X}^4$, which may be taken as the cores of 2-handles H^2_+, H^2_- attached to $\tilde{N}(\tilde{C})$ along $\tilde{C}'_1, \tilde{C}'_2$ -- with H^2_+ having framing $m \in \mathbb{Z}$ as before. Hence $\tilde{N}(\tilde{C}) \cup H^2_+ \cup H^2_-$ is diffeomorphic with $S^2 \times D^2$, since H^2_- is attached with framing $-m$. We may assume that H^2_+, H^2_- are interchanged by the covering transformation of $\tilde{X}^4$, thus projecting to a 2-handle H^2 attached to $N(C)$ with core D^2. The "framing" for H^2 is only defined mod 2, and is given partly by m.

There is a locally flat embedded $\mathbb{RP}^2$ in X^4 -- the union of D^2 and a Möbius band M^2 in $N(C)$ bounded by C'. Now there are only two non-orientable S^1-bundles over $\mathbb{RP}^2$ -- $S^1 \times \mathbb{RP}^2$ and $S^2 \tilde{\times} S^1$; in fact removing on S^1 bundle over D^2, i.e. a solid torus, $S^1 \times D^2$, leaves an S^1 bundle over M^2, with boundary $S^1 \times S^1$. This latter bundle is obtained by identifying the ends of $S^1 \times I \times I$, i.e. $S^1 \times I \times \{-1\}$ with $S^1 \times I \times \{1\}$, by some orientation-reversing diffeomorphism of $S^1 \times I$, and is thus a twisted line bundle over a torus T (as the boundary is connected). If a,b are generators of $\pi_1(T)$, then $\pi_1(\partial(S^1 \times D^2))$ can be assumed to have generators a,b^2. The boundary of the meridian disk of $S^1 \times D^2$ must be of the form b^2a^m, since a is the homotopy class of the fiber. Therefore for the bundle, $\pi_1 = \mathbb{Z} \times \mathbb{Z}_2$ with generators $a, ba^{m/2}$ if m is even, $\pi_1 = \mathbb{Z}$ with generator $ba^{(m-1)/2}$ if m is odd. Note that to obtain $S^2 \tilde{\times} S^1 = \partial N(C)$, we must take m odd.

We illustrate with the manifold $\mathbb{RP}^2 \times D^2$: begin with $S^2 \times D^2$, obtained by adding 2-handles H^2_+, H^2_- to $S^1 \times B^3$ along curves C'_1, C'_2 in $(S^1 \times B^3)$ isotopic to the generator $S^1 \times \{x\}$ of $\pi_1(S^1 \times S^2)$. H^2_+ is added with framing $+1$, H^2_- with framing -1. We may assume that the attaching tubes of H^2_+, H^2_-

are interchanged by the involution $\eta: S^1 \times B^3 \to S^1 \times B^3$

$$\eta(x,y) = (-x,-y) \quad \forall (x,y) \in S^1 \times B^3 .$$

η extends to an involution of $S^2 \times D^2$, interchanging H^2_+ and H^2_-, and whose quotient is the union of $B^3 \underset{\sim}{\times} S^1$ and a 2-handle H^2 attached along the curve $C' \subset S^2 \underset{\sim}{\times} S^1$ with odd framing (needed to change from m odd initially to m even). The disc bundle over $\mathbb{RP}^2$ so obtained may be identified by reducing the structure group to the orthogonal group -- equivalently, by identification of the boundary, an S^1 bundle over $\mathbb{RP}^2$. We obtain $\mathbb{RP}^2 \times S^1$ in this case, by taking $(S^2 \times S^1 - C_1' \times D^2 - C_2' \times D^2)/\eta = S^1 \times S^1 \times I/\eta$, and gluing on a solid torus with meridian identified with the curve C' in Figure 25a.

If H^2_+, H^2_- are each attached with 0-framing, extending η in the essentially unique way gives $\mathbb{RP}^2 \underset{\sim}{\times} D^2$ as quotient -- hence $S^2 \underset{\sim}{\times} S^1$ is obtained by gluing a solid torus onto $S^1 \times S^1 \times I/\eta$ with meridian along the curve C'' in Figure 25b.

For an arbitrary such non-orientable 4-manifold X^4, it is interesting to

<u>CONJECTURE</u>. <u>The curve</u> $C' \subset S^2 \underset{\sim}{\times} S^1 \subset X^4$ <u>always bounds a locally flat embedded disc in</u> $X^4 - \operatorname{int} N(C)$ -- <u>equivalently</u>, C <u>lies on a locally flat embedded</u> $\mathbb{RP}^2 \subset X^4$, <u>where</u> $[C]$ <u>gives an orientation-reversing element of order</u> 2 <u>in</u> $\pi_1(X)$.

If our conjecture is true, X^4 contains either $\mathbb{RP}^2 \times D^2$ or $\mathbb{RP}^2 \underset{\sim}{\times} D^2$ as an embedded submanifold, according to H^2 is attached with odd or even framing. We have thus proved

<u>THEOREM 6.2</u>. 1. <u>If</u> $C \subset \mathbb{RP}^2 \times D^2 \subset X^4$, <u>then</u>

$$X^4_{ik} = (X^4 - \mathbb{RP}^2 \times D^2) \cup_{id} (\mathbb{RP}^2 \times D^2)_{ik} \quad \underline{\text{and}} \quad \tilde{X}^4_{ik} \cong \tilde{X}^4 .$$

2. <u>If</u> $C \subset \mathbb{RP}^2 \underset{\sim}{\times} D^2 \subset X^4$, <u>then</u>

$$X^4_{ik} = (X^4 - \mathbb{RP}^2 \underset{\sim}{\times} D^2) \cup_{id} (\mathbb{RP}^2 \underset{\sim}{\times} D^2)_{ik} \quad \underline{\text{and}} \quad \tilde{X}^4_{ik} \cong \tilde{X}^4 \# \tilde{Q}_{i0} .$$

Akbulut (A1) has shown that the construction on $\mathbb{RP}^2 \times D^2$ gives $\mathbb{RP}^2 \times D^2$ back again. Hence

<u>COROLLARY 6.3</u>. <u>If</u> $C \subset \mathbb{RP}^2 \times D^2 \subset X^4$, <u>then</u> $X^4_{ik} \cong X^4$.

<u>REMARK</u>. Changing the framing of the 2-handle H^2 in X by $+1$ affects a change in the framings of the 2-handles H^2_+ and H^2_- in $\tilde{X}$ by $+2$. We therefore require a diffeomorphism from $\tilde{X}$ to $\tilde{X}$ which commutes with the covering transformation and achieves this change in framing. The map of $S^1 \times S^2 = \partial\tilde{M}_{ik}$ to itself, which is two complete rotations of S^2 in traversing the S^1 factor, gives the result of framing but it is difficult and probably impossible in general to extend this to an equivariant diffeomorphism of $\tilde{X}$.

The structure of $\tilde{Q}_{i0} = H^0 \cup H^1 \cup H^2_{ik} \cup \bar{H}^2_{ik} \cup H^3 \cup H^4$ is completely determined ((Mo)) by $U_{ik} = H^0 \ H^1 \cup H^2_{ik} \cup H^2_{ik}$. By codimension 3 isotopic unknotting, removing $H^3 \cup H^4 \cong S^1 \times B^3$ from $\tilde{Q}_{i0}$ is equivalent to removing $C_G \times B^3$, a tubular neighborhood of C_G. Hence there are three commuting involutions on U_{ik} (a homotopy $S^2 \times D^2$), obtained by restricting each of G,H and HG. The restrictions of G and H are free, whereas GH has S_{GH} as a knotted 2-sphere of fixed points. On $C_G \times B^3$, the involutions are

$$G:(x,y) \longrightarrow (x,-y) \quad \forall (x,y) \in S^1 \times B^3 \cong C_G \times B^3$$

$$H:(x,y) \longrightarrow (-x,-y)$$

$$GH:(x,y) \longrightarrow (-x,y) .$$

The situation is depicted in Figure 26: $U_{i0}/GH \cong S^2 \times D^2$, $U_{i1}/GH = S^2 \times D^2 \# \tilde{\Sigma}$, $\bar{U}_{ik} = U_{ik}/G$ is a homotopy $\mathbb{R}P^2 \times D^2$, and $\bar{\bar{U}}_{ik} = Q_{ik} - S^1 \underset{\sim}{\times} B^3 = (\mathbb{R}P^2 \underset{\sim}{\times} D^2)_{ik}$. $U^*_{ik} = \bar{U}_{ik}/H_1 = \bar{U}_{ik}/G_2$ is a homotopy $\mathbb{R}P^2 \times D^2$.

<u>PROBLEM</u>. How are the manifolds U_{ik}, $\bar{U}_{ik}$, U^*_{ik} and $S^2 \times D^2$ related? In particular, (i) Is $\bar{U}_{ik}$ standard? This would give $\tilde{Q}_{i0} \cong S^4$ $i = 0,1$, and thus there would be an exotic involution of S^4 (which is known to be possible (FS)).

(ii) Is $\bar{U}_{ik}$ diffeomorphic with $\mathbb{R}P^2 \times D^2$? -- in which case we again obtain $\tilde{Q}_{i0} = S^4$. If not, we obtain an exotic $\mathbb{R}P^2 \times D^2$, which is not constructed by Cappell and Shaneson's methods.

Fukuhara (F) has investigated involutions on homotopy 4-spheres with a circle of fixed points. His examples arise by gluing together two contractible 4-manifolds by an involution on the boundary, which represents the boundary as the 2-fold cover of S^3 branched over a knot K. Removing the circle of fixed points gives a free involution on a homotopy $S^2 \times D^2$, with a natural homotopy equivalence of the quotient to $\mathbb{R}P^2 \times D^2$. Fukuhara shows that an obstruction to homotoping this homotopy equivalence to a diffeomorphism is given by the signature of the knot K.

Hence we have obtained many more examples of such involutions, and moreover on the standard 4-sphere. Fukuhara constructs an exotic homotopy equivalence to $RP^2 \times D^2$ using the Brieskorn sphere (2,3,13), but does not prove that the homotopy sphere constructed is S^4. This can be shown to be the case using the link calculus, an exercise we leave to the reader.

The signatures of the knots in our examples are computable, but we have not carried out the computation. Recall that $RP^2 \times D^2$ does admit exotic self-homotopy equivalences (see Akbulut's paper in these proceedings).

Cappell and Shaneson's construction may also be applied to 4-manifolds W^4 such that $\pi_1(W^4)$ contains an element x of order 2, which is orientation-preserving in W^4. Let C be an embedded circle such that $[C] = x$ -- then

$N(C) \cong S^1 \times B^3$, and there is an embedded circle $C' \subset \partial N(C) \cong S^1 \times S^2$ bounding a Möbius band in $N(C)$, and such that $[C'] = x^2 = 1$. Remove $N(C)$ and replace with a punctured 3-torus bundle, the mapping torus corresponding to any matrix $A \in SL(3,\mathbb{Z})$ such that $\det(A-1) = \pm 1$, and $\det(A+1) = \pm 1$: the only possibilities for A are D_0 or D_0^{-1} and D_1 or D_1^{-1}. Gluing $T^3 - R' \times_{g\phi_{ik}} S^1$ into $W^4 - \operatorname{int} N(C)$, there are four possible outcomes as before, denoted W^4_{ik}, $i = 0,1$, $k = 0,1$. Assume now that x induces a non-zero element in $H_1(W^4)$. If $p: \tilde{W}^4 \to W^4$ is a double covering projection, with $p^{-1}(C)$ connected and null homotopic in $\tilde{W}^4$, then as before $p^{-1}(C')$ has two connected components, each of which bounds a pl-locally flat embedded disc D^2 in $\tilde{W}^4 - \operatorname{int} p^{-1}(N(C)) \cong \tilde{W}^4 - \operatorname{int} S^1 \times B^3$. D^2 may be considered the core of a 2-handle attached to $p^{-1}(N(C))$ with some framing $m \in \mathbb{Z}$. We again obtain $\tilde{W}^4_{ik} \cong \tilde{W}^4$ or $\tilde{W}^4 \# Q_{i0}$ respectively as m is odd or m is even -- and again it is interesting to conjecture that C lies in a p.l. locally-flat embedded $\mathbb{RP}^2 \subset W^4$.

Examples for such a W^4 are all the orientable D^2 bundles over $\mathbb{RP}^2$, of which there are infinitely many.

Exotic behavior may also arise from this alternative construction. Cappell and Shaneson have been considering their modification on $Q^3 \times I$, where Q^3 is quaternionic space (arising for example as the boundary of a neighborhood of an embedded $\mathbb{RP}^2$ in S^4). Note that the center of $\pi_1(Q^3)$ is an element of order 2.

The 4-fold cover of $Q^3 \times I$, corresponding to the epimorphism

$$\pi_1(Q^3 \times I) \longrightarrow \mathbb{Z}_2 \times \mathbb{Z}_2$$

is $\mathbb{RP}^3 \times I$. As $\mathbb{RP}^3$ contains a 1-sided $\mathbb{RP}^2$, there are four intersecting copies of $\mathbb{RP}^2$ in $\mathbb{RP}^3 \times \{\frac{1}{2}\}$ permuted by the action of $\mathbb{Z}_2 \times \mathbb{Z}_2$ as covering transformation group, and whose orientation-reversing curves are projected to a single embedded curve $C \subset Q^3 \times \{\frac{1}{2}\}$. Using C to modify $Q^3 \times I$ by Cappell and Shaneson's technique, any manifold X^4_{ik} so constructed has a 4-fold covering space the manifold obtained from $\mathbb{RP}^3 \times I$ by carrying out the modification on the four curves in the pre-image of C. However, a neighborhood of $\mathbb{RP}^2$ in $\mathbb{RP}^3$ is a twisted line bundle, whose product with the unit interval is an orientable D^2 bundle over $\mathbb{RP}^2$ with boundary $\mathbb{RP}^3 \# \mathbb{R}^3$. The double cover of the boundary is thus $S^1 \times S^2$, and thus is obtained from $S^1 \times B^3$ by adding two 2-handles along curves $S^1 \times \{x\}$, $S^1 \times \{y\}$ with framing $m=0$. (Note that here the 2-handles both have the same framing m, since the covering transformation is orientation-preserving.)

So the 8-fold cover of X^4_{ik} is $S^3 \times I \#_4 \tilde{Q}_{i0}$, i.e. $\#_4 \tilde{Q}_{i0}$ with two open 4-cells removed. Hence if $\tilde{Q}_{i0}$ is <u>not</u> diffeomorphic to S^4, we find that X^4_{ik} is another counterexample to the 4-dimensional s-cobordism theorem. Note that the Cappell-Shaneson construction applied to $\mathbb{RP}^3 \times I$ will also give such a counterexample in this case.

On the other hand, if $\tilde{Q}_{i0}$ is diffeomorphic to S^4 but X^4_{ij} is not diffeomorphic to $Q^3 \times I$, then there is an exotic free action of $\pi_1(Q^3)$ on $S^3 \times I$, i.e. which is not smoothly equivalent to the standard orthogonal action. This can be extended to an action of $\pi_1(Q^3)$ on S^4 with two fixed points.

CHARACTERISTIC SUBMANIFOLDS

Given a free involution σ on a closed 4-manifold V^4, a characteristic submanifold for σ is a submanifold $M^4 \subset V^4$ with ∂M^4 connected, such that $M^4 \cap \sigma(M^4) = \partial M^4$ (hence $\sigma|_{\partial M^4}$ is a free involution on ∂M^4) and $V^4 = M^4 \cup_\sigma \sigma(M^4)$. The free involutions on the homotopy 4-spheres considered in the construction of Q_{ik} and $\mathbb{RP}^4$ all have characteristic submanifolds M^4 built with handles of index ≤ 2. This decomposes the quotient as the union of M^4 and he mapping cylinder of the involution restricted to ∂M^4.

Let W^4 be any closed 4-manifold with $H_1(W^4, \mathbb{Z}_2) \neq 0$ -- for example, any non-orientable closed 4-manifold.

THEOREM 6.4. *$W^4 = M^4 \cup N(M^3)$, where M^4 has a handle decomposition consisting entirely of 0-,1- and 2-handles, and $N(M^3)$ is a neighborhood of M^3, a closed connected 3-manifold 1-sided in W^4. ($N(M^3)$ is the mapping cylinder of a free involution $\sigma: \partial M^4 \longrightarrow \partial M^4$, and $M^3 = \partial M^4/\sigma$).*

COROLLARY 6.5. *If V^4 is a closed 4-manifold with free involution σ, then $V^4 = M^4 \cup_\sigma M^4$, where M^4 has handles of index ≤ 2 only, and σ induces a free involution on ∂M^4 which is connected. M^4 is a characteristic submanifold for σ.*

PROOF OF THEOREM. 1. If $H_1(W^4, \mathbb{Z}_2) \neq \{0\}$, there is a continuous map $f: W^4 \to \mathbb{RP}^5$ such that $f_*: H_1(W^4, \mathbb{Z}_2) \to H_1(\mathbb{RP}^5, \mathbb{Z}_2)$ is onto. This follows from obstruction theory, since $\pi_i(\mathbb{RP}^5) = 0, 2 \leq i \leq 4$. There is thus no obstruction to extending an appropriate map of the 1-skeleton of W^4, skeleton by skeleton.

We may suppose that f is transverse to $RP^4 \subset \mathbb{R}^5$, in which case $f^{-1}(\mathbb{RP}^4)$ is the union of closed 3-manifolds in W^4. If there is more than one component in $f^{-1}(\mathbb{RP}^4)$, let M^3_1, M^3_2 be two such, and join these in W^4 by an arc λ with $\operatorname{int}\lambda \cap f^{-1}(\mathbb{RP}^4) = \emptyset$. The ends of $f(\lambda)$ lie in $\mathbb{RP}^4$, and thus $f(\lambda)$ may be homotoped into $\mathbb{RP}^4$. Surgery on a neighborhood of λ enables this homotopy to be realized, i.e. $M^3_1 \cup M^3_1 \subset f^{-1}(\mathbb{RP}^4)$ is replaced by $M^3_1 \#_{\partial N(\lambda)} M^3_2$. Thus by suitably homotoping f, we may assume that $f^{-1}(\mathbb{RP}^4) = M^3$ is connected.

2. M^3 is one-sided in W^4: for suppose M^3 is two sided. Then any loop in $N(M^3)$ meets M^3 an even number of times. However, $N(M^3) = f^{-1}(N(\mathbb{RP}^4))$ since f_* is onto we may assume without loss of generality that some loop in

W^4 is mapped to a loop in $N(\mathbb{R}P^4)$ meeting $\mathbb{R}P^4$ an odd number of times. This is a contradiction.

3. Let $Q^4 = W^4 - \text{int}\, N(M^3)$, and suppose H^3 is a 3-handle of Q^4. Then $H^3 \cap \partial N(M^3) = B^3_{11} \cup B^3_{21}$, where B^3_{11} and B^3_{21} may be assumed disjoint and project respectively to disjoint balls $p(B^3_{11}) = \bar{B}^3_1$, $p(B^3_{21}) = \bar{B}^3_2$, $p: \partial N(M^3) \to M^3$ being the projection map. Hence $p^{-1}(\bar{B}^3_1) = B^3_{11} \cup B^3_{12}$, $p^{-1}(\bar{B}^3_2) = B^3_{21} \cup B^3_{22}$ whence B^3_{11}, B^3_{12}, B^3_{21}, B^3_{22} are mutually disjoint 3-balls in $\partial N(M^3)$ (Figure 27a).

Thus there are embeddings $\phi_i : B^3 \times [-1,1] \to W^4$, $i=1,2$ such that

$$\phi_i(B^3 \times \{-1\}) = B^3_{i2} \; , \; \phi_i(B^3 \times \{0\}) = \bar{B}^3_i \; , \; \phi_i(B^3 \times \{1\}) = B^3_{i1} \; .$$

Let $\bar{M}^3 = M^3 \cup \partial(H^3 \cup \phi_1(B^3 \times [0,1]) \cup \phi_2(B^3 \times [0,1])) - \text{int}\,\bar{B}^3_1 - \text{int}\,\bar{B}^3_2 \cong M^3 \# S^2 \times S^1$ since $H_* = H^3 \cup \phi_1(B^3 \times [0,1]) \cup \phi_2(B^3 \times [0,1])$ is an embedded 4-ball with boundary $\bar{B}^3_1 \cup \bar{B}^3_2 \cup S^2 \times I$. Now let U^4 be a neighborhood in Q^4 of $(\partial H^3 - \text{int}\, B^3_{11} - \text{int}\, B^3_{12}) \cong S^2 \times I$, and let $V = H^3 - \text{int}\, U^4$ (Figure 27c). Thus $V \cap \partial N(M^3) = E^3_{12} \cup E^3_{12}$ where $E^3_{11} \subset \text{int}\, B^3_{11}$, $E^3_{12} \subset \text{int}\, B^3_{12}$ and E^3_{11}, E^3_{12} are embedded 3 balls. We may assume without loss of generality that there is a 3-ball $\bar{B}^3 \subset \text{int}\, B^3$ such that $\phi_1(\bar{B}^3 \times \{1\}) = E^3_{11}$, $\phi_2(\bar{B}^3 \times \{1\}) = E^3_{21}$. Finally, take $H^1 = V \cup \phi_1(\bar{B}^3 \times [-1,1]) \cup \phi_2(\bar{B}^3 \times [-1,1])$, and let

$$\bar{Q}^4 = (Q^4 \cup H^1) - \text{int}\, U^4 - \text{int}(U^4 \cap \partial Q^4) \; . \qquad \text{(Figure 27d)}$$

Clearly $N(\bar{M}^3) = W^4 - \text{int}\, \bar{Q}^4$ is a neighborhood of the 1-sided 3-manifold $\bar{M}^3$, and $\bar{Q}^4$ has exactly the handle decomposition of Q^4 except that the 3-handle H^3 has been removed and replaced with the 1-handle H^1. Continuing in this way, we arrive at a decomposition of W^4 satisfying the properties desired for the theorem.

Similarly, one can prove the following.

THEOREM. Suppose M^4 is a smooth closed 4-manifold, and $P^3 \subset M^4$ is a smoothly embedded two-sided submanifold which is non-separating. Let $Q^4 = M^4 - \text{int}\, N(P^3)$. Then we can modify P^3 to $\bar{P}^3$, Q^4 to $\bar{Q}^4$ such that $\bar{Q}^4$ has a handle decomposition with 0-, 1-, and 2-handles only.

APPENDIX

Conjugacy in $SL(3,\mathbb{Z})$

1. Let $X = \begin{bmatrix} x & a & b \\ d & y & c \\ e & f & z \end{bmatrix} \in SL(3,\mathbb{Z})$.

$$C_X(t) = \text{Det}(tI - X) = t^3 - At^2 + Bt - 1$$

$$\mu = \text{g.c.d. } \{a, b, c, d, e, f, x-y, y-z\} \; .$$

THEOREM (A1): 1. *If* ± 1 *is an eigenvalue of* X, *then* X *is conjugate to*

$$\begin{bmatrix} \pm 1 & 0 & 0 \\ d' & y' & c' \\ e' & f' & z' \end{bmatrix}$$

2. *If* ± 1 *is not an eigenvalue of* X *then* X *is conjugate to* $\lambda I + \mu Y$ *where*

$$Y = \begin{bmatrix} 0 & 0 & 1 \\ m & 0 & 0 \\ n & p & q \end{bmatrix}.$$

Hence $C_X(t) \equiv C_{\lambda I}(t) \equiv (t-\lambda)^3 \bmod \mu$. This implies $3\lambda \equiv A$, $3\lambda^2 \equiv B$, $\lambda^3 \equiv 1 \bmod \mu$, and in particular $A^2 \equiv 3B \bmod \mu$, i.e. μ divides $A^2 - 3B$. Therefore there are only finitely many choices for μ, relative to $C_X(t)$ fixed.

PROOF: Case 1. Assume $a,b \neq 0$ and let $\lambda = \text{g.c.d.}\ \{a,b\}$, with $\delta a + \varepsilon b = \gamma$.

$$\begin{bmatrix} 0 & \frac{a}{\gamma} & \frac{b}{\gamma} \\ 0 & -\varepsilon & \delta \\ 1 & 0 & 0 \end{bmatrix} \begin{bmatrix} x & a & b \\ d & y & c \\ e & f & z \end{bmatrix} \begin{bmatrix} 0 & 0 & 1 \\ \delta & -\frac{b}{\gamma} & 0 \\ \varepsilon & \frac{a}{\gamma} & 0 \end{bmatrix} = \begin{bmatrix} 0 & \frac{a}{\gamma} & \frac{b}{\gamma} \\ 0 & -\varepsilon & \delta \\ 1 & 0 & 0 \end{bmatrix} \begin{bmatrix} \gamma & 0 & x \\ * & * & * \\ * & * & * \end{bmatrix} = \begin{bmatrix} * & * & * \\ * & * & * \\ \gamma & 0 & x \end{bmatrix}$$

Now
$$\begin{bmatrix} 0 & 0 & -1 \\ 0 & -1 & 0 \\ -1 & 0 & 0 \end{bmatrix} \begin{bmatrix} * & * & * \\ * & * & * \\ \gamma & 0 & x \end{bmatrix} \begin{bmatrix} 0 & 0 & -1 \\ 0 & -1 & 0 \\ -1 & 0 & 0 \end{bmatrix} = \begin{bmatrix} x & 0 & \gamma \\ * & * & * \\ * & * & * \end{bmatrix}$$

Case 2. $a = 0 = b$. Then $x = \pm 1$ since $\det X = 1$ and ± 1 is an eigenvalue of X.

Case 3. $b = 0$, $a \neq 0$.

$$\begin{bmatrix} -1 & 0 & 0 \\ 0 & 0 & -1 \\ 0 & -1 & 0 \end{bmatrix} \begin{bmatrix} x & a & 0 \\ d & y & c \\ e & f & z \end{bmatrix} \begin{bmatrix} -1 & 0 & 0 \\ 0 & 0 & -1 \\ 0 & -1 & 0 \end{bmatrix} = \begin{bmatrix} x & 0 & z \\ e & z & f \\ d & c & y \end{bmatrix}.$$

So in all cases we are reduced to

Case 4. $a = 0$, $b \neq 0$. If b does not divide c, then $\text{g.c.d}\ \{b,c\} < |b|$. Exactly as in Case 1 we can replace b by $\text{g.c.d.}\ \{b,c\}$. Eventually we obtain that b divides c (e.g. if $b = \pm 1$). In this case

$$\begin{bmatrix} 1 & 0 & 0 \\ -\frac{c}{b} & 1 & 0 \\ 0 & 0 & 1 \end{bmatrix} \begin{bmatrix} x & 0 & b \\ d & y & c \\ e & f & z \end{bmatrix} \begin{bmatrix} 1 & 0 & 0 \\ \frac{c}{b} & 1 & 0 \\ 0 & 0 & 1 \end{bmatrix} = \begin{bmatrix} x & 0 & b \\ d' & y & 0 \\ e' & f & z \end{bmatrix} .$$

So we can assume that $a = c = 0$. Next

$$\begin{bmatrix} 1 & 0 & 0 \\ 0 & 1 & 0 \\ 0 & 1 & 1 \end{bmatrix} \begin{bmatrix} x & 0 & b \\ d & y & 0 \\ e & f & z \end{bmatrix} \begin{bmatrix} 1 & 0 & 0 \\ 0 & 1 & 0 \\ 0 & -1 & 1 \end{bmatrix} = \begin{bmatrix} x & -b & b \\ d & y & 0 \\ d+e & f+y-z & z \end{bmatrix}$$

and
$$\begin{bmatrix} 1 & 0 & 0 \\ 0 & 0 & 1 \\ 0 & -1 & 0 \end{bmatrix} \begin{bmatrix} x & -b & b \\ d & y & 0 \\ d+e & f+y-z & z \end{bmatrix} \begin{bmatrix} 1 & 0 & 0 \\ 0 & 0 & -1 \\ 0 & 1 & 0 \end{bmatrix} = \begin{bmatrix} x & b & b \\ d+e & z & z-f-y \\ -d & 0 & y \end{bmatrix}$$

By the usual argument we can replace b by g.c.d. $\{b, z-f-y\}$ unless b divides $z-f-y$. Similarly

$$\begin{bmatrix} 1 & 0 & 0 \\ 0 & 1 & 1 \\ 0 & 0 & 1 \end{bmatrix} \begin{bmatrix} x & 0 & b \\ d & y & 0 \\ e & f & z \end{bmatrix} \begin{bmatrix} 1 & 0 & 0 \\ 0 & 1 & -1 \\ 0 & 0 & 1 \end{bmatrix} = \begin{bmatrix} x & 0 & b \\ e+d & f+y & z-f-y \\ e & f & z-f \end{bmatrix} .$$

As $b \mid z-f-y$,

$$\begin{bmatrix} 1 & 0 & 0 \\ -\frac{(z-f-y)}{b} & 1 & 0 \\ 0 & 0 & 1 \end{bmatrix} \begin{bmatrix} x & 0 & b \\ e+d & f-y & z-f-y \\ e & f & z-f \end{bmatrix} \begin{bmatrix} 1 & 0 & 0 \\ \frac{z-f-y}{b} & 1 & 0 \\ 0 & 0 & 1 \end{bmatrix}$$

$$= \begin{bmatrix} x & 0 & b \\ e+d+\frac{(f+y-x)(z-f-y)}{b} & f+y & 0 \\ e+\frac{(z-f-y)}{b} f & f & z-f \end{bmatrix} \quad (+)$$

By the argument above, we can replace b by g.c.d. $\{b, (z-f)-f-(f+y)\}$ = g.c.d. $\{b, z-3f-y\}$. So without loss of generality, $b \mid z-3f-y$ and $b \mid z-f-y$. Therefore $b \mid 2f$, $b \mid 2(z-y)$ follows. Dually, using x, d, y instead of z, f, y with the matrix X in the form with $a = c = 0$, we obtain that $b \mid 2d$, $b \mid 2(x-y)$ without loss of generality. Finally using $(+)$, we obtain that b can be replaced by g.c.d.

$\{b, x - \left(e + d + \frac{(f+y-x)(z-f-y)}{b}\right) - (f+y)\}$. So without loss of generality,

$b|(x-f-y) - e - d - \frac{(f+y-x)(z-f-y)}{b}$. But $b|2f$, $2(x-y)$, $2d$, $z-f-y$. Hence $b|2e$ follows.

Case 5. Suppose we eventually obtain ± 1 -- then we may assume $b=1$, since

$$\begin{bmatrix} -1 & 0 & 0 \\ 0 & -1 & 0 \\ 0 & 0 & 1 \end{bmatrix} \begin{bmatrix} x & 0 & b \\ d & y & 0 \\ e & f & z \end{bmatrix} \begin{bmatrix} -1 & 0 & 0 \\ 0 & -1 & 0 \\ 0 & 0 & 1 \end{bmatrix} = \begin{bmatrix} x & 0 & -b \\ d & y & 0 \\ -e & -f & z \end{bmatrix}.$$

Then
$$\begin{bmatrix} 1 & 0 & 0 \\ 0 & 1 & 0 \\ x-y & 0 & 1 \end{bmatrix} \begin{bmatrix} x & 0 & 1 \\ d & y & 0 \\ e & f & z \end{bmatrix} \begin{bmatrix} 1 & 0 & 0 \\ 0 & 1 & 0 \\ y-x & 0 & 1 \end{bmatrix} = \begin{bmatrix} y & 0 & 1 \\ d & y & 0 \\ e' & f & z+x-y \end{bmatrix}$$

$= \lambda I + \mu Y$ as desired, with $\lambda = y$, $\mu = 1$.

Case 6. Assume $b > 1$ and $b|2(x-y)$, $2(y-z)$, $2d$, $2f$, $2e$. If b is odd, then $b|(x-y)$, $(y-z)$, d, f, e. Hence

$$\begin{bmatrix} x & 0 & b \\ d & y & 0 \\ e & f & z \end{bmatrix} = \begin{bmatrix} y+rb & 0 & b \\ mb & y & 0 \\ mb & pb & y+qb \end{bmatrix}$$

$$\begin{bmatrix} 1 & 0 & 0 \\ 0 & 1 & 0 \\ r & 0 & 1 \end{bmatrix} \begin{bmatrix} y+rb & 0 & b \\ mb & y & 0 \\ n'b & pb & y+qb \end{bmatrix} \begin{bmatrix} 1 & 0 & 0 \\ 0 & 1 & 0 \\ -r & 0 & 1 \end{bmatrix} = \begin{bmatrix} y & 0 & b \\ mb & y & 0 \\ n'b & pb & y+(q-r)b \end{bmatrix}$$

$= \lambda I + \mu Y$ where $\lambda = y$ and $\mu = b$.

If b is even and $b|(x-y)$, $(y-z)$, d, f, e, we also obtain the result. So we can assume $b \nmid (x-y)$ or $b \nmid d$ say. Since $b|(x-y+d)$, if $b \nmid (x-y)$ then $b \nmid d$. So without loss of generality, either $b \nmid d$, e, or f. Suppose say $b \nmid d$. We now use g.c.d. $\{d,e\}$ to find a new value of b and a new matrix.

Eventually we reach $\begin{bmatrix} x' & 0 & b' \\ d' & y' & 0 \\ e' & f' & z' \end{bmatrix}$ where b' d, and $b'|2d'$, $2e'$, $2f'$,

$2(x'-y')$, $2(y'-z')$. Again we can suppose b' is even, say $b' = 2b''$. Then

$$\begin{bmatrix} x' & 0 & b' \\ d' & y' & 0 \\ e' & f' & z' \end{bmatrix} \equiv \begin{bmatrix} x' & 0 & 0 \\ 0 & x' & 0 \\ 0 & 0 & x' \end{bmatrix} \bmod b''$$

Therefore $\begin{bmatrix} x & 0 & b \\ d & y & 0 \\ e & f & z \end{bmatrix} \equiv \begin{bmatrix} x' & 0 & 0 \\ 0 & x' & 0 \\ 0 & 0 & x' \end{bmatrix} \bmod b'' \Rightarrow b''|b$. But

$2b''|d$ and $(b,d) = \frac{b}{2}$. We conclude that $b''|\frac{b}{2} \Rightarrow 2b''|b$. As $2b''|d$, $2b'' \neq b$ and so $b' < b$. Eventually the procedure must terminate -- in fact, reversing the argument shows $\frac{b}{2}|b' = 2b''$ and so $b' = \frac{b}{2}$ must be true. Hence $b'|d', e', f', x'-y', y'-z'$ follows immediately.

REMARKS. 1. Since X is conjugate to $\lambda I + \mu Y$, $X \equiv \lambda I \bmod \mu$ and so $\mu|a, b, c, d, e, f, x-y, y-z$. On the other hand, if $\nu|a, b, c, d, e, f, x-y, y-z$, then $X \equiv \gamma I \bmod \nu$. Hence $\lambda I + \mu Y \equiv \gamma I \bmod \nu$, i.e. $\mu Y \equiv (\gamma - \lambda) I$. Hence $\mu \equiv 0 \bmod \nu \Rightarrow \nu|\mu$. So $\mu = \text{g.c.d.}\{a, b, c, d, e, f, x-y, y-z\}$.

We now set $X = \lambda I + \mu Y = \begin{bmatrix} \lambda & 0 & \mu \\ m\mu & \lambda & 0 \\ n\mu & p\mu & \lambda+q\mu \end{bmatrix}$. Note that μ and $\lambda \pmod \mu$ are invariants of the conjugacy class of X.

2. If $p=1$, then

$$\begin{bmatrix} \lambda & 0 & \mu \\ m\mu & \lambda & 0 \\ n\mu & \mu & \lambda+q\mu \end{bmatrix} \sim \begin{bmatrix} \lambda & 0 & \mu \\ (m+n)\mu & \lambda+\mu & (q-1)\mu \\ n\mu & \mu & \lambda+(q-1)\mu \end{bmatrix}$$

$$\sim \begin{bmatrix} \lambda+\mu & 0 & \mu \\ (m+n+q-1)\mu & \lambda+\mu & (q-1)\mu \\ (n+q-2)\mu & \mu & \lambda+(q-2)\mu \end{bmatrix}$$

$$\sim \begin{bmatrix} \lambda+\mu & 0 & \mu \\ m'\mu & \lambda+\mu & 0 \\ n'\mu & \mu & \lambda+(q-2)\mu \end{bmatrix}$$

Hence we can obtain $0 \leq \lambda \leq \mu - 1$. In this case q, m, n are completely determined by the characteristic polynomial, and so all such matrices are conjugate, with possibly a finite number of choices for λ with $0 \leq \lambda \leq \mu - 1$.

3. $C_X(t) = \det(tI - X) = \det \begin{bmatrix} t-\lambda & 0 & -\mu \\ -m\mu & t-\lambda & 0 \\ -n\mu & -p\mu & t-\lambda-q\mu \end{bmatrix}$

$$= (t-\lambda)^3 - q\mu(t-\lambda)^2 - mp\mu^3 - n\mu^2(t-\lambda)$$

$$= t^3 - at^2 + bt - 1 .$$

Thus

$$a = 3\lambda + q\mu$$

$$b = 3\lambda^2 + 2q\mu\lambda - n\mu^2$$

$$1 = \lambda^3 + q\mu\lambda^2 - n\mu^2\lambda + mp\mu^3 .$$

NOTE. Given λ,μ then q,n and mp are determined in terms of a,b. The difficulty is that m,p are only given as a factorization of the numbers mp, not explicitly.

If A $SL(3;\mathbb{Z})$ has characteristic polynomial $f_a(x) = x^3 - ax^2 + (a-1)x - 1$, we can proceed further:

By Theorem A1, we may assume A is conjugate in $SL(3,\mathbb{Z})$ to a matrix of the form

$$A' = \begin{bmatrix} \lambda & 0 & \mu \\ m\mu & \lambda & 0 \\ n\mu & p\mu & \lambda+q \end{bmatrix} , \quad \mu > 0 .$$

LEMMA A2: In A', we may assume $\mu=1$.

PROOF:

$$\lambda^3 + \lambda^2 q\mu + mp\mu^3 - n\mu^2 = 1 \qquad \text{(i)}$$

$$3\lambda + q\mu = a \qquad \text{(ii)}$$

$$3\lambda^2 + 2\lambda q\mu - n\mu^2 = a - 1 \qquad \text{(iii)}$$

Suppose μ is even. Then $\lambda^3 \equiv 1 \pmod 2 \Rightarrow \lambda \equiv 1 \pmod 2$ from (i)

$\Rightarrow a \equiv 3\lambda \equiv 1 \pmod 2$ from (ii)

$\Rightarrow a - 1 \equiv 3\lambda^2 \equiv 1 \pmod 2$ from (iii)

which is a contradiction. Thus μ is odd. Taking congruences $\bmod \mu$ gives

$$\lambda^3 \equiv 1 \pmod \mu , \quad 3\lambda \equiv a \pmod \mu , \quad 3\lambda^2 \equiv (a-1) \pmod \mu$$

$$\Rightarrow 3\lambda^2 \equiv 3\lambda - 1 \pmod \mu$$

$$3 \equiv 3\lambda^3 \equiv (3\lambda-1)\lambda \equiv 3\lambda^2 - \lambda \pmod \mu$$

$$\Rightarrow 3\lambda^2 \equiv 3\lambda - 1 \equiv 3 + \lambda \pmod \mu$$

$$\Rightarrow 2\lambda \equiv 4 \pmod \mu$$

$$\Rightarrow 2 \equiv 2\lambda^3 \equiv \lambda^2 \cdot 2\lambda \equiv 4\lambda^2 \equiv (2\lambda)^2 \equiv 16 \pmod \mu$$

$$\Rightarrow 14 \equiv 0 \bmod \mu , \text{ and } \mu = 1 \text{ or } 7 .$$

If $\mu = 7$, $\lambda \equiv 2 \pmod \mu \Rightarrow \lambda = 7s + 2$ for some $s \in \mathbb{Z}$,

Then

$$a = 3(7s+2) + 7q = 21s + 6 + 7q$$

$$a - 1 = 3(7s+2)^2 + 2 \cdot 7q(7s+2) - 7^2 n$$

$$1 = (7s+2)^3 + (7s+2)^2 \cdot 7q + 7^3 mp - 7^2 n(7s+2)$$

$$\Rightarrow a \equiv 21s + 6 + 7q \pmod{49} \qquad \text{(a)}$$

$$a - 1 \equiv 84s + 12 + 28q \pmod{49} \qquad \text{(b)}$$

$$1 \equiv 84s + 8 + 28q \pmod{49} \qquad \text{(c)}$$

$$\Rightarrow a - 2 \equiv 4 \pmod{49} \quad \text{by} \quad \text{(b)} - \text{(c)}$$

$$\Rightarrow 3a + 1 \equiv 12 \pmod{49} \quad \text{by} \quad 4\text{(a)} - \text{(b)}$$

$$\Rightarrow 19 \equiv 12 \pmod{49}, \quad \text{a contradiction .}$$

Thus $\mu=1$, giving the following possible representatives of a given class:

$$\alpha_1 \begin{bmatrix} 0 & 0 & 1 \\ m & \lambda & 0 \\ n+\lambda^2-\lambda(2\lambda+q) & p & 2\lambda+q \end{bmatrix} = \begin{bmatrix} 1 & 0 & 0 \\ 0 & 1 & 0 \\ \lambda & 0 & 1 \end{bmatrix} \begin{bmatrix} \lambda & 0 & 1 \\ m & \lambda & 0 \\ n & p & q+\lambda \end{bmatrix} \begin{bmatrix} 1 & 0 & 0 \\ 0 & 1 & 0 \\ -\lambda & 0 & 1 \end{bmatrix}$$

$$\alpha_2 \begin{bmatrix} 0 & 0 & 1 \\ m+n & \lambda+p & (a-\lambda)-(\lambda+p) \\ n & p & (a-\lambda)-p \end{bmatrix} = \begin{bmatrix} 1 & 0 & 0 \\ 1 & 1 & 0 \\ 0 & 0 & 1 \end{bmatrix} \begin{bmatrix} 0 & 0 & 1 \\ m & \lambda & 0 \\ n & p & a-\lambda \end{bmatrix} \begin{bmatrix} 1 & 0 & 0 \\ -1 & 1 & 0 \\ 0 & 0 & 1 \end{bmatrix}$$

NOTE. We have replaced q by $a = q - 3\lambda$ so trace $= a$.

$$\alpha_3 \begin{bmatrix} 0 & 0 & 1 \\ (m+n)+(a-2\lambda+p)(\lambda+p) & \lambda+p & 0 \\ n+(a-2\lambda+p) & p & a-(\lambda+p) \end{bmatrix}$$

$$= \begin{bmatrix} 1 & 0 & 0 \\ -(a-2\lambda+p) & 1 & 0 \\ 0 & 0 & 1 \end{bmatrix} \begin{bmatrix} 0 & 0 & 1 \\ m+n & \lambda+p & (a-2\lambda+p) \\ n & p & a-(\lambda+p) \end{bmatrix} \begin{bmatrix} 1 & 0 & 0 \\ (a-2\lambda+p) & 1 & 0 \\ 0 & 0 & 1 \end{bmatrix} .$$

THEOREM A3. Suppose A $SL(3,\mathbb{Z})$ has characteristic polynomial $f_a(x) = x^3 - ax^2 + (a-1)x - 1$. Then A is conjugate to a matrix of form

$$A_{a,\lambda,p} = \begin{bmatrix} 0 & 0 & 1 \\ m & \lambda & 0 \\ n & p & a-\lambda \end{bmatrix} \qquad \begin{array}{l} n = \lambda(a-\lambda) - (a-1) \\ \\ mp = 1 + n\lambda \end{array}$$

Further, by α_3 we may assume $0 \leq \lambda < p$ (it is easy to obtain $p \geq 0$).

REMARK. We could alternatively arrange $0 \leq \lambda < m$. That $m \neq 0 \neq p$ follows from

LEMMA A4. $f(x) = x^3 - ax^2 + bx - 1$, $a,b \in \mathbb{Z}$, is irreducible over $\mathbb{Q}$ if $(a-b) \equiv 1 \pmod 2$.

PROOF: Define $\phi: \mathbb{Z} \times \mathbb{Z} \to \mathbb{Z}$ by $\phi(p,q) = p^3 - ap^2q + bq^2p - q^3$. Then $\phi(p,q) \equiv 0 \pmod 2$ iff $p \equiv 0 \equiv q \pmod 2$. Suppose p,q are nonzero integers satisfying $f(\frac{p}{q}) = 0$ -- without loss of generality we may assume p and q are coprime. Thus $\phi(p,q) = q^3 f(\frac{p}{q}) = 0$, a contradiction.

COROLLARY A5. $m \neq 0 \neq p$ in Theorem A3.

PROOF:

$$n\lambda + 1 = 0 = \lambda(\lambda(a-\lambda) - (a-1)) + 1 = 0$$

$$\Rightarrow -\lambda^3 + a\lambda^2 - (a-1)\lambda + 1 = 0$$

and thus $\lambda \notin \mathbb{Z}$ by the Lemma, a contradiction.

By the remark above, if $|m| = 1$ or $|p| = 1$, we may assume $\lambda = 0$, and we obtain $\begin{bmatrix} 0 & 0 & 1 \\ 1 & 0 & 0 \\ -(a-1) & 1 & a \end{bmatrix}$. Alternatively, if $\lambda = 1$, then $n = 0$, and we have

$$\begin{bmatrix} 0 & 0 & 1 \\ 1 & 1 & 0 \\ 0 & 1 & a-1 \end{bmatrix} = \begin{bmatrix} 1 & 0 & 0 \\ 1-a & 1 & 1 \\ 0 & 0 & 1 \end{bmatrix} \begin{bmatrix} 0 & 0 & 1 \\ 1 & 0 & 0 \\ -(a-1) & 1 & a \end{bmatrix} \begin{bmatrix} 1 & 0 & 0 \\ a-1 & 1 & -1 \\ 0 & 0 & 1 \end{bmatrix} \quad *_1$$

$$= \begin{bmatrix} 0 & -1 & 0 \\ -1 & a-1 & -1 \\ 0 & 0 & -1 \end{bmatrix} \begin{bmatrix} 0 & 1 & 0 \\ 0 & 0 & 1 \\ 1 & -(a-1) & a \end{bmatrix} \begin{bmatrix} 1-a & -1 & 1 \\ -1 & 0 & 0 \\ 0 & 0 & -1 \end{bmatrix} \quad *_2$$

$$= \begin{bmatrix} 1 & 0 & 0 \\ 0 & 1 & 1 \\ 0 & 0 & 1 \end{bmatrix} \begin{bmatrix} 0 & 0 & 1 \\ 1 & 0 & -(a-1) \\ 0 & 1 & a \end{bmatrix} \begin{bmatrix} 1 & 0 & 0 \\ 0 & 1 & -1 \\ 0 & 0 & 1 \end{bmatrix} \quad *_3$$

The conjugation $*_1$ shows that considerable degeneracy occurs in this characterization of matrices in $SL(3,\mathbb{Z})$. Matrices explicitly given in Cappell and Shaneson (CS1) are those conjugated in $*_2$. The last equation shows that the choice $\lambda = 1$ is equivalent to the rational canonical form for the characteristic polynomial $f_a(x)$. We chose $\lambda = 1$ in the topological construction previously.

The class number of $f_a(x)$ is the number of distinct conjugacy classes of matrices in $SL(3,\mathbb{Z})$ with $f_a(x)$ as characteristic polynomial. We may compute these class numbers using standard techniques of algebraic number theory; as for example in Janusz (J), Borevich and Shafarevich (BS).

We may define an equivalence relation on ideals S,T of a commutative ring R by $[S] = [T] \Leftrightarrow \exists \alpha,\beta \in R$ such that $\alpha S = \beta T$. Clearly any two principal ideals are equivalent. The ideal class group $C(R)$ of R is the abelian group generated by the ideal classes, with composition given by $[ST] = [S][T]$, and identity the class of principal ideals. The following theorem is taken from Newman (Ne).

THEOREM. There is a 1:1 correspondence between similarity classes in $GL(n,\mathbb{Z})$ of matrices A such that $f(A) = 0$, and the elements of the ideal class group $C(R)$ of the ring $R = \mathbb{Z}[\theta]$, where $f(x)$ is a monic polynomial with integer coefficients irreducible over $\mathbb{Q}$ and θ is a root of $f(x) = 0$.

In the particular case $n = 3$, the correspondence is obtained by taking the basis $\{1,\theta,\theta^2\}$ for $\mathbb{Q}[\theta]$, and considering A as the matrix of a $\mathbb{Q}$-linear transformation $\mathbb{Q}[\theta] \to \mathbb{Q}[\theta]$. We may thus choose an eigenvector $[x_1, x_2, x_3]^T \in (\mathbb{Q}(\theta))^3$ corresponding to the eigenvalue θ of A, such that $x_i \in \mathbb{Z}[\theta]$, $i = 1,\ldots,3$. As representative for the ideal class corresponding to A we take the ideal $\mathcal{U}_A = \langle x_1, x_2, x_3\rangle \subset \mathbb{Z}[\theta]$.

Given $x \in \mathbb{Q}[\theta]$, there is a well-defined $\mathbb{Q}$-linear transformation

$$r_x : \mathbb{Q}[\theta] \to \mathbb{Q}[\theta] \quad , \quad r_x(y) = xy \quad \forall\, y \in \mathbb{Q}(\theta) .$$

The discriminant $\Delta(a_1, a_2, a_3)$ of $\{a_1, a_2, a_3\} \subset \mathbb{Q}(\theta)$ is defined by

$$\Delta(a_1, a_2, a_3) \overset{\text{def}}{\equiv} \det(\operatorname{trace}\ (r_{a_i a_j})) .$$

For $f(x) = a_0x^3 + a_1x^2 + a_2x + a_3$, $a_i \in \mathbb{Z}$ $i = 0,\ldots,3$, and $f(x)$ irreducible over $\mathbb{Q}$, the discriminant $\Delta(f)$ of the field $\mathbb{Q}[\theta]$ is defined to be $\Delta(f) \equiv \Delta(1,\theta,\theta^2)$, where θ is a root of $f(x) = 0$. By van der Waerden (V)

$$\Delta(f) = a_1^2a_2^2 - 4a_0a_2^3 - 4a_1^3a_3 - 27a_0^2a_3^2 + 18a_0a_1a_2a_3 .$$

Now let R' denote the integral closure of $\mathbb{Z}$ in $\mathbb{Q}[\theta]$, i.e. the subring of $\mathbb{Q}[\theta]$ consisting of all elements which are roots of monic polynomials with integral coefficients. The class number $|C(R')|$ of R' may be calculated fairly easily in some cases, and related to $|C(R)|$ by the following considerations:

If $\{y_1, y_2, y_3\}$ is an integral basis for R', a free rank 3 $\mathbb{Z}$-module, we may define $\Delta_{R'} \equiv \Delta(y_1, y_2, y_3)$ which is independent of the basis chosen. Necessarily $|\Delta_{R'}| \neq 1$. Now for some m, $0 \leq m \in \mathbb{Z}$, $\Delta(f) = m^2\Delta_{R'}$, and we have the important result that $R' = R$ if $m = 1$.

When $f(x)$ is totally real, i.e. has three real roots, we may read off the class number $|C(R')|$ from Table 7, p. 428 of Borevich and Shafarevich (BS), in case $|\Delta(f)| < 20{,}000$. $f(x)$ is totally real iff $\Delta(f) \geq 0$.

For the polynomial $f_a(x)$, irreducible over $\mathbb{Q}$ by Lemma A4, we obtain

$$\Delta(f_a) = a(f-2)(a-3)(a-5) - 23 = \Delta(f_{5-a})$$

and thus values for the discriminant occur in pairs. $f_a(x)$ is totally real when $a < 0$, $a \geq 6$. This may be seen directly: Note that $f_a(1) = -1$, and thus $f_a(x)$ is totally real if $\exists y \in \mathbb{R}$ such that $y < 1$ and $f_a(y) \geq 0$. Observe that

$$f_a(\tfrac{1}{2}) = \frac{1}{8} - \frac{a}{4} + \frac{a}{2} - \tfrac{1}{2} - 1 = \frac{1}{8}(2a - 11)$$

$$f_a(-\tfrac{1}{2}) = -\frac{1}{8} - \frac{a}{4} - \frac{a}{2} + \tfrac{1}{2} - 1 = \frac{1}{8}(-6a - 5).$$

We may thus fill in the last column of the following Table 1 directly from Table 7, p. 428 of Borevich and Shafarevich:

TABLE 1

a	$\Delta(f_a)$	Prime factors	R=R'?	$\|C(R')\|$
6, -1	49	7, 7	YES	1
7, -2	257	257	YES	1
8, -3	697	17, 41	YES	1
9, -4	1489	1489	YES	1
10, -5	2777	2777	YES	2
11, -6	4729	4729	YES	1
12, -7	7537	7537	YES	2
13, -8	11417	7, 7, 233	?	3
14, -9	16609	17,977	YES	2
0, 5	-23	23	YES	(1)
1, 4	-31	31	YES	(1)
2, 3	-23	23	YES	(1)

The last three entries in the last column must be calculated directly.

The norm $N(x)$ of $x \in \mathbb{Q}[\Theta]$ is defined to be the determinant of r_x. Given an ideal $\mathfrak{A} \in R'$, the norm $N(\mathfrak{A})$ of $\mathfrak{A}$ is the ideal in $\mathbb{Z}$ generated by all $N(x)$, $x \in \mathfrak{A}$. Since $\mathbb{Z}$ is principal, we may define the absolute norm $\mathcal{N}(\mathfrak{A})$ of $\mathfrak{A}$ by $|\mathcal{N}(\mathfrak{A})| = m > 0$, where $N(\mathfrak{A}) = m\mathbb{Z}$. The absolute norm is multiplicative on ideals, i.e. $\mathcal{N}(\mathfrak{A}\mathfrak{B}) = \mathcal{N}(\mathfrak{A})\mathcal{N}(\mathfrak{B})$ for ideals $\mathfrak{A}, \mathfrak{B} \subset R'$.

An ideal B of R' is prime if $ab \in B \Rightarrow a \in B$, or $b \in B$. For some prime $p \in \mathbb{Z}$, $B \cap \mathbb{Z} = p\mathbb{Z}$, and $\mathcal{N}(B) = p^k$ for some $k \in \mathbb{N}$. Every ideal in R' has a unique factorization as a product of prime ideals, and thus the ideal class group C(R') is generated by classes of prime ideals.

The Minkowski Bound states that C(R') is generated by classes of ideals $\mathcal{U}_i \subset R'$ with $|\mathcal{N}(\mathcal{U}_i)| \leq \frac{3!}{3^3}\left(\frac{4}{\pi}\right)^s \sqrt{|\Delta_{R'}|}$, where 2s is the number of complex roots of $f(x) = 0$.

Let θ_a be a root of $f_a(x) = 0$, R'_a denote the integral closure of $\mathbb{Z}$ in $\mathbb{Q}[\theta_a]$. For $0 \leq a \leq 5$, $\Delta(f_a)$ is prime, and $|\Delta(f_a)| < 36$. Thus $\Delta_{R'_a} = \Delta(f_a)$, $R'_a = \mathbb{Z}[\theta_a]$ and since $s = 1$ the Minkowski Bound gives $C(R'_a)$ generated by classes $[\mathcal{U}_i]$ with

$$|\mathcal{N}(\mathcal{U}_i)| \leq \frac{3^2}{3^3} \cdot \left(\frac{4}{\pi}\right) \cdot \sqrt{36} < 2 .$$

Hence $C(R'_a)$ is trivial in each of these cases, i.e. $|C(R'_a)| = 1$.

Suppose $R'_a = R_a = \mathbb{Z}[\theta_a]$: Any element $[\mathcal{U}]$ of $C(R_a)$ is represented by an ideal $\mathcal{U} = \langle x_1, x_2, x_3\rangle$ where $[x_1, x_2, x_3]^T$ is an eigenvector with eigenvalue θ_a of a matrix A representing the similarity class corresponding to $[\mathcal{U}]$. We take A of the form given in Theorem A3; thus

$$\begin{bmatrix} 0 & 0 & 1 \\ m & \lambda & 0 \\ n & p & a-\lambda \end{bmatrix} \begin{bmatrix} x_1 \\ x_2 \\ x_3 \end{bmatrix} = \theta_a \begin{bmatrix} x_1 \\ x_2 \\ x_3 \end{bmatrix}$$

$$\Rightarrow x_3 = \theta_a x_1 \quad , \quad x_2(\theta_a - \lambda) = m x_1 \, ,$$

and thus we may take $[x_1, x_2, x_3]^T = [\theta_a - \lambda, m, \theta_a(\theta_a - \lambda)]^T$. Hence

$$\mathcal{U} = \langle \theta_a - \lambda, m, \theta_a(\theta_a - \lambda)\rangle = \langle m, \theta_a - \lambda\rangle \quad . \qquad (*)$$

is a representative of the ideal class corresponding to the conjugacy class of A.

Since θ_a satisfies $f_a(\theta_a) = 0$, $(\theta_a - \lambda)$ satisfies $f((\theta_a - \lambda) + \lambda)$, giving

$$\begin{aligned} N(\theta_a - \lambda) = f(\lambda) &= \lambda^3 - a\lambda^2 + (a-1)\lambda - 1 \\ &= -(1 + \lambda^2(a-\lambda) - \lambda(a-1)) \\ &= -(1 + n\lambda) \\ &= -mp \, . \end{aligned}$$

In particular, if $(m,p) = 1$, $\mathcal{N}(\langle\theta_a - \lambda, m\rangle)$ divides m. If $\langle m, \theta_a - \lambda\rangle$ is prime, $p \mid \mathcal{N}(\langle\theta_a - \lambda, m\rangle) \Rightarrow p \mid m$.

To apply this to the simplest case of non-trivial ideal class group corresponding to $a = -5$, we require some more results from algebraic number theory. By Janusz (J), factorization of the principal ideal pR', p any prime number, is achieved by reducing the coefficients of $f(x)$ modulo p, and factorizing the resulting polynomial over $\mathbb{Z}/p\mathbb{Z}$. To each irreducible factor $h_i(x)$ of degree k there corresponds a prime ideal B_i with $\mathcal{N}(B_i) = p^{k_i}$. We may thus obtain all generating classes for $C(R')$ by factorization of all ideals pR', $p \leq \delta$, where δ is the maximum value for absolute norms allowed by the Minkowski Bound.

EXAMPLE. $a = -5$. Let θ be a root of $x^3 + 5x^2 - 6x - 1 = 0$. We determine the structure of the ideal class group $C(\mathbb{Z}[\theta])$. Since $\Delta(f_a) = 2777$, which is a prime, $R' = R = \mathbb{Z}[\theta]$, and $C(R)$ is generated by classes of ideals B_i with

$$|\mathcal{N}(B_i)| < \frac{3!}{3^3}\sqrt{2777} < 12 .$$

Factorizing the ideals 2R, 3R, 5R, 7R, 11R, we obtain

p = 2: $(x^3 + 5x^2 - 6x - 1) \pmod 2 \equiv x^3 + x^2 - 1 \pmod 2$ is irreducible

$\Rightarrow 2R = B_2$, a prime, principal ideal with $\mathcal{N}(B_2) = 2^3$.

p = 3: $x^3 + 5x^2 - 6x - 1 \equiv x^3 + 2x - 1 \equiv (x-2)(x^2 + 2x - 1) \pmod 3$

$3R = B_3B_3'$, where $\mathcal{N}(B_3) = 3$, $\mathcal{N}(B_3') = 3^2$

$1 = [3R] = [B_3][B_3'] \Rightarrow [B_3'] = -B_3]^{-1}$.

Since we know $|C(R)| = 2$ by Table 1, $[B_3'] = [B_3]$.

p = 5: $x^3 + 5x^2 - 6x - 1 \equiv x^3 - x - 1 \equiv (x-2)(x^2 + 2x - 2) \pmod 5$

$\Rightarrow 5R = B_5B_5'$, $\mathcal{N}(B_5) = 5$, $\mathcal{N}(B_5') = 5^2 > 12$

and $1 = [5R] = [B_5][B_5'] = [B_5]^2$.

p = 7: $x^3 + 5x^2 - 6x - 1 \equiv (x-4)(x^2 + 2x - 5) \pmod 7$

$\Rightarrow 7R = B_7B_7'$. $\mathcal{N}(B_7) = 7$. $\mathcal{N}(B_7') = 7^2 > 12$

and $1 = [7R] = [B_6][B_7'] = [B_7]^2$.

p = 11: $x^3 + 5x^2 - 6x - 1$ is irreducible $\pmod{11}$

$\Rightarrow 11R = B_{11}$, principal, and $\mathcal{N}(B_{11}) = 11^3 > 12$

Hence $C(R) = \langle [B_3], [B_5], [B_7] \rangle$.

λ	-3	-2	-1	0	1	2
$f(\lambda)$	$35 = 5.7$	23	$9 = 3.3$	-1	-1	$15 = 3.5$

Thus $(\theta-2)R$ is divisible by B_3, and since B_3 divides $3R$, B_3 divides $(\theta+1)R = ((\theta-2)+3)R$. Hence $(\theta+1)R = B_3^2$, and $(\theta-2)R = B_3B_5$, $(\theta+3)R = B_5B_7$ giving $[B_3] = [B_5] = [B_7]$.

We thus consider the ideals $\langle 3, \theta-2\rangle$, $\langle 5, \theta+3\rangle$, $\langle 7, \theta+3\rangle$.

I. $\qquad \langle 3, \theta+1\rangle\langle 3, \theta+1\rangle = \langle 9, 3(\theta+1), (\theta+1)^2\rangle$

$$= \langle 9, 3(\theta+1), \theta^2+2\theta+1\rangle .$$

Now $\theta(\theta+1)^2 = \theta^3+2\theta^2+\theta = 6\theta-5\theta^2+1+2\theta^2+\theta = 7\theta-3\theta^2+1$

$$(7\theta-3\theta^2+1) + 3(\theta+1)^2 = 13\theta+4 = (\theta+1) + 4(3\theta+3) - 9$$

$\Rightarrow (\theta+1) \in \langle 3, \theta+1\rangle^2$. But $\theta^3+\theta^2 = 6\theta-5\theta^2+1+\theta^2$

$$= 6\theta-4\theta^2+1 = \theta^2(\theta+1)$$

and $9 = 6(\theta+1) - (6\theta-3) = 6(\theta+1) - \{6\theta-4\theta^2+1)$

$$+ 4(\theta-1)(\theta+1)\} \in \langle\theta+1\rangle .$$

II. $\qquad \langle 5, \theta+3\rangle\langle 3, \theta-2\rangle = \langle 5, \theta-2\rangle\langle 3, \theta-2\rangle$

$$= \langle 15, 5(\theta-2), 3(\theta-2), (\theta-2)^2\rangle = \langle 15, \theta-2\rangle$$

$$\theta^2(\theta-2) = \theta^3 - 2\theta^2 = 6\theta-5\theta^2+1-2\theta^2 = 6\theta-7\theta^2+1 .$$

Further, $\theta(\theta-2)^2 = \theta^3-4\theta^2+4\theta = 6\theta-5\theta^2+1-4\theta^2+4\theta$

$$= 10\theta-9\theta^2+1$$

$$\Rightarrow (10\theta-9\theta^2+1) - (6\theta-7\theta^2+1) = 4\theta-2\theta^2 \in \langle\theta-2\rangle$$

$$\Rightarrow \theta(4\theta-2\theta^2) = 4\theta^2-12\theta+10\theta^2-2 = 14\theta-12\theta-2 \in \langle\theta-2\rangle$$

$$\Rightarrow \theta(4\theta-2\theta^2) = 4\theta^2-12\theta+10\theta^2-2 = 14\theta^2-12\theta-2 \in \langle\theta-2\rangle$$

$$\Rightarrow (14\theta^2-12\theta-2) + (10\theta-9\theta^2+1) = 5\theta^2-2\theta-1 \in \langle\theta-2\rangle$$

$$\Rightarrow 2(5\theta^2-2\theta-1) = 10\theta^2-4\theta-2 \in \langle\theta-2\rangle$$

$$(10\theta^2-4\theta-2) + (10^2\theta-9\theta+1) = \theta^2+6\theta-1$$

$$(\theta^2+6\theta-1) - (\theta-2)^2 = 10\theta-5$$

$$(10\theta-5) - 10(\theta-2) = 15 \in \langle\theta-2\rangle$$

Hence $\langle\theta-2\rangle = \langle 15, \theta-2\rangle = \langle 5, \theta+3\rangle\langle 3, \theta-2\rangle$.

Thus we must have $B_3 = \langle 3, \theta-2\rangle$, $B_5 = \langle 5, \theta+3\rangle$.

III. $\langle 5, \theta+3\rangle\langle 7, \theta+3\rangle = \langle 35, 5(\theta+3), 7(\theta+3), (\theta+3)^2\rangle = \langle 35, (\theta+3)\rangle$

$$\theta(\theta+3)^2 = \theta^3 + 6\theta^2 + 9\theta = 6\theta - 5\theta^2 + 1 + 6\theta^2 + 9\theta = \theta^2 + 15\theta + 1$$

$$\theta^2(\theta+3) = \theta^3 + 3\theta^2 = 6\theta - 5\theta^2 + 1 + 3\theta^2 = 6\theta - 2\theta^2 + 1$$

$$\theta(6\theta - 2\theta^2 + 1) = 6\theta^2 - 2\theta^3 + \theta = 6\theta^2 - 12\theta + 10\theta^2 - 2 + \theta = 16\theta^2 - 11\theta - 2$$

$$16\theta(\theta+3) - (16\theta^2 - 11\theta + 2) = 59\theta + 2$$

$$16(\theta^2 + 15\theta + 1) - (16\theta^2 - 11\theta - 2) = 251\theta + 18$$

$$(251\theta + 18) - 4(59\theta + 2) = 15\theta + 10\theta \Rightarrow 35 = 15(\theta+3) - (15\theta + 10) \in \langle\theta + 3\rangle$$

$$\Rightarrow \langle\theta + 3\rangle = \langle 35, \theta + 3\rangle = \langle 5, \theta + 3\rangle\langle 7, \theta + 3\rangle .$$

By uniqueness of prime factorization and multiplicativity of the absolute norm,

$$\mathcal{N}(\langle 3, \theta+1\rangle) = 3 , \mathcal{N}(\langle 5, \theta+3\rangle) = 5 , \mathcal{N}(\langle 7, \theta+3\rangle) = 7$$

and so $B_3 = \langle 3, \theta+1\rangle$, $B_5 = \langle 5, \theta+3\rangle$, $B_7 = \langle 7, \theta+3\rangle$.

As a representative matrix for the similarity class corresponding to $[B_3] = [B_5] = [B_7]$, we take

$$\begin{bmatrix} 0 & 0 & 1 \\ -5 & 2 & 0 \\ -8 & 3 & -7 \end{bmatrix} ,$$

since by (*) the ideal class generated by this matrix is $[\langle -5, \theta - 2\rangle] = [\langle 5, \theta + 3\rangle]$.

BIBLIOGRAPHY

1. Akbulut, S. These proceedings.

2. Akbulut, S. and Kirby, R. "An Exotic Involution of S^4", Topology 18 (1979) 75-81, and correction to appear.

3. Akbulut, S. and Kirby, R. "Mazur Manifolds", Mich. Math. J.

4. Andrews, J. J. and Curtis, M. L. "Free Groups and Handlebodies", Proc. Amer. Math. Soc. 16 (1965) 192-195.

5. Birman, J., Gonzales-Acuna, F. and Montesinos, J. M. "Heegard Splittings of Prime 3-manifolds are not Unique", Mich. Math. J. 23 (1976) 97-103.

6. Borevich, Z. K. and Shafarevich, I. R. "Number Theory", Academic Press 1966.

7. Bredon, G. "Compact Transformation Groups", Academic Press 1972.

8. Cappell, S. and Shaneson, J. "There Exist Inequivalent Knots with the Same Complement", Ann. of Math. 103 (1976) 349-353.

9. Cappell, S. and Shaneson, J. "Some New Four Manifolds", Ann. of Math. 104 (1976) 61-72.

10. Fintushel, R. and Stern, R. "An Exotic Free Involution on S^4", Ann. of Math. (1981) 357-365.

11. Fukuhara, S. "On the Invariant for a Certain Type of Involution on Homology 3-spheres and its Application", J. Math. Japan 30 (1978) 653-665.

12. Gluck, H. "The Embedding of 2-spheres in the 4-sphere", Bull. Amer. Math. Soc. 67 (1961) 586-589.

13. Janusz, . "Algebraic Number Fields", Academic Press.

14. Kirby, R. "Problems in Low-Dimensional Topology", Notes from the Stanford Conference, 1976.

15. Kirby, R. "A Calculus for Framed Links in S^3", Invent. Math. 45 (1978) 35-56.

16. Laudenbach, F. and Poenaru, V. "A Note on 4-Dimensional Handlebodies", Bull. Soc. Math. France 100 (1972) 337-344.

17. Lickorish, W. B. R. "Surgery on Knots", Proc. Amer. Math. Soc. 60 (1976) 296-298.

18. Lopez de Medrano, S. "Involutions on Manifolds", Springer-Verlag 1971.

19. Matumoto, T. and Siebenmann, L. "The Topological s-Cobordism Theorem Fails in Dimensional 4 or 5", Math. Proc. Camb. Phil. Soc. 84, (1978) 85-87.

20. Mazur, B. "A Note on Some Contractible 4-manifolds", Ann. of Math. 73 (1961) 221-228.

21. Montesinos, J. M. "Heegard Diagrams for Closed 4-manifolds", preprint. To appear in "Geometric Topology", Academic Press.

22. Newman, M. "Integral Matrices", Academic Press 1972.

23. Norman, R. A. "Dehn's Lemma for Certain 4-manifolds", Invent. Math. 7 (1969) 143-147.

24. Rourke, C. P. and Sanderson, B. J. "Introduction to Piecewise Linear Topology", Springer-Verlag 1972.

25. Ryder, R. W. "The Class Number Problem", Amer. Math. Monthly 86 (1979) 200-202.

26. Singer, J. "Three Dimensional Manifolds and Their Heegard Diagrams", Trans. Amer. Math. Soc. 35 (1933) 88-111.

27. Van der Waerden, B. L. "Modern Algebra", vol. 1, Springer 1931.

28. Waldhausen, F. "On Irreducible 3-manifolds which are Sufficiently Large", Ann. of Math. 87 (1968) 56-88.

29. Wall, C. T. C. "Free Piecewise-Linear Involutions on Spheres", Bull. Amer. Math. Soc. 74 (1968) 554-558.

30. Yoshida, "On the Browder-Livesay Invariant of Free Involutions on Homology 3-spheres", Math. J. Okayama Univ. 22 (1980), 91-106.

DEPARTMENT OF MATHEMATICS
UNIVERSITY OF CALIFORNIA
BERKELEY, CA 94720

DEPARTMENT OF MATHEMATICS
UNIVERSITY OF MELBOURNE
AUSTRALIA

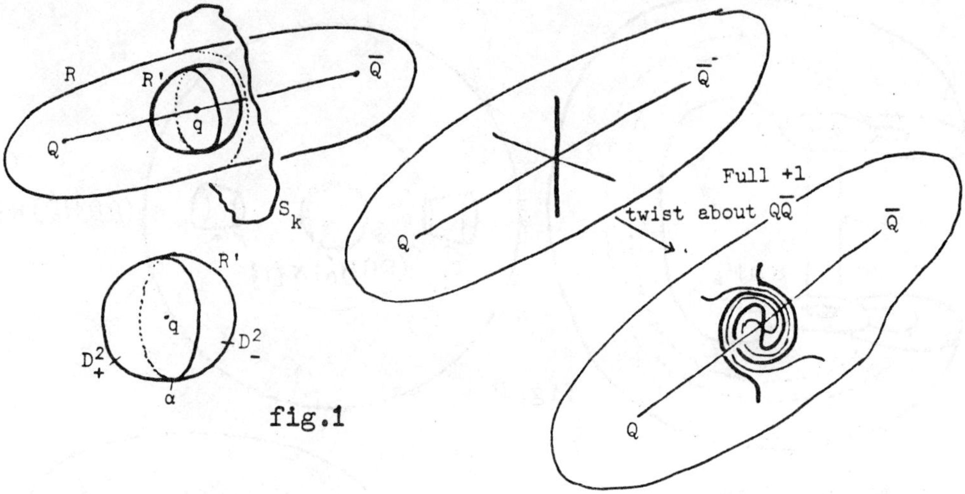
R
R'
q
Q
$\bar{Q}$
S_k
D^2_+
D^2_-
α
Full +1
twist about $Q\bar{Q}$

fig.1

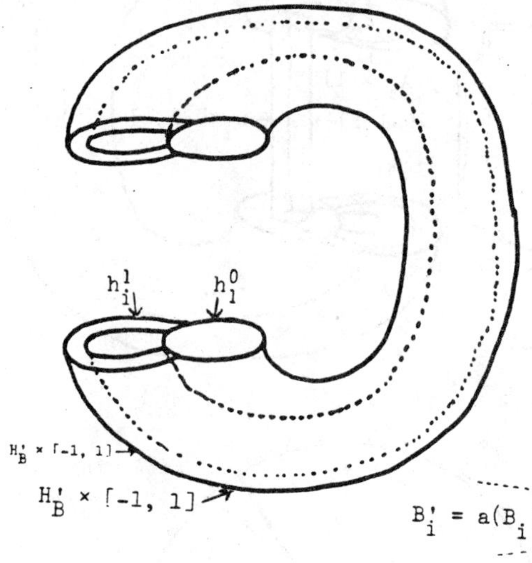
h^1_i
h^0_1
$H'_B \times [-1, 1]$
$H'_B \times [-1, 1]$

fig.2a

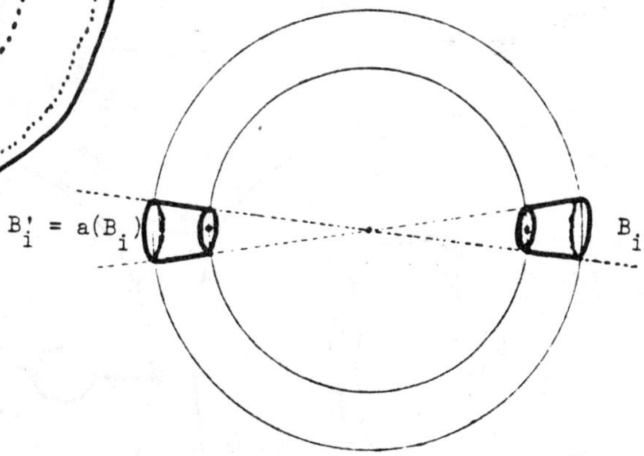
$B'_i = a(B_i)$
B_i

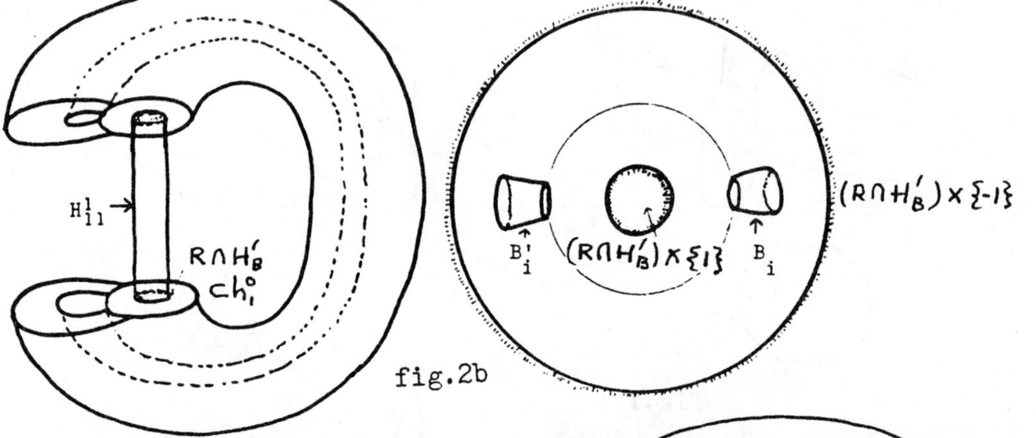

fig.2b

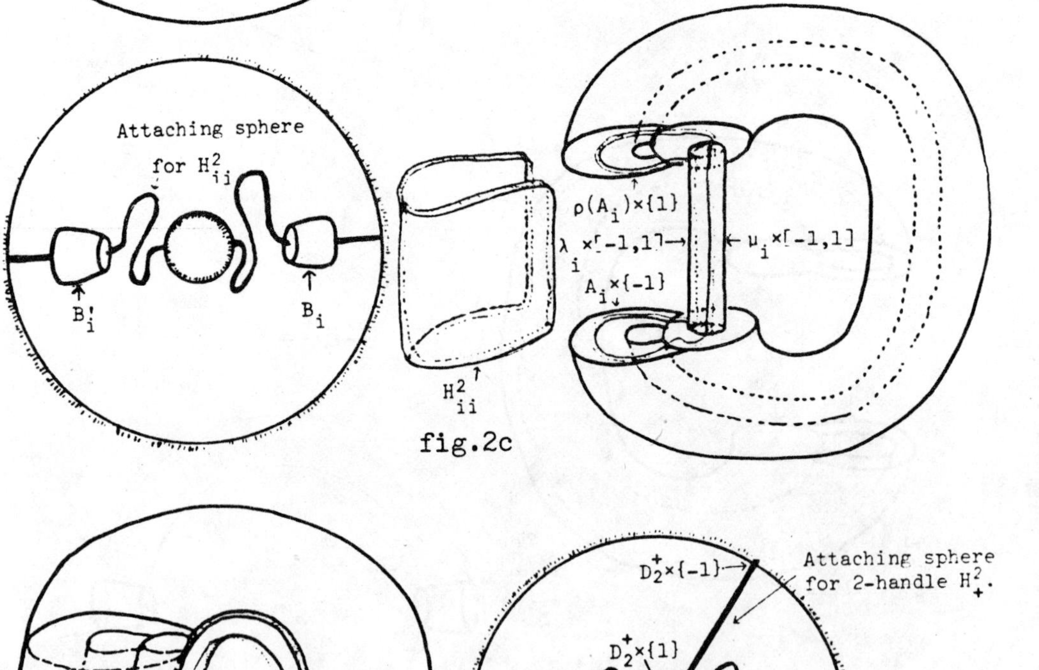

fig.2c

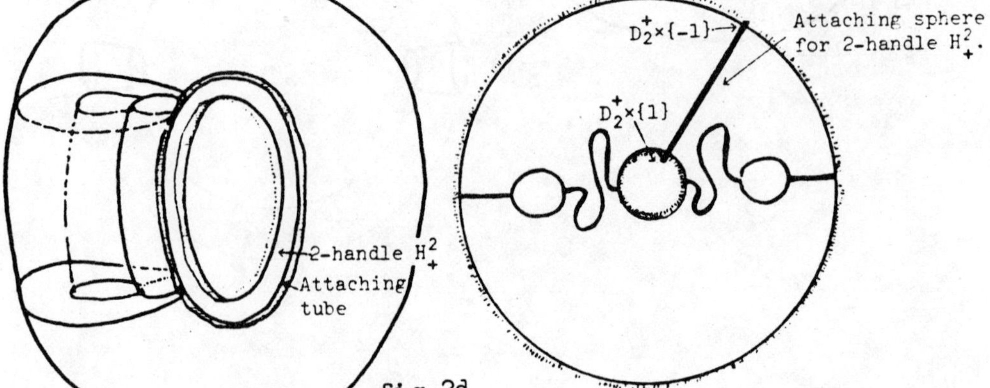

fig.2d

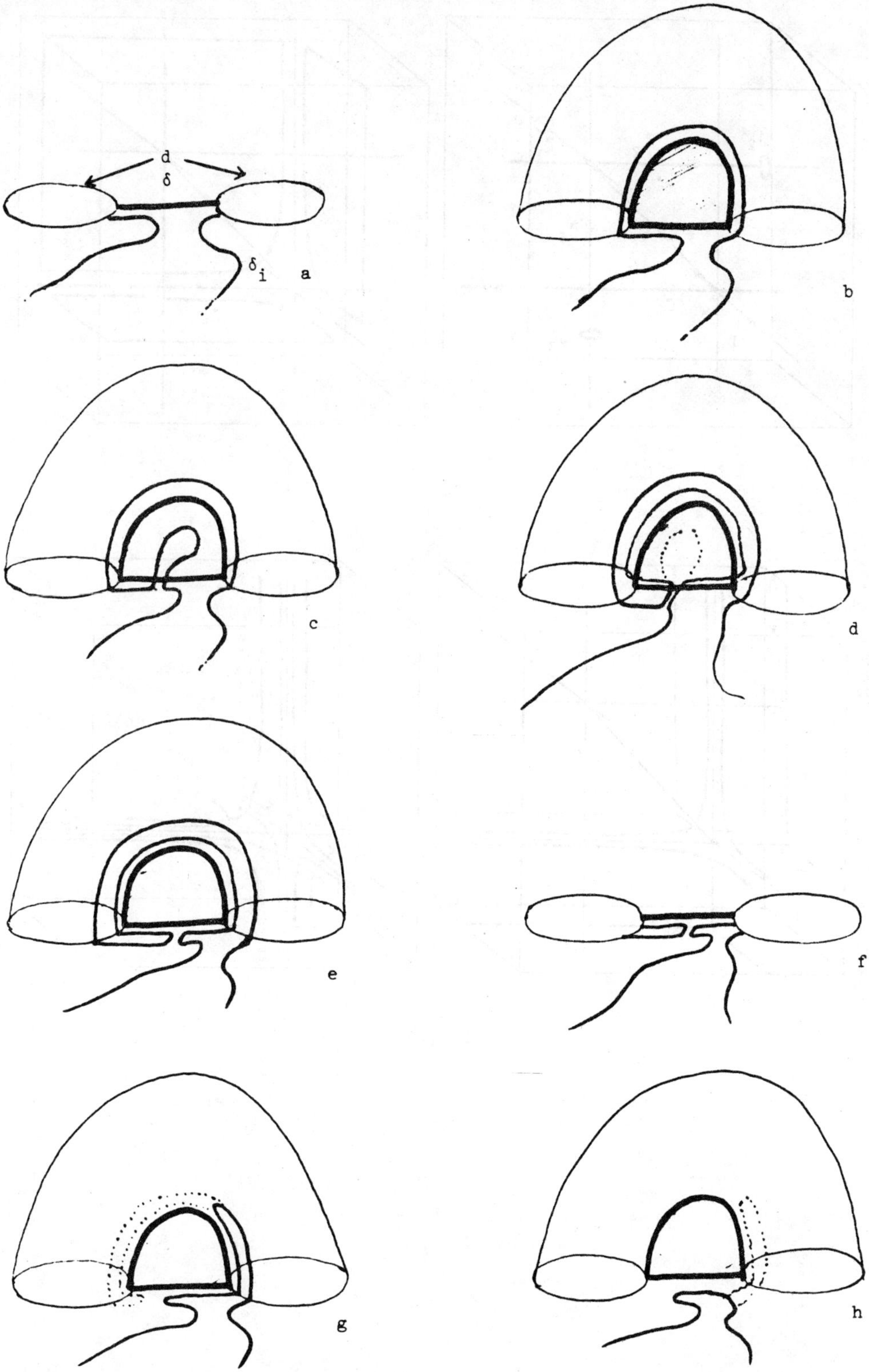

fig.3

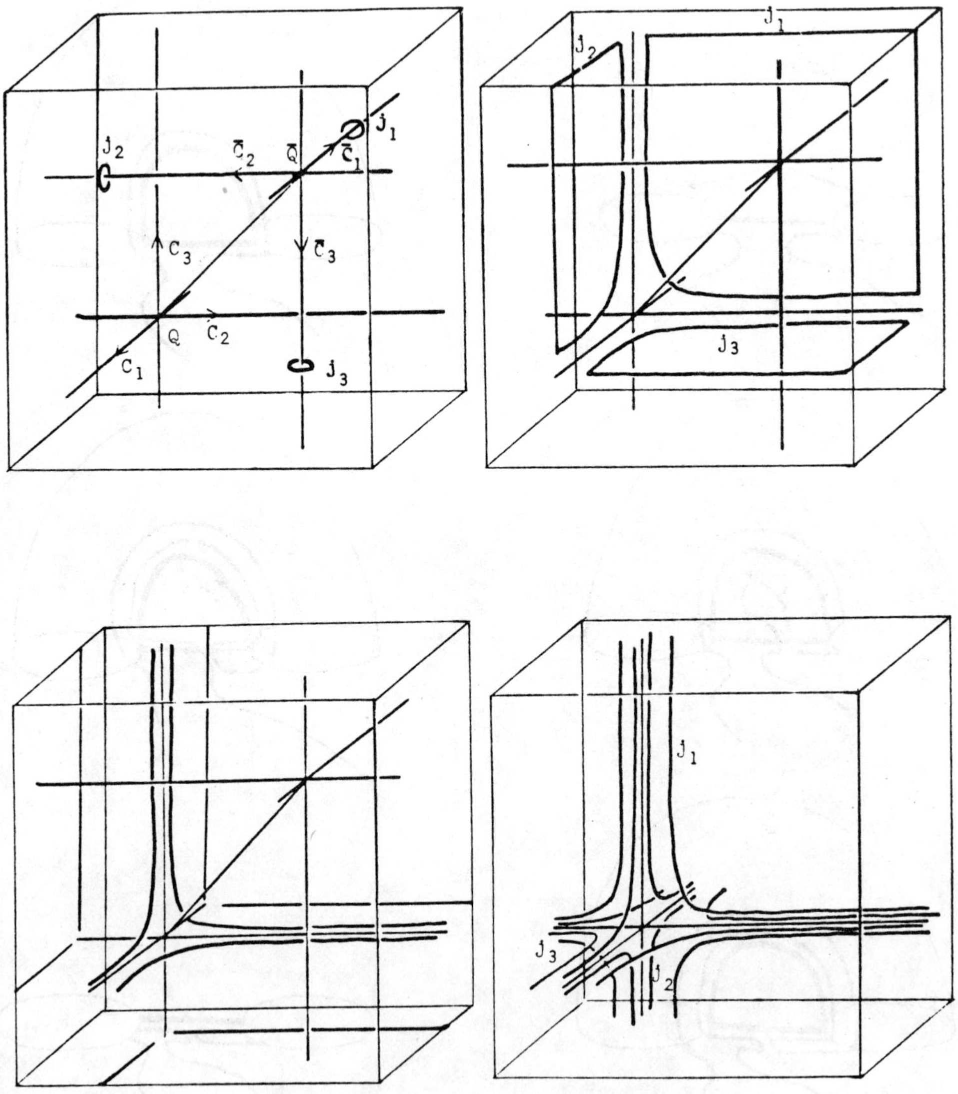

fig.4

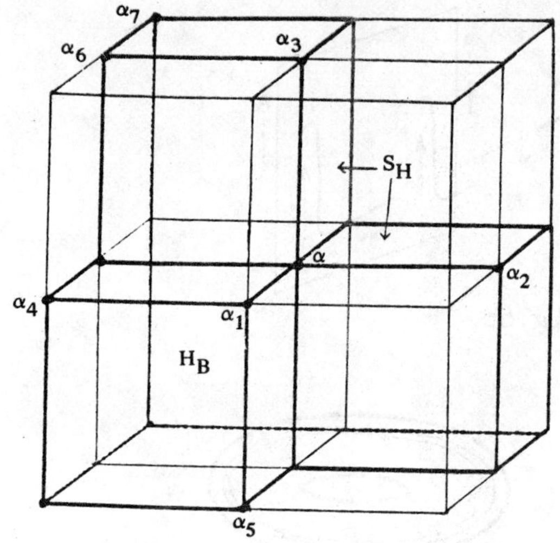

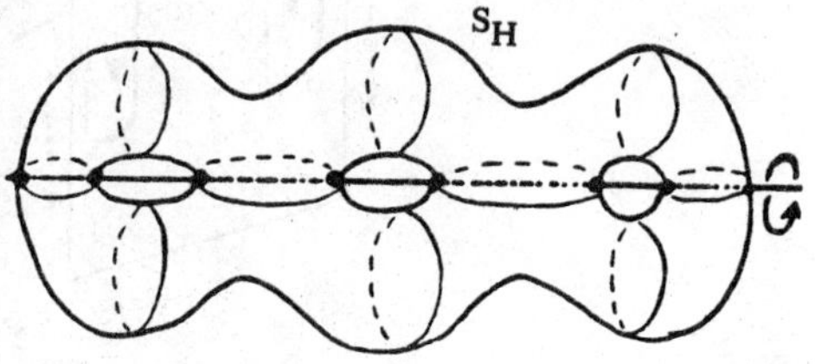

fig.5

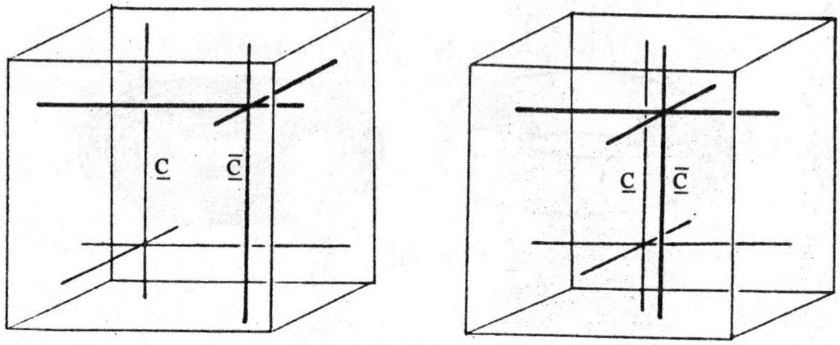

fig.6a

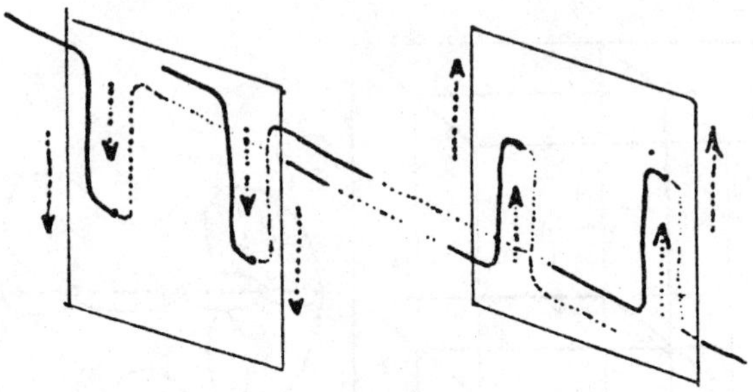

fig.6b

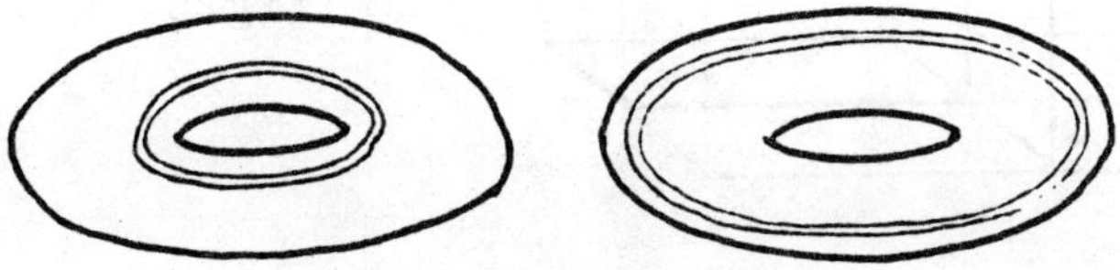

fig.6c

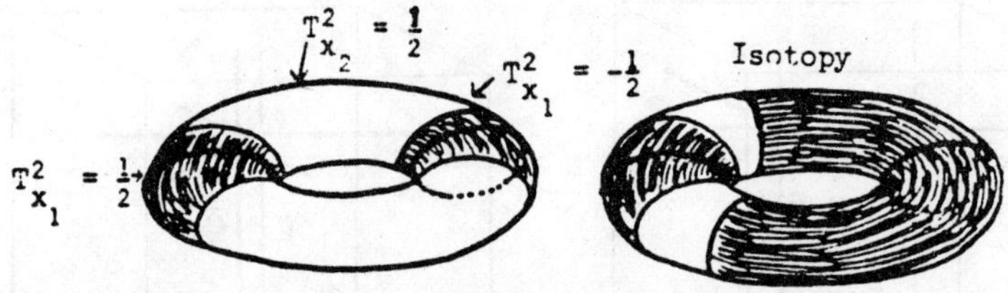

fig.6d

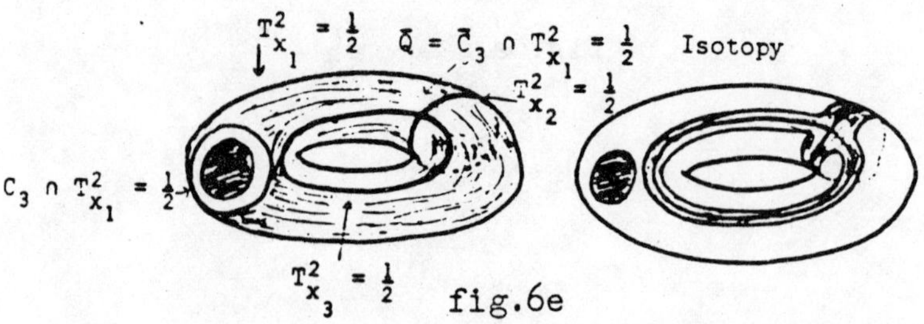

fig.6e

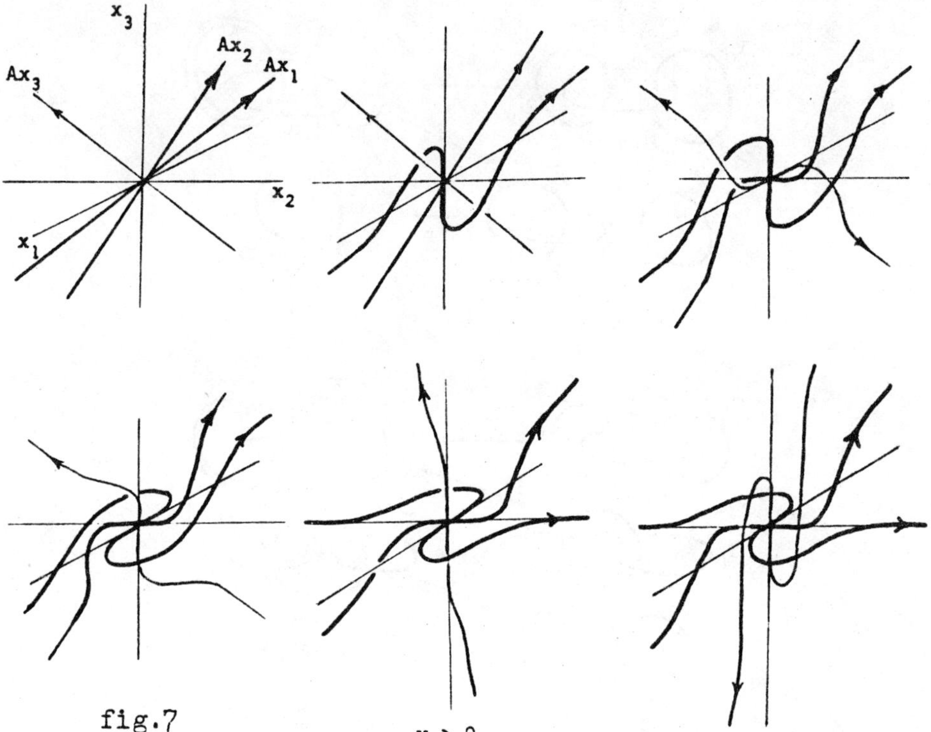

fig.7

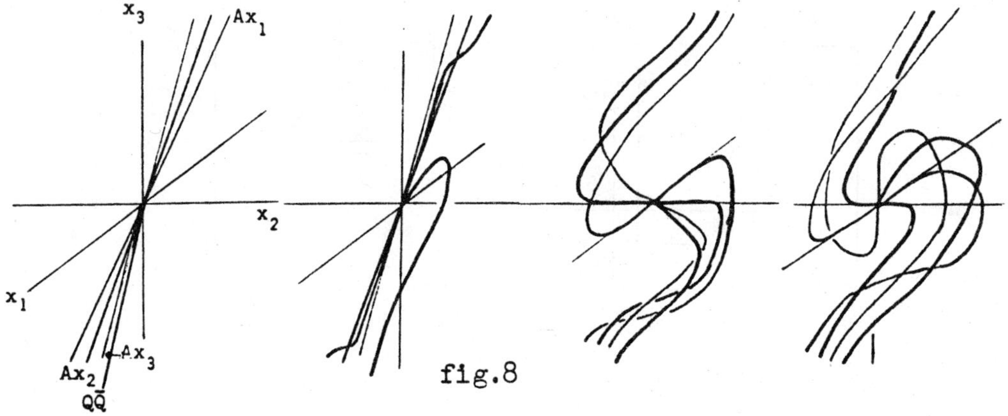

fig.8

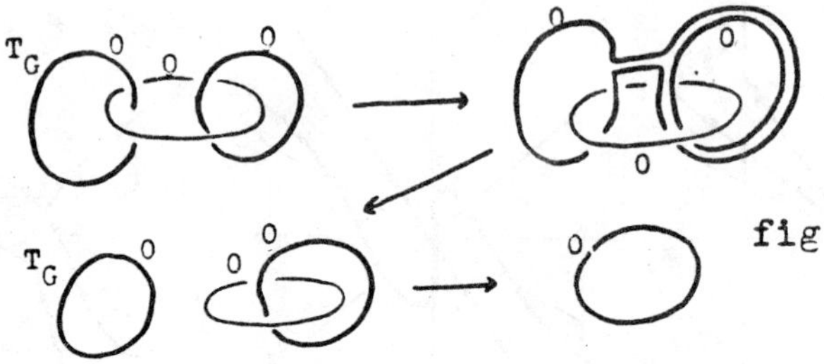

fig.9a

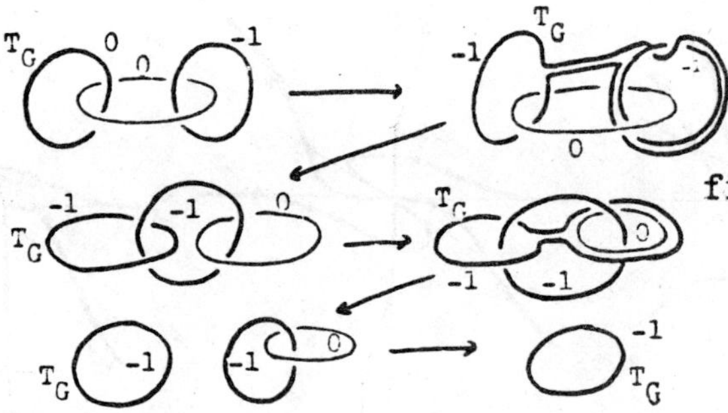

fig.9b

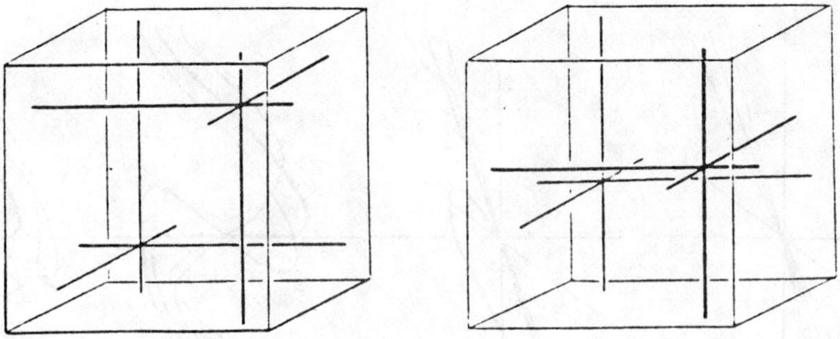

fig.10a

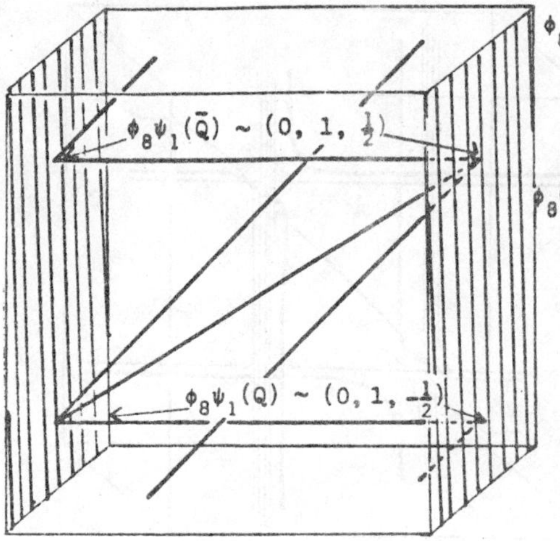

fig.10c

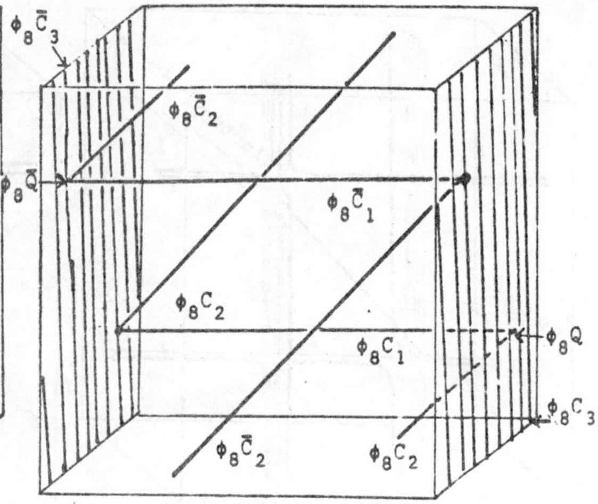

fig.10b

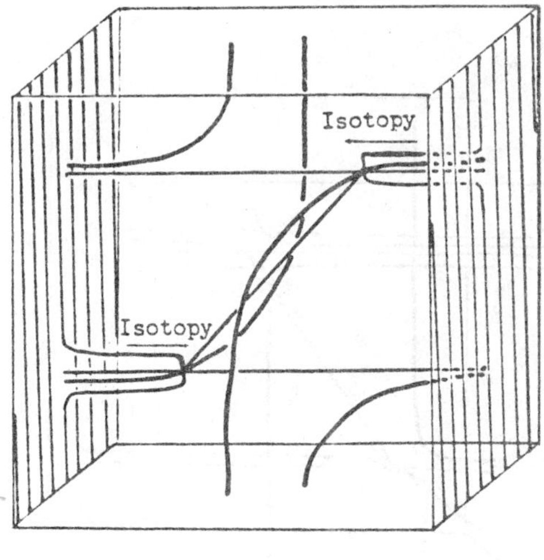

fig.10d

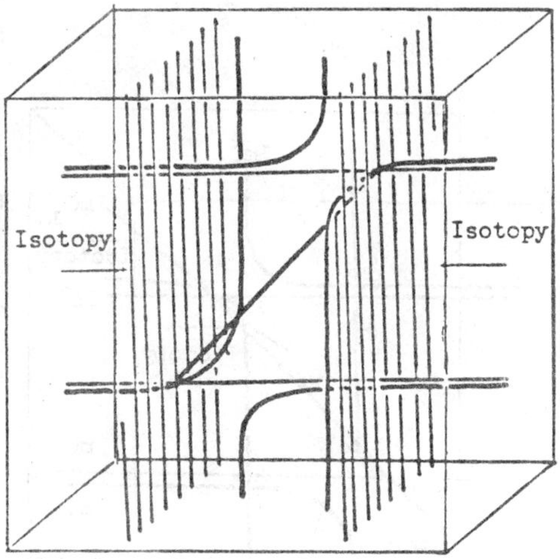

fig.10e

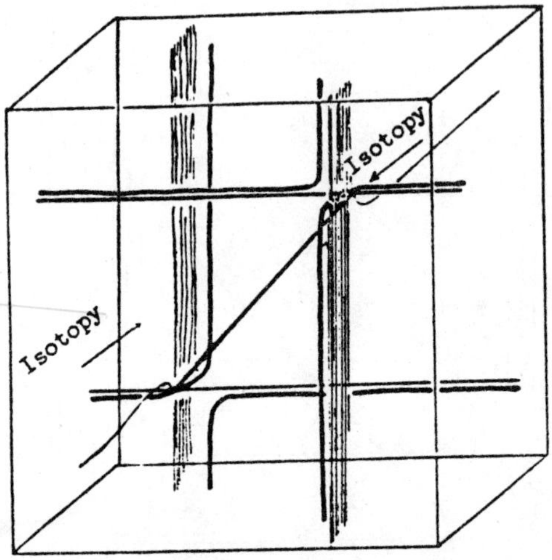

fig.10f

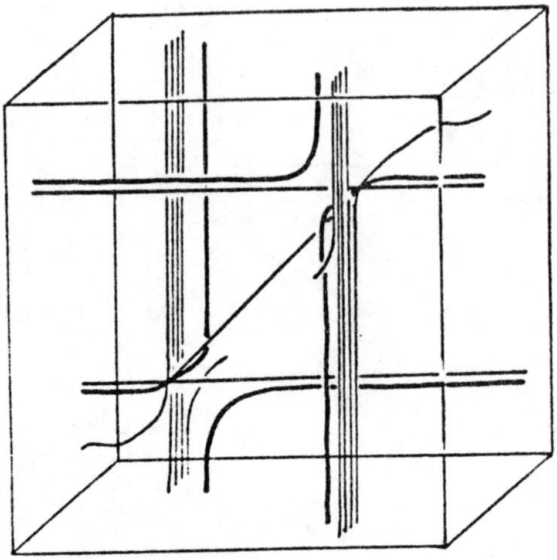

fig.10g

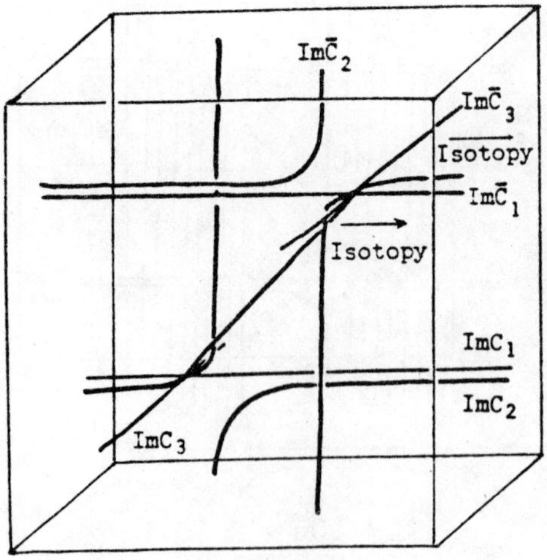

fig.11a

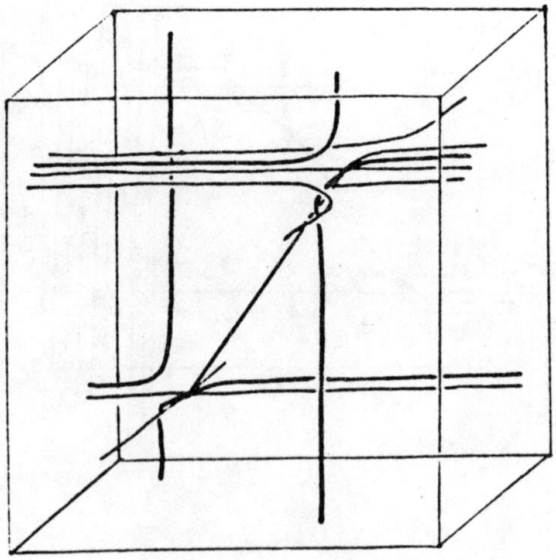

fig.11b

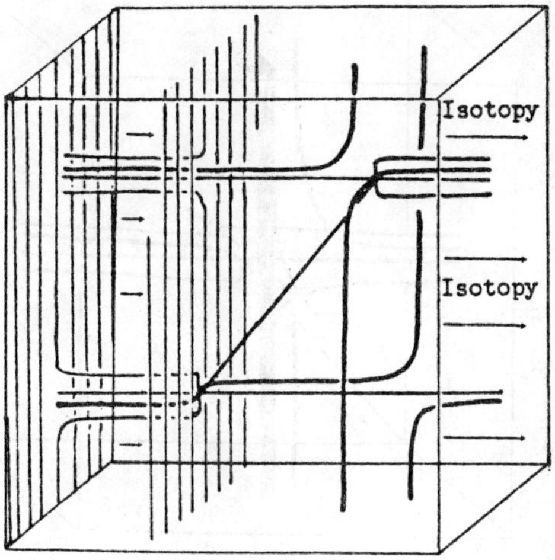

fig.11c

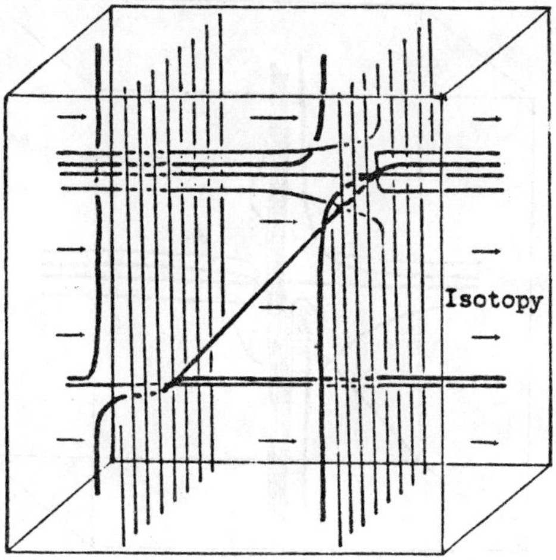

fig.11d

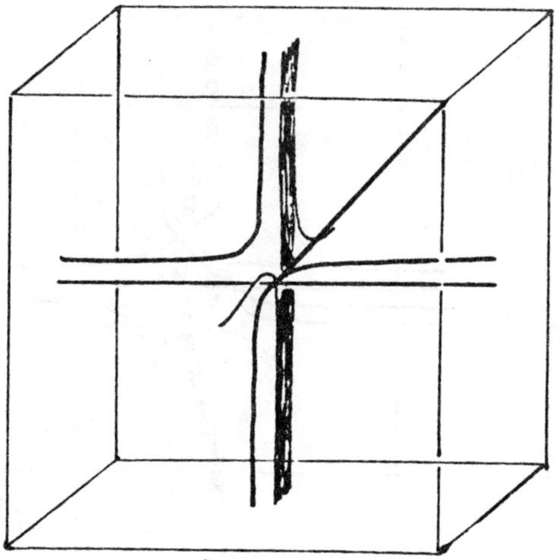

$v > 0$

fig.11e

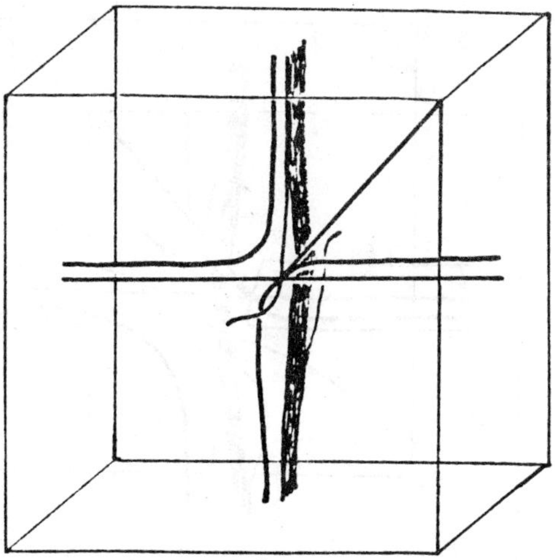

$v < 0$

fig.11f

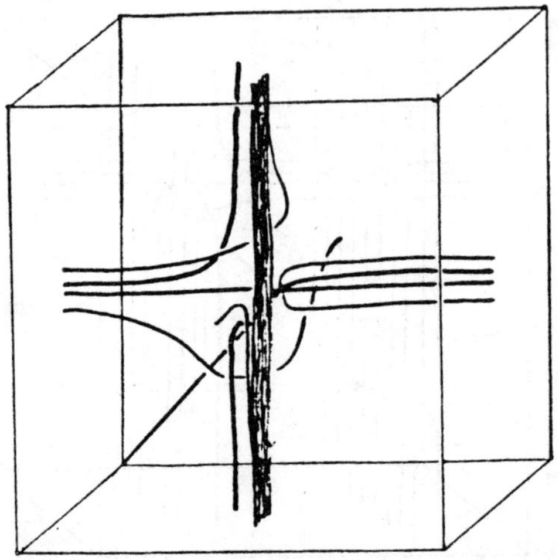

fig.11g

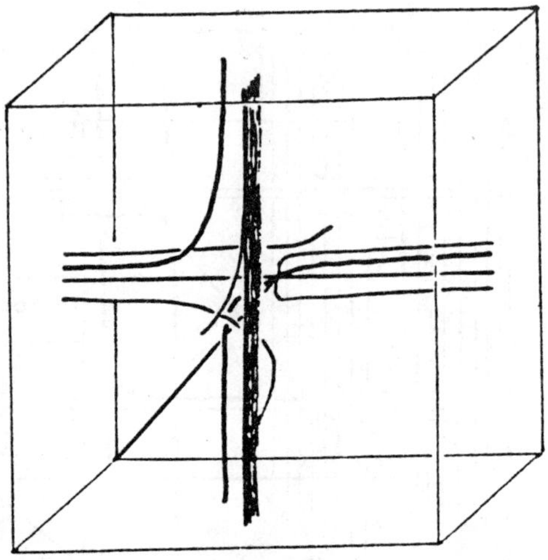

fig.11h

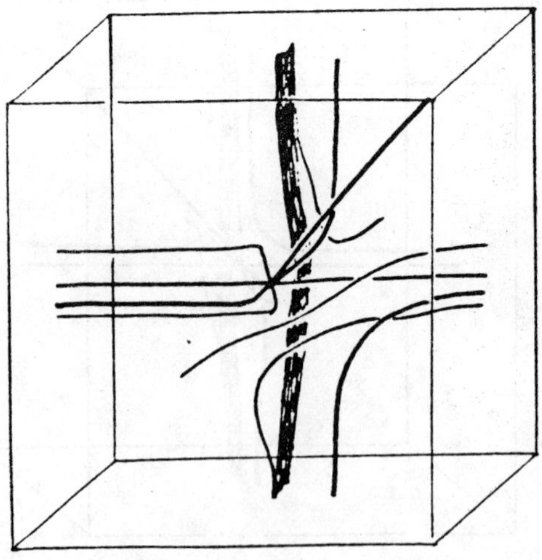

fig.12a

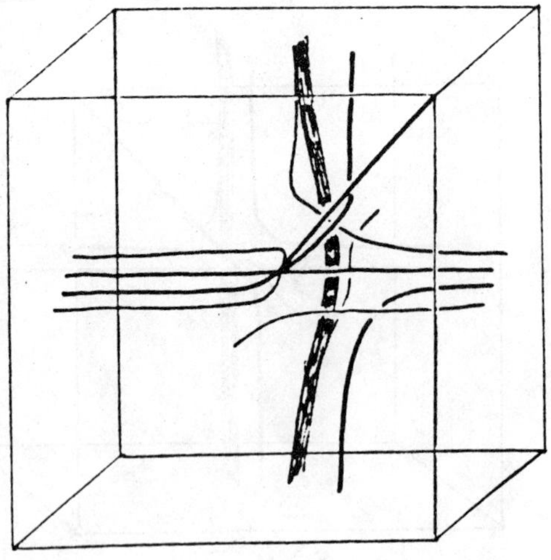

fig.12b

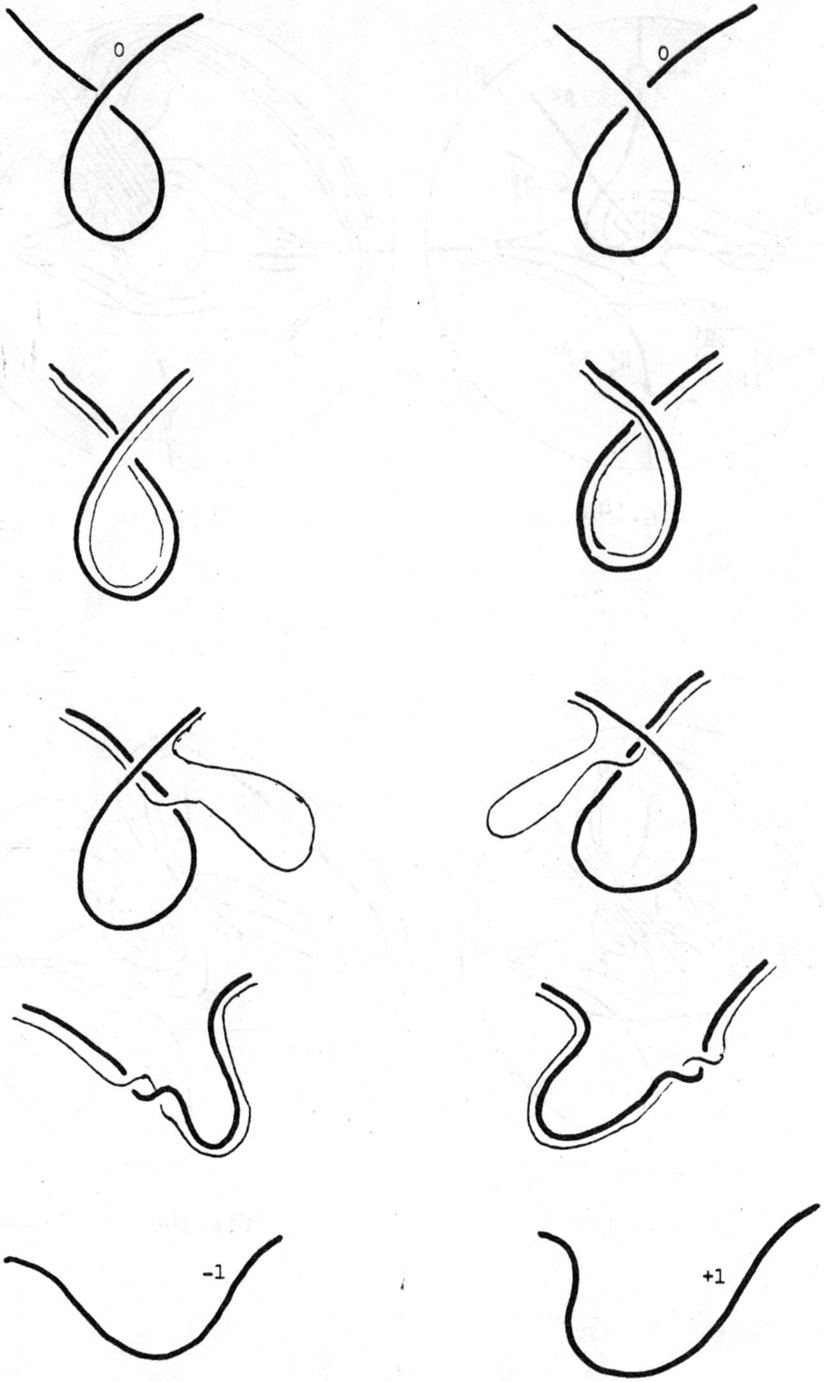

fig.13

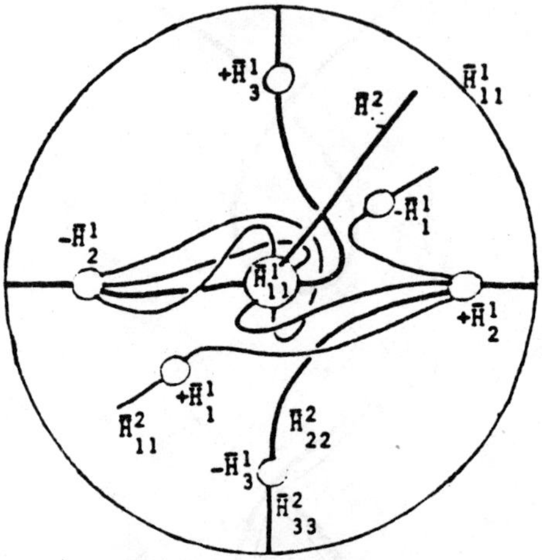

fig.14a

fig.14b

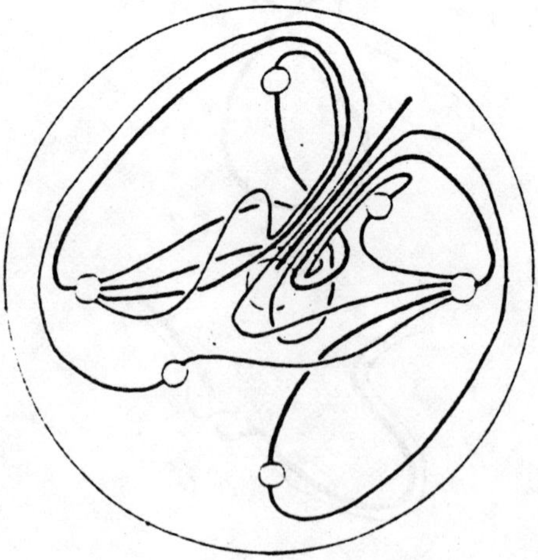

fig.14c

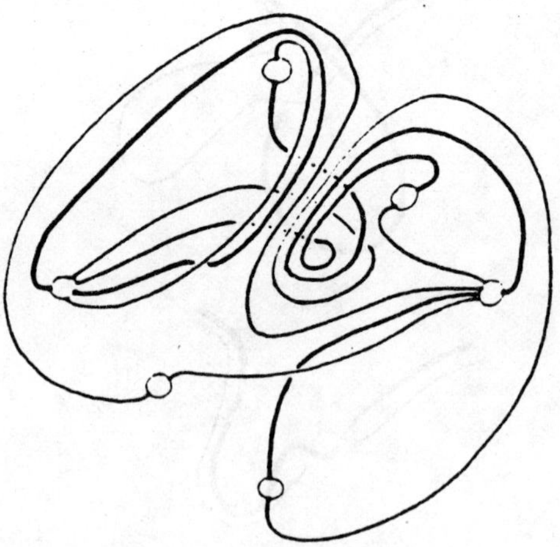

fig.14d

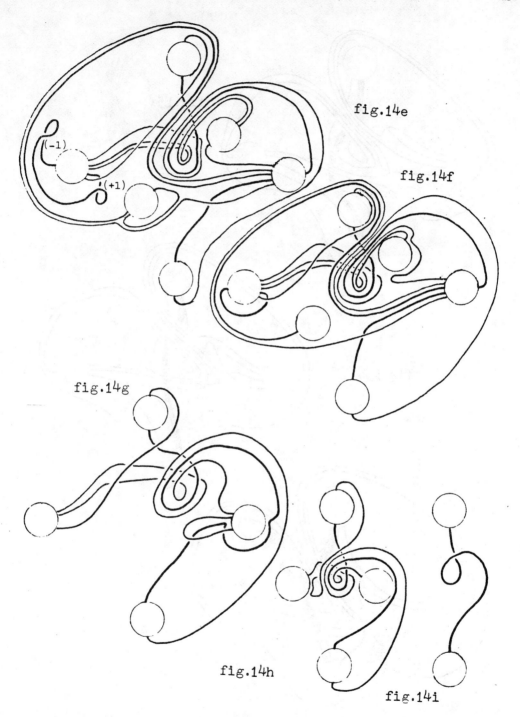

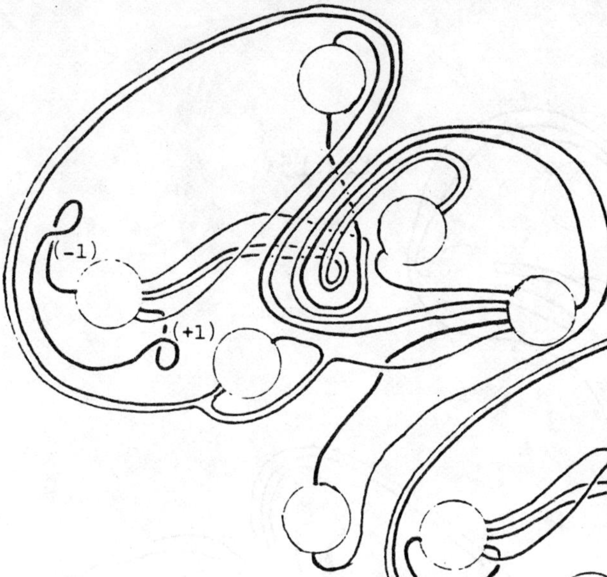

fig.14e

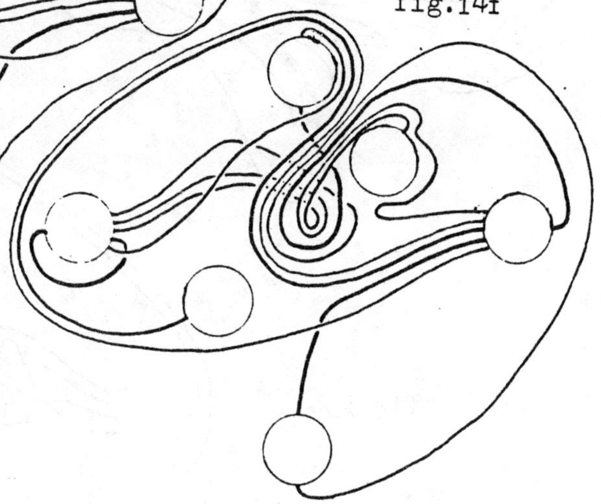

fig.14f

fig.14g

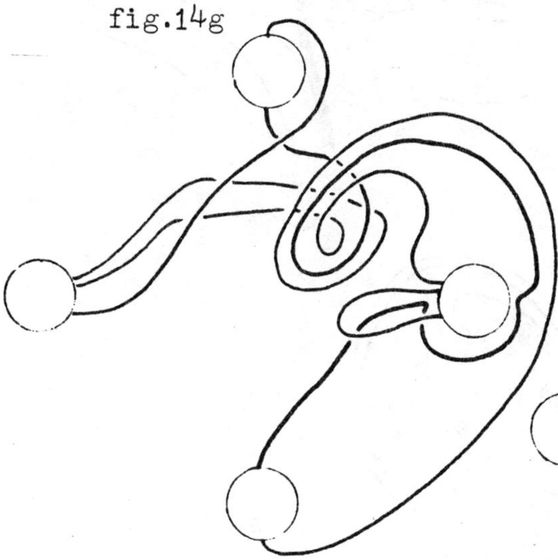

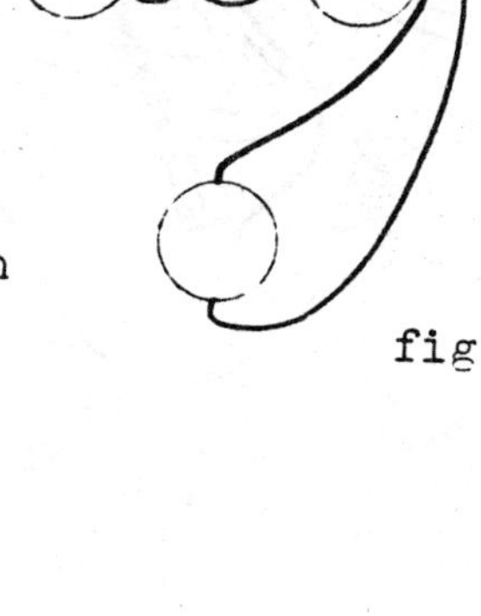

fig.14h

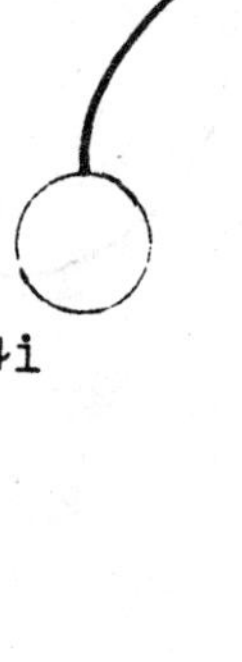

fig.14i

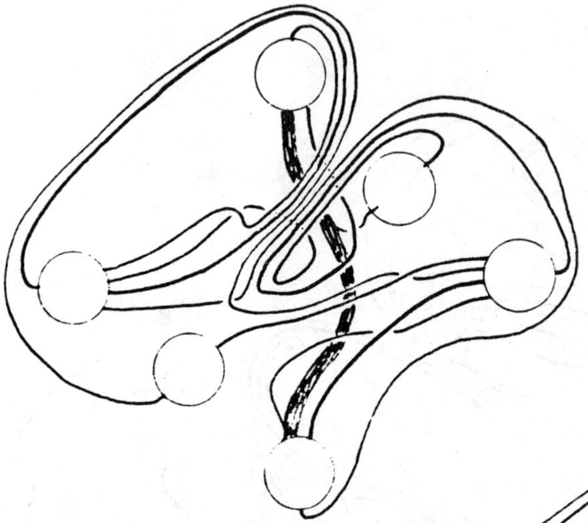

fig.15a

fig.15b

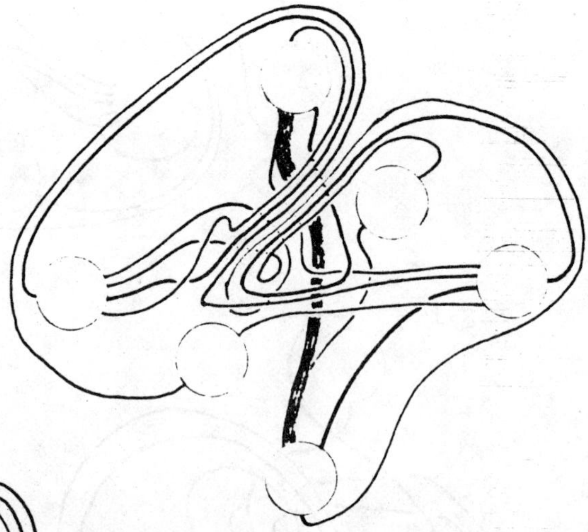

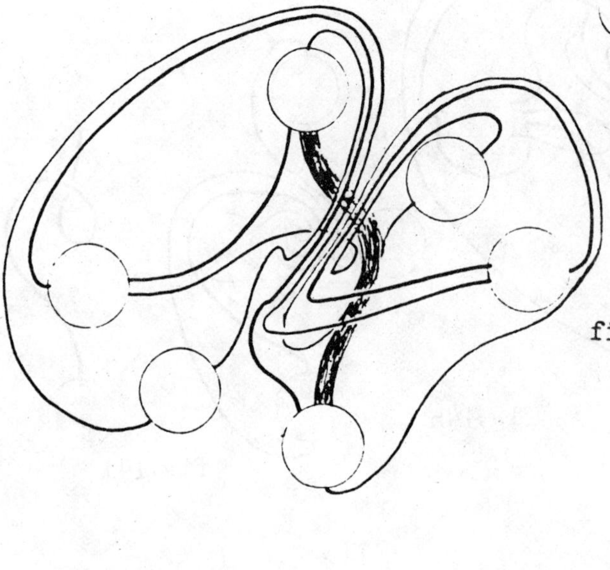

fig.16a

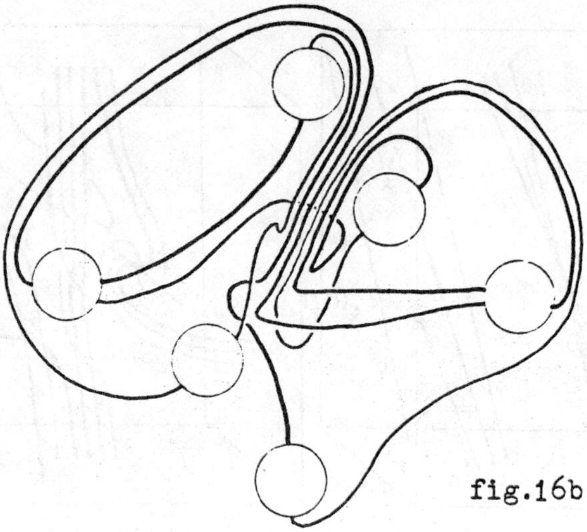

fig.16b

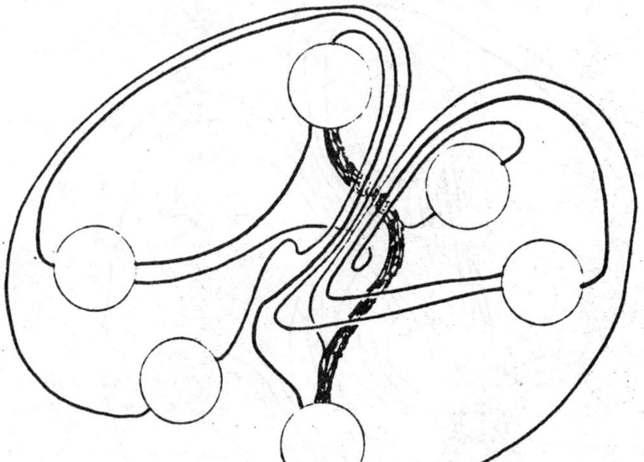

fig.16c

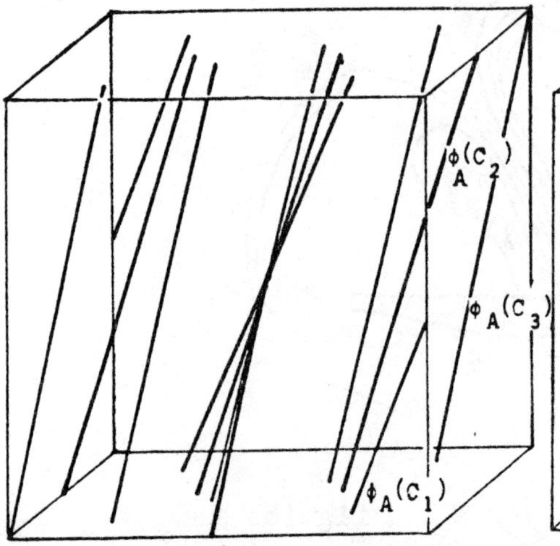

fig.17a

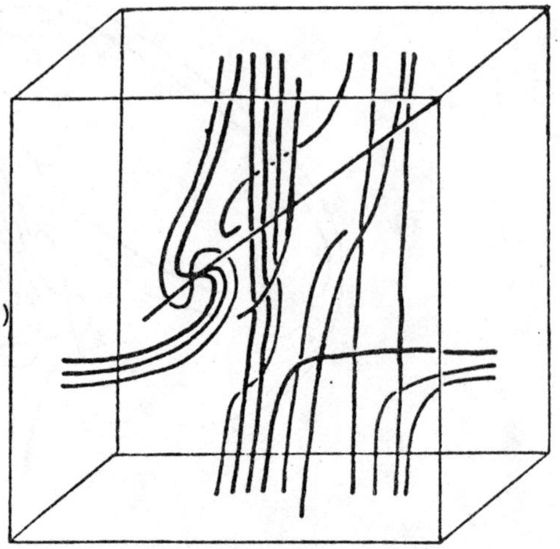

fig.17b

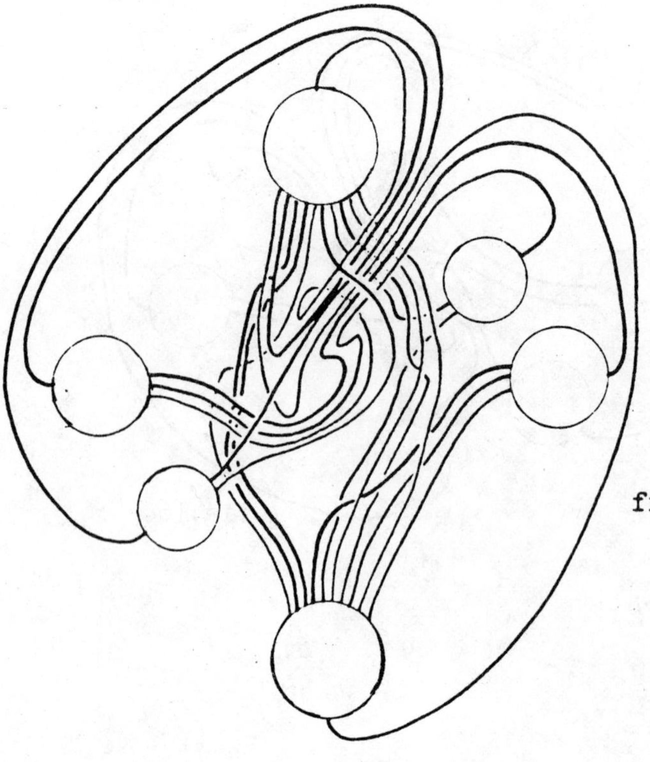

fig.17c

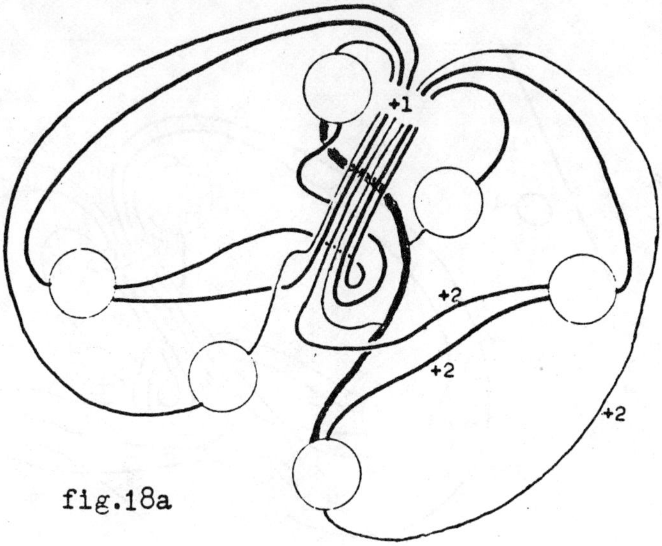

fig.18a

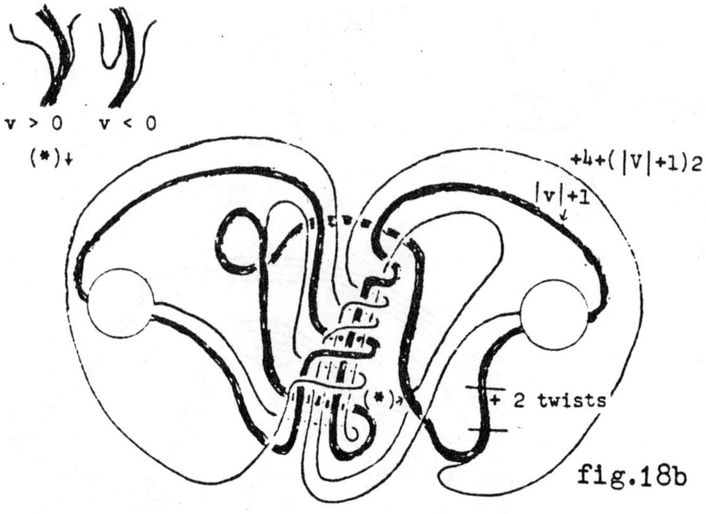

fig.18b

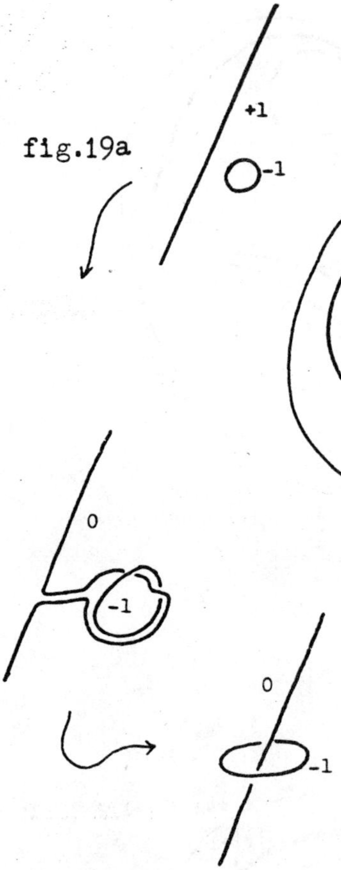

fig.19a

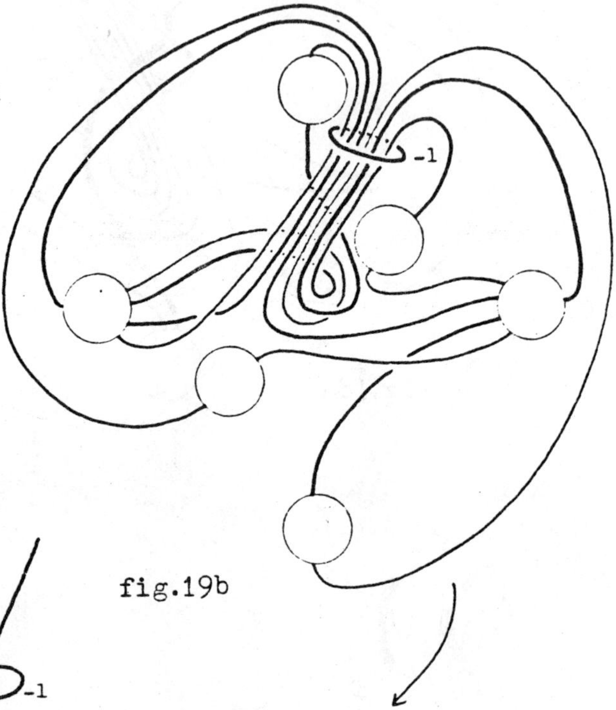

fig.19b

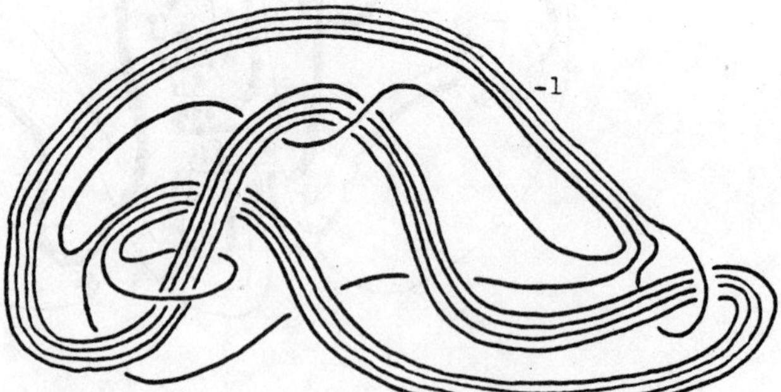

fig.19c

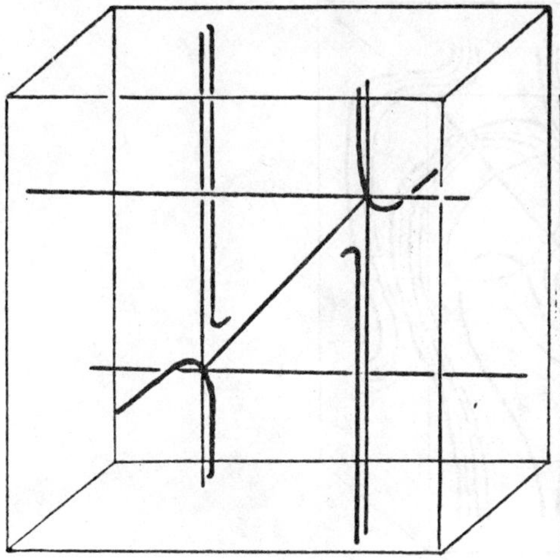

fig.20 B_o

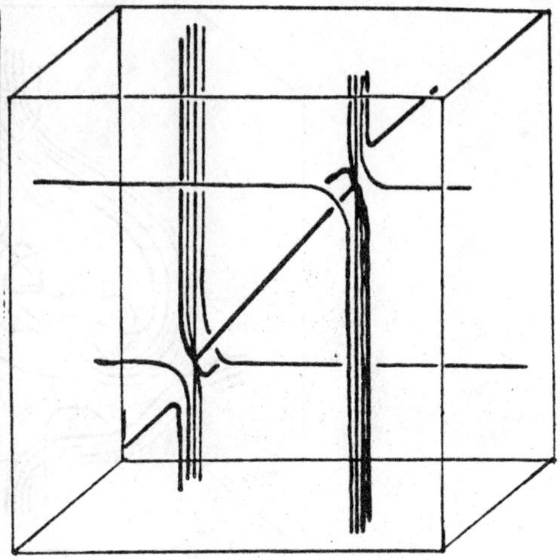

fig.20 B_1

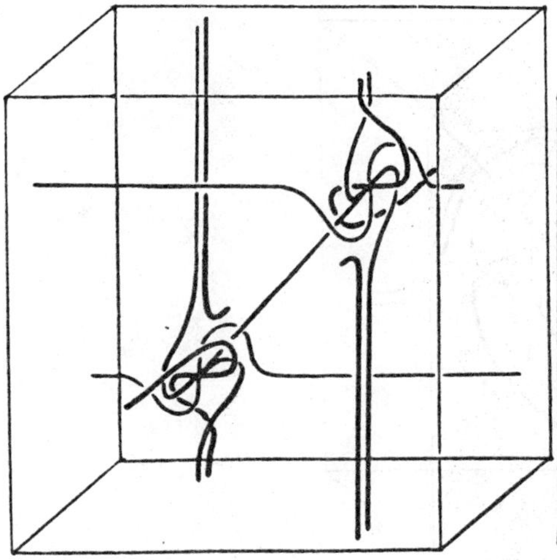

fig.21 B_0

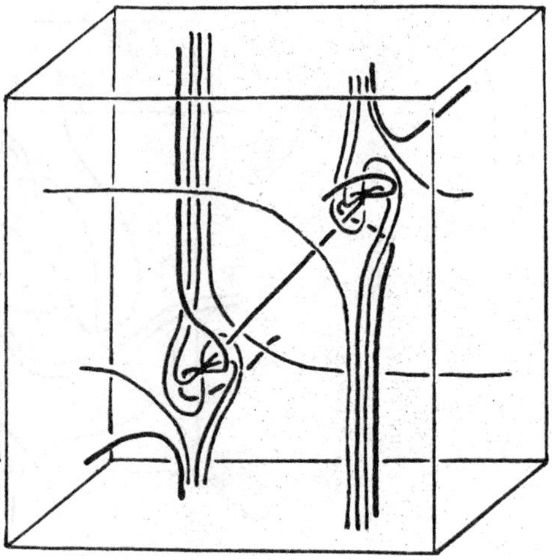

fig.21 B_1

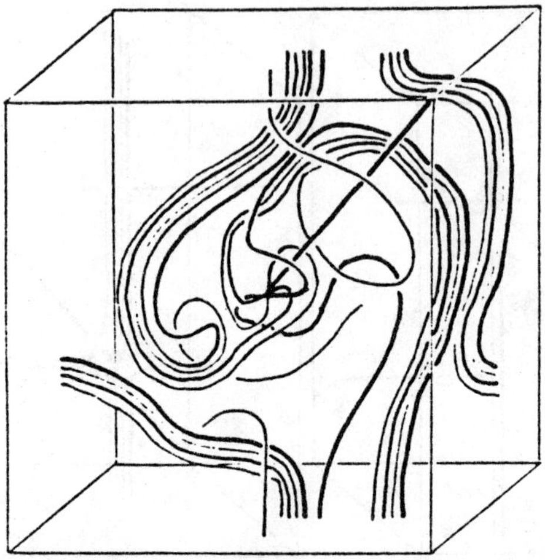

fig.22 B_1

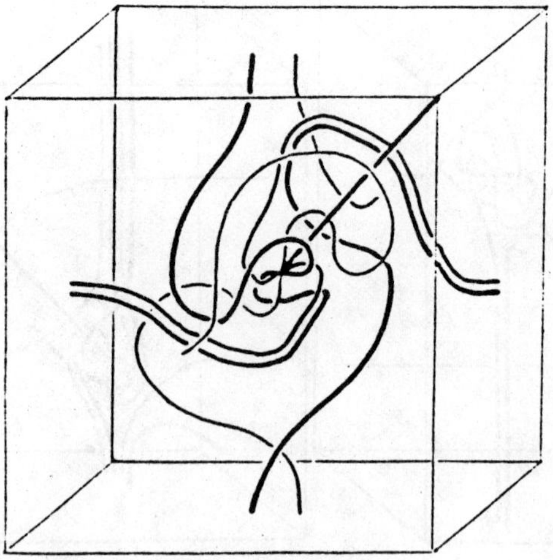

fig.22 B_0

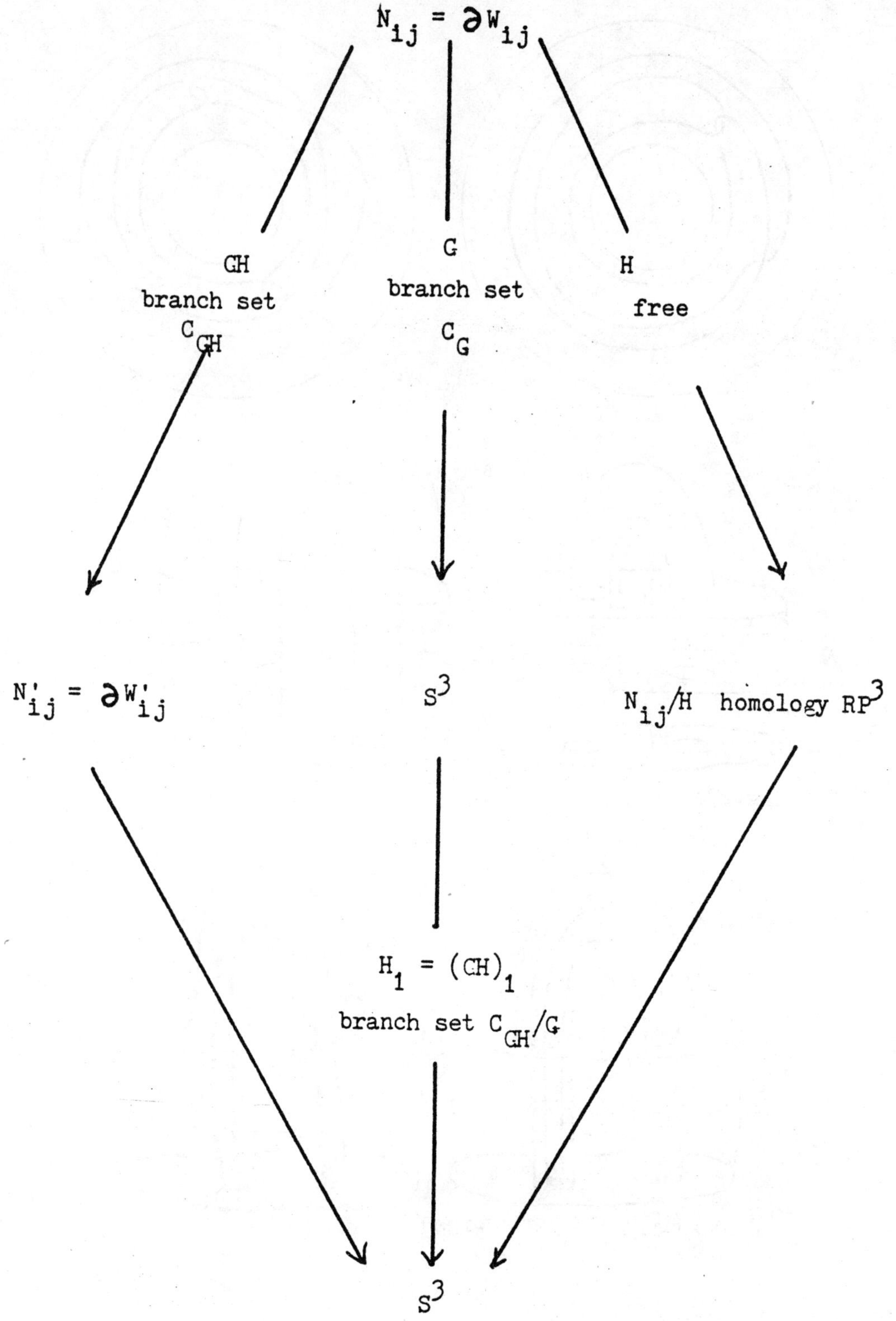
$N_{ij} = \partial W_{ij}$
GH
branch set
C_{GH}
G
branch set
C_G
H
free
$N'_{ij} = \partial W'_{ij}$
S^3
N_{ij}/H homology RP^3
$H_1 = (GH)_1$
branch set C_{GH}/G
S^3

fig.23

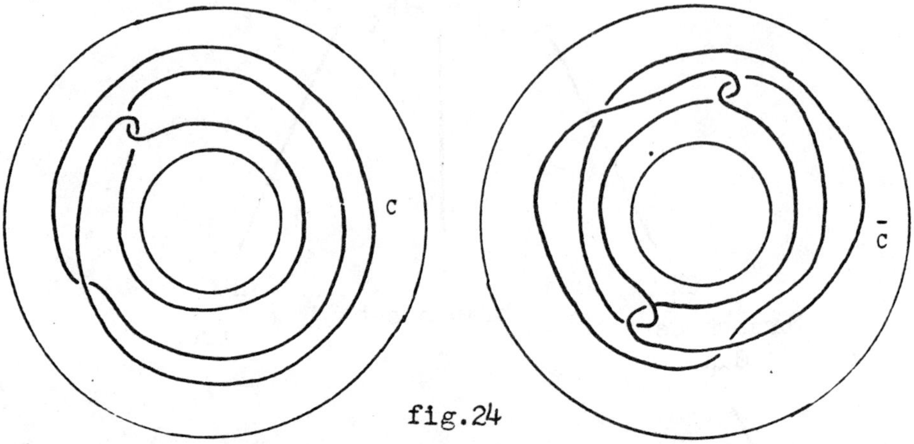

fig.24

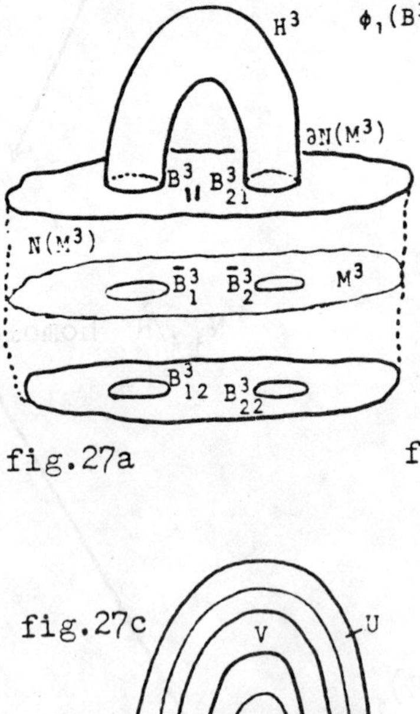

fig.27a

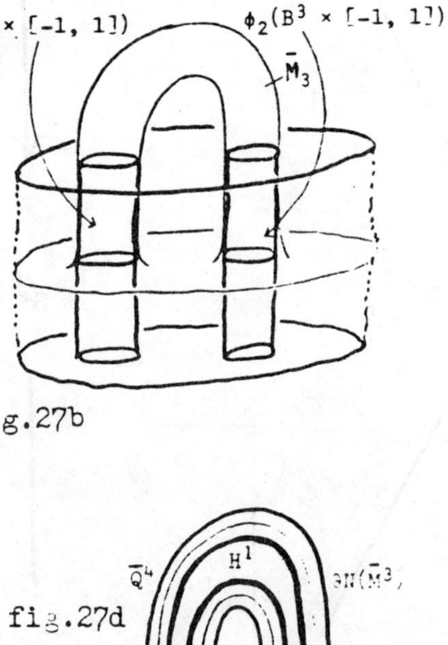

fig.27b

fig.27c

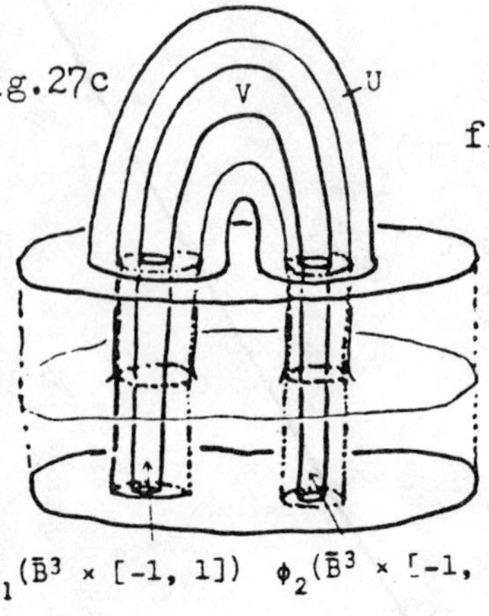

fig.27d

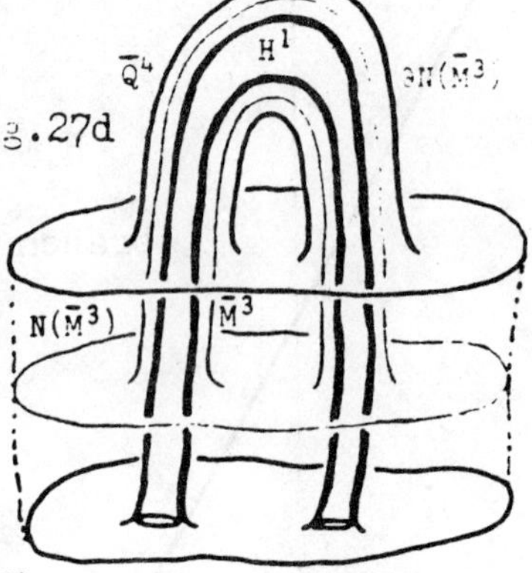

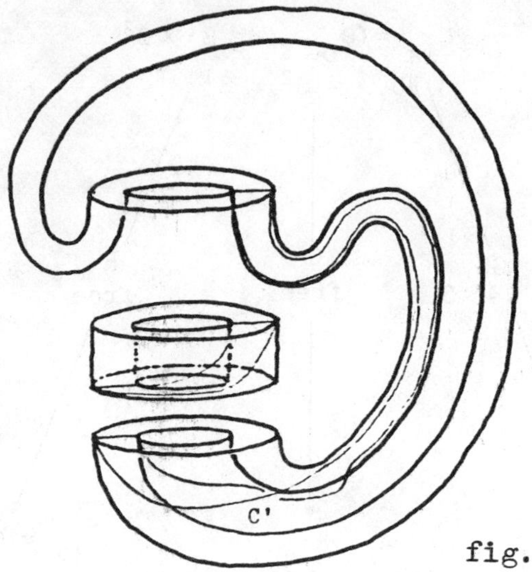

fig.25a

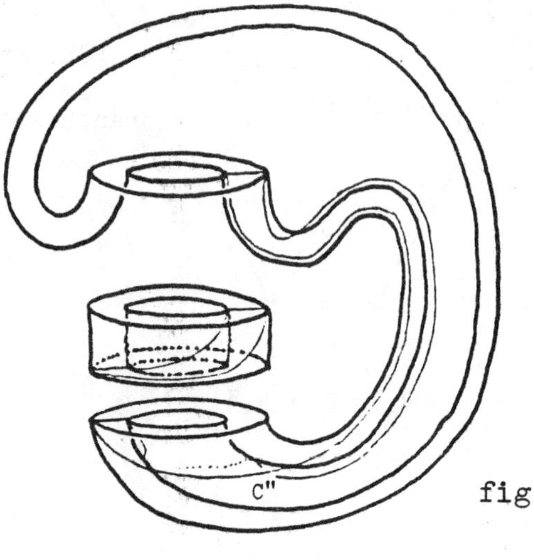

fig.25b

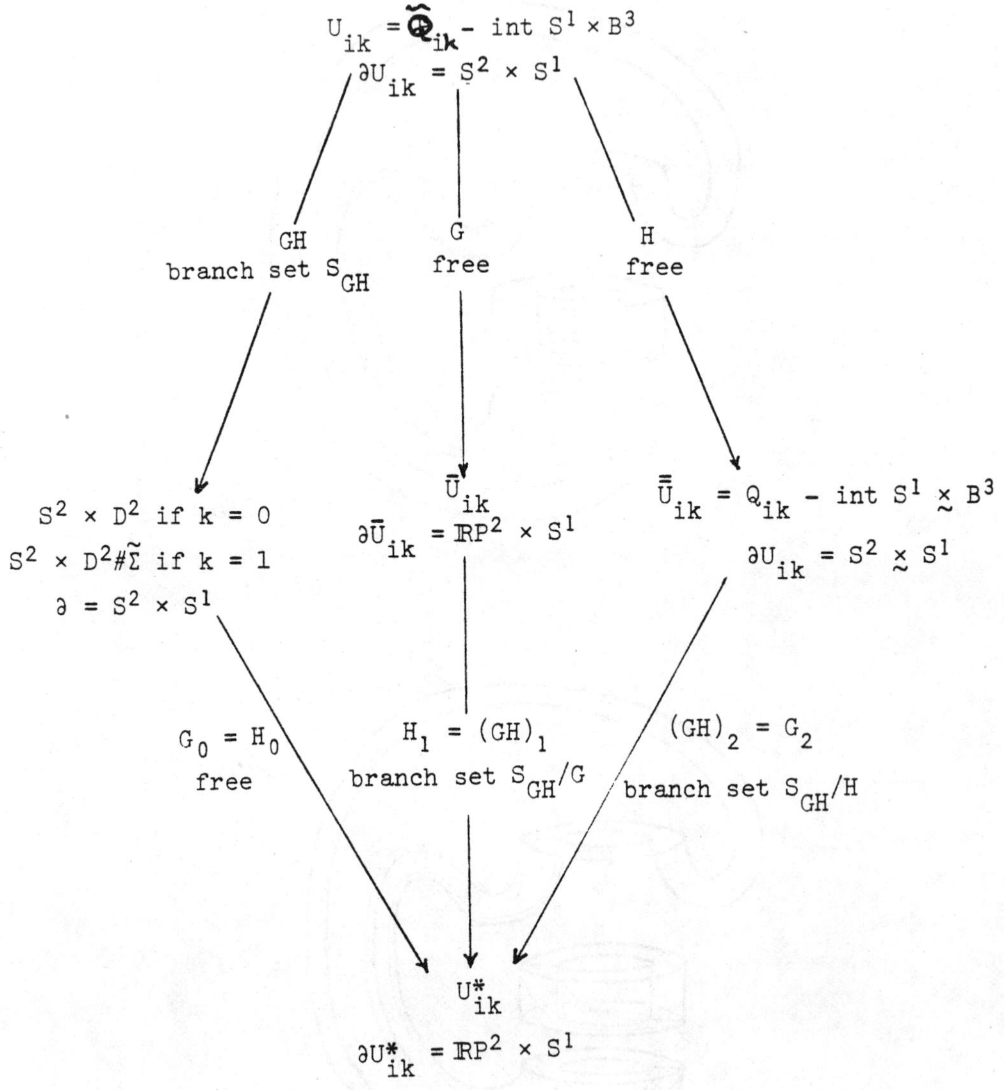
U_ik = Q̃_ik - int S^1 × B^3
∂U_ik = S^2 × S^1
GH
branch set S_GH
G
free
H
free
S^2 × D^2 if k = 0
S^2 × D^2#Σ̃ if k = 1
∂ = S^2 × S^1
Ū_ik
∂Ū_ik = ℝP^2 × S^1
Ū̄_ik = Q_ik - int S^1 ×̰ B^3
∂U_ik = S^2 ×̰ S^1
G_0 = H_0
free
H_1 = (GH)_1
branch set S_GH/G
(GH)_2 = G_2
branch set S_GH/H
U*_ik
∂U*_ik = ℝP^2 × S^1

fig.26

Contemporary Mathematics
Volume 35, 1984

A FAKE 4-MANIFOLD

Selman Akbulut[1]

In this paper we study 4-dimensional fake manifolds; mainly the fake $\mathbb{RP}^4$ which was constructed by Cappell and Shaneson [CS]. This is a smooth closed manifold Q^4 which is simple homotopy equivalent to $\mathbb{RP}^4$ but not diffeomorphic to $\mathbb{RP}^4$. From Q^4 we construct a new 4-manifold:

THEOREM 1: There exists a closed smooth manifold M^4 which is simple homotopy equivalent to $S^3 \tilde{\times} S^1 \# S^2 \times S^2$ but not diffeomorphic to it.

Here $S^3 \tilde{\times} S^1$ denotes the twisted S^3 bundle over S^1. Figure 4.6 is a handlebody of M^4. This handlebody is surprisingly simple, namely: $M^4 = B^3 \tilde{\times} S^1 \cup_\partial$ (two 2-handles) $\cup_\partial B^3 \tilde{\times} S^1$. From this, it easily follows that if $M_0^4 = M^4 - \mathrm{int}(B^3 \tilde{\times} S^1)$ then M_0^4 is a fake $B^3 \tilde{\times} S^1 \# S^2 \times S^2$ and furthermore:

COROLLARY: $M_0^4 \times I \approx (B^3 \tilde{\times} S^1 \# S^2 \times S^2) \times I$

where $\approx$ denotes a differmorphism. Along the way we prove that Q^4 is stably trivial.

THEOREM 2: $Q^4 \# \mathbb{CP}^2 \approx \mathbb{RP}^4 \# \mathbb{CP}^2$

This is interesting because the connected sum of Q^4 with arbitrarily many copies of $S^2 \times S^2$ is not diffeomorphic to the connected sum of $\mathbb{RP}^4$ with arbitrarily many copies of $S^2 \times S^2$ [CS].

In Section 2 we prove a structure theorem for Q^4 similar to the 2-fold cover of Q^4 [AK_4], namely we demonstrate a properly imbedded 2-disk $\Delta^2 \subset D^2 \times \mathbb{RP}^2$ (in fact a ribbon disk) with $\partial\Delta^2 = S^1 \times$ (a point) such that Q^4 is obtained by twisting $D^2 \times \mathbb{RP}^2$ along Δ^2 (Gluck construction) and taking a union with $B^3 \tilde{\times} S^1$. Figure 2.11 is the picture of Δ^2. From this we obtain a solution to a problem of Cappell and Shaneson ([K_1], problem 4.14-B); namely removing the tubular neighborhood of the nontrivial circle in $D^2 \times \mathbb{RP}^2$ and replacing with a certain $(T^3 - B^3)$-bundle over S^1 does not yield a fake $D^2 \times \mathbb{RP}^2$ but it gives a fake self homotopy equivalence of $D^2 \times \mathbb{RP}^2$. In [AK_3] the structure of the 2-fold cover $\tilde{Q}$ of Q was studied and it was shown that $\tilde{Q}$ is an invertible homotopy sphere (in particular it is homeomorphic to S^4), and $\tilde{Q}$ is obtained from S^4 by removing a tubular neighborhood of a knotted S^2 and sewing it back (Gluck construction). Therefore comparing this paper to

[1] Supported in part by N.S.F. grant MCS-8116915

[AK_3] and [AK_4] at times could be useful. We would like to thank Larry Taylor for many helpful discussions on 4-manifold surgery. We also want to thank R. Kirby for a happy collaboration in [AK_3] and [AK_4] which led to this paper.

0. PRELIMINARIES

Throughout the paper we use $\approx$ to denote a diffeomorphism. In this section we discuss handlebodies of 4-manifolds. This presentation is similar to that of [AK_1] and [AK_2], except here 4-manifolds can be nonorientable. Recall that we can present any 2-manifold as a line (a local view of the boundary of the 0-handle) along with the attaching arcs of 1-handles and attaching circles of 2-handles. For example T^2 is

which is a shorthand for:

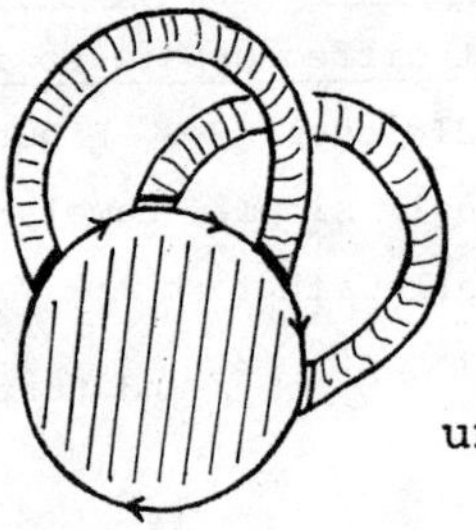

union the 2-handle

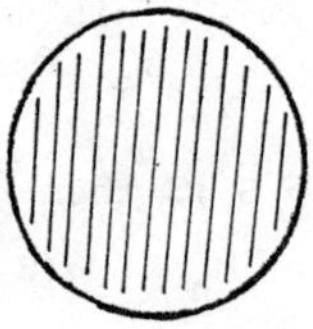

Similarly any 3-manifold can be represented by a plane (a local view of the boundary of the 0-handle) along with attaching discs of 1-handles and attaching circles of 2-handles and a 3-handle. This corresponds to the Heegaard presentation. For example the punctured 3-torus is:

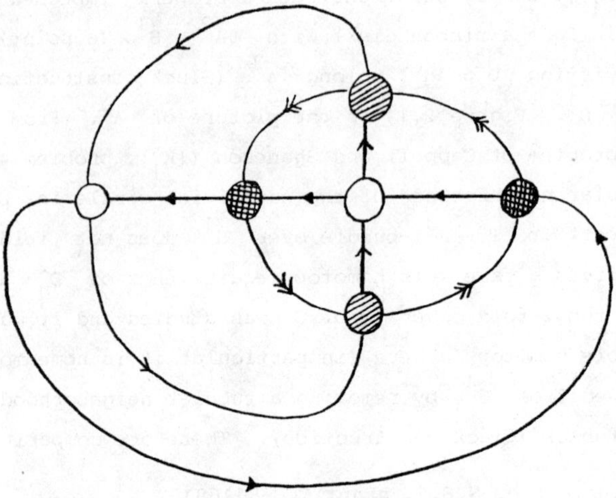

For a given 4-manifold M^4 we draw the handlebody picture of M^4 in the similar way. Namely we will view M^4 from the boundary of the 0-handle $(=S^3)$ and draw the attaching balls of the 1-handles and the attaching circles of the 2-handles in S^3. We will not indicate three and four handles in our pictures.

A pair of balls indicate an attaching $S^0 \times B^3$ of an oriented 1-handle. If we imagine coordinate axes in the centers of these balls the 1-handle identifies the boundaries of these balls by the map $(x,y,z) \rightsquigarrow (x,-y,z)$

This is well defined because the axis, which is reflected, is the axis given by connecting the centers of these balls. In case of orientation reversing handles we put an arc on the centers of these balls which indicates the identification $(x,y,z) \rightsquigarrow (x,-y,-z)$

These arcs indicate the normal direction to the plane where the reflection is performed to the oriented 1-handle to get this handle. We can put one of the balls B^3_- of the 1-handle at the point of ∞, in which case we just draw the other ball B^3_+

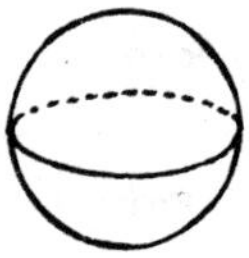

In the case of oriented 1-handle the boundary of B^3_+ is identified with the boundary of B^3_- by identity (i.e. the radial map taking ∂B^3_+ to ∂B^3_-). In the case of nonoriented 1-handle we either draw B^3_+ as

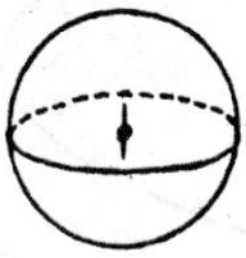

which means ∂B^3_+ is first reflected across the plane perpendicular to the arc then identified with ∂B^3_- by identity; or we draw:

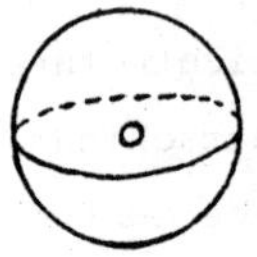

which means we first perform the antipodal map to ∂B^3_+ before identifying with ∂B^3_- by identity. We also denote an oriented 1-handle by an unknotted circle with a dot on it (see [A] and also [AK_1]). The dotted circle means that we delete the thickened unknotted disc the unknot bounds in B^4 obtaining $S^1 \times B^3$. In other words anything that goes through the dotted circle is going over the 1-handle.

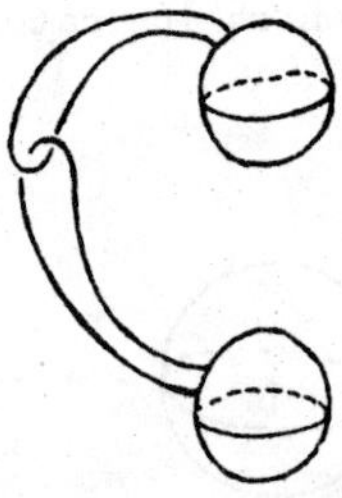

is the same as

Replacing dot by a zero on the dotted circle corresponds surgering $S^1 \times B^3$ to $S^2 \times B^2$; and the vice-versa. We also use dotted ribbon knots which means that we delete the thickened ribbon disc from B^4 (also see [AK_2]). Since a ribbon knot may not bound a unique ribbon disc in B^4 we shade the particular ribbon to indicate the deleted ribbon disc.

If we don't specify the framing on the attaching knot of a two handle, it is the one coming from the normal vector field on the surface of the paper. If we put integers on the knot such as ... (n) ... it means that we add n-full twist to the above framing. This makes the framings well defined even in the presence of orientation reversing 1-handles. For example

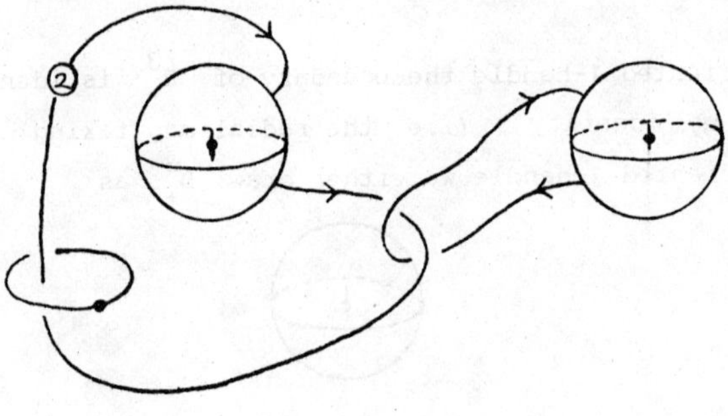

is the same as:

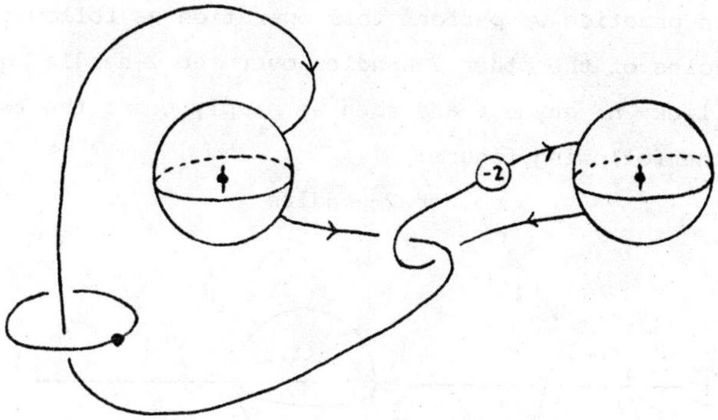

Because 2 twist becomes -2 twist having gone across the orientation reversing 1-handle.

Here is an example of a 4-manifold $M^4(n,m)$:

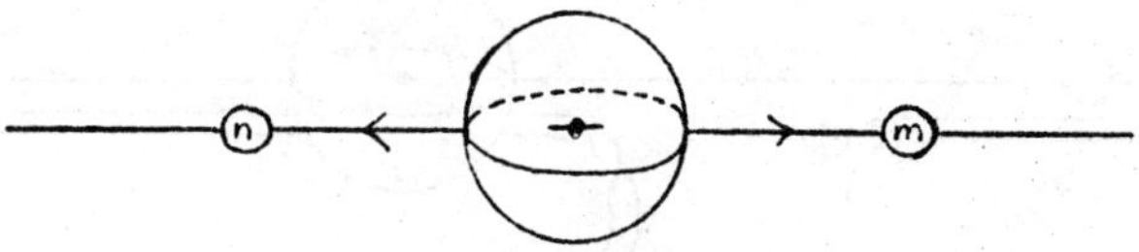

First of all by rotation of the ball B^3_+ 360 ° around the y-axis we get $M(n,m) \approx M(n+1,m-1)$, and by transferring twists across the 1-handle we get $M(n,m) \approx M(n-m,0)$.

Hence $$M(n,m) \approx \begin{cases} M(0,0) & \text{if } n+m \text{ even} \\ M(1,0) & \text{if } n+m \text{ odd} \end{cases}$$

$M^4(0,0)$ is just $D^2 \times \mathbb{R}P^2$ because it is the 4-dimensional trivial thickening of the handlebody of $\mathbb{R}P^2$ which is

(the other attaching arc of the 1-handle is at ∞). Hence $M^4(1,0)$ is $D^2 \tilde{\times} \mathbb{R}P^2$ (the nontrivial D^2-bundle over $\mathbb{R}P^2$) which is the nontrivial thickening of the handlebody of $\mathbb{R}P^2$.

Recall $\mathbb{R}P^4 = D^2 \tilde{\times} \mathbb{R}P^2 \underset{\partial}{\cup} B^3 \tilde{\times} S^1$ hence $\partial(D^2 \tilde{\times} \mathbb{R}P^2) \approx \partial(B^3 \tilde{\times} S^1)$. For a given 4-manifold M^4 containing $D^2 \tilde{\times} \mathbb{R}P^2$ we call the operation:

$$M^4 \rightsquigarrow \hat{M}^4 = (M - D^2 \tilde{\times} \mathbb{R}P^2) \underset{\partial}{\cup} B^3 \tilde{\times} S^1$$

blowing down the $\mathbb{RP}^2$. This is similar to the "blowing down $\mathbb{CP}^2$" operation of [K_2]. In practice we perform this operation as follows: We slide the attaching circles of the other 2-handles over the 2-handle h of $D^2 \tilde{\times} \mathbb{RP}^2$ until they don't link h anymore and then we simply erase the two handle of $D^2 \tilde{\times} \mathbb{RP}^2$ as in the following figure:

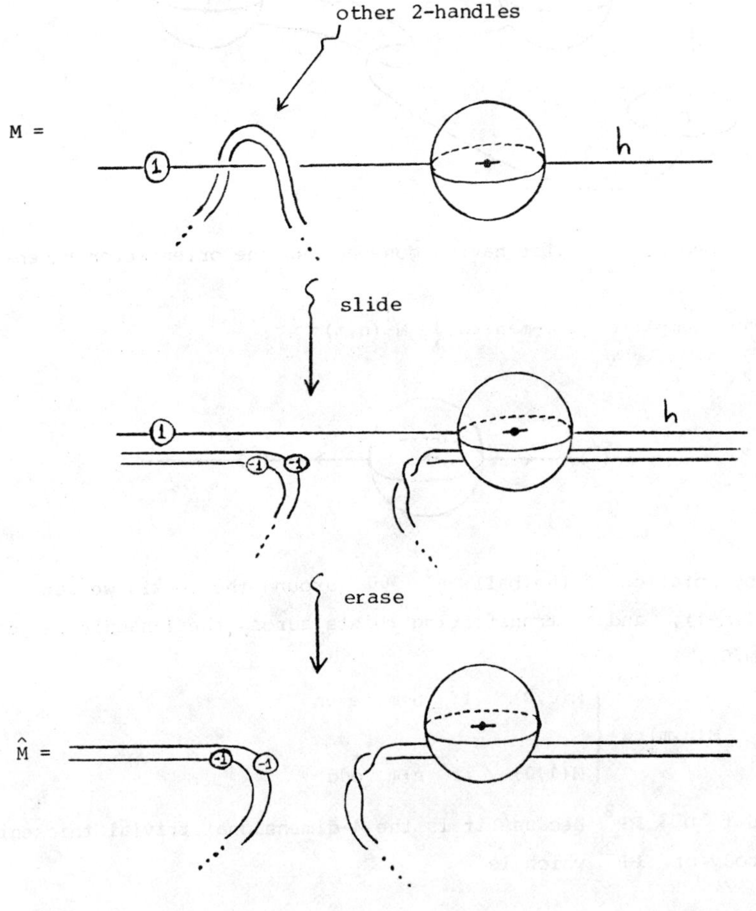

We leave the verification, that this process corresponds to the blowing down operation, as an exercise to the reader. We call the inverse of this operation blowing up an $\mathbb{RP}^2$.

If in a given 4-manifold an attaching circle of a two handle goes through an oriented 1-handle geometrically once we can cancel this pair of 1 and 2 handles by simply erasing them from the picture. The attaching circles of other two handles which go through the 1-handle has to be modified as follows (see [AK_3])

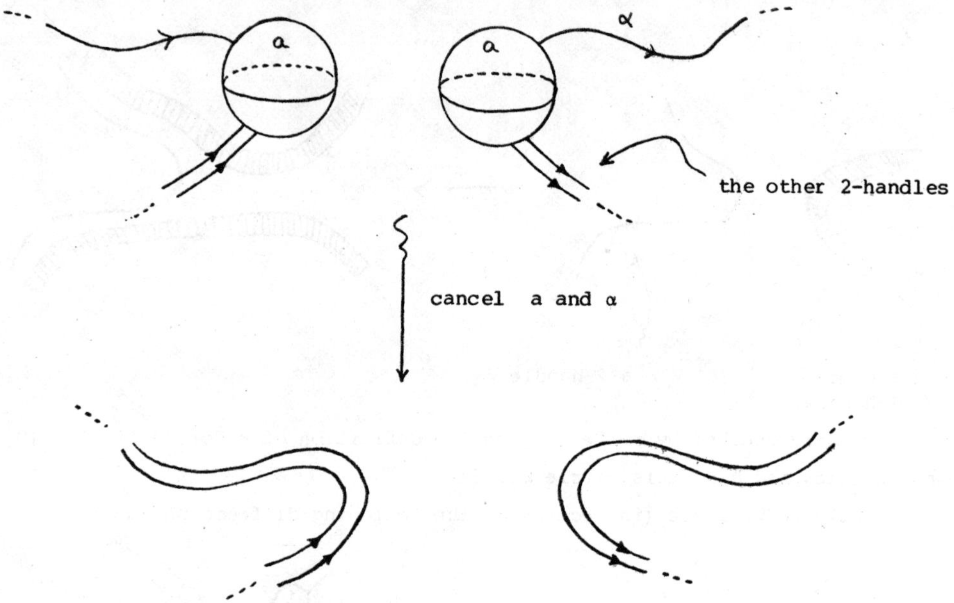

Also if we have a trivial 2-handle 0^0 and a 3-handle attached onto this in the obvious way (i.e. along $S^2 \subset S^2 \times S^1 = \partial(0^0)$) we can cancel them by simply erasing 0^0 and forgetting the 3-handle.

In our figures we use arrows such as

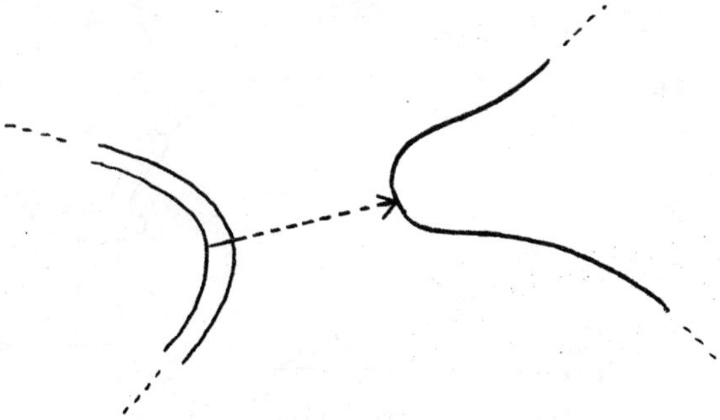

we ignored them, unless it indicated that we do a handle slide as shown by the arrow in which case it means slide two handles over each other i.e.

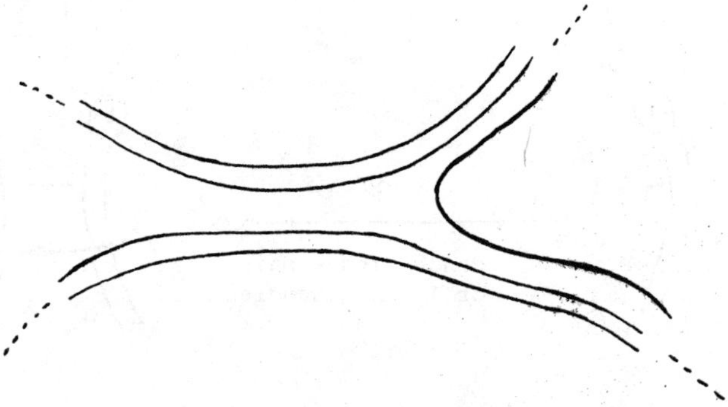

Sometimes 1-handles can be slid over 2-handles such as:

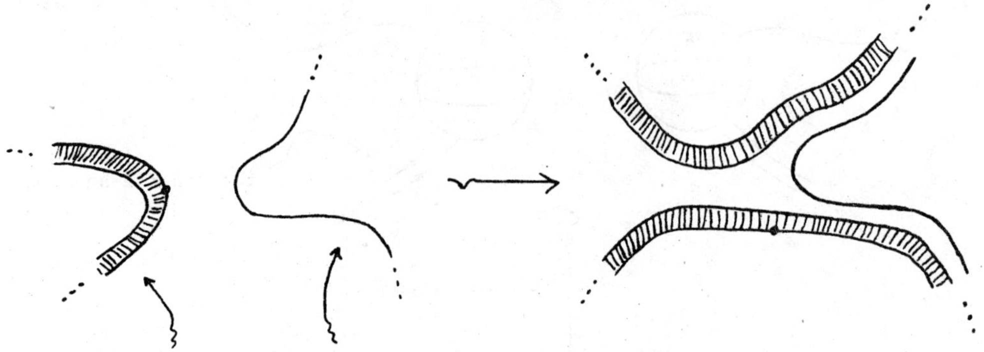

This is easily checked by reflecting on the definition of a dotted circle (B^4 minus a thickened disc this circle bounds).

Finally we will use (in Section 4) the following diffeomorphism

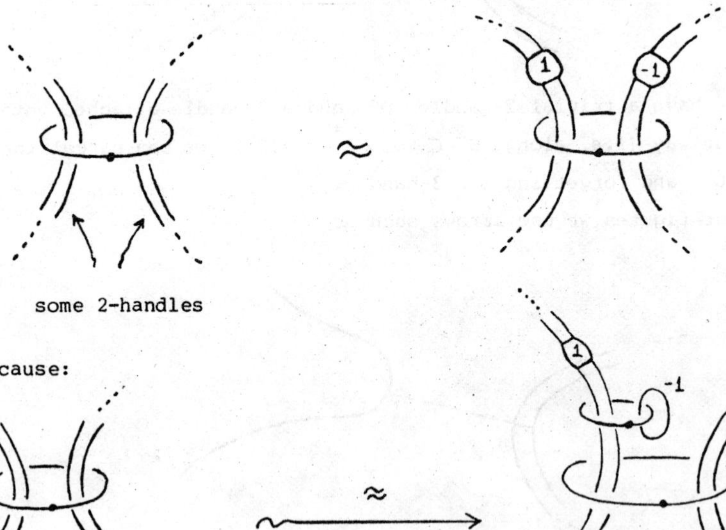

This is because:

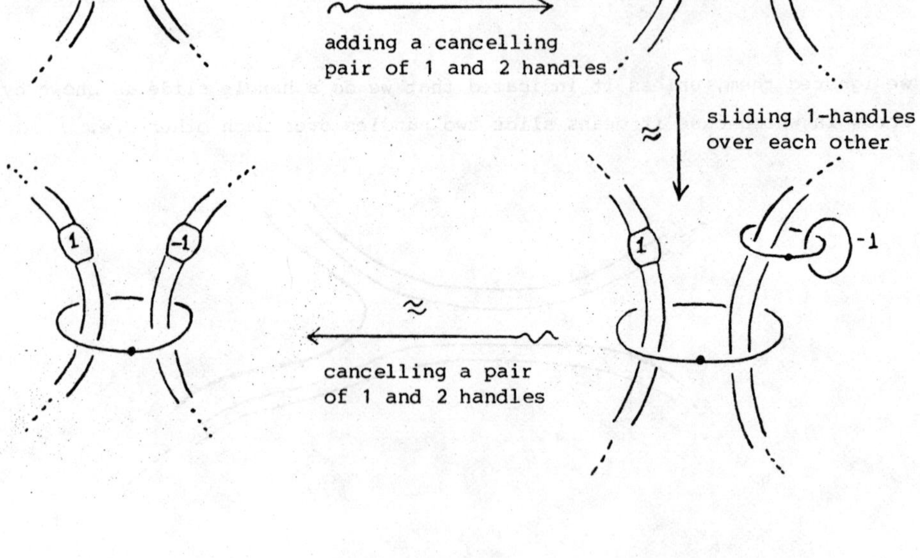

1. STRUCTURE OF Q^4

Let $A = \begin{pmatrix} 0 & 1 & 0 \\ 0 & 0 & 1 \\ -1 & 1 & 0 \end{pmatrix}$ then A induces an orientation reversing linear map $A: \mathbb{R}^3 \to \mathbb{R}^3$. Since A preserves the integral lattice in $\mathbb{R}^3$ it induces a self diffeomorphism of T^3 where $T^3 = \mathbb{R}^3/\mathbb{Z}^3$. Since $A(0) = 0$, A in fact induces a diffeomorphism $\tilde{A}: T_0^3 \to T_0^3$ where $T_0^3 = T^3 - \text{interior}(D^3)$ and D^3 is an imbedded disc in T^3. Then since $\tilde{A}|\partial T_0^3$ is an orientation reversing diffeomorphism of S^2 after a small isotopy we can assume that $\tilde{A}|\partial T_0^3$ is the antipodal map of S^2. Let $C = T_0^3 \times I/(x,0) \sim (\tilde{A}(x),1)$ be the mapping torus of $\tilde{A}$. Then clearly ∂C is the twisted S^2 bundle over S^1 which we denote by $S^2 \tilde{\times} S^1$. Let $D^2 \tilde{\times} \mathbb{RP}^2$ be the twisted D^2 bundle over $\mathbb{RP}^2$ (it is the tubular neighborhood of $\mathbb{RP}^2$ in $\mathbb{RP}^4$). Since $\partial(D^2 \tilde{\times} \mathbb{RP}^2) = S^2 \tilde{\times} S^1$ we can construct: $Q^4 = C \cup_\partial (D^2 \tilde{\times} \mathbb{RP}^2)$. This is the Cappell and Shaneson's construction of a fake $\mathbb{RP}^4$ [CS]. We draw a handlebody picture of Q^4 by the method of [AK_3]: Figure 1.1 is the handlebody picture of $T_0^3 \times I$. We isotop $\tilde{A}$ so that

(i) $\tilde{A}$ is the antipodal map on the small ball centered at the origin.

(ii) $\tilde{A}$ takes the 1-handles of T_0^3 to itself.

Figure 1.2 indicates this isotopy. The first picture in this figure is the images of the coordinates axis under $\tilde{A}$. Since the opposite faces of the cube is identified the coordinate axes are the cores of the 1-handles. So the isotopy moves the end points of the arcs to the centers of the sides of the cube (hence into the 1-handles).

So the handlebody of C is obtained from the handlebody of $T_0^3 \times I$ by identifying with $\tilde{A}$. This identification adds a $k+1$ handle to $T_0^3 \times I$ for every k handle of $T_0^3 \times I$. Hence we add one 1-handle three 2-handles and three 3-handles to get C. Figure 1.3 is the handlebody picture of C except the three handles are not drawn even though they are there. The new 1-handle is a nonoriented 1-handle (because of (i)) attached along the ball at the origin (as indicated in the figure) and the ball at ∞ (hence not seen in the figure).

Figure 1.4 is the same as Figure 1.3 except the two handles α_2, α_3 are not drawn, and the 1-handle a_1 is cancelled by the 2-handles which goes over a_1 once. We get Figure 1.5 by cancelling the 1-handle a_3 by the 2-handle which goes over a_3 once. Figure 1.6 is the same as Figure 1.5 except the 1-handle a_2 is indicated as a dotted circle. By further isotopies we get Figures 1.7 and 1.8. By rotating the B^3 at ∞ (where the one end of the 1-handle attached) by 180 degrees we get Figure 1.9. Hence the notation on the 1-handle in Figure 1.9 is changed. We claim that the boundary of the manifold in Figure 1.9 is $S^2 \tilde{\times} S^1 \# S^2 \times S^1$. To see this surger the 1-handle (i.e. replace the dot with a zero), and then surger the two handle (i.e. put a dot on the attaching circle of this handle) as in Figure 1.10. A further isotopy gives

Figure 1.11. If we now cancel the new 1-handle with the obvious 2-handle (the one corresponding to the circle going through the 1-handle once) we get Figure 1.12. The boundary is obviously $S^2 \tilde{\times} S^1 \# S^1 \times S^2$.

Now here comes an important point! Recall starting with Figure 1.4 we ignored to draw the 2-handles α_2, α_3. If we draw these handles and carry them along the processes of Figure 1.4 through Figure 1.12 α_2, α_3 will end up being two unknotted circles in Figure 1.12 (check). This means that α_2, α_3 are attached to trivial circles on the boundary of the Figure 1.4 and therefore two of the three 3-handles of Figure 1.3 must be cancelling the 2-handles α_2, α_3. In other words we are justified in ignoring α_2, α_3 from the picture along with two 3-handles. So Figure 1.9 along with one three handle is the picture of C^4.

Q^4 is obtained by gluing $D^2 \tilde{\times} \mathbb{R}P^2$ to C^4. Hence to get Q^4 we must add a 2-handle, a 3-handle and a 4-handle (upside down $D^2 \tilde{\times} \mathbb{R}P^2$) to C^4 along ∂C^4. Since we don't draw 3 and 4-handles we only indicate the attaching circle of the 2-handle γ. γ is attached along the standard circle which goes twice around $S^2 \tilde{\times} S^1 = \partial C$ (see Section 0). In fact γ is attached as in Figure 1.13. To check this we apply the diffeomorphism $\partial(C^4) \approx S^2 \tilde{\times} S^1$ of Figures 1.9 - 1.12 to Figure 1.13; and we see that this diffeomorphism takes γ to the 'right' circle in Figure 1.12. The framing on γ is any odd number; so we assign $+1$ framing as indicated in the figure. Hence Figure 1.13 along with two 3-handles and a 4-handle is the handlebody of Q^4.

By doing the indicated handle slides to Figures 1.13 and 1.14 we get Figures 1.14 and 1.15 respectively. Notice the 2-handle δ in Figure 1.15 goes over the 1-handle a_2 geometrically once (after an isotopy), hence it cancels it. After this cancellation we will have one 1-handle, two 2-handles (α_1 and γ) along with two 3-handles and a four handle left. We want to turn this handlebody upside down; i.e. we want to draw it as two 1-handles, two 2-handles and one 3-handle and a 4-handle. To do this we draw the dual 2-handles σ and τ, then we change the interior of the handlebody to $B^3 \tilde{\times} S^1 \# B^3 \times S^1$ via surgeries and handle slides, while carrying σ and τ. Then $B^3 \tilde{\times} S^1 \# B^3 \times S^1$ and the 2-handles σ, τ (and a three and a four handle) will be what we want.

To do this we go back to Figure 1.14 carrying along σ and τ. We then replace the dots on the handles α_1 and a_2 (i.e. surgery) and by an isotopy we get Figure 1.16 (this is similar to going from Figure 1.9 to Figure 1.11). After performing the obvious handle cancellation as in Figure 1.17 we arrive at Figure 1.18. Then by doing the indicated handle slides to this figure we get Figure 1.19. If we ignore σ, τ Figure 1.19 becomes just $D^2 \tilde{\times} \mathbb{R}P^2 \# S^2 \times D^2$. Hence in order to change the interior to $B^3 \tilde{\times} S^1 \# B^3 \times S^1$ we have to

(1) Surger the imbedded S^2

(2) Blow down the $\mathbb{R}P^2$

We surger S^2 by putting dot on the unknotted 2-handle as in Figure 1.19. We will blow down $\mathbb{R}P^2$ a little later (Figure 1.23 and Figure 1.24). By sliding σ over τ twice in the obvious way we get Figure 1.20. We continue to call the slid 2-handle by σ. By isotopies we get to Figures 1.21, 1.22. By a further isotopy (this time pulling the 1-handle around) we get Figure 1.23. By blowing down $\mathbb{R}P^2$ (i.e. γ) in Figure 1.23 we get Figure 1.24. After performing the indicated handle slides and pulling the 1-handle to the standard position we get Figure 1.25. After isotoping the ball at ∞ into the picture we get Figure 1.26 which is $Q^4 - \text{int}(B^3 \tilde{\times} S^1)$.

Figures 1.27 through 1.33 give even a simpler handlebody for $Q^4 - \text{int}(B^3 \tilde{\times} S^1)$. We go to Figure 1.27 from Figure 1.24 by an isotopy, then by the indicated handle slides we get Figure 1.28 and then Figure 1.29. After isotopies and the indicated handle slides we get Figures 1.30 through 1.33. Figure 1.33 is $Q^4 - \text{int}(B^3 \tilde{\times} S^1)$.

2. THE RIBBON IN $D^2 \times \mathbb{R}P^2$

Let $N^4 = Q^4 - \text{int}(B^3 \tilde{\times} S^1)$; N^4 is a fake $D^2 \tilde{\times} \mathbb{R}P^2$. Figure 1.14 with one 3-handle is the handlebody of N^4. This is because Figure 1.14 along with two 3-handles and a 4-handle is Q^4. Figure 2.1 is N^4. This is because if we cancel b with ϑ we get Figure 1.14 back.

Let $V^4 = N^4$ minus the handles $b \cup \vartheta$, then $V^4 = C^4 \cup$ the 2-handle γ attached by 0-framing, but $D^2 \times \mathbb{R}P^2 = B^3 \tilde{\times} S^1 \cup$ the 2-handle γ attached by 0-framing (Section 0). Hence V^4 is obtained from $D^2 \times \mathbb{R}P^2$ by replacing a tubular neighborhood of the orientation reversing circle with C^4. We will show that V^4 is diffeomorphic to $D^2 \times \mathbb{R}P^2$. This answers a question of Cappell and Shaneson ([K_1], problem 4.14-B).

To get V^4 we ignore b and ϑ from Figure 2.1 and add one 3-handle. Then we do a handle slide (as indicated in Figure 2.1) to get Figure 2.2. By another handle slide we get Figure 2.3. By cancelling the obvious pair of handles from Figure 2.3 we get Figure 2.4 which is $D^2 \times \mathbb{R}P^2 \# S^2 \times D^2$. The three handle cancels $S^2 \times D^2$ (check) and we end up with $D^2 \times \mathbb{R}P^2$. Hence we have shown that $V^4 = D^2 \times \mathbb{R}P^2$.

Now we go back to N^4; i.e. we add back the handles b, ϑ to Figure 2.1. If we carry along the handles b, ϑ during the diffeomorphism $V^4 \approx D^2 \times \mathbb{R}P^2$ (as in Figures 2.1-2.4) we get Figures 2.5-2.8. Along the way we slide the 1-handle b over a 2-handle as indicated in Figure 2.5. By isotoping the B^3 at ∞ into the picture we get Figure 2.9. After a handle slide and an isotopy we get Figures 2.10 and 2.11. In Figure 2.11 the shaded ribbon disc is the ribbon 1-handle which $D^2 \times \mathbb{R}P^2$ is twisted along to get N^4. Reader can verify that the 2-fold cover of Figure 2.11 gives the ribbon 2-sphere in S^4 which is discussed in [AK_4].

3. $Q^4 \# \mathbb{C}P^2 \approx \mathbb{R}P^4 \# \mathbb{C}P^2$

Recall Figure 1.19 after blowing down $\mathbb{R}P^2$ (i.e. γ) gives $N^4 = Q^4 - B^3 \tilde{\times} S^1$. Because in Section 1 we have seen that the blown down Figure 1.19 along with a 3-handle and a 4-handle gives Q^4. To prove $Q^4 \# \mathbb{C}P^2 \approx \mathbb{R}P^4 \# \mathbb{C}P^2$ it suffices to show that $N^4 \# \mathbb{C}P^2 \approx (D^2 \tilde{\times} \mathbb{R}P^2) \# \mathbb{C}P^2$.

We claim the loop ρ in Figure 3.1 is the trivial loop on the boundary. This can be seen by going back to Figure 1.18 and sliding ρ over τ and then going back to Figure 1.15 and carrying ρ along. In Figure 1.15 ρ becomes the trivial dual circle to δ. Since σ and τ have zero framings we turn them into 1-handles; they then cancel α_1 and γ. After cancelling δ with a_2 ρ becomes an unknot in $\partial(B^3 \tilde{\times} S^1)$.

Hence if we add a 2-handle to Figure 3.1 along ρ with +! framing it corresponds connected summing with $\mathbb{C}P^2$. We do this; and then by sliding σ over ρ we get Figure 3.2. An isotopy gives Figure 3.3. By a handle slide we obtain Figure 3.4. After cancelling the obvious pair of one and two handles we get Figure 3.5. We slide +1 framed handle over the 0-framed handle it becomes free. Then we blow down $\mathbb{R}P^2$ and obtain Figure 3.6 which is $(D^2 \tilde{\times} \mathbb{R}P^2) \# \mathbb{C}P^2$ we are done.

4. A FAKE $S^3 \tilde{\times} S^1 \# S^2 \times S^2$

Recall Figure 1.24 is $N^4 = Q^4 - \mathrm{int}(B^3 \tilde{\times} S^1)$. By performing only one of the indicated handle slides (the arrow pointing up) to Figure 1.24 we get Figure 4.1. By a diffeomorphism (see end of Section 0) we get Figure 4.2. By surgering Figure 4.2 (i.e. removing the dot) and then blowing down the obvious $\mathbb{R}P^2$ we get Figure 4.3. By isotopies we get Figures 4.4 and 4.5. By isotoping the 1-handle we get Figure 4.6 which we call M_0^4. Since the surgery (removing the dot) to Figure 4.2 is performed to a null homotopic loop, it corresponds to taking connected sum with $S^2 \times S^2$. Therefore $N^4 \# S^2 \times S^2 = (D^2 \tilde{\times} \mathbb{R}P^2) \cup M_0^4$. Since $Q^4 \# S^2 \times S^2$ is fake [CS] so is $N \# S^2 \times S^2$. This implies that M_0^4 has to be a fake $B^3 \tilde{\times} S^1 \# S^2 \times S^2$, since any self-diffeomorphism of $S^2 \tilde{\times} S^1$ extends to $B^3 \tilde{\times} S^1 \# S^2 \times S^2$.

$\partial(M_0 \times I)$ is the double of M_0^4. This is standard because it is obtained from Figure 4.6 by attaching two trivial (dual) 2-handles with 0-framings (i.e. an unknotted circle for each 2-handle which links it geometrically once). By sliding the 2-handles of M_0^4 over the new 2-handles we get

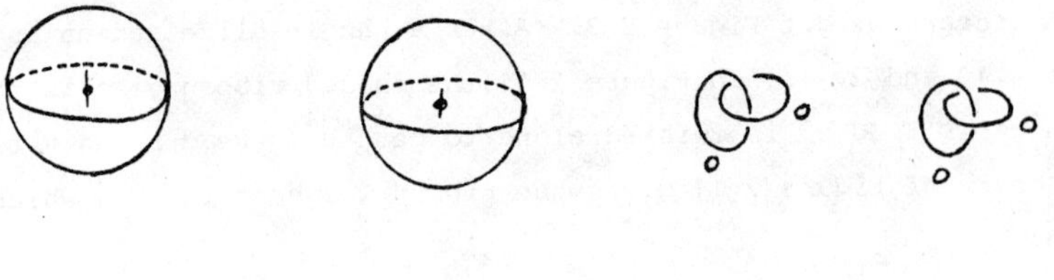

which is (along with a 3-handle and a 4-handle) $S^3 \tilde{\times} S^1 \overset{2}{\#} S^2 \times S^2$.

The fact that $M_0^4 \times I \approx (B^3 \tilde{\times} S^1 \# S^2 \times S^2) \times I$ follows from 5-dimensional surgery exact sequence:

$$L_6(\mathbb{Z},-) \to \mathcal{S}(X \times I, \partial) \to [X \times I/\partial; G/PL] \to \underset{\overset{\parallel}{0}}{L_5(\mathbb{Z},-)}$$

where $X = B^3 \tilde{\times} S^1 \# S^2 \times S^2$. The first map is zero map (check) and $[X \times I/\partial; G/PL] = 0$ so $\mathcal{S}(X \times I, \partial) = 0$ and the claim follows (see [W]).

BIBLIOGRAPHY

[A] S. Akbulut, On 2-dimensional homology classes of 4-manifolds, Math. Proc. Camb. Phil. Soc. (1977), 82, 99-106.

[AK_1] S. Akbulut and Robion Kirby, Mazur manifolds, Mich. Math. J. 26 (1979), 259-284.

[AK_2] ———————————, Branched covers of surfaces in 4-manifolds, Math. Ann. 252, (1980), 111-131.

[AK_3] ———————————, An exotic involution of S^4, Topology, Vol. 18, (1979), 75-81.

[AK_4] ———————————, Homotopy 4-spheres and a correction to "An exotic involution of S^4" (to appear).

[CS] S. Cappell and J. Shaneson, Some new 4-manifolds, Ann. of Math. 104 (1976) 61-72.

[K_1] R. Kirby, Problems in low dimensional manifold theory, Proc. Sym. Pure Math., Vol. 32 (1978) 273-312.

]K_2] R. Kirby, A calculus of framed links in S^3, Invent. Math. 45 (1978), 35-56.

[W] C. T. C. Wall, Surgery on compact manifolds, Academic Press, London. Math. Soc. Mon. No. 1 (1970).

DEPARTMENT OF MATEHMATICS
MICHIGAN STATE UNIVERSITY
EAST LANSING, MI 48824

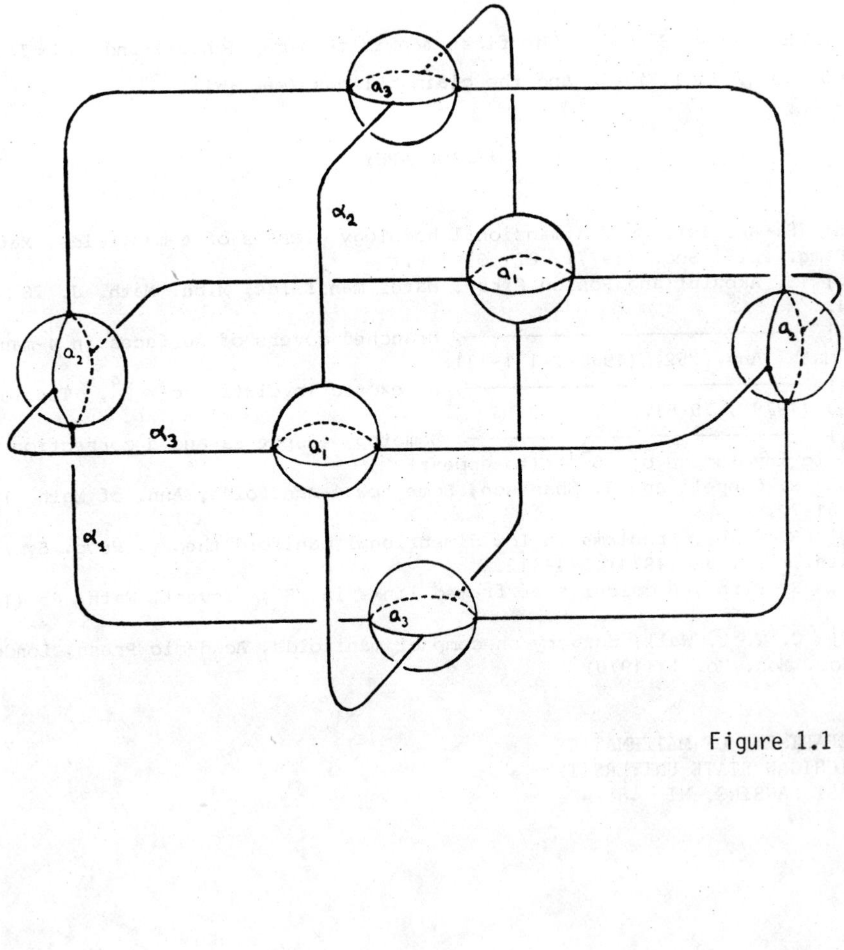

Figure 1.1

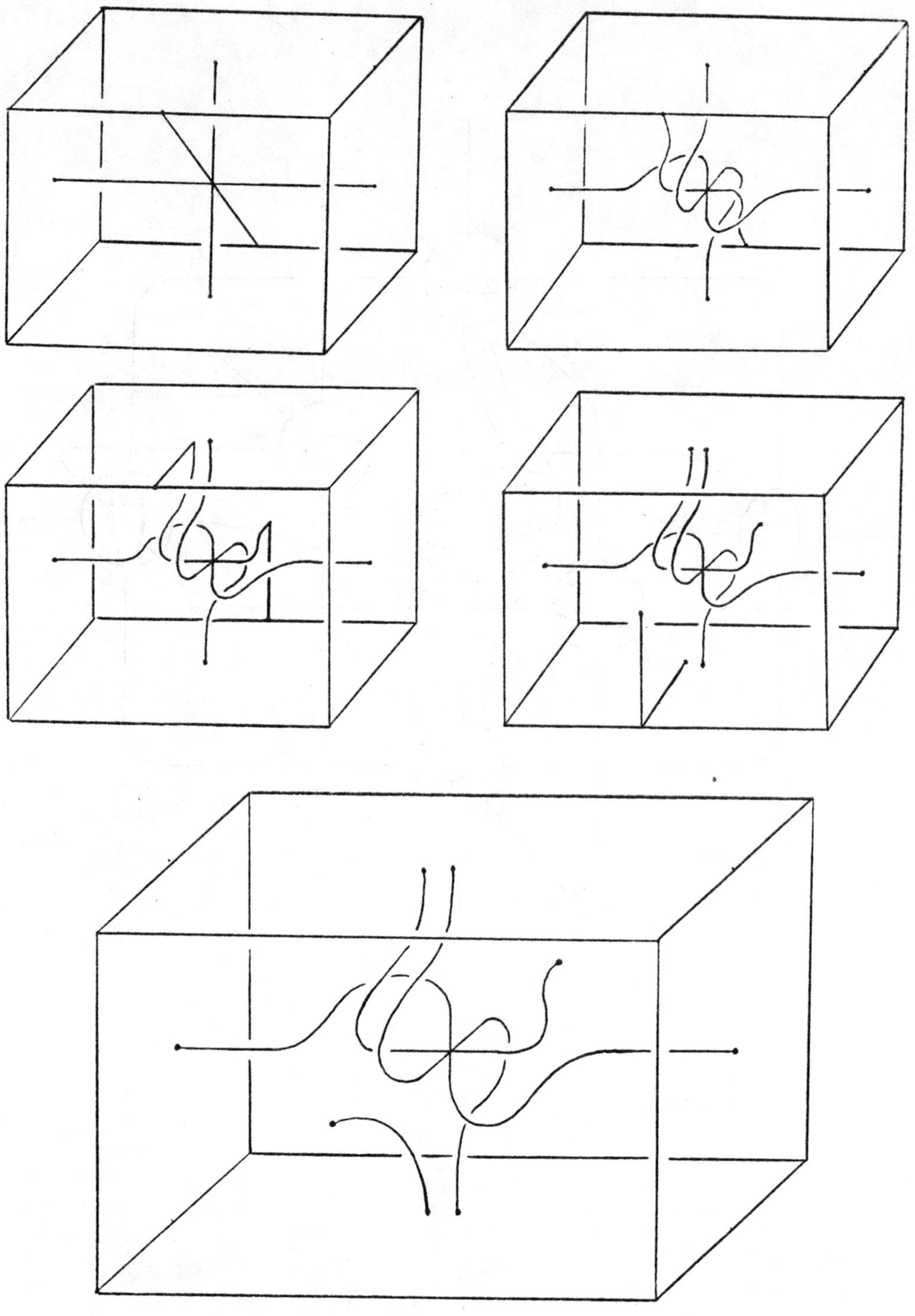

Figure 1.2

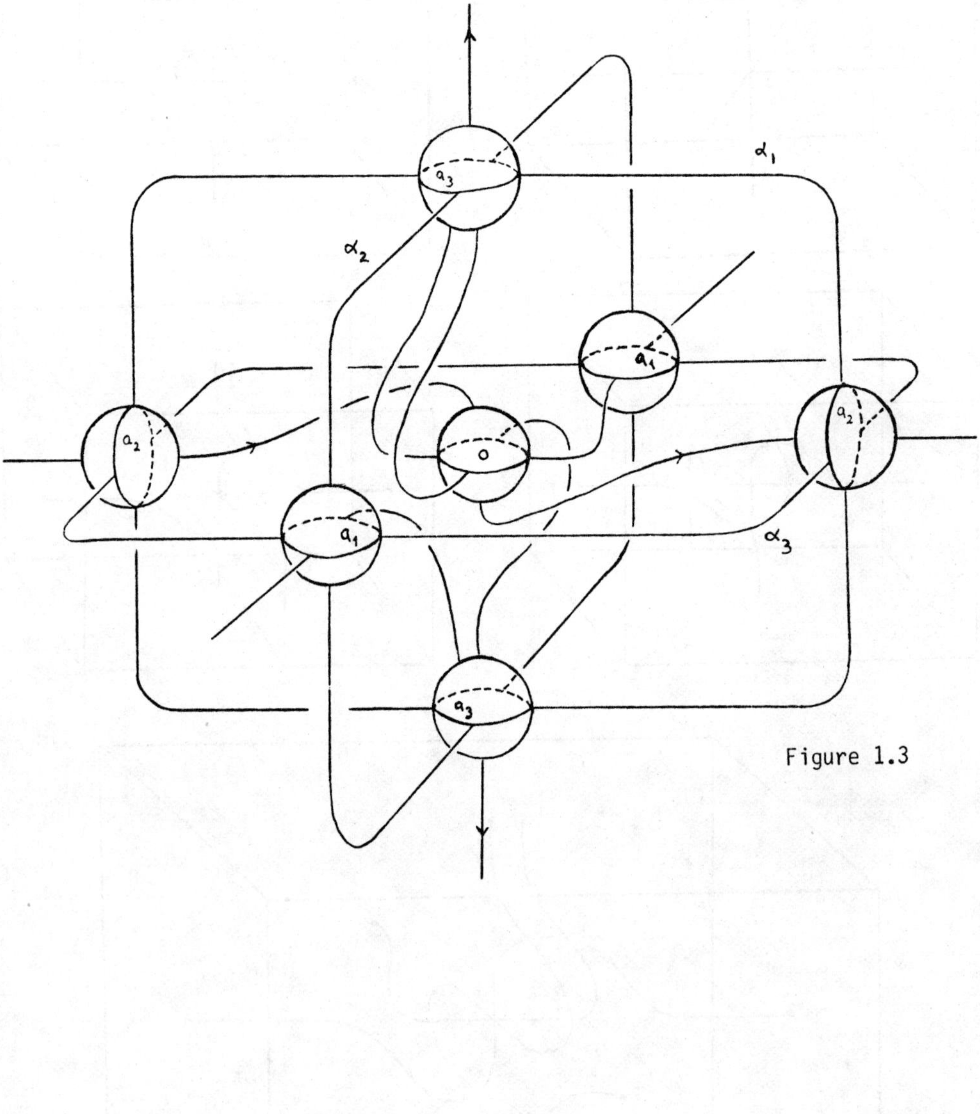

Figure 1.3

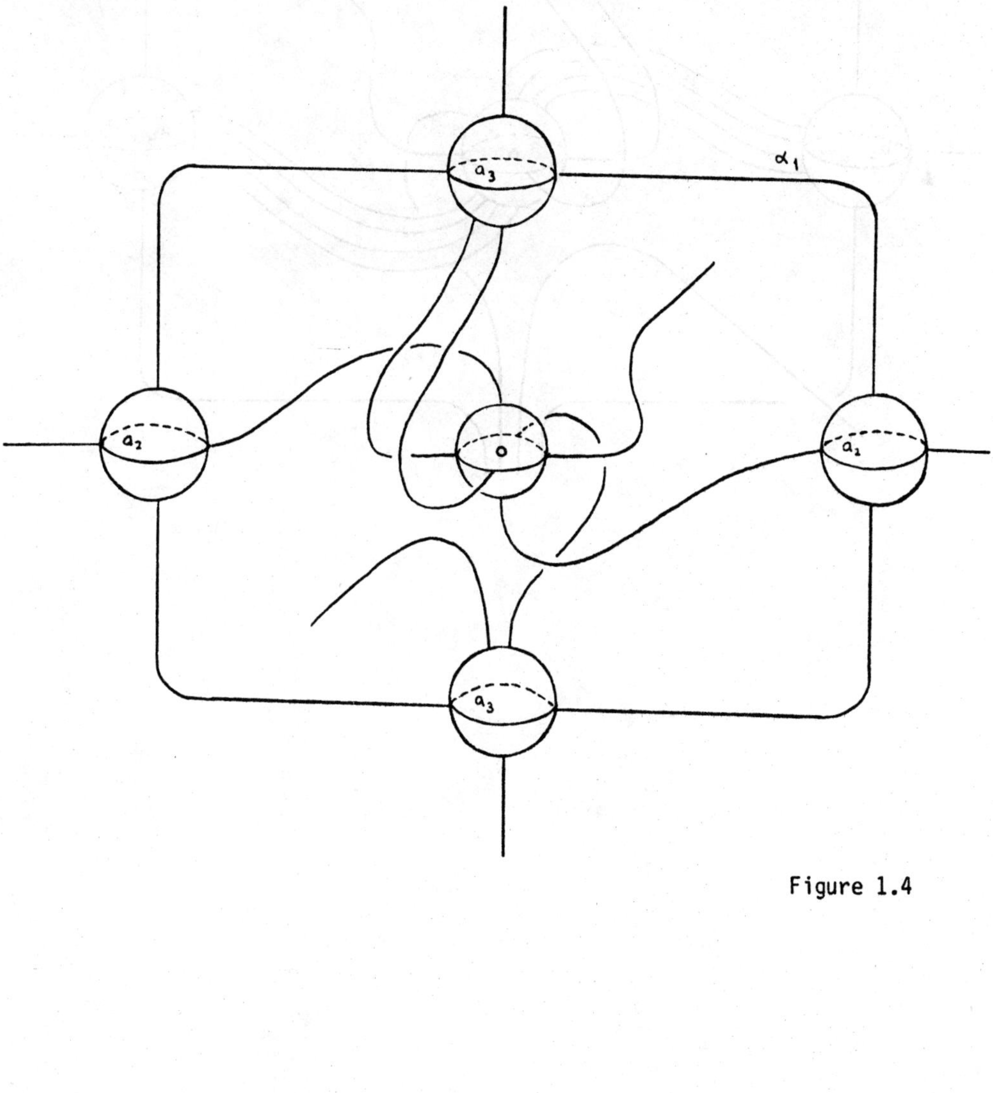

Figure 1.4

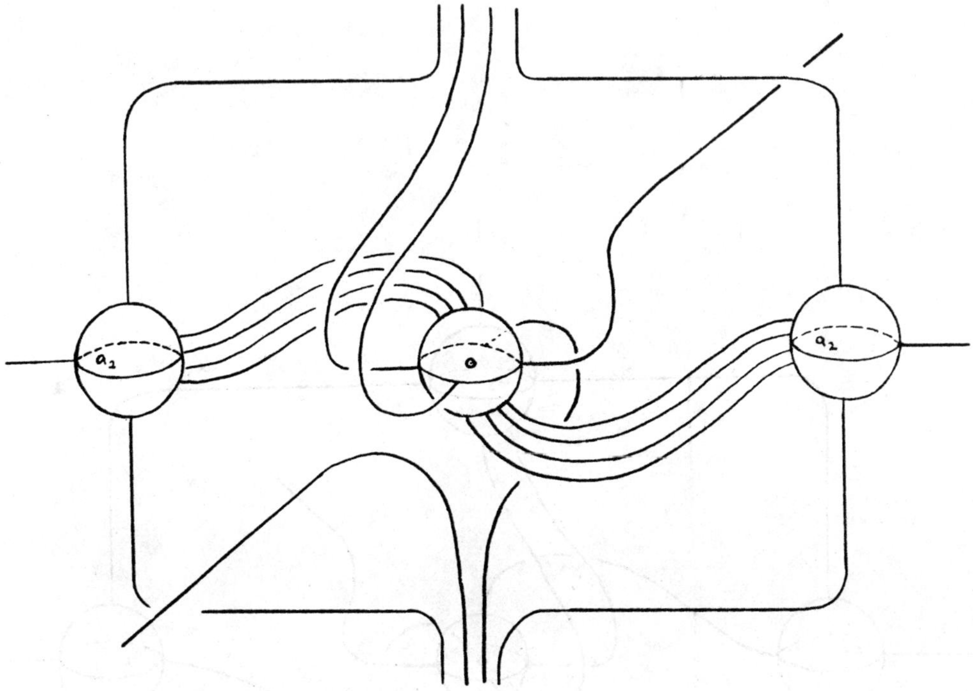

Figure 1.5

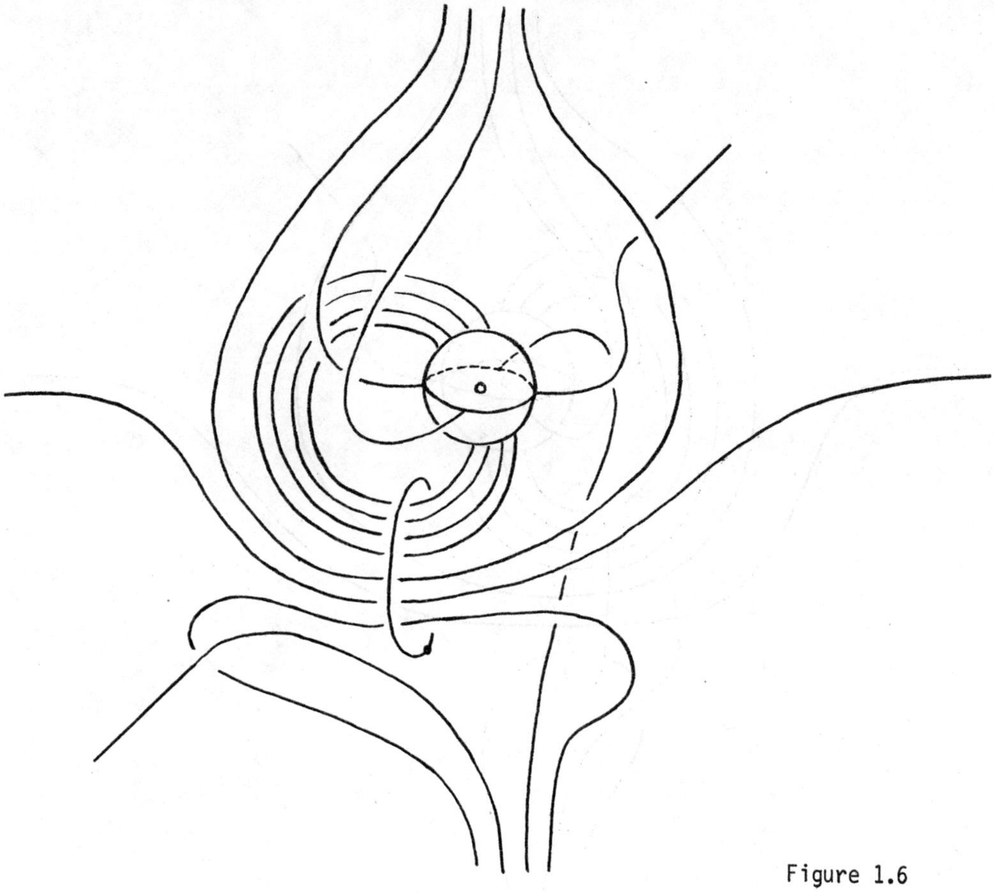

Figure 1.6

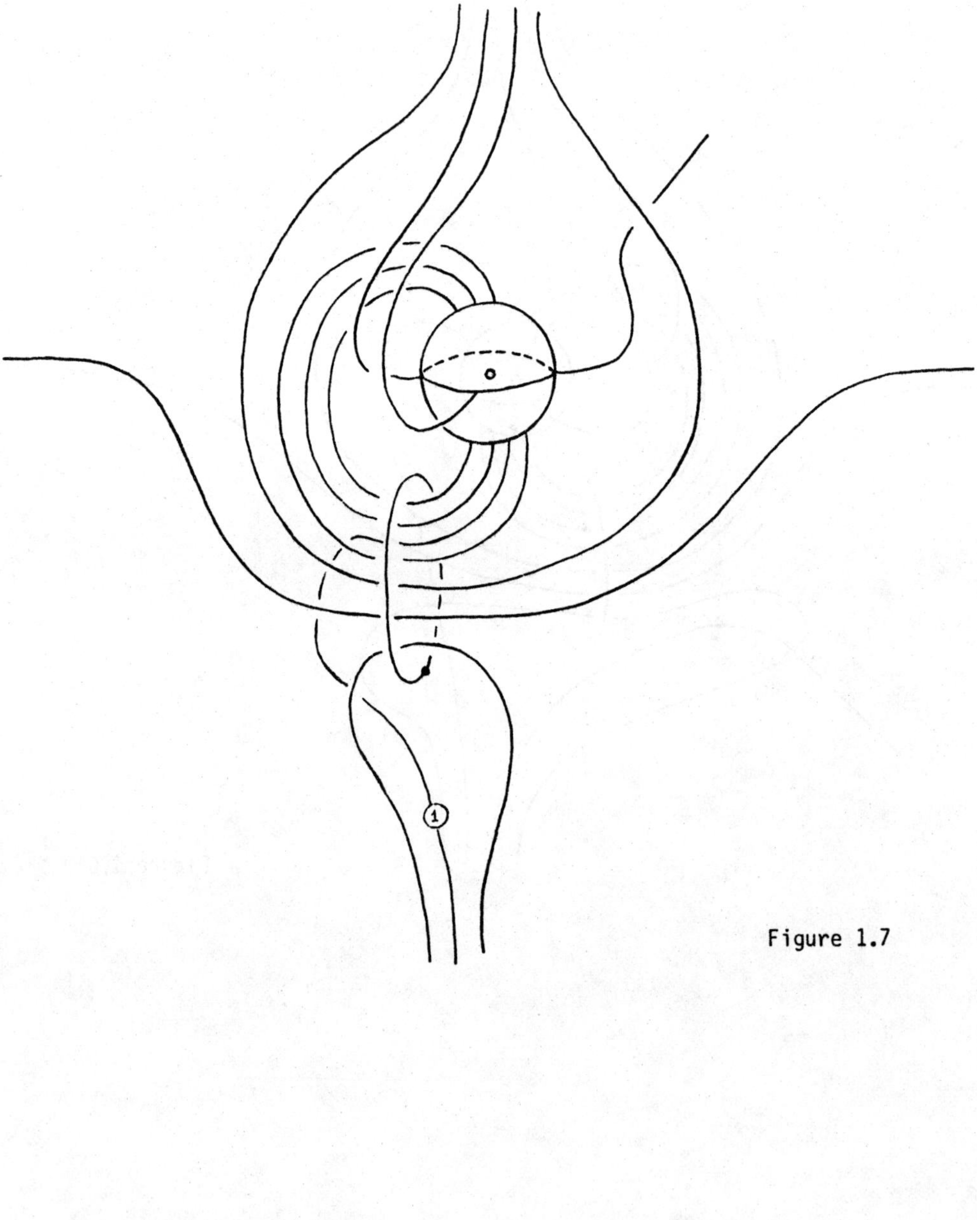

Figure 1.7

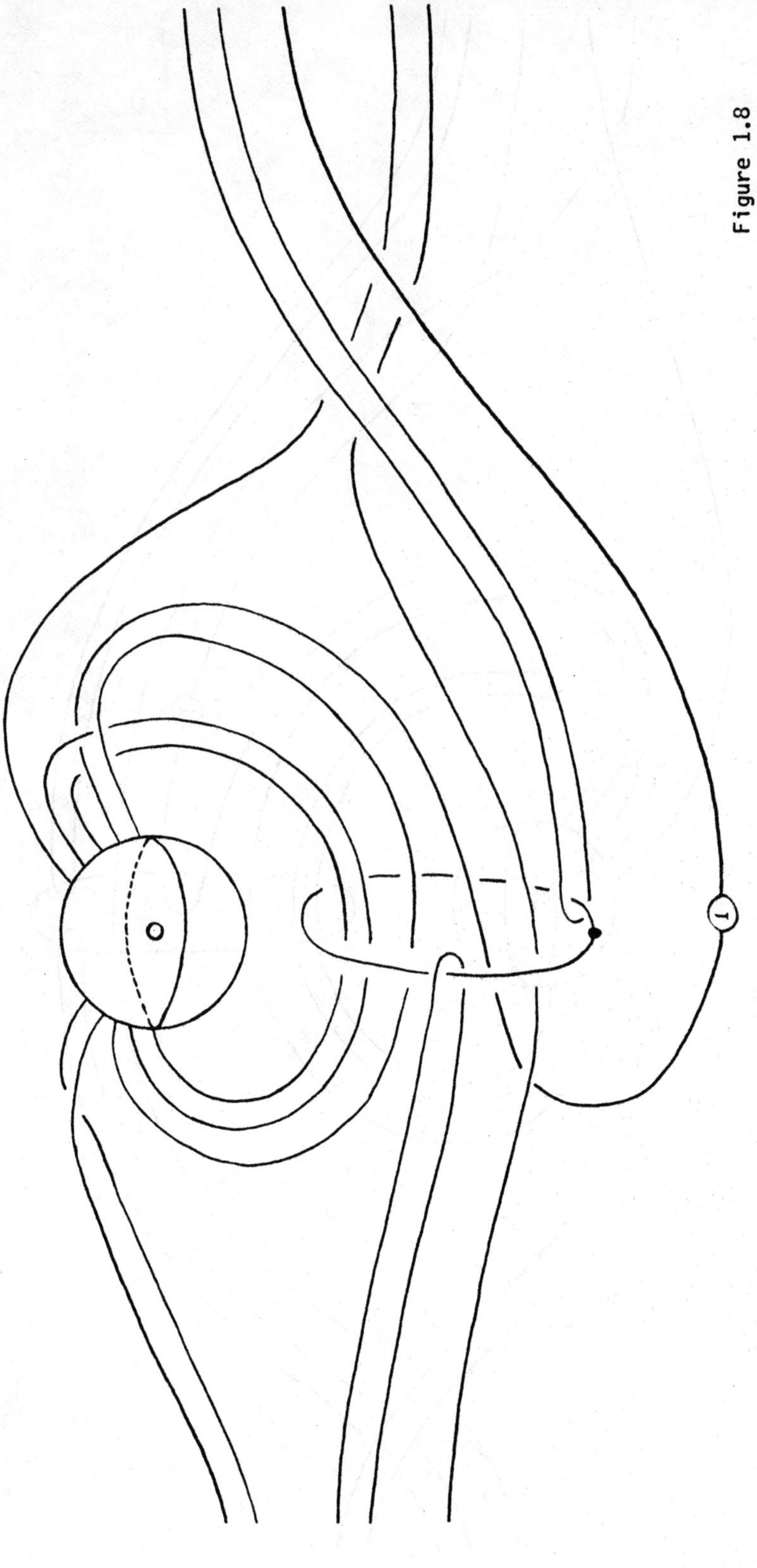

Figure 1.8

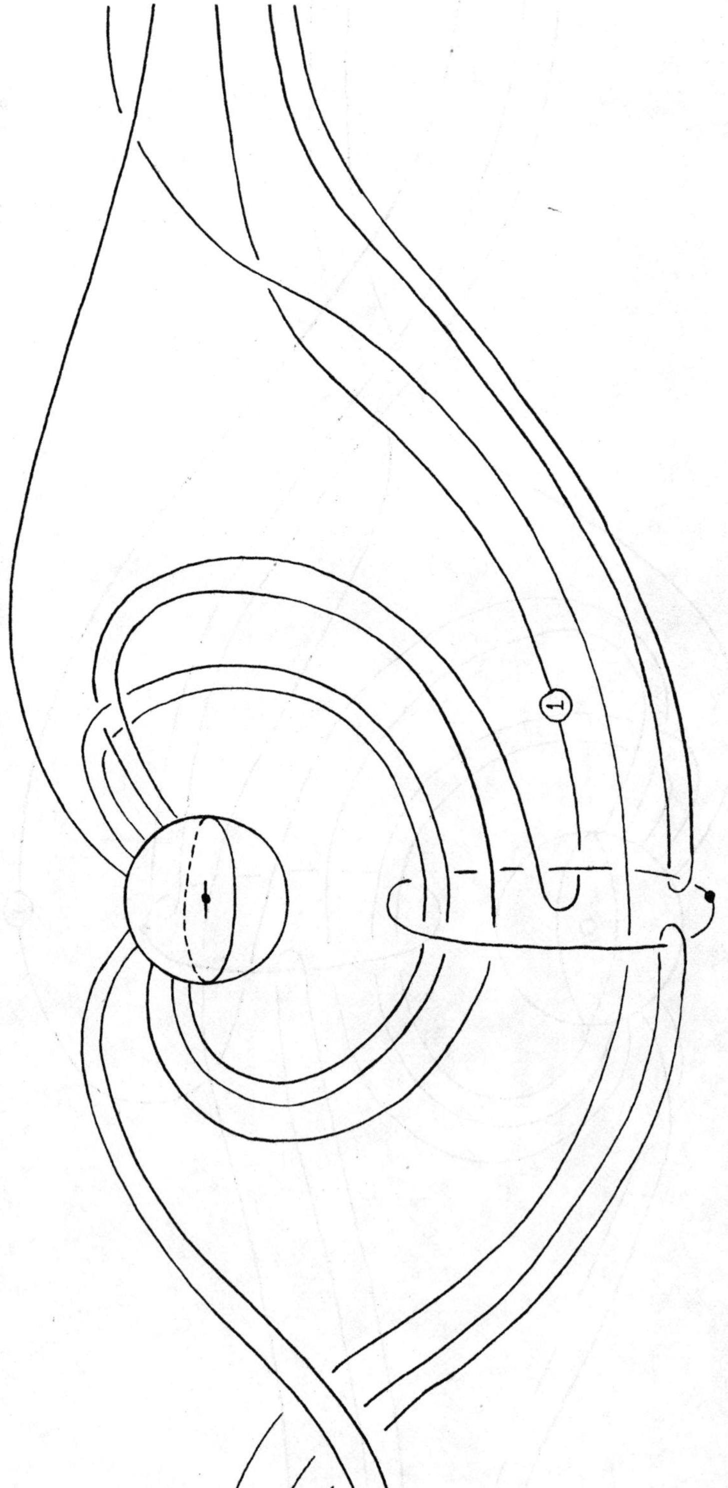

Figure 1.9

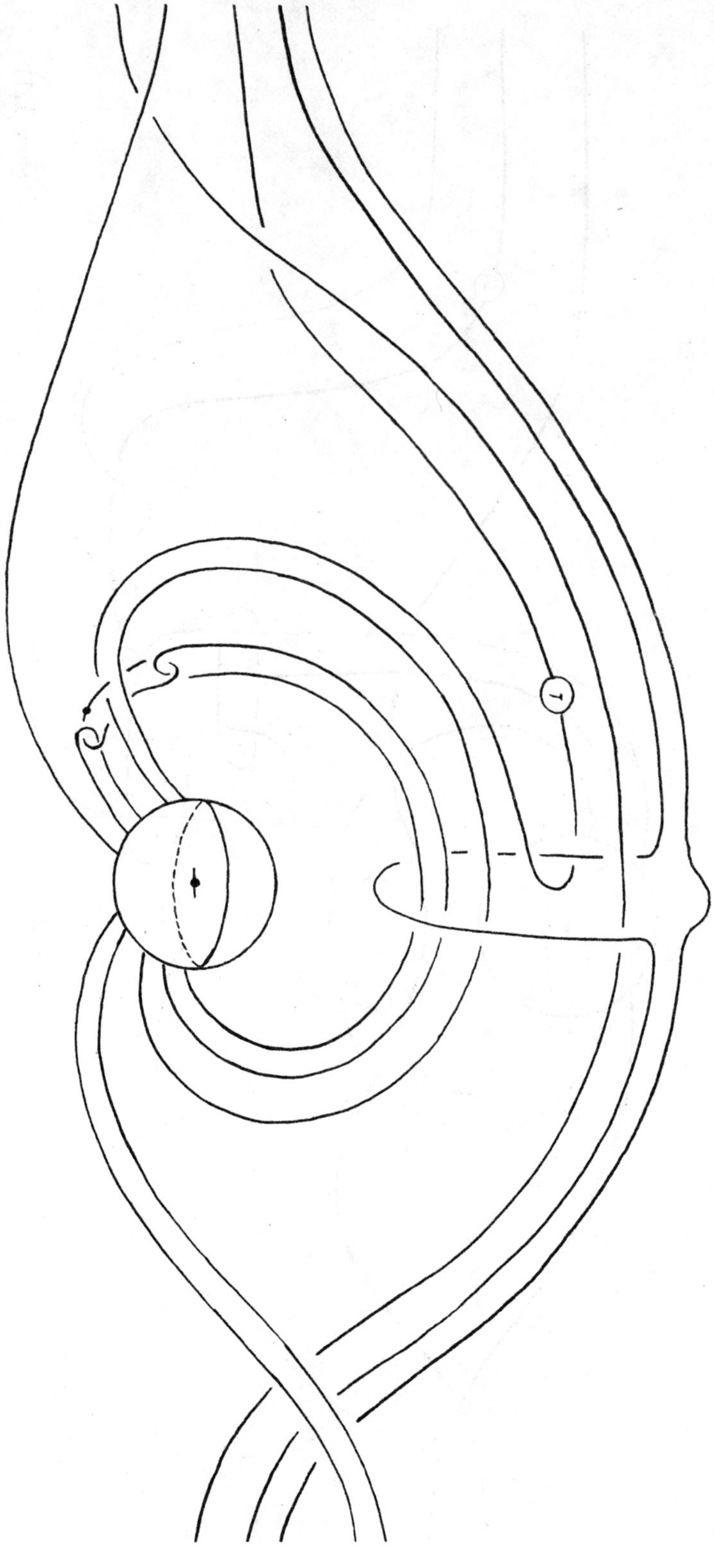

Figure 1.10

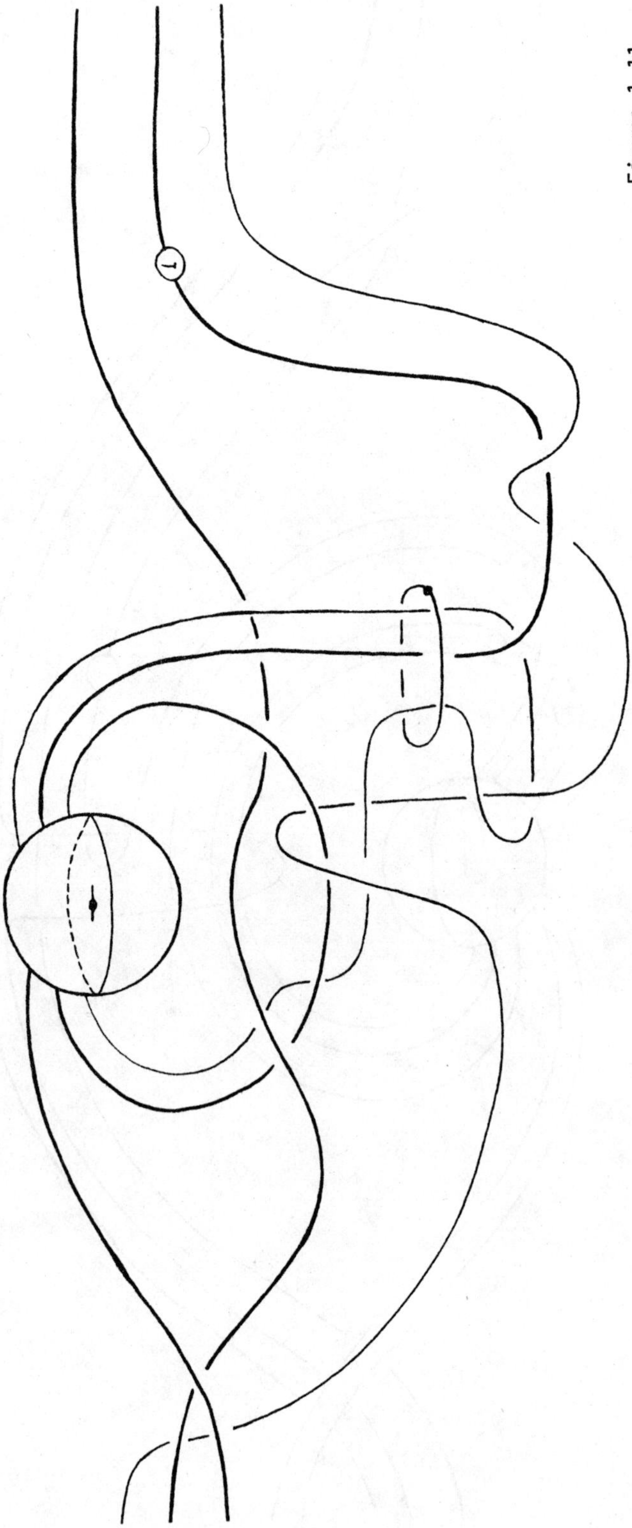

Figure 1.11

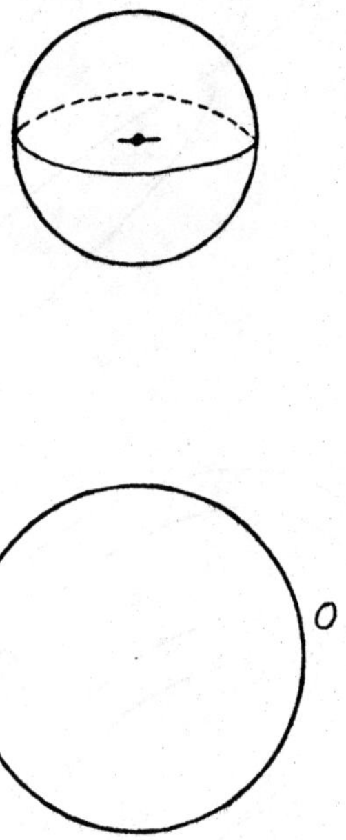

Figure 1.12

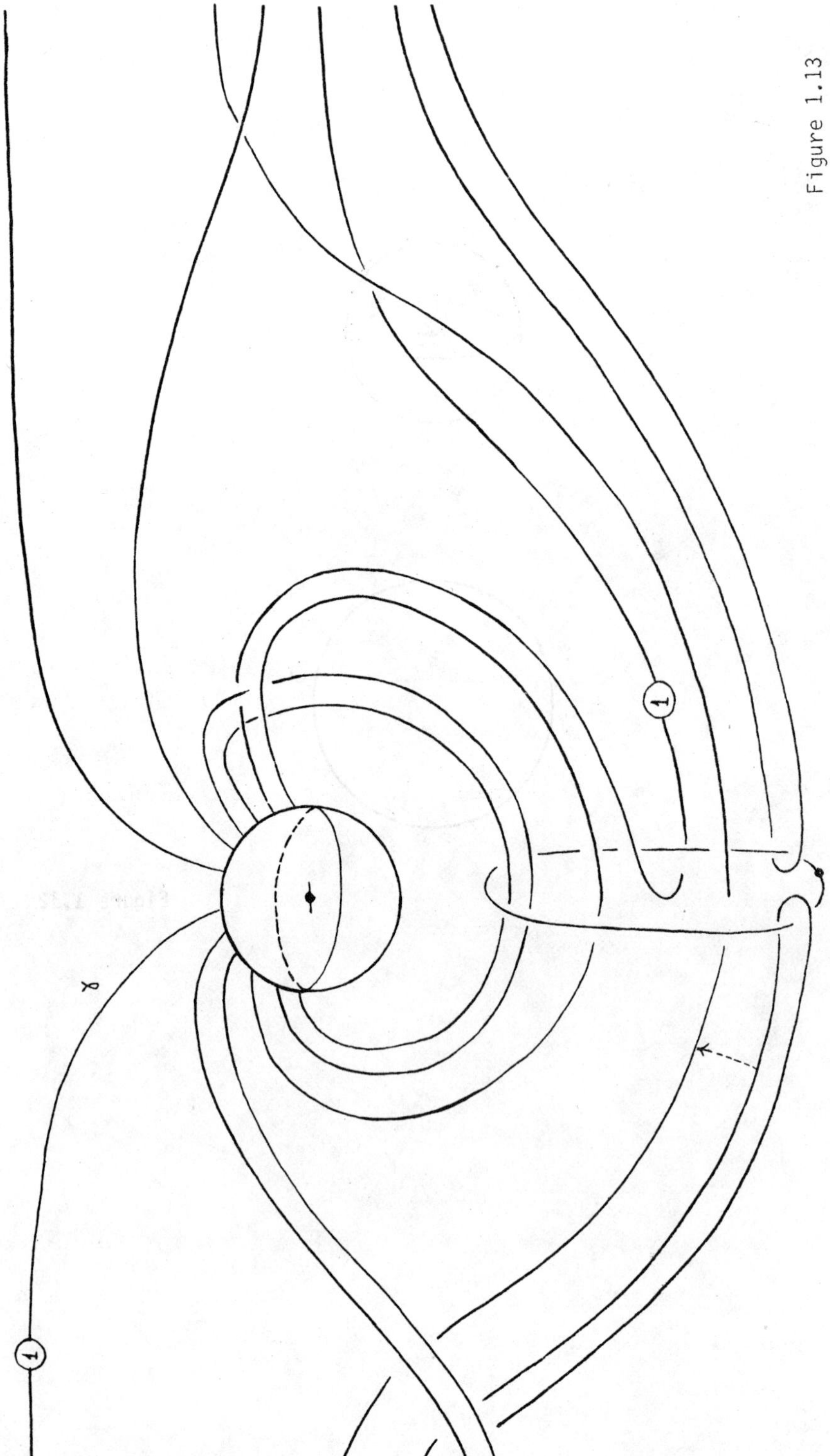

Figure 1.13

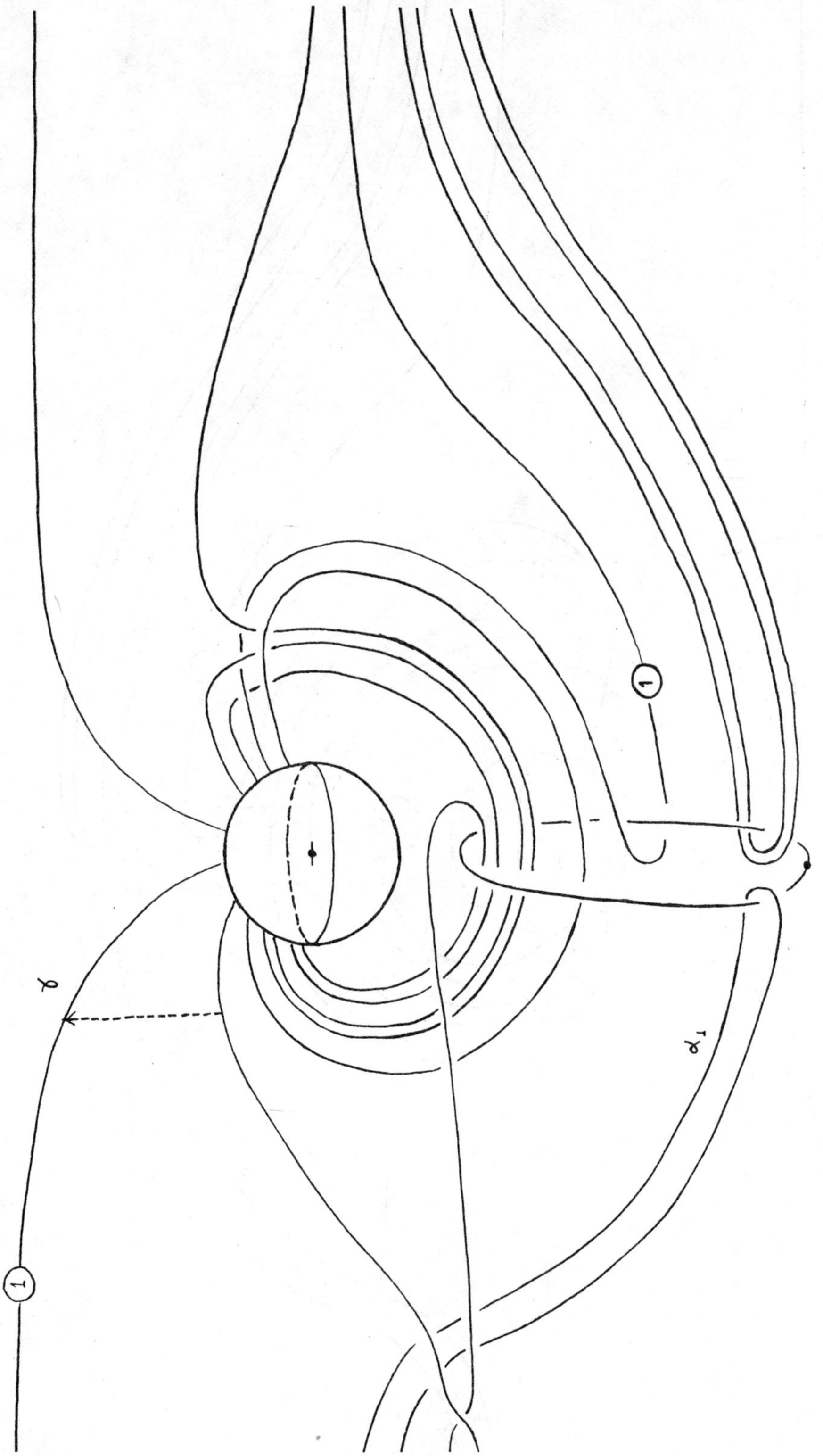

Figure 1.14

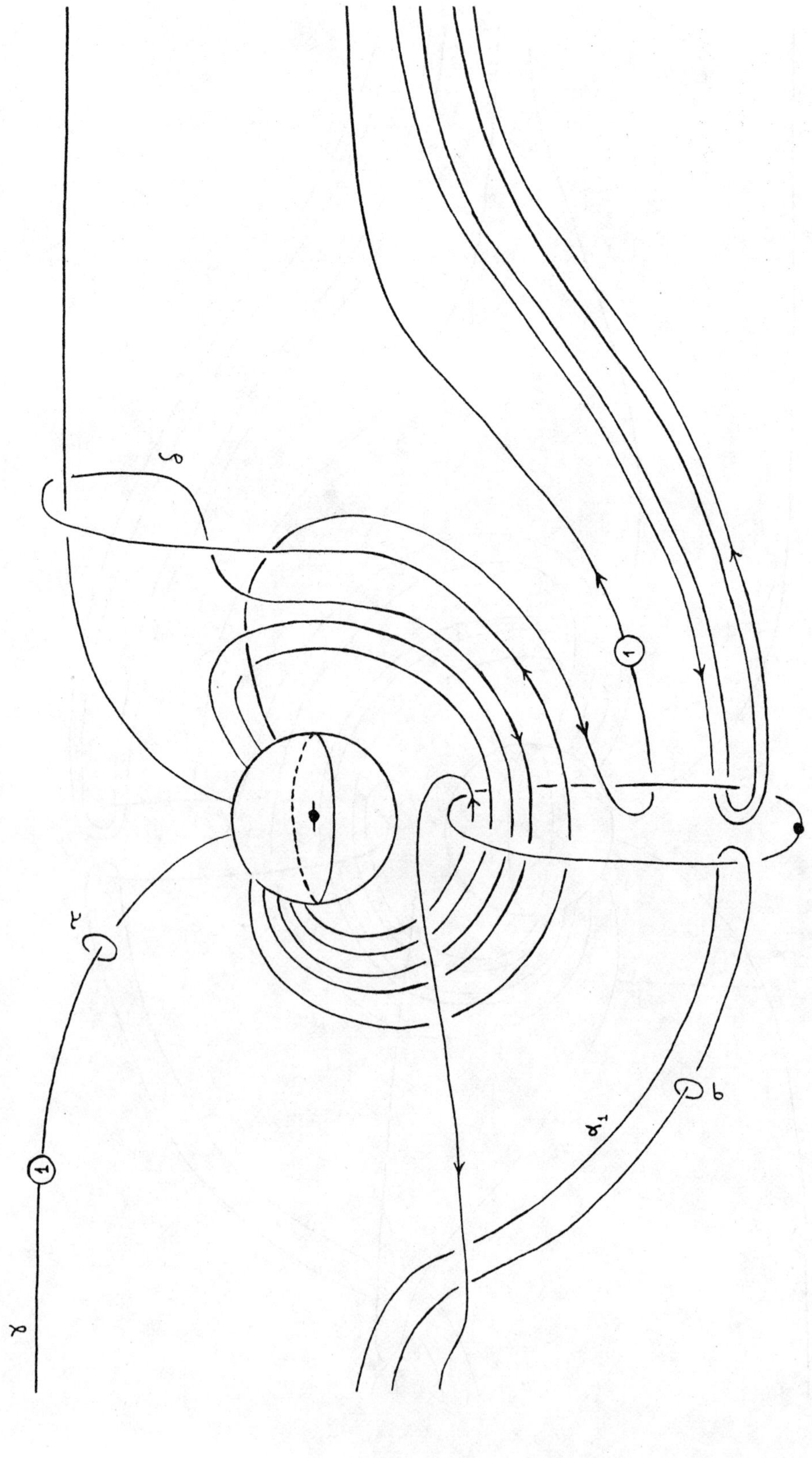

Figure 1.15

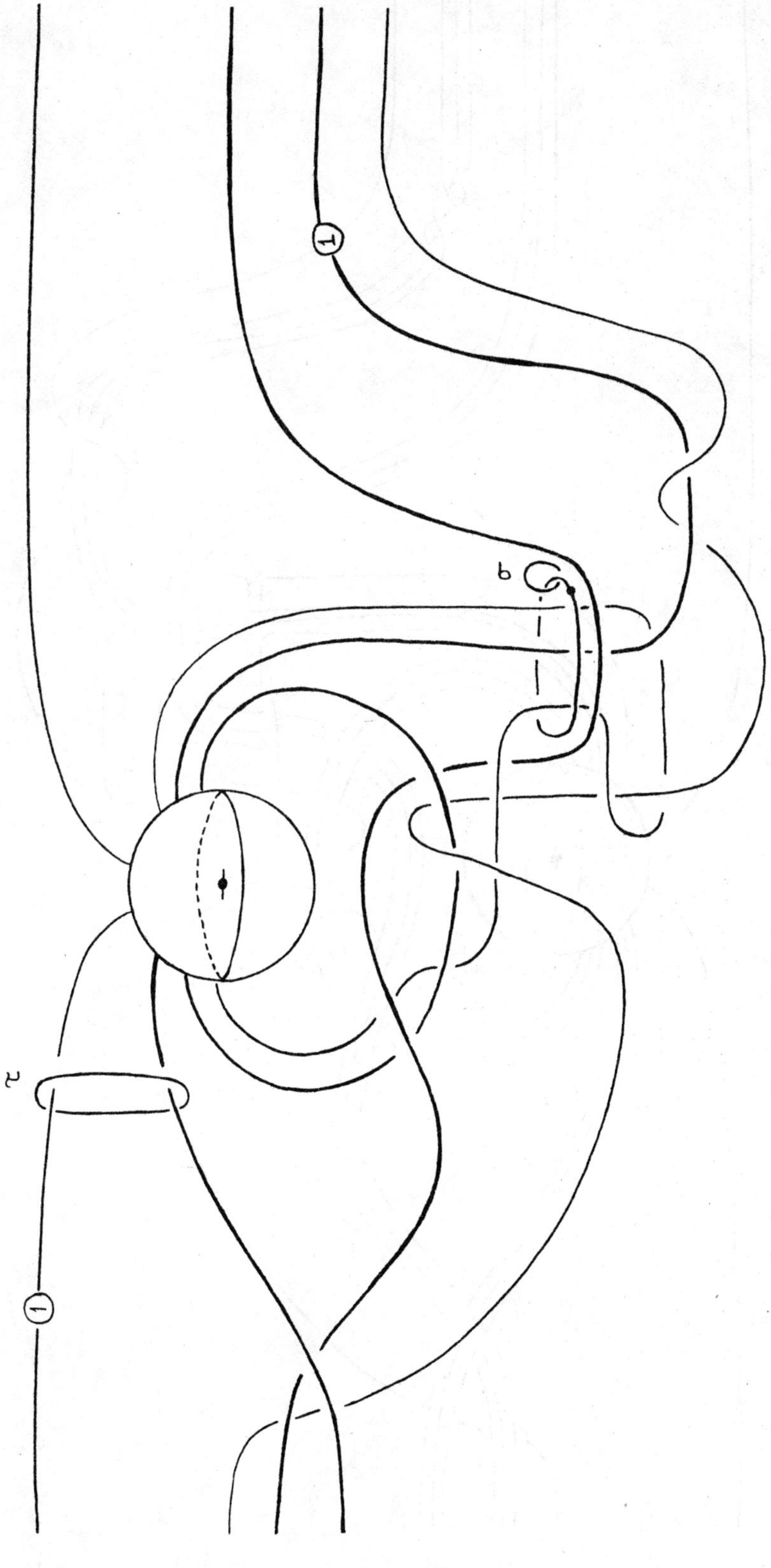

Figure 1.16

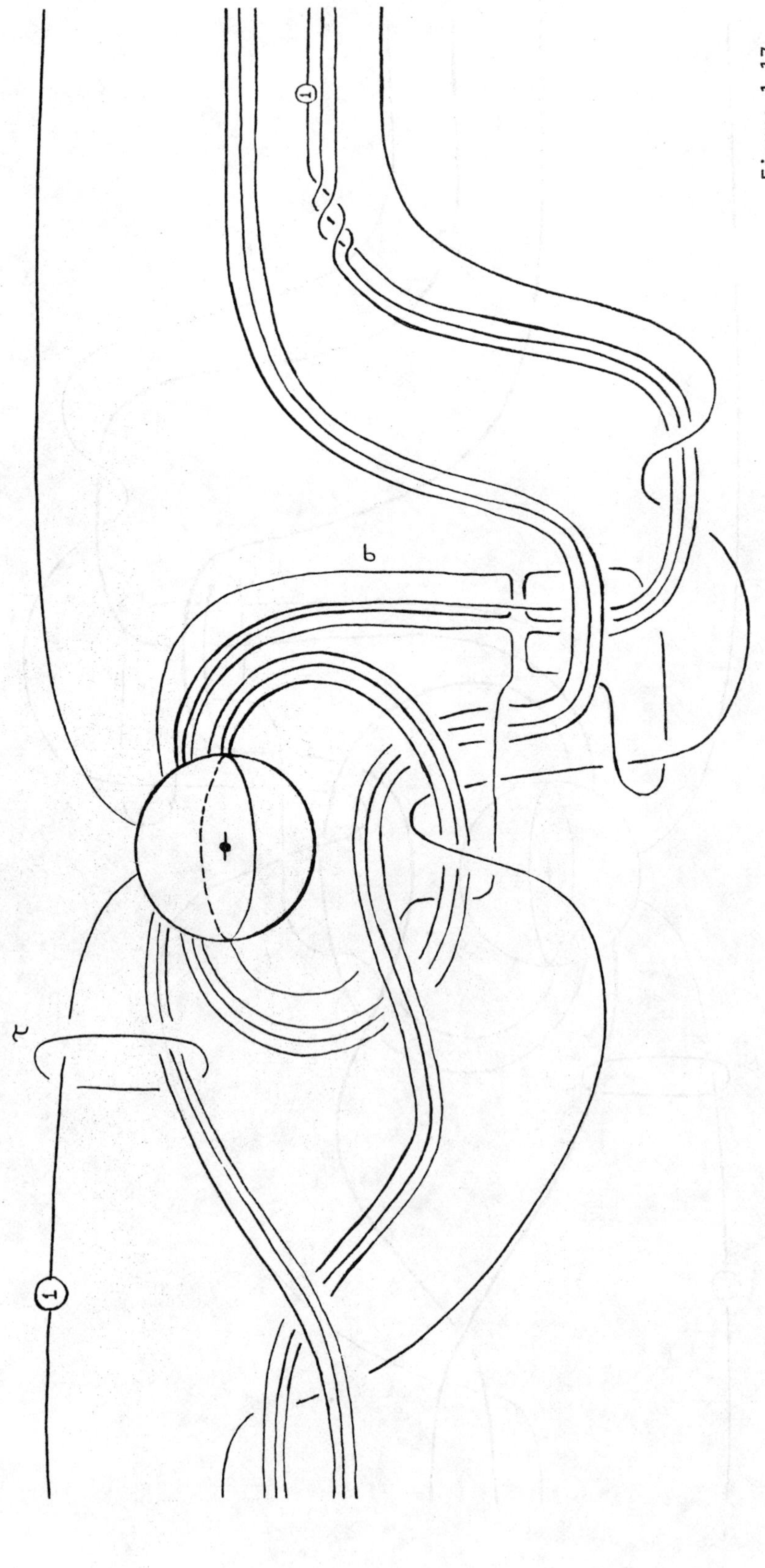

Figure 1.17

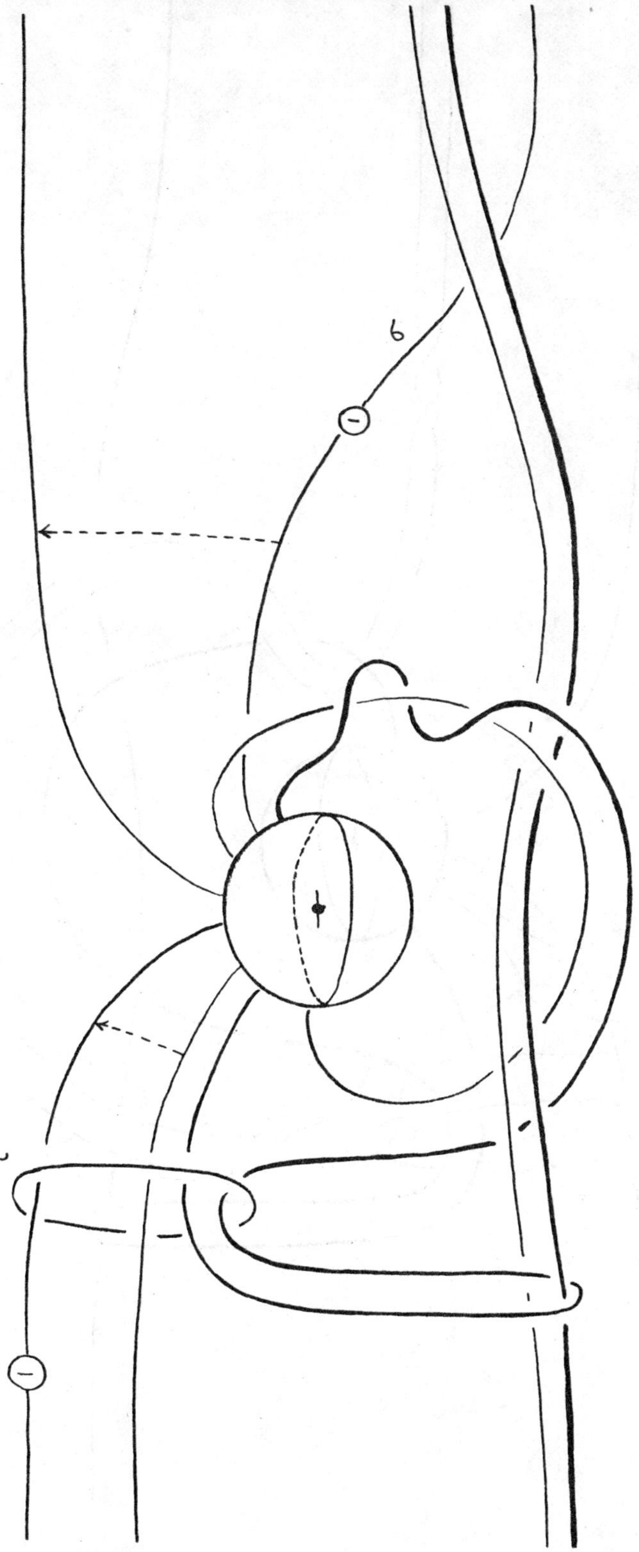

Figure 1.18

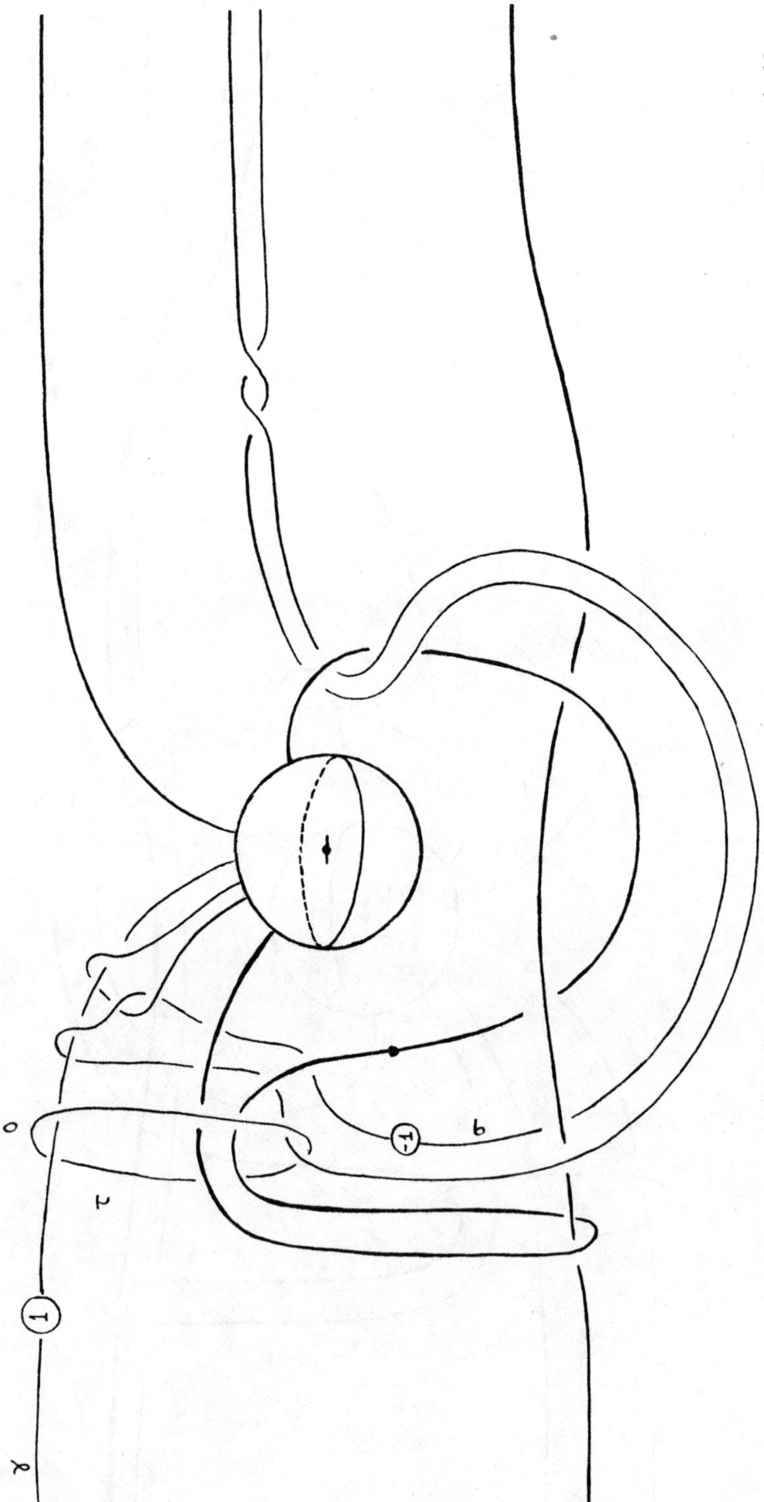

Figure 1.19

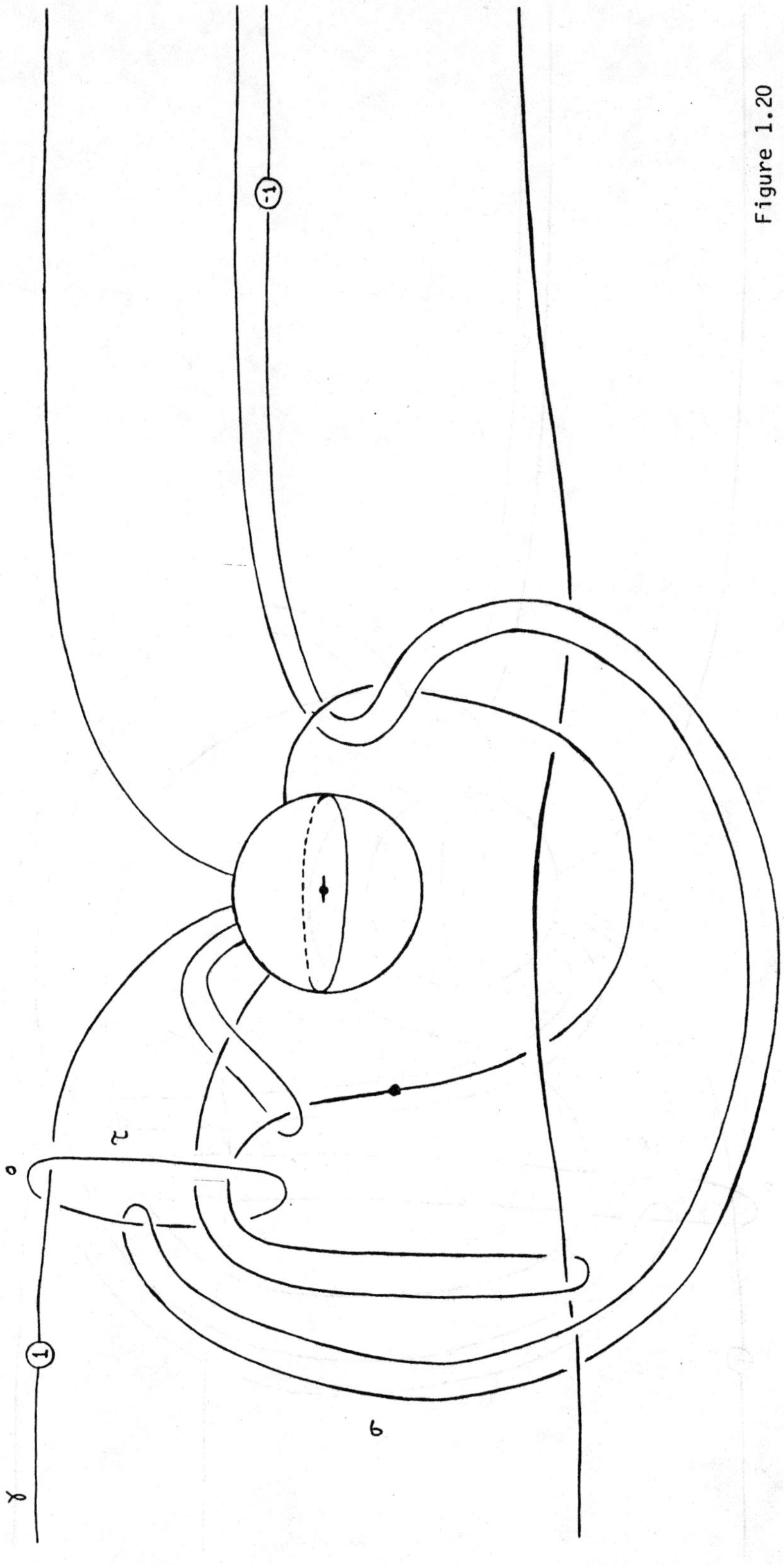

Figure 1.20

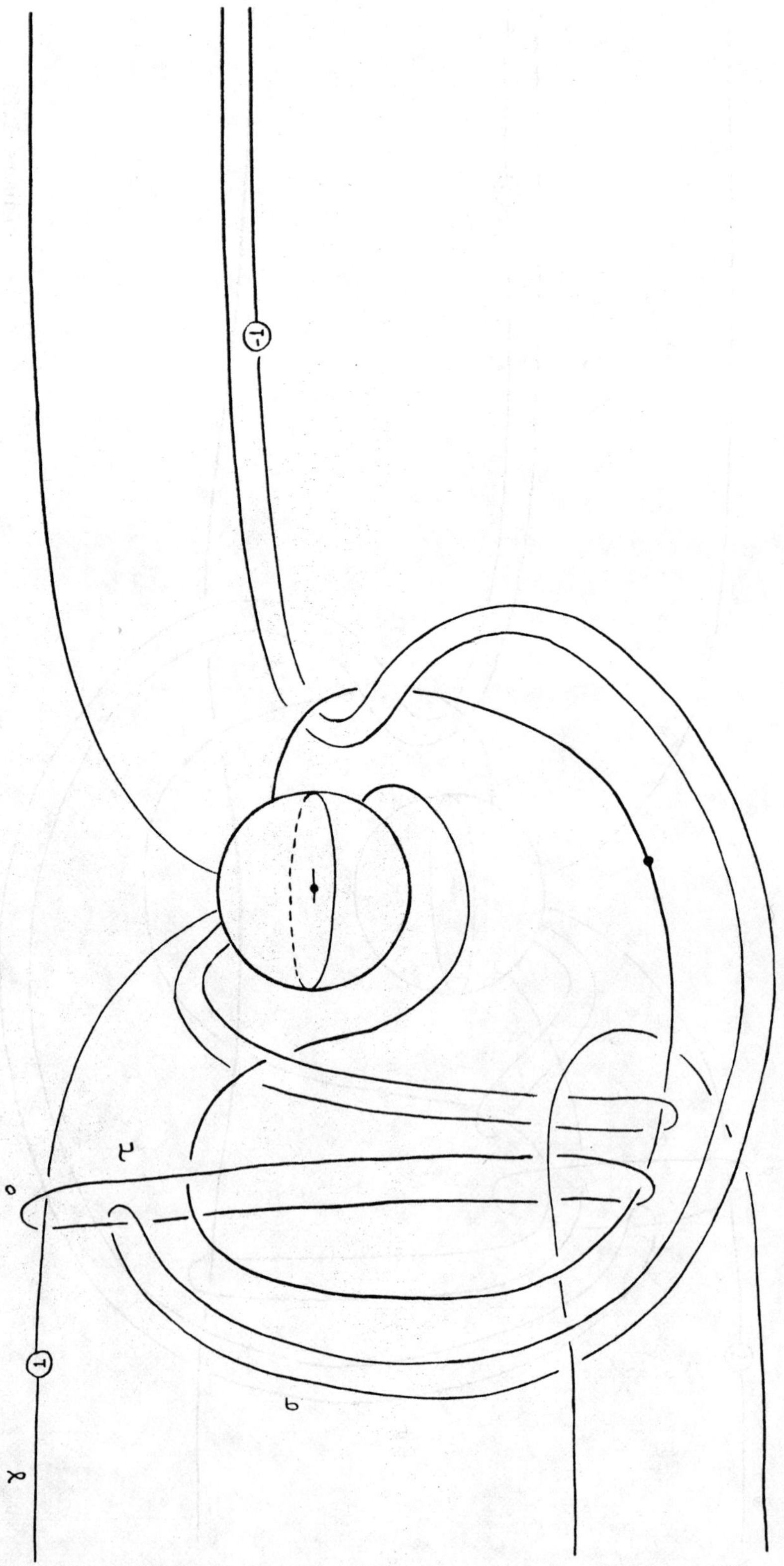

Figure 1.21

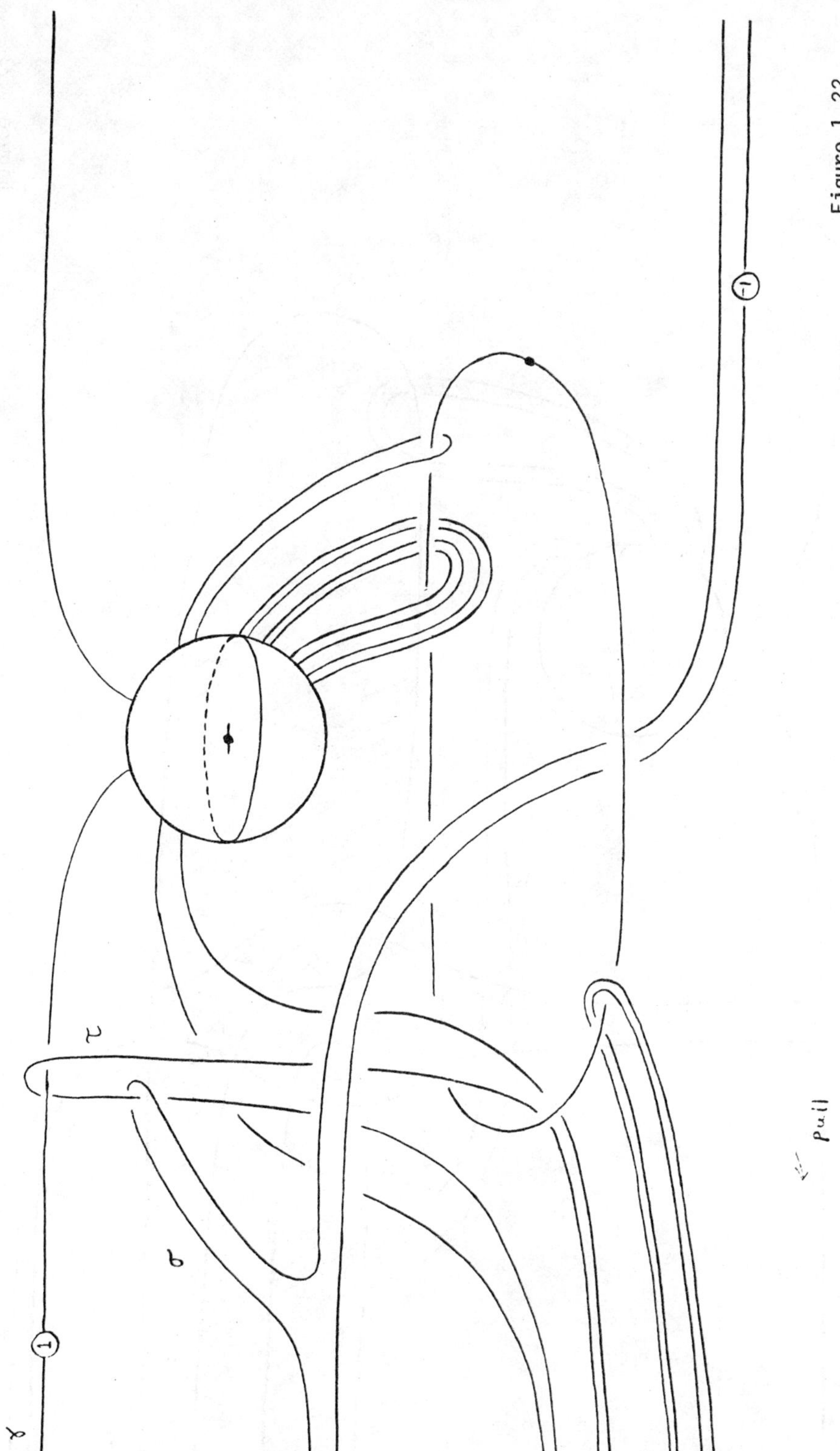

Figure 1.22

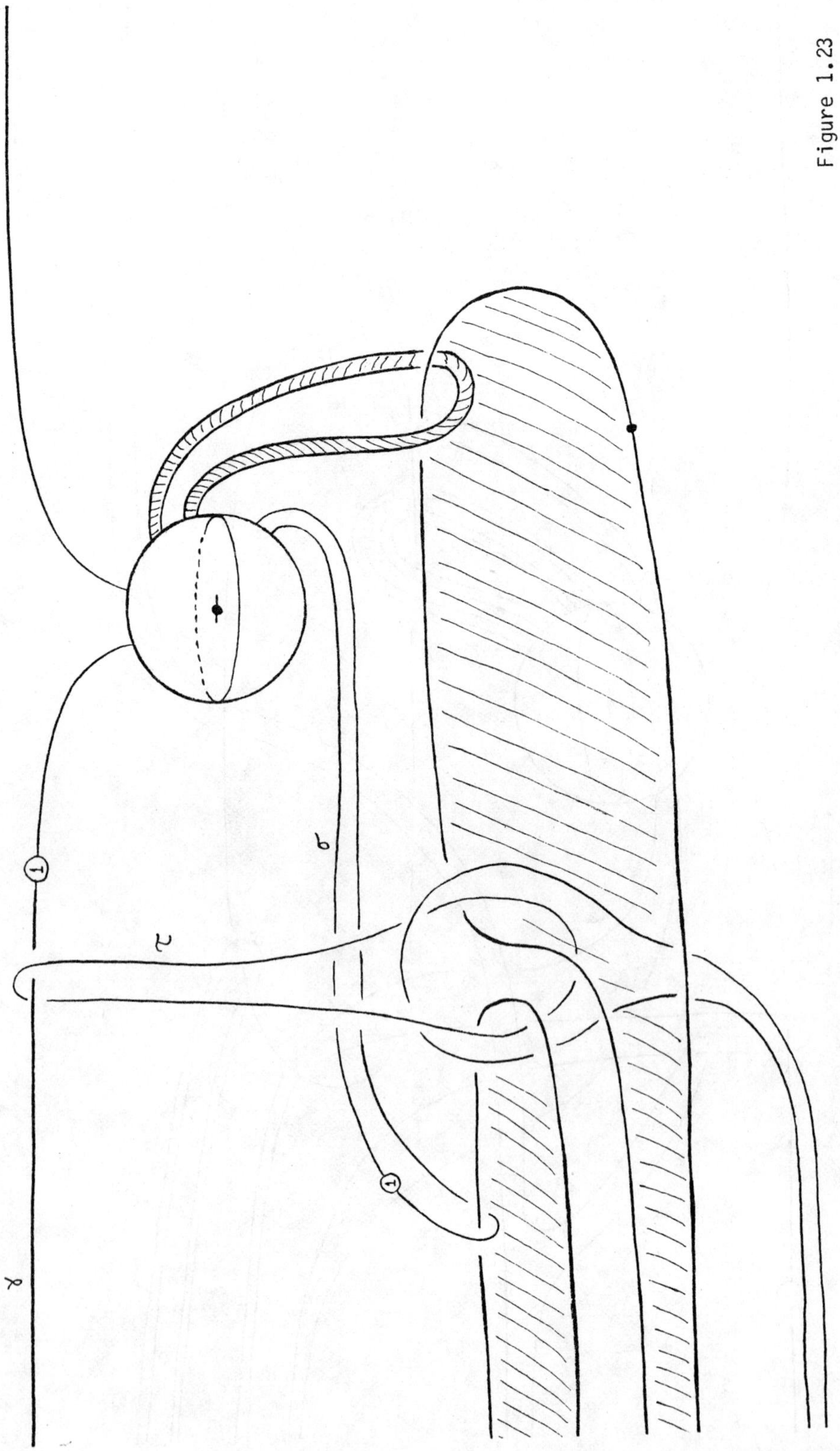

Figure 1.23

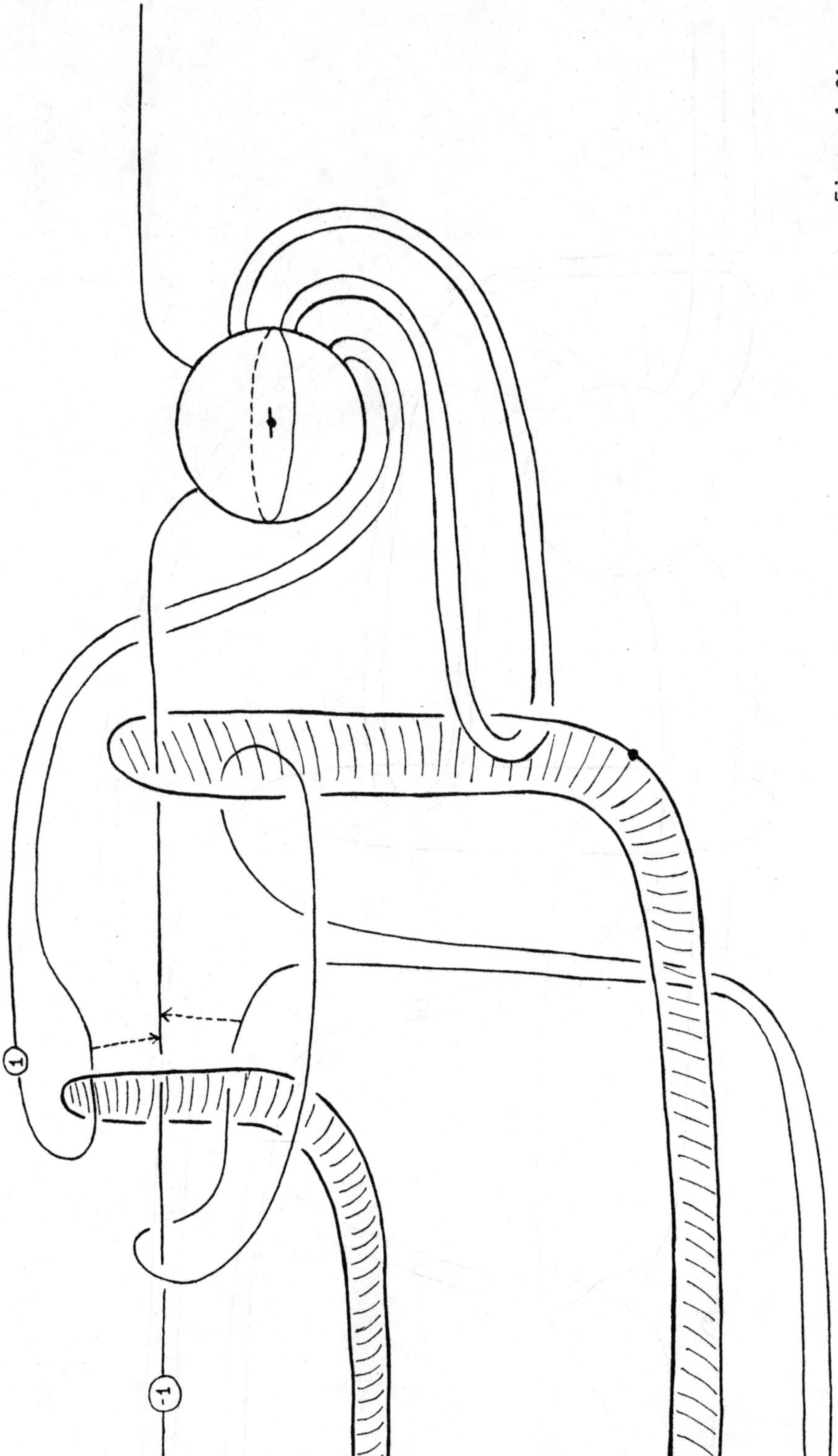

Figure 1.24

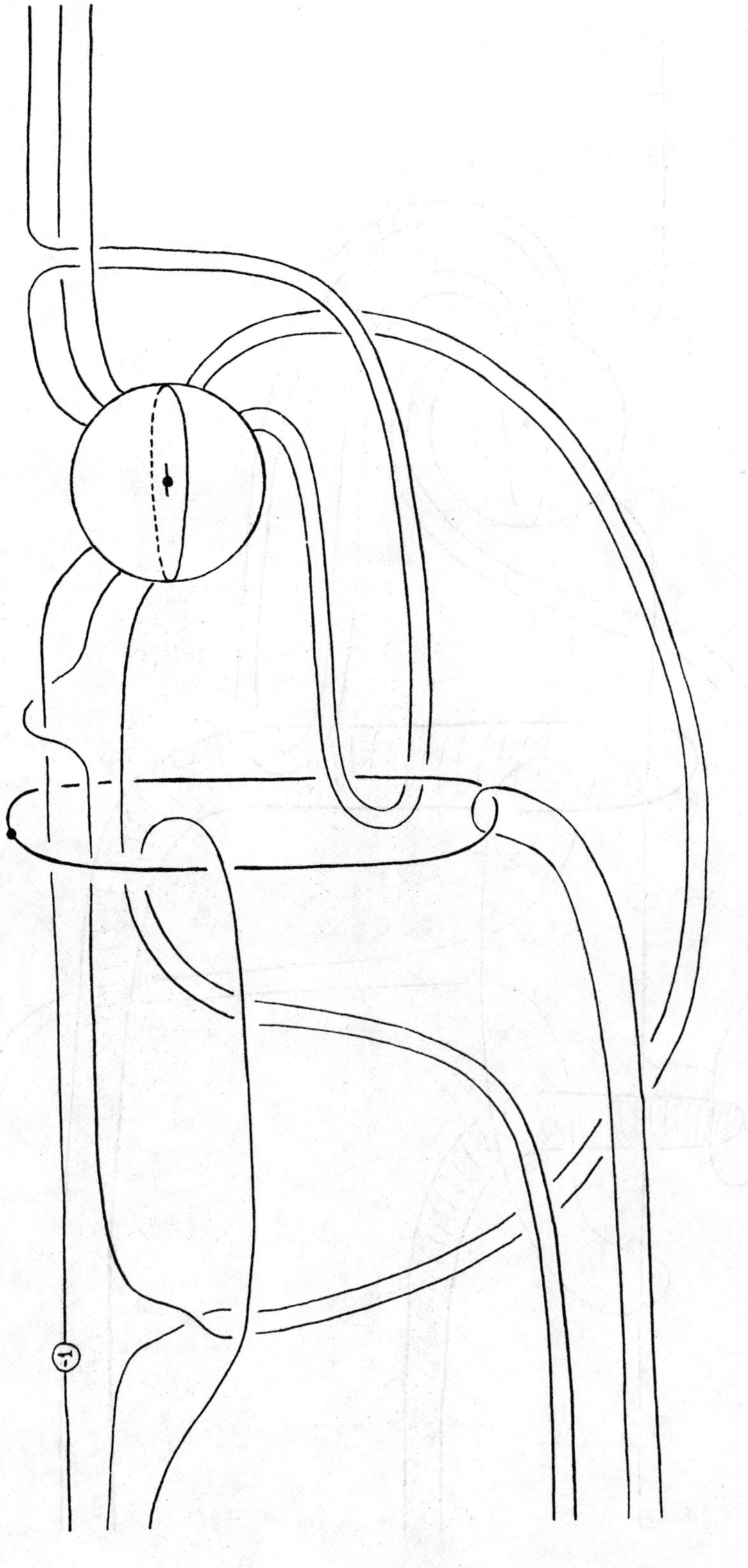

Figure 1.25

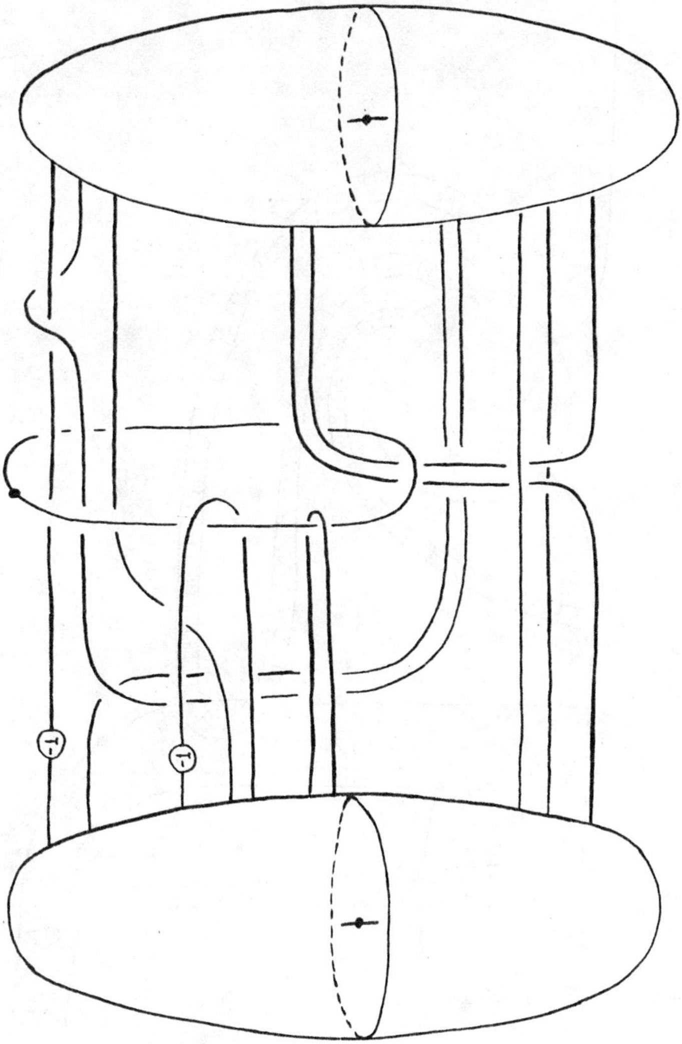

Figure 1.26

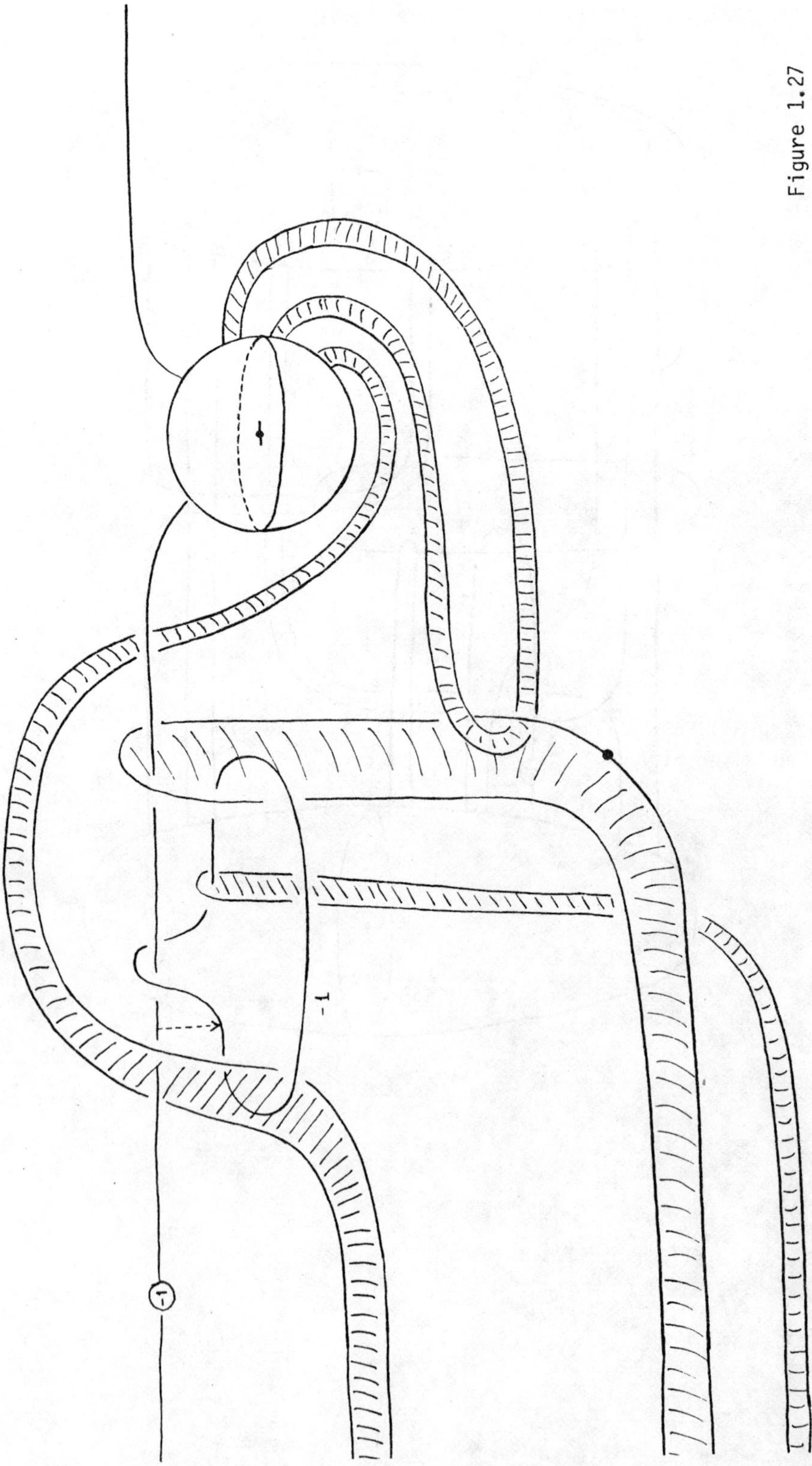

Figure 1.27

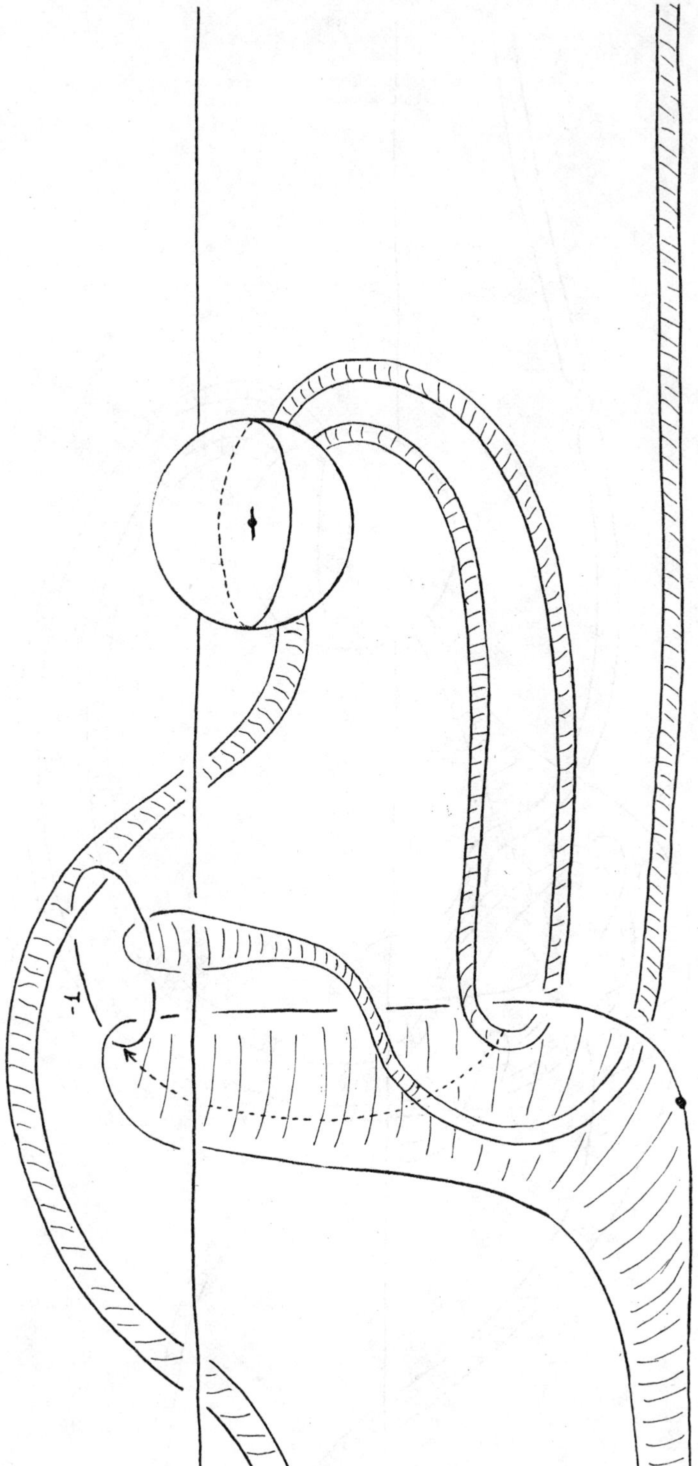

Figure 1.28

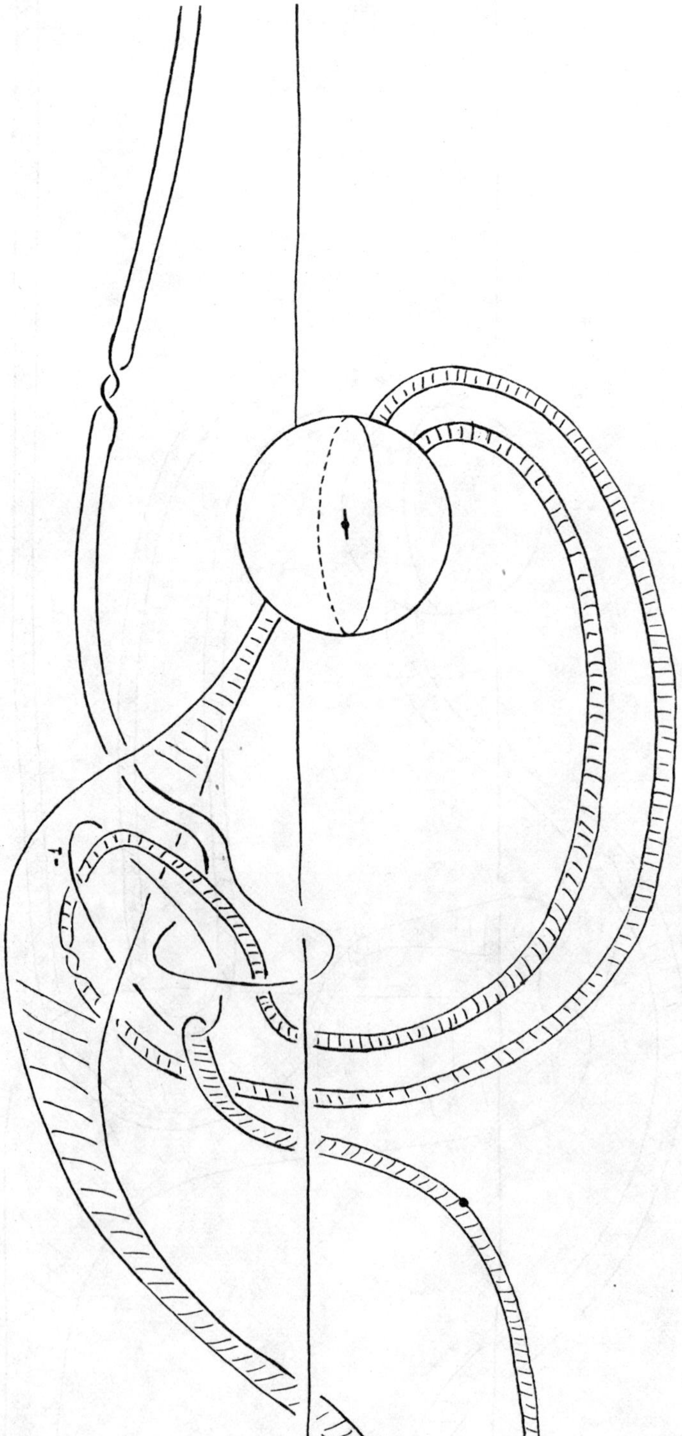

Figure 1.29

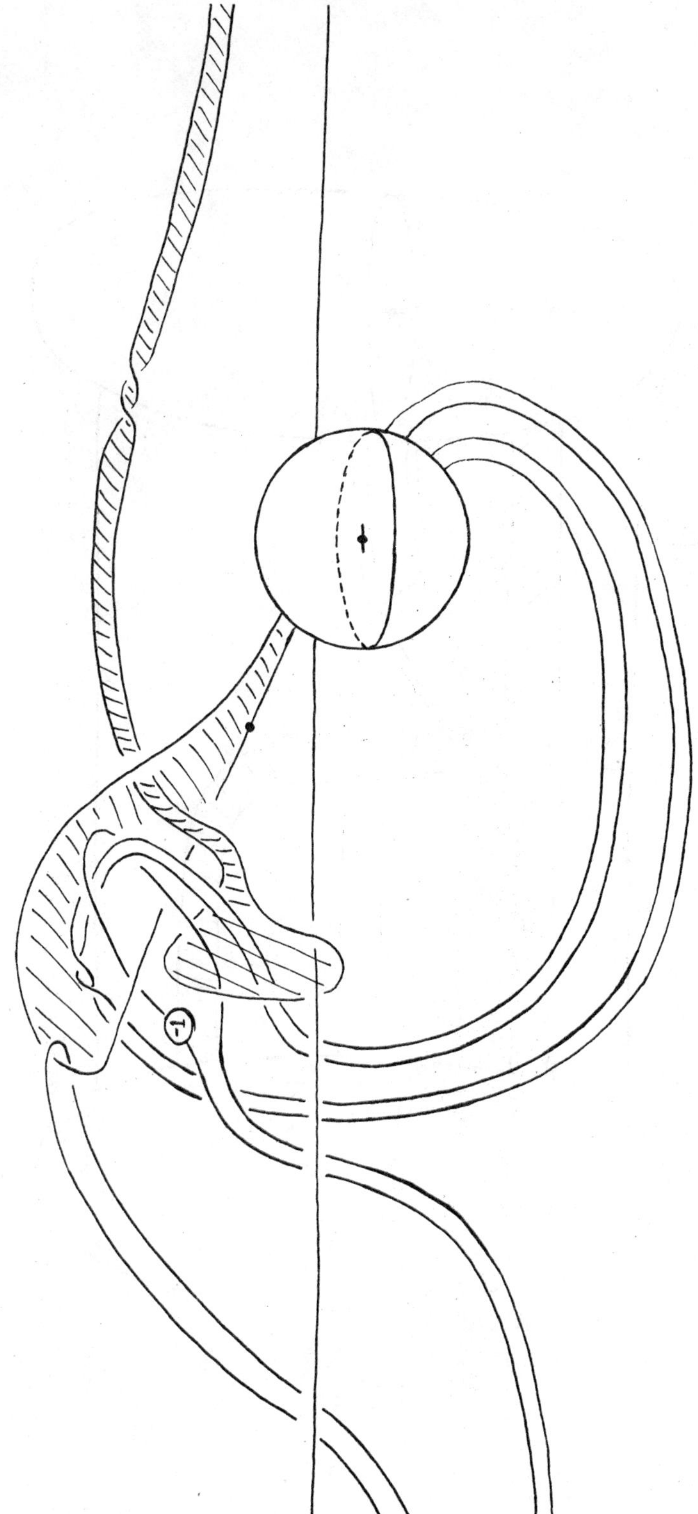

Figure 1.30

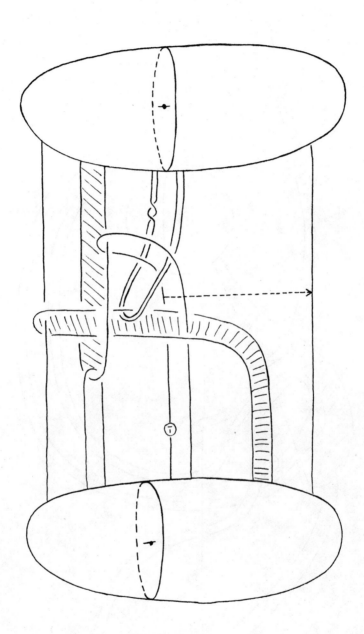

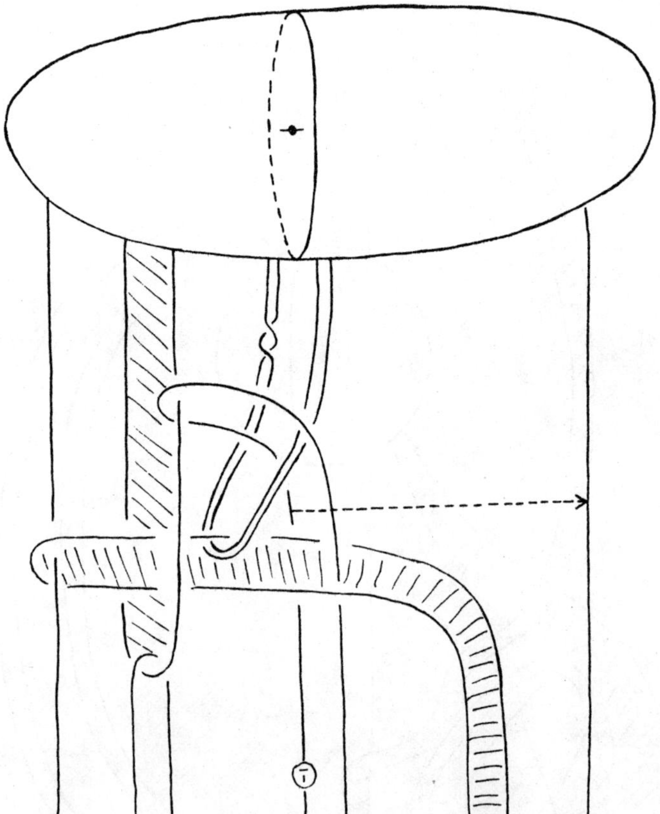

Figure 1.31

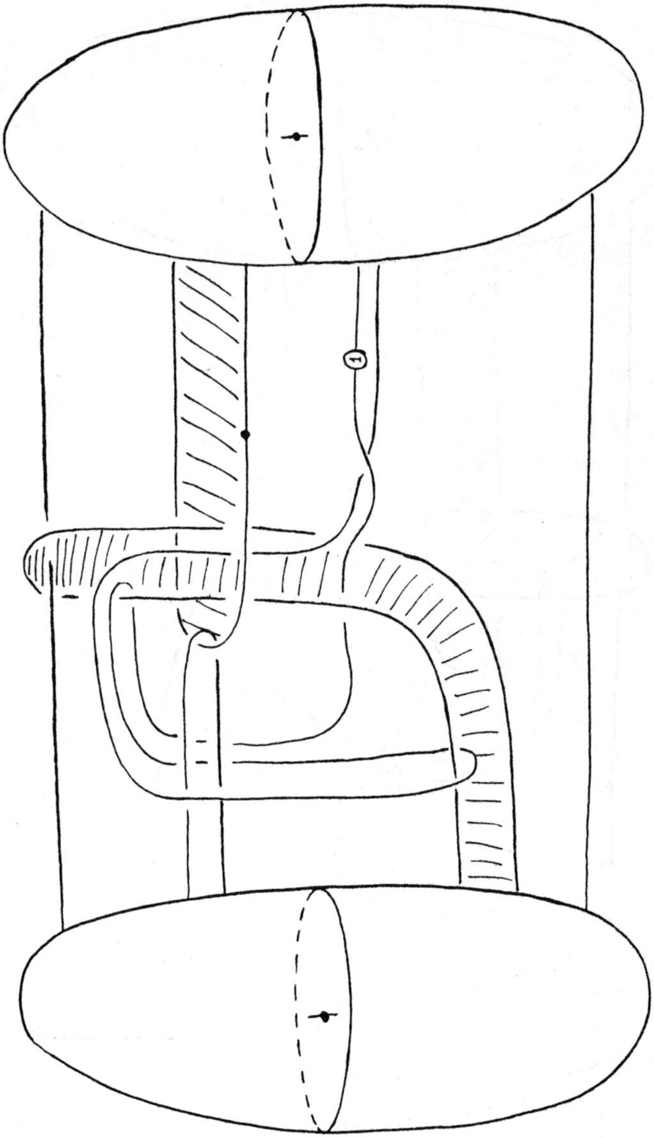

Figure 1.32

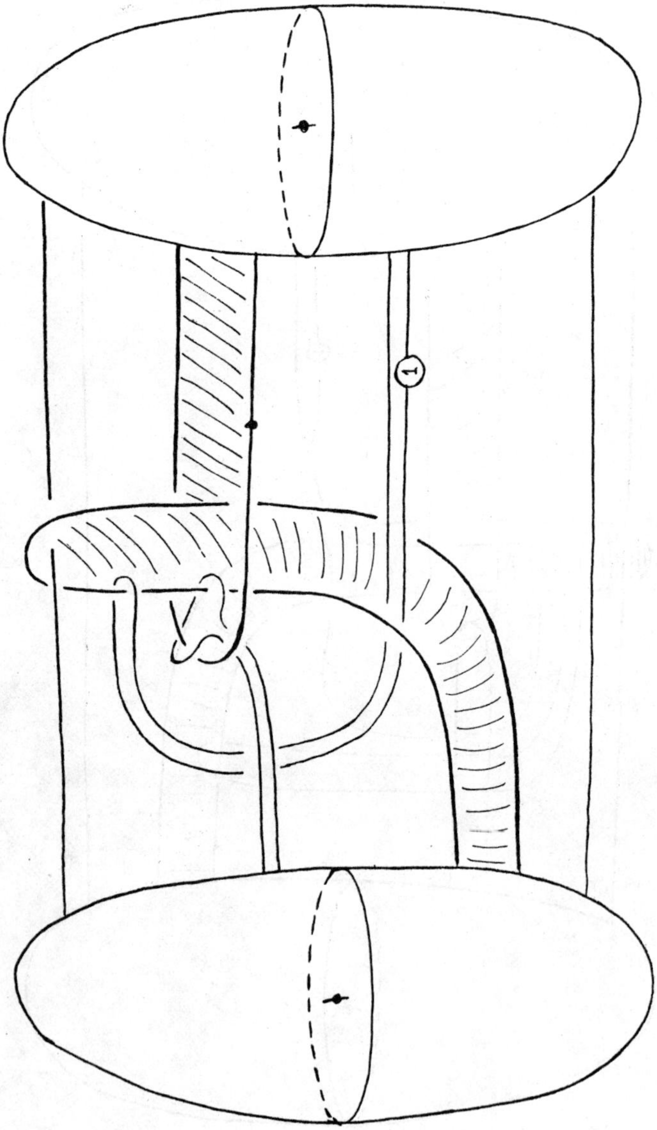

Figure 1.33

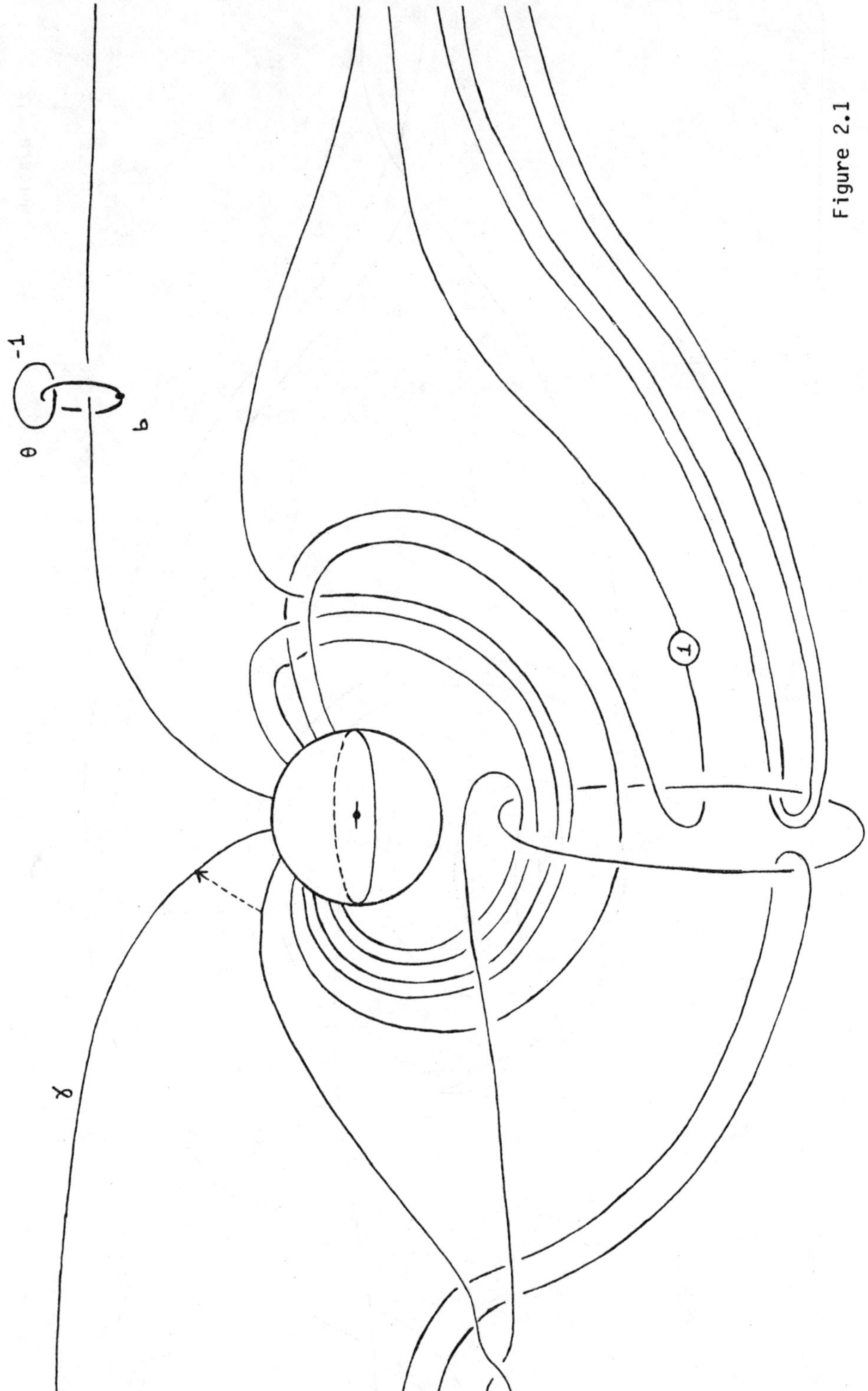

Figure 2.1

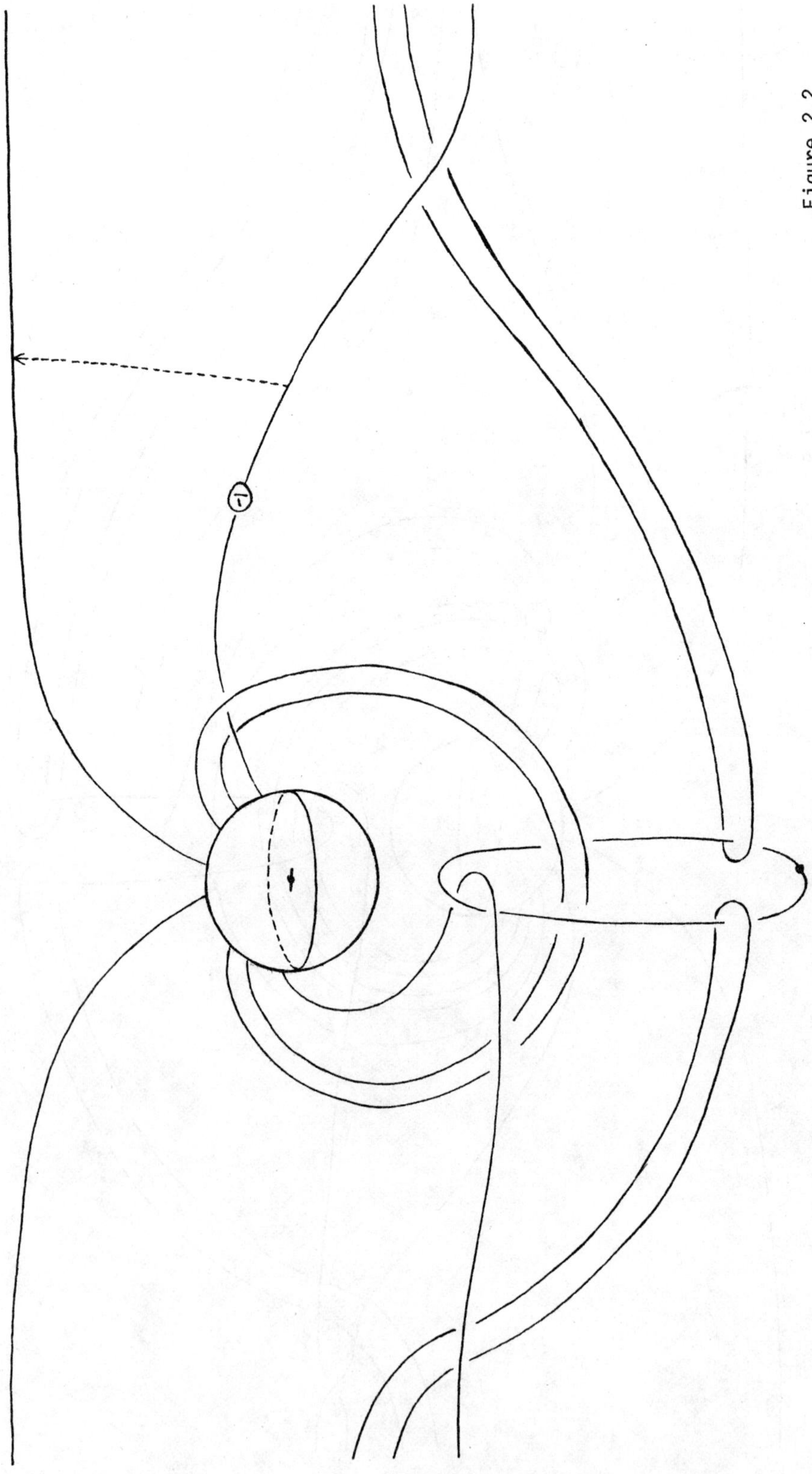

Figure 2.2

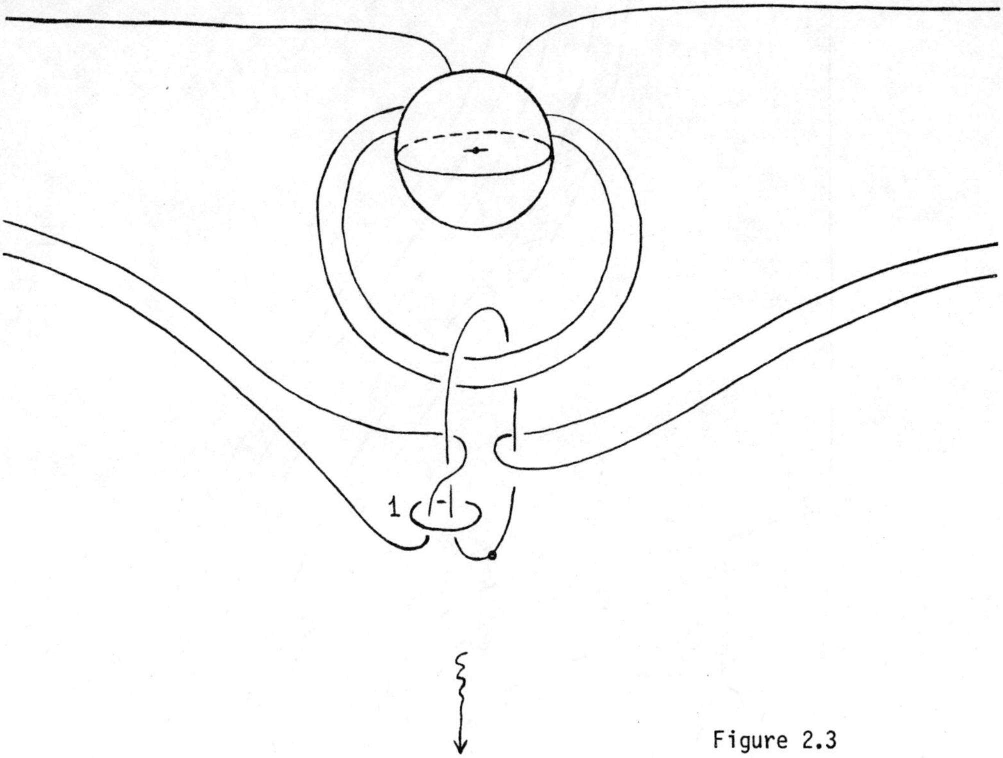

Figure 2.3

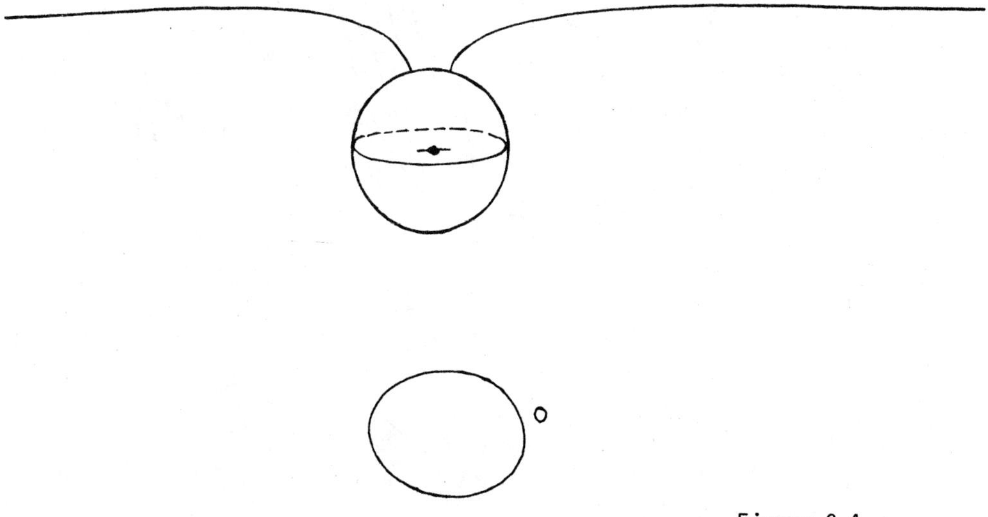

Figure 2.4

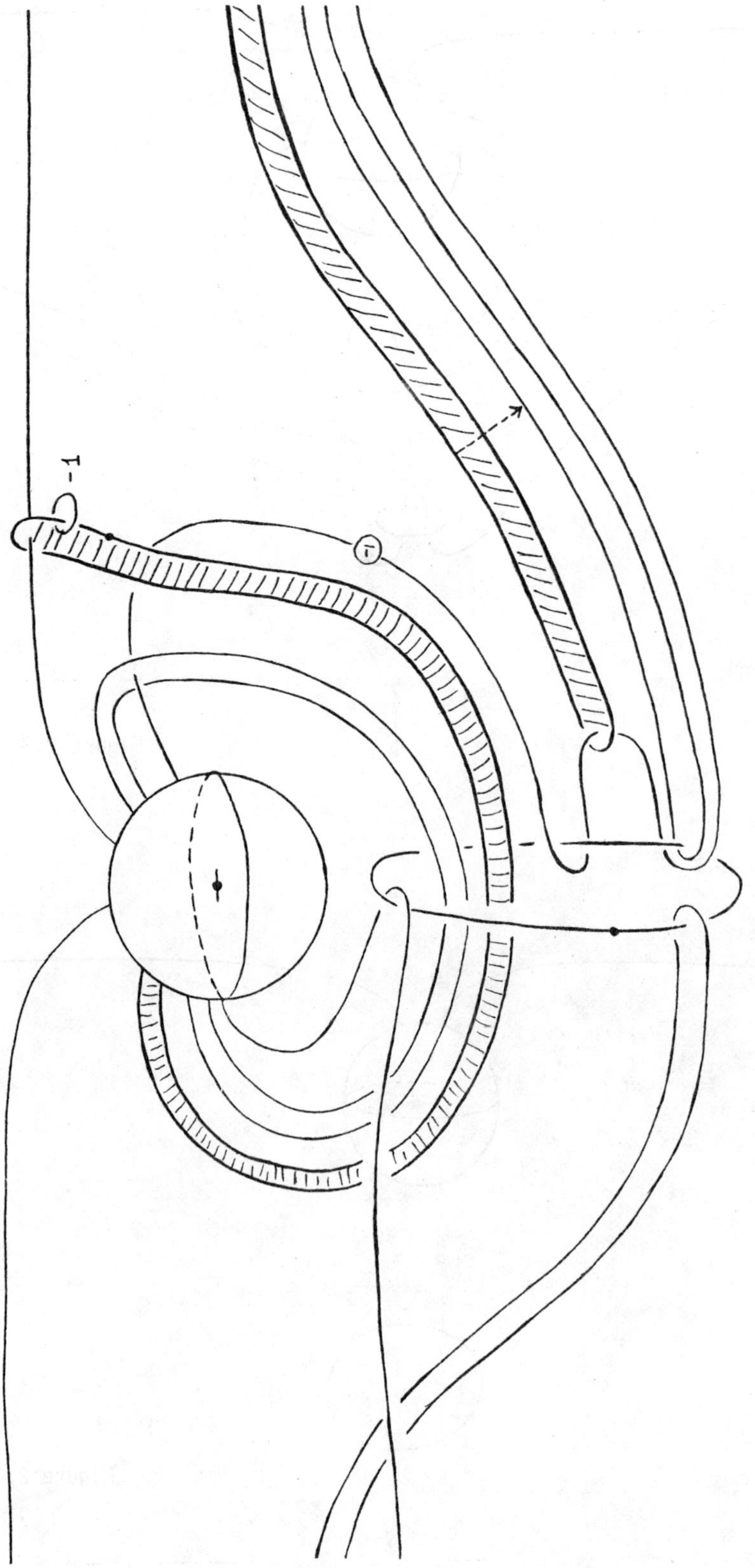

Figure 2.5

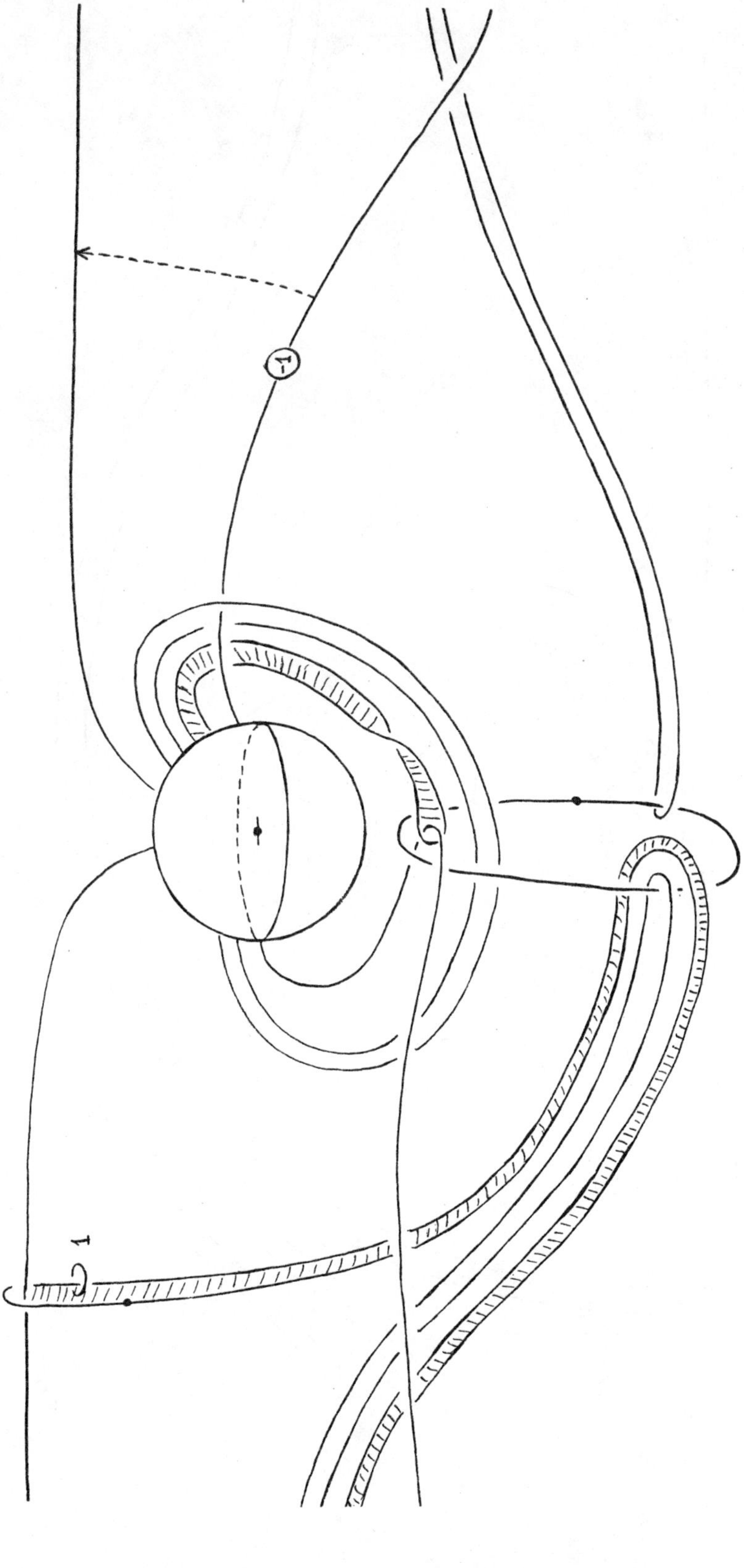

Figure 2.6

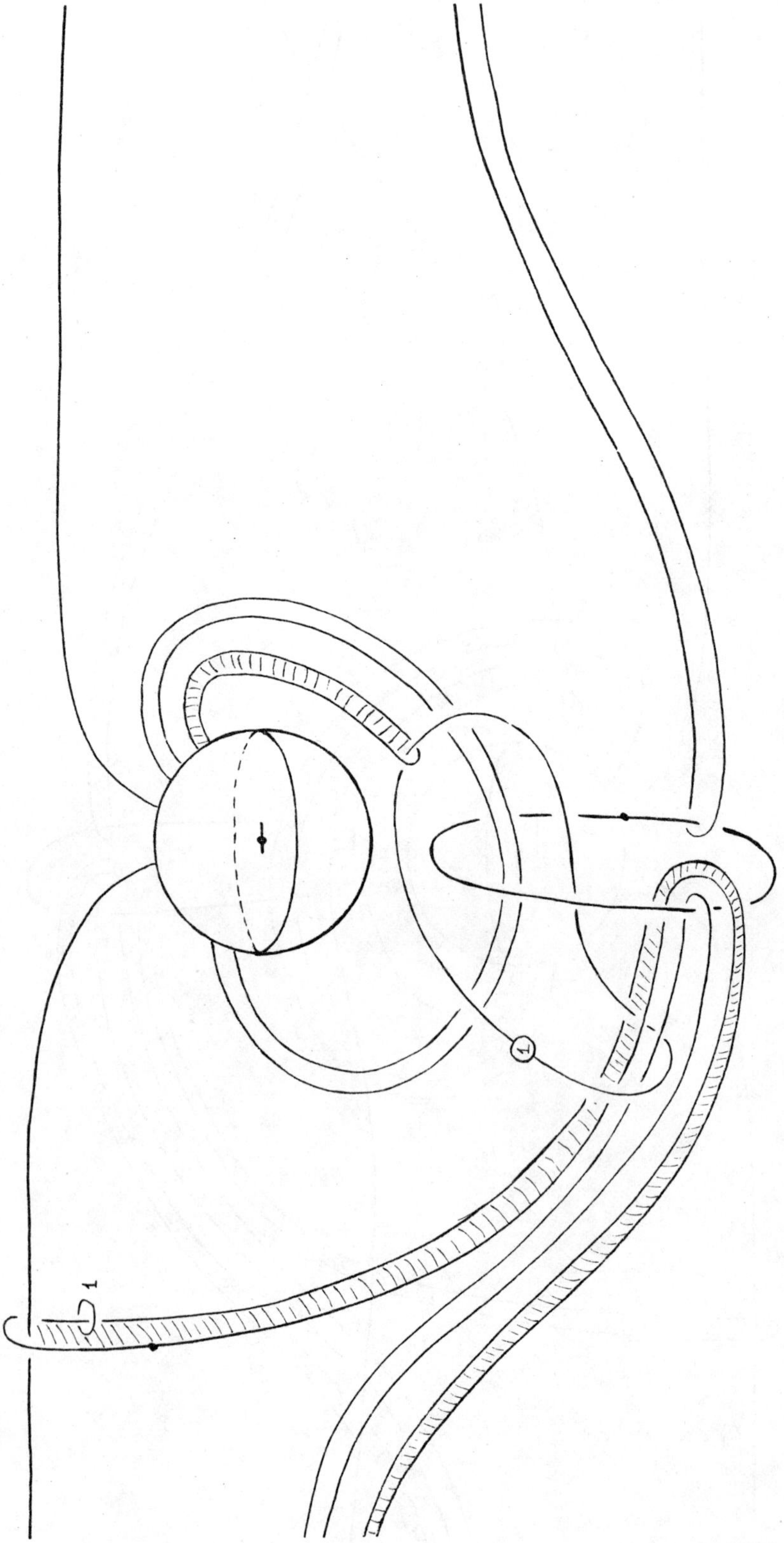

Figure 2.7

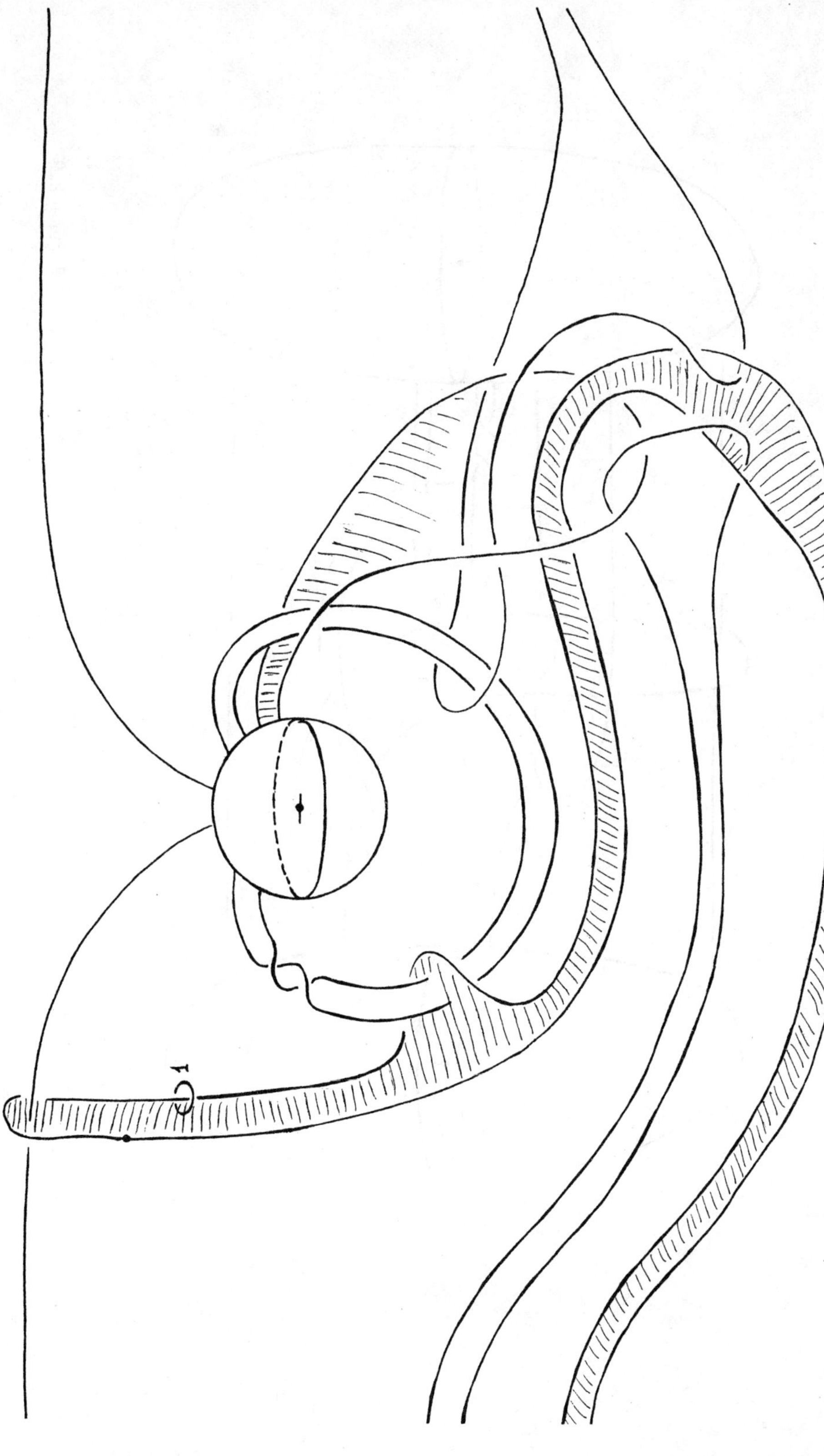

Figure 2.8

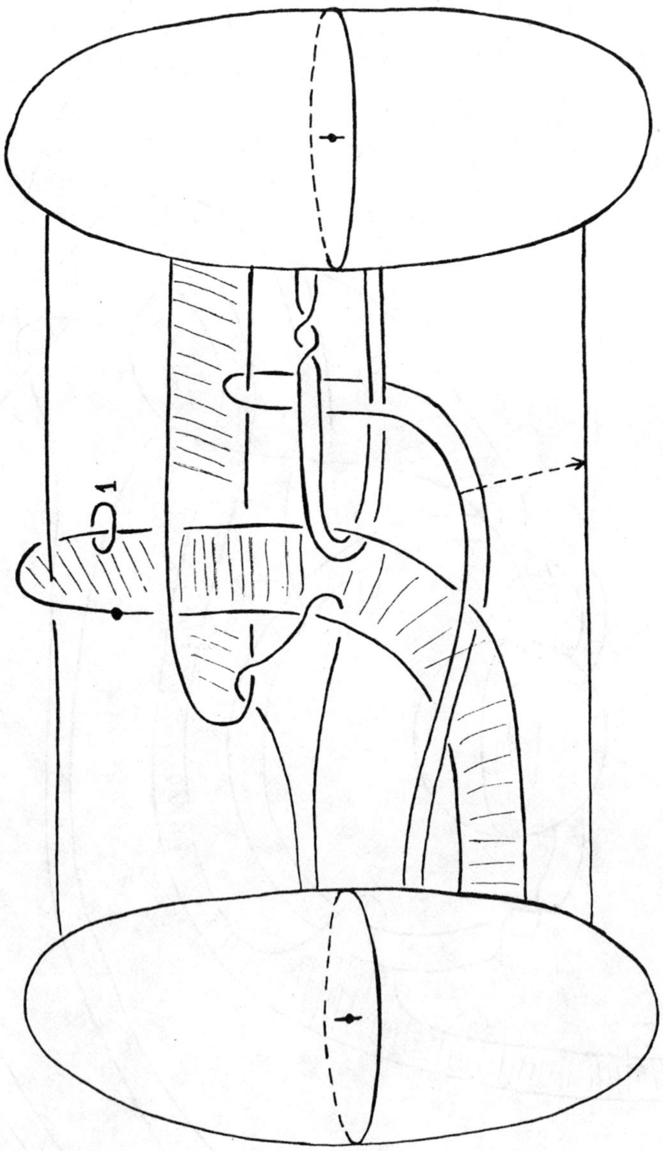

Figure 2.9

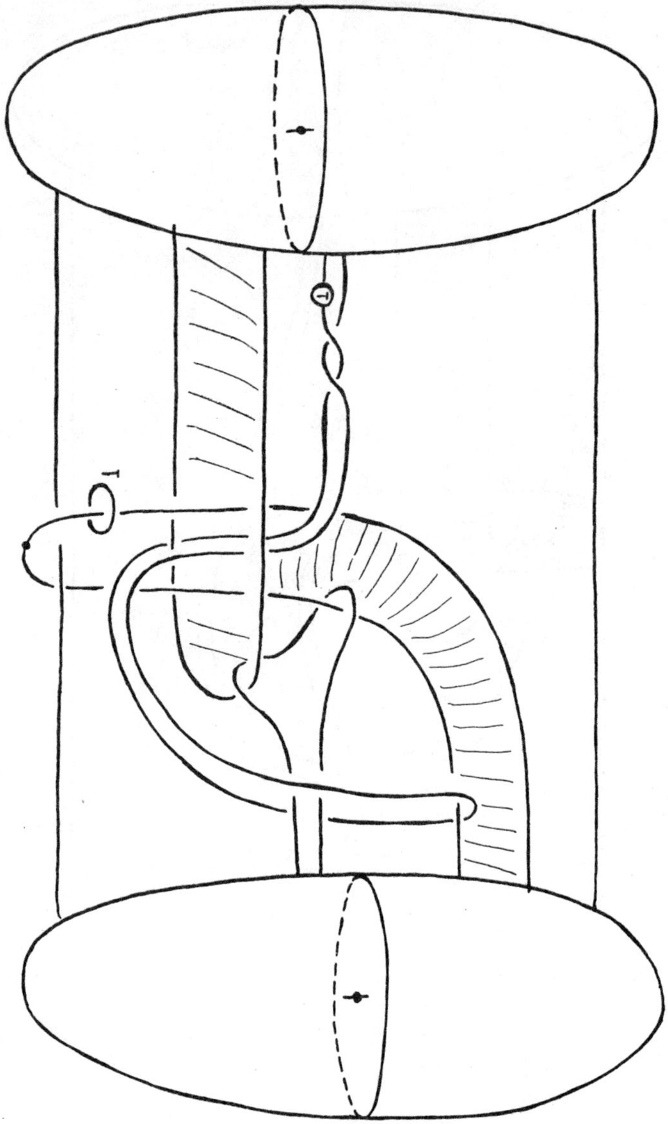

Figure 2.10

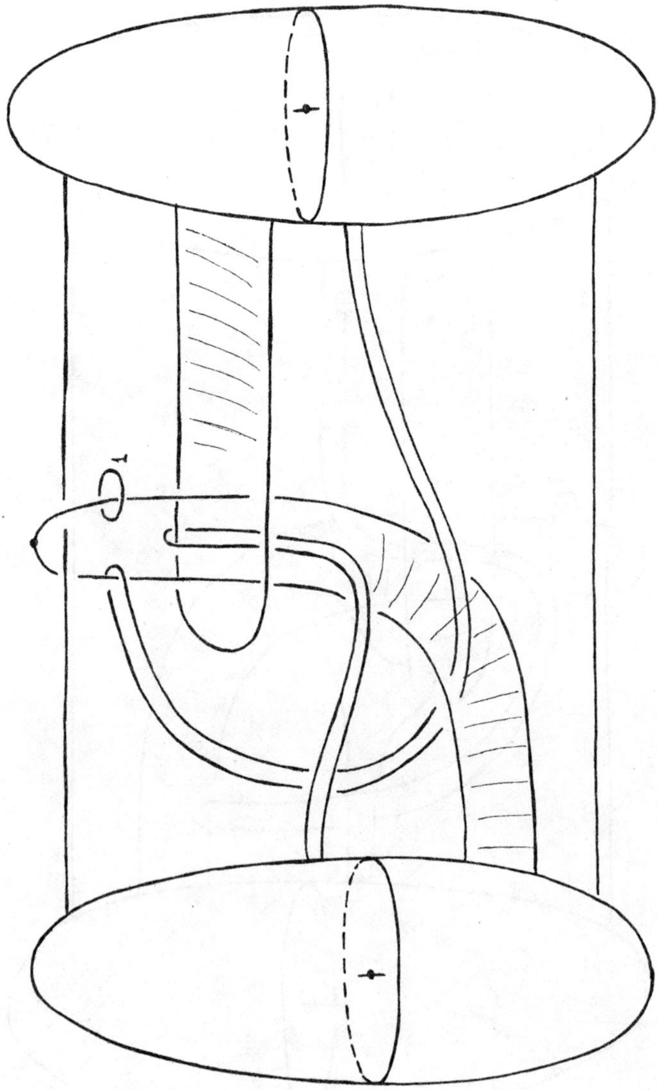

Figure 2.11

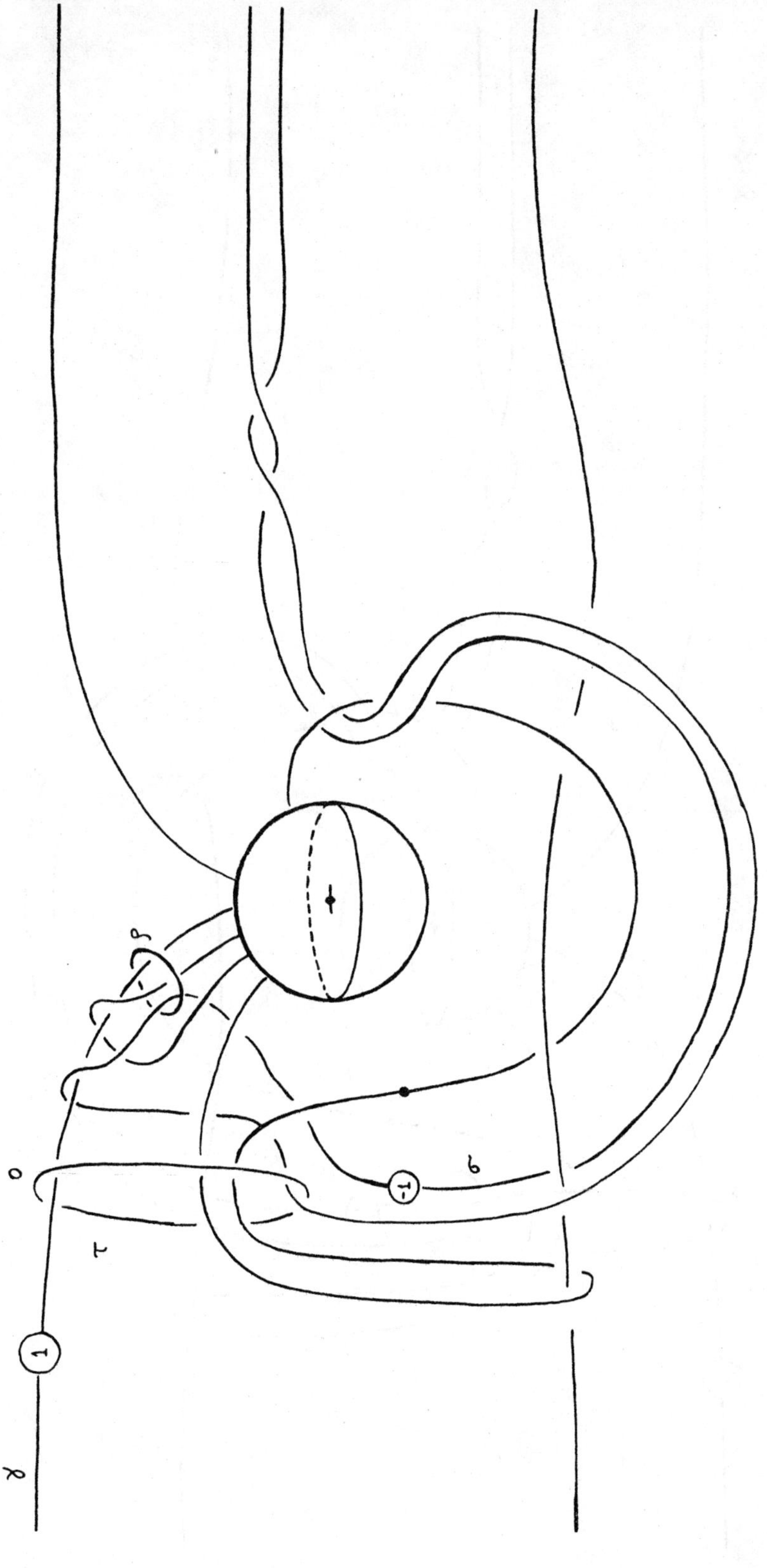

Figure 3.1

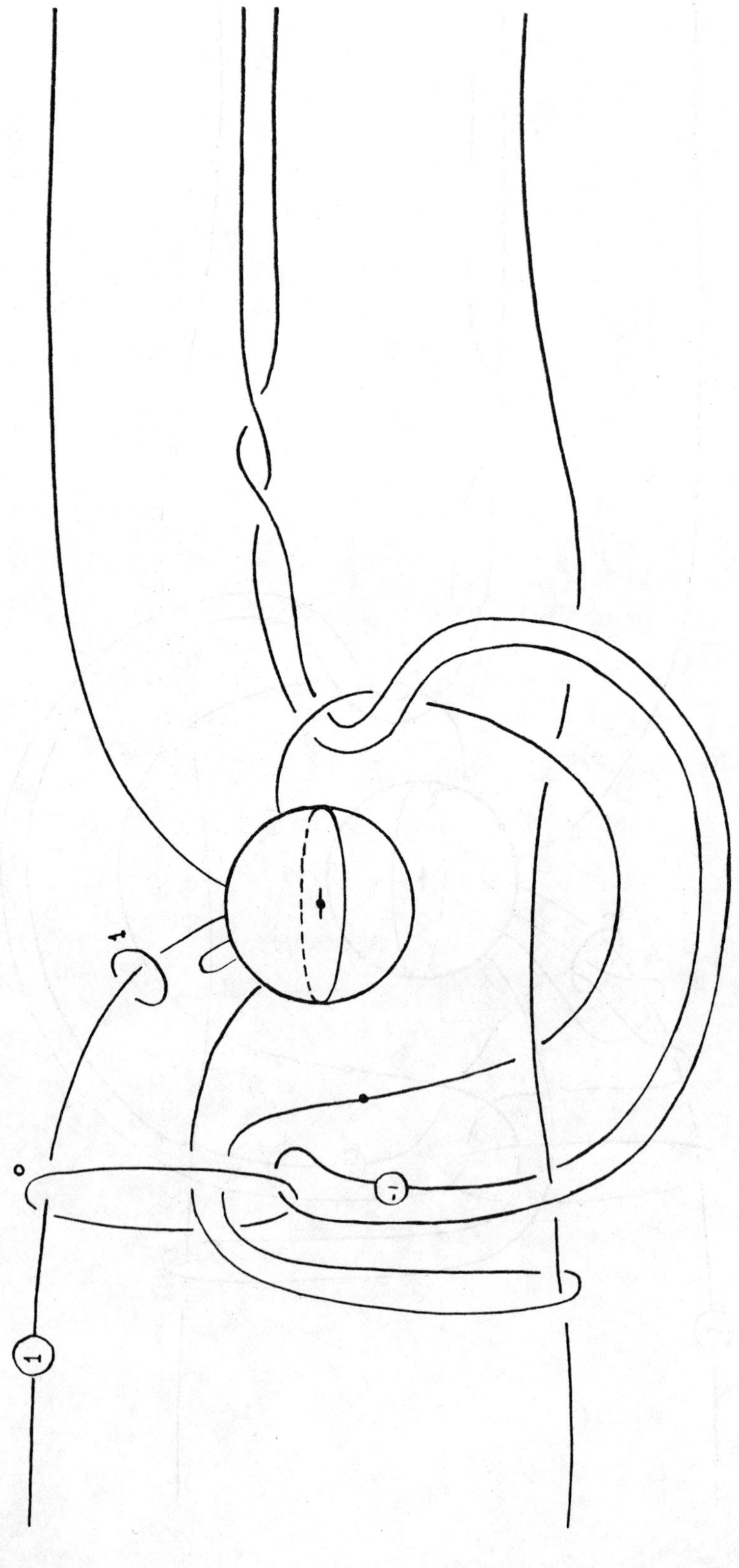

Figure 3.2

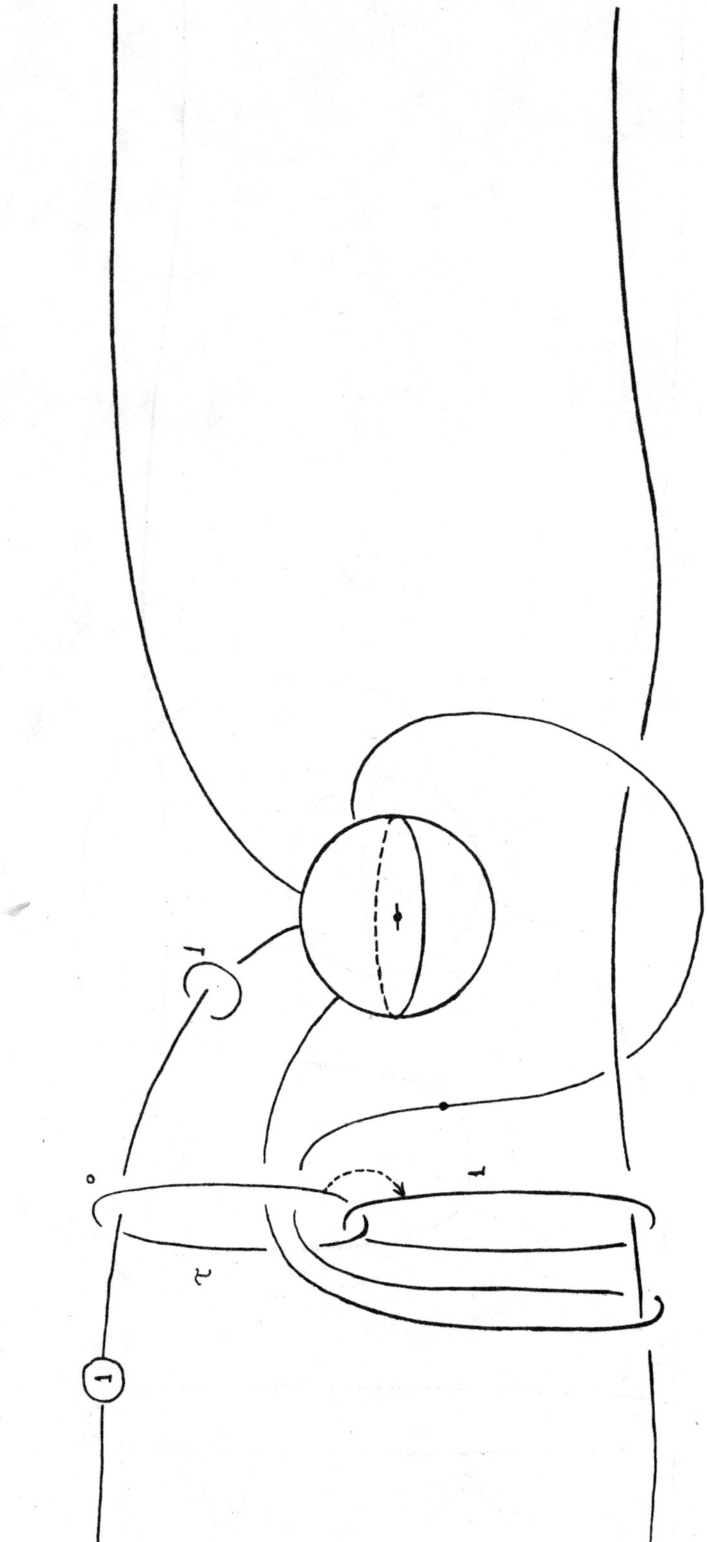

Figure 3.3

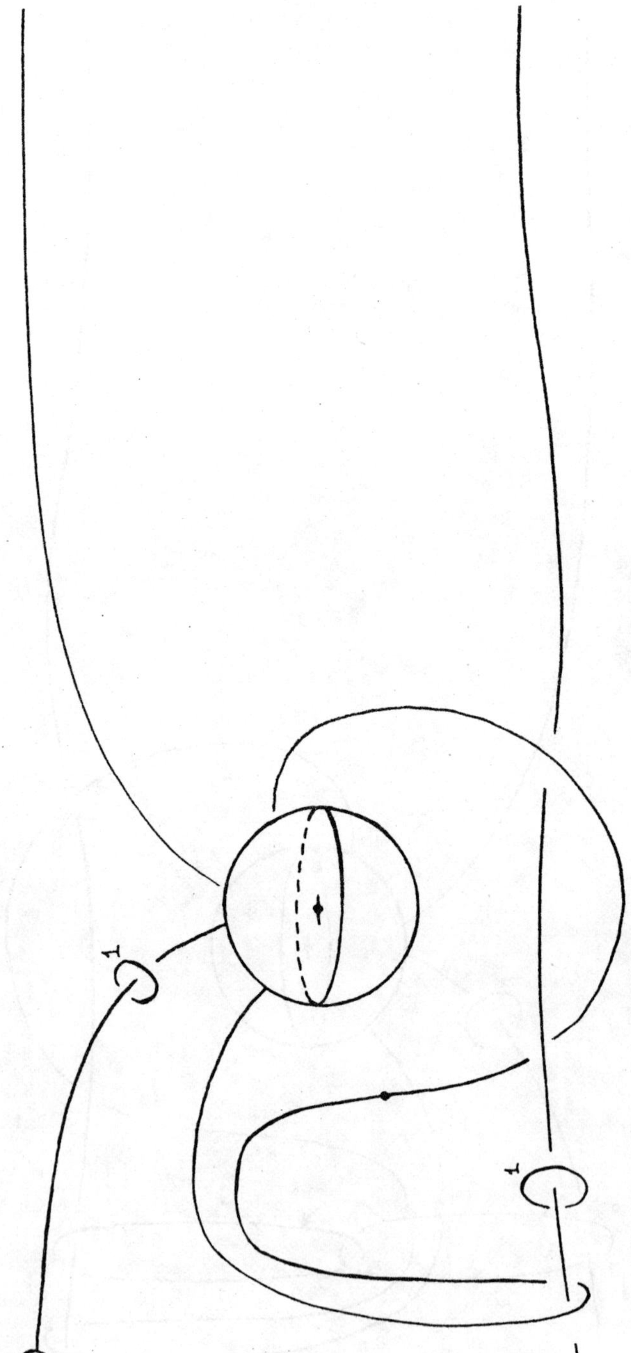

Figure 3.4

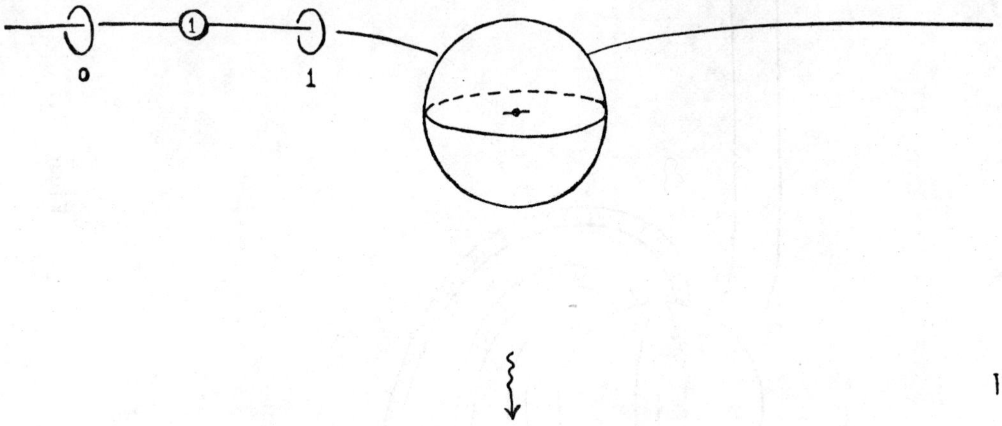

Figure 3.5

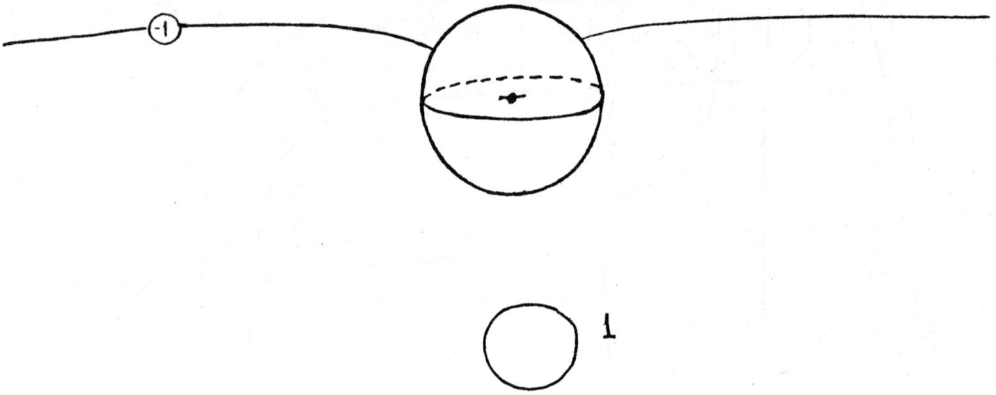

Figure 3.6

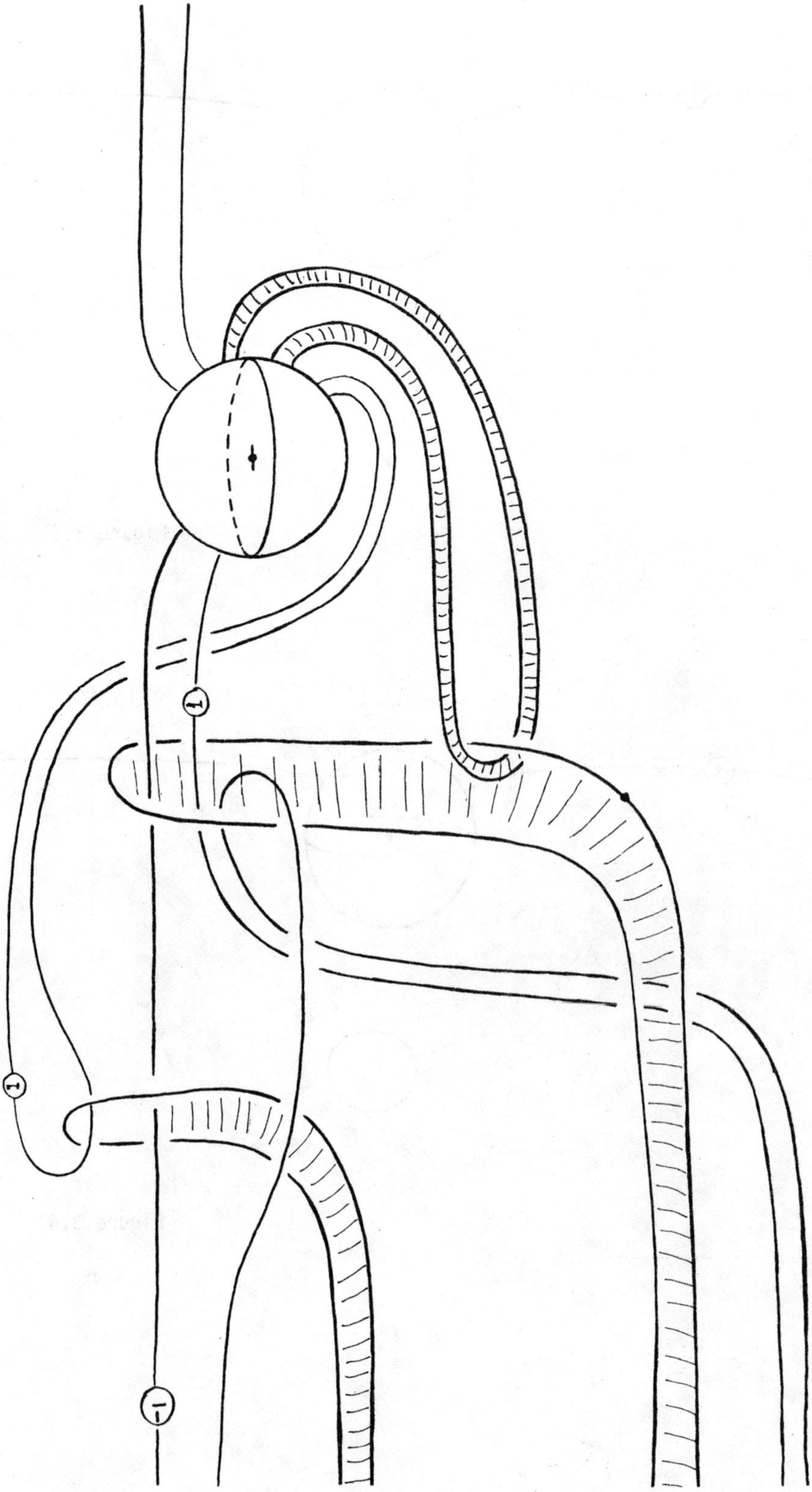

Figure 4.1

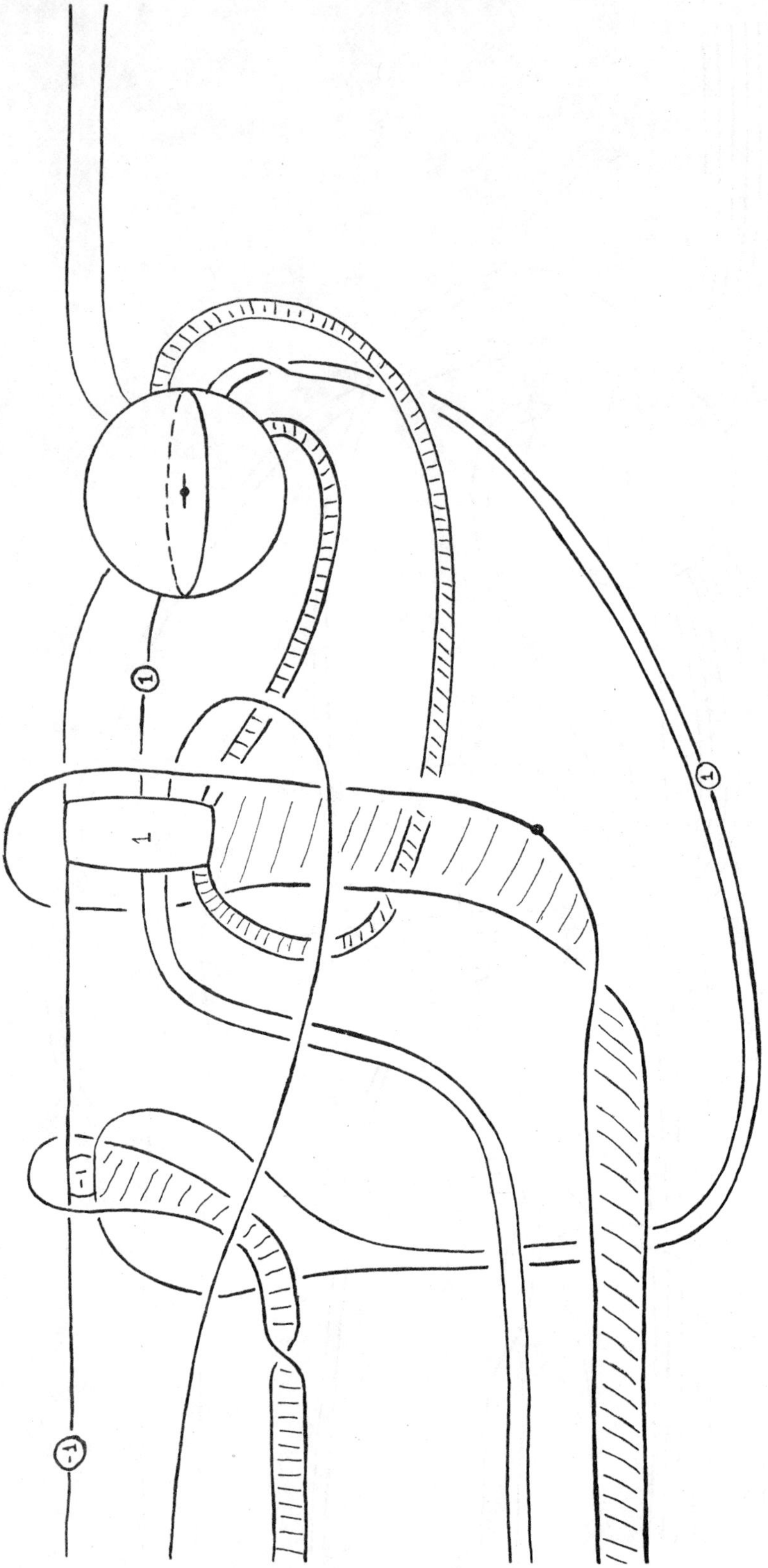

Figure 4.2

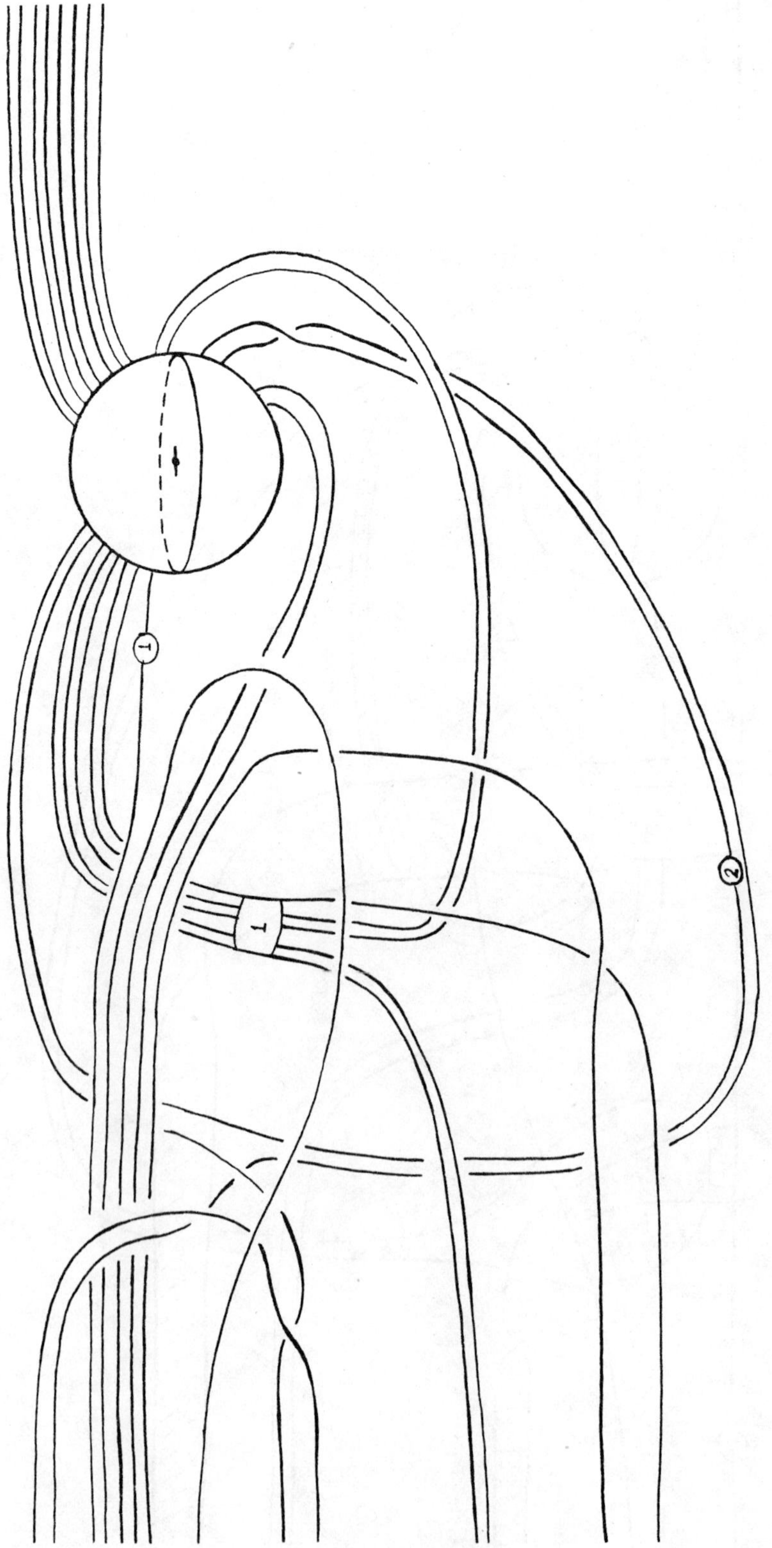

Figure 4.3

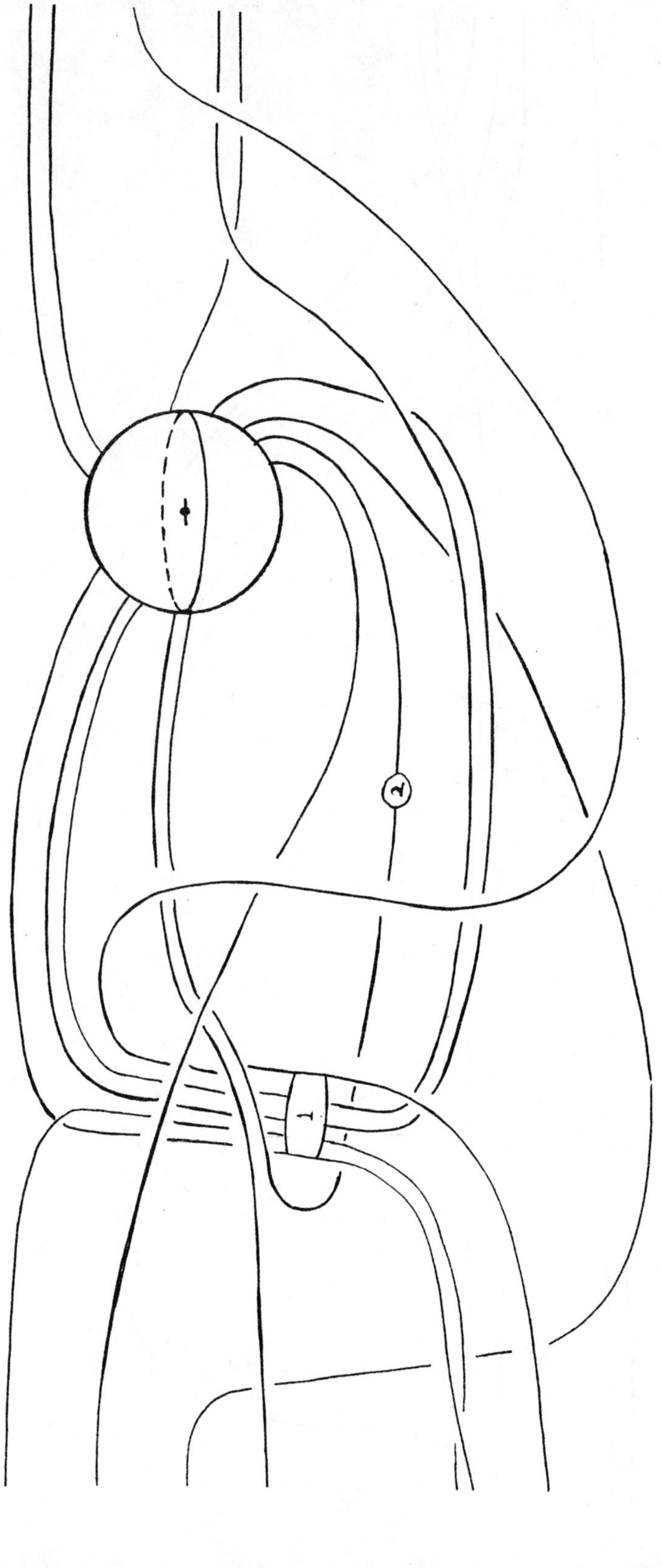

Figure 4.4

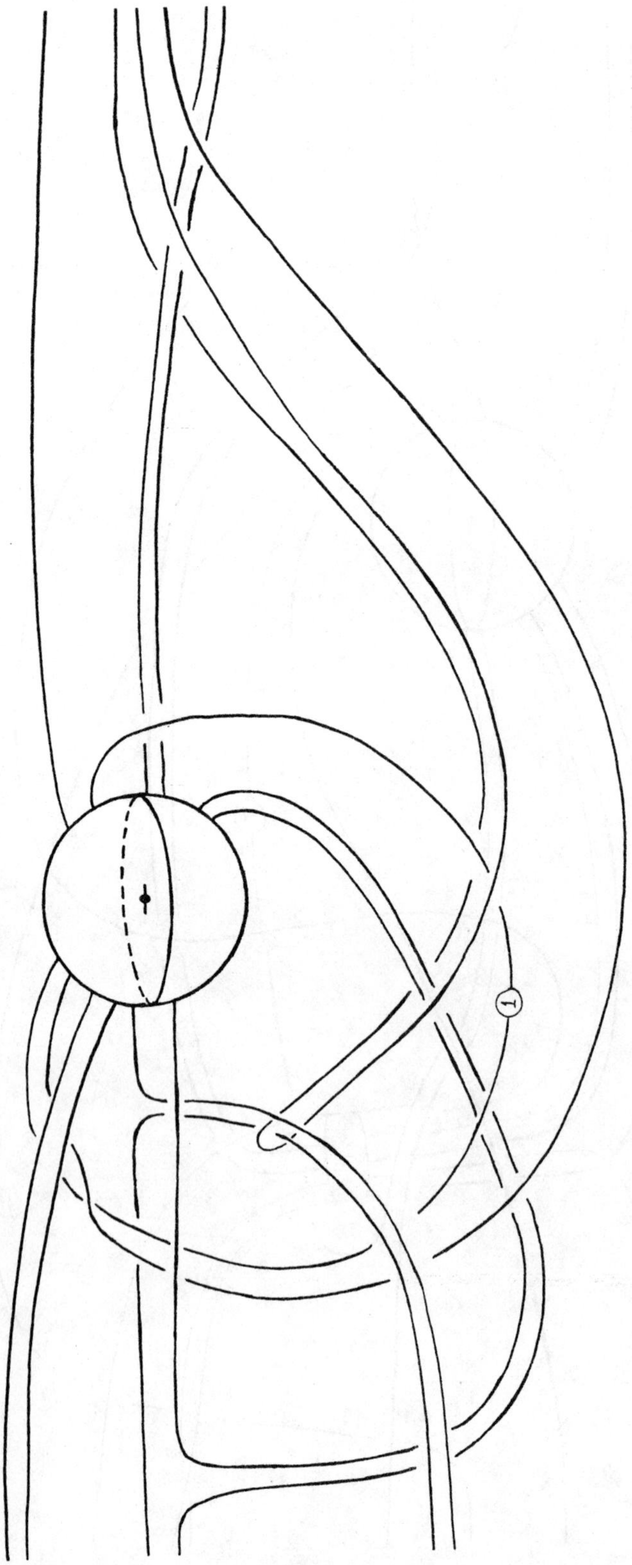

Figure 4.5

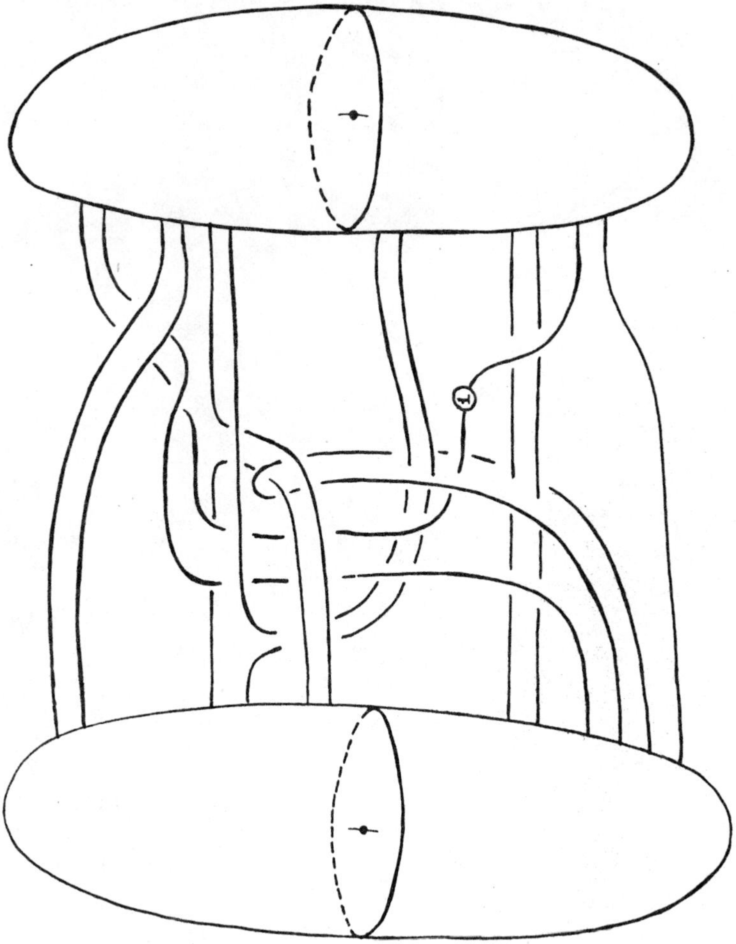

Figure 4.6

Contemporary Mathematics
Volume 35, 1984

APPROXIMATING CELL-LIKE MAPS OF S^4 BY HOMEOMORPHISMS

Fredric D. Ancel

ABSTRACT. We present a proof of FREEDMAN'S APPROXIMATION THEOREM: A surjective map $f:S^n \to S^n$ can be approximated by homeomorphisms if (1) $S(f) = \{y \in S^n : \operatorname{diam} f^{-1}(y) > 0\}$ is a nowhere dense subset of S^n, and (2) $\{f^{-1}(y) : y \in S(f)\}$ is a null collection (for every $\varepsilon > 0$, $\{y \in S(f) : \operatorname{diam} f^{-1}(y) \geq \varepsilon\}$ is a finite set). We then show that these hypotheses can be weakened as follows. A suggestion of R. D. Edwards allows us to replace (1) by: f has a bald spot (there is a non-empty open subset U of S^n such that $f|f^{-1}U : f^{-1}U \to U$ is a homeomorphism). (2) can be replaced by: $S(f)$ is a tame zero-dimensional subset of S^n (each point of $S(f)$ has arbitrarily small collared n-cell neighborhoods whose boundaries miss $S(f)$).

1. INTRODUCTION

Let X and Y be compact metric spaces, and let $f:X \to Y$ be a map. For $\varepsilon > 0$, a map $g:X \to Y$ is <u>within</u> <u>ε</u> of f if $d(f(x),g(x)) < \varepsilon$ for every $x \in X$. f <u>can be approximated by homeomorphisms</u> if for every $\varepsilon > 0$, there is a homeomorphism from X to Y which is within ε of f. If the space Y is locally contractible (for instance, if Y is a manifold), then an easily verified necessary condition for f to be approximable by homeomorphisms is that for each $y \in Y$, $f^{-1}(y)$ contracts to a point in each of its neighborhoods in X. This leads us to the following definition. A subset of X is <u>cell-like</u> if it contracts to a point in each of its neighborhoods in X. The Whitehead continuum is a cell-like (but not contractible) subset of S^3 of great renown. The map $f:X \to Y$ is <u>cell-like</u> if $f^{-1}(y)$ is a cell-like subset of X for each $y \in Y$. We shall consider the question of whether a given cell-like map between spheres can be approximated by homeomorphisms.

For $n \neq 4$, the approximation theorems of [A] and [S] imply that any cell-like map $f:S^n \to S^n$ can be approximated by homeomorphisms. The proofs of these results depend on techniques which are specific to dimension 3 or to high dimensions, and which until recently had no analogues in dimension 4. M. Freedman's August, 1981 construction of topological 2-handles in 4-manifilds [F] changed this situation dramatically. Indeed, in July, 1982 (during the conference whose Proceedings these are), F. Quinn used Freedman's work to obtain a general theorem [Q] which has as a corollary that any cell-like map between

0271-4132/84 $1.00 + $.25 per page

4-spheres can be approximated by homeomorphisms. Freedman's construction depends crucially on the fact that certain special types of cell-like maps between spheres can be approximated by homeomorphisms. We call this fact Freedman's Approximation Theorem. Its ingenious proof (which works in all dimensions) is expounded below, along with proofs of several extensions. Thus the general result that any cell-like map between 4-spheres can be approximated by homeomorphisms follows from Quinn's work which in turn depends on the special case established by Freedman's Approximation Theorem.

To state Freedman's Approximation Theorem and its extensions, we require the following definitions. Again let X and Y be compact metric spaces, and let $f:X\to Y$ be a map. The singular set of f, denoted $S(f)$, is the set $\{y\in Y: f^{-1}(y)$ contains more than one point$\}$. Observe that for every $\varepsilon>0$, the set $\{y\in Y:\operatorname{diam} f^{-1}(y)\geq\varepsilon\}$ is compact. Since $S(f)=\bigcup_{i=1}^{\infty}\{y\in Y:\operatorname{diam} f^{-1}(y)\geq 1/i\}$, we conclude that $S(f)$ is σ-compact. A subset of Y is nowhere dense if its closure has empty interior. A collection $\mathcal{C}$ of subsets of X is a null collection if for every $\varepsilon>0$, $\{C\in\mathcal{C}:\operatorname{diam} C\geq\varepsilon\}$ is a finite set. Thus, if $\{f^{-1}(y):y\in S(f)\}$ is a null collection of subsets of X, then $S(f)$ is a countable set. f has a bald spot if there is a non-empty open subset U of Y such that $f|f^{-1}U:f^{-1}U\to U$ is a homeomorphism. Thus, f has a bald spot if it is surjective and if $c\ell\,(f)\neq Y$.

Let M be an n-manifold. An n-cell C in $\operatorname{int} M$ is collared if there is an embedding of $\partial C\times[0,1]$ in $M-\operatorname{int} C$ which takes $\partial C\times\{0\}$ onto ∂C. A σ-compact subset S of $\operatorname{int} M$ is tame zero-dimensional in M if each point of S has arbitrarily small collared n-cell neighborhoods whose boundaries miss S (in other words, for every $y\in S$ and every neighborhood U of y in M, there is a collared n-cell C in M such that $y\in\operatorname{int} C$, $C\subset U$ and $(\partial C)\cap S=\emptyset$).

We shall present proofs of the following theorems.

THEOREM 1: FREEDMAN'S APPROXIMATION THEOREM. *A surjective map $f:S^n\to S^n$ can be approximated by homeomorphisms if* (1) $S(f)$ *is a nowhere dense subset of* S^n *and* (2) $\{f^{-1}(y) : y\in S(f)\}$ *is a null collection*.

A suggestion of R. D. Edwards for reorganizing Freedman's proof of Theorem 1 leads to a proof of:

THEOREM 2. *A map $f:S^n\to S^n$ can be approximated by homeomorphisms if* (1) f *has a bald spot and* (2) $S(f)$ *is a countable subset of* S^n.

Finally an "amalgamation procedure" combines with a shrinking principle due to R. H. Bing to yield:

THEOREM 3. *A map $f:S^n\to S^n$ can be approximated by homeomorphisms if* (1) f *has a bald spot and* $S(f)$ *is a tame zero-dimensional subset of* S^n.

Before embarking on the proofs of these theorems, we make several remarks.

First, we note that the surjectivity hypothesis in Theorem 1 would be redundant in Theorems 2 and 3, because the bald spot hypothesis implies that f is degree 1 and, thus, surjective.

Second, we note that although the hypotheses of these three theorems do not explicitly state that f is a cell-like map, they easily imply that it is. For let $y \in S(f)$. The tame zero-dimensionality of $S(f)$ implies that $f^{-1}(y)$ has arbitrarily tight closed neighborhoods whose frontiers are (n-1)-spheres. These neighborhoods must be contractible. Hence $f^{-1}(y)$ is cell-like.

In his construction of topological 2-handles in 4-manifolds, Freedman applies Theorem 1 at a crucial point to a map $f: S^4 \to S^4$. The validity of this application depends on $S(f)$ being nowhere dense in S^4. In Freedman's context, $S(f)$ is nowhere dense because its closure is a 1-dimensional subset of S^4.

We close this section with some comments about the proof of Theorem 1, including a comparison to M. Brown's proof of the Generalized Schoenflies Theorem.

Freedman's Approximation Theorem might be regarded as a generalization of [Br], because Brown's method of proof implicitly establishes the following:

THEOREM 0. <u>A surjective map</u> $f: S^n \to S^n$ <u>can be approximated by homeomorphisms if</u> $S(f)$ <u>is a finite set.</u>

There is a superficial resemblance between the techniques used by Brown to prove Theorem 0 and those used by Freedman for Theorem 1. We find it instructive to review the outline of Brown's argument for Theorem 0, to contrast the two methods of proof, and to focus on the difficulties that must be overcome by any proof of Theorem 1 which don't arise in the proof of Theorem 0.

To review Brown's proof of Theorem 0, consider a surjective map $f: S^n \to S^n$ with a finite singular set. First, one argues by induction on the number of points in $S(f)$ that for each $y \in S(f)$, $f^{-1}(y)$ is a cellular subset of S^n ($f^{-1}(y)$ has arbitrarily tight n-cell neighborhoods in S^n). (This is a slight oversimplification; in the actual proof, one must work with a map $f: B^n \to S^n$ such that $S(f)$ is finite and disjoint from $f(\partial B^n)$.) Second, one uses the cellularity of the preimages of the points of $S(f)$ to "shrink" these sets independently to produce a homeomorphism approximating f. Neither of these steps is possible under the hypotheses of Freedman's Approximation Theorem. First, since $S(f)$ may be countably infinite, no induction argument will establish the cellularity of the preimages of the points of $S(f)$. Second, even if the cellularity of the preimages of the points in $S(f)$ is given in advance, they cannot be shrunk independently. The problem is that a motion which shrinks the larger preimage sets small may necessarily stretch some of the smaller sets. The classic example of this phenomenon is Bing's null cellular decomposition of S^3 [B2] whose quotient map is not approximable by homeomorphisms because its quotient space is not S^3.

In Freedman's proof of Theorem 1, the cellularity of the preimages of the points of $S(f)$ is never established in the course of the argument. It follows only after the proof is finished as a consequence of the conclusion of the theorem.

Freedman's proof is not a traditional "shrinking argument" in the sense of decomposition space theory. It has a more complex logical structure. Instead of shrinking the large point inverses of f, it uses a replication device which makes the large point images of f disappear at the cost of complicating the logical framework of the argument. Specifically, the replication device forces the use of relations which are neither maps nor their inverses. In fact, the approximating homeomorphism which is the goal of the proof arises as the limit of such relations. For this reason, simple techniques for manipulating relations appear.

2. TWO LEMMAS

We introduce some terminology and establish two lemmas which find use in the proofs of Theorems 1 and 2.

The first lemma is a general position property of countable subsets of manifolds. The following remarks about the homeomorphism group of a compactum are included to simplify its proof.

Suppose X is a compact space with metric ρ. Let $\mathcal{H}(X)$ denote the space of homeomorphisms of X with the compact-open topology. (One basis for the compact-open topology on $\mathcal{H}(X)$ consists of all sets of the form $\{h \in \mathcal{H}(X) : h \subset O\}$ where O varies over the open subsets of $X \times X$.) The compact-open topology on $\mathcal{H}(X)$ is induced by the "supremum metric" σ which is defined by $\sigma(g,h) = \sup\{\rho(g(x),h(x)) : x \in X\}$. Although σ is generally not a complete metric on $\mathcal{H}(X)$, a complete metric τ on $\mathcal{H}(X)$ is easily produced in terms of σ by the formula $\tau(g,h) = \sigma(g,h) + \sigma(g^{-1},h^{-1})$. For a subset A of X, define $\mathcal{H}(X,A) = \{h \in \mathcal{H}(X) : h|A = 1|A\}$. If $A \subset X$, then $\mathcal{H}(X,A)$ is a closed subset of $\mathcal{H}(X)$; hence, the complete metric τ on $\mathcal{H}(X)$ restricts to a complete metric on $\mathcal{H}(X,A)$.

Two subsets S and T of a metric space X are <u>separated</u> in X if $(c\ell S) \cap T = \emptyset = S \cap (c\ell T)$ (or equivalently if there are disjoint open subsets U and V of X such that $S \subset U$ and $T \subset V$).

LEMMA 1. <u>Let</u> M <u>be a compact manifold</u>.

(1) <u>If</u> S <u>is a countable subset of</u> $\operatorname{int} M$ <u>and</u> T <u>is the union of a countable number of nowhere dense subsets of</u> M, <u>then</u> $1|M$ <u>can be approximated by homeomorphisms</u> h <u>of</u> M <u>such that</u> $h(S) \cap T = \emptyset$ <u>and</u> $h|\partial M = 1|\partial M$.

(2) <u>If</u> S <u>and</u> T <u>are countable nowhere dense subsets of</u> $\operatorname{int} M$, <u>then</u> $1|M$ <u>can be approximated by homeomorphisms</u> h <u>of</u> M <u>such that</u> $h(S)$ <u>and</u> T <u>are</u>

separated in M and $h|\partial M = 1|\partial M$.

PROOF OF (1). Let $S=\{s_i\}$, and let $T=\cup_{j=1}^{\infty}T_j$ where each T_j is a nowhere dense subset of M. For each $i \geq 1$, let $U_{i,j} = \{h \in \mathcal{H}(M,\partial M)\colon h(s_i) \notin c\ell T_j\}$. It is easily seen that each $U_{i,j}$ is a dense open subset of $\mathcal{H}(H,\partial M)$. Since $\mathcal{H}(M,\partial M)$ has a complete metric, we conclude via the Baire Category Theorem that $\cap_{i=1}^{\infty}{}_{j=1}^{\infty}U_{i,j}$ is a dense subset of $\mathcal{H}(M,\partial M)$. Statement (1) now follows because $1|M$ can be approximated by elements of $\cap_{i=1}^{\infty}{}_{j=1}^{\infty}U_{i,j}$.

PROOF OF (2). Assume $S=\{s_i\}$ and $T=\{t_j\}$ are countable nowhere dense subsets of $\operatorname{int} M$. For each $i \geq 1$, let $U_i = \{h \in \mathcal{H}(M,\partial M)\colon h(s_i) \notin c\ell T\}$ and let $V_i = \{h \in \mathcal{H}(M,\partial M)\colon t_i \notin h(c\ell S)\}$. It is easily seen that each U_i and each V_i are dense open subsets of $\mathcal{H}(M,\partial M)$. As above, since $\mathcal{H}(M,\partial M)$ has a complete metric, the Baire Category Theorems implies that $\cap_{i=1}^{\infty}(U_i \cap V_i)$ is a dense subset of $\mathcal{H}(M,\partial M)$. Statement (2) now follows because $1|M$ can be approximated by elements of $\cap_{i=1}^{\infty}(U_i \cap V_i)$. ▥

The second lemma concerns relations. It is used in the proofs of Theorems 1 and 2 to guarantee that the sequences of relations which are produced in these proofs converge to homeomorphisms. In order to streamline the next lemma and the proofs of Theorems 1 and 2, we now establish some notation for relations which generalizes the usual functional notation.

Let $R \subset X \times Y$; i.e., R is a relation from the set X to the set Y. Define

$$R^{-1} = \{(y,x) \in Y \times X\colon (x,y) \in R\} .$$

If $S \subset Y \times Z$, define

$$S \circ R = \{(x,z) \in X \times Z\colon (x,y) \in R \text{ and } (y,z) \in S \text{ for some } y \in Y\} .$$

If $x \in X$, define $R(x) = \{y \in Y\colon (x,y) \in R\}$. Thus for $y \in Y$, $R^{-1}(y) = \{x \in X\colon (x,y) \in R\}$. If $x \in X$, then $R(x)$ is called a <u>point image</u> of R; and if $y \in Y$, then $R^{-1}(y)$ is called a <u>point inverse</u> of R. If $A \subset X$, define $R(A) = \cup\{R(x)\colon x \in A\}$ and define $R|A = R \cap (A \times Y)$.

LEMMA 2. *Let R be a closed subset of $X \times Y$ where X and Y are compact metric spaces. Suppose T is a closed subset of X, $\varepsilon > 0$ and $\operatorname{diam} R(x) < \varepsilon$ for every $x \in X-T$. Then there is a closed subset N of $X \times Y$ such that $R|X-T \subset \operatorname{int} N$, $\operatorname{diam} N(x) < \varepsilon$ for every $x \in X-T$, and $N|T = R|T$.*

PROOF. Let $M_1 \supset M_2 \supset M_3 \supset \cdots$ be a decreasing sequence of closed neighborhoods of R in $X \times Y$ such that $\cap_{i=1}^{\infty}M_i = R$. We assert that if A is a compact subset of $X-T$, then for some $i \geq 1$, $\operatorname{diam} M_i(x) < \varepsilon$ for every $x \in A$. For otherwise, there are sequences $\{(x_i,y_i)\}$ and $\{(x_i,z_i)\}$ in $A \times Y$ such that for each $i \geq 1$, (x_i,y_i) and (x_i,z_i) lie in M_i and $\operatorname{diam}\{y_i,z_i\} \geq \varepsilon$. Since A and Y are compact, then by passing to subsequences, we can assume that the sequence $\{x_i\}$ converges to a point x in A, and that the sequences $\{y_i\}$

and $\{z_i\}$ converge to points y and z, respectively, in Y. Consequently, $\text{diam}\{y,z\} \geq \varepsilon$. Also since $R = \bigcap_{i=1}^{\infty} M_i$, it follows that (x,y) and (x,z) belong to R. Hence y and z belong to $R(x)$. Since $\text{diam}\, R(x) < \varepsilon$, we have a contradiction. Our assertion follows.

Let $\{A_i\}$ be a sequence of compact subsets of $X-T$ such that $A_i \subset \text{int}\, A_{i+1}$ for each $i \geq 1$, and $\bigcup_{i=1}^{\infty} A_i = X-T$. Set $A_0 = \emptyset$. The above assertion implies that by passing to an appropriate subsequence of $\{M_i\}$, we obtain a decreasing sequence $N_1 \supset N_2 \supset N_3 \supset \cdots$ of closed neighborhoods of R such that $\bigcap_{i=1}^{\infty} N_i = R$ and for each $i \geq 1$, $\text{diam}\, N_i(x) < \varepsilon$ for every $x \in A_i$. Set $N = (\bigcup_{i=1}^{\infty} N_i | A_i) \cup (R|T)$. We find it convenient to define, for each $i \geq 1$, a closed neighborhood P_i of R in $X \times Y$ by setting $P_i = (\bigcup_{j=1}^{i-1} N_j | A_j) \cup N_i$. Then $N = \bigcap_{i=1}^{\infty} P_i$; so N is a closed subset of $X \times Y$. For each $i \geq 1$, since $P_i | A_i = N | A_i$, then $R|\text{int}\, A_i \subset \text{int}\, P_i | \text{int}\, A_i \subset \text{int}\, N$; it follows that $R|X-T \subset \text{int}\, N$. For each $i \geq 1$, if $x \in A_i - A_{i-1}$, then $\text{diam}\, N(x) = \text{diam}\, N_i(x) < \varepsilon$; hence $\text{diam}\, N(x) < \varepsilon$ for every $x \in X-T$. Clearly $N|T = R|T$. ▩

3. FREEDMAN'S APPROXIMATION THEOREM

A map $f: B^n \to B^n$ is <u>admissible</u> if $f|\partial B^n = 1|\partial B^n$, $S(f)$ is a nowhere dense subset of B^n, $c\ell S(f) \subset \text{int}\, B^n$, and $\{f^{-1}(y): y \in S(f)\}$ is a null collection.

We shall now argue that Freedman's Approximation Theorem reduces to:

THEOREM 1A. <u>Every admissible map</u> $f: B^n \to B^n$ <u>can be approximated by homeomorphisms</u>.

PROOF OF FREEDMAN'S APPROXIMATION THEOREM FROM THEOREM 1A. Assume Theorem 1A. Suppose $f: S^n \to S^n$ is a map with a nowhere dense singular set whose point inverses form a null collection. Let $\varepsilon > 0$. Since $S(f)$ is nowhere dense, there is a collared n-cell C in $S^n - c\ell S(f)$ of diameter $< \varepsilon$. The Generalized Schoenflies Theorem [Br] produces homeomorphisms $\varphi: B^n \to c\ell(S^n - f^{-1}C)$ and $\psi: B^n \to c\ell(S^n - C)$; furthermore ψ can be adjusted so that $\psi|\partial B^n = f \circ \varphi|\partial B^n$. Then $\psi^{-1} \circ f \circ \varphi: B^n \to B^n$ is an admissible map. The uniform continuity of ψ provides a $\delta > 0$ so that ψ carries any set of diameter $< \delta$ to a set of diameter $< \varepsilon$. Theorem 1A gives us a homeomorphism $g: B^n \to B^n$ which is within δ of $\psi^{-1} \circ f \circ \varphi$. It follows that $\psi \circ g \circ \varphi^{-1}: c\ell(S^n - f^{-1}C) \to c\ell(S^n - C)$ is a homeomorphism which is within ε of $f|c\ell(S^n - f^{-1}C)$. Since $\psi \circ g \circ \varphi^{-1}$ maps $f^{-1}(\partial C)$ homeomorphically onto ∂C, and since $\text{diam}\, C < \varepsilon$, then $\psi \circ g \circ \varphi^{-1}$ extends to a homeomorphism of S^n which is within ε of f.

The geometric idea lying at the heart of the proof of Freedman's Approximation Theorem is a very simple replication device which is crystallized in the following lemma. In this lemma, the pre-image pattern of the given admissible map φ on $\varphi^{-1}D$ is replicated by a new admissible map ψ on $\psi^{-1}D$; and the replication is witnessed by a homemorphism $\chi: \varphi^{-1}D \to \psi^{-1}D$ such that

$\psi \circ \chi = \varphi|\varphi^{-1}D$. We foreshadow the proof of the theorem to the extent of remarking that this replication allows us to replace the map φ by a relation R which equals χ on $\varphi^{-1}D$ and which equals $\psi^{-1} \circ \varphi$ on $B^n - \varphi^{-1}D$. R represents an improvement over φ in that it has no non-trivial point inverses in $\varphi^{-1}D$. The apparent disadvantage of this procedure is that it exchanges a map for a relation.

We need the following terminology for the lemma. Let $| \ |$ denote the Euclidean norm on $\mathbb{R}^n$; i.e., $|x| = (\Sigma_{i=1}^n x_i^2)^{\frac{1}{2}}$ for $x = (x_1, \ldots, x_n) \in \mathbb{R}^n$. An n-cell C in $\mathbb{R}^n$ is <u>round</u> if there is a point x in $\mathbb{R}^n$ called the <u>center</u> of C and a positive number r called the <u>radius</u> of C such that $C = \{y \in \mathbb{R}^n : |x-y| \leq r\}$. Note that if C is a round n-cell in B^n and D is a compactum in $\operatorname{int} C$, then a homeomorphism $\sigma: C \to B^n$ such that $\sigma|D = 1|D$ is easily obtained by sliding along the radial structure emanating from the center of C.

LEMMA 3 (THE REPLICATION DEVICE). *Suppose $\varphi: B^n \to B^n$ is an admissible map, C and D are each the union of a finite number of disjoint round n-cells in $\operatorname{int} B^n$, $D \subset \operatorname{int} C$ and $S(\varphi) \cap \partial D = \emptyset$. Then there is an admissible map $\psi: B^n \to B^n$ and a homeomorphism $\chi: \varphi^{-1}D \to \psi^{-1}D$ such that $\psi \circ \chi = \varphi|\varphi^{-1}D$, $\psi(\operatorname{int} C) = \operatorname{int} C$, ψ restricts to the identity on $B^n - \operatorname{int} C$, $S(\psi) \cap \partial D = \emptyset$, and $S(\varphi) - D$ and $S(\psi) - D$ are separated.*

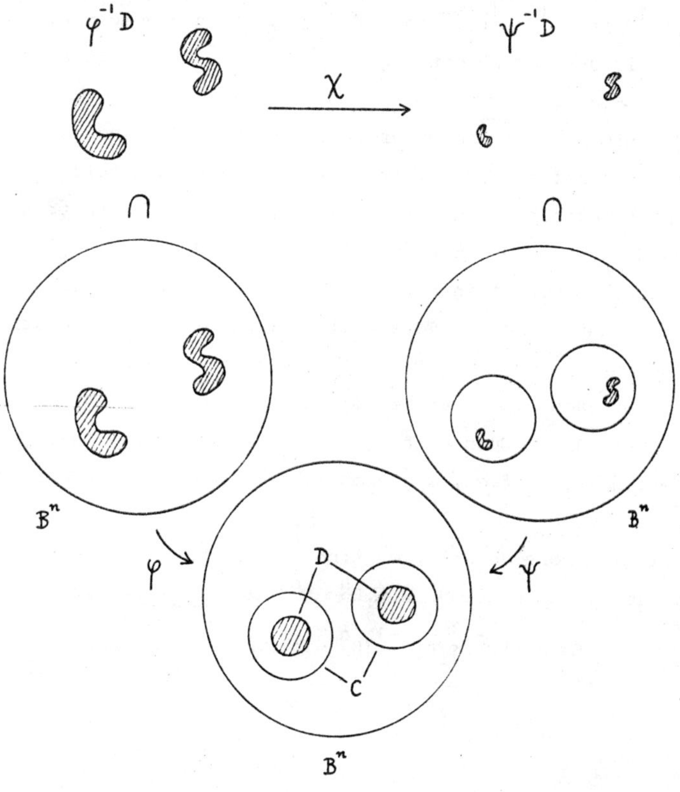

PROOF. $C = \bigcup_{i=1}^{k} C_i$ where each C_i is a round n-cell in $\operatorname{int} B^n$. We shall define ψ so that for each i, $1 \leq i \leq k$, $\psi | C: C_i \to C_i$ is a minaturized replica of φ.

Let $1 \leq i \leq k$. Let $D_i = D \cap C_i$. We shall construct a homeomorphism $\tau_i: C_i \to B^n$ such that $\tau_i | D_i = 1 | D_i$, and $S(\varphi) - D_i$ and $\tau_i^{-1}(S(\varphi)) - D_i$ are separated. To begin, there is a homeomorphism $\sigma_i: C_i \to B^n$ such that $\sigma_i | D_i = 1 | D_i$. Since $S(\varphi)$ and $\sigma_i^{-1}(S(\varphi))$ are countable and nowhere dense, then we can apply Lemma 1 in $C_i - \operatorname{int} D_i$ to obtain a homeomorphism λ_i of C_i which restricts to the identity on $D_i \cup (\partial C_i)$ such that $\lambda_i(\sigma_i^{-1}(S(\varphi)) - D_i$ and $S(\varphi) \cap (\operatorname{int} C_i - D_i)$ are separated. Since $c\ell S(\varphi) \subset \operatorname{int} B^n$, then $c\ell(\lambda_i \circ \sigma_i^{-1}(S(\varphi)) \subset \operatorname{int} C_i$. It follows that $\lambda_i \circ \sigma_i^{-1}(S(\varphi)) - D_i$ and $S(\varphi) - D_i$ are separated. The desired homeomorphism τ_i is obtained by setting $\tau_i = \sigma_i \circ \lambda_i^{-1}$.

Define the map $\psi: B^n \to B^n$ by

$$\psi = \begin{cases} \tau_i^{-1} \circ \varphi \circ \tau_i & \text{on } C_i \quad \text{for } 1 \leq i \leq k \\ 1 \quad \text{on} & B^n - \operatorname{int} C \end{cases}$$

Since $S(\psi) = \bigcup_{i=1}^{k} \tau_i^{-1}(S(\varphi))$, it is easily verified that ψ is an admissible map, $S(\psi) \cap \partial D = \emptyset$, and $S(\varphi) - D$ and $S(\psi) - D$ are separated.

Since $\psi^{-1} D_i = \tau_i^{-1}(\varphi^{-1} D_i)$ for $1 \leq i \leq k$, then a homeomorphism $\chi: \varphi^{-1} D \to \psi^{-1} D$ is defined by setting $\chi | \varphi^{-1} D_i = \tau_i^{-1} | \varphi^{-1} D_i$ for $1 \leq i \varepsilon k$. Clearly $\psi \circ \chi = \varphi | \varphi^{-1} D$. ⁞⁞⁞

PROOF OF THEOREM 1A. The proof is inductive. The induction step, which has a rather technical statement, is isolated in Lemma 4 below.

We begin by describing the strategy of the proof. Let $f: B^n \to B^n$ be an admissible map. Let $\varepsilon > 0$. Set $N_0 = \{(x,y) \varepsilon B^n \times B^n: |f(x) - y| \leq \varepsilon\}$. N_0 is a closed neighborhood of f in $B^n \times B^n$. Our goal is to produce a homeomorphism $h: B^n \to B^n$ such that $h \subset N_0$. This will be accomplished by constructing a decreasing sequence $N_0 \supset N_1 \supset N_2 \supset \cdots$ of closed subsets of $B^n \times B^n$ with the property that for each $i \geq 1$ and every $x \varepsilon B^n$, $N_i(x)$ and $N_i^{-1}(x)$ are non-empty sets of diameter $< 1/i$. Upon setting $h = \bigcap_{i=0}^{\infty} N_i$, we see that $h: B^n \to B^n$ is a bijection which is, in fact, a homeomorphism because it is a closed subset of $B^n \times B^n$.

Before we give more details, we find it convenient to introduce one more bit of terminology. A relation $R \subset B^n \times B^n$ is admissible if

$$R = h \cup g^{-1} \circ f | f^{-1}(B^n - \operatorname{int} A)$$

where

(1) $f: B^n \to B^n$ and $g: B^n \to B^n$ are admissible maps,

(2) A is the union of a finite number of disjoint round n-cells in $\operatorname{int} B^n$ such that $(S(f) \cup S(g)) \cap \partial A = \emptyset$ and $S(f) - A$ and $S(g) - A$ are separated, and

(3) $h: f^{-1}A \to g^{-1}A$ is a homeomorphism such that $g \circ h = f|f^{-1}A$.

$$\begin{array}{ccc} f^{-1}A & \xrightarrow{h} & g^{-1}A \\ \cap & & \cap \\ B^n & \xrightarrow{R} & B^n \\ & {\scriptstyle f}\searrow \quad \swarrow{\scriptstyle g} & \\ & B^n & \end{array}$$

Let $R = h \cup g^{-1} \circ f|f^{-1}(B^n - \operatorname{int} A)$ be an admissible relation in $B^n \times B^n$, where f, g, h and A are as prescribed above. We observe that R is a closed subset of $B^n \times B^n$. This is a consequence of two statements. First f, g and h are compact because each is a continuous function with compact domain and range. Second, the operations of inversion, composition and restriction over a closed set all transform compact relations into compact relations. We also observe that the inverse of an admissible relation is admissible.

We now give the details of the proof of Theorem 4. Set $R_0 = f$; then R_0 is an admissible relation (with $g = 1|B^n$, $A = \emptyset$ and $h = \emptyset$). The closed neighborhood N_0 of R_0 has already been defined. We shall construct a sequence $\{R_i\}$ of admissible relations in $B^n \times B^n$ and a sequence $\{N_i\}$ of closed subsets of $B^n \times B^n$ such that for each $i \geq 1$ the following conditions hold.

(1_i) $R_i \subset \operatorname{int} N_{i-1}$, $\operatorname{diam} R_i^{-1}(y) < 1/i+1$ for every $y \in B^n$ when i is odd, and $\operatorname{diam} R_i(x) < 1/i+1$ for every $x \in B^n$ when i is even.

(2_i) N_i is a closed neighborhood of R_i in $B^n \times B^n$ such that $N_i \subset N_{i-1}$, $\operatorname{diam} N_i^{-1}(y) < 1/i+1$ for every $y \in B^n$ when i is odd, and $\operatorname{diam} N_i(x) < 1/i+1$ for every $x \in B^n$ when i is even.

R_0 and N_0 are already in hand. We proceed inductively. Let $i \geq 1$ and assume we have an admissible relation R_{i-1} and a closed neighborhood N_{i-1} of R_{i-1} in $B^n \times B^n$. We obtain R_i satisfying (1_i) via Lemma 4 below. When i is <u>odd</u>: we apply Lemma 4 by substituting $(R_{i-1}, 1/i+1, N_{i-1})$ for (R, ε, N); then Lemma 4 produces R_*, and we set $R_i = R_*$. When i is <u>even</u>: we apply Lemma 4 by substituting $(R_{i-1}^{-1}, 1/i+1, N_{i-1}^{-1})$ for (R, ε, N); then Lemma 4 produces R_*, and we set $R_i = R_*^{-1}$.

Next we use Lemma 2 to obtain N_i satisfying (2_i). When i is <u>odd</u>: we apply Lemma 2 by substituting $(B^n, B^n, R_i^{-1}, \emptyset, 1/i+1)$ for $(X, Y, R, T, \varepsilon)$; then Lemma 2 produces N, and we set $N_i = N^{-1} \cap N_{i-1}$. When i is <u>even</u>: we apply Lemma 2 by substituting $(B^n, B^n, R_i, \emptyset, 1/i+1)$ for $(X, Y, R, T, \varepsilon)$; then Lemma 2 produces N, and we set $N_i = N \cap N_{i-1}$.

Let $i \geq 2$. Since R_i is admissible, then $R_i(x)$ and $R_i^{-1}(x)$ are non-empty for every $x \in B^n$. Since $R_i \subset N_i \subset N_{i-1}$, then (2_{i-1}) and (2_i) imply that $N_i(x)$ and $N_i^{-1}(x)$ are non-empty sets of diameter $< 1/i$ for every $x \in B^n$. ⦀

LEMMA 4. _If_ $R \subset B^n \times B^n$ _is an admissible relation_, $\varepsilon > 0$, _and_ N _is a closed neighborhood of_ R _in_ $B^n \times B^n$, _then there is an admissible relation_ $R_* \subset B^n \times B^n$ _such that_ $\operatorname{diam} R_*^{-1}(y) < \varepsilon$ _for every_ $y \in B^n$ _and_ $R_* \subset \operatorname{int} N$.

PROOF. Since R is admissible, then $R = h \cup g^{-1} \circ f | f^{-1}(B^n - \operatorname{int} A)$, where f,g,A and h are as prescribed in the definition of "admissible relation". Let $Z = \{z \in S(f) : \operatorname{diam} f^{-1}(z) \geq \varepsilon\} - A$. Z is a finite subset of $\operatorname{int} B^n$ because f is an admissible map. The significance of Z is that $\{f^{-1}(z) : z \in Z\} = \{R^{-1}(y) : y \in B^n$ and $\operatorname{diam} R^{-1}(y) \geq \varepsilon\}$, and the latter set is precisely the set of point inverses of R whose diameters must be reduced.

Here is a rough idea of how we proceed. We enclose Z in the union D of a finite number of small disjoint round n-cells in $\operatorname{int} B^n$. Then we use the Replication Device (Lemma 3) to modify the map g so that the preimage pattern of f on $f^{-1}D$ is replicated by g on $g^{-1}D$. This allows us to redefine R on $f^{-1}D$ so that it carries $f^{-1}D$ _homeomorphically_ onto $g^{-1}D$. In this way, the large point inverses of R simply vanish at the expense of complicating the structure of the map g.

There is a finite collection $C_1, C_2, \ldots, C_k$ of disjoint round n-cells in $\operatorname{int} B^n$ such that if $C = \bigcup_{i=1}^k C_i$, then $Z \subset \operatorname{int} C$, $C \cap (A \cup c\ell(S(g))) = \emptyset$, and $f^{-1}C_i \times g^{-1}C_i \subset \operatorname{int} N$ for $1 \leq i \leq k$. The second condition can be achieved because $S(f) - A$ and $S(g) - A$ are separated, and Z is a finite subset of $S(f) - A$. The third condition holds automatically for C_i's of sufficiently small diameter because for each $z \in Z$, $f^{-1}(z) \times g^{-1}(z) = R | f^{-1}(z) \subset \operatorname{int} N$. (The third condition will be used to insure that $R_* \subset \operatorname{int} N$.) Since $S(f)$ is a countable set, then for each i, $1 \leq i \leq k$, there is a round n-cell D_i such that $D_i \subset \operatorname{int} C_i$ and if $D = \bigcup_{i-1}^k D_i$, then $Z \subset \operatorname{int} D$ and $S(f) \cap \partial D = \emptyset$.

We now apply Lemma 3 with f in the role of φ, to obtain an admissible map $\psi : B^n \to B^n$ and a homeomorphism $\chi : f^{-1}D \to \psi^{-1}D$ such that $\psi \circ \chi = f | f^{-1}D$, $\psi(\operatorname{int} C) = \operatorname{int} C$, $\psi = 1$ on $B^n - \operatorname{int} C$, $S(\psi) \cap \partial D = \emptyset$, and $S(f) - D$ and $S(\psi) - D$ are separated.

We define the map $g_* : B^n \to B^n$ by $g_* = \psi\ g$. Since $S(\psi) \subset C$ and $C \cap c\ell S(g) = \emptyset$, then evidently $S(g_*) = S(\psi) \cup S(g)$ and g_* is an admissible map.

We set $A_* = A \cup D$. Then A_* is the union of a finite number of disjoint round n-cells in $\operatorname{int} B^n$. It is easily verified that $(S(f) \cup S(g_*)) \cap \partial A_* = \emptyset$ and that $S(f) - A_*$ and $S(g_*) - A_*$ are separated.

Since $C \cap (A \cup c\ell(S(g))) = \emptyset$ and $\psi^{-1}D \subset C$, then $g_*^{-1}A_* = g^{-1}A \cup g^{-1}(\psi^{-1}D)$ and $g^{-1}|\psi^{-1}D$ is a homeomorphism. Hence a homeomorphism $h_*: f^{-1}A_* \to g_*^{-1}A_*$ is defined by setting $h_*|f^{-1}A = h$ and $h_*|f^{-1}D = g^{-1} \circ \chi$. It follows easily that $g_* \circ h_* = f|f^{-1}A_*$.

Finally, an admissible relation $R_* \subset B^n \times B^n$ is defined by setting $R_* = h_* \cup g_*^{-1} \circ f|f^{-1}(B^n - \operatorname{int} A_*)$.

Note that $R_*^{-1} = h_*^{-1} \cup f^{-1} \circ g_*|g_*^{-1}(B^n - \operatorname{int} A_*)$. Hence, if $y \in B^n$ and $\operatorname{diam} R_*^{-1}(y) > 0$ then $y \in g_*^{-1}(B^n - \operatorname{int} A_*)$ and $R_*^{-1}(y) = f^{-1}(g_*(y))$. Z,D and A_* are chosen to guarantee that $\{z \in S(f): \operatorname{diam} f^{-1}(z) \geq \varepsilon\} \subset \operatorname{int} A_*$. Since $g_*(y) \notin \operatorname{int} A_*$, it follows that $\operatorname{diam} f^{-1}(g_*(y)) < \varepsilon$. Thus $\operatorname{diam} R_*^{-1}(y) < \varepsilon$.

Lastly, we demonstrate that $R_* \subset \operatorname{int} N$. First, since $g_*^{-1} = g^{-1}$ on $B^n - \operatorname{int} C$ and $h_* = h$ on $f^{-1}A$, it follows that $R_*|f^{-1}(B^n - \operatorname{int} C) = R|f^{-1}(B^n - \operatorname{int} C) \subset \operatorname{int} N$. Second, we use the equation $g_* \circ h_* = f|f^{-1}A_*$ to deduce that $h_* \subset g_*^{-1} \circ f$; therefore, $R_* \subset g_*^{-1} \circ f$. For $1 \leq i \leq k$, since $\psi(C_i) = C_i$, then $g_*^{-1}(C_i) = g^{-1}(C_i)$. Therefore, for $1 \leq i \leq k$,

$$R_*|f^{-1}C_i \subset g_*^{-1} \circ f|f^{-1}C_i \subset f^{-1}(C_i) \times g_*^{-1}(C_i) = f^{-1}(C_i) \times g^{-1}(C_i) \subset \operatorname{int} N.$$

Consequently, $R_*|f^{-1}C \subset \operatorname{int} N$. It is now evident that $R_* \subset \operatorname{int} N$. ⁞⁞⁞

4. MAPS WITH A BALD SPOT

The proofs of Theorems 1 and 2 are quite similar, and we rely on the reader's familiarity with the proof of Theorem 1 at several points in the proof of Theorem 2. We feel the reader may be aided, if we pause here to draw some comparisons between the two proofs.

The proof of Theorem 1 produces a homeomorphism by an infinite process which alternates between excising point images and point inverses of an admissible relation. Successive steps in this process apply the replication device to "opposite sides" of the relation. The ability to "switch sides" repeatedly depends on the point images and point inverses of the relation being separated (when viewed in the appropriate space). Disjointness alone is not sufficient. This separation can be achieved only because the singular set of the original map is nowhere dense.

When the singular set of the original map is countable but not necessarily nowhere dense (as in Theorem 2), then the replication device yields relations whose point images and point inverses can be made disjoint but can't necessarily be separated. This injects serious complications into the plan to produce a homeomorphism by a process which deals alternatively with point images and point inverses. Fortuitously, we find that we need not focus on approximating the original map by a homeomorphism. Instead, as is shown below, in the reduction of Theorem 2 to Theorem 2B, it suffices to approximate the inverse of the original map by a special kind of map, called an "acceptable" map. As a result, we can concentrate on eliminating point inverses, and we can ignore point images. Our inability to separate point images and point inverses will not hamper us, because we shall apply the replication device (repeatedly) on "one side" only. (Since we wish to excise point inverses, we apply the replication device on the left or domain-side of the relation.) (We shall find it necessary to preserve the disjointness of the point images and point inverses for technical reasons, to insure that the map which is the limit of infinitely many left-sided applications of the replication device is acceptable.) Thus, at the expense of adding another reduction step to the proof, we are able to get by with repeated applications of the replication device on one side only, and we avoid having to separate point images and point inverses. The observation that infinitely many left-sided applications of the replication device lead to a map approximating the inverse of the original map is due to R. D. Edwards. It is this observation which makes it possible to replace the hypothesis that the singular set of the original map is nowhere dense by the bald spot hypothesis.

In Theorem 2, we have replaced the hypothesis that $\{f^{-1}(y): y \in S(f)\}$ be a null collection by the weaker hypothesis that $S(f)$ be countable. This is an advantage, because the countability of $S(f)$ is the easier of the two hypothesis to detect and to preserve throughout the inductive process of the proof. Furthermore, the weaker hypothesis poses no additional difficulty in the proof for the following reason. Let $f: X \to Y$ be a map between compact spaces, let $\varepsilon > 0$, and consider the compact set $\{y \in S(f): \operatorname{diam} f^{-1}(y) \geq \varepsilon\}$. Under the stronger hypothesis, this set is finite; while under the weaker hypothesis, this set is compact and countable. We must deal with such a set in the proof of the Replication Lemma, where we must enclose it in the union of a finite number of small disjoint round n-cells. Fortunately, this can be accomplished for a compact countable set almost as easily as it can for a finite set.

The notions of "acceptable map" and "acceptable relation" appear in the proof of Theorem 2 in roles corresponding to those played by "admissible map" and "admissible relation" in the proof of Theorem 1. A map $f: B^n \to B^n$ is <u>acceptable</u> if $f|\partial B^n = 1|\partial B^n$ and $S(f)$ is a countable subset of $\operatorname{int} B^n$.

Theorem 2 reduces to:

THEOREM 2A. <u>Every acceptable map</u> $f: B^n \to B^n$ <u>can be approximated by homeomorphisms</u>.

PROOF THAT THEOREM 2A IMPLIES THEOREM 2. This proof is essentially the same as the proof that Theorem 1A implies Theorem 1. In this case, to locate a small collared n-cell C in the complement of the closure of the singular set, one uses the bald spot hypothesis rather than the nowhere density of the singular set. ⁞⁞⁞

Theorem 2A, in turn, reduces to:

THEOREM 2B. *If* $f: B^n \to B^n$ *is an acceptable map and* N *is a neighborhood of* f *in* $B^n \times B^n$, *then there is an acceptable map* $g: B^n \to B^n$ *such that* $g^{-1} \subset N$.

Theorem 2A is proved by repeated application of Theorem 2B, the output of Theorem 2B at one stage being used as the input at the next. Thus, the essential property of the map g produced by Theorem 2B is that it is acceptable. Indeed, general principles tell us that since the acceptable map $f: B^n \to B^n$ is cell-like, it is a fine homotopy equivalence [H] and automatically gives rise to a map $g: B^n \to B^n$ such that $g^{-1} \subset N$. However, this information is of no use in proving Theorem 2A unless g is known to be acceptable.

PROOF OF THEOREM 2A FROM THEOREM 2B. Assume Theorem 2B. Let $f: B^n \to B^n$ be an acceptable map. Let $\varepsilon > 0$. Set $f_0 = f$ and $N_0 = \{(x,y) \in B^n \times B^n: |f(x) - y| \leq \varepsilon\}$. N_0 is a closed neighborhood of f_0 in $B^n \times B^n$. We seek a homeomorphism $h: B^n \to B^n$ such that $h \subset N_0$. To this end, we shall construct a sequence $\{f_i\}$ of acceptable maps from B^n to itself, and a sequence $\{N_i\}$ of closed subsets of $B^n \times B^n$ such that the following conditions hold.

(1_i) $f_i^{-1} \subset \operatorname{int} N_{i-1}$.

(2_i) N_i is a closed neighborhood of f_i in $B^n \times B^n$ such that $N_i \subset N_{i-1}^{-1}$ and $\operatorname{diam} N_i(x) < 1/i+1$ for every $x \in B^n$.

We already have f_0 and N_0. We proceed inductively. Let $i \geq 1$ and assume we have an acceptable map $f_{i-1}: B^n \to B^n$ and a closed neighborhood N_{i-1} of f_{i-1} in $B^n \times B^n$. We apply Theorem 2B to obtain an acceptable map $f_i: B^n \to B^n$ such that $f_i^{-1} \subset \operatorname{int} N_{i-1}$. Since $\operatorname{diam} f_i(x) = 0$ for every $x \in B^n$, then Lemma 2 provides a closed neighborhood N of f_i in $B^n \times B^n$ such that $\operatorname{diam} N(x) < 1/i+1$ for every $x \in B^n$. Set $N_i = N \cap (N_{i-1}^{-1})$. Then f_i and N_i satisfy (1_i) and (2_i).

Clearly $N_0 \supset N_2 \supset N_4 \supset \cdots$ is a decreasing sequence of closed subsets of $B^n \times B^n$. Also for every $i \geq 2$ and every $x \in B^n$, since $f_i(x)$ and $f_i^{-1}(x)$ are non-empty, then (2_i) implies that $N_i(x)$ and $N_i^{-1}(x)$ are non-empty subsets of diameter $< 1/i$. It follows that $h = \bigcap_{i=0}^{\infty} N_{2i}$ is a homeomorphism of B^n which lies in N_0. ⁞⁞⁞

As the discussion at the beginning of this section suggests, the central geometric idea of the proof of Theorem 2 is, as before, a replication device. This device is codified by the following lemma. Notice that the direction of the homeomorphism χ is the opposite of its direction in Lemma 3.

LEMMA 5 (THE REPLICATION DEVICE). Suppose $\varphi: B^n \to B^n$ is an acceptable map, C and D are each the union of a finite number of disjoint round n-cells in $\operatorname{int} B^n$, and T is a countable subset of $\operatorname{int} C$ such that $D \subset \operatorname{int} C$ and $S(\varphi) \cap \partial D = \emptyset$. Then there is an acceptable map $\psi: B^n \to B^n$ and a homeomorphism $\chi: \psi^{-1}D \to \varphi^{-1}D$ such that $\varphi \circ \chi = \psi|\psi^{-1}D$, $\psi(\operatorname{int} C) = \operatorname{int} C$, ψ restricts to the identity on $B^n - \operatorname{int} C$, and $[S(\psi) \cup \psi(T)] \cap [\partial D \cup (S(\varphi) - D)] = \emptyset$.

PROOF. $C = \bigcup_{i=1} C_i$ where each C_i is a round n-cell in $\operatorname{int} B^n$. As in the proof of Lemma 3, for each i, $1 \le i \le k$, $\psi|C_i: C_i \to C_i$ will be a miniaturized replica of ψ.

Let $1 \le i \le k$. Let $D_i = D \cap C_i$ and $T_i = T \cap C_i$. We begin with a homeomorphsim $\sigma_i: C_i \to B^n$ such that $\sigma_i|D_i = 1|D_i$. Since $S(\varphi)$ and $\sigma_i^{-1}(S(\varphi))$ are countable, then we can apply Lemma 1 in $C_i - \operatorname{int} D_i$ to obtain a homeomorphism λ_i of C_i which restricts to the identity on $D_i \cup (\partial C_i)$ such that $\lambda_i(\sigma_i^{-1}(S(\varphi))) \cap (S(\varphi) - D) = \emptyset$. Then $\lambda_i(\sigma_i^{-1}(S(\varphi))) \cap \partial D = \emptyset$, because $S(\varphi) \cap \partial D = \emptyset$ and σ_i and λ_i fix ∂D. We now define the homeomorphism $\tau_i: C_i \to B^n$ by $\tau_i = \sigma_i \circ \lambda_i^{-1}$. Then $\tau_i|D_i = 1|D_i$ and $\tau_i^{-1}(S(\varphi)) \cap [\partial D \cup (S(\varphi) - D] = \emptyset$. Since $S(\tau_t^{-1} \circ \varphi \circ \tau_i) = \tau_i^{-1}(S(\varphi))$, it follows that $(\tau_i^{-1} \circ \varphi \circ \tau_i)^{-1}[\partial D \cup (S(\varphi) - D)]$ is the union of a finite number of (n-1)-spheres and a countable set. Hence we can apply Lemma 1 in C_i to obtain a homeomorphism μ_i of C_i which restricts to the identity on ∂C_i such that $\mu_i(T_i)$ is disjoint from $(\tau_i^{-1} \circ \varphi \circ \tau_i)^{-1}[\partial D \cup (S(\varphi) - D]$. Consequently, $(\tau_i^{-1} \circ \varphi \circ \tau_i \circ \mu_i)(T_i) \cap [\partial D \cup (S(\varphi) - D)] = \emptyset$.

Define the map $\psi: B^n \to B^n$ by

$$\psi = \begin{cases} \tau_i^{-1} \circ \varphi \circ \tau_i \quad \mu_i & \text{on } C_i \text{ for } 1 \le i \le k \\ 1 \text{ on } B^n - \operatorname{int} C \end{cases}$$

Since $S(\psi) = \bigcup_{i=1} \tau_i^{-1}(S(\varphi))$, it is easily verified that ψ is an acceptable map, and that $[S(\psi) \cup \psi(T)] \cap [\partial D \cup (S(\varphi) - D] = \emptyset$.

Since $\tau_i \circ \mu_i(\psi^{-1}D_i) = \varphi^{-1}D_i$ for $1 \le i \le k$, then a homeomorphism $\chi: \psi^{-1}D \to \varphi^{-1}D$ is defined by setting $\chi|\psi^{-1}D_i = \tau_i \circ \mu_i|\psi^{-1}D_i$ for $1 \le i \le k$. Clearly $\varphi \circ \chi = \psi|\psi^{-1}D$. ∎

PROOF OF THEOREM 2B. The proof is inductive, and the induction step is isolated in Lemma 6 below.

We first describe the strategy of the proof. Let $f: B^n \to B^n$ be an acceptable map, and let N_0 be a closed neighborhood of f in $B^n \times B^n$. We seek an acceptable map $g: B^n \to B^n$ whose inverse lies in N_0. To obtain g, we first construct a decreasing sequence $N_0 \supset N_1 \supset N_2 \supset \cdots$ of closed subsets of $B^n \times B^n$ such that for each $i \ge 1$:

(1) $N_i|\partial B^n = 1|\partial B^n$,

(2) $N_i^{-1}(y)$ is a non-empty set of diameter $< 1/i$ for every $y \in B^n$, and

(3) $\{x \in B^n: \operatorname{diam} N_i(x) \geq 1/i\}$ is a countable set.

Then we set $g = (\bigcup_{i=0} N_i)^{-1}$. Condition (2) forces g to be a function from B^n to itself. g is continuous because it is a closed subset of $B^n \times B^n$. Condition (1) implies that $g|\partial B^n = 1|\partial B^n$ and $S(g) \subset \operatorname{int} B^n$. Since $S(g) \subset \bigcup_{i=1}\{x \in B^n: \operatorname{diam} N_i(x) \geq 1/i\}$, then condition (3) forces $S(g)$ to be a countable set. We conclude that g is an acceptable map. Obviously $g^{-1} \subset N_0$.

Before proceeding with the details of the proof we establish the definition of "acceptable relation" and several other convenient bits of notation. Notice that in passing from admissible relations to acceptable relations, h changes from a homeomorphism to a map and its direction is reversed. A relation $R \subset B^n \times B^n$ is acceptable if

$$R = h^{-1} \cup g^{-1} \circ f|f^{-1}(B^n - \operatorname{int} A)$$

where

(1) $f: B^n \to B^n$ and $g: B^n \to B^n$ are acceptable maps,

(2) A is the union of a finite number of disjoint round n-cells in $\operatorname{int} B^n$ such that $(S(f) \cup S(g)) \cap \partial A = \emptyset$ and $(S(f) - A) \cap (S(g) - A) = \emptyset$, and

(3) $h: g^{-1}A \to f^{-1}A$ is a map such that $f \circ h = g|g^{-1}A$ and $S(h)$ is a countable subset of $f^{-1}(\operatorname{int} A)$.

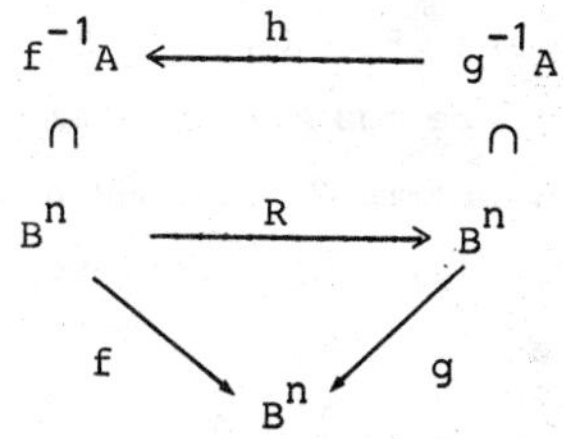

Let $R \subset X \times Y$ be a relation. Define

$$\sigma(R) = \bigcup\{R^{-1}(y): y \in Y \text{ and } R^{-1}(y) \text{ contains more than one point}\}$$

and define

$$\tau(R) = \{x \in X: R(x) \text{ contains more than one point}\}.$$

Now let $R \subset B^n \times B^n$ be an acceptable relation. Then $R = h^{-1} \cup g^{-1} \circ f|f^{-1}(B^n - \operatorname{int} A)$ where f, g, h and A are as prescribed in the direction of "acceptable relation". We make four observations.

(1) R is a closed subset of $B^n \times B^n$

(2) $R|\partial B^n = 1|\partial B^n$

(3) $\sigma(R)$, $\tau(R)$ and ∂B^n are all disjoint.

(4) For each $\varepsilon > 0$, $\{x \in B^n : \operatorname{diam} R(x) \geq \varepsilon\}$ is a compact countable set.

The first observation is valid for the same reason that an admissible relation is a closed set. Observation (2) is clear. The third observation follows from the equations: $\sigma(R) = f^{-1}(S(f) - A)$ and $\tau(R) = S(h) \cup f^{-1}(S(g) - A)$. It follows that $\sigma(R) \cup \tau(R) \subset \operatorname{int} B^n$. Also since $(S(f) - A) \cap (S(g) - A) = \emptyset$ and $S(h) \subset f^{-1}(A)$, it is clear that $\sigma(R) \cap \tau(R) = \emptyset$. To prove observation (4), note that $\{x \in B^n : \operatorname{diam} R(x) \geq \varepsilon\}$ is the union of the two sets $\{x \in S(h) : \operatorname{diam} h(x) \geq \varepsilon\}$ and $f^{-1}(\{z \in S(g) : \operatorname{diam} g^{-1}(z) \geq \varepsilon\} - \operatorname{int} A)$. These two sets are compact and countable because $S(h)$ and $S(g)$ are countable and $(S(f) - A) \cap (S(g) - A) = \emptyset$.

We now give the details of the proof. Set $R_0 = f$; then R_0 is an acceptable relation (with $g = 1|B^n$, $A = \emptyset$ and $h = \emptyset$). The closed neighborhood N_0 of R_0 is given. We shall construct a sequence $\{R_i\}$ of acceptable relations in $B^n \times B^n$ and a sequence $\{N_i\}$ of closed subsets of $B^n \times B^n$ such that for each $i \geq 1$, the following conditions hold.

(1_i) $R_i \subset N_{i-1}$, $R_i|\sigma(R_i) \subset \operatorname{int} N_{i-1}$ and $\operatorname{diam} R_i^{-1}(y) < 1/i$ for every $y \in B^n$.

(2_i) $R_i \subset N_i \subset N_{i-1}$, $R_i|\sigma(R_i) \subset \operatorname{int} N_i$, $N_i|\partial B^n = 1|\partial B^n$, $\operatorname{diam} N_i^{-1}(y) < 1/i$ for every $y \in B^n$, and $\{x \in B^n : \operatorname{diam} N_i(x) \geq 1/i\}$ is a countable set.

R_0 and N_0 are given. We proceed inductively. Let $i \geq 1$ and assume we have an acceptable relation R_{i-1} and a closed subset N_{i-1} of $B^n \times B^n$ such that $R_{i-1} \subset N_{i-1}$ and $R_{i-1}|\sigma(R_{i-1}) \subset \operatorname{int} N_{i-1}$. We apply Lemma 6 below to obtain R_i satisfying (1_i), by substituting $(R_{i-1}, 1/i, N_{i-1})$ for (R, ε, N). Then Lemma 6 produces R_*, and we set $R_i = R_*$.

To obtain N_i satisfying (2_i), we must apply Lemma 2 twice. First, since $\operatorname{diam} R_i^{-1}(y) < 1/i$ for every $y \in B^n$, Lemma 2 provides a closed neighborhood L of R_i^{-1} in $B^n \times B^n$ such that $\operatorname{diam} L(y) < 1/i$ for every $y \in B^n$. For the second application of Lemma 2, we set

$$T = \{x \in B^n : \operatorname{diam} R_i(x) \geq 1/i\} \cup \partial B^n .$$

Since $\{x \in B^n : \operatorname{diam} R_i(x) \geq 1/i\}$ is compact, then T is a closed subset of B^n. Also $\sigma(R_i) \subset B^n - T$ because $T \subset \tau(R_i) \cup \partial B^n$. Lemma 2 now provides a closed subset M of $B^n \times B^n$ such that $R_i|B^n - T \subset \operatorname{int} M$, $\operatorname{diam} M(x) < 1/i$ for every $x \in B^n - T$, and $M|T = R_i|T$. It follows that $R_i|\sigma(R_i) \subset \operatorname{int} M$ because $\sigma(R_i) \subset B^n - T$, and that $M|\partial B^n = 1|\partial B^n$ because $\partial B^n \subset T$ and $R_i|\partial B^n = 1|\partial B^n$. Thus $\{x \in B^n : \operatorname{diam} M(x) \geq 1/i\}$ coincides with the countable set $\{x \in B^n : \operatorname{diam} R_i(x) \geq 1/i\}$. We conclude that ($2_i$) is satisfied if we set $N_i = L^{-1} \cap M \cap N_{i-1}$.

Let $i \geq 1$. Note that $R_i^{-1}(y) \neq \emptyset$ for every $y \in B^n$ because R_i is acceptable. Thus, (2_i) implies that $N_i^{-1}(y)$ is non-empty and of diameter $< 1/i$

for every $y \in B^n$. Also $N_i \subset N_{i-1}$, $N_i|\partial B^n = 1|\partial B^n$, and $\{x \in B^n : \operatorname{diam} N_i(x) \geq 1/i\}$ is a countable set. Now, as we argued earlier, an acceptable map $g: B^n \to B^n$ such that $g^{-1} \subset N_0$ is specified by $g = (\bigcap_{i=0}^{\infty} N_i)^{-1}$. ▥

LEMMA 6. <u>If</u> $R \subset B^n \times B^n$ <u>is an acceptable relation</u>, $\varepsilon > 0$ <u>and</u> N <u>is a closed subset of</u> $B^n \times B^n$ <u>such that</u> $R \subset N$ <u>and</u> $R|\sigma(R) \subset \operatorname{int} N$, <u>then there is an acceptable relation</u> $R_* \subset B^n \times B^n$ <u>such that</u> $\operatorname{diam} R_*^{-1}(y) < \varepsilon$ <u>for every</u> $y \in B^n$, $R_* \subset N$ <u>and</u> $R_*|\sigma(R_*) \subset \operatorname{int} N$.

PROOF. Since R is acceptable then $R = h^{-1} \cup g^{-1} \circ f|f^{-1}(B^n - \operatorname{int} A)$ where f, g, h and A are as prescribed in the definition of "acceptable relation". Let $Z = \{z \in S(f) : \operatorname{diam} f^{-1}(z) \geq \varepsilon\} - A$. Z is a compact countable subset of $\operatorname{int} B^n - A$ because $S(f)$ is a countable subset of $\operatorname{int} B^n - \partial A$. The significance of Z is that $\{f^{-1}(z) : z \in Z\} = \{R^{-1}(y) : y \in B^n \text{ and } \operatorname{diam} R^{-1}(y) \geq \varepsilon\}$, and the latter set is precisely the set of point inverses of R whose diameter must be reduced.

We proceed as we did in the proof of Lemma 4. We enclose Z in the union D of a finite number of small disjoint round n-cells in $\operatorname{int} B^n$. Then we use the Replication Device (Lemma 5) to modify g so that there is a natural map from $g^{-1}D$ to $f^{-1}D$. We can then alter R on $f^{-1}D$ so that $R|f^{-1}D$ is the inverse of this map, thereby eliminating all the non-trivial point inverses of R arising from points of $g^{-1}D$. In particular, this eliminates all point inverses of R of diameter $\geq \varepsilon$.

For each $z \in Z$, since $f^{-1}(z) \times g^{-1}(z) = R|f^{-1}(z) \subset R|\sigma(R) \subset \operatorname{int} N$, then z has a neighborhood U_z in $\operatorname{int} B^n - A$ such that $f^{-1}U_z \times g^{-1}U_z \subset \operatorname{int} N$. We now begin choosing a sequence $C_1, C_2, C_3, \ldots$ of disjoint round n-cells in $\operatorname{int} B^n$ such that for each $i \geq 1$, $\partial C_i \cap Z = \emptyset$ and $z \in \operatorname{int} C_i \subset C_i \subset U_z$ for some $z \in Z$. Since Z is countable, we can continue to choose C_i's for as long as some points of Z remain uncovered. However, since Z is compact, this process must terminate after a finite number of choices, yielding a finite collection $C_1, C_2, \ldots, C_k$ of disjoint round n-cells in $\operatorname{int} B^n$ such that if $C = \bigcup_{i=1} C_i$, then $Z \subset \operatorname{int} C$, $C \cap A = \emptyset$ and $f^{-1}C_i \times g^{-1}C_i \subset \operatorname{int} N$ for $1 \leq i \leq k$. (The third condition will be used to insure that $R_* \subset N$ and $R_*|\sigma(R_*) \subset \operatorname{int} N$.) Since $S(f)$ is a countable set, then for each i, $1 \leq i \leq k$, there is a round n-cell D_i such that $D_i \subset \operatorname{int} C_i$, and if $D = \bigcup_{i=1} D_i$, then $Z \subset \operatorname{int} D$ and $S(f) \cap \partial D = \emptyset$.

We now apply Lemma 5 with f in the role of φ and $S(g) \cap \operatorname{int} C$ in the role of T. We obtain an acceptable map $\psi: B^n \to B^n$ and a homeomorphism $\chi: \psi^{-1}D \to f^{-1}D$ such that $f \circ \chi = \psi|\psi^{-1}D$, $\psi(\operatorname{int} C) = \operatorname{int} C$, $\psi = 1$ on $B^n - \operatorname{int} C$, and $[S(\psi) \cup \psi(S(g) \cap \operatorname{int} C)] \cup [\partial D \cup (S(f) - D)] = \emptyset$. At this point, it is convenient to observe that since $\psi(S(g) - \operatorname{int} C) = S(g) - \operatorname{int} C$, and the latter set is disjoint from both ∂A and $S(f) - A$, then $S(\psi) \cup \psi(S(g))$ is disjoint from both $\partial(A \cup D)$ and $S(f) - (A \cup D)$. Also note that $S(f) \cap \partial(A \cup D) = \emptyset$.

We define the map $g_*:B^n \to B^n$ by $g_* = \psi \circ g$. Since $S(g_*) = S(\psi) \cup \psi(S(g))$, then g_* is evidently an acceptable map.

We set $A_* = A \cup D$. Then A_* is the union of a finite number of disjoint round n-cells in $\operatorname{int} B^n$. It follows from our observations above that $(S(f) \cup S(g_*)) \cap \partial A_* = \emptyset$ and $(S(f) - A_*) \cap (S(g_*) - A_*) = \emptyset$.

Since $g_*^{-1}A_* = g^{-1}A \cup g^{-1}(\psi^{-1}D)$, then a map $h_*:g_*^{-1}A_* \to f^{-1}A_*$ is defined by setting $h_*|g^{-1}A = h$ and $h_*|g^{-1}(\psi^{-1}D) = \chi \circ g|g^{-1}(\psi^{-1}D)$. It is easy to check that $f \circ h_* = g_*|g_*^{-1}A_*$. Since $S(h_*) = S(h) \cup \chi(S(g|g^{-1}(\psi^{-1}D)$ and $\psi(S(g)) \cap \partial D = \emptyset$, then $S(h_*)$ is a countable subset of $f^{-1}(\operatorname{int} A_*)$.

Now we can define an acceptable relation $R_* \subset B^n \times B^n$ by the formula $R_* = h_*^{-1} \cup g_*^{-1} \circ f|f^{-1}(B^n - \operatorname{int} A_*)$.

It follows that $R_*^{-1} = h_* \cup f^{-1} \circ g_*|g_*^{-1}(B^n - \operatorname{int} A_*)$. Now suppose $y \in B^n$ and $\operatorname{diam} R_*^{-1}(y) > 0$. Then $y \in g_*^{-1}(B^n - \operatorname{int} A_*)$ and $R_*^{-1}(y) = f^{-1}(g_*(y))$. Z, D and A_* are chosen so that $\{z \in S(f) : \operatorname{diam} f^{-1}(z) \geq \varepsilon\} \subset \operatorname{int} A_*$. Since $g_*(y) \notin \operatorname{int} A_*$, it follows that $\operatorname{diam} f^{-1}(g_*(y)) < \varepsilon$. Thus $\operatorname{diam} R_*^{-1}(y) < \varepsilon$.

Lastly, we demonstrate that $R_* \subset N$ and $R_*|\sigma(R_*) \subset \operatorname{int} N$. Since $g_*^{-1} = g^{-1}$ on $B^n - \operatorname{int} C$ and $h_*^{-1} = h^{-1}$ on $f^{-1}A$, it follows that $R_*|f^{-1}(B^n - \operatorname{int} C) = R|f^{-1}(B^n - \operatorname{int} C) \subset N$. Also the equation $f \circ h_* = g_*|g_*^{-1}A_*$ implies that $h_*^{-1} \subset g_*^{-1} \circ f$, from which we deduce that $R_* \subset g_*^{-1} \circ f$. For $1 \leq i \leq k$, since $\psi(C_i) = C_i$, then $g_*^{-1}(C_i) = g^{-1}(C_i)$. Therefore, for $1 \leq i \leq k$,

$$R_*|f^{-1}C_i \subset g_*^{-1} \circ f|f^{-1}C_i \subset f^{-1}C_i \times g_*^{-1}C_i = f^{-1}C_i \times g^{-1}C_i \subset \operatorname{int} N .$$

Consequently, $R_*|f^{-1}C \subset \operatorname{int} N$. It is now evident that $R_* \subset N$. Since $\sigma(R) = f^{-1}(S(f) - A)$ and $\sigma(R_*) = f^{-1}(S(f) - A_*)$, then $\sigma(R_*) \subset \sigma(R)$. Thus,

$$R_*|\sigma(R_*) - f^{-1}C = R|\sigma(R_*) - f^{-1}C \subset R|\sigma(R) \subset \operatorname{int} N.$$

Since $R_*|\sigma(R_*) \cap f^{-1}(C) \subset R_*|f^{-1}(C) \subset \operatorname{int} N$, we conclude that $R_*|\sigma(R_*) \subset \operatorname{int} N$. ⁞⁞

5. TAME ZERO-DIMENSIONAL SINGULAR SETS

We shall deduce Theorem 3 from Theorem 2 by passing from a map with a tame zero-dimensional singular set to a map with a countable singular set. This transformation requires two propositions. The first is that any σ-compact tame zero-dimensional set can be enclosed in a null collection of small disjoint collared n-cells. This fact is established below in Lemma 7. The second is a fundamental decomposition shrinking principle which originates in the work of R. H. Bing, and is known as "the Null Star-like Equivalent Shrinking Principle". It applies here to show that a decomposition of an n-manifold determined by a null collection of disjoint collared n-cells is shrinkable. We describe this principle in more detail below.

Lemma 7 captures the fundamental properties of tame zero-dimensional sets. Before presenting this lemma, we feel it appropriate to comment on the definition of "tame zero-dimensionality". Let M be a compact n-manifold. One of the classical definitions of zero-dimensionality implies that a subset S of M is zero-dimensional if every point of S has arbitrarily small neighborhoods in M whose frontiers miss S. The definition of tame zero-dimensionality applies only to σ-compact subsets of $\text{int}\, M$; recall that it states that a σ-compact subset S of $\text{int}\, M$ is tame zero-dimensional if each point of S has arbitrarily small collared n-cell neighborhoods in M whose boundaries miss S. Clearly, the definition of tame zero-dimensionality makes sense for arbitrary (not just σ-compact) subsets of $\text{int}\, M$, and comparison with the above classical definition of zero-dimensionality tempts us to drop the restriction to σ-compacta. We resist this temptation for the following reason. Originally a subset of manifold was called "tame" if it behaved like a piecewise linearly embedded polyhedron of the same dimension. Thus, a tame zero-dimensional subset should behave in some sense like a finite set of points. As the level of understanding of tame sets rose, it was recognized that the specific properties which tame sets share with piecewise linearly embedded polyhedra of the same dimension are their general position properties. For a tame zero-dimensional set, the appropriate general position property is expressed below in statement (2) of Lemma 7. This general position property can be proved for tame zero-dimensional σ-compacta. However, it is not necessarily valid for arbitrary subsets of $\text{int}\, M$ which satisfy the definition of tame zero-dimensionality. An illustration of this phenomenon is given in the next paragraph. For this reason, we do not use the term "tame zero-dimensional" outside the class of σ-compacta.

Let $J = \{(x,y,z) \in \mathbb{R}^3 : x, y \text{ and } z \text{ are irrational}\}$. J is not σ-compact. However J satisfies the definition of tame zero-dimensionality, because any prism of the form $[a,b] \times [c,d] \times [e,f]$ where a,b,c,d,e and f are rational, is a collared 3-cell whose boundary misses J. Let A be the Cantor set in $\mathbb{R}^3$ known as Antoine's necklace. A is a compact wild (= not tame) zero-dimensional nowhere dense subset of $\mathbb{R}^3$ with the following property. Every non-empty open subset of A contains a wild Cantor set - in fact, a smaller copy of A. We assert that no homeomorphism of $\mathbb{R}^3$ carries J off A. Thus J does not possess the general position property which characterizes tame zero-dimensional σ-compacta. For a simple proof by contradiction, suppose h is a homeomorphism of $\mathbb{R}^3$ such that $h(J) \cap A = \emptyset$. Then $h^{-1}A \subset \mathbb{R}^3 - J$. Since $\mathbb{R}^3 - J$ is the union of countably many flat 2-dimensional planes, the Baire Category Theorem implies that some non-empty open subset U of $h^{-1}A$ must lie in one of these planes. Since any Cantor set which lies in a flat 2-dimensional plane is tame in $\mathbb{R}^3$, then U contains no wild Cantor sets. Hence, hU is a non-empty open subset of A which contains no wild Cantor sets.

LEMMA 7. Let S be a σ-compact subset of the interior of a compact manifold M. The following three statements are equivalent.

(1) S is tame zero-dimensional.

(2) If T is the union of a countable number of nowhere dense subsets of M, then $1|M$ can be approximated by homeomorphisms h of M such that $h(S) \cap T = \emptyset$ and $h|\partial M = 1|\partial M$.

(3) For every $\varepsilon > 0$, there is a null collection $\{C_i\}$ of disjoint collared n-cells of diameter $< \varepsilon$ in $\operatorname{int} M$ such that $S \subset \bigcup_{i=1} \operatorname{int} C_i$.

PROOF. (1) implies (2). Assume statement (1). We first establish statement (2) in the special case that S is compact and T is nowhere dense.

Let $\varepsilon > 0$. Since S is compact, it is covered by a finite collection $\{K_i : 1 \le i \le p\}$ of collared n-cells of diameter $< \varepsilon$ in $\operatorname{int} M$ such that $S \cap \partial K_i = \emptyset$ for $1 \le i \le p$. For $1 \le i \le p$, let $L_i = K_i - \bigcup_{j<i} \operatorname{int} K_j$. Then $\{\operatorname{int} L_i : 1 \le i \le p\}$ is a cover of S by disjoint open sets of diameter $< \varepsilon$.

Let $1 \le i \le p$. Set $S_i = S \cap L_i$. S_i is a compact subset of $\operatorname{int} L_i$. Hence, S_i is covered by a finite collection $\{C_{i,j} : 1 \le j \le q(i)\}$ of collared n-cells in $\operatorname{int} L_i$ such that $S_i \cap \partial C_{i,j} = \emptyset$ for $1 \le j \le q(i)$, and $\{C_{i,j} : 1 \le j \le q(i)\}$ is irreducible in the sense that no proper subcollection covers S_i. For each j, $1 \le j \le q(i)$, there are collared n-cells $D_{i,j}$ and $E_{i,j}$ and a homeomorphism $h_{i,j}$ of M such that

(a) $E_{i,j} \subset \operatorname{int} D_{i,j} \subset D_{i,j} \subset \operatorname{int} C_{i,j}$,

(b) $S_i \cap (C_{i,j} - \operatorname{int} D_{i,j}) = \emptyset$,

(c) $E_{i,j}$ is disjoint from $C_{i,k}$ whenever $k \ne j$ for $1 \le k \le q(i)$, and $E_{i,j} \cap T = \emptyset$.

(d) $h_{i,j}(D_{i,j}) = E_{i,j}$ and $h_{i,j}|M - \operatorname{int} C_{i,j} = 1|M - \operatorname{int} C_{i,j}$.

Define the homeomorphism h_i of M by $h_i = h_{i,q(i)} \circ \cdots \circ h_{i,2} \circ h_{i,1}$. Then $h_i|M - \operatorname{int} L_i = 1|M - \operatorname{int} L_i$; so h_i is within ε of $1|M$. Also we assert that $h_i(S_i) \subset \bigcup_{j=1}^{q(i)} E_{i,j}$. To prove this, let $x \in S_i$. Choose j, $1 \le j \le q(i)$, so that $x \in C_{i,j}$ and $x \notin C_{i,k}$ for $1 \le k < j$. Then $h_{i,k}$ fixes x for $1 \le k < j$. Also $x \in D_{i,j}$, so that $h_{i,j}(x) \in E_{i,j}$. Consequently, $h_{i,k}$ fixes $h_{i,j}(x)$ for $j < k \le q(i)$. It follows that $h_i(x) = h_{i,j}(x) \in E_{i,j}$. Since each $E_{i,j}$ misses T, we have that $h_i(S_i) \cap T = \emptyset$.

Now we define the homeomorphism h of M by setting $h|L_i = h_i|L_i$ for $i \le i \le p$ and setting $h|M - \bigcup_{i=1} \operatorname{int} L_i = 1|M - \bigcup_{i=1} \operatorname{int} L_i$. Then h is within ε of $1|M$, $h(S) \cap T = \emptyset$ and $h|\partial M = 1|\partial M$. This finishes the proof of statement (2) in the special case.

To prove statement (2) in the general case, we write $S = \bigcup_{i=1}^{\infty} S_i$ and $T = \bigcup_{j=1}^{\infty} T_j$ where each S_i is compact and each T_j is nowhere dense.

For each $i \geq 1$ and $j \geq 1$, let $U_{i,j} = \{h \in \mathcal{H}(M, \partial M) : h(S_i) \cap c\ell T_j = \emptyset\}$. Since S is tame zero-dimensional, so is each S_i; hence $h(S_i)$ is tame zero-dimensional for each $i \geq 1$ and every $h \in \mathcal{H}(M, \partial M)$. Since each T_j is nowhere dense, so is each $c\ell T_j$. Therefore, we can deduce from the special case of statement (2) proved above, that each $U_{i,j}$ is a dense subset of $\mathcal{H}(M, \partial M)$. Also each $U_{i,j}$ is evidently an open subset of $\mathcal{H}(M, \partial M)$. Since $\mathcal{H}(M, \partial M)$ has a complete metric, we conclude via the Baire Category Theorem that $\bigcap_{i=1}^{\infty} \bigcap_{j=1}^{\infty} U_{i,j}$ is a dense subset of $H(M, \partial M)$. Statement (2) now follows because $1|M$ can be approximated by elements of $\bigcap_{i=1}^{\infty} \bigcap_{j=1}^{\infty} U_{i,j}$.

(2) implies (3). Assume statement (2). One can easily choose a null collection $\{C_i\}$ of disjoint collared n-cells of diameter $< \varepsilon/3$ in $\operatorname{int} M$ such that $\bigcup_{i=1} \operatorname{int} C_i$ is a dense subset of M. Then $M - \bigcup_{i=1} \operatorname{int} C_i$ is nowhere dense in M. Statement (2) provides a homeomorphism h of M within $\varepsilon/3$ of $1|M$ such that $h(S) \cap (M - \bigcup_{i=1} \operatorname{int} C_i) = \emptyset$ and $h|\partial M = 1|\partial M$. It follows that $\{h^{-1} C_i\}$ is a null collection of disjoint collared n-cells of diameter $< \varepsilon$ in $\operatorname{int} M$ whose interiors cover S.

(3) implies (1). Assume statement (3). Let $x \in S$ and let U be an open neighborhood of x in M. Choose $\varepsilon > 0$ so that ε is less than the distance from x to M-U. Statement (3) provides a null collection $\{C_i\}$ of disjoint collared n-cells of diameter $< \varepsilon$ in $\operatorname{int} M$ whose interiors cover S. Hence, $x \in \operatorname{int} C_i$ for some $i \geq 1$. Also $\partial C_i \cap S = \emptyset$. Since $\operatorname{diam} C_i < \varepsilon$, then $C_i \subset U$. This proves S is tame zero-dimensional. ∎

Perhaps the fundamental geometric tool of decomposition space theory is the Null Star-like Equivalent Shrinking Principle. A compact subset F of $\mathbb{R}^n$ is <u>star-like</u> if there is a point p in F such that every ray in $\mathbb{R}^n$ emanating from p intersects F in a connected set. A compact subset F of the interior of an n-manifold M is <u>star-like equivalent</u> if there is a neighborhood U of F in M and an embedding $e: U \to \mathbb{R}^n$ such that $e(F)$ is star-like. Observe that any collared n-cell in an n-manifold is star-like equivalent.

THE NULL STAR-LIKE EQUIVALENT SHRINKING PRINCIPLE. <u>Suppose</u> $f: M \to X$ <u>is a surjective map from a compact boundaryless manifold</u> M <u>to a compact metric space</u> X. <u>If</u> $\{f^{-1}(y) : y \in S(f)\}$ <u>is a null collection of star-like equivalent sets, then</u> f <u>can be approximated by homeomorphisms</u>.

This principle has manifested itself in many forms, apparently originating in [B1], and playing major roles in a number of significant results including [C], [E] and [F].

PROOF OF THEOREM 3. Let $f: S^n \to S^n$ be a map with a bald spot and a tame zero-dimensional singular set. Let $\varepsilon > 0$. Then there is a collared n-cell D in S^n disjoint from $S(f)$, and Lemma 7 provides a null collection $\{C_i\}$ of disjoint collared n-cells of diameter $< \varepsilon$ in $S^n - D$ such that $S(f) \subset \bigcup_{i=1} \operatorname{int} C_i$.

Let $X = \{C_i : i \geq 1\} \cup \{\{y\} : y \in S^n - \cup_{i=1} C_i\}$; i.e., X is the quotient space obtained from S^n by identifying each C_i to a point. Let $\pi : S^n \to X$ denote the quotient map; thus $y \in \pi(y)$ for every $y \in S^n$. We endow X with the quotient topology. This makes $\pi : S^n \to X$ continuous and makes X a compact metric space. Notice that since $\{\pi^{-1}(x) : x \in S(\pi)\} = \{C_i : i \geq 1\}$, then the Null Star-like Equivalent Shrinking Principle asserts that $\pi : S^n \to X$ can be approximated by homeomorphisms. Consequently, X is homeomorphic to S^n.

Consider the map $\pi \circ f : S^n \to X$. Its singular set is the countable set $\{\pi(C_i) : i \geq 1\}$. Also it has a bald spot because $f|f^{-1}(\text{int}\, D)$ and $\pi|\text{int}\, D$ are homeomorphisms. Since X is homeomorphic to S^n, Theorem 2 implies that $\pi \circ f : S^n \to X$ can be approximated by homeomorphisms. (This procedure, which encloses $S(f)$ in the null collection $\{C_i\}$ to yield a map $\pi \circ f$ with a countable singular set, is called "amalgamation".)

Let d denote the given metric on S^n, and let d' be a metric on X. Since $\text{diam}\, C_i < \varepsilon$ for each $i \geq 1$, then there is a $\delta > 0$ such that for all $y, z \in S^n$, if $d'(\pi(y), \pi(z)) < \delta$, then $d(y,z) < \varepsilon$. Let $g : S^n \to X$ and $h : S^n \to X$ be homeomorphisms such that g is within $\delta/2$ of π, and h is within $\delta/2$ of $\pi \circ f$. We assert that the homeomorphism $g^{-1} \circ h : S^n \to S^n$ is within ε of f. To see this, let $y \in S^n$. Then $d'(\pi \circ f(y), h(y)) < \delta/2$ and $d'(\pi(g^{-1} \circ h(y)), g(g^{-1} \circ h(y))) < \delta/2$. Hence $d'(\pi(f(y)), \pi(g^{-1} \circ h(y)) < \delta$. Therefore, the choice of δ insures that $d(f(y), g^{-1} \circ h(y)) < \varepsilon$. ▥

BIBLIOGRAPHY

[A] S. Armentrout, Cellular decompositions of 3-manifolds that yield 3-manifolds, Memoir Amer. Math. Soc. 107 (1971).

[B1] R. H. Bing, Upper semicontinuous decompositions of E^3, Annals of Math. 65 (1957), 363-374.

[B2] ——————, Point-like decompositions of E^3, Fund. Math. 50 (1962), 431-453.

[Br] M. Brown, A proof of the generalized Schoenflies theorem, Bull. Amer. Math.Soc. 66 (1961), 74-76.

[C] J. W. Cannon, Shrinking cell-like decompositions of manifolds. Codimension three, Annals of Math. 110 (1979), 83-112.

[E] R. D. Edwards, Approximating certain cell-like maps by homeomorphisms, preprint (1977).

[F] M. H. Freedman, The topology of four-dimensional manifolds, J. of Diff. Geom. 17 (1982), 357-453.

[H] W. E. Haver, Mappings between ANR's that are fine homotopy equivalences, Pacific J. Math. 58 (1975), 457-461.

[Q] F. Quinn, Ends of maps III: dimensions 4 and 5, J. of Diff. Geom. 17(1982), 503-521.

[S] L. C. Siebenmann, Approximating cellular maps by homeomorphisms, Topology 11 (1972), 271-294.

Contemporary Mathematics
Volume 35, 1984

LINKING NUMBERS IN BRANCHED COVERS

Sylvain E. Cappell* and Julius L. Shaneson*

INTRODUCTION

Let $\alpha: S^1 \to N^3$ be a knot in a 3-dimensional manifold and let $f: \hat{N} \to N$ denote a branched covering space of N branched along α. This note sketches a method based on a 4-dimensional construction for studying invariants of $\hat{N}$ and of the branch set $f^{-1}(\alpha) \subset N$. Our method gives a way of relating a noncyclic branched cover of α to a branched cyclic cover of a different associated knot β, which we call a characteristic knot for α. Here our results will be discussed only for $N = S^3$ and f an (irregular) dihedral branched covering set; the invariant studied in the present note will be the linking numbers of the components of the branch set $f^{-1}(\alpha)$. The method can be used to study other invariants, or other branched covers, as well. The 4-dimensional construction itself was announced and described some 10 years ago in [CS2].

The particular interest of dihedral covering space lies in their extraordinary simplicity and generality. Classically, it was studied as the simplest "non-abelian" cover of a knot and thus gave rise to the simplest "non-abelian" (i.e. not obtained from the cyclic covers) invariants of knots [Re]. More recently, M. Hilden and J. Montesinos showed that every oriented 3-manifold is such a 3-fold dihedral branched covering space of S^3 branched along a knot [Hi], [Mo]. In [CS2] we announced a formula for the Rohlin μ-invariant of any mod 2 3-dimensional homology sphere presented as a 3-fold dihedral covering space. That formula, in terms of various linking numbers, could be extended to all dihedral covers, provided that a certain conjecture on the linking numbers of the components of branch sets, a conjecture apparently long familiar to students of this subject, were verified. That conjecture is the theorem of the present note.

Our study of Rohlin μ-invariants of dihedral branched covers will be presented elsewhere. For certain special classes of knots, e.g. ribbon knots,

*Work supported by NSF Grants

this formula simplifies. As we noted in [CS2] this can be used to show that various (algebraically slice) knots are not ribbon. As noted in [CS2] these methods can also be used to compute Atiyah-Singer invariants used by Casson and Gordon [CG] in their study of ribbon and slice knots. An extensive study of that has been made by Litherland [Li].

Precisely, let $\alpha: S^1 \to S^3$ be a knot and $\rho: G \to D_{2p}$ a homomorphism of the knot group $G = \pi_1(S^3 - \alpha(S^1))$ onto the dihedral group of order $2p$, p odd. The p-fold irregular (respectively: regular) dihedral cover of α is the branched cover of S^3, branched along α, associated to the subgroup $\rho^{-1}(Z_2)$ (resp., $\rho^{-1}(e)$) of G, for $e \in Z_2 \subset D_{2p}$. Let $f: M_\alpha \to S^3$ (resp., $\hat{f}: \hat{M}_\alpha \to S^3$) denote this covering space of degree p (resp. $2p$). Consideration of the diamond of subgroups of D_{2p}

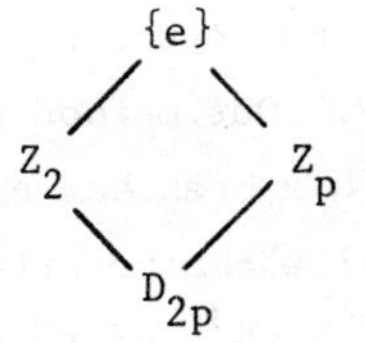

gives a corresponding diamond of covering spaces,

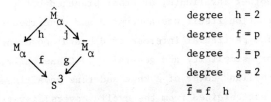

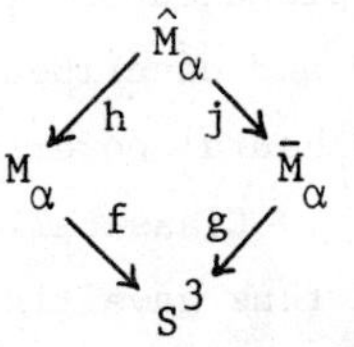

degree $h = 2$
degree $f = p$
degree $j = p$
degree $g = 2$
$\hat{f} = f\ h$

where $\bar{M}_\alpha \to S^3$ is the 2-fold cyclic cover of S^3 branched along α, $\hat{M}_\alpha \to \bar{M}_\alpha$ is a p-fold cyclic unbranched covering space, M_α is the quotient of a lift to $\hat{M}_\alpha$ of the covering translation of period 2 of $\bar{M}_\alpha$.[1]

---

[1] Classically, one sees from this that dihedral covers of S^3 branched along α correspond to elements of order p in $H^1(\bar{M}_\alpha; \mathbf{Z}_p)$ producing $\pi_1(\bar{M}_\alpha) \to \mathbf{Z}_p$ and the associated cover $\hat{M} \to \bar{M}$. Recall that the order of $H_1(\bar{M}_\alpha; \mathbf{Z})$ is just $\Delta_\alpha(-1)$, for $\Delta_\alpha(t)$ the Alexander polynomial of α. (A conceptual explanation of $|H_1(\bar{M}_\alpha; \mathbf{Z})| = |\Delta(-1)|$ was provided using 4-manifolds in [CS1].) Thus one concludes classically that for p odd and square-free, α has a p-fold dihedral covering space if and only if $\Delta_\alpha(-1) \equiv 0 \pmod{p}$; for p prime there is a <u>unique</u> such cover if $\Delta_\alpha(-1) \not\equiv 0 \pmod{p^2}$ (cf. [F1]).

From this, we read off easily a description of the branch set, the inverse image of α, in each of these covers. Clearly $g^{-1}(\alpha)$ is a single circle of branching index 2. Hence, $\bar{f}^{-1}(\alpha) = j^{-1}g^{-1}(\alpha)$ consists of p circles $\hat{\alpha}_0, \hat{\alpha}_1, \ldots, \hat{\alpha}_{p-1}$ each of branching index 2; here these circles are indexed by the convention $T^i\alpha_0 = \alpha_i$ for T a fixed choice of a generator of the covering translation group Z_p of the map $j: \hat{M}_\alpha \to \bar{M}_\alpha$. The covering translation of period 2, $\phi: \hat{M}_\alpha \to \hat{M}_\alpha$, is associated to the 2-fold covering space $h: \hat{M}_\alpha \to M_\alpha$. Notice that ϕ and T are just generators for the dihedral group D_{2p} acting as covering translation on $\hat{M}_\alpha$; the action of D_{2p} on the components of $\bar{f}^{-1}(\alpha)$ is equivalent to that of D_{2p} on the p verticies of a polygon with p sides. As $M_\alpha = \hat{M}_\alpha$/action of ϕ, in M_α, $f^{-1}(\alpha)$ consists of one circle α_0, of branching index 1, and $(p-1)/2$ circles of branching index 2, $\alpha_1, \ldots, \alpha_{(p-1)/2}$; thus, $\hat{M}_\alpha$ can be viewed as a 2-fold covering space of M_α branched along α_0 with $h^{-1}(\alpha_i) = \hat{\alpha}_i \cup \hat{\alpha}_{p-1}$, $1 \leq i \leq (p-1)/2$.

Fixing an orientation for S^3 and α, the covers M_α and $\hat{M}_\alpha$ are correspondingly oriented, as are the branch curves α_i and $\hat{\alpha}_i$. When M_α is a rational homology sphere, let $v_{i,j}$ denote the linking number of α_i with α_j, $i \neq j$, $0 \leq i,j \leq (p-1)/2$; when M_α is a mod 2 homology sphere these $v_{i,j}$ are rational numbers with odd denominator.

The study of the behavior of these numbers is one of the oldest topics in topology. This is partially because these are the simplest "non-abelian invariants" that can be used to distinguish knots. Calculations of them for this purpose were used by Reidemeister [Re]. An early paper of Bankwitz and Schumann [BS] stated that if α is a 2-bridge knot, then $v_{0,i} = \pm 2$; their proof is difficult to reconstruct; clear and more precise modern proofs of this were given by Perko [Pe1] and by Burde [B1]. Note that if α is a 2-bridge knot, by considering its Heegard genus it is easy to show that then M_α is actually S^3 [B2]. While it is not hard to develop methods for calculating the $v_{i,j}$ (cf. [Re], [F3]) really efficient general algorithms were developed by K. Perko [Pe2] and further studied by Hartley and Murasugi [HM].

The following was perhaps conjectured by everyone who has thought about linking numbers in branched covers; it generalizes to all knots the classical result for 2-bridge knots and is suggested by calculating examples. It is, moreover, needed in understanding other invariants (e.g. μ-invariants) of branched covers.

THEOREM I. <u>If the</u> p-<u>fold dihedral branched covering space</u> M_α <u>is a</u> mod 2 <u>homology sphere, then the linking numbers of the branch curves satisfy</u> $v_{i,0} \equiv 2 \pmod 4$, $1 \leq i \leq (p-1)/2$ and $v_{i,j} \equiv 0 \pmod 2$, $1 \leq i,j \leq (p-1)/2$, $i \neq j$.

Counterexamples to the converse of this theorem are provided, according to calculations of Ken Perko, by some 10-crossing knots with $p=3$ [Pe1].

Actually, as noted by Perko, the numbers $v_{i,0}$, $1 \leq i \leq (p-1)/2$ determine all the $v_{i,j}$. This follows readily from the following transfer argument. First of all, note that as $\hat{M}_\alpha$ is a 2-fold branched cyclic cover of M_α, $\hat{M}_\alpha$ is a mod 2 homology sphere if and only if M_α is. (The homology of a 2-fold cyclic branched cover is given by the Alexander polynomial at (-1); cf. [CS1].) Let u_j denote the linking number of $\hat{\alpha}_i$ with $\hat{\alpha}_{i+j}$, $1 \leq j \leq p-1$; this is independent of i, $0 \leq i \leq p-1$, as the $\hat{\alpha}_i$ are permuted by the covering translations. For the same reason, $u_i = u_{p-i}$.

Standard transfer considerations show, as noted by Perko [Pe1] that:

$$v_{i,j} = u_{i+j} + u_{|i-j|}, \quad 0 \leq i,j \leq \frac{p-1}{2}, \quad i \neq j$$

and, in particular, $v_{i,0} = 2u_i$, so that

$$v_{i,j} = \frac{1}{2}(v_{\min(i+j,p-i-j),0} + v_{|i-j|,0}) \; ; \; 1 \leq i,j \leq \frac{p-1}{2} \, .$$

Hence the numbers $v_{i,0}$ determine all the $v_{i,j}$ and the main theorem of this note will follow from:

THEOREM II. *If the 2p-fold (regular) dihedral branched covering space $\hat{M}_\alpha$ is a mod 2 homology sphere, then the linking numbers of the branch curves satisfy*

$$u_i \equiv 1 \pmod 2), \quad 1 \leq i \leq (p-1) \, .$$

Outline of Method

Here is a summary of our approach to this and related problems on branch covers.

Step 1. An effective method for studying 3-manifolds M described as branched covers of S^3 along a knot α is to utilize a 4-manifold W^4, with $\partial W = M$ obtained by letting W be a branched cover of D^4 along K^2, where $K \cap S^3 = \alpha(S^1)$. It is easy to do this for cyclic covers; just let K be a Seifert surface of α pushed into D^4; this method of studying cyclic covers was introduced by us in [CS2] and independently by L. Kauffman. However, it will not work for more general branched covers as the fundamental group of D^4-{pushed in Seifert surface} is Z and thus has no nonabelian covers.

For noncylcic covers, we employ instead for K a certain (non-manifold) 2-complex. This works at least for all metacyclic covers; in particular, for dihedral covers the resulting W^4 is a manifold even though K^2 is not. (In other settings, the singularity which arises is readily understood and can be resolved.)

Step 2. We relate questions about linking numbers of branch curves in $M^3 = \partial W^4$ to intersection numbers of parts of 2-dimensional surfaces in the branch set of W^4.

Step 3. We get information on these intersection numbers by relating these 2-dimensional surfaces to a kind of equivariant second Stiefel-Whitney class of W^4 and then get our result from an equivariant version of the standard fact that in an oriented 4-manifold, w_2^2 is just the Euler characteristic mod2.

Of course, this method can be used to study many other invariants of such branched covers. An interesting way to view the geometrical procedure outlined in Step 1, and carried out in Section 1 below for dihedral covers, is that it reveals a close relationship between a dihedral (or metacyclic) cover of S^3 branched along α and a cyclic cover of a characteristic knot β associated below to α.

1. Characteristic knots and a cobordism construction.

Fix an orientation of S^3 and adopt the unique conventions so that the circles in Figure 1 have linking number +1.

Fig. 1

If α is a (smooth or P.L. locally flat) knot in S^3, let $\Delta_\alpha(t)$ denote its Alexander polynomial.

Definition. Let α and β be (oriented) knots in S^3. Then β is called a mod p characteristic knot for α if there exists an oriented Seifert surface of α, V, $\partial V = \alpha$, so that $\beta \subset \mathring{V}$ represents a nonzero (primitive) class $[\beta]$ of $H_1(V)$ and so that

$$(L_V + L_V')\beta \equiv 0 \pmod p) .$$

L the linking pairing of V in S^3. More precisely, $L_V(x,y) = \ell(f_+x,y)$, where f_+ is induced by pushing V off itself using a positive normal, and ℓ denotes linking numbers and $L_V(x,\beta) + L_V(\beta,x) \equiv 0 \pmod p$, all $x \in H_1(V)$.

Note: If α is a nontrivial knot with Seifert surface V with p square-free, and if $p|\Delta_\alpha(-1)$, then α has a mod p characteristic knot embedded in V. (Proof: Note that $\Delta_\alpha(-1) = \pm\det(L_V + L_V')$ and use the well-known fact that a primitive class in $H_1(V)$ is represented by an embedded circle.)

Suppose α is a knot with Seifert surface V and $\beta \subset \mathring{V}$ is a mod p characteristic knot of α. We proceed to construct a cobordism relating the dihedral covering spaces of S^3 with branch sets α to the cylcic cover of S^3 with branch set β.

Let $\pi: \Sigma(\beta,p) \to S^3$ be the p-fold cyclic branched cover of S^3, branched along β. If $x \in H_1(V-\beta)$, then the intersection number on V, $x\cdot\beta = 0$; hence

$$L_V(x,\beta) - L_V(\beta,x) = x(L_V - L_V')\beta = 0 \ .$$

Since $(L+L')\beta \equiv 0 \pmod p$, it follows that $2L_V(\beta,x) \equiv 0 \pmod p$. Since $\det(L+L;) \equiv \det(L-L') \pmod 2$, and since $\det(L-L') = \pm 1$ by Poincare duality, p is odd. Hence $L_V(\beta,x) \equiv 0 \pmod p$. Therefore

$$\pi^{-1}(V) = V_0 \cup V_1 \cup \ldots \cup V_{p-1} ,$$

$\pi|V_i : V_i \to V$ a P.L. homeomorphism and

$$V_i \cap V_j = \pi^{-1}(\beta) \quad , \quad i \neq j \ .$$

Let $T: \Sigma \to \Sigma$ be a generator of the group of covering translations corresponding to a positively oriented meridional circle of β in S^3 (i.e. $T|$fiber of a neighborhood of $\pi^{-1}\beta$ is rotation by $2\pi/p$). Assume the indices have been chosen so that $TV_i = V_{i+1}$, $0 \leq i \leq p-2$, and $TV_{p-1} = V_0$.

Let $V \times [-1,1] \subset S^3$ be a neighborhood of $V = V \times 0$, and let $h(x,t) = (x,-t)$ for $x \in V$ and $t \in [-1,1]$. Then

$$\pi^{-1}(V \times [-1,1]) = J_0 \cup \ldots \cup J_{p-1} = J ,$$

with $\pi|J_i : J_i \to V \times [-1,1]$ a P.L. homeomorphism and with $V_i \subset J_i$. Clearly, $\pi^{-1}(V \times [-1,1]) = J$ is the p-fold branched cyclic cover of $V \times [-1,1]$ along β. Let

$$\bar{h}: J \to J$$

be a lift of h, i.e. $\pi\bar{h} = h$ $(\pi|J)$, with $\bar{h}(V_0) \subset V_0$. Then $\bar{h}(J_i) \subset J_{p-1}$, $1 \leq i \leq p-1$, $\bar{h}(J_0) = J_0$, and $\bar{h}$ fixes precisely V_0.

Let $\Sigma = \Sigma(\beta,p)$ and let

$$Y = \Sigma \times [0,1] \ / \ \{(x,1) = (\bar{h}(x),1) \ \text{ for } \ x \in J\}$$

the space obtained by identifying $(x,1)$ and $(\bar{h}(x),1)$ in $\Sigma \times I$. Let $\pi': Y \to S^3 \times I/\{(x,t) = (x,t) = (x,-t),\ x \in V\} \cong S^3 \times I$ be induced by $\pi \times 1_{[0,1]}$. Y is evidently an orientable (smooth) cobordism of Σ to a closed manifold, $M_{\alpha,\beta}$, say, and π' is a branched covering projection with branching set B the image of $V \times 0 \cup \beta \times I$ in $S^3 \times I/\{(x,t) = (x,-t),\ x \in V\}$, which is canonically P.L. homeomorphic to V. Orient Y so that π' has positive degree. (Usual convention: $\partial[S^3 \times I] = [S^3 \times 1] - [S^3 \times 0]$, and thus $\partial Y = [M] - [\Sigma]$.)

Clearly the tuple

$$(S^3 - \mathrm{Int}(V \times [-1,1]), \partial(V \times [-1,1]), \alpha)/\{(x,t) = (x,t) \mid (x,t \in \partial(V \times [-1,1])$$

is canonically P.L. homeomorphic to (S^3, V, α). Hence the restriction ω of π' to $M_{\alpha,\beta}$ is a branched covering of S^3 along α. Note that $\omega^{-1}(\alpha)$ has $(p+1)/2$ components, with branching index 2 on $(p-1)/2$ of them. In fact, if V_i' denotes the image of $V_i \times 1$ in Y (so $V_i' = V_{p-1}'$, $1 \le i \le p-1$), then $\omega^{-1}(\alpha) = \partial V_0' \cup \ldots \cup V_{(p-1)/2}'$, and $\partial V_0'$ is the component with branching index 1. Write $\alpha_i = \partial V_i'$, $0 \le i \le \frac{p-1}{2}$.

<u>Proposition 1.1.</u> <u>$M_{\alpha,\beta} \overset{\omega}{\to} S^3$ is a dihedral metacylcic, branched covering space of S^3 along</u> α.

<u>Proof</u>: Let $D_p = \{u,\tau \mid \tau^2 = 1,\ u^p = 1,\ \tau u = u^{-1}\tau\}$. The group $\pi_1(S^3-\alpha)$ has the form (Higman-Neumann-Neumann construction)

$$Z * G/\{t i_+(x) t^{-1} = i_-(x),\ x \in H\}$$

where G is the fundamental group of $S^3 - V$, H that of V, t is a generator of the infinite cyclic group, represented by a meridian m of α, and i_+ and i_- are induced by pushing V into its complement along positive and negative normal vectors, respectively.

Define $\rho: G \to D_p$ by

$$\rho(\xi) = u^{\ell(\xi,\beta)} ,$$

and let $\rho(t) = \tau$. Since $L_V(x,\beta) \equiv -L_V(\beta,x) \pmod p$, these definitions determine a homeomorphism

$$\rho: \pi_1(S^3 - \alpha) \to D_p .$$

Assuming $M_{\alpha,\beta}$ is connected, the fundamental group of the unbranched covering $M_{\alpha,\beta} - \omega^{-1}(\alpha)$ also has an (HNN)-representation. In particular, using Van Kampen's theorem (and a base point near α_0), $\pi_1(M_{\alpha,\beta}-\omega^{-1}(\alpha))$ is generated by a meridian m_0 of α_0 with $\omega(m_0) = m$ and elements in the image of $\pi_1(M_{\alpha,\beta} - \omega^{-1}(V))$. Let $\omega' = \omega \mid M_{\alpha,\beta}-\omega^{-1}(\alpha)$. Since, by construction, $\omega \mid M_{\alpha,\beta}-\omega^{-1}V$ is the cylcic cover $\pi \mid \Sigma-\pi^{-1}(\mathrm{Int}\,V \times [-1,1])$, of $S^3 - \mathrm{Int}(V\times[-1,1]) = S^3 - V$, it follows that $M_{\alpha,\beta}-\omega^{-1}(\alpha)$ is connected and that $\ell(\omega_*'\eta,\beta) \equiv 0 \pmod p$ for $\eta \in \pi_1(M_{\alpha,\beta}-\omega^{-1}(V))$, and so $\rho(\omega_*'\eta)$ is the trivial element. Clearly $\rho(\omega_*'[m_0]) = \rho([m]) = \rho(t) = \tau$. Thus the image of $\rho \circ \omega_*'$ is $\{\tau,1\}$, which proves the result; in particular we have the following:

Proposition 1.2. Dihedral p-fold branched covers of α are in 1 to 1 correspondence to equivalence classes of characteristic knots viewed as representing elements of order p in the kernel of the mod p reduction of $(L_V + L_V^t)$, modulo the action of $\mathbb{Z}_p^*$.

Let F_0 be a stable framing of the tangent bundle of S^3, compatible with the orientation. Let $N_1 = N(V_0' \cup \ldots \cup V'_{(p-1)/2} \cup \pi^{-1}(\beta) \times I)$ be a regular neighborhood of $(\pi')^{-1}(B)$, meeting the boundary regularly. Clearly N_1 may be chosen so that the restriction of π' to a neighborhood $V_0' - V_0' \cap N_1$ is a homeomorphism. Therefore the stable framing of $Y - \mathring{N}_1$ induced from $F_0 \times I$ via the unbranched covering $\pi'|Y-\mathring{N}_1$ extends to a framing F' of $Y-\mathring{N}_2$, N_2 a regular neighborhood of $V_1' \cup \ldots \cup V'_{(p-1)/2} \cup \pi^{-1}(\beta) \times I$.

Recall that given a q-fold covering map $S^1 \to S^1$ and a stable framing of S^1 that extends over D^2, the induced framing extends over D^2 iff q is odd. Therefore $F'|(\Sigma - \mathring{N}_1 \cap \Sigma)$ extends to a fiber of the tubular neighborhood $N_1 \cap \Sigma$ of β in Σ; i.e. to the complement of a cell in Σ. Hence, as $\pi_2(SO) = 0$, it extends to all of Σ. It follows easily (recall $\pi' = \pi \times id_{[0,\varepsilon)}$ near Σ) that F' extends to a stable framing F of $Y - V_1' \cup \ldots \cup V'_{(p-1)/2}$. The sole obstruction to extending $F|\Sigma$ to all of Y is an element

$$\theta(F|\Sigma) \in H^2(Y;\Sigma;\mathbb{Z}_2)$$

Proposition 1.3. Let $D\colon H^2(Y;\Sigma;\mathbb{Z}_2) \to H_2(Y,M;\mathbb{Z}_2)$ be the Poincare duality isomorphism. Then

$$D(\theta(F|\Sigma)) = [V_1']_2 + \cdots + [V'_{(p-1)/2}]_2 ,$$

where $[V_i']_2$ is the element of $H_2(Y,M;\mathbb{Z}_2)$ represented by $(V_i', \partial V_i')$.

Proof: $\theta(F|\Sigma)$ is the restriction of $\theta(F) \in H^2(Y;Y-V_1' \cup \ldots \cup V'_{(p-1)/2};\mathbb{Z}_2)$. Hence, by Poincare duality, $D(\theta(F|\Sigma))$ is the image of

$$D(\theta(F)) \in H_2(V_1' \cup \ldots \cup V'_{(p-1)/2} \cup M,M;\mathbb{Z}_2)$$

$$\cong H_2(V_1' \cup \quad \cup V'_{(p-1)/2};\alpha_1 \cup \quad \cup \alpha_{(p-1)/2};\mathbb{Z}_2) .$$

Using Meyer-Vietoris, the right side is of course just $\bigoplus_{i=1}^{(p-1)/2} H_2(V_i';\alpha_i)$. Hence $D(\theta(F|\Sigma))$ is a linear combination of the classes $[V_1'],\ldots,[V'_{(p-1)/2}]_2$.

Now let (D_i^2, S_i^1), $i=1,\ldots,\frac{p-1}{2}$ be the disjoint fibers of the normal tubes of $V_1',\ldots,V'_{(p-1)/2}$, respectively. Clearly $\pi'|S_i^1$ is a two-fold covering map; hence, as noted above, $F|S_i^1$ does not extend to D_i^2. It follows (e.g. represent any element of $H_2(Y;\mathbb{Z}_2)$ by a 2-manifold transverse to all V_i' and consider the obstruction to framing a neighborhood of this 2-manifold that $D(\theta(F))$ is as stated.

Remark. This argument could be reformulated as an instance of the general principle that if $P^4 \xrightarrow{f} Q^4$ is a branched covering space of orientable 4-manifolds, then $D(w_2(P)) = D(f^* w_2(Q)) + [S^2]$, where S^2 is the subset of the branching set in Y consisting of points of even branching degree. (This follows from the familiar simplicial formula for Stiefel-Whitney classes.)

Corollary 1.4. The image of $[V_1']_2 + \cdots + [V'_{(p-1)/2}]_2$ in $H_2(Y,\partial Y;\mathbb{Z}_2)$ is precisely $Dw_2(Y)$, $w_2(Y) \in H^2(Y;\mathbb{Z}_2)$ the second Stiefel Whitney class of Y.

Now having constructed a cobordism Y^4 of M to Σ it is easy to further produce a compact manifold with Σ on the boundary. In fact, we just observe that $\Sigma = \partial P^4$ where $\phi: P^4 \to D^4$ is obtained as in [CS1] as the branched cyclic cover of D^4 along E^2, a Seifert surface, of the characteristic knot β, whose interior has been pushed into the interior of D^4. See [CS1] for details. Then set $W = Y \cup_\Sigma W$; clearly $\partial W = (\partial Y) - M$.

This 4-manifold W^4 can be described directly as a branched covering space as follows. The maps constructed above $\pi': Y \to S^3 \times I$ and $\phi: P^4 \to D^4$ can be glued together to get a map $\Phi: W = Y \cup_\Sigma P^4 \to S^3 \times I \cup_{S^3} D^4 = D^4$. This map Φ is then seen to be a branched dihedral covering space. The total branching set in D^4 is a 2-complex K^2 obtained by attaching to V^2 a Seifert surface E^2 of β glued to V along $\beta \times \frac{1}{2}$.

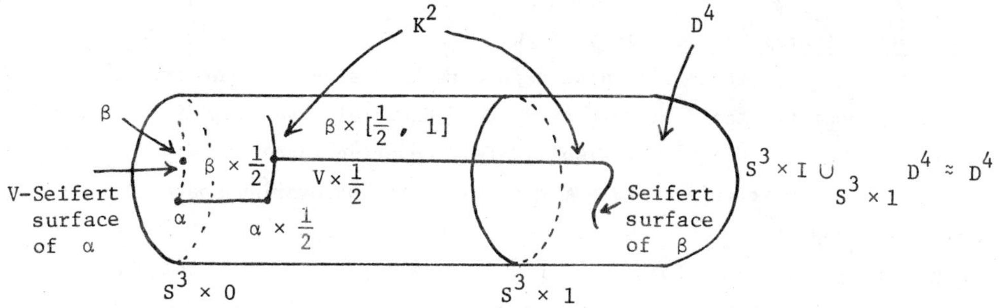

Fig. 2

This branching set in $D^4 = S^3 \times I \cup_{S^3 \times 1} D^4$ fails to be a manifold around the circle $\beta \times \frac{1}{2}$. Nevertheless, as we have seen W^4, the corresponding branched dihedral covering space is a manifold.

Remark. As the branching set in D^4 is not a manifold, it may seem surprising that the branched cover W^4 if a manifold; we explain this directly from another perspective. Consider again the branched dihedral cover W of D^4 along K^2. This is clearly a manifold except in a neighborhood of the inverse of the singularity circle $\beta \times \frac{1}{2}$ lying on K^2. See Figure 2. Now in a neighborhood of $\beta \times \frac{1}{2}$ the pair (D^4,K^2) looks like $S^1 \times (D^3,Q)$ where Q denotes a "figure Y", as can be seen near $\beta \times \frac{1}{2}$ in Figure 2. The dihedral cover of this neighborhood of the circle $\beta \times \frac{1}{2}$ would then be just $S^1 \times \{$a dihedral cover of D^3 branched along $Q\}$. As D^3 = cone on S^2, this cover will be just $S^1 \times \{$cone on the branched cover of S^2 along $S^2 \cap Q\}$. Now $S^2 \cap Q$ = 3 points.

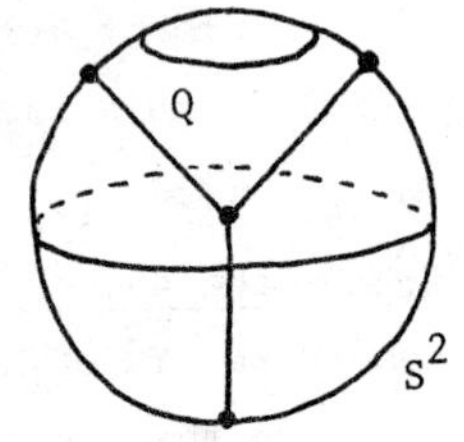

Fig. 3

Last, note that the branched dihedral cover of S^2 along these three points is again S^2. This follows by calculating its Euler characteristic using the fact that there the meridian about each point represents an element of order 2 in the dihedral group. Hence W has no singularity and is a P.L. manifold.

The same geometrical methods used above can be used to extend Proposition 1.4 to the following:

Proposition 1.5. *Let* $D: H^2(W;\mathbb{Z}_2) \to H_2(W,M;\mathbb{Z}_2)$ *be the Poincare duality isomorphism. Then*

$$[V_1']_2 + \cdots + [V'_{(p-1)/2}]_2 = D(w_2(W)),$$ *where* $w_2(W)$ *is the second Stiefel-Whitney class of* W.

Also, we can repeat all these arguments used above for the irregular p-fold dihedral cover for the full regular 2p-fold dihedral cover. In fact this 2p covering space of $\hat{W} \to D^4$ is also a 2-fold covering space of W^4 branched along V_0'; in particular, $\partial\hat{W} = \hat{M} \to M$ is a 2-fold covering space of M branched along α_0. Thus, the branch set in $\hat{M}$ (resp., $\hat{W}$) is the union of p circles (resp., 2-manifolds) $\hat{\alpha}_0,\ldots,\hat{\alpha}_p$ (resp., $\hat{V}_0,\ldots,\hat{V}_p$) which are disjoint (resp., intersect in one common circle $\hat{\beta}$). Here the circles are indexed by the convention $T^i\hat{\alpha}_0 = \hat{\alpha}_i$, where T is the generator of $Z_p \subset D_{2p}$ regarded as the group of covering translations.

Remark 1.6. Notice that in the regular 2p-fold dihedral covering space $\hat{W}$ has $w_2(\hat{W}) = 0$; for, it is a p-fold cyclic (branched) cover of a manifold with zero second Stiefel-Whitney class, the 2-fold cyclic cover of D^4 branched along the Seifert surface V (see [CS1]). On the other hand arguments similar to those used above show that the Poincare dual of $w_2(\hat{W}^4)$ is given by $\sum_{i=0}^{p-1} [\hat{V}_i] \in H_2(\hat{W},\partial\hat{W};\mathbb{Z}_2)$ which hence equals zero. (This can be checked in other ways.)

2. Linking numbers and characteristic classes

Note that for $M_\alpha = \partial W^4$ (resp. $\hat{M}_\alpha = \partial \hat{W}^4$) the irregular (resp., regular) p-fold (resp., 2p-fold) dihedral cover of S^3 as above, as $\hat{M}_\alpha \to \hat{M}_\alpha$ is a 2-fold cyclic branched cover [CS1], $H_1(M_\alpha;\mathbb{Z}_2) = 0$ if and only if $H_1(\hat{M}_\alpha;\mathbb{Z}_2) = 0$. In this section, we assume $H_1(M_\alpha;\mathbb{Z}_2) = 0$; hence the intersection form

$$\begin{array}{ccc} H_2(\hat{W};\mathbb{Z}_2) \times H_2(\hat{W};\mathbb{Z}_2) & \to & \mathbb{Z}_2 \\ (x,y) & \to & [x,y] \end{array}$$

is, by standard Poincare duality, a nonsingular symmetric bilinear pairing. Letting T (resp. ϕ) denote, as before, an element of order p (resp., 2) in D_{2p}, the covering translation group of $\hat{M} \to S^3$, we introduce a new bilinear pairing on $H_2(\hat{W};\mathbb{Z}_2)$ with values in $\mathbb{Z}_2[\mathbb{Z}_p]$:

$$\langle x,y\rangle = \sum_{i=0}^{p-1} [T^{-i}x,\phi y]T^i$$

<u>Lemma 2.1</u>. <u>This pairing is bilinear over $\mathbb{Z}_2[\mathbb{Z}_p]$ and symmetric over $\mathbb{Z}_2[\mathbb{Z}_p]$ and nondegenerate</u>.

<u>Proof</u>: To see that it is <u>symmetric</u> (not Hermitian) note that

$$\begin{aligned} [T^{-i}x,\phi y] &= [\phi y,T^{-i}x] \\ &= [y,\phi T^{-i}x] \\ &= [y,T^{i}\phi x] \\ &= [T^{-i}y,\phi x] \end{aligned}$$

To check bilinarity note that

$$\begin{aligned} \langle Tx,y\rangle &= \sum [T^{-i}(Tx),\phi y]T^i \\ &= \sum [T^{-(i-1)}x,\phi y]T^i \\ &= \sum ([T^{-j}x,\phi y]T^j)T \\ &= \langle x,y\rangle T \end{aligned}$$

The nondegeneracy of this pairing follows from that of the intersection pairing.

Now we need some facts about symmetric forms over $\mathbb{Z}_2[\mathbb{Z}_p]$, p odd.

<u>Proposition 2.2</u>. <u>Let</u> $p \times p \xrightarrow{\langle\ ,\ \rangle} \mathbb{Z}_2[\mathbb{Z}_p]$ <u>be a nonsingular symmetric bilinear pairing on the finitely generated $\mathbb{Z}_2[\mathbb{Z}_p]$ module</u> P. <u>Then there is a unique element</u> $\alpha \in P$ <u>satisfying</u>

$$\langle x,\alpha\rangle^2 = \langle x,x\rangle \ , \ x \in P \ .$$

<u>Notation</u>. α is called the characteristic element of P.

Proof: Consider $L(x) = \langle x,x\rangle^{1/2}$, $x \varepsilon P$. This is well-defined as $\mathbb{Z}_2[\mathbb{Z}_p]$ is a product of finite fields of characteristic 2. Moreover, as $L\colon P \to \mathbb{Z}_2[\mathbb{Z}_p]$ is easily seen to be linear, there is a unique $\alpha \varepsilon P$ with

$$L(x) = \langle x,\alpha\rangle \quad , \quad x \varepsilon P \ .$$

Recall that $\mathbb{Z}_2[\mathbb{Z}_p] = \oplus F_j$ where each F_j is a field of characteristic 2 and $F_0 \cong \mathbb{Z}_2$. Let e_j denote the multiplication identity of F_j; note that $e_0 = 1 + T^1 + T^2 + \cdots + T^{p-1}$ in $\mathbb{Z}_2[\mathbb{Z}_p]$. Correspondingly, a $\mathbb{Z}_2[\mathbb{Z}_p]$ module P decomposes naturally as

$$P = \oplus(P \otimes_{\mathbb{Z}_2[\mathbb{Z}_p]} F_j) \ .$$

Proposition 2.3. For $\alpha \varepsilon P$, the characteristic element of a symmetric bilinear form on the finitely generated $\mathbb{Z}_2[\mathbb{Z}_p]$ module P

$$\langle\alpha,\alpha\rangle = \sum e_j \operatorname{rank}_{F_j} (P \otimes_{\mathbb{Z}_2[\mathbb{Z}_p]} F_j) \ .$$

This follows immediately from the corresponding fact over each field F_j, which is easy as such forms decompose into 1-dimensional forms.

We use this to study $\hat{W}^4$. Let $A = [\hat{V}_0] \varepsilon H_2(\hat{W},\hat{M};\mathbb{Z}_2) \cong H_2(\hat{W};\mathbb{Z}_2)$.

Proposition 2.4. $A \varepsilon H_2(\hat{W};\mathbb{Z}_2)$ is the characteristic element of the pairing $\langle x,y\rangle$.

Lemma 2.5.

$$[x,\phi T^j x] = [T^{i/2}x,A] \ .$$

Proof of Proposition 2.4:

$$\begin{aligned} \langle A,x\rangle^2 &= \left(\sum[T^{-i/2}x,A]T^{i/2}\right)^2 \\ &= \left(\sum[x,\phi T^{-i}x]T^{i/2}\right)^2 , \quad \text{by the lemma} \\ &= \sum[x,\phi T^{-i}x]T^i \qquad \text{in } \mathbb{Z}_2[\mathbb{Z}_p] \\ &= \sum[T^i x,\phi x]T^i \\ &= \langle x,x\rangle \end{aligned}$$

Proof of Lemma 2.5: Consider the 2-fold covering maps $g_i\colon \hat{W} \to \hat{W}/\phi T^i$; for $i=0$, write $g = g_0\colon \hat{W} \to (\hat{W}/\phi T^0) = W$. As in the dihedral group D_{2p}, $\phi T^{i/2} = T^{i/2}\phi T^i$, there is a commutative diagram:

$$\begin{array}{ccc} \hat{W} & \xrightarrow{T^{i/2}} & \hat{W} \\ {\scriptstyle \phi T^i}\downarrow & & \downarrow{\scriptstyle \phi} \\ \hat{W} & \xrightarrow{T^{i/2}} & \hat{W} \end{array}$$

From this there is a homeomorphism $h_i : \hat{W}/\phi T^i \to \hat{W}/\phi$ and a commutative diagram

$$\begin{array}{ccc} \hat{W} & \xrightarrow{T^{i/2}} & \hat{W} \\ g_i \downarrow & & \downarrow g \\ \hat{W}/\phi T^i & \xrightarrow{h_i} & \hat{W}/\phi \end{array}$$

Now, as noted above, $w_2(\hat{W}) = 0$ and hence,

$$[x,x] = 0$$

and thus

$$[x,\phi T^i x] = [x,(i + \phi T^i)x]$$

and using transfers,

$$\begin{aligned} [x,\phi T^i x] &= [g_{i_*}(x), g_i(x)] \\ &= [g_*(T^{i/2}x) , g_*(T^{i/2}x)] \\ &= [g_*(T^{i/2}x), \sum_{i=1}^{(p-1)/2} [V_i]] , \end{aligned}$$

as $w_2(W) = \sum_{i=1}^{(p-1)/2} [V_i]$ by Proposition 1.6. Thus,

$$\begin{aligned} [x,\phi T^i x] &= [T^{i/2}x, (1+\phi) \sum_{i=1}^{(p-1)/2} [\hat{V}_i]] \\ &= [T^{i/2}x, \sum_{i=1}^{p-1} T^i A] \end{aligned}$$

But as noted in Remark 1.7, $\sum_{i=0}^{p-1} T^i A = 0$. Hence,

$$[x,\phi T^i x] = [T^{i/2}x,A] .$$

For P a module over $\mathbb{Z}_2[\mathbb{Z}_p]$, let [P] denote the class represented by P in $R(\mathbb{Z}_p)$, the representation ring of $\mathbb{Z}_p$ over the field $\mathbb{Z}_2$. Notice that $G = \hat{W} \cup_{\partial\hat{W}} \{\text{cone on } \partial\hat{W}\}$ has a natural action of D_{2p} with one fixed point, the cone point, and satisfies, as M is a mod 2-homology sphere, mod 2 Poincare duality.

Hence, $[H_2(\hat{W};\mathbb{Z}_2)] = \sum_{i=0}^{4} [H_i(G;\mathbb{Z}_2)]$ in $R(\mathbb{Z}_p) \otimes \mathbb{Z}_2$. Moreover $\sum_{i=0}^{4} [H_i(G;\mathbb{Z}_2)] = \sum_{i=0}^{4} [C_i(G;\mathbb{Z}_2)]$ in $R(\mathbb{Z}_p) \otimes \mathbb{Z}_2$ for $C_i(G;\mathbb{Z}_2)$ the cellular chain groups of a cellular decomposition of G. However, the action of $\mathbb{Z}_p$ on G is free outside a 2-manifold [1] in $\hat{W}$ and the cone point. Hence in $R(\mathbb{Z}_p) \otimes \mathbb{Z}_2$

$$[H_2(\hat{W};\mathbb{Z}_2)] = k[\mathbb{Z}_2[\mathbb{Z}_p]] \oplus [\mathbb{Z}_2] \text{ , some } k \text{ .}$$

Moreover, as $\hat{W}$ is a 2-fold branched cover of W along V_0, $\chi(\hat{W}) = 2\chi(W) - \chi(V)$ is odd, and hence $\chi(G) \equiv 0 \pmod 2$. Hence, $[H_2(\hat{W};\mathbb{Z}_2)] = [\mathbb{Z}_2[\mathbb{Z}_p]] \oplus [\mathbb{Z}_2]$ in $R(\mathbb{Z}_p) \otimes \mathbb{Z}_2$. Thus, from Propositions 2.4 and 2.3 we conclude that

$$\begin{aligned} \langle A,A\rangle &= 1 + e_0 \\ &= 1 + (1 + T + \cdots + T^{p-1}) \\ &= T + T^2 + \cdots + T^{p-1} \text{ .} \end{aligned}$$

Going back to the definition of the pairing $\langle A,A\rangle$ this says:

Corollary 2.7. *In $H_2(\hat{W},\partial\hat{W};\mathbb{Z}_2)$ the intersection number of $[\hat{V}_0]$ with $[\hat{V}_i]$ is odd.*

Proof of Theorem II: As $\hat{V}_i$ and $\hat{V}_0$ intersect just in the circle β, and this circle of intersections can be removed by pushing one class away from the other, the intersection of $[\hat{V}_i]$ and $[\hat{V}_0]$ is evidently given by the linking numbers of $\partial\hat{V}_i = \hat{\alpha}_i$ and $\partial\hat{V}_0 = \hat{\alpha}_0$ in W. Thus Theorem II, and also Theorem I, follow from Corollary 2.7.

BIBLIOGRAPHY

[BS] Bankwitz, C., and Schumann, H. G., Uber Viergeflechte. Abh. Math. Sem. Univ. Hamburg 10, 263-284 (1934).

[B1] Burde, G., Verschlingungsinvarianten von Knoten und Verkettungen mit zwei Brucken, Math. Zeit., 145, 235-242 (1975).

[B2] ———, On branched coverings of S^3, Canad. J. Math. 23, 84-89 (1971).

[CS1] Cappell, S. E., and Shaneson, J. L., Cyclic branched covering spaces, Knots groups and 3-manifolds; Papers dedicated to the memory of R. H. Fox, ed. by L. P. Neuwirth, Ann. of Math. Studies 84, pp. 165-173, Princeton Univ. Press, 1975.

[CS2] ———, Invariants of 3-manifolds, Bull. Amer. Math. Soc. 81, 559-562 (1975).

[CG] Casson, A., and Gordon, C., Cobordism of classical knots, mimeographed notes, Orsay, 1975.

---

[1] This 2-manifold looks like 2 copies of the Seifert surface of joined together along their boundary. To see this pass first to the 2-fold cover of D^4 and then up to $\hat{W}$.

[F1] Fox, R. H., A quick trip through knot theory, Topology of 3-Manifolds and Related Topics (Proc. Univ. of Georgia Conf. 1961), ed. by M. K. Fort, Jr., Prentice-Hall (1962), 120-167.

[F2] ———, Construction of simply connected 3-manifolds, Topology of 3-Manifolds and Related Topics (Proc. Univ. of Georgia Conf. 1961), ed. by M. K. Fort Jr., Prentice-Hall, (1962), 213-216.

[F3] ———, Metacyclic invariants of knots and links, Canad. J. Math. 22, 193-201 (1970).

[Hi] Hilden, H., Every closed orientable 3-manifold is a 3-fold branched covering space of S^3, Bull. Amer. Math. Soc. 80, 1243-1244 (1974).

[HM] Hartley, R., and Murasugi, K., Covering linkage invariants, Canad. Jour. Math. 29, 1312-1339 (1977).

[H] Hartley, R., Homology invariants, Canad. Jour. Math. 30, 655-670 (1978).

[Li] Litherland, R. A., A formula for the Casson-Gordon invariants of a knot, to appear.

[Mo] Montesinos, M., A representation of closed, orientable 3-manifolds as 3-fold branched covers of S^3, Bull. Amer. Math. Soc. 80, 845-846 (1974).

[Pe1] Perko, K., On dihedral covering spaces of knots, Inventiones Math. 34, 77-82 (1976).

[Pe2] ———, On covering spaces of knots, Glasnik Mat. 9, 141-145 (1974).

[Re] Reidemeister, K., Knoten Theorie, Springer-Verlag, 1932.

COURANT INSTITUTE OF MATHEMATICAL SCIENCES
NYU
NEW YORK, NY 10012

DEPARTMENT OF MATHEMATICS
RUTGERS UNIVERSITY
NEW BRUNSWICK, NJ 08903

[illegible] Fox, R. H., A quick trip through knot theory, Topology of 3-manifolds and related topics (Proc. Univ. of Georgia Inst., 1961), ed. by M. K. Fort, Prentice-Hall (1962), 120-167.

[illegible] ———, [illegible] manifolds, [illegible] (1962) [illegible]

[illegible] ———, [illegible] (1969).

[illegible] 3-fold [illegible] (1974).

[illegible] Covering linkage invariants, [illegible] (1977).

[illegible]

[illegible] the Casson-Gordon invariants [illegible] to appear.

[illegible] representation of closed, orientable 3-manifolds as 3-fold branched coverings of S^3, Bull. Amer. Math. Soc. 80, 845-846 (1974).

[illegible] knots, Invent. Math. [illegible]

[illegible] Knots, [illegible] (1975).

[illegible] Publish or Perish, [illegible] 1976.

COURANT INSTITUTE OF MATHEMATICAL SCIENCES
NEW YORK, NY 10012

DEPARTMENT OF MATHEMATICS
RUTGERS UNIVERSITY
NEW BRUNSWICK, NJ 08903

Contemporary Mathematics
Volume 35, 1984

ATOMIC SURGERY PROBLEMS

Andrew Casson and Michael Freedman*

ABSTRACT. The surgery sequence is the central theorem in manifold theory. In dimension four it is a giant, if improbably, conjecture which would imply almost everything from the four dimensional Poincaré conjecture to "knots with Alexander polynomial equal one are slice". We have reduced the conjecture to an investigation of certain "atomic" surgery problems. This leads to an equivalent reformulation of the conjecture in terms of the classical theory of links in the three sphere.

REVIEW

This is a preliminary draft, written and abandoned in 1976 (or 1977). Andrew had come to visit me; I put him to work on the non-simply-connected version of his theory of flexible handle-bodies. This writeup explores the finite version, though we considered, but could find no use for the non-compact limit (recently considered in Dimonski's Ph.D. thesis). This paper is included in the proceedings at the request of the editors, as an historical relic. Two recent ideas which we suffered in ignorance of were: 1. It is possible (even when $\pi_1 \neq 0$) to concentrate on complexes which serve as substitutes for a disk rather than ones substituting for a wedge of 2-spheres. And the related observation - 2. The more symmetrical grope construction can replace the "1/2-towers" created here. (Bob Edwards was influential in the development of both these ideas - a fact, I am glad to record.) The second deficit greatly complicates our discussion of the s-cobordism theorem. This draft was never proofed by Andrew, has not been updated, and is probably replete with speling erors!

Michael H. Freedman August 1983

0. INTRODUCTION AND PRELIMINARIES

Most efforts to construct smooth four dimensional manifolds can be regarded as an attempt to solve some particular surgery problem with vanishing obstruction. No general theory exists for compact four dimensional surgery problems (although progress has recently been made in the non-compact case, see [F1],

*Partially supported by NSF Grants

0271-4132/84 $1.00 + $.25 per page

[F2], [FQ], and [S]) and the history of effort expended on special cases is discouraging. The only notable success, here, is a technique (See [CS1]) for altering the normal invariants of certain non-orientable 4-manifolds such as RP^4.

One is lead to suspect that many of these surgery problems do not in fact have solutions; for if they did admit solutions why should these always be so difficult to find? However, it is noteworthy that no counterexample is known to the all-encompassing conjecture $A(A^+)$: The surgery-exact-sequence for (oriented) simple Poincaré pairs (X,∂) is exact when $\dim[X,\partial] \geq 4$. It is our purpose to shed some light on this conjecture by reducing the vast diversity of unobstructed four dimensional problems to a smaller collection of "atomic" surgery problems.

In the orientable case a close relationship is developed between atomic problems and certain link slicing problems. This leads to Theorem 2, an equivalent reformulation of conjecture A^+ purely in terms of the classical theory of links in S^3.

In the cases we will consider, the Wall group surgery obstruction vanishes. So, for us, a <u>problem</u> will be a degree one normal map $f:(M^4,\partial) \to (X,\partial)$ from a smooth 4-manifold to simple Poincare space with $\sigma(f) = 0 \in L_4^s(\pi_1 X)$. In the case that the boundaries are non-empty $f|\partial:\partial M \to \partial X$ may not be a homotopy equivalence but is required to induce an isomorphism on $H_*(\ ;\mathbb{Z}[\pi_1 X])$ the homology induced from the universal cover $\tilde{X}$. This requirement implies that the intersection pairing on the kernel $K_2(M^4) \otimes K_2(M^4) \to \mathbb{Z}[\pi_1 X]$ is nonsingular, the necessary condition to define $\sigma(f)$. A solution will mean a normal bordism (rel ∂) to a simple homotopy equivalence.

The choice of generality in this definition has been carefully made. We remark that the problem of h-slicing a knot with Alexander polynomials $\Delta(t) = 1$ (so that π_1(homotopy D^4 - slice) $\cong \mathbb{Z}$) gives rise to a bounded problem f where $f|\partial$ is a $\mathbb{Z}[\pi_1 X] = \mathbb{Z}[\mathbb{Z}]$ equivalence but not, usually, a homotopy equivalence. Also this is the generality in which the stable $(\#n(S^2 \times S^2))$ theory of Shaneson and Cappell [CS2] applies. On the other hand, if $f|\partial$ were required only to be an integral homology equivalence it is known ([CG]) through the study of dihedral signatures that the vanishing of the appropriate "surgery obstruction" (this time lying in a Γ-group [CS3]) is not sufficient to complete surgery up to integral equivalence.

Unless specified to the contrary, constructions are to be carried out in the smooth category; when corners arise they are understood to be rounded in the usual way.

An important notion for us will be: a problem f <u>"reduces to"</u> a problem g, written $f \to g$, this corresponds to finding g inside f, more precisely: We write $f \to g$ iff $f:(M,\partial) \to (X,\partial)$ is normally coborant (rel. ∂) to an $f':(M;\partial) \to (X,\partial)$ such that:

1) There is a (not necessarily connected) codimension-0 smooth submanifold $(N,\partial) \subset$ interior (M), and a simple relative homotopy equivalence: $h:(X,\partial X) \to (X',\partial X)$ such that $g = h \circ f' | (X,\partial): (N,\partial) \to h \circ f'(N,\partial) = (Y,\partial)$ is a problem whose target is a collared Poincare imbedding $(Y,\partial) \subset X'$.

2) $h \circ f'$ is a map of quadruples:

$$h \circ f'(M', N, \overline{M'-N}, \partial M') \to (X, Y, \overline{X-Y}, \partial X)$$

3) The surgery kernel is concentrated in N, i.e. $h \circ f' | \overline{M'-N}: \overline{M'-N} \to \overline{X-Y}$ is a simple absolute homotopy equivalence.

We say f' contains g and use script letters to denote sets of problems. We will write $\mathcal{F} \to \mathcal{G}$ if for each $f \in \mathcal{F}$ there exist $g_1, \ldots, g_n$ such that $f \to g_1 \coprod \cdots \coprod g_n$. Set $\mathcal{Z}(\mathcal{Z}^+)$ = the collection of all (orientable) problems. $\mathcal{A}(\mathcal{A}^+)$ will be the <u>atomic</u> problems (<u>oriented atomic</u> problems). We have given a recipe for constructing a general $a \in \mathcal{A}$ (or $a^+ \in \mathcal{A}^+$).

Consider the four ways of constructing <u>self-plumbings</u> of $S^2 \times D^2$. If i_0 and i_1 are two disjoint product-preserving imbeddings $(D^2 \times D^2) \hookrightarrow S^2 \times D^2$ we may identify: 1) $i_0(a,b) \sim i_1(b,a)$, 2) $i_0(a,b) \sim i_1(\bar{b},\bar{a})$, 3) $i_0(a,b) \sim i_1(\bar{b},a)$, or 4) $i_0(a,b) \sim i_1(b,\bar{a})$ to allow (i_0,i_1) to determine a self-plumbing in one of four possible ways. The first two are <u>oriented self-plumbings</u>. Let $N_2^{j_0,j_1,k_0,k_1}$, or just N_2, denote the 4-manifold with boundary obtained by taking two copies of $S^2 \times D^2$, $(S^2 \times D^2)_0$ and $(S^2 \times D^2)_1$, and performing $(j_\varepsilon, j_\varepsilon, k_\varepsilon$ and $k_\varepsilon)$ self-plumbing of types (1,2,3 and 4) on $(S^2 \times D^2)_\varepsilon$, $\varepsilon = 0$ or 1, and then joining the two copies by a single self-plumbing of type 1. If both $k_0 = k_1 = 0$ we denote the manifold by N_2^+. N_2 collapses to a wedge of singular (immersed) 2-spheres, $N_2 \searrow A \vee B = (S^2 \times 0)_0 / \text{self-plumbings} \vee (S^2 \times 0)_1 / \text{self-plumbings}$.

Suppose that (i_0, i_1) and (j_0, j_1) determine self-plumbings of type 1 and 2 respectively (or 3 and 4 respectively). This pair of self-plumbings determines an imbedded loop $\gamma \subset \partial N_2$ as follows: Let γ_0' and γ_1' be disjointly imbedded arcs in $[S^2 - \text{int}(i_0(D^2 \times 0) \coprod i_1(D^2 \times 0) \coprod j_0(D^2 \times 0) \coprod j_1(D^2 \times 0))]$ with the endpoints $\gamma_0'(0) = i_0(1,0)$, $\gamma_0'(1) = j_0(1,0)$, $\gamma_1'(0) = j_1(1,0)$, and $\gamma_1'(1) = i_1(1,0)$. γ is defined by $\gamma = (\gamma_0', 1) \cup (\gamma_1', 1) \subset \partial N_2$. Evidently there are different choices possible for γ_0' and γ_1' and therefore γ. Let $\gamma_1, \ldots, \gamma_{j_0+j_1+k_0+k_1} = \gamma_\ell$ be a disjoint collection of such γ's for N^2; call this a <u>standard basis</u>.

Here a <u>Kinky handle</u> (<u>oriented kinky handle</u>) will be a 2-handle $D^2 \times D^2$ with interior self-plumbings (of types 1 and 2) with an equal number of types 1 and 2 and of types 3 and 4. (When both numbers are zero we will not call this a kinky handle.)

The symbol $N_{4,i}$ or N_4 will be reserved to denote any 4-manifold with boundary obtained by attaching kinky handles to any N_2 along an appropriately framed standard basis.

The framing is to be determined as follows. Let $\bar{\gamma}$ be the image of γ under the collapse $N_2 \searrow A \vee B$. Let $\mathcal{M}$ be a closed regular neighborhood of $\bar{\gamma}$ in N_2 containing γ in its boundary. $\partial\mathcal{M}(\cong S^1 \times S^2$ since both self-plumbings have the same orientation) meets $A \vee B$ in a pair of circles $c_1 \amalg c_2$ as shown in diagram 1.

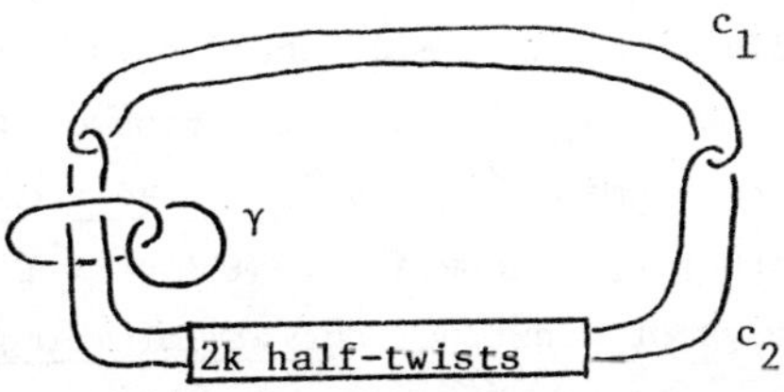

Diagram 1

The circle bearing the dot represents the 1-handle in $S^1 \times S^2$. The number of half twists is even because the sum of the signs of the two self-plumbs are opposite. The appropriate framing for γ is $-k$. N_4 will denote the result of attaching (with 0-framing) oriented kinky handles to an N_2 along an appropriately framed standard basis.

Let s be the total number of type 1 self-plumbings and t be the total number of type 3 self-plumbing in the kinky handles attached to N_2 to form N_4.

CLAIM 1: There is a degree 1-normal map $a:(N_4, \partial) \to ((B \cup s$ (oriented 1-handles) $\cup\, t$(unorientable 1-handles), $\partial) = (Y, \partial)$.

PROOF: This claim corresponds to Lemma 3 [F]; the proof there applies with little modification. ▥

N_4 is simple homotopy equivalent to $S^2\, S^2\, S^1$'s. The inclusion map of kernel modules $K_2(N_4; \mathbb{Z}[\pi_1 Y]) \to K_2(N_4, \partial; \mathbb{Z}[\pi_1 Y])$ is given by the intersection pairing λ on $K_2(N_4; \mathbb{Z}[\pi_1 Y])$. It is easy to see geometrically the two free generators and check that λ is represented by $\begin{array}{c} \\ \alpha \\ \beta \end{array}\!\!\begin{bmatrix} \alpha & \beta \\ 0 & 1 \\ 1 & 0 \end{bmatrix}$ (the kinky handles cancel all self-intersections over the group ring). $\pi_1(N_4) \to \pi_1(Y)$ is an isomorphism so $K_1(N_4; \mathbb{Z}[\pi_1 Y]) \cong 0$; from the long exact sequence of kernel modules $K_1(\partial N_4; \mathbb{Z}[\pi_1 Y]) \cong 0$. It follows from a standard duality argument that:

CLAIM 2: $a|\partial : \partial N_4 \to \partial Y$ is a simple $\mathbb{Z}[\pi_1 Y]$ - equivalence. ▥

Furthermore the self-intersection pairing μ is also made standard ($\mu(\alpha) = \mu(\beta) = 0$) by kinky handles. Thus the surgery obstruction $\sigma(f) \in L_4(\pi_1 Y) = L_4(\text{Free group}) \cong \mathbb{Z}$ vanishes so a is a problem.

We define $\mathcal{A}$ to be the set of all the a's we have just constructed and $\mathcal{A}^+$ to be the set of all orientable a's, $a^+:(N_{4+},\partial) \to (Y,\partial)$.

1. THE REDUCTION TO ATOMIC PROBLEMS

THEOREM 1: $\mathcal{Z} \to \mathcal{A}$ and $\mathcal{Z}^+ \to \mathcal{A}^+$.

PROOF: Let $(f:(M,\partial) \to (X,\partial)) \in \mathcal{Z}$. Preliminary 0 and 1-surgeries may be made to normally cobord f (rel ∂) to $f':(M;\partial) \to (X,\partial)$ with $f'_\#$ an isomorphism on π_1 and $K_*(M',\mathbb{Z}[\pi_1 X]) = K_* = 0$ for $* \neq 2$.

We would like to represent a preferred basis for K_2 by an imbedding of $\coprod N_2$'s $\subset M$. This may be done as follows: Let $(\alpha_1,\ldots,\alpha_n, \beta_1,\ldots,\beta_n)$ be a symplectic basis for K_2. Represent α_1 by a normal framed immersion $a_1:S^2 \longrightarrow M$ with $\mu(a_1) = 0$. Using Casson's Lemma ([F1]) we may arrange that $\pi_1(M-a_1(S^2)) \xrightarrow{\text{inc.}\#} \pi_1(M)$ is an isomorphism. Now we can represent β_1 by a normal framed immersion $b_1:S^2 \longrightarrow M$ with $\mu(b_1) = 0$ and b_1 meeting a_1 in one (transverse) point. Again Casson's Lemma allows us to arrange $\pi_1(M-(a_1(S^2) \cup b_1(S^2))) \xrightarrow{\text{inc.}\#} \pi_1(M)$ to be an isomorphism. A closed regular neighborhood $\mathcal{N}(a_1(S^2) \cup b_1(S^2)) = \mathcal{N}_1 \subset M$ is an imbedded N_2. Proceeding by induction we can represent the hyperbolic pairs $(\alpha_1,\beta_1),\ldots,(\alpha_n,\beta_n)$ by disjoint imbeddings $\coprod^n \mathcal{N}_i \subset M$ with the additional property:

$$\pi_1(M - \coprod_{i=1}^n \mathcal{N}_i) \xrightarrow{\text{inc.}\#} \pi_1(M) \text{ is an isomorphism.}$$

Given an $\subset M$ and a $\bar{\gamma} \subset \mathcal{N}$ as above we must find an appropriately framed γ, with $\gamma \searrow \bar{\gamma}$ to which a kinky handle $(k,\partial) \subset (M - \coprod_{i=1}^n \mathcal{N}_i, \partial \coprod_{i=1}^n \mathcal{N}_i)$ ambiently attaches.

Recall that $h \searrow A \vee B$ and assume that it is self-plumbings of A (say) which are paired by $\bar{\gamma}$. Let B' be a framed immersed 2-sphere which meets $(A_i \vee B_i)$ only in a single transverse point $p \in A$. Consider two γ's, γ^1 and γ^2 as constructed in Section 0 (with $\gamma^1 \searrow \bar{\gamma}$ and $\gamma^2 \searrow \bar{\gamma}$) which differ by one full turn around $i_0(1,0)$ in the choice of γ'_0. If $d^1:(D^2,\partial) \to (M - \coprod_{i=1}^n \mathcal{N}_i, \partial \coprod_{i=1}^n \mathcal{N}_i)$ is a normally immersed null homotopy of γ^1 there is a normally immersed null homotopy d^2 of γ^2 of the form $d^2 = d^1 \# B^1$. Since B^1 is framed d^1 and d^2 induce the same framing on $\mathcal{T}(\mathcal{M})$. Since γ^1 and γ^2 differ by a full twist if $(\partial\mathcal{M};c_1,c_2)$ is put by a diffeomorphism in the form of diagram 2, k will be even in one case and odd in the other. However the framings induced on γ^1 and γ^2 by d^1 and d^2 are equal, so by selecting the correct curve, say γ^1, we ensure that when the framing induced on γ^1 by d^1 is used to trivialize $\partial(\mathcal{M})$ the number of full twists K becomes even. Adding trivial self-intersection in a chart enables us to change the framing d^1 induces on γ by any even number, thus we may assume that $k=0$.

A neighborhood of d^1 cannot yet be used as the desired kinky handle since the number of self-plumbing of types 1 and 2 (and types 3 and 4) may not be equal. A relative Wall form $\mu(d') \in \mathbb{Z}(\pi_1 M)/I$ is defined; it would be sufficient to alter d^1 (without altering the induced framing on γ^1) so that $\mu(d^1)=0$. To do this it is sufficient to find an immersed framed sphere $S \subset (M - \coprod_{i=1}^{n} \mathcal{N}_i)$ with $\mu(S) = \lambda(S) = 0$ and S meeting d^1 in a single transverse point; for then one could set $d^1_{new} = d^1_{old} \# -\mu(d^1_{old})(S)$. There is a distinguished torus (see [F] for definitions) $T \subset \partial\mathcal{N}$ which meets d^1 transversely in a single point. $\pi_1(T) \xrightarrow{\text{inc.}\#} \pi_1(M - \coprod_{i=1}^{n} \mathcal{N}_i)$ is the zero map. It follows immediately that $[T] \in H_2(M - \coprod_{i=1}^{n} \mathcal{N}_i; \mathbb{Z}[\pi_1 X])$ and that $\lambda(T,T)=0$. Geometrically T may be converted into an immersed 2-sphere S by an ambient surger along an immersed 2-disk whose boundary is the meridian (or longitude) of T. As before, $\lambda(S,S)=0$; also counting up self-crossings over $\mathbb{Z}[\pi_1 X]$ (see Diagram 2) shows $\mu(S) = 0$

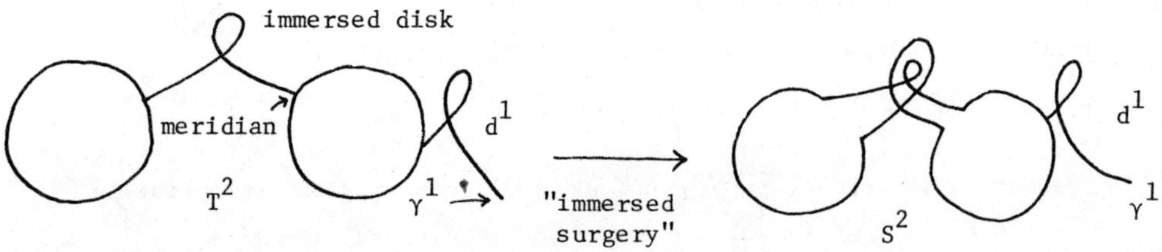

Diagram 2.

Again after a regular homotopy of d^1 we have $\pi_1(M - (\coprod_{i=1}^{n} \mathcal{N}_i \cup d^1)) \to \pi_1(M)$ is an isomorphism.

We can describe a regular neighborhood of $(\mathcal{N} \cup d^1)$ as $\mathcal{N}$ union a kinky handle attached to an appropriately framed γ. Proceeding by induction we are able to prove:

<u>LEMMA 1</u>: <u>The hyperbolic pairs</u> (a_i, b_1) <u>in the kernel group of</u> f', $K_2(M'; \mathbb{Z}[\pi_1 X])$, <u>are represented by disjointly imbedded</u> 4-<u>manifolds</u> $N_{4,i}$ <u>with</u> $\pi_1(M - \coprod_{i=1}^{n} N_{4,i}) \longrightarrow \pi_1(M)$ <u>an isomorphism</u>. ⦀

Let $a_i : (N_{4,i}, \partial) \longrightarrow (Y_i, \partial)$ be the atomic problems with domains $N_{4,1} \cdots N_{4,n}$ which we constructed in Section 0. Set $(a: N \to Y) = \coprod_{i=1}^{n} a_i$. By

Lemma 1 N is a codimension 0 submanifold of M. Lemma 2.8 of [W] and the remark which follows it allows us to find a manifold 1-skeleton for X, that is a simple homotopy equivalence: $h:(X,\partial X) \longrightarrow (X',\partial X)$ where $X' = (\partial X \cup \text{1-cells} \cup \text{2-cells} \cup \text{3-cells}) \cup_{\partial H} (H)$ where H is a smooth manifold with boundary obtained by attaching 1-handles (possibly unoriented) to the 4-ball; $\pi_1(H)$ generates $\pi_1(X')$.

An imbedding: $Y_i \xrightarrow{i} H$, unique up to isotopy, is determined by the map $\pi_1(Y_i) \xrightarrow{a_{i\#}^{-1}} \pi_1(N_i) \xrightarrow{h\circ f|} \pi_1(X')$. After a homotopy of $h \circ f| : N \to X$ is merely the composition $i \circ a : N \to Y \to X$. By alignment (Lemma 4'[F] with the statement generalized slightly to permit nonorientable 1-handles) $h\ f$ is homotopic (rel ∂) to a map of quadruples: $h\circ f:(M, \overline{M-N}, N, \partial M) \longrightarrow (X', \overline{X'-Y}, Y, \partial X)$. Since $K_2(N;\mathbb{Z}[\pi_1 X] \cong K_2(M;\mathbb{Z}[\pi_1 X])$ a Mayer-Vietoris argument shows that $h\circ f|\overline{M-N}$ is a simple $\mathbb{Z}[\pi_1 X]$-equivalence. Furthermore $\pi_1(\overline{M-N}) \xrightarrow{\text{inc.}\#} \pi_1(M) \xrightarrow{f\#} \pi_1(X) \xrightarrow{\text{inc.}\#} \pi_1(\overline{X'-Y})$ are all isomorphisms so $h\ f\ \overline{M-N} \longrightarrow \overline{X'-Y}$ is actually a simple homotopy equivalence. Thus $f \to a$.

If f is an orientable problem f^+ then the a we have constructed is a^+, also orientable. So we have also shown that $f^+ \longrightarrow a^+$. ⁞⁞

<u>COROLLARY 1</u>: <u>If the problems in $\mathcal{A}$ ($\mathcal{A}^+$) all admit solutions then all problems (problems in $\mathcal{Z}^+$) admit solutions.</u>

<u>PROOF</u>. Let $f \in \mathcal{Z}$. By the theorem there is a normal bordism (rel. ∂) B from f to f' with f' containing some $a \in \mathcal{A}$. If B' be some solution to a then it is easily checked that $B \cup_{\text{domain } a} B'$ is a solution to f. ⁞⁞

<u>REMARK 1</u>: Theorem 1 shows that problems over free groups are sufficiently general to capture any surgical phenomena that may be peculiar to dimension four.

<u>REMARK 2</u>: Since there are few people who believe that all problems admit solutions it is worth noting that the implication in Corollary 1 holds for the weaker notion of Λ-solution. Let Λ be a functor from groups with an augmentation into Z_2 to algebras with augmentation. A Λ-solution to a problem is just a normal bordism (rel. ∂) to a simple Λ-homology isomorphism.

<u>REMARK 3</u>: There is some arbitrariness in our choice of $\mathcal{A}$ and $\mathcal{A}^+$. A little of the work of theorem one would be saved if we had settled for larger class in which we abandon the framing assumption on the attached kinky handles and the requirement that the self-plumbings of a kinky handle be paired. Our choice of $\mathcal{A}$ and $\mathcal{A}^+$ is motivated by the simplicity of the corresponding link-diagrams (see [F1]). Another potentially useful feature of the N_4's as we have constructed them is that they are amenable to a continued extension to N_6's, N_8's etc... analogous to the $M_4, M_6, \ldots$ of [F1]).

<u>REMARK 4</u>: A smaller (but still countably infinite) sub class of $\mathscr{A}(\mathscr{A}^+)$, $\mathscr{A}_-(\mathscr{A}_-^+)$ can be used in place of $\mathscr{A}(\mathscr{A}^+)$. A typical problem $a_- \in \mathscr{A}_-$ has its domain an N_4 constructed as follows: Begin with either $N_2^{0,j_1,1,K_1}$ or $N_2^{1,j_1,0,k_1}$. Form N_4 by attaching kinky handles which have only a single pair of self-plumbings to a standard basis $\gamma_1,\ldots,\gamma_{1+j_1+k_1}$. We outline this reduction.

As in the proof of Theorem 1 bord f to f' and represent K_2 by arbitrary N_4's. In a given N_4 as long as some kinky handle has more than a single pair of double points, a pair of double points on either A or B can be created in such a way that one sees a new collection of kinky handles on a new standard basis with the total number of self-plumbings in the kinky handle unchanged. Diagram 3 illustrates this "un-Whitney move ".

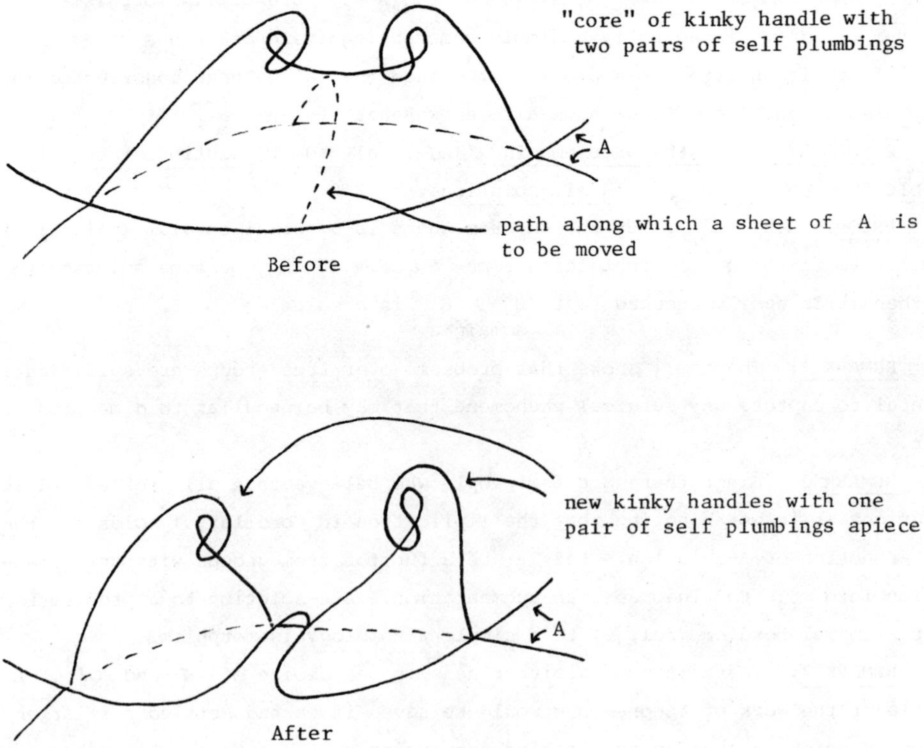

Diagram 3

By induction we end with an N_4 in which all kinky handles have a single pair of self-plumbings.

To achieve $j_0 = 0$, and $k_0 = 1$ (or $j_0 = 1$ and $k_0 = 0$) it is necessary to further bord f' by additional 1-surgeries. A representative surgery and its effect on K_2 is illustrated in Diagram 4.

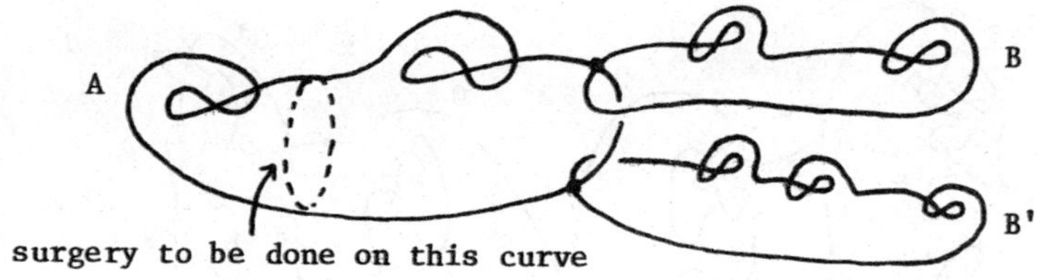

Before

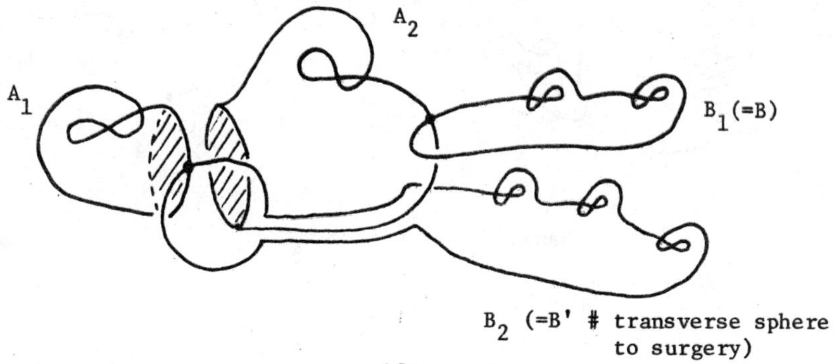

After

Diagram 4

2. THE LINK SLICING PROBLEMS ASSOCIATED TO $\mathcal{A}^+$

Henceforth $N_{4,a}$ (or just N_a) denotes the domain of a problem $a \in \mathcal{A}^+$. To understand the relation between a and link slicing problems it will be necessary to produce a handle decomposition for N_4 analogous to the handle decomposition of M_4 given in Section 1 of [F1]. The situations are quite similar so the handle diagram for N_4 is given below without further justification.

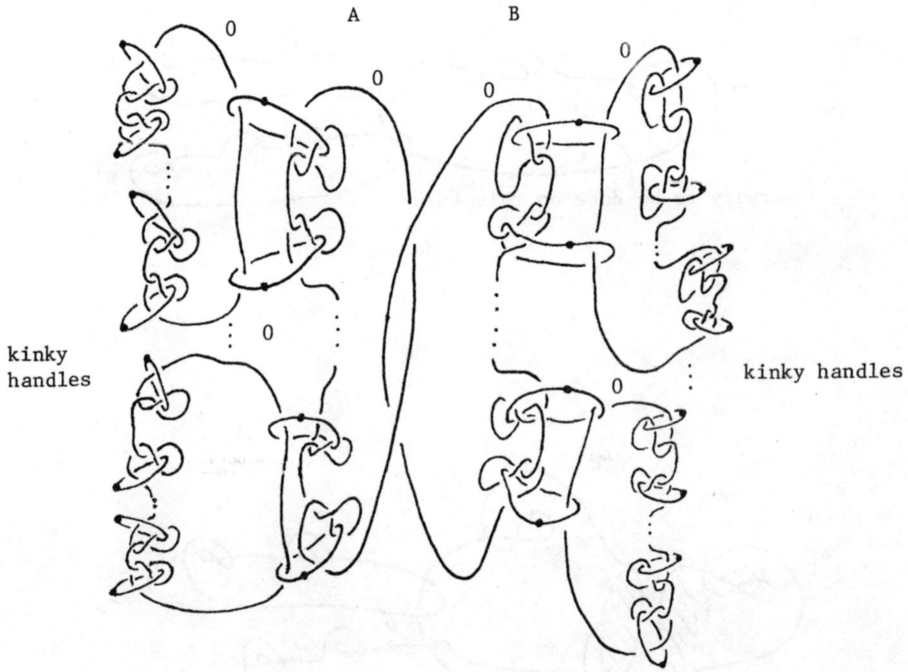

Handle Diagram for N_4

Diagram 5

Henceforth the possible repititions (represented by dots in Diagram 5) will be omitted from the illustrations. Two diagrams, or the links which constitute them will be called similar if they differ only by the admission or omission of such repititions. Using this conversion and after passing 1-handles Diagram 5 becomes Diagram 6.

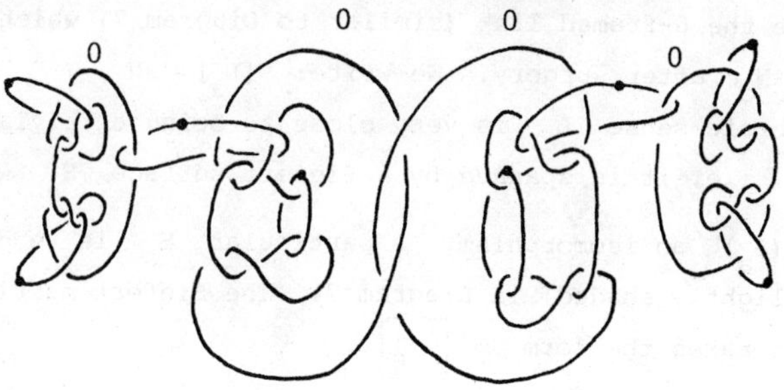

Diagram 6

Changing 1-handles to 2-handles and then manipulating 2-handles (see [K] for the rules of handle calculus) we get Diagram 7 representing ∂N_a.

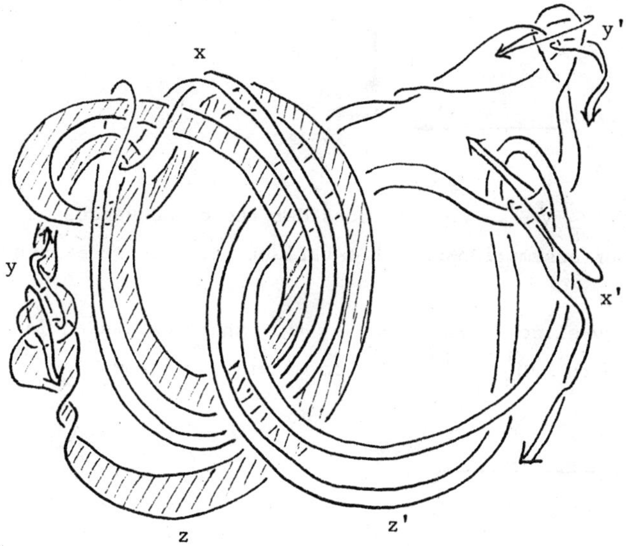

0-framed surgery on L_a, $[L_a] = \partial N_a$
(all up to similarity)

Diagram 7

Let L_a be the 0-framed link (similar to Diagram 7) which gives rise to $\partial(\text{domain }(a)) = \partial N_a$ after surgery. We write: $[L_a] = \partial N_a$.

In an algebraic sense L_a is very close to being a trivial link. L_a is a boundary link, i.e. it is spanned by a Siefert surface S_a with $\pi_0(L_a) \xrightarrow{\text{inc.}_\#} \pi_0(S_a)$ an isomorphism. A particular S_a is readily visible (half of it is lightly shaded) in Diagram 7. The Siefert matrix for S_a is trivial, i.e. it takes the form $\oplus \begin{bmatrix} 0 & 1 \\ 0 & 0 \end{bmatrix}$.

Thus L_a is "algebraically slice" in the strong sense; in the higher dimensions $(\coprod S^{4k+1} \xrightarrow{\bar{L}_a} S^{4k+3},\ k \geq 1)$ this data would imply that there was a slice s_a, i.e. a commutative diagram:

$$\begin{array}{ccc}
\coprod S^{4k+1} & \xrightarrow{L_a} & S^{4k+3} \\
\partial \downarrow & & \partial \downarrow \\
\coprod D^{4k+2} & \xrightarrow{s_a} & D^{4k+4}
\end{array}$$

with $[D^{4k+4} - \bar{s}_a(\coprod D^{4k+2})] \simeq \bigvee S^1$. We will consider the consequences of three progressively stronger assumptions, each a variant of: "There exists an s_a slicing L_a".

ASSUMPTION 1 (homology-slice): For each $a \in \mathcal{A}^+$ there exists an integral homology 4-ball D^4_H with $\partial D^4_H = S^3$ and a commutative diagram

$$\begin{array}{ccc}
\coprod S^1 & \xrightarrow{L_a} & S^3 \\
\partial \downarrow & & \partial \downarrow \\
\coprod D^2 & \xrightarrow{s_a} & D^4_H
\end{array}$$

ASSUMPTION 2 (Algebraically ribbon): In addition to Assumption 1, we require that the inclusion map $[L_a] = (\partial \bar{\mathcal{N}}(S^3 \cup s_a(\coprod D^2)) \longrightarrow (D^4_H - s_a(\coprod D^2))$ induces an epimorphism on fundamental groups.

ASSUMPTION 3 (strongly slice): In addition to Assumption 2, we require that D^4_H is diffeomorphic to D^4 and that s_a restricts to the standard slice on the unknotted and unlinked collection of components $\{x, y, x', y',$ and all similar components$\}$ as shown in Diagram 7.

PROPOSITION 1: Assumption 1 implies every oriented problem has a Λ-solution where $\Lambda(G)$ is the integers with the trivial action of G.

PROOF: This follows from Remark 2 once $(D^4_H - \mathcal{N}(S^3 \cup S_a(\coprod D^2))$, $\partial = [L_a] = \partial N_a) = (Q,\partial)$ is identified as the upper boundary of some B', a $\Lambda = \mathbb{Z}$-solution of a. Given a framing $\mathcal{F}$ for Q. The obstruction $\mathcal{O}$ to constructing a B' with $\partial B' = N \underset{\partial N \cong \partial Q}{\cup} Q$, the isomorphism being the canonical one given by the passage from Diagram 5 to Diagram 7, lies in the 4-dimensional framed bordism of a wedge of circles $F_4(\vee^n_{i=1} S^1_i) \cong \bigoplus_{i=1}^n \pi_3^{\text{stable}}$. We may arrange $\mathcal{O} = 0$ by rechoosing $\mathcal{F}$ near an imbedded wedge of circles in Q. This constructs B' and completes the proof. ⦙⦙

PROPOSITION 2: Assumption 2 implies every oriented problem has a solution.

PROOF: First construct a degree one normal map from $(Q,\partial N)$, a': $(Q,\partial N) \longrightarrow (Y,\partial)$, normally bordant (rel ∂) to a. It follows from Claim 2 (Section 0) that the kernel $k_\partial = \text{kernel } (\pi_1(\partial N) \longrightarrow \pi_1(\partial Y))$ is perfect. Assumption 2 gives commutative diagram

$$\begin{array}{ccc} \pi_1(\partial N) & \longrightarrow & \pi_1(\partial Y) \\ \downarrow \text{epi} & & \downarrow \cong \\ \pi_1(Q) & \longrightarrow & \pi_1(Y) \end{array} .$$

From this we see that $k = \text{kernel } (\pi_1(Q) \longrightarrow \pi_1(Y))$ is a quotient of k_∂ and therefore a perfect group. Thus $K_1(Q;\mathbb{Z}[\pi_1 Y]) \cong 0$. Set $K_2 = K_2(Q;\mathbb{Z}[\pi_1 Y])$.

K_2 is the first nonvanishing kernel group and is therefore stably-free. According to a calculation of Bass [B] $\tilde{K}_0(\mathbb{Z}[\text{free group}]) \cong \underset{\text{copies}}{\oplus} \tilde{K}_0(\mathbb{Z}[\mathbb{Z}]) \cong 0$ so K_2 is actually free. If $\text{rank}_{\mathbb{Z}[\pi_1 Y]} K_2 > 0$ there would be a generator x and an element y, $x,y \in K_2$, with $\lambda(x,y) = 1$ (by nonsingularity of λ). But all intersection numbers must be zero when reduced to $\mathbb{Z}$ since $H_2(Q;\mathbb{Z}) = 0$; so $K_2 \cong 0$.

By duality all the kernel groups vanish; a' is a Λ-solution to a where $\Lambda(G) = \mathbb{Z}[G]$. Remark 2 may now be used to find a Λ-solution $f'': (\bar{M}' - N \cup Q),\partial) \longrightarrow (X,\partial)$ to any $f \in \mathcal{Z}^+$. The following Van Kampen diagram computes kernel $[f''_\#: \pi_1(M'') \longrightarrow \pi_1(X)]$:

$$\begin{array}{ccc} \ker(\pi_1 \partial N) & \longrightarrow & \ker(\pi_1 Q) \\ \downarrow & & \downarrow \\ 0 \cong \ker(\pi_1(\overline{M'-N}) & \longrightarrow & \ker(\pi_1 M'') \end{array}$$

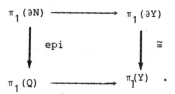

Part 3 in the definition of $f \to g$ yields $\ker(\pi_1(\overline{M'-N}) \cong 0$; Assumption 2 forces the top arrow to be an epimorphism. The pushout $\ker(M'')$ is necessarily trivial. It follows that f'' is actually a solution to f. ⦙⦙

3. THE EQUIVALENCE OF ASSUMPTION 3 AND CONJECTURE A^+

It is an open question of some interest whether there is a smooth knot $k: S^2 \longrightarrow \text{homotopy}(S^4) = S^4_h$ with $\pi_1(S^4_h - k(S^2)) \cong \mathbb{Z}$ and Rochlin invariant $(k) = 1 \in \mathbb{Z}_2$.

<u>OBSERVATION 1:</u> <u>Assumption 2 implies the existence of a knot k with the above properties.</u>

<u>PROOF:</u> Proposition 2 allows us to solve the surgery problem (see [F1]) which constructs a simply connected homology H-cobordism C with $\partial C = \Sigma^3 \coprod -\Sigma^3$ where Σ^3 is the Poincare homology sphere. Let $\bar{C}$ be C with ends identified. Let $\bar{\bar{C}}$ be the result of a framed 1-surgery on the generator of $\pi_1(\bar{C}) \cong \mathbb{Z}$. $\bar{\bar{C}}$ is a homotopy 4-sphere. If the surgered circle is arranged to meet Σ^3 in a single point then its linking 2-sphere is clearly a knot of Rochlin invariant = 1 in $\bar{\bar{C}}$. Furthermore $\pi_1(\bar{\bar{C}}\text{-linking sphere}) \cong \pi_1(\bar{C}\text{-circle}) \cong \pi_1(\bar{C}) \cong \mathbb{Z}$. ⦙⦙

<u>PROPOSITION 3:</u> <u>Assumption 3 implies conjecture</u> A^+: <u>The surgery-exact sequence:</u>

$$L_5^s(\pi_1 X) \xrightarrow{\text{acts}} \mathcal{S}^s(X,\partial) \longrightarrow \mathcal{N}(X,\partial) \xrightarrow{\sigma} L_4^s(\pi_1 X)$$

<u>is exact for oriented simple Poincare pairs</u> (X,∂) <u>when</u> $\dim[X,\partial] = 4$.

It is only necessary to verify exactness at $\mathcal{S}$, the smooth structures (rel. ∂) on X, since Proposition 2 already implies exactness at the normal maps $\mathcal{N}$.

In the proof of Proposition 1, a bordism B' is constructed. With a little care, B' can be constructed to be the domain of a surgery problem: $g: (B', N \cup Q) \longrightarrow (D^3 \times S^2 \natural_{i=1}^{n} (D^4 \times S^1)_i, \partial)$ with g restricted to one singular sphere $g|A$ null homotopic.

By a splitting argument the surgery obstruction $\sigma(g)$ is a collection of signatures. Without changing $\partial B' = \partial N \cup \partial Q$ we are free to vary these signatures by an even integer. If some of these signatures are odd we can change $Q(\text{rel } \partial)$ to make $\sigma(g) = 0$. This is done by altering the slice S_a where necessary by a connected sum of pairs $(D^4_H, s_a(\coprod D^2) \# (S^4_h, k(S^2))$ where k is the knot constructed in Observation 1. Since $\pi_1(S^4_h - k(S^2)) \cong \mathbb{Z}$ our assumption 2, that $\pi_1(\partial Q) \to \pi_1(Q)$ is an epimorphism, is preserved. Thus g has vanishing obstruction. By high dimensional surgery B' is normally bordant (rel ∂) to a simple homotopy equivalence:

$$g': (B'' \; N \cup Q) \longrightarrow (D^3 \times S^2 \mathop{\sqcup}_{i=1}^{n} (D^4 \times S^1)_i, \partial)$$

The proof of Proposition 2 (Section 1) enables us to represent any kernel by

$$\langle a_1, \ldots, a_n, b_1, \ldots, b_n \rangle$$

an imbedded $\coprod N$. Attaching B' to B (as in the proof of Corollary 1) along $\coprod N$ will kill the subkernel basis $\{a_1, \ldots, a_n\}$ and do nothing else on homology with coefficients $\mathbb{Z}[\pi_1 X]$. The usual argument (see Ch. 10 [W]) that L_{4k+1} acts on $\mathscr{S}(X^{4k}, \partial)$ now applies. Thus it follows from Assumption 2 alone implies that the surgery exact sequence is exact if $\mathscr{S}^s(X^4, \partial)$ is interpreted as relative s-cobordism classes of simple homotopy equivalences to X (rel ∂). To complete the proof of Proposition 3 we must prove:

<u>PROPOSITION 3'</u>: <u>Assumption</u> 3 <u>implies the</u> 5-<u>dimensional relative (oriented)</u> s-<u>cobordism theorem.</u>

The proof of the s-cobordism theorem in higher dimensions may be followed until the following difficulty on a mid-level 4-manifold is reached.

<u>PROBLEM</u>: Let M be an oriented 4-manifold (possibly with boundary) and k an integer > 0. Let $\langle a_1, \ldots, a_k \rangle$ and $\langle b_1, \ldots, b_k \rangle$ be two disjointly imbedded collections of spheres (the ascending and descending spheres) with the intersections over $\mathbb{Z}[\pi_1 M]$ given by $\lambda(a_i, b_i) = \delta_{ij}$. We know $\pi_1(M - \coprod_{i=1}^{k} a_i) \longrightarrow \pi_1(M)$ and $\pi_1(M - \coprod_{i=1}^{k} b_i) \longrightarrow \pi_1(M)$ are isomorphisms. The problem is to find an isotopy of $\coprod_{i=1} b_i$ to $\coprod_{i=1} b_i'$ with $a_i \cap b_j' = \emptyset$ for $i \neq j$ and a_i and b_i meeting transversely in one point for all $1 \leq i \leq k$.

In the presence of Assumption 3 we will solve this problem, completing the proof of Proposition 3'. The strategy is to solve the problem in a manifold $\bar{M}$ arrived at by cutting and pasting $M \# k(S^2 \times S^2)$, in the end $\bar{M}$ is identified as the original M. For convenience we have outlined the steps as if $k = 1$.

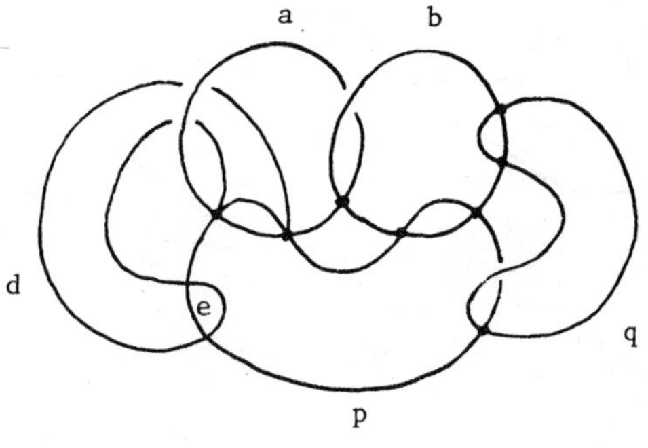

Diagram 8

Step 1: Add a copy of $S^2 \times S^2$ to M. Let p,q be the new transverse spheres in $M \# S^2 \times S^2$. Corollary 2.5 [Q] and the following remark provides an isotopy of $M \# S^2 \times S^2$ which carries b to a sphere b' with $a \cap b' =$ one transverse point. The cost is that b now intersect $p \vee q$.

Step 2: A homological calculation (using $\mathbb{Z}[\pi_1 M]$ coefficients) shows that $\pi_1(M-(a \vee b \cup p \vee q))$ is perfect. "Casson moves" (see added in proof [F1]) creating pairs of double points in $a \cap p$ and $a \cap q$ make $\pi_1(M-(a \vee b \cup p \vee q)) \cong 0$. We show how to eliminate one pair of double points of $a \cap P$ (or $a \cap q$, $b \cap p$, $b \cap q$) without introducing new intersections into these sets.

Step 3: Let d be an immersed (framed) Whitney disk [FQ] pairing an algebraically cancelling pair of intersection in $a \cap p$. Assuming $\text{int}(d) \subset M-(a \vee b \cup p \vee q)$ and (after Casson moves) that its deletion does not change π_1 (complement). Make d imbedded by pushing out self-intersections of d to form pairs of intersections in $d \cap p$. Add a new layer of immersed Whitney disks pairing algebraically cancelling pairs of intersections of d and p. Again to simplify notation, we consider the case of one disk e. As with d we arrange e to be imbedded, $e \cap (a \vee b \cup q)) = \emptyset$, $e \cap p$ consists of cancelling pairs. Also $\text{int}(e) \cap d = \emptyset$ and $\pi_1(M - a \vee b \cup p \vee q \cup d \cup e) \cong 0$. Now make $e \cap p = \emptyset$ by pushing sheets of p off the part of ∂e lying on p. p is now only immersed; call it p'.

Step 4: Push p' by the "Whitney trick" across e. p' will now "link" d in the following sense. The natural Whitney disks (call one f) for killing the double point pairs introduced into p (when it became p') will intersect d.

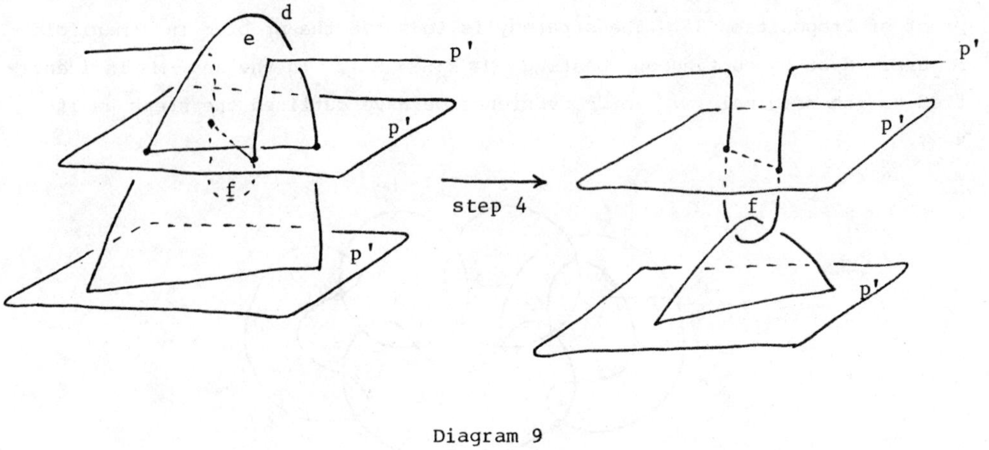

Diagram 9

Step 5: Push p_2' across d by the "Whitney trick". The interior of f lies in $(M - a \vee b \cup q)$.

The intersection $p' \cap \operatorname{int}(f)$ may be arranged in pairs that cancel in the group ring. In fact there is an f' which differs from f only by Casson moves and an isotopy of its interior so that $p' \cap \operatorname{int}(f') = \emptyset$. f' is found by doing a "singular Whitney trick" along an immersed disk. (Each double point of the immersed Whitney disk corresponds to four Casson moves.)

Step 6: A regular neighborhood of $(p' \vee q \cup f')$ is an N_4^+ manifold, $N \subset M - a \vee b$ with $\pi_1(M - (a \vee b \cup N)) \to \pi_1(M)$ an isomorphism). Furthermore we may assume $N \subset (S^2 \times S^2 - D'^4) \natural_{\text{copies}} S^1 \times D^3 = W$, the original $S^2 \times S^2 - \dot{D}^4$ summand union thickened arcs. To see this note that p' is made from p by Casson moves, i.e. pushing along an arc, f' is made from the newly created Whitney disk f by more pushing along arcs. W is merely a regular neighborhood of $(S^2 \times S^2 - D^4 \cup \text{arcs})$. Assume, for the sake of a canonical form, that each arc minus $S^2 \times S^2 - D^4$ is a (nonempty) closed interval.

Step 7: Retrace steps 1 through 6 in the proper generality; pair $a_i \cap p_j$, $a_i \cap q_j$, $b_i \cap p_j$, and $b_i \cap q_j$, $1 \le i, j \le k$, with d's and $d \cap p_i$ and $d \cap q_i$ with e's. The result will be a regular neighborhood, $N = \mathcal{N}(p_i' \vee q_i' \cup f'\text{'s}) \subset M - \coprod_{i=1}^{k} (a \vee b)$. Furthermore we will have an inclusion $N = \coprod_{i=1}^{k} N_i \subset \coprod_{i=1}^{k} [(S^2 \times S^2 - D^4) \natural_{\text{copies}} S^1 \times D^3]_i = \coprod_{i=1}^{k} W_i = W$ analogous to the one constructed above.

Starting with Diagram 7 one may arrive at the following handle-body description of any component of $\overline{W-N}$ $\overline{W_i - N_i} = [\#_{\text{copies}} (S^1 \times S^2) \times [0,1]] \cup_{\partial^+}$ 2-handles. The index set counts the total number of above mentioned arcs or equivalently, I contains an index for each curve of type x, y, x', or y' in Diagram 7. If we let the x, y, x', y', represent the 1-handles of $[\#_{\text{copies}} (S^1 \times S^2)_I] \times 1$ then the 2-handles are attached with 0-framing to the 1-level along the curves of type z and z' in Diagram 7.

Assumption 3 provides standard slices for the x, y, x', and y'. So Q is actually the closed complement of slices for the z and z' in $\#_{\text{copies}} S^1 \times D^3$. The 2-handles of $\overline{W_i - N_i}$ fit into the missing slices so $\overline{W_i - N_i} \underset{\substack{\text{can.} \\ \partial N_i \cong \partial Q_i}}{\cup} Q_i = \#_{\text{copies}} S^1 \times D^3$.

So $\bar{M} = (\overline{M_k - N}) \cup Q = (\overline{M_k - W}) \cup (\overline{W-N} \cup Q) = (\overline{M_k - W}) \cup (\#_{\text{copies}} (S^1 \times D^3)) = (M_k - k(S^2 \times S^2 - D^4)) \cup kD^4 = M$. We have written M_k for $M \# k(S^2 \times S^2)$ and

Q for $\coprod_{i=1}^{k} Q_i$. This completes the proof of Propositions 3' and 3. ⁞⁞

THEOREM 2: The surgery exact sequence of Proposition 3 is exact if and only if Assumption 3 is true.

PROOF: By Proposition 3 it is only necessary to assume exactness and then construct for each a the desired slices $s_a|$ (Components similar to z and z') in $\natural_{\text{copies}} S^1 \times D^3$.

The slice s_a will be found by constructing its closed complement Q. Exactness at $\mathcal{N}(Y,\partial)$ guarantees a solution to $a:(N,\partial) \to (Y,\partial)$. Let $h:(Q,\partial) \to (Y,\partial)$ be the resulting simple homotopy equivalence. Let $T = Q \cup$ 2-handles be obtained by attaching 2-handles to Q along small 0-framed linking circles to the components of types z and z' in Diagram 7. ∂T is now diffeomorphic to $\natural_{\text{copies}} (S^1 \times S^2)$ and the co-cores of the newly attached handles constitute slice $s_a|(\coprod D^2)$ in T for the curves of types z and z' lying on ∂T. $\pi_1(T - s_a(\coprod D^2)) \cong \pi_1(Q) \cong$ Free group so $s_a|$ extends by adding standard 2-handles to T to a slice s_a of $L_a \subset S^3$ which is "algebraically ribbon".

To verify Assumption 3 we must show that (T,∂) can be chosen to be diffeomorphic to $(\natural_{\text{copies}} (S^1 \times D^3),\partial) = (U,\partial)$. h may be extended (by attaching 2-handles to both sides) to a simple homotopy equivalence relative the identity map on ∂, $h':(T,\partial) \to (U,\partial)$. The normal invariant $h(h';\partial) \in \bigoplus_{\text{copies}} \pi_3(G/O) \oplus \pi_4(G/O)$ is signature$(T) = 0$ so h' is normally cobordant relative boundary to the identity. We have $\bar{h}:(B;T,U) \to (U \times [0,1], U \times 0, U \times 1)$, $\bar{h}|T = h' \times 0$, $\bar{h}|U = id_T \times 1$. The surgery obstruction $\sigma(\bar{h}) \in L_5^s$ (free group) on the normal bordism belongs to $H^1(\natural_{i\in I} (S^1 \times D^3); \mathbb{Z})$ and is determined by the signatures of $\bar{h}^{-1}(* \times D^3 \times [0,1])$.

Let C be the manifold in Observation 1. Let $\hat{C} = E^8 \cup_{\partial^+ C} C \cup_{\partial^- C} -E^8$ be C with a copy of the E^8-plumbing glued to its upper and lower boundaries. Let $\hat{\hat{C}}$ be an s-framed 5-manifold $\partial\hat{\hat{C}} = \hat{C}$. Set $H = \hat{\hat{C}}/E^8 \equiv -E^8$. A new normal bordism B' can be formed with $\sigma(\bar{h}) = 0$ by adding copies of H to B. A copy of H is added to B by identifying a framed normal bundle of a circle γ in $\partial^o B = T$ with the framed normal bundle of a generator of $H_1(\hat{C};\mathbb{Z})$. This changes $\sigma(\bar{h})[\gamma]$ by ± 1. $\partial^o B' = T'$ is no longer equal T but results from changing T near a collection of circles. By general position these circles do not meet the slices $s_a|(\coprod D^2)$ and change does not affect any fundamental groups. So assume our slices $s_a|(\coprod D^2)$ lie in T'. Let h'' be the new simple homotopy equivalence.

Since $\sigma(h'')$ is now zero, surgery on h''(rel ∂) produces a relative s-cobordism of $h':(T;\partial) \to (U,\partial)$ to the identity. An application of the s-cobordism theorem completes the proof of Theorem 2. ⁞⁞⁞

REMARK 5: A nonorientable version of Theorem 2 may be proved. However, we would lose the close relationship to links in S^3; the corresponding Assumption 3 would involve the slicing of certain links in nonorientable handle bodies.

This close relationship between surgery and link theory should provide moral support to those who study 4-manifolds through link-diagrams. It might be hoped that information can be recovered in either field by going forward and then backwards along the equivalence. We close with such an example.

EXAMPLE: Suppose that for all a $\partial N_a = \partial Q_a$ for some Q_a with $\pi_1(\partial Q_a) \to \pi_1(Q_a)$ an epimorphism and Q_a $\mathbb{Z}[\pi_1 Y]$ - equivalent to a wedge of circles (i.e. all a are algebraically ribbon). Then all a satisfy $\partial Q_a = \partial Q'_a$ with Q'_a homotopy equivalent to a wedge of circles. The assumption, by Proposition 2 yields exactness at $\mathcal{N}(Y,\partial)$; exactness at $\mathcal{N}(Y,\partial)$ provides Q'_a.

BIBLIOGRAPHY

[B] H. Bass, Projective modules over free groups are free, notes, Columbia University (1964).

[CS1] Sylvain Cappell and Julius Shaneson, Construction of some new four-dimensional manifolds, Bull. AMS. Vol. 82(1976) no. 1, pp. 69-70.

[CS2] ———————— On four-dimensional surgery and applications, Comm. Math. Helv., Vol. 46(1971), pp. 500-528.

[CS3] ———————— The codimension two placement problem and homology equivalent manifolds, Ann. of Math., Vol. 99(1974), pp. 277-348.

[C6] Andrew Casson and Cameron Gordon, Classical knot cobordism, (to appear). Also the first authors address at ICM (1978), Helsinky.

[F1] Michael Freedman, A fake $S^3 \times R$, to appear in Ann. of Math. (1979).

[F2] ———————— A surgery sequence in dimension four, the relations with knot theory, (to appear).

[FQ] ———————— and Frank Quinn, Slightly singular 4-manifolds, (to appear).

[K] Robion Kirby, A calculus for framed links, Invent. Math., Vol. 45(1978), pp. 35-56.

[S] Lawrence Siebenmann, Amorces de la Chirugie en dimension quatre: Un $S^3 \times R$ exotique, Seminare Bourbaki no. 536 (1978-79).

[W] C. T. C. Wall, Surgery on Compact Manifolds, Academic Press 1970.

[Q] Frank Quinn, On the stable topology of 4-manifolds, (to appear).

DEPARTMENT OF MATHEMATICS
UNIVERSITY OF TEXAS
AUSTIN, TX 78712

DEPARTMENT OF MATHEMATICS
UNIVERSITY OF CALIFORNIA, SAN DIEGO
LA JOLLA, CA 92093

Contemporary Mathematics
Volume 35, 1984

SMOOTH 4-MANIFOLDS WITH DEFINITE INTERSECTION FORM

S. K. Donaldson

This is a summary of the lectures at the Durham conference that were devoted to the application of gauge theory to 4-manifold topology, in the form of the following result:

THEOREM. If X is a smooth, compact, simply connected, oriented 4-manifold with positive definite intersection form, then that form is equivalent over the integers to the standard diagonal form.

A detailed account of the proof is now in preparation so I shall attempt here to give an easily accessible presentation of the ideas used. The first six sections give the definitions and basic properties of self-dual connections, and these are used in Section Seven to prove the theorem.

SECTION ONE. SELF-DUALITY

There is a local isomorphism of Lie Groups:

$$SO(4) \approx SO(3) \times SO(3)$$

which gives 4-dimensional Riemannian geometry certain special features. The isomorphism can be realized by the natural decomposition:

$$\Lambda^2\mathbb{R}^4 = \Lambda^2_+\mathbb{R}^4 \oplus \Lambda^2_-\mathbb{R}^4 .$$

into the 3-dimensional eigenspaces of the *-operator, the "self-dual" and "anti self-dual" parts, so that for $\alpha \in \Lambda^2_\pm$:

$$\alpha \wedge \alpha = \pm(\alpha \wedge *\alpha) = \pm|\alpha|^2 \cdot \mathrm{vol}. \qquad (1)$$

If Y is any compact oriented Riemannian 4-manifold this decomposition applies to each of the cotangent spaces to decompose the space of 2-forms:

$$\Omega^2(Y) = \Omega^2_+(Y) \oplus \Omega^2_-(Y).$$

The Laplacian $\Delta = dd^* + d^*d$ commutes with $*$ so we get a similar decomposition of the harmonic 2-forms:

$$\mathcal{H}^2(Y) = \{\alpha \in \Omega^2 \mid \Delta\alpha = 0\} = \mathcal{H}^2_+ \oplus \mathcal{H}^2_- .$$

which, by the defining property (1) and the Hodge Theory (see [8], for this) reflects the signature of the intersection form:

$$H^2(Y;\mathbb{R}) \cong \mathcal{H}^2 \quad ; \quad \tau(Y) = \dim \mathcal{H}_+^2 - \dim \mathcal{H}_-^2 = b_2^+ - b_2^-$$

So if in particular $Y = X$ is the manifold of the theorem, given some metric, then $\mathcal{H}_-^2(X) = 0$. Similarly the first order operator

$$(d^* + d^-) : \Omega^1(Y) \to \Omega^0(Y) \oplus \Omega_-^2(Y)$$

(d^* adjoint to d, d^- the anti self-dual part of d) has

$$\operatorname{index}(d^* + d^-) = b_1 - 1 - b_2^- . \tag{2}$$

SECTION TWO. CONNECTIONS AND CURVATURE

If V is a complex vector bundle over Y, a connection A on V can be defined to be a family of "horizontal subspaces" in the total space, or dually by the associated differential operator

$$\begin{gathered} d_A : \Omega^0(Y;V) \to \Omega^1(Y;V) \\ d_A(f \cdot s) = df \otimes s + f d_A s . \end{gathered} \tag{3}$$

So in any local frame $S = (s_1, \ldots, s_n)$ for V, A is represented by a matrix of 1-forms A^S: $d_A s_i = \sum_{j=1}^{n} A_{ij}^S \otimes s_j$. (This definition follows [3] Appendix C).

If V has extra structure we can define connections that respect that structure.

For any two connections A, B on V the defining property (3) implies that $d_A - d_B$ is an algebraic operator, so $A-B$ is naturally defined as an element of $\Omega^1(Y;\operatorname{End} V)$. Similarly the curvature $F(A)$ of a connection is an element of $\Omega^2(Y;\operatorname{End} V)$ given in the local frame S by the matrix of 2-forms:

$$F^S = dA^S - A^S \wedge A^S \tag{4}$$

There is an obvious notion of isomorphism for connections and we shall eventually be interested in connections up to this equivalence or, more formally, in the quotient of the affine space $\mathcal{A}$ of all connections on V by the "gauge group" $\mathcal{G}$ of automorphisms of V.

The self-dual decomposition extends to bundle valued forms and combines very naturally with the geometry of connections. If A is a connection on a unitary bundle V the "Yang-Mills action" of A is defined to be:

$$\|F\|^2 = \int_Y |F|^2 d\mu = \int_Y |F_+|^2 + |F_-|^2 d\mu$$

(Here $F(A) = F = F_+ + F_-$)

(This was first defined in Mathematical Physics, the associated variational equations are the Yang-Mills equations). Whereas the integral

$$\int_Y |F_+|^2 - |F_-|^2 d\mu = \int_Y \operatorname{Tr}(F \wedge *F)$$

is a characteristic number, a topological invariant of the bundle V. We say that a connection A is <u>self-dual</u> if it has self-dual curvature:

$$F_-(A) = 0 \tag{5}$$

and so for such connections the two integrands above are identical. Henceforth we will be concerned only with bundles of rank 2 and with structure group $SU(2)$. The topological classification of such bundles over the 4-manifold Y is by the integer $c_2 = c_2(V)[Y]$, and for a self-dual connection on V we have:

$$1/8\pi^2 \, \|F\|^2 = -c_2 \geq 0 .$$

If $c_2 = 0$ then any self-dual connection must be a flat connection with vanishing curvature, such are in (1-1) correspondence with their holonomy representation

$$\pi_1(Y) \to SU(2)$$

In particular if $Y = X$ is the simply connected 4-manifold of the theorem then any self-dual connection with $c_2 = 0$ is trivial. We shall study the first interesting case $c_2 = -1$ from now on.

SECTION THREE. THE SPACE OF 1-INSTANTONS

The self-dual connections on a bundle with group $SU(2)$ and $c_2 = -1$ over the standard Riemannian S^4 are all explicitly known and play an important role in the general theory. They illustrate the fact that the self-duality equation (5) is conformally invariant; that is, a conformal transformation $f: S^4 \to S^4$ pulls one self-dual connection back to another. Similarly since $S^4 - \{pt.\}$ is conformally equivalent to $\mathbb{R}^4$ these solutions may also be regarded as self-dual connections or "instantons" on $\mathbb{R}^4$. There is a natural $SO(5)$-invariant self-dual connection coming from the "quaternionic Hopf fibration":

$$\begin{array}{l} S^7 \\ \downarrow \; S^3 \cong SU(2) \\ S^4 \end{array}$$

and the conformal group of S^4 generates all possible solutions from this basic one. Thus the set of equivalence classes of self-dual connections with $c_2 = -1$ on S^4 is parametrized by a <u>moduli space</u>:

$\mathcal{M}(S^4)$ = Conformal Group of $S^4/SO(5) \cong B^5$. Under the standard conformal chart $\mathbb{R}^4 \to S^4 - \{\bar{p}\}$ the conformal transformations corresponding to the segment $\overline{op}$ in B^5 ($o \to p$) are represented by dilations:

$$x \to \lambda \cdot x \qquad 0 < \lambda < 1 \quad , \quad x \in \mathbb{R}^4$$

And on $\mathbb{R}^4$ the instantons represented by this segment in the moduli space, A_λ, say, have curvature densities:

$$|F(A_\lambda)(x)| = \lambda^2/(\lambda^2+|x|^2)^2$$
$$\to 0 \quad \text{as} \quad \lambda \to 0 \quad \text{for} \quad x \neq 0$$
$$\to \infty \quad \text{as} \quad \lambda \to 0 \quad \text{for} \quad x = 0$$

Thus we may compactify the moduli space $\mathcal{M}(S^4)$ intrinsically by adding on S^4 as boundary and saying that a point $p \in S^4$ is the limit of a sequence A_i in $\mathcal{M}$ if $\frac{1}{8\pi^2}|F(A_i)|^2$ tends to the δ-function at p.

The next three sections explain why an analogous moduli space $\mathcal{M}(X)$ for the manifold of the theorem (with some fixed metric) exists. From these we shall deduce the proof of the theorem.

SECTION FOUR. ANALYTICAL PROPERTIES OF SELF-DUAL CONNECTIONS.

A linear elliptic differential operator D defined on some open set $U \subset \mathbb{R}^n$ has the standard property (see [8] for example) that if $C \leq 0$ and if $\{f_i\}$ is a sequence with $Df_i = 0$, $\|f_i\| \leq C$ then some sub-sequence of the f_i converge to a limiting f with $Df = 0$, $\|f\| \leq C$. (For example, if $U \subset \mathbb{C}$ and if $D = \bar{\partial}$ is the Cauchy-Riemann operator then this is the classical theorem of Montel). An immediate consequence is that if D is defined instead on a compact manifold then Ker D is finite dimensional, or equivalently the unit ball in Ker D is compact.

The theorems of K. Uhlenbeck ([6],[7]) extend this standard linear theory to the non-linear self-duality equations, with bounds on the action $\|F\|^2$. Two main differences appear in the local theory for a sequence A_i of self-dual connections defined over $B^4 \subset \mathbb{R}^4$, with $\|F(A_i)\|^2 \leq C$:

(a) Gauge invariance.

In a fixed frame S the A_i^S need have no convergent sub-sequences. For example even if $C = 0$ the A_i could be an infinite sequence of flat connections. This corresponds to the fact that the self-duality equation: $d_-A^S - (A^S \wedge A^S)_- = 0$ (cf. (4)) is not elliptic. One can only hope to find B_i isomorphic to the A_i converging, by fixing the constraint $d{*}B_i^S = 0$ to give an elliptic system.

(b) Non-Linearity.

One will not achieve convergence for all values of the constant C. For less than some fixed C_0 the linear theory extends and the B_i exist and have a convergent sub-sequence. The limit is also a self-dual connection. The Yang-Mills action is conformally invariant in dimension 4 (cf. Section 3) so the same conclusion, with the same constant C_0, applies to balls of arbitrary size.

If now $A_i \in \mathcal{A}$ is a sequence of self-dual connections on the bundle V over the compact manifold Y, so that:

$$\int_Y |F(A_i)|^2 d\mu = -8\pi^2 c_2(V) = 8\pi^2$$

One may apply the local result near to any point $y \in Y$ about which there is a ball $B \quad Y$ with for large i:

$$\int_B |F(A_i)|^2 d\mu \leq C_0$$

On passing to a suitable sub-sequence this is always possible for all but finitely many points in Y (in fact, at most $8\pi^2/C_0$ points) and away from these points the local result gives convergence to a limiting connection A_∞ which is self-dual. By another theorem of Uhlenbeck [7] A_∞ extends over all of Y, but to a connection on a bundle possibly not isomorphic to V: however we can only "lose" curvature in the limiting process, so the only possibility for a new bundle is the trivial one $(c_2 = 0)$ and in this case, by the remarks above in Section 2, if Y is simply connected, for example if $Y = X$, A_∞ is isomorphic to the standard flat connection. Moreover for any ball $B \subset Y$ the value of

$$\frac{1}{8\pi^2} \int_B |F(A_i)|^2 d\mu = \frac{1}{8\pi^2} \int_B \mathrm{Tr}(F \wedge F)$$

is modulo $\mathbb{Z}$ an invariant of $A_i\big|_{\partial B}$ (this is a basic property of characteristic class integrands for manifolds with boundary), so one easily sees that there is at most one point $y \in Y$ over which the curvature of the A_i may gather, and either:

(a) $c_2(A_\infty) = -1$ and $A_i \to A_\infty$ over all of Y,

or (b) A_∞ is the trivial flat connection, and $\frac{1}{8\pi^2} |F(A_i)|^2 \to \delta_y$ for some point $y \in Y$. (Thus the point appears as the Poincaré dual of $-c_2(Y)$.)

SECTION FIVE. THE EXISTENCE OF SELF-DUAL CONNECTIONS

Self-dual connections were first studied [1] under the restriction that the Riemannian base manifold was of a very special type. C. H. Taubes then showed that they exist under more general hypotheses, [5]. His construction may be roughly described thus: given a point $y \in Y$ and a scale $\lambda > 0$, use geodesic co-ordinates to identify a small ball around y with a similar ball in $\mathbb{R}^4$. Under this identification the instanton A_λ goes over to a connection defined in a neighborhood of y which can be extended over all of Y by gluing on to the flat connection. This gives a connection on a bundle over Y with $c_2 = -1$ and with $\| F_- \|^2$ small. Taubes showed that if one took the scale λ very small, so the connection constructed had most of its curvature concentrated around the point y, and if one tried to modify this connection to find a nearby self-dual connection one encountered obstructions in the space $\mathcal{H}^2_-(Y)$ of anti self-dual harmonic 2-forms. By the discussion of Section 1 this space vanishes precisely when the intersection form is positive definite; in particular from Taubes theorem [5] self-dual connections with $c_2 = -1$ exist over the manifold X of the theorem.

This condition on the intersection form for the existence of self-dual connections is definitely necessary in some cases; for example, with suitable Riemannian metrics, self-dual connections exist on a bundle over $-\mathbb{CP}^2$ with $c_2 = -2$ but not with $c_2 = -1$. In general one would hope for a precise result along the lines of the Riemann-Roch theorem for meromorphic functions on a Riemann surface.

SECTION SIX. THE MODULI SPACE OF SELF-DUAL CONNECTIONS

The equivalence classes of self-dual connections on the given bundle V over the manifold Y will be parametrized by a moduli space $\mathcal{M} = \mathcal{M}(Y)$, just as we saw in Section 3 for the case $Y = S^4$. It is easiest to define this abstractly in terms of calculus in Banach spaces.

The infinite dimensional gauge group $\mathcal{G}$ of automorphisms of V acts on $\mathcal{A}$ by conjugation, and by definition the set of equivalence classes of connections is the quotient $\mathcal{B} = \mathcal{A}/\mathcal{G}$. Let us, for simplicity, assume for the moment that $\mathcal{G}/\pm 1$ acts freely on $\mathcal{A}$; then $\mathcal{B}$ is an infinite dimensional manifold with charts defined by transversals in $\mathcal{A}$ to the $\mathcal{G}$-orbits:

$$T_{A,\varepsilon} = \{A + a \mid d_A^* a = 0 \ , \ \|a\| < \varepsilon\}$$

(This is the global version of the constraint $d^*B^S = 0$ in Section 4.) The sub-set $\mathcal{M} \subset \mathcal{B}$ of equivalence classes of self-dual connections is cut out by equations: explicitly in the chart above about a self-dual A these are:

$$\begin{aligned} & d_A^* a = 0 \qquad\qquad (\text{fixing } A + a \in T_{A,\varepsilon}) \\ & d_A^- a - (a \wedge a)_- = 0 \end{aligned} \tag{6}$$

More formally there is an infinite dimensional vector bundle $\begin{smallmatrix}\mathcal{E}\\ \downarrow \\ \mathcal{B}\end{smallmatrix}$ with a canonical section s cutting out $\mathcal{M}$;

$$\mathcal{M} = Z(s) = \{[A] \in \mathcal{B} \mid s[A] = 0\}$$

Suppose, for purposes of comparison, that $\begin{smallmatrix}E\\ \downarrow \!\!\uparrow s\\ B\end{smallmatrix}$, $M = Z(s)$ were analogous finite dimensional objects, then one would have the standard properties:

(a) For generic perturbations $s + \sigma$ of s, $M_\sigma = Z(s + \sigma)$ is a smooth submanifold.

(b) In $KO(M_\sigma)$ $TM_\sigma = E - TB$, hence taking 0 and 1-dimensional components in cohomology: $\mathrm{Dim}(M_\sigma) = \mathrm{Dim}(E - TB)$, $w_1(M_\sigma) = w_1(E - TB)\big|_{M_\sigma}$. So the dimension of M_σ is fixed by the data and if $H^1(B;\mathbb{Z}_2) = 0$ then M_σ is orientable.

Both (a) and (b) extend to our infinite dimensional case. This is because the linearization of the local representation (6) of the equations is the elliptic equation:

$$(d_A^* + d_A^-)a = 0$$

and it is a standard fact that elliptic differential operators over compact manifolds give rise to "Fredholm" operators on Hilbert spaces (cf. Section 4). The usual finite dimensional argument that is used to prove property (a) extends to such Fredholm equations [4] so without loss of generality, we may suppose that $\mathcal{M}$ is a smooth submanifold of $\mathcal{B}$ by making a small perturbation if necessary. (It seems very likely that one can always make this perturbation by varying the metric on the base space Y).

Similarly there is a well defined element in $KO(\mathcal{B})$ which is formally the difference of infinite dimensional spaces $\mathcal{E} - T\mathcal{B}$: the situation is complicated by the $\mathcal{G}$ action, but abusing notation it is the index of the family of operators $(d_A^* + d_A^-)$ in the sense of [2]. The Atiyah-Singer index theorem computes the dimension of this in terms of the original data V,Y: [1]

$$\begin{aligned}\mathrm{Dim}\mathcal{M}(Y) \;=\; \mathrm{index}(d_A^* + d_A^-) &= 8|c_2(V)| + 3\ \mathrm{index}(d^* + d^-)\\ &= 8 - 3(1 - b_1 + b_2^-) \qquad \text{cf. (2)}\end{aligned}$$

In particular for the manifold X of the theorem, $b_1 = b_2^- = 0$, so

$$\mathrm{Dim}\mathcal{M}(X) = 5\ .$$

And a straightforward homotopy calculation shows that, again for our particular X,V, the group $\mathcal{G}/\pm 1$ is connected so, since it acts freely on the contractible space $\mathcal{A}$ the quotient $\mathcal{B}$ is simply connected and $\mathcal{M}(X)$ orientable by the generalization of property (b) above. (I am grateful to Cliff Taubes for a correction on this point.)

Now we see from Section 4 that a sequence in $\mathcal{M}(X)$ can only fail to have convergent sub-sequences if the curvature densities become concentrated over a point of X, and from Section 5 that Taubes constructs solutions of this type depending upon 5 parameters, a point in X and a scale $\lambda > 0$. Thus with a certain amount of effort one proves that there is a collar $\mathcal{U} \subset \mathcal{M}$ with $\mathcal{U} \cong X \times (0,\lambda_0)$ and $\mathcal{M} - \mathcal{U}$ compact, so we may compactify $\mathcal{M}(X)$ to a manifold with boundary X just as in Section 3.

Throughout this section we have assumed that $\mathcal{G}/\pm 1$ acts freely on $\mathcal{A}$, and we should now return and briefly describe the modifications required when this is not the case.

If a connection $A \in \mathcal{A}$ is fixed by a non-trivial $g \in \mathcal{G}$ then g is a co-variant constant $d_A g = 0$. The eigenspaces of $g \in \operatorname{Aut} V$ split V into a direct sum of line bundles $V = L \oplus L^{-1}$ in such a way that the connection A is induced from a connection on L, in the obvious sense.

It is very easy to check that on a simply connected manifold there is a unique connection on a line bundle (up to equivalence) having any prescribed curvature form within the set of representatives of $2\pi i\, c_1(L)$, so on X there is for any line bundle L just one self-dual connection with curvature the harmonic form in $\mathcal{H}^2_+$. Then the condition for such a line bundle L to appear in a splitting of V is:

$$-1 = c_2(V) = -c_1(L)^2$$

Thus the number n of such Abelian reducible connections in the moduli space $\mathcal{M}(X)$ is determined by the intersection form Q:

$$n = n(Q) = \tfrac{1}{2}\#\{\alpha \in H_2(X;\mathbb{Z}) \mid Q(\alpha) = 1\} .$$

(the ½ coming from the choice of factor L, L^{-1}). The stabilizer in $\mathcal{G}$ of one of these reducible connections is a copy of S^1, corresponding to a constant rotation of each factor, and this gives $\mathcal{M}$ a quotient singularity; a neighborhood in $\mathcal{M}$ of any such point has generically the form $\mathbb{C}^3/S^1$ = cone on $\mathbb{CP}^2$.

SECTION SEVEN. PROOF OF THE THEOREM

By using the Gram-Schmidt diagonalization procedure one easily sees that for any positive definite unimodular form Q:

$$n(Q) \leq \operatorname{rank}(Q) \tag{7}$$

with equality if and only if Q is equivalent to the standard diagonal form. But the moduli space $\mathcal{M}(X)$ of Section 6, with its singular points removed, gives an oriented cobordism between X and $n(Q)$ copies of $\mathbb{CP}^2$; hence

$$\operatorname{Rank}(Q) = \tau(X) \leq n(Q)$$

so combining this information with (7) we have $n(Q) = \operatorname{rank}(Q)$ and Q is the standard form.

BIBLIOGRAPHY

[1] M. F. Atiyah, N. J. Hitchin, I. M. Singer, "Self-duality in 4-dimensional Riemannian Geometry" Proc. Roy. Soc. London A362 (1978).

[2] M. F. Atiyah and I. M. Singer, "The index of elliptic operators V" Ann. of Math. 93 (1971).

[3] J. W. Milnor, "Characteristic Classes" Ann. of Math. Studies Princeton 1974.

[4] S. Smale, "An infinite dimensional version of Sards Theorem" Amer. Jour. Math. 87 (1965).

[5] C. H. Taubes, "The existence of self-dual connections" to appear in Journal of Differential Geometry.

[6] K. K. Uhlenbeck, "Connections with L^p-bounds on curvature" Commun. Math. Phys. 3 (1981).

[7] K. K. Uhlenbeck, "Removable Singularities in Yang-Mills Fields" Commun. Math. Phys 3 (1981).

[8] F. W. Warner, "Foundations of Differentiable Manifolds and Lie Groups" Scott Foresman and Co. (1971).

THE MATHEMATICAL INSTITUTE
24-29 St. GILES
OXFORD, ENGLAND

Contemporary Mathematics
Volume 35, 1984

THE SOLUTION OF THE 4-DIMENSIONAL ANNULUS CONJECTURE (AFTER FRANK QUINN)

Robert D. Edwards

After Freedman made his startling breakthrough in August-September 1981, establishing that 6-stage Casson towers contain flat spanning discs, it quickly became clear that in order to derive the most significant consequences using this technique, one should go back to Casson's original tower construction and attempt to refine it a bit, in order to achieve some control on the size of towers. For example, Freedman himself toiled ceaselessly throughout that September, seeking enough control to be able to prove the proper h-cobordism theorem and hence the topological 4-dimensional Poincaré conjecture. He finally succeeded by means of a fairly intricate adaptation of Siebenmann's method for proper h-cobordisms (see [F_2, Sec. 10]).

The sort of control that Freedman achieved can be regarded as 1-dimensional - one merely had to control the wandering of points out toward infinity, or in from infinity. There remained an entirely new layer of results to be established, using techniques already successful in higher dimensions, if only one could impose additional degrees (dimensions) of control. Such control finally was achieved almost a year later by Frank Quinn, who made his clinching discovery and dramatic announcement of success at this conference.

This article is a discussion of that work. In its proper generalized setting, Quinn's work is a maze of ε's, δ's, homotopies, intersection patterns, et cetera, which is as difficult for a novice to wade through as Freedman's original work. But when stripped to its core, Quinn's work becomes much more tractable (but still far from easy). In this article we present this core, using as a target the theorem that is probably the best known corollary of Quinn's work, the 4-dimensional Annulus Theorem. Alternatively, and a bit easier perhaps, one could keep in mind as the target theorem the 5-dimensional proper h-cobordism theorem (say the finite ended case), which originally was proved in [F_2, Sec. 10] by the somewhat specialized argument mentioned above.

*Supported in part by NSF grant MCS80-03571

At its heart, Quinn's work amounts to an ingenious reorganization of Casson's construction, making full use of the great triumvirate of moves in this subject, namely Whitney's, Casson's and Norman's moves. It is remarkable how careful one has to be to achieve even the slenderest amount of control, but once achieved, the payoff is substantial.

This article is written so that a fledgling student of geometric topology should be able to follow most of it. To this end, we have included in Sections 2 and 3 a summary of the more important, previously used constructions that will come up. Hence, anyone already comfortable with Casson's work can proceed immediately to where Quinn's work begins, in Section 4.

There is nothing new in this paper itself, in the sense of new theorems or new constructions that haven't already appeared. However, there is some novelty here in the presentation of this work, primarily Quinn's but also Freedman's, and with this different perspective we achieve some modest economy and (hopefully) clarity.

The sections of this article are:

1. PRELIMINARY MATTERS AND EVER PRESENT HYPOTHESES

All surfaces and manifolds in this article are always assumed to be oriented, and everything is smooth, except where Freedman's work is applied at the end. Since we work in noncompact manifolds, we often deal with infinite collections of data (usually surfaces), but we will always assume that all such data are locally finite. Surfaces are not necessarily connected, but <u>all components are always compact</u> (or perhaps relatively compact), unless clearly indicated otherwise. By <u>component</u> of an immersed surface we mean component in the manifold sense, and so components may intersect. We write $S \cdot T$ to denote the homological intersection number $\in \mathbb{Z}$ between two compact (oriented) surfaces S and T immersed in a 4-manifold. Reference to $S \cdot T$ presumes that $(\partial S \cap T) = \emptyset = (S \cap \partial T)$, or else that there is a preferred way of achieving this. If we write $S \cdot T = 0$ for noncompact surfaces S and T, this means that $S_\alpha \cdot T_\beta = 0$ for all components S_α of S and T_β of T.

All intersections/meetings of arcs, surfaces, etc. in the ambient 4-manifold are assumed to be generically positioned, subject to the constraints imposed by the hypotheses. For example, if a disc is attached to a surface, then near the attaching curve their union locally looks like
$R^2 \times 0 \times 0 \cup R^1 \times 0 \times [0,\infty) \times 0$ in R^4, unless the curve has self-intersections, in which case at such points their union locally looks like
$R^2 \times 0 \times 0 \cup R^1 \times 0 \times [0,\infty) \times 0 \cup 0 \times R^1 \times (-\infty,0] \times 0$ in R^4.

Invariably the union of a collection such as $\{C_\gamma\}$ is denoted by its Roman letter C, and we will abusively make statements like "the collection $C = \cup C_\gamma$," confusing the collection $\{C_\gamma\}$ with its union C. This seems to make statements more readable. A similar abuse occurs when we speak of a "regular homotopy of C_j;" by this we mean a regular homotopy of the (abstract manifold) components of C, and not of the set C itself (so e.g. crossings in C may disappear). Finally, we are constantly moving sets like A,B,C, etc., without renaming them, to keep the notation simple.

The operation of <u>piping</u> is used repeatedly in the subject, to desingularize immersed arcs in surfaces and immersed discs in 4-manifolds:

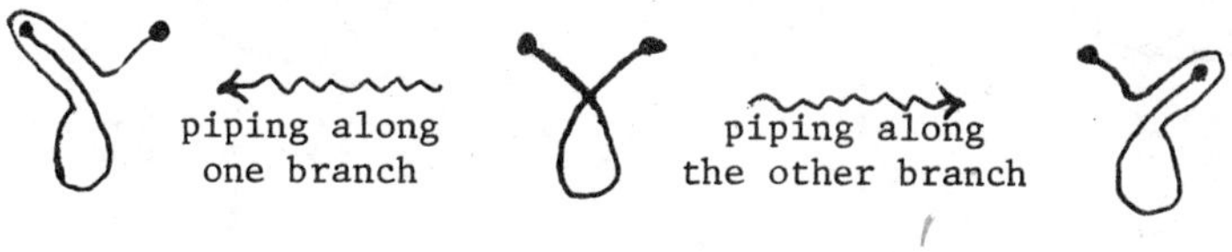

We note that there is always a choice involved in piping, namely the choice of which sheet or branch to pipe. Furthermore, in the case of a disc, there is the choice of which boundary point to pipe toward. In what follows these choices are immaterial, unless explicitly specified. Also, one can often replace piping with its inverse motion, which desingularizes the arc or disc by shrinking it smaller:

However, for custom's and consistency's sake, we maintain the language of the former point of view.

At various points in ensuing discussions, careful attention must be paid to certain framings. However, these discussions are invariably technical, and are definitely peripheral to the central issues; the first-time reader in particular should not dwell on these points.

Definitions in the paper are made where they are first needed. For convenience, we list the major ones, and where they occur:

Section 2: Whitney move, Casson move (= finger move = anti-Whitney move), pre-Whitney loop, pre-Whitney disc, correct framing (as a Whitney disc).

Section 3: tower, correct framing (as the base of a tower).

Section 4: transverse sphere, Norman move (= Norman trick), double surgery.

Section 5: transverse collection of spheres.

2. WHITNEY MOVES, CASSON MOVES AND CASSON'S SURFACE SEPARATION LEMMA

In this section we recall the basic facts about Whitney moves and Casson moves. A convenient model for these moves is as follows. In the 2-cell D^2, consider the intersecting arcs α and β together with the spanning 2-disc W in $\operatorname{int} D^2$ shown below in Figure 2.1a.

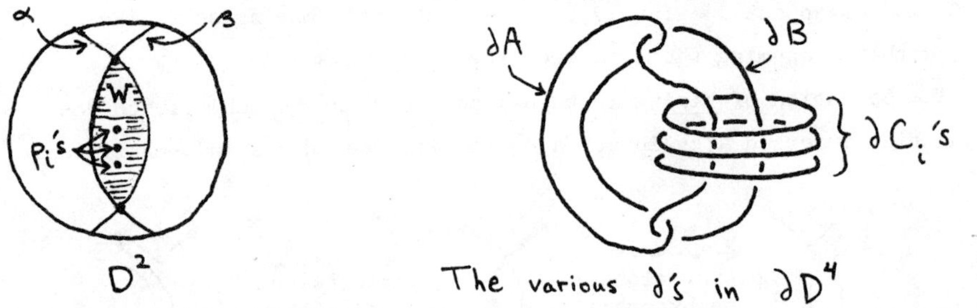

Figure 2.1

In the 4-cell $D^4 = D^2 \times I \times I$, where $I = [-1,1]$, let $A = \alpha \times I \times 0$ and $B = \beta \times 0 \times I$. Then A and B are unknotted 2-cells in D^4 which intersect transversely in two points. They can be isotoped to be disjoint, moving only points close to W ($= W \times 0 \times 0$) in $\operatorname{int} D^4$, by the familiar <u>Whitney move</u> which uses W as a guideway.

Suppose there are additional 2-cells $C_1,\ldots,C_k$ present of the form $C_i = p_i \times I \times I$, where $p_i \in \operatorname{int} W$. (In the model $k=1$ case, the boundary circles ∂A, ∂B and ∂C_1 comprise the familiar Borromean rings in ∂D^4; see Figure 2.1b). Initially the C_i's are disjoint from $A \cup B$, but after the Whitney move there will be intersections. The <u>Casson finger move</u> (<u>or anti-Whitney move</u>) can be described as follows: for each i, before moving $A \cup B$, we isotope C_i off of W by piping $p_i = C_i \cap W$ off of either of the two edges $\partial W \cap A$ or $\partial W \cap B$ of W (which edge depends upon the context), dragging the rest of C_i along, so that it looks as if a finger has been poked into C_i. This makes C_i disjoint from W before $A \cup B$ is moved, at the expense of creating two points of intersection between C_i and either A or B (depending upon the choice of edge above).

This motion of C_i, toward A say, can be regarded as an inverse Whitney move between A and C_i, for one is creating intersections instead of cancelling them. Indeed, one can see appear a Whitney disc W_i for the two newly created intersection points between A and C_i, with $\partial W_i \subset A \cup C_i$ and $W_i \cap W = \partial W_i \cap \partial W$ = one point. (We note for future use that if B is now Whitney-moved across W to free it from A, it will wind up intersecting int W_i in one point; alternatively, if W_i is first moved off of W by piping the arc $\partial W_i \cap A$ along and off the end of the arc $\partial W \cap A$ in A, then this motion creates an intersection point between int W_i and B.) In case the reader has not yet done so, he might find it worthwhile to play with the simple, model Borromean ring situation where there are three discs A, B and $C = C_1$ as described above. The point is, using Whitney moves and Casson moves, any two of these three discs can be made disjoint, but the third disc will always intersect one of the other two.

It was Casson who first showed how to profitably exploit Whitney moves and Casson (finger) moves in combination [C]. These moves occur over and over again in the subsequent work of Freedman and Quinn.

We now describe how the above process typically is applied in the interior of a 4-manifold M. Suppose in M one has surfaces A and B which are connected and imbedded (or, for example, one has imbedded portions of immersed surfaces), having two (transversal) intersection points of opposite sign, p and q say (recall everything is assumed oriented). Let α and β be paths in A and B, respectively, joining p to q, and suppose W is an immersed disc in M whose boundary is attached to $\alpha \cup \beta$. Assuming everything is generically positioned, then W may have self-intersections in its interior and also in its boundary, but not between them. Such a W is called a <u>pre-Whitney disc</u>, and $\alpha \cup \beta$ $(= \partial W)$ is a <u>pre-Whitney loop</u>.

We wish to use W as a Whitney disc to separate A and B. But first we must rectify its possible shortcomings. They are, in the order that they will be dealt with: (1) α and β may have self-intersections, (2) the "framing" of W may be wrong, (3) int W may have self-intersections, and (4) int W may intersect $A \cup B$ (and also int W may have unwanted intersections with some other surface C).

To deal with (1), we simply pipe the self-intersections of α and β off of their respective ends (either ones), keeping α in A and β in B, dragging W along as we do so. This will create additional intersections between int W and A (if β is not embedded) or between int W and B (if α is not imbedded), but they will be dealt with in time (see (4)).

To deal with (2), we first explain what we mean by the framing of W. Since W is immersed, there is an immersion $\pi: D^4 \to M$ of our model 4-ball D^4 (described above) onto a neighborhood of W, carrying the model $\hat{W}$ onto W. (We will for the moment use $\hat{\ }$'s over the model discs.) Then W has the <u>correct framing</u> (as a Whitney disc) if in addition we can make π carry $\hat{A}$ into A and $\hat{B}$ into B. Either one or the other of these containments is easily arranged, but there is a potential obstruction to achieving both simultaneously. For example, if we look at the pre-image circles $\pi^{-1}(A) \cap \partial D^4$ and $\pi^{-1}(B) \cap \partial D^4$, their union may look twisted in ∂D^4, even though both of these circles are unknotted there, and their algebraic linking number is 0 (see Figure 2.2; for future use, we let $t \in \mathbf{Z}$ denote the number of apparent full twists in these pre-image circles).

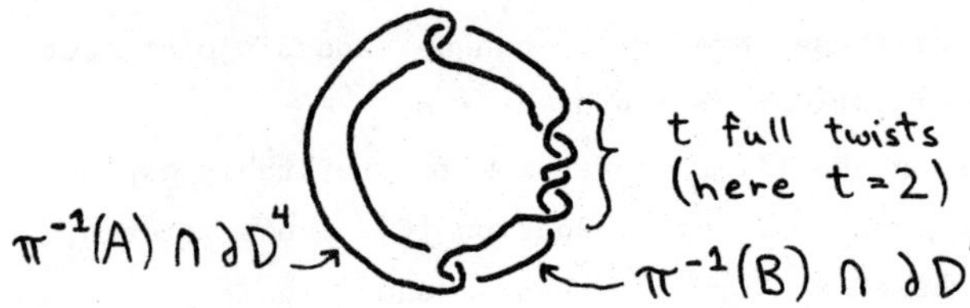

Figure 2.2

To remedy this "framing mismatch", we are forced to alter something. The most convenient change to make is to spin W at (some arbitrary point of) $\partial W - \{p,q\}$, as suggested by the familiar sequence of pictures in Figure 2.3 (here we are working near an arbitrary point $a \in \operatorname{int}\alpha$; we could just as easily work instead near $b \in \operatorname{int}\beta$).

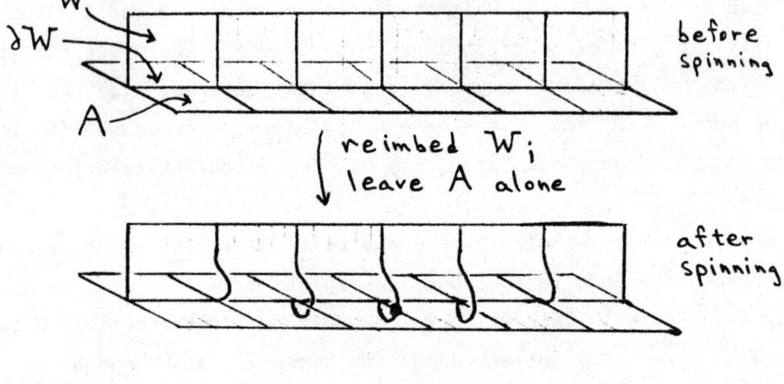

Figure 2.3

This spinning operation is to be regarded as a reimbedding (via isotopy, if you wish) of W rel ∂W, during which A and B are left fixed. Each single spin has the effect of changing the above twisting number t by ± 1, at the expense of introducing a new intersection point between int W and A (or, alternatively, B). Having done such spins we can assume that there is no twisting, and hence that the above immersion π now carries $\hat{A}$ into A and $\hat{B}$ into B, as desired.

To deal with (3), we simply pipe the self-intersection of int W off of its boundary, either at the A side or the B side, at the expense of creating two additional intersection points of int W with A (or B) for each initial self-intersection point of int W. This operation, achieved by regular homotopy of W, does not affect the framing coherence established in (2). (Actually, for many purposes this step (3) is unnecessary, but there is no harm in doing it.)

To deal with (4), we again use piping, this time to move A and B off of int W. Depending upon the particular context at hand, we will either pipe the A intersection points off of the A-edge of W, and the B intersection points off of the B-edge of W, or vice versa. In the former option, which is used most often, we create self-intersections in A and in B, but no intersections between them, whereas in the latter option the reverse is true. As for getting rid of possible unwanted intersections of int W with some other surface C, they can in similar fashion be piped off of either edge of W, at the expense of making C intersect either A or B.

Finally, having rectified (1)-(4) as above as best we can, we can use our newly imbedded Whitney disc W to get rid of the two original intersection points p and q between A and B by moving either A or B across W in the usual manner.

Since the above operation is used so often, we make a formal statement of it, in the generality that we need. The data in the following lemma may well be unbounded, and are subject only to the ever present hypotheses listed in §1.

CASSON'S SURFACE SEPARATION LEMMA. Suppose A and B are surfaces (not necessarily compact or connected) immersed in a 4-manifold, with $A \cdot B = 0$, and suppose W is a union of pre-Whitney discs for all of the intersections between A and B. (As usual, we suppose each disc contains just two points of $A \cap B$, and these points are disjoint from the other discs.) Then there is a regular homotopy of $A \cup B$, supported arbitrarily close to W, which makes A and B disjoint. (However, new self-intersections may be introduced in A and also in B.) Furthermore, if C is any other surface intersecting int W, then C can be kept free of A or B (but not both).

ADDENDUM. The regular homotopy can be bounded (in the distance it moves any point) by the maximum diameter of any individual pre-Whitney disc, plus ε, for some arbitrary $\varepsilon > 0$.

Proof. The proof is just as described above for the single disc case, except that additional care should be taken to make all of the resultant Whitney discs disjoint, which is easily arranged using piping as in operations (1) and (3) above. To achieve the Addendum as stated, strictly speaking one should think in terms of inverse piping instead of piping, as described in Section 1.

3. A FEW WORDS ON TOWERS AND THEIR FRAMINGS

The goal of this article is to construct towers, just as in Casson's work, but this time with control on their ultimate size.

Recall that a (finite) tower is a finite union $C \cup D \cup E \cup \cdots$ of stages of discs immersed in a 4-manifold, where the first stage C is a single disc, the second stage D is a collection of disjoint discs attached to C to kill its fundamental group (which arises from its self-intersections), with each disc of D going through just one crossing point of C; the third stage E is a collection of disjoint discs attached to the discs of D to kill their fundamental groups, etc. See [C] or [F_2] or [G-S]. In this paper, the first stage C will be attached to a union $A \cup B$ of two imbedded surfaces, just as a Whitney disc would be attached, so that $C \cap (A \cup B) = \partial C$. At this point the question of framing arises again, and it is worth a few words, for there is a subtle distinction, often misunderstood, between the situation in the previous section and the situation here.

We wish to describe what it means for C to have the correct framing as the base of a tower. In short, this means that it is possible to attach 2-handles (abstractly) to a regular neighborhood of C to make it into a 4-ball so that in its boundary 3-sphere, the two circles arising from its intersection with $A \cup B$ are geometrically unlinked, i.e., they span disjoint discs there. In terms of framings, this can be described formally as follows.

Let N be a regular neighborhood of C rel ∂C; we think of N as an immersed normal disc bundle over C. Let N_α, N_β and $N_\partial = N_\alpha \cup N_\beta$ denote the induced subbundles over $\alpha = A \cap C$, $\beta = B \cap C$ and $\partial C = \alpha \cup \beta$. We can assume that $A \cap N$ is an interval subbundle of N_α, and similarly for $B \cap N$. These subbundles induce a natural framing on N_∂, i.e., they provide a natural product structure, which we denote by $N_\partial = \partial C \times D^2$. The question of framings here concerns the linking number, call it ℓ, of the circles $\partial C \times 0$ and $\partial C \times *$, $* \in D^2 - 0$, as subsets of ∂N. (Recall that ∂N is homeomorphic to n copies of $S^1 \times S^2$ connect-summed together, where n is the number of self-crossings of C. Both $\partial C \times 0$ and $\partial C \times *$ are null-

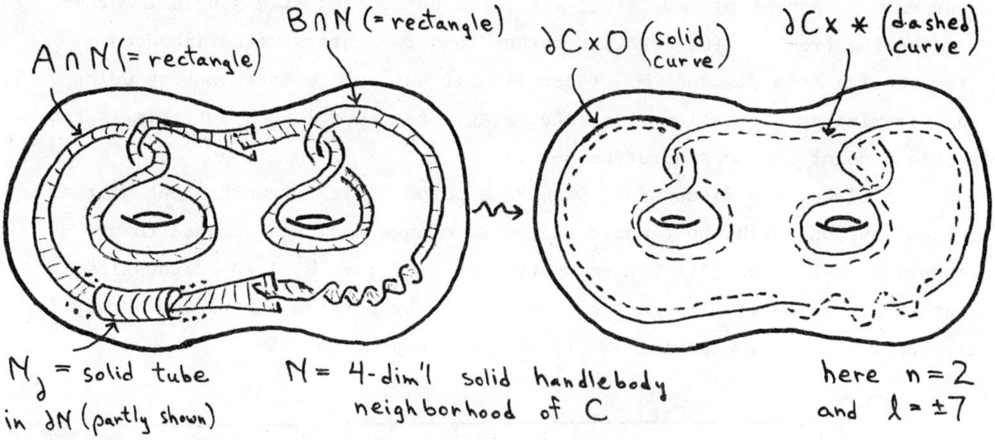

Figure 3.1

homotopic in ∂N.) If $\ell \neq 0$, then no matter how 2-handles are attached to N to make it into a 4-ball, the circles $\partial(A \cap N)$ and $\partial(B \cap N)$ will be geometrically linked in its boundary (even though algebraically unlinked). It turns out that ℓ must be 0 before we can even hope that the tower construction will eventually lead to producing a Whitney disc.

This number ℓ is not to be confused with the twisting number t encountered earlier; t is measuring whether the above product structure on $N_\partial = \partial C \times D^2$ extends to a product structure $C \times D^2$ on N. In fact, we have in effect here an example of the well-known relationship between the euler number t, the homological self-intersection number ℓ and the algebraic number (call it i) of transverse self-intersection points of a closed 2m-dimensional immersed submanifold of an ambient 4m-manifold (everything oriented): $\ell = t + 2i$. We can change t by multiples of 2 by inserting little kinds in int C (pictorially:  ; this is not a regular homotopy operation), but this does not change ℓ. (Nevertheless, it is sometimes convenient to arrange that $t = \ell$ by arranging that $i = 0$ by inserting such kinks; this is a comforting assumption to make throughout all constructions.) On the other hand, the piping operation described earlier, where self-intersections of int C are piped off of the edge of C, changes both ℓ and i, but not t, whereas the spinning operation described earlier changes both ℓ and t, but not i. It is this latter operation, followed by some clean-up motions, which will be used to make $\ell = 0$ during the heart of the proof (Sec. 7).

There is a similar discussion for later stages of the tower, i.e. one must arrange that $\ell = 0$ for each of the immersed discs, collectively called D, that will be attached to C to kill its kinks, and likewise for the layers of discs E, F etc. The only minor difference is that now the framing over each component of ∂D (∂E, etc.) is determined by a single immersed disc coming from C (D, etc.), rather than two intersecting imbedded discs coming from A and B. (Actually, it has been pointed out by Quinn that framing considerations for later stages can be relaxed, but this is too subtle a point for consideration here.)

To close this discussion, we make a parenthetical remark about another method that one might be tempted to use to change framings (called to my attention by Ric Ancel). Returning to Step 2 of the W, A, B discussion in Section 2, one could alternatively change t by putting a kink in A say (or, alternatively, B), leaving W fixed, as shown below:

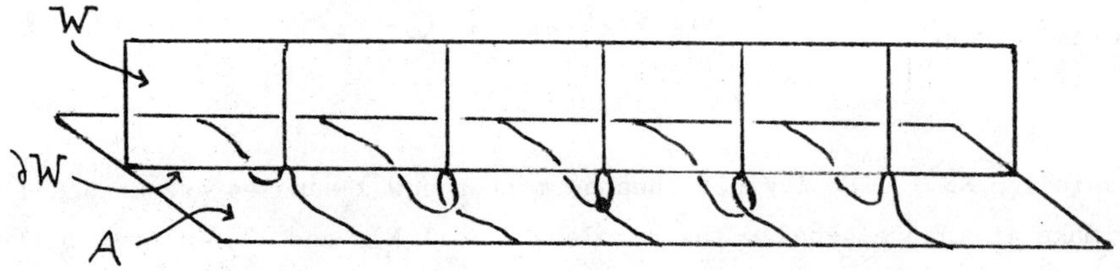

Figure 3.2 (compare to Figure 2.3)

Although this method can sometimes be used, it has the disadvantage of putting a possibly undesirable self-intersection point in A, and also it changes the euler number of the normal bundle of A. One can correct the latter by inserting nearby a kink in A of opposite sign, and then one can go one step further and try to cancel these kinks of A by regular homotopy. Interestingly, if one does so, making A imbedded again, then one is forced to make A intersect $\operatorname{int} W$, and the whole process reduces to the spinning operation described earlier.

4. QUINN'S TRANSVERSE SPHERE LEMMA

In this section, the most basic of the article, we discuss Quinn's fundamental construction. It was first presented in [Q_2, Lemma 3], and a bit more explicitly in [Q_3, Section 3.1]. Our description is intended to be complementary to Quinn's; it will be presented in a symmetrized fashion. In this form Quinn's move bears a striking resemblance to a move used by Štanko twelve years ago in his fundamental taming theorem [St]. As noted at the end of this section, Quinn's construction can be regarded as a variation of Casson's basic π_1-Lemma.

Before beginning, we need the notion of transverse sphere. Suppose C is a connected surface immersed in a 4-manifold. A transverse sphere for C is an immersed sphere $C^{\perp}$ which intersects C transversely in a single point.

It is occasionally required (in Sections 6 and 7; not in Sections 4 and 5) that such a $C^\perp$ have homological self-intersection number 0. This is equivalent to its normal bundle being framed (i.e., a product; perhaps we should really say "framable"), provided that the number of self-intersection points of $C^\perp$ is algebraically 0, a feature which can easily be arranged by adding little kinks in $C^\perp$. On the other hand, every $C^\perp$ produced in this article has homological self-intersection number 0. Hence, for simplicity of exposition we will always assume that transverse spheres have this property (and leave it to aficionados to detect where this hypothesis can be relaxed). We note that Quinn, in his references to "framed immersed S^2's", is tacitly assuming only that such 2-spheres have even homological self-intersection number. Allowing additional kinks, these two hypotheses (framed; even self-intersection) become equivalent, and are really all that is necessary for many applications.

Transverse spheres are useful for getting rid of unwanted intersections. Suppose C is some connected immersed surface in M having a transverse sphere $C^\perp$, and suppose some surface A intersects C transversely (in severalpoints, perhaps). To get rid of these intersections we can pipe them along C over to $C^\perp$, and then connect-sum these resulting fingers of A with copies of $C^\perp$, changing A to $\hat{A} = A \# \underset{n}{\#} C^\perp$, which misses C (where n = the number of intersection points of A with C; see Figure 4.1).

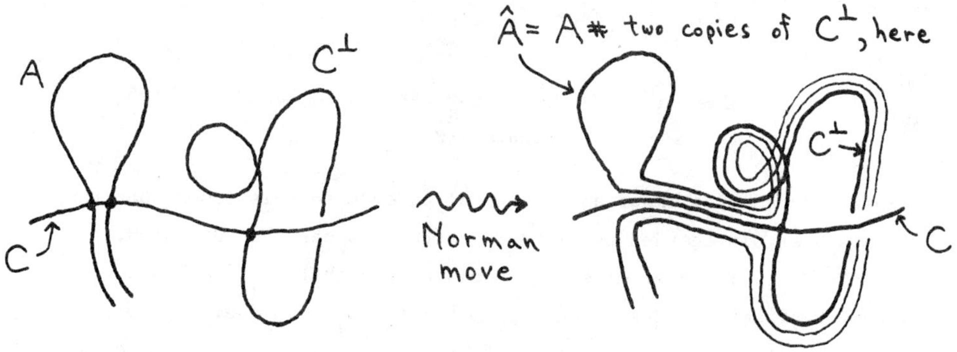

Figure 4.1

This connect-summing operation has come to be known as the Norman trick or Norman move [N]; it is used repeatedly in upcoming sections. Note however that if $C^\perp$ is not imbedded, then the new $\hat{A}$ picks up self-intersections from $C^\perp$.

The following lemma presents the fundamental construction of this article.

QUINN'S TRANSVERSE SPHERE LEMMA. Suppose in the interior of a 4-manifold M one has immersed connected surfaces C_1 and C_2 which meet only transversely, equipped with transverse (immersed) spheres $C_1^\perp$ and $C_2^\perp$, such that $C_1 \cap C_2^\perp = \emptyset = C_2 \cap C_1^\perp$ (hence $C_i \cap C_j^\perp = \delta_{ij}$ points). Suppose at one of the intersection points $p \in C_1 \cap C_2$ one has an imbedded disc W which is attached to $C_1 \cup C_2$ in standard fashion like part of a Whitney disc, with ∂W changing sheets from C_1 to C_2 at p. Suppose that for $i = 1,2$, F_i is a surface, perhaps disconnected, which meets $C_i^{\perp}$ transversely in some finite number of points. For $i = 1,2$, let λ_i be an arc in C_i joining p to $q_i = C_i \cap C_i^{\perp}$. Then, after making finger moves between F_1 and F_2, which are supported arbitrarily close to $X \equiv C_1^{\perp} \cup \lambda_1 \cup \lambda_2 \cup C_2^{\perp}$, but are disjoint from $C_1 \cup C_2$, one can find a transverse 2-sphere $W^{\perp}$ lying arbitrarily close to X, with $W^{\perp} \cdot W^{\perp} = 0$, such that $W^{\perp} \cap (C_1 \cup C_2 \cup F_1 \cup F_2) = \emptyset$. (Recall our convention that F_1 and F_2 denote the <u>repositioned</u> surfaces here.)

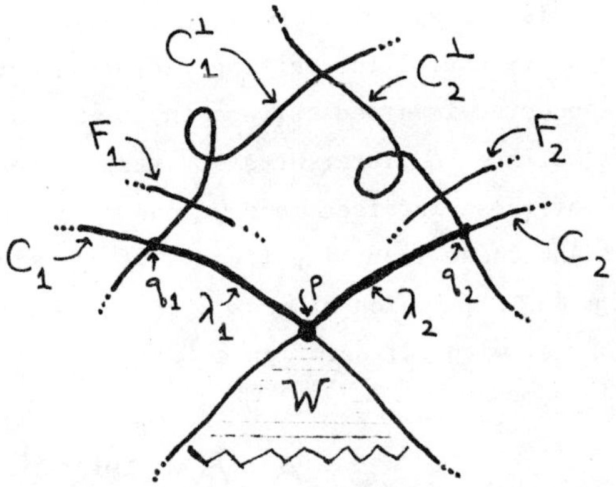

The data for the Transverse Sphere Lemma
Figure 4.2

There are several important technical addenda to this lemma, but they are perhaps best disregarded until the construction has been digested.

ADDENDUM 1. To be precise, we should assume that F_i consists of a finite number of small discs which are normal to $C_i^{\perp}$. Furthermore, we wish to allow some (or all) of these discs of F_i to lie <u>in</u> $C_i^{\perp}$ (and also $C_j^{\perp}$), in case $C_i^{\perp}$ has self-intersections (or intersections with $C_j^{\perp}$), so that we can produce $W^{\perp}$ so that $W^{\perp} \cap (C_1^{\perp} \cup C_2^{\perp}) = \emptyset$ if we wish, at the expense of doing finger moves to $C_1^{\perp} \cup C_2^{\perp}$. (Indeed, the most powerful applications of the Lemma are obtained this way; however, this is definitely not the case to ponder first.) In a similar vein, if $W \cap C_i^{\perp} \neq \emptyset$, we should require that F_i contains $W \cap N(C_i^{\perp})$, where $N(C_i^{\perp})$ is some small neighborhood of $C_i^{\perp}$, in order to be able to produce a $W^{\perp}$ which really does intersect W in only one point. See the remarks on all this at the end of the proof.

ADDENDUM 2. As a special case, one can in fact assume that $C_1 = C_2$ (= C, say), in which case p is a self-intersection point of C, and also one can further assume that $C_1^\perp = C_2^\perp$ (= $C^\perp$). In this case, then, assuming that F_1 and F_2 each contain discs of $C^\perp \cup W$ as in Addendum 1, one can conclude that $W^\perp$ misses $C \cup C^\perp$.

ADDENDUM 3. The finger moves between F_1 and F_2 are in fact supported arbitrarily close to $\xi_1 \cup \lambda_1 \cup \lambda_2 \cup \xi_2$, where ξ_i is a union of arcs in $C_i^\perp$, one for each point of $C_i^\perp \cap F_i$, joining these points to q_i (indeed it seems most natural, and symmetrical, to make the finger moves of F_i be supported arbitrarily close to $\xi_i \cup \lambda_i$, as is done below.) The total number of these double-finger moves performed is the product $|F_1 \cap C_1^\perp| \cdot |F_2 \cap C_2^\perp|$, each move resulting in the creation of two new intersection points between F_1 and F_2. Furthermore, these are the only new intersections created among the given data in the proof, so that for example the intersection $(C_1 \cup C_2) \cap (C_1^\perp \cup C_2^\perp)$ remains two points, even if $C_1^\perp \cup C_2^\perp$ is moved as part of F_i.

Proof of the Lemma. Perhaps one should first note that if either $F_1 = \emptyset$ or $F_2 = \emptyset$ (say for concreteness that $F_1 = \emptyset$), then the proof of the Lemma is easily accomplished, without moving F_2, as follows (see Figure 4.3). One starts with a small "characteristic torus" T for the surfaces C_1 and C_2 lying near p. (Recall that T is the natural torus in $\partial N \approx S^3$ separating the linked circles $C_1 \cap \partial N$ and $C_2 \cap \partial N$, where N is some small 4-ball neighborhood of p; if we write N as $N = B_1 \times B_2$, where B_i is a small 2-cell in C_i centered at p, then $T = \partial B_1 \times \partial B_2 \subset \partial N$.) Let E_1 be a natural spanning disc for T in ∂N on the C_1 side of T, such that $E_1 \cap T = \partial E_1$ (hence $E_1 \cap C_2 = \emptyset$, but $E_1 \cap C_1$ = one point; in the above model, we can take E_1 as $E_1 = *_1 \times B_2$, where $*_1$ is some point in $\partial B_1 \subset C_1$). Let

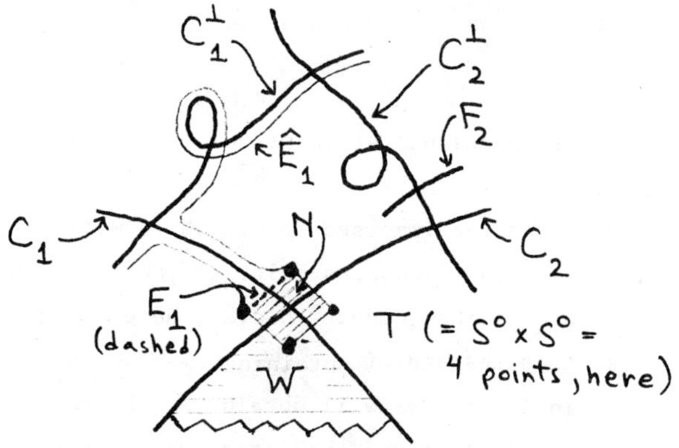

Constructing $W^\perp$ if $F_1 = \emptyset$

Figure 4.3

$\hat{E}_1 = E_1 \# C_1^\perp$ denote a disc gotten from E_1 by applying the Norman trick to E_1, using $C_1^\perp$. That is, $\hat{E}_1$ is gotten from E_1 by tubing E_1 over to $C_1^\perp$ following along the route of λ_1, and then connect-summing E_1 to $C_1^\perp$ via this tube, so that $\hat{E}_1 \cap C_1 = \emptyset$. (Technical note: strictly speaking, for future considerations, we should connect-sum E_1 to a parallel copy of $C_1^\perp$; see the technical point near the end of the proof.) Finally, let $W^\perp$ be gotten from T by doing surgery on T using $\hat{E}_1$. In other words, $W^\perp$ consists of T minus a small band about $\partial E_1 (= \partial \hat{E}_1)$, plus two parallel copies of $\hat{E}_1$ whose boundary circles are glued to the two boundary circles resulting from this discarded band. This sphere $W^\perp$ is immersed, and has 4k self-intersection points arising from the k self-intersection points of $C_1^\perp$. Also, $W^\perp \cdot W^\perp = T \cdot T = 0$. This completes the trivial case.

The simplest nontrivial case of the Lemma is the case where all of the individual surfaces $C_1, C_2, C_1^\perp$ and $C_2^\perp$ are imbedded (actually, one always has without loss that C_1 and C_2 are imbedded, since one only needs subsurfaces of them containing the arcs λ_1 and λ_2), and furthermore $C_1^\perp \cap C_2^\perp = \emptyset$, and each F_i is a single disc meeting $C_i^\perp$ transversally. Schematically, these data are summarized:

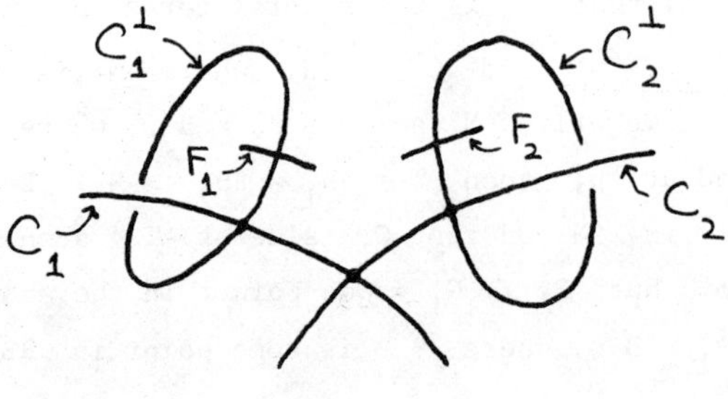

The data for the simplest nontrivial case of the Lemma

Figure 4.4

Understanding this case represents at least 90% of the proof, so we will concentrate on it.

To begin, we describe the immersed sphere which will serve as $W^\perp$ in this model setting. This $W^\perp$ is somewhat similar to the $W^\perp$ constructed above for the trivial $F_1 = \emptyset$ case of the Lemma. However, here we are going to symmetrize the construction, so that instead of obtaining $W^\perp$ by doing a single surgery along the curve ∂E_1 in T, now we will obtain $W^\perp$ by doing a sort of <u>double</u> surgery to T, simultaneously along the intersecting curves ∂E_1 and ∂E_2 in T, where E_2 (and likewise $\hat{E}_2 = E_2 \# C_2^\perp$) denote discs constructed exactly as E_1 and $\hat{E}_1$ were constructed earlier, replacing the subscript 1 by 2 everywhere. Note that E_1 and E_2 lie on opposite sides of T in ∂N. If $W^\perp$ were constructed by using E_1 and E_2 to surger T, instead of $\hat{E}_1$ and $\hat{E}_2$, it would look as in Figure 4.5. After doing these intersecting surgeries, the part of T which remains to become part of $W^\perp$ is a union $P = P_a \cup P_b$ of two squares glued together at their corners, where $P_a = A_1 \cap A_2$ and $P_b = T^2 - \text{int}(A_1 \cup A_2)$, and where in turn A_i is an annular (band) neighborhood of ∂E_i in T. As far as T itself goes, we think of the first surgery (along ∂E_1 say) as removing $\text{int } A_1$ from T, and the second surgery (along ∂E_2) as removing $\text{int } A_2$, but replacing the surgery overlap $A_1 \cap A_2$. (In our model, where $N = B_1 \times B_2$, we can let $A_1 = I_1 \times \partial B_2$ and $A_2 = \partial B_1 \times I_2$, where I_i is a small interval neighborhood of $*_i$ in ∂B_i. Then $P_a = I_1 \times I_2$ and $P_b = (\partial B_1 - \text{int } I_1) \times (\partial B_2 - \text{int } I_2)$.) So we can write $W^\perp$ as $W^\perp = P_a \cup P_b \cup \hat{E}_{1,1} \cup \hat{E}_{1,2} \cup \hat{E}_{2,1} \cup \hat{E}_{2,2}$, where $\hat{E}_{i,1}$ and $\hat{E}_{i,2}$ denote two parallel copies of $\hat{E}_i$ whose boundaries coincide with ∂A_i. Note that $W^\perp$ can be constructed arbitrarily close to $C_1^\perp \cup \lambda_1 \cup \lambda_2 \cup C_2^\perp$, and that $W^\perp \cap (C_1 \cup C_2) = \emptyset$. (Also note that under our present trivializing assumptions on C_1 and C_2, $W^\perp$ is imbedded.)

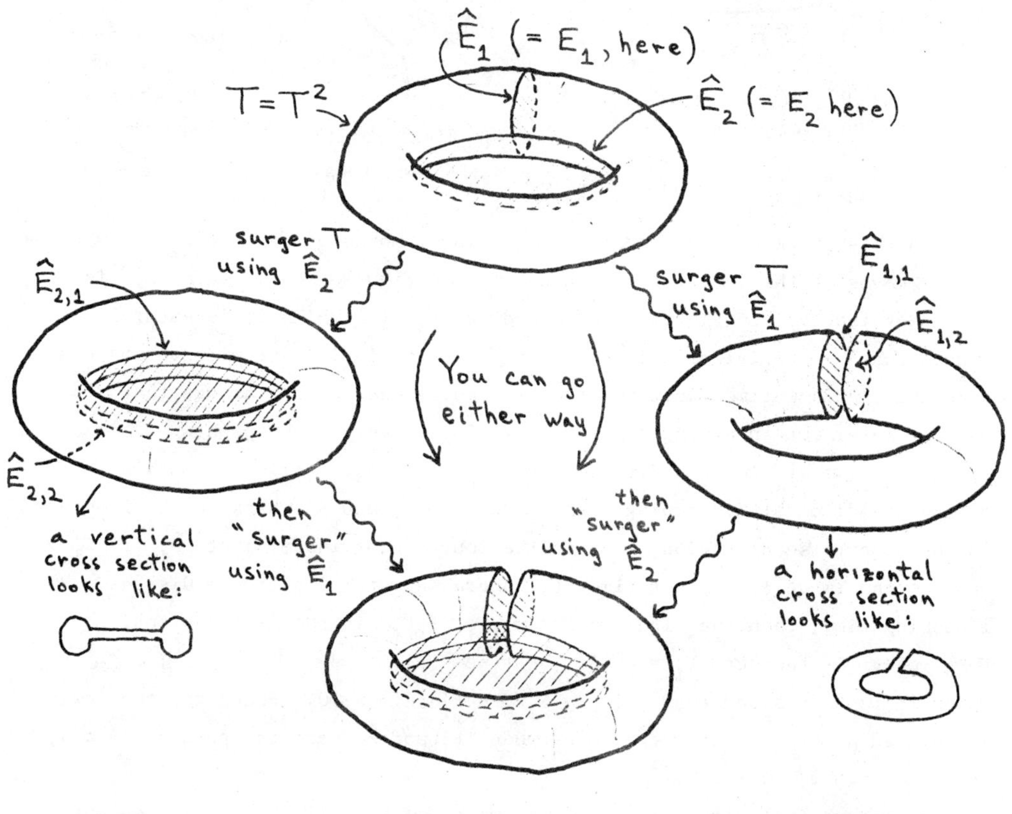

Doing double surgery on the torus T to get the 2-sphere $W^\perp$

Figure 4.5

We now consider how to move $F_1 \cup F_2$ off of $W^{\perp}$ by finger moves of F_1 and F_2. These moves can most easily be described using a 3-dimensional slice H of M, chosen to contain almost all of the relevant data, but disjoint from $C_1 \cup C_2$. This slice H, obtained from a neighborhood of $T^2 \cup E_1 \cup E_2$ in ∂N by rerouting it to follow $\hat{E}_1$ and $\hat{E}_2$ instead of E_1 and E_2 (so $H \approx S^3$ - two points, say), is shown in Figure 4.6a; it contains all of T, $\hat{E}_1$, $\hat{E}_2$ and $W^{\perp}$, and it contains 1-dimensional slices of F_1, F_2 and W, denoted F_1', F_2' and W'. Producting H with an interval $(-\varepsilon,\varepsilon)$ produces an open subset of M, with each subproduct $F_i' \times (-\varepsilon,\varepsilon)$ becoming an open subset of F_i, and likewise for $W' \times (-\varepsilon,\varepsilon) \subset W$.

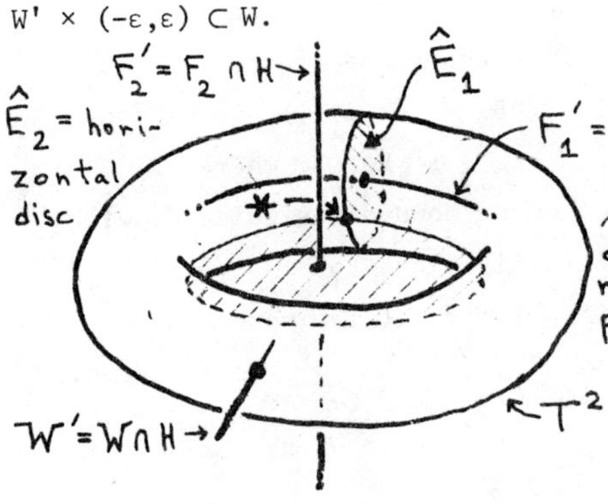

The slice H^3 of M^4 containing the data shown

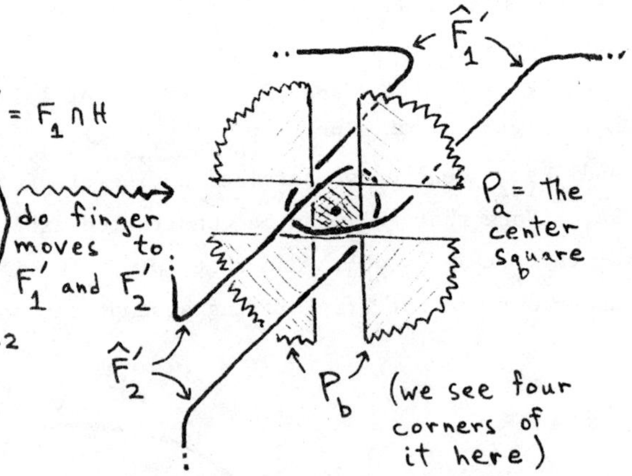

The final positions $\hat{F}_1'$ and $\hat{F}_2'$ of the 1-dimensional slices F_1' and F_2'

Figure 4.6

In this 3-dimensional model H the finger moves of F_1 and F_2 show up as finger moves of the intervals F_1' and F_2' (see Figure 4.6b). Each F_i' is moved (by isotopy) in the plane of the disc $\hat{E}_j$ (in which it is natural to assume that F_i' originally lies; here $j = 2$ if $i = 1$, and vice versa). At some intermediate time, just before these moves end, we can assume that F_1' and F_2' intersect in a single point, say (for symmetry) the point $* = (*_1, *_2) = \partial E_1 \cap \partial E_2 \in T$. The moves of F_1 and F_2 arise from this 3-dimensional model by damping these moves of F_1' and F_2' back to the identity in the transverse direction, i.e. in the fourth coordinate. That is, if $F_{i,t}'$ denotes the image of F_i' at time t, where say starting time $= 0 \le t \le \varepsilon/2 =$ finishing time, then the final position $\hat{F}_i$ of F_i in the 4-dimensional model $H \times (-\varepsilon,\varepsilon)$ is the set $\hat{F}_i = \cup \{F_{i,s}' \times t \mid -\varepsilon < t < \varepsilon,\ s = \varepsilon/2 - t\} \subset H \times (-\varepsilon,\varepsilon)$, where it is understood that $F_{i,s}' = F_{i,0}' = F_i'$ if $s < 0$. Hence the two newly introduced points of intersection between $\hat{F}_1$ and $\hat{F}_2$ are the points $(*,\pm\delta)$, for some δ, $0 < \delta < \varepsilon/2$.

To ensure that, after these moves, we have $W^{\perp} \cap (\hat{F}_1 \cup \hat{F}_2) = \emptyset$, we must require a certain modest compatibility relation between the construction of $W^{\perp}$

and the construction of $\hat{F}_1$ and $\hat{F}_2$. Namely, we must assume that the bands A_1 and A_2 in T (in the construction of $W^{\perp}$) have been chosen sufficiently thin, or reciprocally we must assume that the final 1-dimensional fingers $\hat{F}'_1$ $(= F'_{1,\varepsilon/2})$ and $\hat{F}'_2$ $(= F'_{2,\varepsilon/2})$ have been chosen sufficiently thick, i.e. wide (as opposed to long), so that $\hat{F}'_i \cap A_i = \emptyset$ (see Figure 4.6b). Hence $\hat{F}'_1 \cap T$ (= two points) $\subset \operatorname{int} A_2 - A_1$, and similarly $\hat{F}'_2 \cap T$ (= two points) $\subset \operatorname{int} A_1 - A_2$. Consequently $(\hat{F}'_1 \cup \hat{F}'_2) \cap (P_a \cup P_b) = \emptyset$, and so $(\hat{F}_1 \cup \hat{F}_2) \cap W^{\perp} = \emptyset$. This completes the discussion of this most elementary non-trivial case.

In one sense, the above operation is an elaboration of the following familiar process. Let G be a small 2-cell neighborhood in T of the point $T \cap W$, and let Q be a 4-cell regular neighborhood of the contractible set $(T - \operatorname{int} G) \cup \hat{E}_1 \cup \hat{E}_2 \cup F_1 \cup F_2$ rel $\partial G \cup \partial F_1 \cup \partial F_2$, thinking of the F_i's as discs. Then in the 3-sphere ∂Q, the three boundary circles look like:

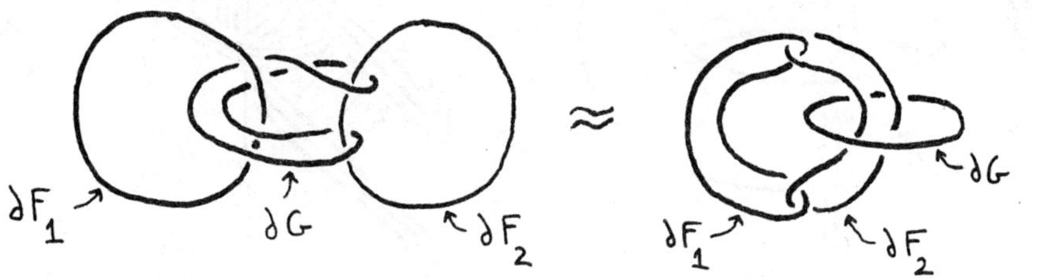

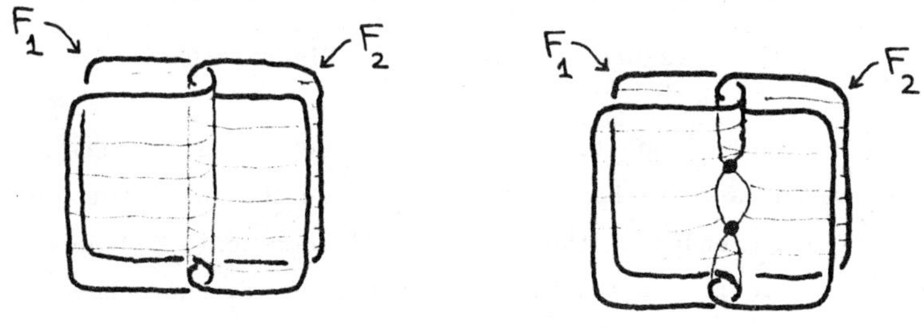

Before finger move; $F_1 \cap F_2 = \emptyset$ After finger move; $F_1 \cap F_2 = 2$ points

$F_1 \cup F_2$ in the 4-cell Q

Figure 4.7

That is, ∂G is a commutator in the complement of the unlink $\partial F_1 \cup \partial F_2$. In order to be able to span ∂G with a disc in Q which misses $F_1 \cup F_2$, we do finger moves to $F_1 \cup F_2$ to make them intersect so that the fundamental group of their complement becomes abelian. This disc spanning ∂G, unioned with G itself, becomes the 2-sphere $W^{\perp}$.

For the general case of the proof, where $C_1^{\perp}$ and $C_2^{\perp}$ (and hence the discs $\hat{E}_1$ and $\hat{E}_2$) have self-intersections and mutual intersections, we still have a model slice H and product $H \times (-\varepsilon,\varepsilon)$ as above, but these sets are no longer imbedded in our 4-manifold, only immersed. Nevertheless, the motions described above still make sense, because they can be transferred from the model to its immersed image; in fact, the finger motions of F_1 and F_2 can each be isotopies, since the path of each finger move can be chosen to avoid the double-point patches in the immersed 4-dimensional model. (Aside: the paragraph after next is relevant to this assertion.)

In the case where each F_i consists of several disjoint discs (recall that in effect this is the most general F_i), we choose the model slice H so that $F_i \cap H$ shows up as several parallel copies of our originally described interval F_i', all lying without loss in the plane of E_j (see Figure 4.8; $j = i\pm1$).

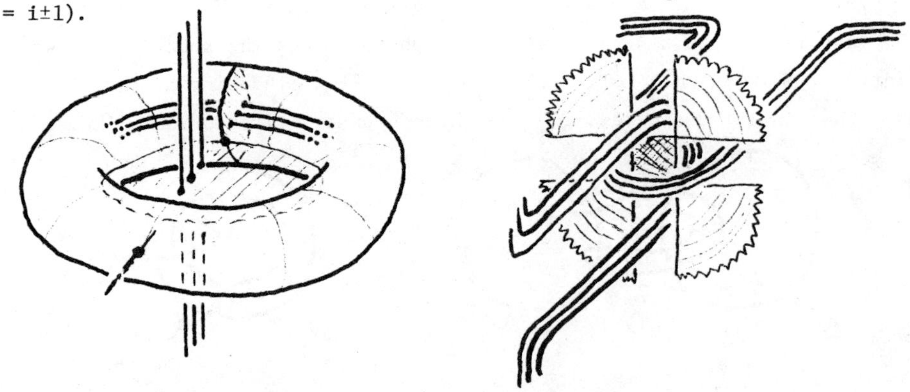

The picture when F_1 and F_2 have several sheets.

Figure 4.8 (compare to Figure 4.6)

The motions originally done to F_i' are now done to all of these parallel copies as a bunch, making sure as before that when done the resultant parallel copies of $\hat{F}_i'$ miss A_i, ensuring thereby that all components of the newly positioned $\hat{F}_i$ miss $W^\perp$. Note that all of the components of F_1 have been made to intersect all of the components of F_2.

Before we finish, there is an important technical point to be made about positioning $\hat{E}_1$ and $\hat{E}_2$ above, and how this relates to Addenda 1 and 2, in which we allow F_i to contain subdiscs of $C_1^\perp \cup C_2^\perp$ in order that $C_1^\perp \cup C_2^\perp$ winds up disjoint from $W^\perp$. Actually, to properly deal with this situation, one should in the construction above, take each $\hat{E}_i$ to be $E_i \# \overline{\overline{C}}{}_i^\perp$ instead of $E_i \# C_i^\perp$, where $\overline{\overline{C}}{}_i^\perp$ denotes a parallel copy of $C_i^\perp$, meaning a copy of $C_i^\perp$ which has been general positioned with respect to $C_i^\perp$. Hence, this new $\hat{E}_i$ meets $C_i^\perp$ transversally (as well as F_i), so it makes sense to allow that some of the discs in F_i be subdiscs of $C_1^\perp \cup C_2^\perp$. We note that the finger moves of F_i are a result of intersections of F_i with $\overline{\overline{C}}{}_i^\perp$ (not with $C_i^\perp$), and that these moves take place arbitrarily close to arcs lying in $\hat{E}_i$ (not arcs in $C_i \cup C_i^\perp$). Hence, for example, if none of the discs of F_i lie in $C_1^\perp \cup C_2^\perp$ for both $i = 1,2$, then in fact all finger motion during the construction is bounded away from $C_1^\perp \cup C_2^\perp$ (but of course it takes place nearby).

In closing, here are two simple illuminating cases to ponder, in which one wants to construct $W^\perp$ so that $W^\perp \cap (C_1^\perp \cup C_2^\perp) = \emptyset$.

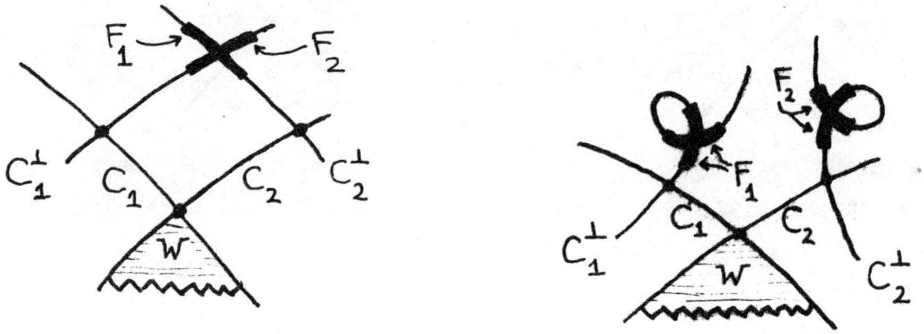

Two simple cases to ponder

Figure 4.9

Case 1 (see Figure 4.9a): Suppose each $C_i^\perp$ is imbedded, but $C_1^\perp \cap C_2^\perp$ = one point. Let each F_i be a disc in $C_{i+1}^\perp$ containing the intersection point. Here the resultant finger moves of $C_1^\perp$ and $C_2^\perp$ will leave them each imbedded, but will create two additional intersection points between them. The resultant $W^\perp$ will have four self-intersection points.

Case 2 (see Figure 4.9b): Suppose $C_1^\perp \cap C_2^\perp = \emptyset$, but suppose each $C_i^\perp$ has a single self-intersection point. Let each F_i consist of two subdiscs of $C_i^\perp$ centered at the crossing point, lying in the different sheets (c.f. the preceding technical point). Here the resultant $C_1^\perp$ and $C_2^\perp$ will each be left with a single self-intersection point (the original ones), but the finger moves done on $C_1^\perp \cup C_2^\perp$, namely two fingers being pushed from each, will create eight intersection points between $C_1^\perp$ and $C_2^\perp$, situated near the point $C_1 \cap C_2$. The resultant $W^\perp$ will have eight self-intersection points, in two groups of four, each group lying near one of the self-intersection points of $C_1^\perp$ or $C_2^\perp$.

This completes the proof of the Lemma.

Wistful note: One could get carried away with Addenda 1 and 2, and ask why in fact could one not allow $C_i^\perp$ to have additional intersections with C_i, hoping nevertheless to construct $W^\perp$ missing C_i, by making C_i part of F_i. But this seems to lead nowhere useful. For example, if one lets $C_1 = R^2 \times 0 \subset R^4 \supset 0 \times R^2 = C_2$, and $W = 0 \times [0,\infty)^2 \times 0$, and one lets $C_i^\perp$ be a small 2-sphere intersecting C_i in two points, then the construction leads to

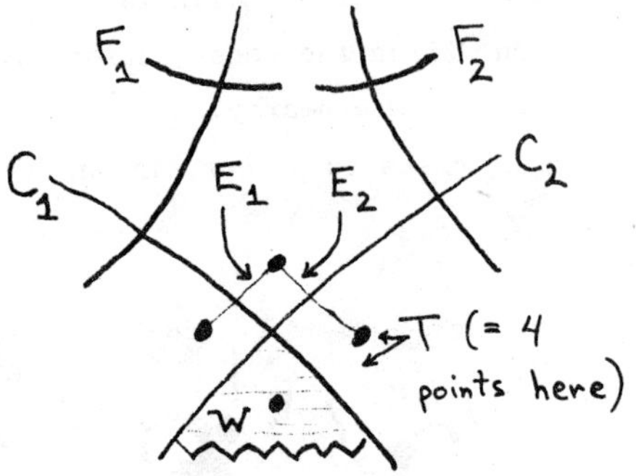

1. The initial setup with the dual torus T and spanning discs E_1 and E_2.

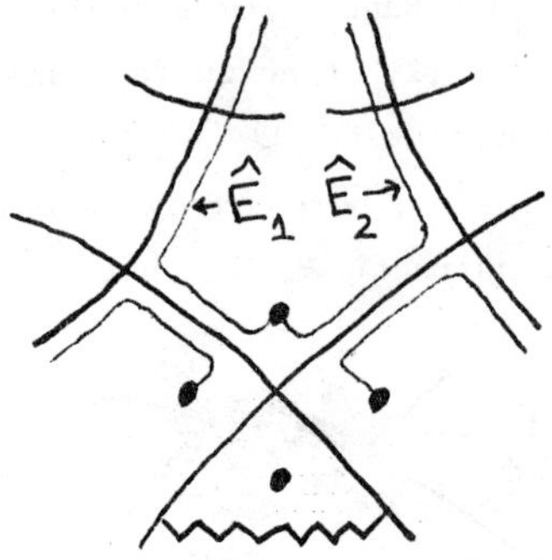

2. Let $\hat{E}_1 = E_1 \# \overline{\overline{C}}{}_1^{\perp}$ and $\hat{E}_2 = E_2 \# \overline{\overline{C}}{}_2^{\perp}$.

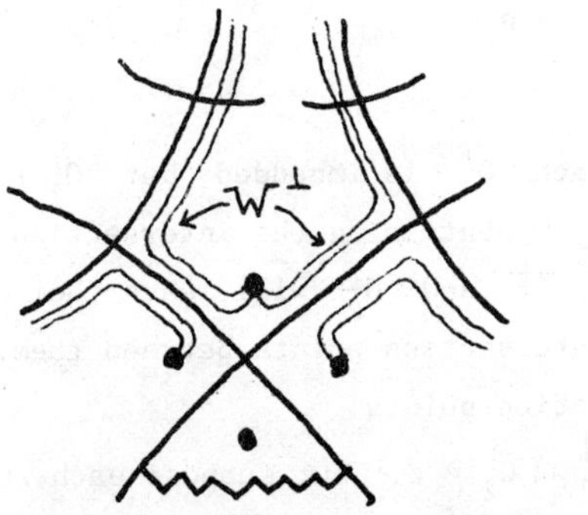

3. Do double surgery to T to get $W^{\perp}$.

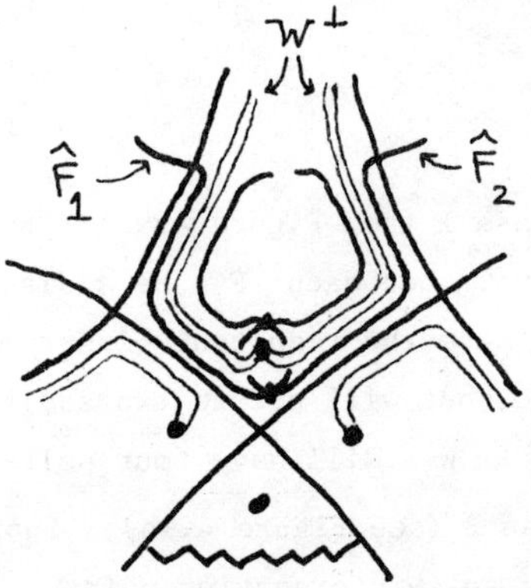

4. Do finger moves to F_1 and F_2 to get them off of $W^{\perp}$.

Schematic summary of the proof of the
Transverse Sphere Lemma
Figure 4.10

the following unproductive situation, in which one has constructed an imbedded transverse sphere $W^{\perp}$ at the expense of making two additional points of intersection between C_1 and C_2 (see Figure 4.11).

To close this section, we note that Quinn's Lemma above can be regarded as a geometrized version of Casson's original π_1-Lemma, applied in a special context. Recall that Casson's π_1-Lemma ([C, p. 3]; see [G-S, Lemma 2.1.1]) asserts that if S is a surface immersed in a 1-connected 4-manifold M_0 and if S has an algebraicly dual class, i.e., there is a surface S^d such that $S \cdot S^d = 1$, then one can do finger moves to S to make $M_0 - S$ 1-connected.

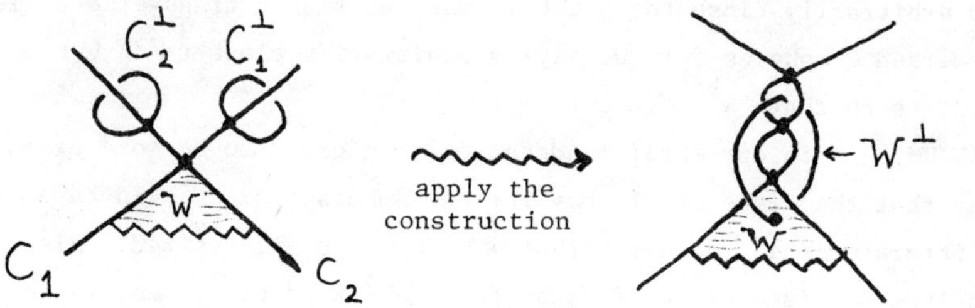

Figure 4.11

The main point is, if some π_1 element ω in $M_0 - S$ (in this case a meridian μ of S) can be expressed as a product of commutators of conjugates of meridians of surfaces (in this case the meridians are all the same μ), then, ω can be killed by doing finger moves to the surfaces (in this case S). Now in Quinn's setup (above), letting the ambient manifold be $M_0 = M - C_1 \cup C_2$ and letting S be $W \cup F_1 \cup F_2$, then the linking circle ω of W is the commutator of the linking circles of C_1 and C_2 (that is evidenced by the torus T), which in turn are products of (conjugates of) the meridians of F_1 and F_2 (evidenced by $C_1^\perp$ and $C_2^\perp$). Hence, Casson's π_1-Lemma asserts that by doing finger moves between F_1 and F_2, one can make μ null-homotopic in $M_0 - S$, which immediately provides the desired complementary sphere $W^\perp$.

5. MULTI-APPLICATIONS OF QUINN'S LEMMA.

In this section we state the Transverse Sphere Lemma in the actual form that it will be used. Since we no longer wish to distinguish the separate surfaces C_1 and C_2, or $C_1^\perp$ and $C_2^\perp$, or F_1 and F_2, we are combining them to become $C, C^\perp$ and F. Hence, C may be an unbounded (but locally finite) immersed surface of many (manifold) components (for us the components will always be compact, of uniformly bounded size, either cells or spheres). The collection $C^\perp$ will always be understood to be a <u>transverse collection of spheres</u> for C, which means that for each component C_γ of C there is an (immersed) sphere component $C_\gamma^\perp$ of $C^\perp$ whose intersection with all of C is just a single point $q_\gamma \in C_\gamma$. Thus $C \cap C^\perp = \bigcup_\gamma (C_\gamma \cap C_\gamma^\perp) = \{q_\gamma\}$.

QUINN'S TRANSVERSE SPHERES LEMMA. Suppose C, $C^\perp$ and F are surfaces (not necessarily connected) immersed in a 4-manifold M, where $C^\perp$ is a transverse collection of (immersed) 2-spheres for C, and F is a collection of discs normal to $C^\perp$. Suppose W is a union of Whitney-like discs attached to C, each disc W_μ associated to (at least) one distinct crossing point of C, say ρ_μ, at which ∂W_μ changes sheets of C, with no other disc of W passing through ρ_μ. Suppose $\Lambda = \bigcup \Lambda_\gamma$, where each Λ_γ is a union of paths in C_γ joining the point $C_\gamma \cap C_\gamma^\perp$ to all of the points of $\{\rho_\mu\}$ which lie in C_γ. Then, after doing finger moves to F (to create self-intersections), which are supported arbitrarily close to $\Lambda \cup C^{\perp}$, one can find a transverse collection $W^{\perp}$ of immersed 2-spheres for W, lying arbitrarily close to $\Lambda \cup C^{\perp}$, with $W^{\perp} \cdot W^{\perp} = 0$, such that $W^{\perp} \cap (C \cup F) = \emptyset$.

ADDENDUM. As in our earlier Addendum 1, in order to be more exact, we should say that the discs of F may include subdiscs of $C^{\perp}$ in case that $C^{\perp}$ has self-intersections, to ensure that $W^{\perp} \cap C^{\perp} = \emptyset$ if desired. Also, F should include subdiscs of W in case that $W \cap C^{\perp} \neq \emptyset$, to ensure that $W^{\perp}$ has no unwanted extra intersections with W. The other comments of the previous Addenda 1,2 and 3 also apply here, suitably adapted.

Note: It is possible, and indeed likely, that the different components of $W^{\perp}$ will intersect each other. However, if we iterate the Lemma, to produce a sequence $W_1^{\perp}, W_2^{\perp}, \ldots$ of transverse collections of spheres, then this sequence will be disjoint, provided that with each iteration F is chosen appropriately. Namely, with each application F should contain subdiscs of (both sheets of) $C^{\perp}$ containing the self-intersection points of $C^{\perp}$, so that $W_i^{\perp}$ will miss $C^{\perp}$, and also as usual F should contain subdiscs of W containing its intersection points with $C^{\perp}$. (Aside: If the reader is perplexed by the choice of the words "containing" here, he should ponder the technical point near the end of the proof in Section 4.) Hence, under this iteration, $C^{\perp}$ and W are constantly being moved, but each $W_i^{\perp}$ so produced is disjoint from $C \cup C^{\perp}$, and $W_i^{\perp}$ need not be moved when the subsequent $W_j^{\perp}$'s are produced.

The most powerful applications of the Lemma are obtained in this manner.

Proof of Lemma. The proof is the same as before, except that now one works on all of the discs in W simultaneously, keeping all of the data generically positioned as much as possible. The only motions required in the construction are the finger moves, which are supported arbitrarily close to 1-dimensional sets which can be chosen disjoint from each other and from other 2-dimensional data. Hence the finger moves can be done disjointly, without disturbing other data. As noted above, the resultant spheres of $W^{\perp}$ certainly may intersect each other.

6. A PRELIMINARY SEPARATION PROPOSITION

The basic problem which confronted Quinn was to find a way to maintain bounded control in Casson's construction when working in a noncompact manifold. As part of his analysis, Quinn had to determine exactly what sort of geometric input was required to accomplish a certain separation step in Casson's work. One result was the following Proposition (implicit in [Q_3, Section 3.2], and referred to there as the Group Separation Statement). Although it is finite in nature, it plays a key role in the noncompact main construction in the next section.

SEPARATION PROPOSITION. Suppose $C = C_1 \cup \ldots \cup C_n$ is a union of compact connected surfaces immersed in a 4-manifold M such that $C_i \cdot C_j = 0$ for $i \neq j$, and suppose that $C^\perp = C_1^\perp \cup \ldots \cup C_n^\perp$ is a transverse collection of (immersed) 2-spheres for C (as in Section 5). Suppose W is a union of pre-Whitney discs for all of the intersections between all pairs $C_i, C_j, i \neq j$ (i.e. the intersection points are paired, and there is one disc for each pair). Then the C_i's can be made disjoint, by regular homotopies which are supported arbitrarily close to $W \cup \Lambda \cup C^\perp$, where $\Lambda = \bigcup_{i=1}^n \Lambda_i$, and Λ_i is any union of paths in C_i joining the point $C_i \cap C_i^\perp$ to all the points of $C_i \cap \bigcup_{j \neq i} C_j$.

ADDENDUM. Furthermore, the newly positioned C can be provided with a (newly positioned) transverse collection $C^\perp$ which lies arbitrarily close to the original union $\Lambda \cup C^\perp$.

We note that the discs of W initially may intersect $C \cup C^\perp$ and each other in many unspecified points. Dealing with these unwanted intersections is the core of the Proposition.

Proof of Proposition. Let W_{ij}, $i < j$, denote the union of the discs in W which are associated to intersections between C_i and C_j.

First note that if $n = 2$, the proof is easy; it is a direct application of the Surface Separation Lemma (Section 2), and we don't even need $C_1^\perp$ and $C_2^\perp$.

If $n \geq 3$, the goal in effect is to reduce this general situation to a collection of disjoint $n = 2$ situations, which then can be separately finished off as above. That is, our primary aim is to achieve the following

Goal: For each $i, j (i < j)$, we wish to arrange that
$W_{ij} \cap C_k = \emptyset$ unless $k = i$ or $k = j$, and also that
$W_{ij} \cap W_{k\ell} = \emptyset$ unless $(i,j) = (k,\ell)$.

In other words, we want each W_{ij} to intersect only C_i and C_j among all of the C_k's, and we want all $n(n-1)/2$ of the W_{ij}'s to be disjoint. One might observe that the second condition is easy to arrange at the expense of the first by means of piping intersections among the W_{ij}'s off of the edges of the W_{ij}'s, but this turns out to be the wrong way to proceed.

Instead, as a preliminary step toward the Goal we first use the Norman trick to get rid of all intersections between $\text{int}\, W$ and C, using $C^\perp$ to reroute the various discs of W. Thus, we can assume that $\text{int}\, W \cap C = \emptyset$. Also, we assume that the W_{ij}'s have been repositioned (via piping along $\partial W \subset C$ as in Section 2) so that their boundaries are all disjoint.

The remainder of the Goal, namely getting the $\text{int}\, W_{ij}$'s disjoint, is achieved using the Transverse Spheres Lemma in Section 5. By means of $n(n-1)/2$ successive applications of it (see the Note there), each time letting F be all of (the possibly repositioned) $C^\perp \cup W$, say, we can find a sequence $W_{1,2}^\perp, W_{1,3}^\perp, \ldots, W_{1,n}^\perp, W_{2,3}^\perp, \ldots, W_{2,n}^\perp, \ldots, W_{n-1,n}^\perp$ of $n(n-1)/2$ disjoint transverse collections of spheres for W. (Actually, each collection $W^{\perp}_{ij}$ need only be a transverse collection for W_{ij}, consisting therefore only of one sphere for each disc in W_{ij}. But there is no profit in trying to be economical here.) The spheres in each collection $W^{\perp}_{ij}$ may intersect each other, but no $W^{\perp}_{ij}$ intersects (the finally positioned) $C \cup C^{\perp}$. Note that during the finger moves required for all of this, new self-intersections are created in $C^{\perp} \cup W$, but no new intersections are created between $C^{\perp} \cup W$ and C, and also C is not moved. (For a mild variation here, see (2) below.)

Now we use the $W^{\perp}_{ij}$'s to achieve the Goal (we no longer need $C^{\perp}$). For each distinct pair $(i,j) < (k,\ell)$ (lexicographic order, say), we use the Norman trick to reposition $W_{k\ell}$ to miss W_{ij}, by using $W^{\perp}_{ij}$ to reroute $W_{k\ell}$. The newly positioned $W_{k\ell}$'s may have additional self-intersections, but they no longer intersect each other. Hence we have achieved the Goal (and in fact we also have that $\operatorname{int} W_{ij} \cap (C_i \cup C_j) = \emptyset$, but this has no significance).

At this point, the discussion in the first two paragraphs of the proof applies to finish the proof.

As for the Addendum, we note that it was not automatically achieved by the above construction; it may well be that the (finally positioned) W_{ij}'s intersect $C^{\perp}$, and hence that when the C_i's are separated, they are made to intersect $C^{\perp}$ in additional points. There are two natural ways to remedy this, both involving constructing one additional layer $W^{\perp}_{*}$:

1) One could carry the construction of the $W^{\perp}_{ij}$'s one step further, producing a last collection $W^{\perp}_{*}$ which is transverse to all of W, and then at the end of the proof one could use this final collection to get rid of intersections of $C^{\perp}$ with W by means of the Norman trick.

2) Alternatively, at the start of the proof, right after the preliminary step, one could make a preliminary application of the Transverse Spheres Lemma (Section 5) to produce an initial transverse collection $W^{\perp}_{*}$, finger-moving $C^{\perp} \cup W$ to do so, and then one could use $W^{\perp}_{*}$ to get rid of the intersections of $C^{\perp}$ with W via the Norman trick. Now when the subsequent collections $\{W^{\perp}_{ij}\}$ are produced, they do not require moving W, and so W remains disjoint from $C^{\perp}$, and so $C^{\perp}$ remains geometrically complementary to C.

This method (2) is used at several points in Section 7.

It is interesting to note that, although the preceding Proposition will be instrumental in achieving control of motions in the next section, nevertheless during the proof above a point may wind up being moved the full diameter of $C \cup C^{\perp} \cup W$.

7. QUINN'S CONSTRUCTION

In this section we will present Quinn's full construction in a specific context, to make it more concrete and more digestible. It will be the situation that arises, for example, in the proof of the 4-dimensional Annulus Conjecture, as explained in the next section. Or, changing a phrase here and there, it is the situation that arises in showing that a manifold proper homotopy equivalent to R^4 is in fact homeomorphic to R^4 (which in turn trivially yields the topological 4-dimensional Poincaré conjecture). As for the appropriate generalized setting for this section, which is more complicated only in appearance, we refer the reader to the relevant parts of $[Q_3]$ and $[Q_1]$.

Everything in this section is smooth, except for the brief discussion surrounding $(*_1)$ and $(*_2)$ below, where we invoke Freedman's work.

We assume in this section that we are presented with a certain smooth noncompact 4-manifold $M = R^4 \#_\alpha (S^2 \times S^2)_\alpha$, that is, M is gotten from R^4 by connect-summing R^4 with some locally finite collection of $S^2 \times S^2$'s. Given in M are four distinguished locally finite, transversally intersecting collections of disjoint imbedded 2-spheres $\{A_\alpha\}$, $\{A^d_\alpha\}$, $\{B_\alpha\}$ and $\{B^d_\alpha\}$. This M will be like the middle level of a 5-dimensional proper h-cobordism built on R^4, in which there are only handles of index 2 and 3, which have been paired in some appropriate controlled manner, with the B_α's (respectively, the A_α's) representing the belt, i.e. ascending 2-spheres of the 2-handles (respectively, the attaching, i.e. descending 2-spheres of the 3-handles), and with the B^d_α's and A^d_α's being respective duals for them. In Section 8 we describe exactly how such an M arises in the proof of the 4-dimensional Annulus Conjecture.

Presenting our hypotheses on M more carefully, let $\{D^4_\alpha\}$ be a locally finite collection of small round balls in R^4 each of diameter < 1, say, and let M be gotten from R^4 by connect-summing R^4 with a collection $\{(S^2 \times S^2)_\alpha\}$ of $S^2 \times S^2$'s at the D^4_α's (for purposes below, we regard that $D^4_\alpha \subset (S^2 \times S^2)_\alpha$ also). Since we will want to talk about boundedness in M, which ultimately is to be related to boundedness in R^4, we assume that M is provided with a (topological) metric which on $R^4 - \cup_\alpha \operatorname{int} D^4_\alpha \subset M$ agrees with the euclidean metric, and such that under this metric each subset $(S^2 \times S^2)_\alpha - \operatorname{int} D^4_\alpha$ of M has uniformly bounded diameter, say < 2. (For example, one could build M from $R^4 = R^4 \times 0$ working in R^5, and take the inherited metric.)

The collections of spheres listed above are as follows. For each α, $A_\alpha \cup A^d_\alpha$ is a spine of $(S^2 \times S^2)_\alpha - \operatorname{int} D^4_\alpha$. So A_α and A^d_α intersect once, transversally, and each has a product normal bundle neighborhood. Similarly, for each α we assume that the spheres B_α and B^d_α intersect once, transversally, and each has a product normal bundle neighborhood. In addition, we assume that the B_α's and B^d_α's have uniformly bounded diameter (but the bound may be huge), and that for each pair α,β, we have $A_\alpha \cdot B_\beta = \delta_{\alpha\beta}$ (kronecker δ). Consequently, each pair $B_\alpha \vee B_\alpha^d$ lies within some uniformly bounded distance of $A_\alpha \vee A_\alpha^d$ (although it does not necessarily lie in $(S^2 \times S^2)_\alpha - D_\alpha^4$). The model situation is that each $B_\alpha \vee B_\alpha^d$ is a "parallel" copy of $A_\alpha \vee A_\alpha^d$, but with the order of the factors reversed, as in Figure 7.1.

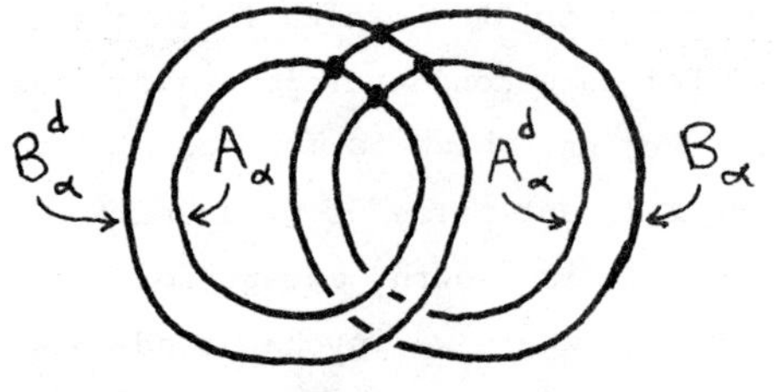

Figure 7.1

As noted already, we explain in Section 8 how such an M might arise in practice; suffice it to say here that if we take the union along M of the two surgery traces (i.e., the 5-dimensional cobordisms gotten by doing two sets of surgeries to M, one set of surgeries on the A_α's ("to one side of M"), and the other set on the B_α's ("to the other side of M"; see Figure 7.2), then this 5-dimensional union is a proper h-cobordism having only 2 and 3-handles, with its middle level being M, one end being R^4, and the other end being a proper-homotopy R^4 (in fact it can be any preassigned, possibly exotic, smooth proper-homotopy R^4).

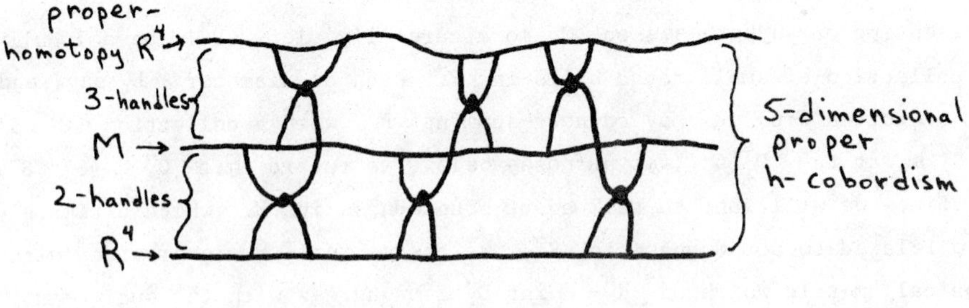

Figure 7.2

It will turn out that this cobordism is topologically a product, based on the work to be done in this section.

The goal of this section is to establish:

$(*_1)$: There is a (uniformly) bounded (topological) ambient isotopy of M, which remains the identity on some open subset of M, and which repositions $B = \cup B_\alpha$ so as to remove all excess intersections between it and $A = \cup A_\alpha$ (i.e. it achieves $A_\alpha \cap B_\beta = \delta_{\alpha\beta}$ points).

The isotopy which gets rid of these excess intersections will be gotten by means of the usual Whitney process, once suitable Whitney discs have been found. Recall that Freedman has shown that 6-stage towers are as good as Whitney discs, in that neighborhoods of such towers contain topologically flat spanning discs. Hence, it suffices to establish:

$(*_2)$: There is a locally finite, disjoint collection of (smooth) 6-stage towers in M, of uniformly bounded size, their bases attached to $A \cup B$ in Whitney-like fashion, one tower for each pair of excess intersections between A and B.

(Actually, before achieving $(*_2)$, a preliminary isotopy of B may be required, which creates additional pairs of intersections with A, but it is understood that $(*_2)$ is providing towers for these, too.)

The remainder of this section is devoted to Quinn's proof of $(*_2)$. From this point on, everything is smooth. Furthermore, all isotopies and regular homotopies of M and its subsets will be assumed bounded, even if not explicitly so stated. Of course, as usual all maps, subsets, etc. will be assumed to be generically positioned, subject to the restrictions at hand.

For convenience at this point, we list all of the intersection properties (algebraic and geometric) of the four locally finite collections of imbedded spheres that we will use.

1) For each pair α,β, $A_\alpha \cap A_\beta = \emptyset = B_\alpha \cap B_\beta$ (if $\alpha \neq \beta$), and $A_\alpha \cdot A_\alpha = 0 = B_\alpha \cdot B_\alpha$ and $A_\alpha \cdot B_\beta = \delta_{\alpha\beta}$.

2) For each pair α,β, $A^d_\alpha \cap A_\beta = \delta_{\alpha\beta}$ points, and similarly $B^d_\alpha \cap B_\beta = \delta_{\alpha\beta}$ points.

However, we note that for all α,β, the intersection numbers $A^d_\alpha \cdot A^d_\beta$, $B^d_\alpha \cdot B^d_\beta$, $A^d_\alpha \cdot B^d_\beta$, $A^d_\alpha \cdot B_\beta$ and $A_\alpha \cdot B^d_\beta$ are immaterial.

Several of the steps which follow are, as one might expect, similar to those used by Casson and Freedman in their analyses of h-cobordisms. Among these is the

Preliminary Setup. In this step, after perhaps isotoping the B_α's, we find new transverse collections $A^\perp = \cup A^\perp_\alpha$ and $B^\perp = \cup B^\perp_\alpha$ of immersed spheres of uniformly bounded size, such that the combined collection $A^\perp \cup B^\perp$ is transverse to the combined collection $A \cup B$ (i.e., for each α, $A^\perp_\alpha \cap (A \cup B) = A^\perp_\alpha \cap A_\alpha$ = one point, and similarly for each $B^\perp_\alpha$; see Section 5), and such that for each α we have $A^\perp_\alpha \cdot A^\perp_\alpha = 0 = B^\perp_\alpha \cdot B^\perp_\alpha$. (In fact, we get $A^\perp \cdot A^\perp = 0 = B^\perp \cdot B^\perp$, but that is not needed.)

We show how to produce $A^\perp$. (Technical note: the construction which follows is a mild variation on the one used in [F_2, Lemma 10.1] (derived in turn from [C, III, Lemma 1]), for we make $A^\perp_\alpha$ from a parallel copy of B_α, not from A^d_α.)

The construction which follows amounts to an application of Casson's Surface Separation Lemma (Section 2). Let P be a parallel copy of B, so that $P \cap B = \emptyset$ (recall $B \cdot B = 0$; here then $P = \cup P_\alpha$ is a locally finite union of disjoint imbedded spheres). Since $P_\alpha \cdot A_\beta = \delta_{\alpha\beta}$, we can pair off the excess intersection points between P and A and choose for them a union W of pre-Whitney discs of uniformly bounded size. After getting ∂W imbedded, we can arrange that $\mathrm{int} W \cap A = \emptyset$ by doing the Norman trick to $\mathrm{int} W$, using A^d. Still, $\mathrm{int} W$ may intersect B (but its other intersections, for example those with P, A^d, B^d and itself, we don't care about). These points of $\mathrm{int} W \cap B$ are piped off of the A-edge of W by isotopy of B (creating new pairs of intersections between A and B, but that is acceptable). Now we can use W as in the Surface Separation Lemma to regularly homotope P to get rid of its excess intersections with A (making sure that if we spin W to correct framings, we do so at its P edge; here intersections of $\mathrm{int} W$ with P and with itself lead only to self-intersections in P). The newly positioned P is our desired $A^\perp$. Note that $A^\perp \cdot A^\perp = P \cdot P = 0$, since $A^\perp$ has been obtained from P (hence B) by regular homotopy.

In a similar fashion, starting with a parallel copy Q of A, and interchanging the roles of A, B, etc. above, one produces the desired $B^\perp$. (Actually, one can get $B^\perp$ more quickly simply by starting with B^d and getting rid of its intersections with A by means of the Norman trick, using for this the new collection $A^\perp$. This requires appealing to some of the "immaterial" intersection properties of B^d mentioned after (2) above.) ///

At this point, having completed the preliminary setup, we are ready to begin what amounts to an induction, which we will cycle through six times, constructing one tower layer each time (more on this later). Before beginning, however, we wish to relabel our new collections of spheres, and make some of them discs, to make this first quasi-induction step more like the later ones. So, for each α, choose a distinguished point $p_\alpha \in A_\alpha \cap B_\alpha$, and remove a small open round ball from M centered at p_α. Calling the resultant manifold M_0, we henceforth denote by $\{\Delta_\sigma\}$ the entire resultant collection of (properly imbedded) discs in M_0, i.e. the holed A_α's and B_α's, and we denote by $\{\Delta_\sigma^\perp\}$ the entire transverse collection $\{A_\alpha^\perp\} \cup \{B_\alpha^\perp\}$ of (immersed) spheres in M_0 produced in the Preliminary Setup (thus the indexing set for σ is two copies of the index set for α). Now the goal $(*_2)$ can be restated thus:

$(*_3)$ There is a locally finite disjoint collection of (smooth) 6-stage towers in M_0, of uniformly bounded size, one tower for each pair of intersections between the Δ_σ's.

For convenience, we list the intersection properties of the imbedded discs

$\Delta = \cup\Delta_\sigma$ and the immersed spheres $\Delta^\perp = \cup\Delta_\sigma^\perp$ which are used henceforth.

(1_Δ) For each pair σ,τ, with $\sigma \neq \tau$, we have $\Delta_\sigma \cdot \Delta_\tau = 0$, and also $\Delta_\sigma^\perp \cdot \Delta_\sigma^\perp = 0$ (<u>even</u> here would do, instead of 0).

(2_Δ) The collections are complementary (transverse), i.e., for each σ, $\Delta_\sigma^\perp \cap \Delta = \Delta_\sigma^\perp \cap \Delta_\sigma = 1$ point.

In particular, then, the intersection numbers $\Delta_\sigma^\perp \cdot \Delta_\tau^\perp$ are immaterial for $\sigma \neq \tau$.

We break the inductive part of the proof (i.e. achieving $(*_3)$) into eight bite-sized steps, which we will cycle through a total of six times. We call attention to the summary table which appears later in this section.

The overall purpose of this first induction round (= the first eight steps) is to construct a collection $C = \cup C_\gamma$ of immersed discs attached to Δ which will serve as the bases of our desired towers. Thus, the C_γ's are to be disjoint, with $\text{int}\, C$ $(= \cup\, \text{int}\, C_\gamma)$ disjoint from Δ (which, incidentally, is never moved here) and with ∂C $(= \cup\, \partial C_\gamma)$ imbedded and properly framed.

<u>Step 1</u>. <u>Selecting and initializing the</u> C_γ's. Here we find a collection $C = \cup C_\gamma$ of (immersed) pre-Whitney discs of uniformly bounded size, for all of the (paired) intersections among the Δ_σ's. These C_γ's will become the bases of our towers. We want the C_γ's to have the following properties (for now):

(i) the boundaries of the C_γ's are imbedded and disjoint,

(ii) the framings of the C_γ's are correct as bases of towers (relative to the way they are attached to Δ), as explained in Section 3,

(iii) for each γ, $\text{int}\, C_\gamma \cap \Delta = \phi$, and

(iv) for each pair γ,δ, $\gamma \neq \delta$, we have $C_\gamma \cdot C_\delta = 0$ (one can interpret (ii) as saying that $C_\gamma \cdot C_\gamma = 0$).

To begin the construction, suppose the intersections among the Δ_σ's have been paired, and let $C = \cup C_\gamma$ be any collection of pre-Whitney discs for these pairs. The discs may be assumed to be of uniformly bounded diameter, from the geometry of M_0 and the boundedness of the components of Δ. This property will be maintained throughout, and nothing further will be said about it. As we modify these discs to produce the desired collection of Step 1, we will continue to denote them by $C = \cup C_\gamma$, to minimize notational proliferation.

The C_γ's initially may have none of the desired properties (i)-(iv) above. We will work to correct these defects, much as in the proof of the Casson Lemma (Section 2), but here we must work a bit harder, since the desired repositioning is a bit more delicate. The method to be followed here is basically Casson's (from [C, Lecture I]), with some minor variations to make it more geometric, as presented for example in [F_2, Section 3].

To arrange property (i), one works just as in Section 2, desingularizing the various attaching paths by piping their intersections off of their ends.

Regarding property (ii), initially we arrange it to hold only modulo 2. That is, we arrange that

(ii_e) the framings of the C_γ's are correct modulo 2, i.e., the framing mismatch is even.

This is achieved as usual by spinning C_γ at ∂C_γ, as explained in Sections 2,3 (as earlier, we may spin at either arc of ∂C_γ; it doesn't matter which. Of course, at most one spin is required).

Next, property (iii) is arranged, by means of the Norman trick, using $\Delta^\perp$ to get $\operatorname{int} C$ off of Δ. In effect each C_γ is replaced by some linear combination $C_\gamma \# \Sigma n_{\gamma,\sigma} \Delta^\perp_\sigma$. The self-intersections of these new C_γ's agree mod 2 with the original self-intersections, because $\Delta^\perp_\sigma \cdot \Delta^\perp_\sigma = 0$ (of course even would do here), and so property (ii_e) is maintained.

Before achieving property (iv), and property (ii) on the nose, we note the existence of a certain collection $C^d = \cup C^d_\gamma$ of immersed spheres, with $C^d \cdot C^d = 0$, which are algebraically dual to the collection C (i.e. $C^d_\gamma \cdot C_\delta = \delta_{\gamma\delta}$) and are disjoint from Δ. To get them, start with a collection of (small imbedded disjoint framed) tori $T^d = \cup T^d_\gamma$ dual to C, as explained at the start of the proof in Section 4 (the distinguished tori, in the language of [F_2]). Do (single) surgery to each T^d_γ to turn it into a homologous immersed sphere C^d_γ, using $\Delta^\perp$ and the Norman trick to avert intersections with Δ. As T^d had the desired algebraic properties, so does C^d. (However, we note that $C^d_\gamma \cap C_\delta$ may have extra pairs of points. One could use Section 5 here to construct $C^\perp_\gamma$'s, but they would be of no additional help at this point.)

Now, returning to properties (ii) and (iv), observe that they can be achieved by connect-summing each C_γ with appropriately many copies of the various $C^\perp_\delta$'s. For example, assuming the subscripts $\{\gamma\}$ are ordered in some sequence, we can replace each C_γ by the multifold connect-sum $C_\gamma \# \sum_{\delta<\gamma} (-C_\delta \cdot C_\gamma) C^d_\delta \; \# \; (-C_\gamma \cdot C_\gamma/2) C^d_\gamma$. This completes Step 1. ///

We pause a moment here to remark on how Quinn's construction is about to depart from Casson's. The goal for Casson/Quinn at this point is to get the C_γ's disjoint from each other. For Casson, everything was finite, and so he was able to proceed one C_γ at a time, making it disjoint from all of the previous ones, before proceeding to the next C_γ. In the infinite case, however, this process breaks down, for the usual reason that a point may wind up getting moved an infinite number of times, right out to infinity. Thus the natural question was, how could one reorganize Casson's procedure for the infinite case into infinitely many disjoint collections of finite procedures. This is what Quinn is doing in the steps which follow.

Quinn's idea at this point is to partition the C_γ's into finitely many groups, each group itself consisting of infinitely many disjoint finite subgroups of C_γ's. Within any given group, the subgroups are to be quite isolated, separated from each other by some distance much larger than the size of any disc or sphere yet encountered in the proof. The motivating analogy is, if we think of the C_γ's as tiny, microscopic cells in 4-space, then we want to take a medium-sized handle decomposition of 4-space, and let all of the C_γ's which intersect the 0-handles comprise one group, with the subgroups being engendered by the individual 0-handles; let all of the remaining C_γ's which intersect the (disjoint) 1-handles comprise the next group, etc. In the next step, this is formalized (and should be skipped by those familiar with such details).

<u>Step 2</u>. <u>Partitioning the C_γ's into groups</u>. To begin, we need a covering of R^4 by some finite number, p say, of closed subsets $K_1,\ldots,K_p$ (thinking of each K_i as an infinite, disjoint union of cubes), where the components of each K_i are uniformly bounded in size, and yet the distance between any two components of any single K_i is at least ℓ, where ℓ is some number much larger than the size of any sphere or disc yet encountered in the proof. (The smallest possible choice for p is 5. However, for $p=2^4$ one has the natural checkerboard collection of cubes obtained by letting, for each subset $\varphi \subset \{1,2,3,4\}$, K_φ be the union of cubes in R^4 with edges parallel to the axes, of edge length $\ell \gg 0$, centered at points of the form $(\ell z_1,\ldots,\ell z_4)$ where $z_i \in 2\mathbf{Z}$ or $z_i \in \mathbf{Z}-2\mathbf{Z}$ according as $i \in \varphi$ or $i \notin \varphi$.) These subsets $\{K_i\}$ of R^4 give rise to a collection of subsets of M, still denoted $\{K_i\}$, having the same sort of properties, say by assigning each $(S^2\times S^2)_\alpha$ in the definition of M to the lowest indexed K_i that D^4_α intersects. Using these K_i's we can partition the C_γ's. Let C_1 be the union of those C_γ's which intersect K_1, and in general, let C_i be the union of those C_γ's which intersect K_i but not any earlier K_j's. ///

The individual C_γ's are going to be separated in two steps. First the different groups produced in Step 2 will be separated (Steps 3 and 4). The motion here will be small compared to the large distance between individual subgroups of any given group, and so these subgroups will remain bounded far apart. Next, one prepares to separate the C_γ's within the individual groups. This requires some auxiliary data, namely some pre-Whitney discs W and some transverse spheres $C^\perp$, which are to be constructed for each subgroup (Steps

5-7). Finally, in Step 8 the individual disjoint groups-plus-auxiliary-data can be worked on separately, and the individual C_γ's made disjoint.

As we will see in upcoming steps, $\Delta^\perp$ is sort of the backbone of the proof, for we are always returning to use it to produce lots of different layers of $C^\perp$'s. To this end, the following step is an example of a useful general principle: It is desirable to keep $\Delta^\perp$ separated from as much of the other data as possible, so that when it is to be used again, these other data needn't be moved.

Step 3. Getting the $\Delta_\sigma^\perp$'s off of the C_γ's. The goal here is to arrange that $\Delta^\perp \cap C = \emptyset$. This will be achieved by regular homotopies of C and $\Delta^\perp$.

To begin, apply the Transverse Spheres Lemma of Section 5 to obtain a transverse collection $C^\perp = \cup C_\gamma^\perp$ of immersed spheres for C, so that $C^\perp \cap \Delta = \emptyset$. Here we are applying the Lemma with the sets C, $C^\perp$, W and F of the Lemma being respectively the sets $\Delta, \Delta^\perp$, C and subdiscs of C here, and so the proof entails moving C by regular homotopy. (Aside: This unfortunate mismatch of notation was bound to occur somewhere in the proof, inasmuch as the Lemma is applied in several different places.)

Having produced $C^\perp$, we can now use the Norman trick to get rid of the intersections between $\Delta^\perp$ and C, by connect-summing the $\Delta_\sigma^\perp$'s with appropriate $C_\gamma^\perp$'s, as needed. (This happens to be a regular homotopy of $\Delta^\perp$, as each $C_\gamma^\perp$ is regularly null-homotopic by construction. After this, this $C^\perp$ is no longer of any use, although fresh ones will be needed later.) ///

Step 4. Getting the groups of C_γ's disjoint. First we construct p-1 disjoint transverse collections of spheres $C_1^\perp, C_2^\perp, \ldots, C_{p-1}^\perp$ for C so that $C_i^\perp \cap (\Delta \cup \Delta^\perp) = \emptyset$. This is achieved by p-1 successive applications of the Transverse Spheres Lemma (Section 5; see its Note), with the sets C, $C^\perp$, W and F of the Lemma being $\Delta, \Delta^\perp$, C and subdiscs of $\Delta^\perp$ here, for $i=1$ to $p-1$, producing $C_i^\perp$ at the ith step, making sure at each step that one stays away from the previously constructed $C_j^\perp$'s. Note that C (as well as Δ) needn't be moved here, only $\Delta^\perp$ (repeatedly).

Now we can use these $C_i^\perp$'s to make the different groups $\{C_i\}$ disjoint. Starting with C_1, using $C_1^\perp$ to move each C_j off of C_1, $j > 1$, via the Norman trick. Next, use $C_2^\perp$ to move each (newly positioned) C_k off of (the new) C_2, $k > 2$. Continue in this manner. Note that when done, the initialization accomplished in Step 1 still holds, and similarly it remains true that $\Delta^\perp \cap C = \emptyset$. The above $C_i^\perp$'s, being useful no longer, are discarded. ///

Having separated the groups of C_γ's, we now must provide the individual groups with some additional data. The first of these are some pre-Whitney discs.

Step 5. Selecting W, and making $W \cap (\Delta \cup \Delta^{\perp}) = \emptyset$. For each group C_i, let W_i be a collection of pre-Whitney discs for all of the intersections between different cells of C_i (which, recall, only occur between cells of the same subgroup). Let $W = \cup W_i$ (we needn't initialize the framings of W in any manner, at this point).

Using $\Delta^{\perp}$, get W off of Δ by means of the Norman trick.

We next arrange that $\Delta^{\perp} \cap W = \emptyset$. To do so, construct a transverse collection $C^{\perp}$ to C which misses $\Delta \cup W$ (but there is no need to make it miss $\Delta^{\perp}$), by using the Transverse Spheres Lemma (Section 5) with the sets $C, C^{\perp}$, W and F of the Lemma being $\Delta, \Delta^{\perp}, C$ and subsets of W here; this may require regularly homotoping W, but not C (nor Δ and $\Delta^{\perp}$), as $\Delta^{\perp} \cap C = \emptyset$. Given $C^{\perp}$, we can move $\Delta^{\perp}$ off of W first by piping $\Delta^{\perp}$ off of the edges of W, and then getting rid of the resultant intersections of $\Delta^{\perp}$ with C by using the Norman trick with respect to $C^{\perp}$. So now we have $W \cap (\Delta \cup \Delta^{\perp}) = \emptyset$, and we have retained that $\Delta^{\perp} \cap C = \emptyset$. The above $C^{\perp}$ is no longer needed. ///

At this point, we must separate the members of W which are attached to distinct groups of C.

Step 6. Separating the groups $\{W_i\}$. As before, this is accomplished using layers of $C^{\perp}$'s. Using the Transverse Spheres Lemma $p(p-1)$ times in succession, construct $p(p-1)$ disjoint collections of immersed 2-spheres $C^{\perp}_{i,j}$, $1 \leq i,j \leq p$, $i \neq j$, where for each i, $C^{\perp}_{i,j}$ is transverse to C_i, and $C^{\perp}_{i,j} \cap (\Delta \cup \Delta^{\perp} \cup W) = \emptyset$ (the sets $C, C^{\perp}$, W and F of the Lemma are $\Delta, \Delta^{\perp}$, C_i and subdiscs of $\Delta^{\perp}$ here, making sure as usual that when constructing $C^{\perp}_{i,j}$, one stays away from other C_k's, and away from previously constructed $C^{\perp}_{k,\ell}$'s). Only $\Delta^{\perp}$ need to be moved here (repeatedly).

Now, for each i,j, to get W_j off of $C_i \cup W_i$ $(j \neq i)$, first move W_j off of W_i by piping it off of W_i edges, and then move W_j off of C_i by means of the Norman trick, using $C^{\perp}_{i,j}$. When done, we have $(C_i \cup W_i) \cap (C_j \cup W_j) = \emptyset$ for all $i \neq j$. Since these motions are small with respect to the distance between subgroups of groups, we now have that any individual C_γ or W_μ intersects any other individual C_δ or W_ν (four possibilities here) only if these two intersecting cells belong to the same subgroup of the same group. Finally, we note that all of the properties arranged in Steps 1 through 5 remain true. The above $C^{\perp}_{i,j}$'s are no longer needed. ///

The last data to be provided for the subgroups are some complementary spheres for the C's.

Step 7. Providing groups of $C^{\perp}$'s. Once again we appeal to the Transverse

Spheres Lemma, this time constructing p disjoint collections $C_1^\perp,\ldots,C_p^\perp$ of immersed 2-spheres, where $C_i^\perp$ is transverse to C_i, and $C_i^\perp \cap C_j = \emptyset$ for $i \neq j$, and $C_i^\perp \cap (\Delta \cup \Delta^\perp \cup W) = \emptyset$. This is accomplished just as in Step 6, moving only $\Delta^\perp$ (repeatedly). As before, it follows from distance considerations that any two different cells in this entire collection $C^\perp = \cup C_i^\perp$ will intersect only if their mates (i.e. duals) in C belong to the same subgroup of the same group. ///

Finally, we are in a position to complete our separation of the individual C_γ's.

Step 8. Separating within the subgroups. At this point, each subgroup consists of a finite number of C_γ's, their associated $C_\gamma^\perp$'s produced in Step 7, and a union W_* of pre-Whitney discs produced in Steps 5 and 6, one disc for each pair of intersections between distinct C_γ's of the subgroup. Furthermore, all of these data for distinct subgroups are disjoint. Hence we can apply the Separation Proposition (Section 6) separately to each subgroup of each group, as we have just the data we need. Consequently, we can make all of the C_γ's disjoint. As noted in the Addendum (Section 6), we can leave ourselves with a transverse collection $C^\perp$ for C, for use in the next induction round (or we could just make $C^\perp$ using $\Delta^\perp$). ///

The C_γ's are now positioned to serve as bases of towers. That is, they are disjoint from each other, with imbedded boundaries, and their interiors are disjoint from Δ, and they are correctly framed. Furthermore, they are equipped with a transverse collection of spheres $C^\perp$ (whose members, however, may intersect quite badly, but as usual are bounded). The next round of induction proceeds to construct a disjoint collection D of discs to serve as the second stages of the towers, just as C was constructed above. Hence, in this second round, C and $C^\perp$ play the role of Δ and $\Delta^\perp$ in the first round (note that they satisfy the analogues of properties (1_Δ) and (2_Δ) listed earlier, the only properties used). The only difference is that now the individual D_λ's are attached to individual C_γ's instead of to pairs of C_γ's, so this requires changing a few words in Step 1, but otherwise all the operations remain the same.

Concerning distances, note that in Step 8, there is no bound on how far an individual C_γ may move within an individual subgroup, other than it stays close to the subgroup. Hence, if the diameters of these subgroups are bounded by some constant d_1, then this number serves (approximately) as a bound for all motions of the first induction round. For the second round, then, we must greatly enlarge our scale, for example grouping the D's so that individual subgroups are much further apart than distance d_1 (but still $p=2^4$,

SUMMARY TABLE OF THE EIGHT INDUCTIVE STEPS OF SECTION 7

Abbreviations (used either as nouns or adjectives): c.s. = connect-sum; r.h. = regular homotopy; N.t. = Norman trick

	Δ	$\Delta^{\perp}$	$C = \bigcup C_{\gamma}$	layers of $C^{\perp}$'s (always transverse to C)	W
At the start	initial data never moved	initial data			
Step 1: Selecting and initializing the C_{γ}'s.		not moved	Select them and initialize them: (i) get their ∂'s imbedded (ii) make their framings correct modulo 2 (iii) make int C miss Δ (iv) finish correcting framings, and make $C_{\gamma} \cdot C_{\delta} = 0$		
Step 2: Grouping the C_{γ}'s.		not moved	Assign the C_{γ}'s to p different groups, whose unions are denoted $C_1, C_2, \ldots, C_p$.		
Step 3: Making $\Delta^{\perp} \cap C = \emptyset$.		3b. Apply the N.t., c.s.'ing $\Delta^{\perp}$ to $C^{\perp}$ to make $\Delta^{\perp} \cap C = \emptyset$.	R.h.'d during 3a.	3a. Construct a layer $C^{\perp}$ disjoint from Δ, and use it for 3b. Then discard it.	
Step 4: Making the groups $C_1, \ldots, C_p$ disjoint.		R.h.'d when making the layers of $C^{\perp}$'s. Keep $\Delta^{\perp} \cap C = \emptyset$.	4b. Make the different groups $C_1, \ldots, C_p$ disjoint by applying the N.t., c.s.'ing the C_{γ}'s of different groups to different layers of $C^{\perp}$'s.	4a. Construct p − 1 disjoint layers of $C^{\perp}$'s, disjoint from $\Delta \cup \Delta^{\perp}$, and use them to accomplish 4b. Then discard them.	

	$\Delta^\perp$	$C = \bigcup C_\gamma$	layers of $C^\perp$'s	W
Step 5: Selecting $W = W_1 \cup \cdots \cup W_p$, and making $W \cap (\Delta \cup \Delta^\perp) = \emptyset$.	5c. Apply the N.t., c.s.'ing $\Delta^\perp$ to $C^\perp$ to make $\Delta^\perp \cap W = \emptyset$, keeping $\Delta^\perp \cap C = \emptyset$ and $\Delta \cap W = \emptyset$.	not moved	5b. Construct a collection $C^\perp$ disjoint from $\Delta \cup W$, and use it in 5c. Then discard it.	5a. Select $W_1,\ldots,W_p$, making them miss Δ by using $\Delta^\perp$. They are r.h.'d in 5b when $C^\perp$ is constructed.
Step 6: Making the groups $\{W_i\}$ disjoint.	R.h.'d when making the layers of $C^\perp$'s. Keep $\Delta^\perp \cap (C \cup W) = \emptyset$.	not moved	6a. Construct $p(p - 1)$ disjoint layers of $C^\perp$'s, disjoint from $\Delta \cup \Delta^\perp \cup W$, and use them in 6b. Then discard them.	6b. Apply the N.t., c.s.'ing the W_i's to the different layers of $C^\perp$'s to separate the groups of W_i's.
Step 7: Constructing disjoint groups of $C^\perp$'s.	R.h.'d when making the layers of $C^\perp$'s. Keep $\Delta^\perp \cap (C \cup W) = \emptyset$.	not moved	Construct p disjoint layers of $C^\perp$'s, one for each group disjoint from $\Delta \cup \Delta^\perp \cup W$.	not moved
Step 8: Separating within the subgroups.	not moved and no longer needed	Completely separate the individual C_γ's by applying the Separation Proposition (§6) within each subgroup.	Used during the separation process and then discarded. A final layer of $C^\perp$'s is provided for the next induction round.	Used during the separation process and then discarded.

Now cycle through Steps 1-8 five more times, as explained in the text, to build disjoint 6 stage towers.

or even $p=5$ groups will suffice). Nevertheless, everything remains bounded, and in fact the bound is simple function of the original diameters of sets.

Cycling through the induction process six times, producing 6-stage towers $T = C \cup D \cup E \cup F \cup G \cup H$, completes the goal $(*_3)$ of this section.

It is interesting to note that as one builds successive layers of the towers, the duals of these layers spread out further and further, intersecting more and more distant duals.

We note in Appendix 3 a few technical differences between the above construction and Quinn's.

8. THE PROOF OF THE 4-DIMENSIONAL ANNULUS THEOREM

The n-dimensional Annulus Conjecture (AC_n) asserts that for any homeomorphism $h:R^n \to R^n$ such that $h(B^n) \subset \operatorname{int} B^n$, the closed difference $B^n - h(\operatorname{int} B^n)$ is homeomorphic to the annulus $S^{n-1} \times I$. This conjecture is intimately related to the n-dimensional Stable Homeomorphism Conjecture (SH_n), which asserts that any orientation preserving (= o.p.) homeomorphism $h:R^n \to R^n$ is stable, that is, can be written as a finite composition of homeomorphisms $h = h_m \ldots h_2 h_1$ where each h_i is the identity on some open set. There is a rich collection of facts and consequences surrounding these conjectures, which we will not go into here; see for example [B-G]. But for our present purposes we recall:

1) $SH_n \Rightarrow AC_n$ for any given n.
2) An o.p.-homeomorphism h of R^n is stable if
   a) it is a composition of stable homeomorphisms, or
   b) it is differentiable at some point, with non-singular derivative there, or
   c) it is uniformly bounded, i.e. $\{\|h(x)-x\| \mid x \in R^n\}$ is bounded.
3) SH_n is true for all $n \neq 4$ (classical for $n=1$, from [R] for $n=2$, from [M] for $n=3$, and [K] for $n \geq 5$).

We will discuss the following

THEOREM (Quinn): SH_4 (hence AC_4) is true.

Consequently, the Stable Homeomorphism and Annulus Conjectures are finally established for all dimensions.

The proof of this theorem follows the lines of a remarkably prescient proposal of Connell and Hollingsworth [C-H, pp. 161,179]. In short, they noted that given an o.p.-homeomorphism $h:R^n \to R^n$, if one knew that $h \times id: R^n \times R^1 \to R^n \times R^1$ were stable, and if one could establish a certain sort of controlled (n+1)-dimensional h-cobordism theorem, then one could deduce that h was stable. This they presented as one possible application of many that would follow if one could carry to conclusion their ideas and conjectures about "geometric groups" set forth in [C-H].

When in 1968 Kirby established the stable homeomorphism conjecture for dimensions ≥ 5, but not 4, then the Connell-Hollingsworth proposal grew in credibleness. Finally in 1977-78 Quinn succeeded in supplying the missing algebraic details of the Connell-Hollingsworth program. But as expected, the resultant controlled h-cobordism theorem (for example) could be established only for dimensions ≥ 6. The critical 5-dimensional case eluded Quinn, for the usual 4-dimensional reasons prevailing in the middle level of the cobordism (see below). However, Freedman's work offered new prospects, and a year after Freedman's breakthrough, Quinn succeeded in establishing the 5-dimensional controlled h-cobordism theorem, obtaining the Annulus Theorem as one particular consequence (of many). This is what we have been aiming toward in this exposition.

We give now the details of the Connell-Hollingsworth-Quinn program that reduce the 4-dimensional Stable Homeomorphism Conjecture to Quinn's result established in Section 7. (We note that as an alternative to this route, one could instead apply Kirby's original argument directly in dimension 4, using Quinn's work to complete the discussion of 4-dimensional homotopy tori, but there seems to be no clear advantage in proceeding that way.)

Suppose, then, that $h:R^4 \to R^4$ is an o.p.-homeomorphism. The idea will be to express h as a composition $h = gf$ of two homeomorphisms, where f is bounded (defined above) and g is a diffeomorphism on some open set. Hence, h will be stable (see Fact (2) above).

To start, we use h to put a certain, possibly nonstandard, smooth structure on $R^4 \times [0,1]$, as follows (we use $|\ |$ here to emphasize the underlying topological space). It will be described in terms of two coordinate charts (φ_0, U_0) and (φ_1, U_1), where as usual U_i is an open subset of $|R^4 \times [0,1]|$ and $\varphi_i : U_i \overset{\to}{\simeq} \varphi(U_i) \subset R^5$. Let $U_0 = |R^4 \times [0,2/3)| \subset |R^4 \times [0,1]|$ and let φ_0 = inclusion: $U_0 \subset R^4 \times R^1 = R^5$. Let $U_1 = |R^4 \times (1/3,1]| \subset |R^4 \times [0,1]|$. We will choose φ_1 so that $\varphi_1 | |R^4 \times 1| = h: R^4 \times 1 \to R^4 \times 1$ and (see Figure 8.1) $\varphi_1 \varphi_0^{-1} | R^4 \times (1/3,2/3)$ is smooth, i.e., $\varphi_1 | R^4 \times (1/3,2/3)$ is smooth. To get φ_1, we apply Kirby's 5-dimensional Stable Homeomorphism Thereom [K], together with Connell's Smooth Approximation Theorem [Co] (as established in dimension 5 by Bing [B]), to find a diffeomorphism $\psi: R^4 \times (1/3,2/3) \to R^4 \times (1/3,2/3)$ which approximates $h \times id$ as close as we like, even in the majorant sense, i.e.,

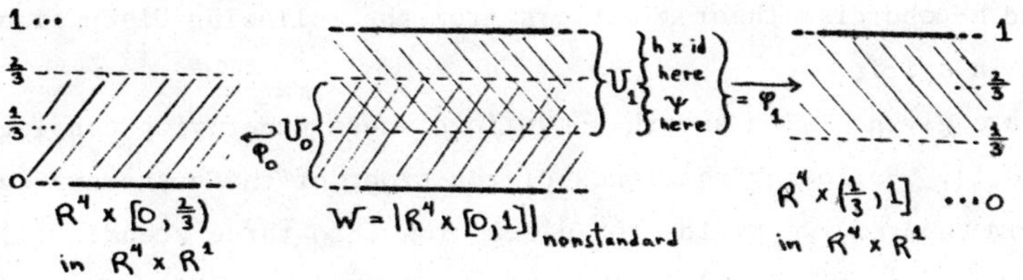

Figure 8.1: Defining W

$\|\psi(x) - (h \times id)(x)\| \to 0$ as fast as we like for $x \to$ end in $R^4 \times (1/3,2/3)$. Then we can define φ_1 to be $h \times$ inclusion on $|R^4 \times [2/3,1]| \subset U_1$, and ψ on $|R^4 \times (1/3,2/3)| \subset U_1$.

We denote by W this new smooth manifold whose underlying space is $R^4 \times [0,1]$. Clearly W is an h-cobordism. If we could show that W is smoothly a product in some reasonably well controlled sense, then we would be done. For example, suppose we could establish

(*) For some $k > 0$, there is a diffeomorphism $G: R^4 \times [0,1] \to W$ such that πG is k-close to π, where $\pi: R^4 \times [0,1] = |W| \to R^4$ is vertical projection.

Granted that (*) holds, and assuming without loss that $G(R^4 \times 1) = \partial_+ W =$ the 1-end of W, then let $g = \varphi_1 G|R^4 \times 1 : R^4 \times 1 \to R^4 \times 1$ and let $f = (G|R^4 \times 1)^{-1}$. Regarding $|\partial_+ W|$, $R^4 \times 1$ and R^4 as being identified in the obvious manner (to avoid a clutter of maps), we get that $gf = h$, where f is bounded and g is smooth, and hence h is stable, as noted above.

Unfortunately it turns out that (*) not only is unknown, it is in fact false for arbitrary smooth structures on $R^4 \times [0,1]$, because of the existence of exotic structures on R^4. However, Quinn establishes the following weaker statement, which is sufficient for his needs here.

(**) For some $k > 0$, there is a homeomorphism $G: R^4 \times [0,1] \to W$, with $G|U \times [0,1]$ a diffeomorphism for some open set $U \subset R^4$, such that πG is k-close to π (π as above).

Granted that (**) holds, then the argument that h is stable is the same as above. So it remains to discuss (**).

To prove (**), one attempts to prove (*) using the methods that do in fact work in higher dimensions, and finds that by using the work presented in Section 7 together with Freedman's work, one can at least deduce (**).

We offer an outline of this argument. We will confine ourselves to the specific contest of (**), although it should be remarked that the full-blown controlled h-cobordism theorem differs from the following discussion only detail, not in spirit.

We are given W, a smooth 5-manifold whose underlying topological space is $R^4 \times [0,1]$. Following the lines of the proof of the customary compact h-cobordism theorem, we divide the discussion into three steps: (1) establishing the existence of a bounded handlebody structure on $(W, \partial_- W)$, (2) trading 0 and 5-handles for 2 and 3-handles, and trading 1 and 4-handles for 3 and 2-handles, respectively, and (3) cancelling the 2 and 3-handles, all the time maintaining boundedness control (or even ε-control, if you wish, but that isn't required here).

We elaborate these steps. We emphasize that everything is smooth here.

Step 1: Imposing a bounded handlebody structure on $(W, \partial_- W)$. Letting $\partial_- W = R^4 \times 0 \subset W$, a handlebody structure on $(W, \partial_- W)$ is a filtration $W_{-1} \subset W_0 \subset \cdots \subset W_5 = W$ of W by 5-dimensional submanifolds of W, closed as subsets, such that W_{-1} is a collar of $\partial_- W$, and W_i is obtained from W_{i-1} by attaching (disjoint) i-handles to $\partial_+ W_{i-1} \equiv \partial W_{i-1} - \partial_- W$. The collection of such handles may be infinite, but it is presumed to be locally finite. This handlebody structure is bounded if all handles and all fibers of the collar are uniformly bounded in size (using say the standard metric on $R^4 \times [0,1]$).

It is a routine matter to get such a bounded handlebody structure: simply take an ordinary handlebody structure which starts with a thin collar, and then subdivide the handles to make the new handles small, isotoping the attaching maps as required in order to make handles be attached only to unions of handles of lower index.

Step 2: Trading handles into the middle dimension. One does the following argument first for 0-handles and then 1-handles, and dually for 5-handles and then 4-handles. Let i be 0 or 1, and assume that the $i = 0$ case has been done if $i = 1$. In particular, we can assume that any i-handle is attached to $\partial_+ W_{-1}$ (which, when $i = 0$, means nothing). Focusing on an individual i-handle H, it is traded for an $(i + 2)$-handle $\hat{H}$ by introducing a trivial $i + 1$, $i + 2$ (complementary) handle pair G_{i+1}, G_{i+2} and then isotoping G_{i+1} to be in complementary position to H, so that it and H can be cancelled, leaving behind the repositioned $G_{i+2} = \hat{H}$. In more detail, the topological product structure on W provides a predictably bounded homotopy of core H rel its attaching region into $\partial_+ W_{-1}$. After some general positioning we can assume that this homotopy hits no handles of index $> i + 1$, i.e. lies in W_{i+1} (this motion is bounded because the handles are bounded). After some further general positioning, we can move the homotopy off of

$\partial_- W \cup$ the cores of handles of index $\leq i+1$ and hence into $\partial_+ W_{i+1}$, so it becomes a homotopy in $\partial_+ W_{i+1}$ carrying a parallel copy of core H (lying in the belt region of ∂H) rel its attaching region into $\partial_+ W_{-1} \cap \partial_+ W_{i+1}$. Now introduce a small trivial $i+1$, $i+2$ handle pair G_{i+1}, G_{i+2} attached to $\partial_+ W_{i+1}$ but missing the i and $i+1$ handles (so in fact it is attached to $\partial_+ W_{-1}$), and lying somewhere near the image of the homotopy. Using the homotopy, and the fact that for 0 and 1-dimensional submanifolds of a 4-manifold, homotopy gives rise to isotopy, one verifies that the attaching map of G_{i+1} can be isotoped in $\partial_+ W_{i+1}$ so as to put G_{i+1} in complementary position to H, as asserted. All motions here come from bounded homotopies, and so are uniformly bounded. (For higher dimensional cobordisms, this argument must be presented a bit more carefully; see [Q_1, Thm. 6.1] following [W].)

Step 3: Cancelling 2 and 3-handles. At the end of Step 2, having gotten rid of all of the 0,1,4 and 5-handles, we can write our cobordism W as $W = \partial_- W \times [0,1/3] \cup \bigcup\{H_{2,\beta}\} \cup \bigcup\{H_{3,\alpha}\} \cup \partial_+ W \times [2/3,1]$, where $\partial_- W \times [0,1/3]$ and (for symmetry's sake) $\partial_+ W \times [2/3,1]$ are collars for the two boundary components of W, having fibers of uniformly bounded size, and $\{H_{2,\beta}\}$ and $\{H_{3,\alpha}\}$ are locally finite collections of 2-handles and 3-handles, all of uniformly bounded size.

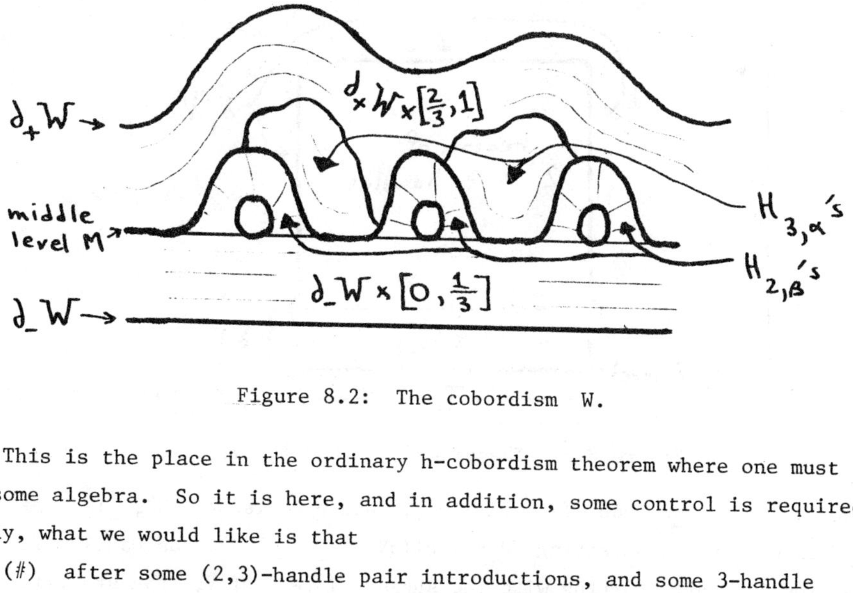

Figure 8.2: The cobordism W.

This is the place in the ordinary h-cobordism theorem where one must use some algebra. So it is here, and in addition, some control is required. Namely, what we would like is that

(#) after some (2,3)-handle pair introductions, and some 3-handle slides of uniformly bounded size, the 3-handles $\{H_{3,\alpha}\}$ can be put in 1-1 correspondence with the 2-handles $\{H_{2,\beta}\}$, so that for any pair α, β, $A_\alpha \cdot B_\beta = \delta_{\alpha\beta}$ (kronecker δ), where A_α is the attaching (descending) sphere of $H_{3,\alpha}$ and B_β is the belt (ascending) sphere of $H_{2,\beta}$, and this intersection is taking place in the middle 4-manifold level $M = \partial_+ W_2$.

Forgeting size for the moment, in the finite simply-connected case, establishing (#) requires only elementary algebra, being nothing more than the fact that an integer matrix of determinant 1 is reducible to the identity matrix by (say) row operations. But in this infinite controlled setting this is another matter, and in fact this is the problem addressed in the earlier work of Connell and Hollingsworth. Inasmuch as that program was successfully brought to conclusion by Quinn in 1977-78, we consequently can presume that (#) above holds. (Interestingly, Quinn [Q_1] deduces his main results, namely the Connell-Hollingsworth conjectures, by starting from the fact that when suitably cast in a manifold setting, they can be established by using the torus trick, in the same spirit as Kirby's original work).

We take the liberty at this point of describing a variant manner of establishing (#), which amounts to a geometrization of Quinn's argument in [Q_1], making direct use of the previously known theorems upon which Quinn modeled his proof. The key point is, (#) is a condition which lends itself to stabilization-destabilization (of dimension W). To be precise, let V be the 6-dimensional relative cobordism gotten by crossing W with I (see Figure 8.3),

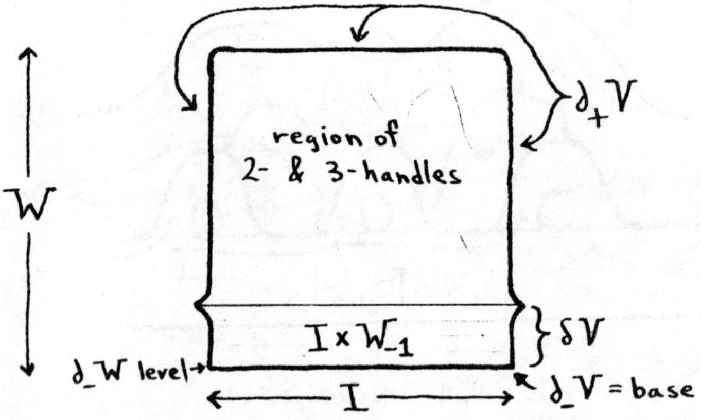

Figure 8.3. The cobordism $V = I \times W$.

letting $\partial_- V = I \times \partial_- W$ and $\delta V = \partial I \times W_{-1}$, where we recall W_{-1} is some collar of $\partial_- W$ in W, and then letting $\partial_+ V = \mathrm{cl}(\partial V - (\partial_- V \cup \delta V))$, as usual. The key observation is that, starting with the subset $I \times W_{-1}$ of V, the stabilized handles $\{I \times H_{2,\alpha}\}$ and $\{I \times H_{3,\beta}\}$ provide a handle decomposition of V based on $I \times W_{-1}$. These handles are of index 2 and 3, as earlier. Now, in dimension 6 we know, using the Product Structure Theorem of Kirby and Siebenmann [K-S], that V is a smooth product, and in fact one can perturb the topological structure an arbitrarily small amount to make it smooth. Consequently, the above handle structure can be changed by the usual handle operations (births, deaths, and slides) to become trivial (so, for example, thinking Morse-theoretically, one can construct a Cerf diagram). Furthermore, since the Product Structure Theorem can be applied locally, in bounded patches, one can argue that all of these handle operations can be done in bounded fashion.

Now, if only handles of index 2 and 3 appear during this transition, then this would immediately provide a solution to (#), for that is exactly what (#) is saying. In general, however, handles of other indices may be appearing, disappearing and sliding over each other. Nevertheless, one can make a Cerf-theoretic argument that all handles not of index 2 or 3 can be traded for handles of index 2 or 3 (as for example in [H-W]). Thus (#) is established.

In short, we are saying that (#) can be achieved in dimension 5 because it can be achieved, geometrically in fact, in dimension 6.

Assuming now that (#) holds, we return to our discussion of the cobordism W, showing how to establish (**) by using the work of Section 7. By our description of W, we see that the middle level $M = \partial_+ W_2$ is obtained from $\partial_- W = R^4$ by performing a locally finite collection of 1-surgeries. Since the surgery circles in R^4 are necessarily unknotted and unlinked and uniformly bounded in size, we can regard M as being obtained from R^4 by connect-summing with infinitely many copies of $S^2 \times S^2$ at a locally finite collection $\{D^4_\alpha\}$ of uniformly bounded balls in R^4. In M we make note of each resultant subset $(S^2 \times S^2 - \operatorname{int} D^4)_\alpha$ by labeling a spine of it, say $B_\alpha \cup B^d_\alpha$, consisting of two transverse imbedded 2-spheres, one of them the belt sphere B_α of the handle $H_{2,\alpha}$ and the other some dual B^d_α for it. (Since the 2 and 3 handles have been paired by (#), we now use the same index set $\{\alpha\}$ for both sets of handles.)

We can make the same sort of discussion at the other end of W, to see that M is obtained from $\partial_+ W$ by connect-summing with $S^2 \times S^2$'s at a locally finite collection of uniformly bounded balls in $\partial_+ W$. As above we mark each $S^2 \times S^2$ of this collection in M via a spine $A_\alpha \cup A^d_\alpha$, i.e. a wedge of imbedded 2-spheres, where A_α is the attaching 2-sphere for the 3-handle $H_{3,\alpha}$, and A^d_α is some dual for it in M.

At this point we are ready to apply the discussion in Section 7, where the sets M, A_α, A^d_α, B_α and B^d_α correspond to the sets above. Condition (#) above is exactly what is hypothesized at the start of Section 7, and furthermore all considerations of boundedness prevail. Hence, by Section 7 we can find a disjoint locally finite collection of topological Whitney discs for the excess intersections between the A_α's and the B_α's. These discs can be used to perform (topological) isotopies in M, to be regarded as (topological) isotopies of the attaching maps of the 3-handles of W, to make the 3-handles geometrically complementary to the 2-handles. So the handles can be cancelled, leaving W with a product structure. One routinely checks that, as the only non-smooth part of the proof is the preceding repositioning of the 3-handles and subsequent 2-3 cancellations, there remains an open subset U of $\partial_- W$ (which can be made dense if you wish) over which the product structure can be made to smoothly agree with that of W. Hence (**) above is established, and the 4-dimensional Stable Homeomorphism and Annulus Theorems follow.

APPENDIX 1. Casson's Imbedding Theorem via Quinn's Lemma. The preceding material has imbedded in it a proof of Casson's original theorem, but it may not be apparent. Indeed, if one is willing to grant the Transverse Spheres Lemma (Section 5) and the subsequent Separation Proposition (Section 6), then Casson's construction can be presented quite succinctly. We do this here. We begin by recalling

CASSON'S IMBEDDING THEOREM ([C], mildly paraphrased).

Suppose $C_1,\ldots,C_n$ are immersed 2-discs in a simply-connected 4-manifold M, with the ∂C_i's imbedded disjointly in ∂M, such that $C_i \cdot C_j = 0$ for $i \neq j$. Suppose there exist $\beta_i \in H_2(M)$, $1 \leq i \leq n$, such that $\beta_i \cdot \beta_i$ is even and $\beta_i \cdot C_j = \delta_{ij}$. Then the C_i's can be regularly homotoped to be disjoint, and one can build disjoint (infinite) towers $T_1,\ldots,T_n$ in M whose bases are these separated C_i's.

Proof. Casson's original construction proceeded a disc at a time fixing up the first layer $C = \cup C_i$ of discs for his towers, then a disc at a time through the second layers, etc., all the time having to spend repeated effort to recover the necessary working hypothesis that the complement of the union of most 2-dimensional data at hand be simply connected. However, with the aid of the Transverse Spheres Lemma, basically one can proceed an entire layer at a time. We give the argument in summary fashion, presuming that only experienced hands have gotten this far. As in Casson's proof, we are immediately entering an inductive procedure.

Let $C^d = \cup C_i^d$ be a union of immersed 2-spheres representing the classes $[\beta_i]$, provided by the Hurewicz isomorphism theorem (i.e., surger the surfaces representing the β_i's). To begin, we regularly homotope C^d, as well as C, to make C^d into a transverse collection $C^\perp = \cup C_i^\perp$ of spheres for C. This can be done all at once using Casson's Surface Separation Lemma (Section 2), taking C as A and C^d as B, and finding the necessary pre-Whitney discs W as a consequence of the 1-connectivity of M.

Let W (unrelated to the preceding W) be a collection of pre-Whitney discs for all of the intersections between all pairs C_i, C_j, provided by the 1-connectivity of M and the hypothesis that $C_i \cdot C_j = 0$. Applying the Separation Proposition of Section 6, we see that the C_i's can be regularly homotoped to be disjoint, and furthermore (by the Addendum) the resultant union C can be provided with a transverse collection of spheres, still denoted $C^\perp$. Thus the C_i's are now separated, and we will not need to move them any more.

We now begin to construct the next layers of the towers. Let $D = \cup D_j$ be a collection of immersed discs in M which are attached to the C_i's to "kill their kinks". We wish to make certain preliminary improvements to these discs, just as we did to the C_γ's in Step 1 of Section 7. First (i) we make the ∂D_j's disjointly embedded, and then (ii_e) we spin the D_j's at their boundaries to make their framings correct modulo 2 (as second stages of towers, as in Section 3). Next (iii) we get $\operatorname{int} D$ off of C by means of the Norman Trick, using $C^\perp$. Before proceeding further, we observe as in Step 1of Section 7 that the D_j's have algebraically dual immersed spheres $\{D_j^d\}$ disjoint from C, obtained by surgering the small dual tori of the D_j's, using $C^\perp$ to keep the surgeries off of C (note that $D_j^d \cdot D_k^d = 0$ for all j,k, even if $C^\perp \cdot C^\perp \neq 0$). Now, we can arrange that (ii,iv) $D_j \cdot D_k = 0$ for all j,k by connecting-summing each D_j with appropriate copies of the various D_k^d's.

Now, to formally complete the induction process, we regard the D_j's as being attached to a small regular neighborhood N_C of C. So in the simply-connected manifold $M_D = M - \operatorname{int} N_C$ we are back in the same sort of situation in which we started, now with D_j's in place of C_i's. So we can cycle through the induction again, separating the D_j's and providing a new layer of E_k's, etc. Continuing, one can produce towers $T = C \cup D \cup E \cup \ldots$ of arbitrary length, as desired.

APPENDIX 2. Freedman's Big Reimbedding Theorem via Quinn's Lemma. The first formidable aspect of Freedman's work is his sequence of Reimbedding Theorems, of which the most intricate by far is his 5-stage Reimbedding Theorem. In this appendix, we note that Quinn's Transverse Spheres Lemma (= TSL; see Section 5), once mastered, substantially eases the hard technicalities of Freedman's proof, for example rendering unnecessary any discussion of triangular bases. Quinn himself recognized there was room for improvement in Freedman's argument [Q_2]; our discussion carries this another step further.

Since [G-S] is so close at hand, we refer to it for notation and statements of theorems (notational exception: we leave initial data unsuperscripted, writing e.g. T_3 or C_4 in place of their T_3^0 or C_4^0, but we (as they) do use superscripts for later copies, e.g. T_3^1 or C_4^1). In particular, our discussion is presented in the context of their one-stage improved versions. The relevant Theorems there are 3.3 (= 6.0 = the Little Reimbedding Theorem) and 4.0.0 (= 6.1 = the Big Reimbedding Theorem). Hence our goal here is to describe how, given a 4-stage tower T_4, to reimbed a new 4-stage tower T_4^1 into T_4, with agreement on the first two stages, so that the new imbedding is trivial on π_1, and also is π_1-negligible in the customary manner.

We present the argument in five steps. In brief, the idea is that in Steps 1 (= the Little Reimbedding Theorem) and 4 we work inside of the first three stages T_3, using Quinn's Lemma to produce first one and then lots and lots of disjoint transverse spheres for the third stage (we will assume for simplicity of language throughout that the third stage has only one component, i.e., each of the first two stages has just one kink each). In Step 5 the Norman trick is applied, using these transverse spheres to change the fourth stage kinks into kinks coming from the transverse spheres, which lie in T_3 and hence are null-homotopic in T_4. The intermediate Steps 2 and 3 are necessitated by the fact that in Step 1 kinks were introduced into the original third stage, and so they must be provided with their own fourth stage kinky discs, which need to be correctly positioned, requiring some argument.

In more detail, the five steps are as follows:

<u>Step 1</u>. Do the Little Reimbedding Theorem, i.e., regularly homotope the original third stage C_3 to become C_3^1 so that C_3^1 has a transverse sphere $C_3^\perp \subset T_3$ which misses the first two stages $C_1 \cup C_2$. This gives rise to a transverse sphere $C_2^\perp \subset T_3$ for the second stage.

<u>Some details</u> (originally in [F_1]). Letting τ be a small distinguished dual torus for the third stage located near the self-crossing point of the second stage, Freedman noted in [F_1] that the two natural generating circles of τ, which are meridians of C_2, are null-homotopic in T_3 missing $C_1 \cup C_2$. Hence, if one does double surgery on τ (as in Section 4) using these two immersed null-homotopy discs to produce an immersed sphere $C_3^\perp$ missing $C_1 \cup C_2$, at the same time doing finger moves to the intersections of the original third stage C_3 with these discs as described in Section 4, then we can produce a newly positioned third stage C_3^1 with transverse sphere $C_3^\perp$. (It is observed in [G-S] that the above finger moves need not link the first stage C_1, i.e., $\pi_1(C_3^1) \to \pi_1(T_3 - C_1)$ is trivial; this will be used later, to avoid having to glue on an additional earlier stage.) Using $C_3^\perp$ and Freedman's observation above one can get a transverse sphere $C_2^\perp$ to the second stage so that $C_2^\perp \cap C_{1-3}^1 = C_2^\perp \cap C_2 = 1$ point (recall $C_{1-3}^1 = C_1 \cup C_2 \cup C_3^1$).

Step 2. Exhibit fourth stage discs for all of the third stage kinks, and get their interiors off of the union of the first three stages C^1_{1-3}.

Some details. During Step 1 some new self-intersections (= kinks) arose in constructing the new third stage C^1_3. Since $C^1_3 \subset T_3$, these kinks are null-homotopic in T_4, and furthermore by Gompf's observation (see Step 1), these null-homotopies can be chosen disjoint from the first stage C_1. Using $C^\perp_3$ and $C^\perp_2$ produced in Step 1, we can make these null-homotopies disjoint also from the second and third stages $C_2 \cup C^1_3$. Hence these null-homotopies, together with the original fourth stage C_4, give us a new collection C^1_4 of fourth stage discs (not necessarily disjoint) for all of the kinks of C^1_3, with $\operatorname{int} C^1_4 \cap C^1_{1-3} = \emptyset$.

Step 3. Produce a new transverse sphere $C^\perp_3$ (and from it $C^\perp_2$) lying in T_3 which misses C^1_4 as well as C^1_{1-3}.

Some details. The new fourth stage C^1_4 may well intersect the transverse sphere $C^\perp_2$ produced in Step 1 (Aside: $C^\perp_2$ needn't intersect the original fourth stage C^0_4, but this fact isn't used.) However, we can use Quinn's Transverse Sphere Lemma (Section 5) again, more or less repeating the construction of Step 1, except this time using $C^\perp_2$ to provide the null-homotopies of the second stage meridians, to produce our new $C^\perp_3$. To be precise, we apply the TSL with C, $C^\perp$, W and F there being C_2, $C^\perp_2$, C^1_3 and C^1_4 here. The finger moves of this operation will put extra kinks into C^1_4, turning it into C^2_4 (all of these kinks, both old and new, will be dealt with in Step 5). Finally, use the newly produced $C^\perp_3$ to produce a new transverse sphere $C^\perp_2$ as at the end of Step 1 so that $C^\perp_2 \cap (C_1 \cup C_2 \cup C^1_3 \cup C^2_4) = C^\perp_2 \cap C_2 = 1$ point. To see that one can arrange that $C^\perp_2 \cap C^2_4 = \emptyset$ here requires checking that Freedman's null-homotopy of a meridian of C_2 in $T^1_3 - C_1 \cup C_2$, where T^1_3 is a small neighborhood of $C_1 \cup C_2 \cup C^1_3$, can be chosen to miss the collar $T^1_3 \cap C^2_4$, but this is clear, either by direct inspection, or by observing that any such intersections could be pushed off the attaching boundary of C^2_4, making extra intersections of the null-homotopy with C^1_3, which are then gotten rid of like all of the other intersections by connect-summing with $C^\perp_3$.

Step 4. Construct lots of disjoint transverse spheres $C^\perp_{3,1}, C^\perp_{3,2}, C^\perp_{3,3}, \ldots$, all lying in T_3, with $C^\perp_{3,i} \cap C^2_{1-4} = C^\perp_{3,i} \cap C^1_3 = 1$ point, without moving C^2_{1-4} $(\equiv C_1 \cup C_2 \cup C^1_3 \cup C^2_4)$, using the Transverse Spheres Lemma repeatedly.

Some details. The point is, we new have at our disposal a transverse-sphere making machine, whose basic components are a distinguised torus τ dual to the third stage C^1_3 (just as in Step 1), together with the transverse sphere $C^\perp_2$ produced in Step 3 which can be used to provide null-homotopies of the natural generating circles of τ (= meridians of C_2), these null-homotopies missing C^2_{1-4}. To be precise, we apply the TSL with the sets C, $C^\perp$, W and F there being C_2, $C^\perp_2$, C^1_3 and $C^\perp_2$ here, producing a (the first) transverse sphere $C^\perp_{3,1}$, at the same time finger-moving $C^\perp_2$ so that when done $C^\perp_2 \cap C^\perp_{3,1} = \emptyset$. Then the TSL can be applied again to make a second transverse sphere $C^\perp_{3,2}$ disjoint from the first, again leaving a repositioned $C^\perp_2$ disjoint from it. This process can be reiterated as long as desired. The $C^\perp_{3,i}$'s so produced have more and more kinks as i increases, because each inherits kinks from the current $C^\perp_2$, which itself is getting more and more complicated because of the finger moves repeatedly being performed on it. The total number of $C^\perp_{3,i}$'s required is the total number of crossing in C^2_4, plus one more, as we will see in Step 5.

<u>Step 5</u>. Apply the Norman trick, using the transverse spheres $\{C^\perp_{3,i}\}$ to transform the kinks of the fourth stage C^2_4 into kinks which lie in T_3, and hence are null-homotopic in T_4, thereby producing the desired new $T^1_4 \subset T_4$.

<u>Some details</u>. For each self-crossing of C^2_4, choose one of the sheets and push it along the other sheet and off the edge of C^2_4 in the usual manner, making for the moment two intersections with C^1_3, and then get rid of these intersections via the Norman trick, using one of the $C^\perp_{3,i}$'s (you can use the same one for both intersections). This is the same idea as in Freedman's original argument. Since the $C^\perp_{3,i}$'s lie in T_3, it is clear that the new 4th stage C^3_4 so produced is null-homotopic in T_4. Letting T^1_4 be a small regular neighborhood of $C_1 \cup C_2 \cup C^1_3 \cup C^3_4$, we have our desired reimbedding. Note that T^1_4 is π_1-negligible in the usual desired sense (i.e., $\pi_1(T_4 - T^1_4) \to \pi_1(T_4 - C_1)$ is an isomorphism) because of the last unused transverse sphere $C^\perp_{3,*}$.

This completes the proof of the Big Reimbedding Theorem.

As a variation in the above argument, one could have not bothered producing $C^\perp_2$ from $C^\perp_3$ in Step 3, and could have in Step 4 produced many disjoint transverse spheres to (the various disc-components of) C^2_4, instead of to C^1_3, these transverse spheres lying in T_3, using the TSL with C, $C^\perp$, W and F there being C^1_3, $C^\perp_3$, (various components of) C^2_4 and $C^\perp_3$ here. This variation is perhaps marginally more efficient than the one presented.

APPENDIX 3. <u>Quinn's Disc Deployment Lemma</u>. A significant portion of our exposition above of Quinn's work (primarily Section 7) was tailored to a specific use, namely the proof of the Annulus Conjecture. But in fact the same proof works to yield what Quinn calls the thin h-cobordism theorem (= ε-h-cobordism theorem = controlled h-cobordism theorem, meaning an h-cobordism theorem in which distances are controlled); the only change required concerns the discussion of distances. On the other hand, if one wishes to prove a controlled surgery theorem in dimension four, then the proof in Section 7 requires a mild strengthening, to what Quinn calls the Disc Deployment Lemma. It is to be regarded as a controlled analogue of Casson's Imbedding Theorem. For purposes of discussion, we recall only a special case of the Lemma (see [Q_3, Section 3.2] for the general statement, which requires too many definitions to be given here). It should be compared to Casson's Imbedding Theorem in Appendix 1 above.

<u>Quinn's Disc Deployment Lemma (special case)</u>. Given a 4-manifold M, and given $\varepsilon>0$, there is a $\delta>0$ such that if $C = \cup C_\gamma$ is a locally finite collection of 2-discs immersed in M, with boundaries imbedded disjointly in ∂M, such that $C_\gamma \cdot C_\delta = 0$ for $\gamma \neq \delta$, and if $C^\perp = \cup C_\gamma^\perp$ is a transverse collection of immersed spheres for C, with $C_\gamma^\perp \cdot C_\gamma^\perp = 0$ (<u>even</u> would do), such that each C_γ and each $C_\gamma^\perp$ has diameter $<\delta$, then there is a collection $T = \cup T_\gamma$ of disjoint 6-stage towers in M attached to the curves ∂C, such that each tower T_γ has diameter $<\varepsilon$.

Note: Throughout it is understood that if M is not compact, then ε and δ are continuous functions from M to $(0,\infty)$.

The idea of the proof is just as in Section 7: first one uses a series of steps like those in Section 7 to regularly homotope the C_γ's to be disjoint, so that the C_γ's can serve as the bases of the towers; then one constructs a new layer D of disjoint discs to serve as the second stages of the towers, etc. But there is one important differences between the setup here and the earlier discussion of Section 7: here we are missing the preceding layer Δ and its complement $\Delta^\perp$. Thus, we cannot produce disjoint transverse collections $C_1^\perp$, $C_2^\perp$, etc., whenever we wish.

The manner by which Quinn proceeds amounts to shifting the construction of Section 7 by one notch, so that in effect Δ and $\Delta^\perp$ there become C and $C^\perp$ here, C and $C^\perp$ there become W and $W^\perp$ here, and W there becomes X (say) here. In other words, to prove the above Lemma, one starts by selecting a collection W of pre-Whitney discs for <u>all</u> of the intersections between different cells of C, and one goes through Steps 1 through 8 of Section 7, now working with W in place of C (and C, $C^\perp$ in place of Δ, $\Delta^\perp$), to get $\operatorname{int} W$ off of C and ultimately to get the discs of W disjoint (without ever moving C). After doing so one can use W to regularly homotope the C_γ's to be disjoint, and then one can begin the whole process over to get the second layer D. In applying Steps 1-8 above, one will, for example in Step 2, put the W_μ's into different groups; in Step 4 get the groups disjoint using disjoint transverse collections $W_1^\perp$, $W_2^\perp$, etc.; in Steps 5 and 6 provide disjoint collections X_1, X_2, etc., of pre-Whitney discs for the different groups of W_μ's, and finally in Step 8 separate the individual W_μ's.

We note that the only reason we didn't present Quinn's proof in this fashion in Section 7, where for purposes of concreteness we were interested only in a very specific h-cobordism theorem, was that such a presentation would have called for an additional layer of discs, namely the above X, which seemed unwarranted in a proof which is already taxing enough.

APPENDIX 4. Some Remarks on Non-simply Connected Developments. In Freedman's extension of his work to the nonsimply connected setting, accomplished during the Fall of 1982, the most important consideration was to come to grips with what was happening on the fundamental group level when one did finger moves such as those discussed in Section 4. Here we describe Freedman's key observation, in the context of the constructions presented in Section 4.

Suppose $T_\# = T_0 \cup E_1 \cup E_2$ consists of a punctured torus T_0 (with circle boundary, i.e. $T_0 \approx T^2 - \text{int } B^2$), together with discs E_1 and E_2 attached to a figure eight basis in T_0, as in Section 4. Suppose $f : T_\# \looparrowright M^4$ is a generically positioned immersion of $T_\#$ into a 4-manifold M^4, with $f^{-1}(\partial M^4) = \partial T_\#$, such that f extends to an immersion $\hat{f} : N \to M$ of a regular neighborhood N rel $\partial T_\#$ of $T_\# \subset R^3 \subset R^4$ in R^4. (Aside: this extension condition is not really necessary, but it substantially simplifies the ensuing discussion, and also in applications it can invariably be arranged without loss of generality. We leave it to the reader to ponder the more general situation).

We suppose that the singularities of f lie only in int $\hat{E}_1 \cup$ int $\hat{E}_2$, where $\hat{E}_i = f(E_i)$, so that in particular $\hat{T}_0 \equiv f(T_0)$ is an imbedded copy of T_0. In other words, the image $\hat{T}_\# \equiv \hat{T}_0 \cup \hat{E}_1 \cup \hat{E}_2$ and its regular neighborhood $\hat{N} \equiv \hat{f}(N)$ are in effect obtained from $T_\#$ and N by introducing self-crossings at points of int $E_1 \cup$ int E_2 (note that int $\hat{E}_1$ may intersect int $\hat{E}_2$).

Now, if either $\hat{E}_1$ or $\hat{E}_2$ is by itself imbedded, say $\hat{E}_1$, then one can do surgery on $\hat{T}_0$ using $\hat{E}_1$ to produce an imbedded disc in $\hat{N}$ spanning $\partial\hat{T}_0$.

In general, however, both $\hat{E}_1$ and $\hat{E}_2$ may have self-intersections, as well as mutual intersections. The simplest nontrivial case to consider is where each of $\hat{E}_1$ and $\hat{E}_2$ has just one self-intersection, and there are no mutual intersections. In other words, each disc has just one kink, period. Then $\pi_1(\hat{N})$ is free on two generators ε_1 and ε_2, say, arising from these respective crossing points. (We will suppress discussion of basepoints here, although any proper discussion should address this issue. As Freedman points out, the entire surface $\hat{T}_0$ can be treated as a basepoint, since it is null-homotopic in $\hat{N}$.) As above, one could do surgery to $\hat{T}_0$, using either $\hat{E}_1$ or $\hat{E}_2$, to produce an immersed disc $\hat{D}_1$ or $\hat{D}_2$ spanning $\partial\hat{T}_0$ in $\hat{N}$ (each $\hat{D}_i$ would have four self-intersection points), in which case the image of $\pi_1(\hat{D}_i)$ in $\pi_1(\hat{N})$ would be the infinite cyclic subgroup generated by ε_i. However, Freedman observed that there is a third possible way to proceed. Namely, one can produce an immersed disc $\hat{D}$ in $\hat{N}$ spanning $\partial\hat{T}_0$ by doing <u>double</u> surgery to $\hat{T}_0$ using both $\hat{E}_1$ and $\hat{E}_2$, at the same time doing finger moves to the self-crossings of $\hat{E}_1$ and $\hat{E}_2$ to produce self-intersections in $\hat{D}$ (eight of them, which will occur near the point $\hat{E}_1 \cap \hat{E}_2$), just as described in Section 4. In this case it turns out that <u>the image of $\pi_1(\hat{D})$ in $\pi_1(\hat{N})$ is the infinite cyclic subgroup generated by the product element $\varepsilon_1^*\varepsilon_2^*$ in</u> $\pi_1(\hat{N})$, where each ε_i^* denotes either ε_i or ε_i^{-1} (your choices, for $i = 1$ and $i = 2$) depending upon which sheet at each crossing point is pushed along which sheet. Verifying this key observation is a matter of examining carefully the construction discussed in Section 4.

More generally, suppose $\hat{E}_1$ has $p > 0$ self-intersections, giving rise to elements $\alpha_1,\ldots,\alpha_p$ in $\pi_1(\hat{N})$, and suppose $\hat{E}_2$ has q self-intersections, giving rise to elements $\beta_1,\ldots,\beta_q$ in $\pi_1(\hat{N})$ (hence $\pi_1(\hat{N})$ is freely generated by $\alpha_1,\ldots,\alpha_p$, $\beta_1,\ldots,\beta_q$). If one produced an immersed disc $\hat{D}_1$ or $\hat{D}_2$ as above by doing (single) surgery to $\hat{T}_0$ using $\hat{E}_1$ or $\hat{E}_2$, then the image of $\pi_1(\hat{D}_1)$ in $\pi_1(\hat{N})$ would be the subgroup freely generated by $\alpha_1,\ldots,\alpha_p$, and likewise the image of $\pi_1(\hat{D}_2)$ in $\pi_1(\hat{N})$ would be the subgroup freely generated by $\beta_1,\ldots,\beta_q$. However, if one produced an immersed disc $\hat{D}$ as above by doing <u>double</u> surgery to $\hat{T}_0$ using both $\hat{E}_1$ and $\hat{E}_2$, at the same time doing the usual finger moves, producing $8pq$ self-crossings in $\hat{D}$, then the image of $\pi_1(\hat{D})$ in $\pi_1(\hat{N})$ would be the subgroup generated by all of the products $\alpha_i^*\beta_j^*$, $1 \le i \le p$ and $1 \le j \le q$, where each α_i^* is either always α_i or always α_i^{-1} (i.e., the superscript $*$ on each occurrence of α_i is always the same superscript, independent of j, but possibly varying with i), and similarly each β_j^* is either always β_j or always β_j^{-1} (so there are $p + q$ choices to be made here, again determined by which of the two sheets is finger-pushed at each of the original $p + q$ crossings).

In the most general situation, both $\hat{E}_1$ and $\hat{E}_2$ have mutual intersections, as well as self-intersections. Suppose these r mutual intersections give rise to elements $\gamma_1,\ldots,\gamma_r$ in $\pi_1(\hat{N})$. Now, at each of these intersections, when one does the double surgery and associated finger pushing to form $\hat{D}$, one may finger push either the $\hat{E}_2$ sheet to follow (an arc in) $\hat{E}_1$, or alternatively one may push the $\hat{E}_1$ sheet to follow (an arc in) $\hat{E}_2$. Suppose that at the $\gamma_1,\ldots,\gamma_s$ crossings $(0 \leq s \leq r)$ one does the former type of push, whereas at the $\gamma_{s+1},\ldots,\gamma_r$ crossings one does the latter type of push. Then it turns out that the image of $\pi_1(\hat{D})$ in $\pi_1(\hat{N})$ is the subgroup generated by all products of precisely two elements of $\pi_1(\hat{N})$, where the first element of the product is one of $\{\alpha_1^*,\ldots,\alpha_p^*,\gamma_1,\ldots,\gamma_s\}$, and the second element of the product is one of $\{\beta_1^*,\ldots,\beta_q^*,\gamma_{s+1},\ldots,\gamma_r\}$ (so there are $(p+s)\cdot(q+r-s)$ products of this form), where as earlier the $*$'s are each ± 1 according to choice of sheets.

Finally, it is important to note that, when producing $\hat{D}$ by doing double surgery in this manner, one has the option at each crossing point of doing no finger-pushing at that point. In such a case, that particular fundamental group element (e.g. α_i, or respectively β_j or γ_k) would remain represented in the image of $\pi_1(\hat{D})$, but it would not appear as the first term (respectively, the second term or either term) in any of the product elements described above. All in all, then, this gives one a lot of options in deciding exactly which elements of $\pi_1(\hat{N})$ are to be represented in the image of $\pi_1(\hat{D})$.

To illustrate Freedman's application of these ideas, we outline briefly how he used this construction in the case where the ambient 4-manifold M had finite (nontrivial) fundamental group. Consider first the model case where $\pi_1(M) \approx \mathbb{Z}/2$. Suppose that in the immersed image $\hat{T} \subset \hat{N} \subset \hat{M}$, that when one forms the immersed disc $\hat{D}$, one does no finger pushing at crossings of $\hat{T}$ representing $0 \in \mathbb{Z}/2$, but one does finger pushes (along either sheet) at all crossings representing $1 \in \mathbb{Z}/2$. Then it turns out that all the resultant self-crossings of $\hat{D}$ will represent $0 \in \mathbb{Z}/2$, because in the product elements mentioned above, both the first and second factors will represent $1 \in \mathbb{Z}/2$, and hence their product will represent $0 \in \mathbb{Z}/2$.

More generally, one could proceed as follows. Suppose $g \in \pi_1(M)$ is any preselected, fixed nontrivial element. In the immersed image $\hat{T} \subset \hat{N} \subset M$, choose labels so that $\alpha_1,\ldots,\alpha_\ell$ $(0 \leq \ell \leq p)$; $\beta_1,\ldots,\beta_m$ $(0 \leq m \leq q)$ and $\gamma_1,\ldots,\gamma_n$ $(0 \leq n \leq r)$ all represent $g \in \pi_1(M)$, whereas all the other α_i's, β_j's and γ_k's represent elements of $\pi_1(M) - \{g\}$. When producing $\hat{D}$, suppose one does finger pushes only at these crossings representing g, choosing sheets so that one gets all products of the form $\alpha_i\beta_j^{-1}$ and $\gamma_k\beta_j^{-1}$, where $1 \leq i \leq \ell$, $1 \leq j \leq m$ and $1 \leq k \leq n$ (so in particular at the γ_k points one pushes the $\hat{E}_2$ sheet along the $\hat{E}_1$ sheet). Then all of these products are of the form gg^{-1} and hence are trivial. Hence the image of $\pi_1(\hat{D})$ lies in $\pi_1(M) - \{g\} \subset \pi_1(M)$.

If $\pi_1(M)$ is finite, it turns out that one can use this idea repeatedly to ultimately produce a disc $\hat{D}$ such that $\pi_1(\hat{D})$ represents trivially in

$\pi_1(M)$. One must, in order to do this, start with a finite 2-sided tower of imbedded surfaces, capped off with a single layer of immersed, possibly intersecting discs, with the tower being of height at least $|\pi_1(M)| - 1$ (in the above discussion, the height was 1). (In any 4-dimensional surgery or 5-dimensional s-cobordism problem, Freedman has shown that one can construct such towers of arbitrary finite height, using constructions from the elementary side of the theory.) Then, one applies the preceding construction for each nontrivial element of $\pi_1(M)$ in turn, each time sacrificing one layer of the tower, producing a new tower which no longer carries that element. In the end, then, one has produced an immersed disc D carrying no nontrivial elements of $\pi_1(M)$. In this manner Freedman was able to extend all of the appropriate simply-connected theorems to the corresponding finite-π_1 settings (and with another clever idea or two, to the poly- (finite or cyclic) settings).

BIBLIOGRAPHY

R. H. Bing, Radial engulfing, in: Conference on the Topology of Manifolds (held at Michigan State University, 1967), J. G. Hocking, ed., 1-18.

M. Brown and H. Gluck, Stable structures on manifolds I, II and III, Annals of Math. 79(1974), 1-58. Announced in Bull. Amer. Math. Soc. 69(1963), 51-58.

A. Casson, Three lectures on new infinite constructions in 4-dimensional manifolds, lecture notes from Université de Paris-Sud (Orsay), France, 1980.

E. H. Connell, Approximating stable homeomorphisms by piecewise linear ones, Annals of Math. 78(1963), 326-338. Announced in Bull. Amer. Math. Soc. 69(1963), 87-90.

E. H. Connell and J. Hollingsworth, Geometric groups and Whitehead torsion, Trans. Amer. Math. Soc. 140(1969), 161-181.

M. H. Freedman, A fake $S^3 \times R$, Annals of Math. 110(1979), 177-201.

__________, The topology of 4-dimensional manifolds, J. Diff. Geom. 17(1982), 357-453.

R. Gompf and S. Singh, On Freedman's reimbedding theorems, these proceedings.

A. Hatcher and J. Wagoner, Pseudo-isotopies of compact manifolds, Asterisque #6, Société Math. de France, 1973.

R. Kirby, Stable homeomorphisms and the annulus conjecture, Annals of Math. 89(1969), 575-582.

R. Kirby and L. Siebenmann, Foundational essays on topological manifolds, smoothings and triangulations, Annals of Math. Studies #88, Princeton Univ. Press, 1977.

E. E. Moise, Affine structures in 3-manifolds I-V, Annals of Math. 54(1951), 506-533; 55(1952), 172-176, 203-222; 56(1952), 96-114.

R. A. Norman, Dehn's lemma for certain 4-manifolds, Invent. Math. 7(1969), 143-147.

F. Quinn, Ends of Maps I, Annals of Math. 110(1979), 275-331.

______, The embedding theorem for towers, these proceedings.

______, Ends of maps, III: dimensions 4 and 5, J. Diff. Geom. 17(1982), 503-521.

T. Radó, Über den Begriff der Riemannschen Fläche, Acta Univ. Szeged 2(1924-26), 101-121.

L. Siebenmann, Amorces de la chirurgie en dimension 4, un $S^3 \times R$ exotique (d'après A. Casson et M. H. Freedman), Séminaire Bourbaki (1978-79), #536.

__________, La conjecture de Poincaré topologique en dimension 4 (d'après M. H. Freedman), Séminaire Bourbaki (1981-82), #588.

M. A. Stan'ko, Approximation of the imbedding of compacta in a codimension larger than two, Dokl. Akad. Nauk SSSR 198(1971), 783-786 = Soviet Math. Dokl. 12(1971), 906-909.

C. T. C. Wall, Geometrical connectivity I, Jour. London Math. Soc. 3(1971), 597-604.

Contemporary Mathematics
Volume 35, 1984

A μ-INVARIANT ONE HOMOLOGY 3-SPHERE THAT BOUNDS AN ORIENTABLE RATIONAL BALL

Ronald Fintushel[1] and Ronald J. Stern[2]

In this note we show that the Brieskorn homology sphere $\Sigma(2,3,7)$ bounds an orientable rational ball Q. It is known that the μ-invariant of $\Sigma(2,3,7)$ is one as it bounds the plumbed 4-manifold W^4

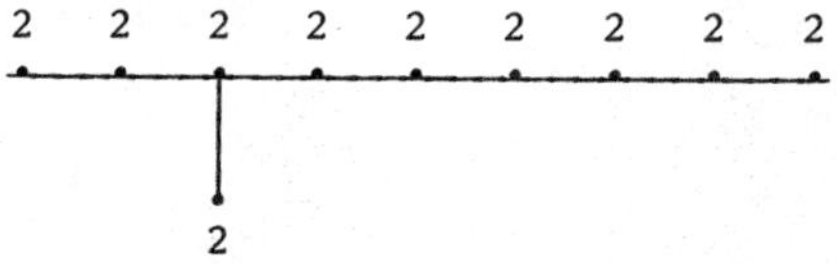

Note that W^4 has an even intersection form with signature $\sigma(W^4) = 8$ and rank 10. Thus $M^4 = Q \cup_\Sigma W^4$ is a closed orientable 4-manifold with even intersection form of signature 8 and rank 10. (Note that M^4 cannot be a spin 4-manifold.) As a corollary we have the following recent theorem of N. Habegger [1]:

COROLLARY. *Every even unimodular symmetric bilinear form* F *with* $\left|{}^{\mathrm{rank}(F)}/_{\sigma(F)}\right| \geq 5/4$ *can be realized as the intersection form of a closed orientable 4-manifold.*

THEOREM. $\Sigma(2,3,7)$ *bounds an orientable rational ball* Q^4.

PROOF. First we attach a 1-handle and a 2-handle to $\Sigma(2,3,7) \times I$ to obtain a rational homology cobordism W_1 between $\Sigma(2,3,7)$ and a 3-manifold K^3 which has the integral homology of $L(4,-1)$. Then we describe an integral homology cobordism W_2 between K^3 and $L(4,-1)$. Since $L(4,-1)$ bounds a rational ball W_3, we let $Q = W_1 \cup W_2 \cup W_3$. This is done as follows.

It is well known that $\Sigma(2,3,7)$ is obtained by $+1$ surgery on the figure eight knot. Attach a 1-handle to $\Sigma(2,3,7) \times I$ to obtain a cobordism from $\Sigma(2,3,7)$ to $\Sigma(2,3,7) \# S^2 \times S^1$:

[1] Supported in part by NSF grant MCS 7900244A01.
[2] Supported in part by NSF grant MCS 8002843A01.

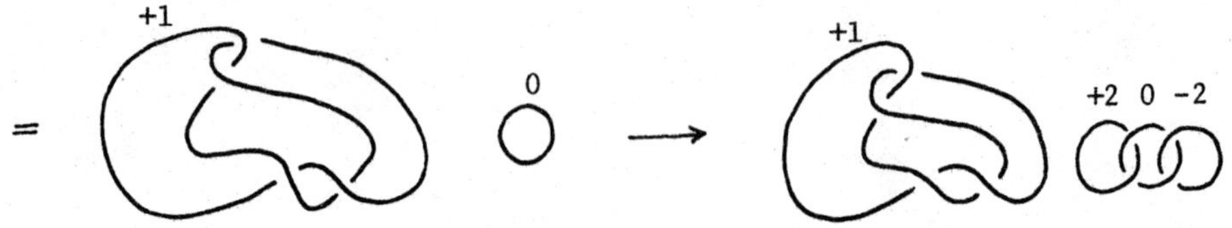

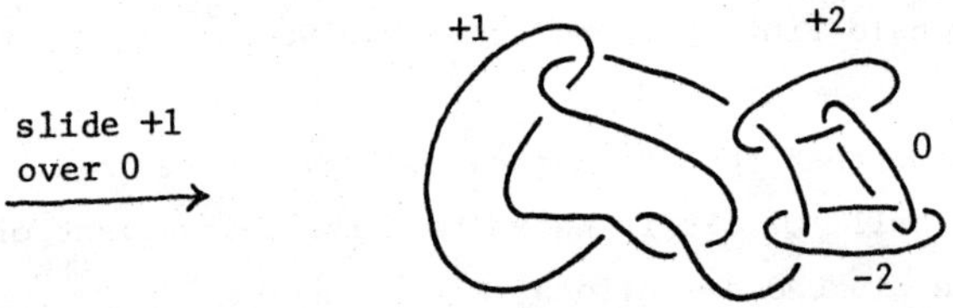

Now attach a 2-handle

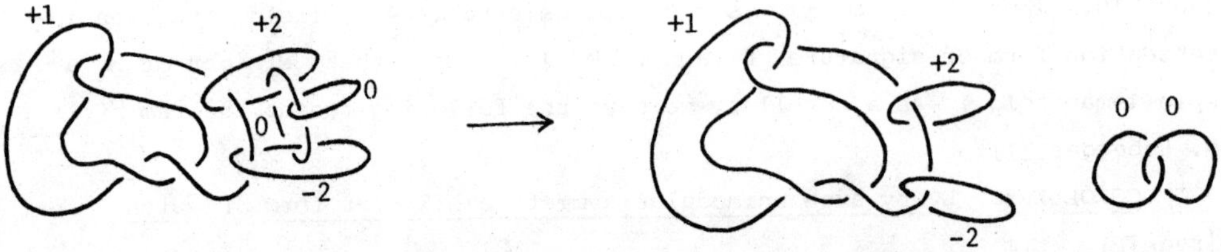

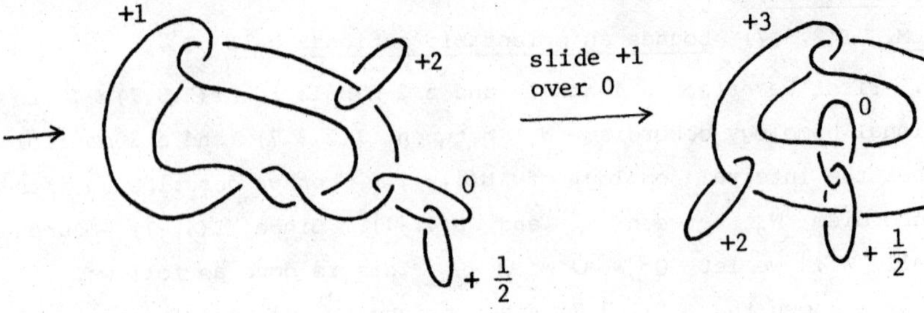

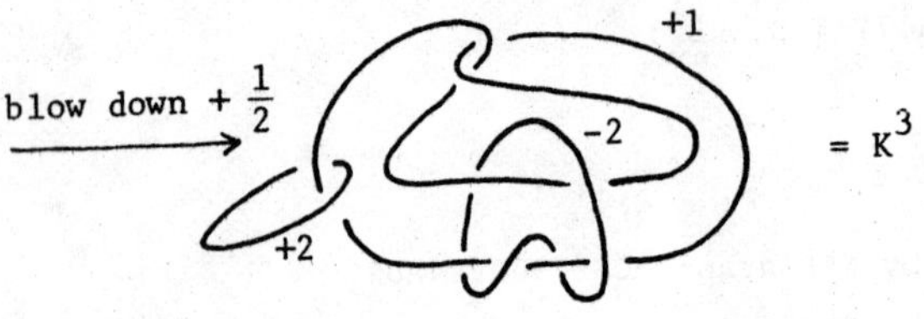

This describes the cobordism W_1. To see that it is a rational homology cobordism note that the attached 2-handle kills 4 times the generator of H_1 which was introduced by the 1-handle.

Now the link

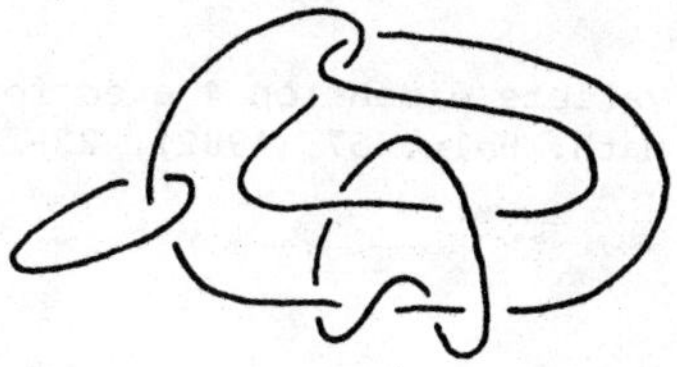

is ribbon concordant to the link

by means of the ribbon

Thus K^3 is integral homology cobordant to

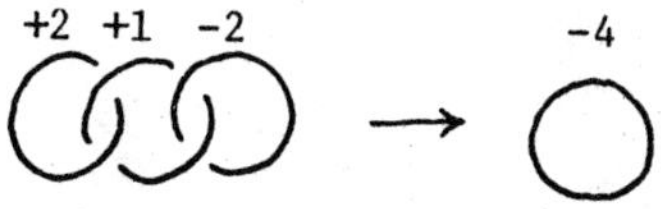

i.e. to $L(4,-1)$. Hence we have W_2.

Finally $L(4,-1)$ bounds a rational ball W_3. To see this attach the following 2-handle to $L(4,-1)$ to obtain $S^2 \times S^1$:

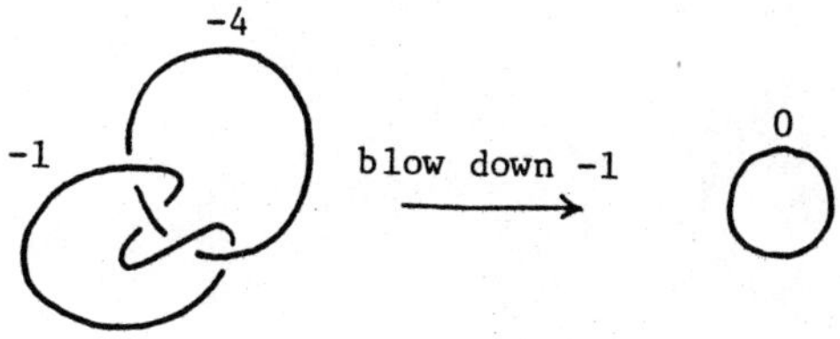

QUESTION. Does there exist a closed orientable 4-manifold with definite even intersection pairing and signature 8?

BIBLIOGRAPHY

1. N. Habegger, Une variete dimension 4 avec form d'intersection paire et signature -8, Comment. Math. Helv. 57 (1982), 22-24.

DEPARTMENT OF MATHEMATICS
TULANE UNIVERSITY
NEW ORLEANS, LA 70118

DEPARTMENT OF MATHEMATICS
UNIVERSITY OF UTAH
SALT LAKE CITY, UT 84112

Contemporary Mathematics
Volume 35, 1984

ANOTHER CONSTRUCTION OF AN EXOTIC $S^1 \underset{\sim}{\times} S^3 \# S^2 \times S^2$

Ronald Fintushel[1] and Ronald J. Stern[2]

This note was motivated by Selman Akbulut's talk at this conference. (See [A].) As Akbulut pointed out, if one could construct an exotic twisted S^3-bundle over S^1, with a homotopy equivalence $g: N^4 \to S^1 \underset{\sim}{\times} S^3$, then if a transverse preimage of an S^3-fiber is a homology sphere H^3, we must have $\mu(H^3) \neq 0$. But splitting N^4 along H^3 yields an acyclic 4-manifold whose boundary is $H^3 \# H^3$. Thus searching for an exotic $S^1 \underset{\sim}{\times} S^3$ is an approach toward finding the long sought after element of order 2 in θ^3_H.

Akbulut's construction is suggested by the fact that the complement of a tubular neighborhood $E(\mathbb{RP}^2)$ of $\mathbb{RP}^2$ in $\mathbb{RP}^4$ is $S^1 \underset{\sim}{\times} B^3$. His idea was to look for an $\mathbb{RP}^2$ in Q^4, Cappell and Shaneson's exotic $\mathbb{RP}^4$ ([CS]), such that $\pi_1(Q^4 - \mathbb{RP}^2) = \mathbb{Z}$, and then form $Q^4 - E(\mathbb{RP}^2) \cup S^1 \underset{\sim}{\times} B^3$. Unable to find such an $\mathbb{RP}^2$ embedded in Q^4, Akbulut was nonetheless able to find an $\mathbb{RP}^2$ in $Q^4 \# S^2 \times S^2$ with $\pi_1(Q^4 \# S^2 \times S^2 - \mathbb{RP}^2) = \mathbb{Z}$ and he was then able to form $Q^4 \# S^2 \times S^2 - E(\mathbb{RP}^2) \cup S^1 \underset{\sim}{\times} B^3$ an exotic $S^1 \underset{\sim}{\times} S^3 \# S^2 \times S^2$.

After seeing Akbulut's talk we decided to see if one could construct an exotic $S^1 \underset{\sim}{\times} S^3 \# S^2 \times S^2$ using the techniques we promoted in [FS_1] and [FS_2]. As we show this is quite simple to do and the invariant ρ of these papers can be used to detect the fact that the construction is exotic. Instead of viewing $S^1 \underset{\sim}{\times} S^3$ as $S^1 \underset{\sim}{\times} B^3 \cup S^1 \underset{\sim}{\times} B^3$, it is more convenient from our point of view to think of $S^1 \underset{\sim}{\times} S^3$ as $S^2 \times MB \cup S^1 \times B^3$ (MB = Mobius band). For our construction we start with K^3 a Seifert-fibered homology $S^2 \times S^1$ obtained by surgering an exceptional fiber of $\Sigma(3,5,19)$ and form X^4, the mapping cylinder of the free involution contained in the S^1-action on K^3. If we could show that K^3 bounded a homotopy $B^3 \times S^1$ with π_1 mapping onto, we could take its union with X^4 and thus construct a fake $S^1 \underset{\sim}{\times} S^3$. We cannot do this, but we are able to show that K^3 bounds a homotopy $B^3 \times S^1 \# S^2 \times S^2$ and thus we are able to form M^4, a homotopy $S^1 \underset{\sim}{\times} S^3 \# S^2 \times S^2$. As in [$FS_1$] we can show that if M^4 were s-cobordant to $S^1 \underset{\sim}{\times} S^3 \# S^2 \times S^2$ then

---

[1] Supported in part by NSF grant MCS 7900244A01.

[2] Supported in part by NSF grant MCS 8002843A01.

0271-4132/84 $1.00 + $.25 per page

$$\mu(K/\mathbb{Z}_2) - \frac{1}{2}\alpha(K,\mathbb{Z}_2) = \rho(M^4) = \rho(S^1 \underset{\sim}{\times} S^3 \# S^2 \times S^2)$$
$$= \mu(S^2 \times S^1) - \frac{1}{2}\alpha(S^2 \times S^1, \mathbb{Z}_2) = 0 \pmod{16}$$

for some almost framing of $K/\mathbb{Z}_2$. However $\alpha(K;\mathbb{Z}_2) = 0$ and the two μ-invariants of $K/\mathbb{Z}_2$ are both $8 \pmod{16}$; so M^4 is exotic. Finally, we are able to show that the double cover $\tilde{M}$ is standard, i.e. $\tilde{M}$ is diffeomorphic to $S^1 \times S^3 \# S^2 \times S^2 \# S^2 \times S^2$.

We now proceed with the construction of M^4. Let K^3 be the homology $S^2 \times S^1$ which is the boundary of the plumbing manifold

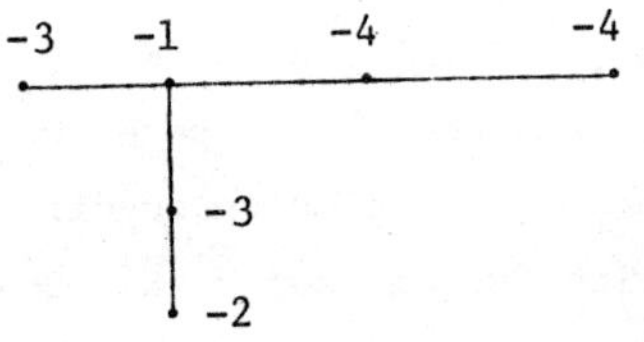

Then K is Seifert fibered with Seifert invariants $((1,1),(3,-1),(5,-2),(15,-4))$; so the involution contained in the S^1-action on K is free. Let X^4 be the mapping cylinder of the orbit map $K \to K/\mathbb{Z}_2$. As was shown in our earlier paper [FS_2, Lemma 3.1] there is a $\mathbb{Z}_2$-equivariant map $K \to S^2 \times S^1$ which induces isomorphisms on homology. (The involution on $S^2 \times S^1$ is identity $\times$ antipodal.) Taking mapping cylinders there is an induced map $f: X \to S^2 \times MB$ which induces isomorphisms on homology.

We have the following Kirby calculus picture for K:

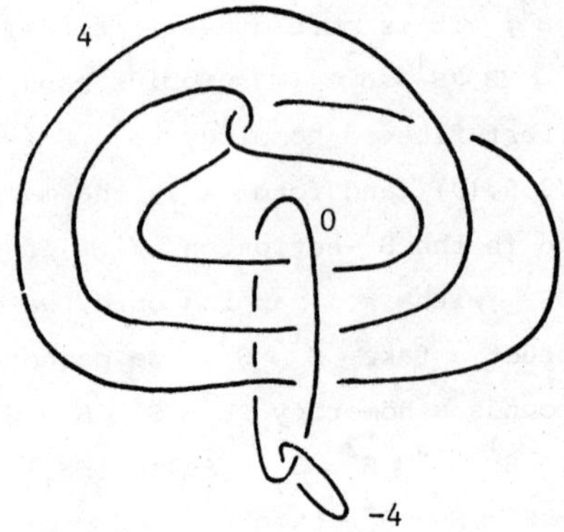

(cf [FS_1; p. 362]).

Now construct a cobordism Y^4 from K to $\partial_+ Y = \hat{K}$ by attaching the following 2-handles to $K \times I$:

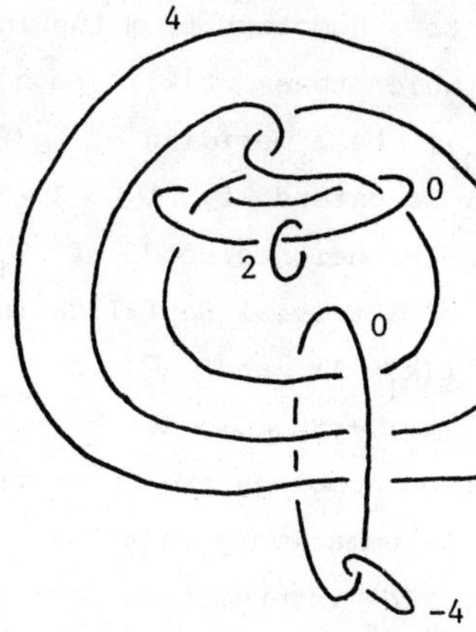

We claim that f extends over these 2-handles to a map:

$$f: X \cup Y \to S^2 \times MB \cup (S^2 \times S^1 \times I \# S^2 \times S^2) \quad .$$

To see this follow the 2-handles back through the Kirby calculus argument in [FS_2; p. 361-362]. The attaching circles are k_0 with 0-framing and k_2 with 2-framing:

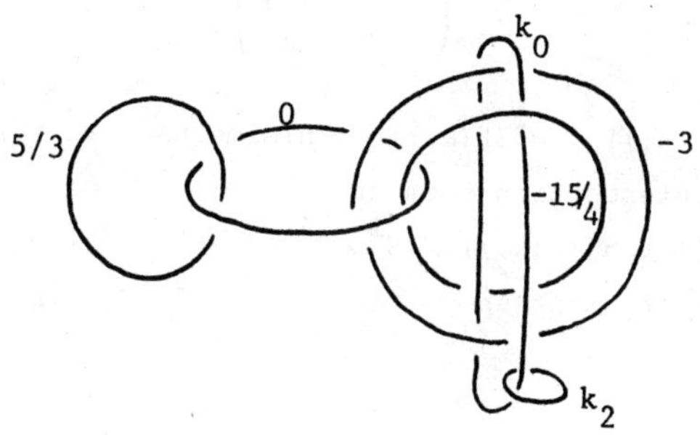

On K-(exceptional fibers), f preserves S^1-fibers and is a 15-fold covering. The image of f (see [FS_2; Lemma 3.1]) is $S^2 \times S^1$:

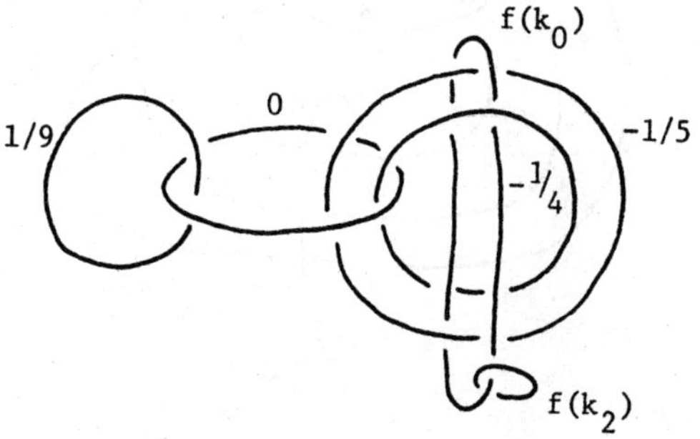

In $S^2 \times S^1$, $f(k_0)$ is nullhomologous (in the above diagram we see that $f(k_0)$ bounds a genus 1 surface) therefore $f(k_0)$ is nullhomotopic in $S^2 \times S^1$. So there is a homotopy in $S^2 \times S^1$ of $f(k_0)$ to a trivial knot. By the homotopy extension property this extends to a homotopy from the identity of $S^2 \times S^1$ to a map g of $S^2 \times S^1$ to itself which takes $f(k_0)$ to a trivial knot. We can also easily arrange that $g(f(k_1))$ be a meridian of $g(f(k_0))$. Composing f with the above ambient homotopy, we extend $f: X \cup K \times I \to S^2 \times S^1 \times I$ so that $f|K \times \{1\} \to S^2 \times S^1 \times \{1\}$ maps tubular neighborhoods of k_1 and k_2 onto tubular neighborhoods of the components of a trivial Hopf link in $S^2 \times S^1 \times \{1\}$.

For some framings a_1 on $f(k_1 \times 1)$ and a_2 on $f(k_2 \times 1)$, f will extend over $Y = K \times I \cup h^2(k_1) \cup h^2(k_2) \to S^2 \times S^1 \times I \cup h^2(f(k_1)) \cup h^2(f(k_2))$. Because $f|K$ induces isomorphisms on homology the naturality of the Mayer-Vietoris sequence and the 5-lemma imply that the intersection form of these two manifolds is the same. The intersection form of Y has matrix

$$\begin{pmatrix} 0 & 1 \\ 1 & 2 \end{pmatrix},$$

and therefore is even unimodular with signature 0. Hence the same is true for the intersection form

$$\begin{pmatrix} a_1 & 1 \\ 1 & a_2 \end{pmatrix}$$

of $S^2 \times S^1 \times I \cup h^2(f(k_1)) \cup h^2(f(k_2))$. This means that this intersection form is the same as the intersection form of $S^2 \times S^2$. Hence $S^2 \times S^1 \times I \cup h^2(f(k_1)) \cup h^2(f(k_2)) \cong S^2 \times S^1 \times I \# S^2 \times S^2$.

Another 5-lemma argument shows that $f|\hat{K} \to S^2 \times S^1$ induces isomorphisms on homology. $\hat{K}$ is:

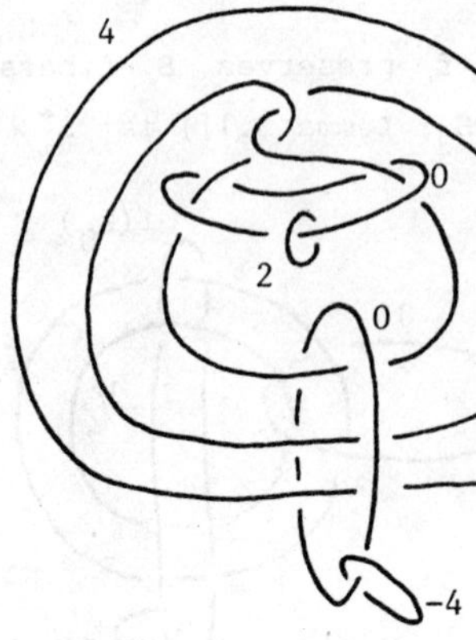

=

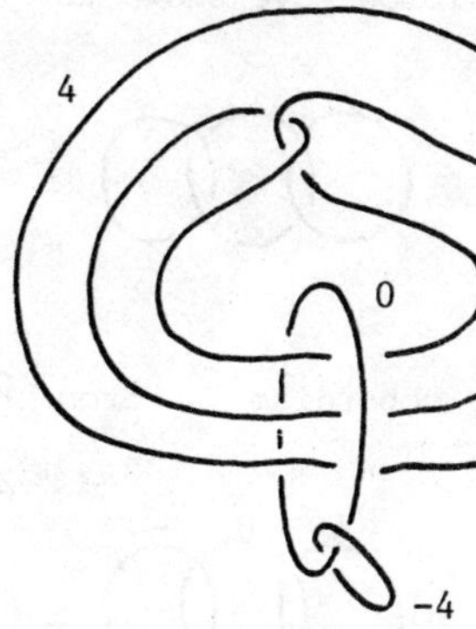

=

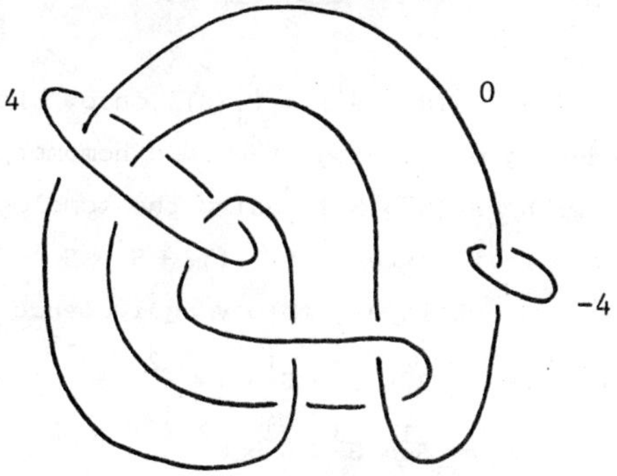

But the link

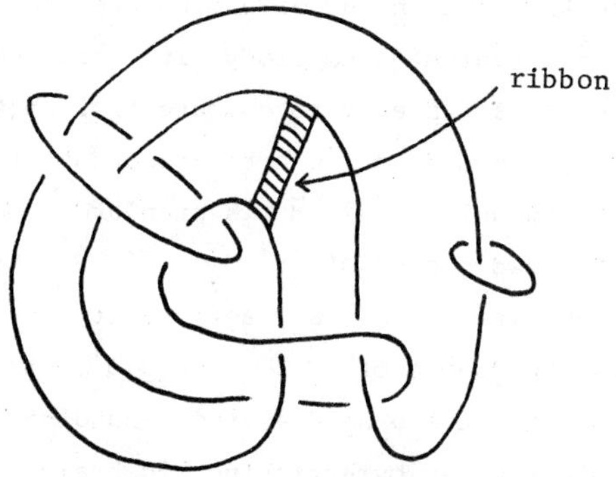

in S^3 is concordant by the ribbon move shown to

Hence there is a homology cobordism Z from $\hat{K}$ to $S^2 \times S^1 =$

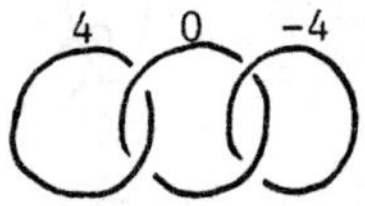

with $\pi_1(\hat{K}) \to \pi_1(Z)$ and $\pi_1(S^2 \times S^1) \to \pi_1(Z)$ onto. Let $\bar{f}: S^2 \times S^1 \to S^2 \times S^1$ be a diffeomorphism inducing on homology the same homomorphism as $(f|\hat{K})_*$. (Here we identify $H_*(\hat{K})$ with $H_*(S^2 \times S^1)$ using the homology cobordism Z.) Then by obstruction theory $f \cup \bar{f}$ extends to $f: Z \to S^2 \times S^1 \times I$. Since $\bar{f}$ extends over $B^3 \times S^1 \to B^3 \times S^1$ we obtain a homology equivalence

$$f: M = X \cup Y \cup Z \cup B^3 \times S^1 \to S^2 \times MB \cup S^2 \times S^1 \times I \# S^2 \times S^2 \cup S^2 \times S^1 \times I \cup B^3 \times S^1$$

$$= S^1 \tilde{\times} S^3 \# S^2 \times S^2 .$$

Using Van Kampen's theorem one checks that $\pi_1(M^4) = \mathbb{Z}$ and hence f induces an isomorphism on fundamental groups. Let $\tilde{f}: \tilde{M} \to S^1 \times S^3 \# S^2 \times S^2 \# S^2 \times S^2$ be the induced map on oriented double covers. As $\tilde{f}$ is degree one, the induced homomorphisms on homology with $\mathbb{Z}[\mathbb{Z}]$ coefficients split [W; Lemma 2.2]. However, all homology groups are free and in any dimension are the same rank, so $\tilde{f}$, hence f_1 induces an isomorphism on homology with local coefficients. So f is a homotopy equivalence. It is easy to compute that $\rho(M) \equiv 8 \pmod{16}$ (see [FS_2; proof of Prop. 5.5]); hence M is not s-cobordant to $S^1 \tilde{\times} S^3 \# S^2 \times S^2$.

We now show that the double cover $\tilde{M}$ is standard. Note that $\tilde{M}$ is obtained by gluing together two copies of $Y \cup Z \cup B^3 \times S^1$ by the involution $t: K \to K$. Since t is contained in an S^1 action, t is isotopic to the identity. Hence $\tilde{M}$ is the double of $Y \cup Z \cup B^3 \times S^1$. A handle decomposition for $Y \cup Z \cup B^3 \times S^1$ consists of a 0-handle, two 1-handles, and three 2-handles. (The cobordism Z is constructed by attaching algebraically cancelling 2 and 3-handles to $\hat{K} \times I$.) So the framed link picture for $\tilde{M}$ is obtained by adding a meridional circle labelled "0" to each circle representing a 2-handle. Using these it is easy to slide 2-handles to obtain

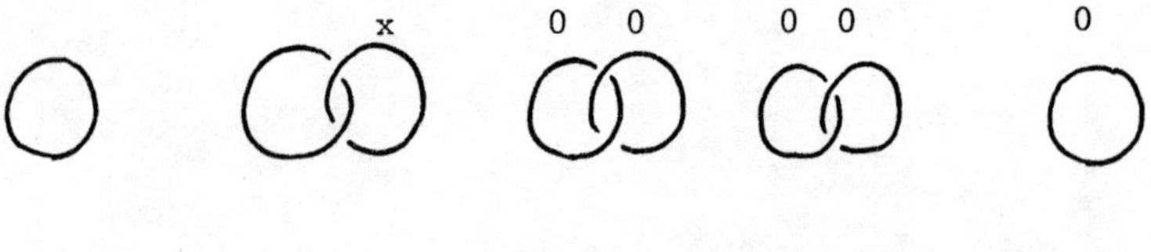

i.e. $\tilde{M} \cong S^3 \times S^1 \# S^2 \times S^2 \# S^2 \times S^2$.

BIBLIOGRAPHY

[A] S. Akbulut, A fake 4-manifold, these proceedings.

[CS] S. Cappell and J. Shaneson, Some new four manifolds, Ann. of Math. 104 (1976), 61-72.

[FS_1] R. Fintushel and R. Stern, An exotic free involution on S^4, Ann. of Math. 113 (1981), 357-365.

[FS_2] ——————————, Seifert fibered 3-manifolds and non-orientable 4-manifolds, to appear in Proceedings of the 1981 AMS special session on low dimensional topology, AMS New Contemporary Mathematics Series.

[W] C. T. C. Wall, Surgery on Compact Manifolds, Academic Press, 1970.

DEPARTMENT OF MATHEMATICS
TULANE UNIVERSITY
NEW ORLEANS, LA 70118

DEPARTMENT OF MATHEMATICS
UNIVERSITY OF UTAH
SALT LAKE CITY, UT 84112

Contemporary Mathematics
Volume 35, 1984

ON FREEDMAN'S REIMBEDDING THEOREMS

Robert E. Gompf and Sukhjit Singh

ABSTRACT. An improvement of Freedman's 3-Stage Reimbedding Theorem is given and its consequences are studied. In particular, an analogue of Freedman's 5-Stage Reimbedding Theorem is proved for 4-stage towers instead of 5-stage towers. Consequently, the second untwisted double of the Whitehead link is TOP slice (by flat disks). This appears to be new. Quite a bit of this paper is devoted to the exposition or the formalization of the techniques involved in proving the various Reimbedding Theorems; this is an elementary and complete account of these theorems.

0. INTRODUCTION. M. Freedman [F1] has proved some remarkable results for topological 4-manifolds. In particular, he has solved the ubiquitous 4-dimensional topological Poincaré conjecture (which roughly states: a homotopy 4-sphere is a sphere). A considerable portion of this work of Freedman (see also [F2]) deals with what he calls, "Reimbedding Theorems". These theorems have proved to be an indispensable tool in the "exploration" of a Casson handle, see [F1]. It appears that these theorems or their variants may also be useful in solving some other problems. It is our impression that these theorems, by themselves, are an important contribution of Freedman [F1,F2].

The main concern of this paper is these reimbedding theorems. The main innovation presented here is an "Improved 3-Stage Reimbedding Theorem" which is due to the first author. Consequently, all the reimbedding theorems of [F1] which come after the "3-Stage Reimbedding Theorem" require one less stage. These theorems are carefully summarized in Section 6 for the convenience of reference. As an application, we observe that every 5-stage tower contains a topological 2-handle with the same attaching curve; see Section 5. Consequently, the second double of the Whitehead link is topologically slice; see Theorem (5.3.2) for a specific statement.

This paper is also written with an intent of exposition, axiomatization, and formalization of the techniques involved in the proof of these theorems.

0271-4132/84 $1.00 + $.25 per page

An effort is made to distill together the best features of [F1] and [F2]. Furthermore, we have tried to complement or supplement the detail or exposition available there; we give alternative proofs, discussions, etc., wherever possible.

We assume familiarity with Casson's important Lecture I of [C] and some general knowledge of Freedman [F2]. The portion on the reimbedding theorems given in [F1] and this paper may be read simultaneously.

We wish to thank both M. Freedman and R. Kirby for their encouragement and other help. We first learned about this subject from Rob Kirby's course [K2]; we thank him again for his inspiring lectures. The second author wishes to thank U. C. Berkeley, Mathematics Department, for their hospitality during 1981-82; in particular, he thanks Emery Thomas.

1. NOTATION, TERMINOLOGY, AND OTHER CONVENTIONS. Most of the notation and terminology is standard; we have followed [C,F1,F2,K1] for these matters whenever convenient. Here is a brief discussion of some other conventions.

The word map should be interpreted as a morphism in a suitable category which will be clear from the context, e.g., a map between groups means a homomorphism. An unlabelled map $\pi_1 A \to \pi_1 X$ will be understood to be induced by the inclusion of A into X (i.e., A is a subspace of X). All homology groups will be with integral coefficients unless otherwise stated.

Suppose A is a subspace of X. We denote by $\text{Int}\,A$ or $\mathring{A}$ the interior of A in X. It is often convenient to abbreviate commonly used words and phrases, e.g., regular neighborhood = reg.nbd., π_1 - negligible or π_1 - negligibility = π_1 - neg., etc. We prefer to write $\pi_1 A$ rather than $\pi_1(A)$, i.e., we get rid of the (cumbersome to type) parentheses whenever this does not cause confusion; we also do this for other functors.

Suppose c is a simple closed curve (circle) contained in a space X. We often think of c representing an element of $\pi_1 X$ as follows: if $h: S^1 \to X$ is an imbedding with $h(S^1) = c$, we consider the homotopy class of h as an element of $\pi_1 X$ (where the base points are appropriately chosen). The choice of h will be either unimportant or clear from the context. Although we explicitly state the general position or transversality condition on subspaces of a manifold, assume this is the case whenever there is any doubt. A closed or open reg.nbd. of a subset in a manifold M, whenever defined, will be denoted by $N(A)$ or $\mathring{N}(A)$, respectively.

Suppose $A_1, A_2, \ldots, A_n$ are subsets of a set X. By $A_i \cap A_j = \delta_{ij}$ we mean that $A_i \cap A_j$ is empty when $i \neq j$ and a singleton set otherwise. The infinite cyclic group will be denoted by Z. The free product of two groups G and H is denoted by $G * H$.

2. PRELIMINARIES.

(2.0) A LEMMA OF CASSON: THE SIMPLY CONNECTED CASE. We assume familiarity with [C] and we follow [C] for notation and terminology whenever convenient. Let W denote a smooth 4-manifold with non-empty boundary ∂W.

(2.0.0) DEFINITION. A map $f: S \to W$, where S is an oriented surface with (possibly empty) boundary, is called a <u>normal immersion</u> if

a) f is a smooth immersion;

b) $f(S)$ meets ∂W in embedded $f(\partial S)$;

c) f is a transverse to ∂W; and

d) all self-intersections are transverse double-points in $\operatorname{Int} W$.

It is often convenient to forget the map and say "$s = f(S)$ is a <u>normally immersed surface</u> in W".

In the sequel, the surface S will usually be either a disk or a disk with one hole (annulus); we refer to s, in this case, by a <u>normally immersed disk or annulus</u>, respectively.

(2.0.1) DEFINITION. Suppose $f: S \to W$ is a normal immersion of an oriented surface S into a simply connected 4-manifold W. An <u>algebraic dual</u> of f is an element β of H_2W such that the intersection number $f \cdot \beta = 1$. (Note that β can be represented by an immersed 2-sphere.)

This definition is interesting only when W is simply connected; see Definition (2.1.0) when W is not simply connected. Although Definition (2.1.0) includes the simply connected case, we have treated the simply connected case separately, which, we hope, will motivate and clarify the non-simply connected case.

(2.0.2) DEFINITION. A map $f: X \to Y$ is called <u>π_1-negligible</u> (abbreviate: <u>π_1-neg.) in</u> Y if the inclusion $Y - f(X) \to Y$ induces an isomorphism on π_1. A subset of Y is <u>π_1-neg. in</u> Y if its inclusion into Y has this property.

With this terminology, we have a lemma from [C]:

(2.0.3) LEMMA. <u>Suppose</u> $f: D^2 \to W$ <u>is a normal immersion of the disk</u> D^2 <u>into a simply connected 4-manifold</u> W. <u>Then:</u> f <u>has an algebraic dual if and only if</u> f <u>can be regularly homotoped rel</u> ∂D^2 <u>to a π_1-neg. normal immersion</u> $g: D^2 \to W$.

(2.0.4) REMARKS. This lemma is proved by altering f (or $d = f(D^2)$) by "Casson moves" or "finger moves" to kill certain commutators in $\pi_1(W-d)$; see [C] for details. <u>Note that Casson moves can be made to miss any preassigned 2-complex.</u> <u>Also, note that the lemma remains true for a normally immersed annulus (or even an arbitrary surface).</u>

(2.1) A LEMMA OF CASSON: THE NON-SIMPLY CONNECTED CASE. The discussion below is essentially what appears in Freedman [F1], see Section 3. Let W denote an oriented smooth 4-manifold with non-empty boundary which may or may not be simply connected (the interesting case is when W fails to be simply connected). Suppose $f: S \to W$ is a normal immersion. Put $s = f(S)$. In this setting, we have the following:

(2.1.0) DEFINITION. An <u>algebraic dual</u> of f (or s) is an immersed 2-sphere z in W meeting s in points $x, x_1, y_1, x_1, y_2, \ldots, x_n, y_n$ such that for each i, $1 \le i \le n$, the points x_i and y_i are paired over W, i.e., x_i and y_i have opposite signs of intersection and there is a <u>Whitney circle</u> c_i, the union of an arc on S and an arc on z joining x_i and y_i, which bounds an immersed disk d_i in W; d_i is called a <u>Whitney disk.</u> An algebraic dual for s (in W) is called a <u>geometric dual</u> if it meets s in exactly one point.

The following lemma of Freedman [F1] replaces Lemma (2.0.3) in the non-simply connected case:

(2.1.1) LEMMA (π_1-LEMMA). <u>The normally immersed surface s (or the normal immersion f) in W can be regularly homotoped rel boundary to a normally immersed surface s' (or a normal immersion f') which is π_1-neg. in W if and only if s (or f) has an algebraic dual.</u>

Since Freedman [F1] gives a (formal) proof, we merely give an informal sketch of a proof. It is useful to carefully understand the basic ideas of this proof, since they are needed later on.

PROOF OF π_1-LEMMA: AN INFORMAL SKETCH. It suffices to prove that f' exists when s has an algebraic dual. Let $z, x, x_1, y_1, \ldots, x_n, y_n, c_1, d_1, \ldots, c_n, d_n$ be as in Definition (2.1.0). The proof is finished if $n = 0$. Suppose $n = 1$ until further notice; see Figure (2.A) for a schematic drawing.

<u>Casson Moves on s to Remove Intersections with $\mathrm{Int}\, d_1$.</u> Suppose s and the Whitney disk d_1 are transverse. Consider the case when the intersection of s and $\mathrm{Int}\, d_1$ is non-empty. For each point p in s and $\mathrm{Int}\, d_1$, use a Casson move to push s along an arc in d_1 from p to a point in $(s \cap d_1)$ until it falls off $\partial d_1 \cap s$; see Figure (2.B). Do these Casson moves simultaneously along disjoint arcs. This introduces new pairs of intersection for s. Observe that we have regularly homotoped f or s to obtain a new immersion f' or s'.

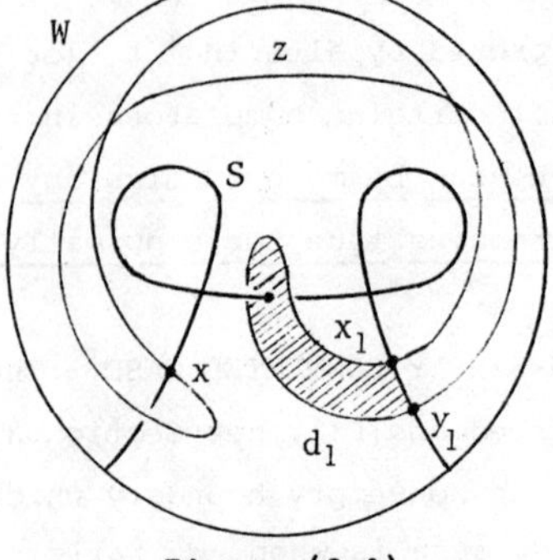

Figure (2.A)

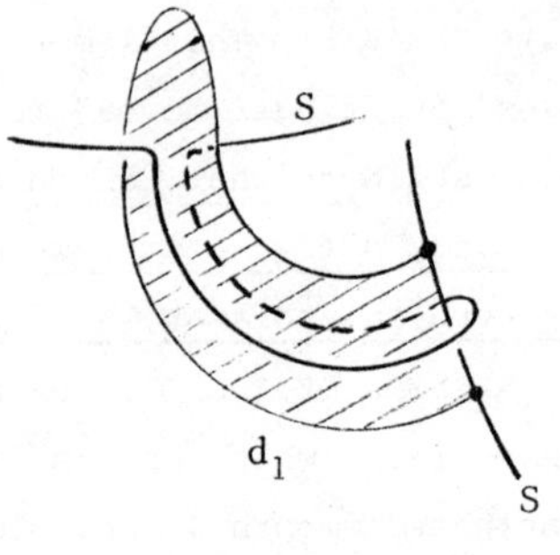

Figure (2.B)

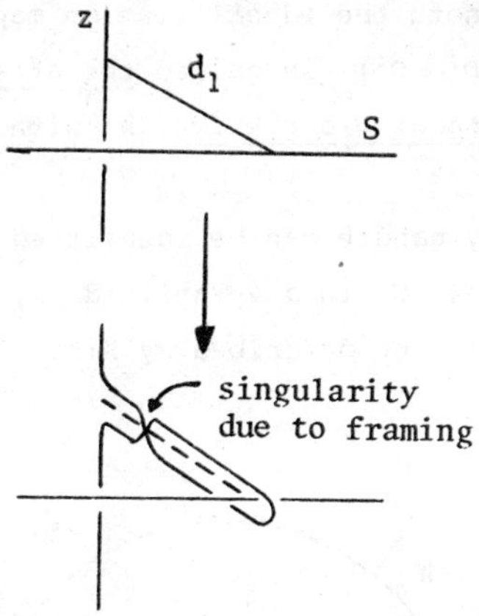

Figure (2.C)

Singular Whitney Trick on z. Use the "singular Whitney trick" to push z across d_1 to cancel the points x_1, y_1. Thus we obtain a new immersed 2-sphere z' which is a geometric dual of s'. For more details see [F1]. Figure (2.C) gives some idea of this push of z. Now s' is π_1-neg. since it has a geometric dual (every meridian bounds a disk). ⦀

(2.2) Kinky Handles. A 2-handle is a pair $(D^2 \times D^2, \partial D^2 \times D^2)$ where $\partial D^2 \times D^2$ is called the attaching region, $D^2 \times \{0\}$ is called the core, $\{0\} \times D^2$ is called the cocore, and $\partial D^2 \times \{0\}$ is called the attaching curve. A kinky handle $(k, \partial^- k)$ is a 2-handle with a finite but nonzero number of self-plumbings. Our Figure (2.D) represents a kinky handle $(k, \partial^- k)$ with one self-plumbing (or one kink). We often write k instead of $(k, \partial^- k)$.

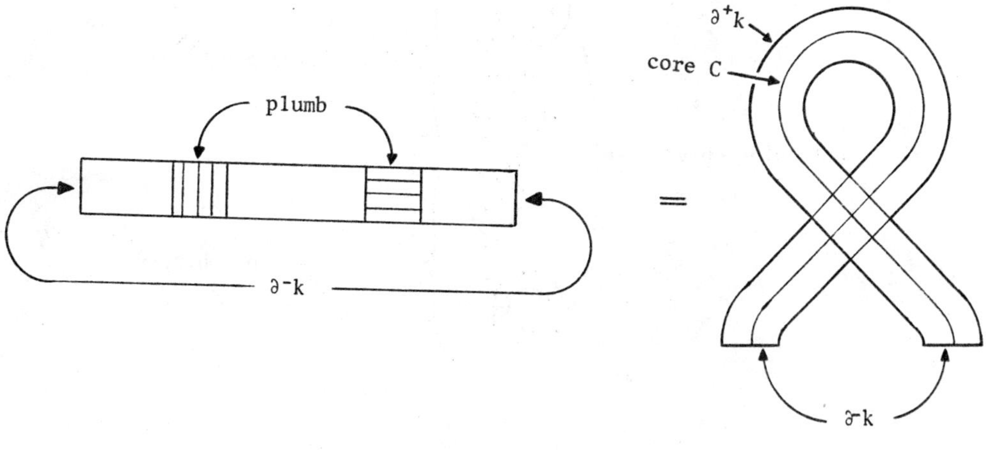

Figure (2.D)

Let $\pi: D^2 \times D^2 \to k$ denote the identification map used to produce a kinky handle k. Now $\partial^- k = \pi(\partial D^2 \times D^2)$ is called the attaching region for k. We call $\pi(\partial D^2 \times \{0\})$ the attaching curve for k with a framing which is discussed in (2.2.2).

Equivalently, a kinky handle can be identified with a regular neighborhood of a normally immersed disk C in a 4-manifold W, see Figure (2.E).

A kinky handle can also be described by Kirby calculus [K1], see Figure (2.F).

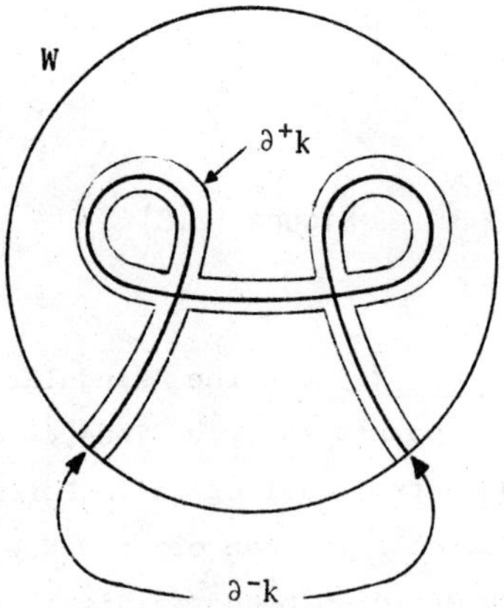

Figure (2.E)

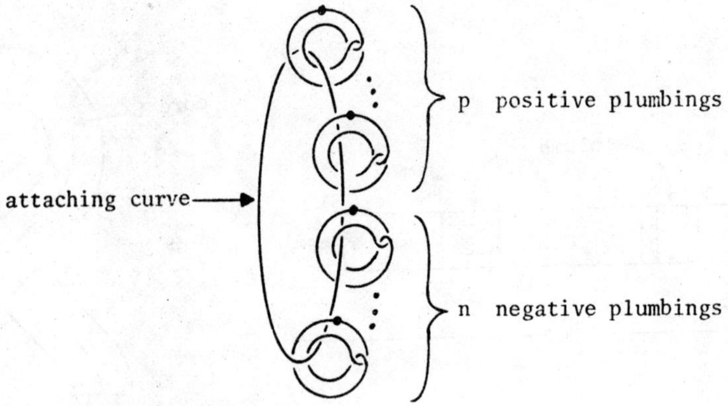

Figure (2.F)

Observe that every self-plumbing of $D^2 \times D^2$ corresponds to a transverse double-point of the core disk $D^2 \times \{0\}$ which inherits a + or - sign when $D^2 \times D^2$ is given the standard orientation. Notice that a kinky handle k has a core C corresponding to the core of $D^2 \times D^2$ i.e., k collapses to C, where C is a normally immersed disk in k and ∂C = the attaching curve for k. For further information see [F1,F2].

(2.2.1) A Standard Family of Curves. Suppose $(k, \partial^- k)$ is a kinky handle with n kinks (= number of self-plumbings). Casson [C, p.6] describes a family of pairwise disjoint framed (curves) circles $\{c_1, \ldots, c_n\}$ in $\partial^+ k$ such that if $\hat{k}$ is constructed by (abstractly) attaching a 2-handle along each of these framed circles, the resulting pair $(\hat{k}, \partial^- k)$ is diffeomorphic to the handle $(D^2 \times D^2, \partial D^2 \times D^2)$. We can draw a link picture of k, as in Figure (2.F), in which these circles appear as zero-framed meridians of the dotted circles. We also require, as in [C], that each c_i meets exactly one distinguished torus in exactly one point. Any family of framed curves $\{c_1, \ldots, c_n\}$ which fits the above description will be called a standard family of curves. There are several different isotopy classes of such families, due to nontrivial self-diffeomorphisms of k fixing $\partial^- k$.

(2.2.2) The Standard Framing for the Attaching Curve. In order to attach a kinky handle $(k, \partial^- k)$ to the boundary of a 4-manifold (as we do with 2-handles), we define a standard framing for $\partial^- k$. Capping off a standard family (as above) turns $(k, \partial^- k)$ into a 2-handle, whose attaching region is also $\partial^- k$. The standard framing for this 2-handle (i.e., the product structure on its attaching region $\partial D^2 \times D^2$) gives us the desired framing for $\partial^- k$. This definition is independent of the choice of standard family, for it is a "homological" invariant in the following sense: Suppose we attach a 2-handle to k along $\partial^- k$ with some framing, obtaining a manifold $\hat{k}$ with $H_2 \hat{k} = Z$. Then the intersection pairing on $H_2 \hat{k}$ is zero if and only if the 2-handle was attached via the standard framing of $\partial^- k$. (This may be verified by capping off a standard family on k $\hat{k}$, obtaining a disk bundle over S^2.) This is equivalent to the following characterization (see [F1]): If we push a "parallel" (normally displaced) copy $(C', \partial C')$ off of the core disk $(C, \partial C)$ in $(k, \partial^- k)$, with $\partial C'$ displaced from ∂C via the standard framing on $\partial^- k$, then the algebraic sum of the (signed) intersections of C and C' is zero.

Note that the standard framing is not the one induced by the normal bundle of C (i.e., the one obtained from $\partial D^2 \times D^2$ via the map π of (2.2.0)). In fact, these framings differ by exactly twice the number (Self C) of self-intersections of C (counted with sign). This is a consequence of the following formula for a closed (oriented) surface F in a 4-manifold: The homological intersection number $[F] \cdot [F] = \chi(\nu) + 2 \text{ Self } F$, where $\chi(\nu)$ is the normal Euler number of F.

We now observe the following important fact: the standard framing is preserved by Casson moves. More precisely, suppose we have an imbedding $(k,\partial^- k) \to (M,\partial M)$ of a kinky handle into a 4-manifold. Suppose that the core C of k is altered (rel boundary) by Casson moves, to obtain a normally immersed disk $\tilde{C}$. Taking a regular neighborhood of $\tilde{C}$, we obtain a new kinky handle $(k',\partial^- k')$, with $\partial^- k' = \partial^- k$. We claim that the two induced framings on $\partial^- k$ will agree. This follows easily from the homological interpretation of these framings. (Glue a 2-handle h onto M along the circle ∂C in ∂M, and note that the immersed spheres $C \cup \text{core}(h)$ and $C' \cup \text{core}(h)$ represent the same (nonzero) homology class.)

(2.2.3) Towers. A 1-stage tower T_1 is a kinky handle k^1. A 2-stage tower T_2 is obtained from T_1 by attaching a kinky handle to T_1 along each member of a standard family of framed curves by matching the framings. Recall that the attaching region for a kinky handle is always framed as in (2.2.2). By the second stage of T_2 we mean either the collection of all the kinky handles attached to T_1 to obtain T_2 or the union of these kinky handles (the precise meaning will always be clear from the context). Suppose an n-stage tower T_n has been constructed. We define a standard family of curves for T_n as the union of a standard family for each nth stage kinky handle. We construct an $(n+1)$-stage tower by attaching a kinky handle to each curve belonging to a standard family for T_n.

(2.2.4) A Casson Handle. Suppose $T_1 \to T_2 \to T_3 \to \dots$ are inclusions of towers constructed as above. Define T_n^- as the union of $\text{Int}\, T_n$ with $\partial^- T_n = \partial^-$(the first stage kinky handle for T_n). A Casson handle CH is the union of the corresponding inclusion of towers $T_1^- \to T_2^- \to T_3^- \to \dots$ with the direct limit topology. It is a deep theorem of Freedman that any $(CH, \partial^- T_1)$ is homeomorphic as a pair to the standard open 2-handle $(D^2 \times \mathring{B}^2, \partial D^2 \times \mathring{B}^2)$.

(2.2.5) Link Pictures for Towers. It is often useful to draw a link picture for a tower T_n. We have already drawn a link picture (see Figure (2.F)), for an arbitrary 1-stage tower T_1. Figure (2.G) is a link picture of a 2-stage tower T_2 having exactly one kinky handle with exactly one kink at each stage.

Figure (2.H) is a link picture of a 3-stage tower T_3. The first stage of T_3 is a kinky handle with two kinks, the second stage has a kinky handle with one kink and a kinky handle with two kinks, and the third stage has two kinky handles each with one kink and a kinky handle with two kinks.

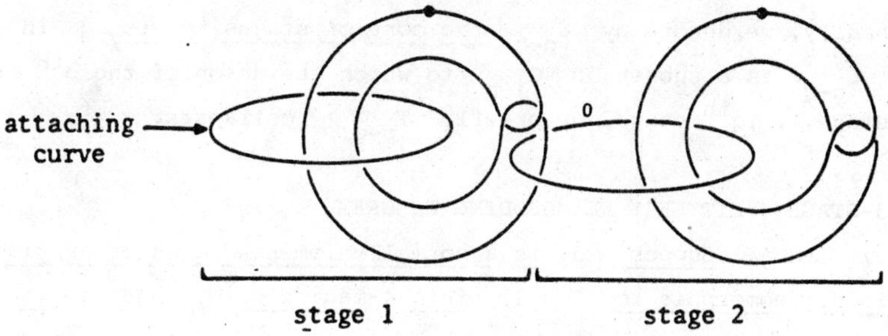

Figure (2.G)

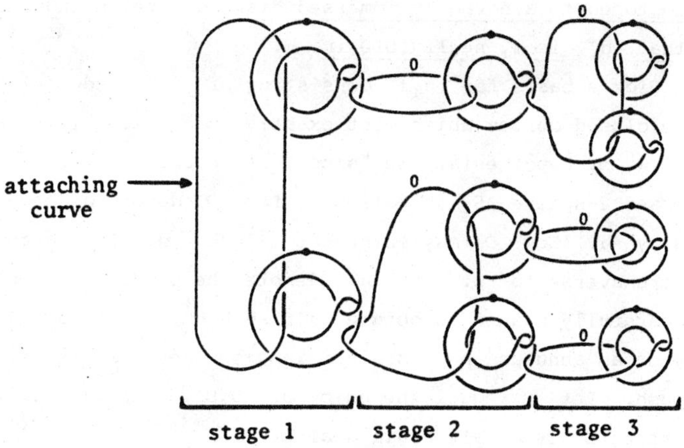

Figure (2.H)

(2.2.6) Cores for Towers. We have already defined core for T_1 or a kinky handle; see (2.2). Observe that every tower T_n has a 2-complex as a strong deformation retract. This can be seen by collapsing each kinky handle to its core C and observing that each boundary ∂C traces an annulus under the collapse of the previous stage: the union of all these cores together with annuli is this 2-complex which is denoted by C_{1-n} and called a core of T_n. Figure (2.I) shows C_{1-2} for a 2-stage tower T_2.

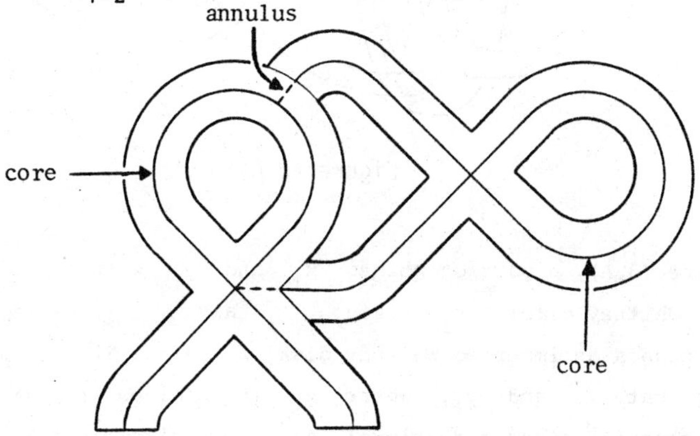

Figure (2.I)

More generally, we denote by C_{p-q} _the core of stages_ p _to_ q _in_ T_n: we identify C_{p-q} as a subset of C_{1-n} to which the union of the pth stage, $(p+1)^{th}$ stage,...,q^{th} stage, or briefly T_{p-q}, collapses.

3. THE 3-STAGE ("LITTLE") REIMBEDDING THEOREM

(3.0) LEMMA. _Suppose_ d _is a normally immersed disk or annulus (an annulus is diffeomorphic to_ $S^1 \times I$) _in a 4-manifold_ W, _and_ τ _is an imbedded torus in_ W _which meets_ d _transversely in exactly one point, such that the inclusion_ $\tau \to W$ _induces the zero map_ $\pi_1\tau \to \pi_1 W$. _Then:_ d _can be regularly homotoped to a normally immersed disk_ d' _rel boundary by Casson moves such that_ d' _is_ π_1_-negligible in_ W.

PROOF. Pick a basis for $\pi_1\tau$ consisting of $[\alpha]$ and $[\beta]$ where α and β are simple closed curves which meet exactly once. Now $[\alpha]$ and $[\beta]$ are trivial in $\pi_1 W$ by hypothesis. We "singularly surger" τ along α to obtain an immersed 2-sphere S as follows. Let D denote a normally immersed disk bounded by α. Now D may intersect τ and d. Push a copy D' off D which is transverse to D. Let α' denote the boundary of D'. We require that α is carefully pushed to obtain α' such that α and α' constitute the boundary of an annulus A contained in τ, where A is traced by α during the push. Let S equal the union of D, D', and $(\tau - \mathrm{Int}\,A)$. Figure (3.A) shows that S is an algebraic dual of d.

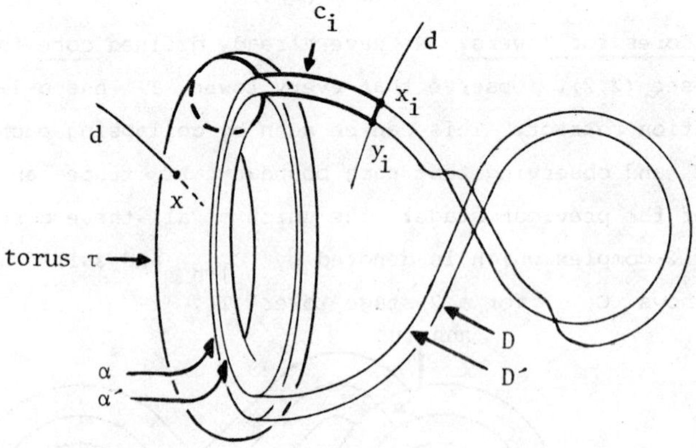

Figure (3.A)

In Figure (3.A), a pair of points x_i and y_i with opposite signs are shown, and a Whitney circle c_i through x_i and y_i is exhibited. The circle c_i bounds an immersed Whitney disk d_i since it is homotopic to β. We emphasize that x_i and y_i, where y_i is obtained from x_i by the push, always have opposite signs and trivial c_i when paired in this way. Our proof is finished by observing that each point of intersection of d with D gives a pair of this type; see Lemma (2.1.1). ▮

(3.1) Some Lemmas. Recall that T_n denotes an n-stage tower and C_{1-m} denotes the core of the first m-stage subtower T_m of T_n. The following is the main lemma of this section:

(3.1.0) LEMMA. *$\pi_1(T_n - C_{1-m}) \approx Z * F$ whenever $1 \leq m < n$, where F is the free group generated by a standard family of curves for T_n (i.e., for the nth stage of T_n), and Z is generated by a meridian of the first stage core C_1. In particular, the inclusion $(T_n - C_{1-m}) \to (T_n - C_1)$ induces an isomorphism of the fundamental groups when $1 \leq m < n$.*

This decomposition $Z * F$ of $\pi_1(T_n - C_{1-m})$ will be called the canonical decomposition.

The following sublemma is a technical prerequisite for the proof of Lemma (3.1.0):

(3.1.1) SUB-LEMMA ("BACKING-UP LEMMA"). *If N is the subgroup of $\pi_1(T_m - C_{1-m})$ normally generated by a standard family of curves for T_m, then the quotient $\pi_1(T_m - C_{1-m})/N$ is isomorphic to Z where Z is generated by the meridian of C_1.*

The following is immediate from the Seifert-Van Kampen Theorem.

(3.1.2) AN OBSERVATION. *Suppose X is a connected space containing a solid torus T. Suppose $(k, \partial^- k)$ is a kinky handle. Construct a space Y by attaching k to X by identifying $\partial^- k$ with T. Then: $\pi_1 Y \approx \pi_1 X/N * \pi_1 k$ where N is the subgroup of $\pi_1 X$ normally generated by the core of the solid torus.* (Use the fact that the core is null-homotopic in k.)

PROOF OF SUBLEMMA (3.1.1). We will demonstrate our method in the special case of the following 3-stage tower.

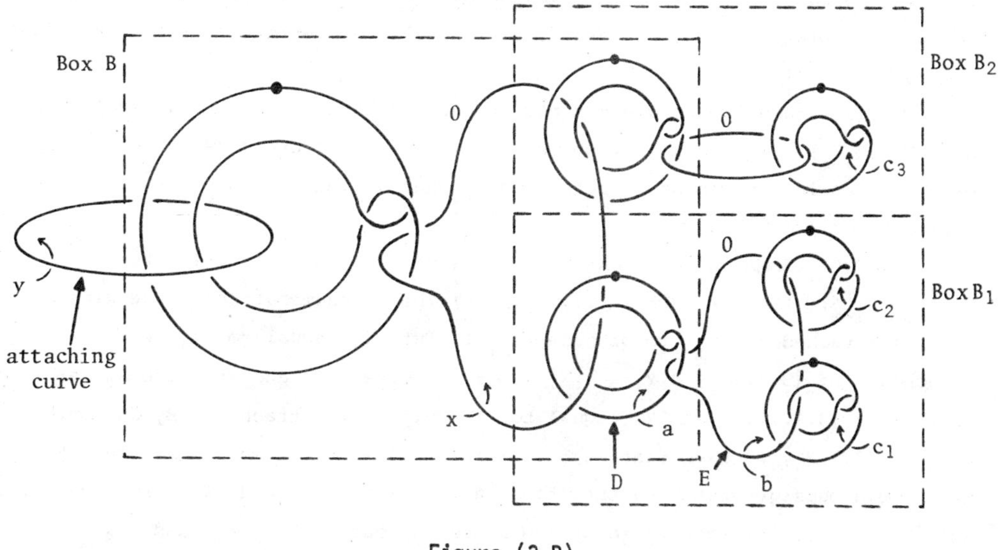

Figure (3.B)

Figure (3.B) represents a tower T_3 with the first stage a kinky handle with one kink, the second stage a kinky handle with two kinks, and the third stage two kinky handles. (The one inside the box B_1 has two kinks and the one inside B_2 has one kink.)

Consider the 3-manifold $M^3 = \partial T_3 - \text{Int}$ (attaching region). The 4-manifold $T_3 - C_{1-3}$ is M^3 up to homotopy (T_3 is a regular neighborhood of C_{1-3}). The link picture (see Figure (3.B)) now describes M^3 as a 0-framed surgery in S^3 on each component of this link other than the attaching curve, followed by the removal of the interior of the attaching region. (Note that dots are replaced by zeros.)

Let L denote the link consisting of all the curves of Figure (3.B). Now $\pi_1 M^3$ is the quotient of $\pi_1(S^3 - L)$ by some "surgery relations" which are in one-to-one correspondence with components of L other than the attaching curve. At any rate, we have a presentation of $\pi_1 M^3$, whose generators are the meridians of curves in L.

We want to prove that this presentation of $\pi_1 M^3$ reduces to $\langle y:\emptyset\rangle$ after adding the relations $c_1 = c_2 = c_3 = 1$. (Recall that by definition a standard family can always be represented by meridians of the top stage dotted circles in an appropriate link picture; see (2.2.1).) Consider the box B_1. We show that after adding the relations $c_1 = c_2 = 1$ we have $a = b = 1$. The equality $a = 1$ follows from the handle relation $a^{\alpha} c_1^{\beta} (c_1^{-1})^{\gamma} c_2^{\delta} (c_2^{-1})^{\varepsilon} = 1$ corresponding to the component E, where $\alpha, \dots, \varepsilon$ are some elements of $\pi_1 M^3$ and a^{α} denotes $\alpha^{-1} a \alpha$. The relation $b(a^{-1})^{\lambda} x (a^{-1})^{\mu} x^{-1} a^2 = 1$ corresponding to D and the relation $a = 1$ implies $b = 1$. Now proceed with box B_2 in a similar manner. This allows us to "back-up" and apply the procedure to the box B: We call this method the <u>backing up technique</u>. It is now clear that adding the relations c_1, c_2 and c_3 kills all the meridians except y. This proves our result.

The backing up technique can now be used to handle the more general case of $(T_m - C_{1-m})$. (Observe that we must apply the procedure given above to all of the boxes in a given stage before backing up to the previous stage.) This finished our proof. ////

PROOF OF LEMMA (3.1.0). We first consider the case $n = m+1$. Put $X_1 = T_m - C_{1-m}$. Let $(k_1, \partial^- k_1), \dots, (k_s, \partial^- k_s)$ be an enumeration of the kinky handles attached to T_m to produce T_{m+1}. Put Y_1 equal to X_1 with k_1 attached. By (3.1.2), $\pi_1 Y_1 \approx \pi_1 X_1 / N_1 * F_1$, where N_1 and F_1 are as described in (3.1.2). Put Y_2 equal to Y_1 with k_2 attached. By an application of (3.1.2), we have that $\pi_1 Y_2 \approx \pi_1 X_1 / N_2 * F_1 * F_2$, where F_1 and F_2 are free groups generated by the standard curves on k_1 and k_2, respectively, and N_2 is normally generated by the attaching curves for k_1 and k_2.

Proceeding in this manner we have $Y_s = T_{m+1} - C_{1-m}$ and $F = F_1 * \cdots * F_s$. The subgroup $N = N_s$ of $\pi_1 X_1$ is normally generated by the standard family of attaching curves for T_m, and $\pi_1 Y_s \approx \pi_1 X_1/N * F$. Since $\pi_1 X_1/N \approx Z$ by Sub-Lemma (3.1.1), our proof is finished for all (m+1)-stage towers.

We now consider the case $n = m+2$. It is similar to the previous case. Observe that by (3.1.2) the free factor generated by the standard family of curves in $(T_{m+1} - C_{1-m})$ vanishes in $T_{m+2} - C_{1-m}$. It is replaced by a free group F' generated by standard families of curves at the $(m+2)^{th}$ stage. Thus $\pi_1(T_{m+2} - C_{1-m}) \approx Z * F'$. The general case when $n = m+\ell$ follows in a similar manner. ⦀

(3.1.3) PROPOSITION. *Let* $(k, \partial^- k)$ *be a kinky handle belonging to the* j^{th} *stage of* T_n *with* $2 \le j \le n$. *Let* $\tau \subset \partial^+ k$ *denote a distinguished torus in* k. *If* $m < n$, *then the map* $\pi_1 \tau \to \pi_1(T_n - C_{1-m})$ *induced by the inclusion is trivial*.

(3.1.4) REMARKS. Observe that the map $\pi_1 \tau \to \pi_1 T_n$ induced by the inclusion is always trivial for τ as above. The conclusion of Proposition (3.1.3) becomes false when $m = n$, or when τ is a first stage distinguished torus.

PROOF OF PROPOSITION (3.1.3). By [C, see page 7], each of the two standard generators of $\pi_1 \tau$ is a meridian of the kinky handle k. By Lemma (3.1.0), such a meridian is nullhomotopic in $T_n - C_{1-m}$ when $m < n$. This finishes our proof. ⦀

We now have all the ingredients to prove the following theorem of Freedman [F1].

(3.2) 3-STAGE REIMBEDDING THEOREM. *Every 3-stage tower* T_3^0 *contains another 3-stage tower* T_3^1 *satisfying*:

a) *(agreement)* $C_{1-2}^0 = C_{1-2}^1$; *and*

b) *(π_1-neg.)* $\pi_1(T_3^0 - T_3^1) \to \pi_1(T_3^0 - C_1^0)$,

or equivalently, $\pi_1(T_3^0 - T_3^1) \to \pi_1(T_3^0 - C_{1-2}^0)$ *is an isomorphism*.

PROOF. Let $(k_1, \partial^- k_1), \ldots, (k_n, \partial^- k_n)$ denote the third stage kinky handles attached to T_2^0 to obtain T_3^0. Let τ_i denote the second stage distinguished torus for k_i, and let $\bar{d}_i$ denote the core of k_i together with the annulus. By Proposition (3.1.3), for each $1 \le i \le n$, the map $\pi_1 \tau_i \to \pi_1(T_3^0 - C_{1-2}^0)$ is trivial.

Put $W = T_3^0 - \mathring{N}(C_{1-2}^0)$, $d = \bar{d}_1 \cap W$, and $\tau = \tau_1$. By Lemma (3.0), find a normally immersed disk d' such that d' is π_1-neg. in W and d' misses $(\bar{d}_2 \cap W), \ldots, (\bar{d}_n \cap W)$. Put $\tilde{d}_1$ equal to d' union $(\bar{d}_1 - W)$. Observe that $\tilde{d}_1$ is π_1-neg. in $(T_3^0 - C_{1-2}^0)$.

Again, put $W = T_3^0 - \mathring{N}(C_{1-2}^0 \cup \tilde{d}_1)$, $d = \bar{d}_2 \cap W$, and $\tau = \tau_1$. Since $\tilde{d}_1$ is π_1-neg. in $(T_3^0 - C_{1-2}^0)$, it follows that the inclusion $\tau \to W$ induces the trivial map on π_1. We proceed as above, to find a normally immersed disk $\tilde{d}_2$ in W satisfying: (a) $\tilde{d}_2$ is π_1-neg. in $T_3^0 - (C_{1-2}^0 \cup \tilde{d}_1)$, and $\tilde{d}_2$ misses $(\bar{d}_3 \cap W), \ldots, (\bar{d}_n \cap W)$. The inductive step is now clear. We continue in this manner and obtain disjoint normally immersed disks $\tilde{d}_1, \ldots, \tilde{d}_n$ whose union is π_1-neg. in $T_3^0 - C_{1-2}^0$.

It is easy to see that a regular neighborhood of $C_{1-2}^0 \cup \tilde{d}_1 \cup \cdots \cup \tilde{d}_n$ is a tower T_3^1 inside T_3^0 satisfying the required properties, since Casson moves do not change the standard framing of a kinky handle (see (2.2.2)). ∎

(3.3) <u>An Improvement of the 3-Stage Reimbedding Theorem</u>. The proof of Theorem (3.2) requires that we make Casson moves on each $\bar{d}_i$ (recall $\bar{d}_i$ is a core of a third stage kinky handle together with annulus). It was discovered by the first author that the required Casson moves on each $\bar{d}_i$ can be made in a controlled manner such that they do not link the core C_1^0 of the first stage kinky handle. More precisely, the following conclusion (c), in addition to (a) and (b) in Theorem 3.2 holds:

c) <u>(no linking C_1^0) The image</u> $\mathrm{Im}[\pi_1 C_3^1 \to \pi_1(T_3^0 - C_1^0)]$ <u>lies in the image</u> $\mathrm{Im}[\pi_1 C_3^0 \to \pi_1(T_3^0 - C_1^0)]$, <u>where C_3^1 denotes the core of the third stage of T_3^1.</u>

Note that the latter image is precisely the factor F of the canonical decomposition $\pi_1(T_3^0 - C_1^0) = Z * F$. The conclusion (c) will prove useful in reducing a stage in subsequent reimbedding theorems of Freedman [F1]; a complete summary of these matters is given in Section 6.

We next prove this "Improved 3-Stage Reimbedding Theorem" by giving an alternate proof of Theorem (3.2) and keeping track of the Casson moves, which we call "careful Casson moves".

PROOF. We begin with the link picture Figure (3.C) which represents a 3-stage tower T_3^0 where the exchange trick has been applied on the third stage. We shall prove our claim for this particular tower; the general case will follow from the pattern of this proof.

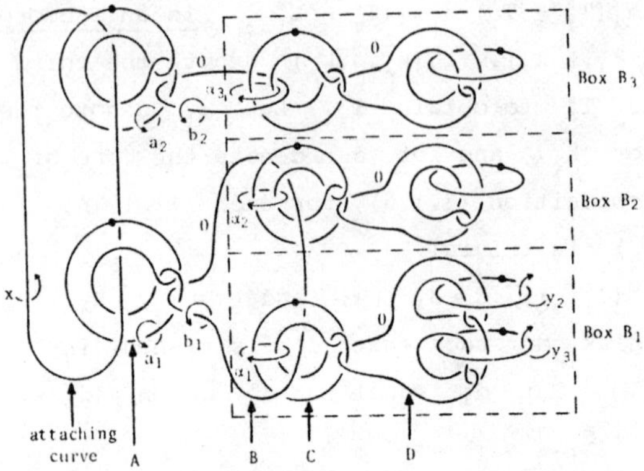

Figure (3.C)

Let L denote the link in S^3 given in Figure (3.C), consisting of all the components except the attaching curve. The boundary of T_3^0 minus an open regular neighborhood of the attaching curve is a 3-manifold M^3. Note that M^3 is also obtained by performing 0-framed surgery on each component of L and deleting a regular neighborhood of the attaching curve. Observe that $M^3 \times I \approx T_3^0 - \mathring{N}(C_{1-3}^0)$. Figure (3.C) will represent M^3 or $M^3 \times I$ in the following discussion.

We note that in $\pi_1(M^3 \times I)$, b_1 equals a product of conjugates of α_1, α_2 and their inverses. This follows from the surgery relation $b_1(a_1^{-1})^{\lambda} x(a_1^{-1})^{\mu} x^{-1}a_1^2 = 1$ (from the link component A), together with the relation that a_1 is conjugate to $\alpha_2\alpha_1$ (from component B). It follows that there is a singular punctured 2-disk K_1 bounded by b_1 and copies of α_1 and α_2. In fact, K_1 is determined by a homotopy between b_1 and conjugates of α_1, α_2, and their inverses. Similarly, we obtain a singular punctured 2-disk K_2 bounded by b_2 and copies of α_3.

Now consider $T_3^0 - C_{1-2}^0$. Let $\bar{d}_1$, $\bar{d}_2$ and $\bar{d}_3$ denote the cores of the third stage kinky handles together with annuli. In the sequel, we will alter $\bar{d}_1$, $\bar{d}_2$ and $\bar{d}_3$ to obtain $\tilde{d}_1$, $\tilde{d}_2$ and $\tilde{d}_3$ by "careful Casson moves" in $\mathrm{Int}(T_3^0 - C_{1-2}^0)$. We will choose these Casson moves to miss the 2-complex K_1 union K_2, and so that each α_i, $i = 1,2,3$, bounds an immersed disk in $X = (T_3^0 - C_{1-2}^0$ minus $\tilde{d}_1$, $\tilde{d}_2$, and $\tilde{d}_3$).

We now indicate why this suffices to prove Theorem (3.2) (conditions (a) and (b)). The loops b_1 and b_2 are nullhomotopic in X, since the punctures in the punctured disks K_1 and K_2 can now be filled. Thus, the generators of the second stage distinguished tori are trivial in $\pi_1 X$, since the loops b_1 and b_2 are meridians of the two second stage kinky handle cores. We singularly surger these tori in X (see proof of Lemma (3.0)) to obtain geometric spherical duals of $\tilde{d}_1$, $\tilde{d}_2$ and $\tilde{d}_3$ which provide specific nullhomotopies (in X) of a meridian for each $\tilde{d}_i$. This proves that the union of $\tilde{d}_1$, $\tilde{d}_2$ and $\tilde{d}_3$ is π_1-neg. in $T_3^0 - C_{1-2}^0$. Hence, if we let T_3^1 be a regular heighborhood of C_{1-2}^0 union with $\tilde{d}_1$, $\tilde{d}_2$ and $\tilde{d}_3$, our proof of Theorem (3.2) is clearly finished.

There are now two tasks remaining. We must construct "careful Casson moves" which allow for each α_i to bound a disk in X, and we must verify that (c) is satisfied. At this juncture, we let the proof bifurcate into a geometric argument given in (3.3.1) or an algebraic argument given in (3.3.2); the reader may choose either of these two depending on his or her taste.

(3.3.1) Continuation of Proof: A Geometric Argument. We consider the circle α_1 and the core (with annulus) $\bar{d}_1$; see Box B_1 of Figure (3.C). Figure (3.D) shows α_1 bounding a punctured disk δ_0 in $W = M^3 \times I \approx T_3^0 - N(C_{1-3}^0)$ with two punctures whose boundaries are denoted by β and β'. Let δ_1 denote a disk with nine holes (four due to framing) as shown in Figure (3.E).

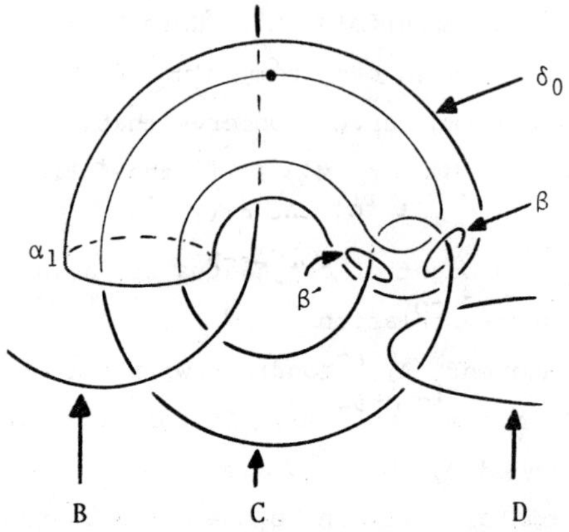

Figure (3.D)

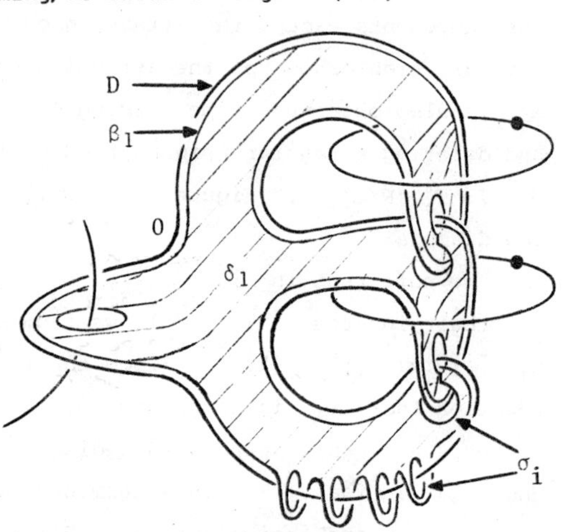

Figure (3.E)

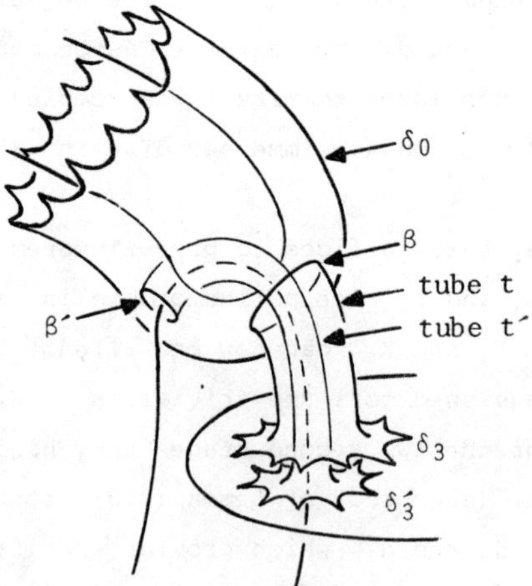

Figure (3.F)

Observe that δ_1 is, as drawn, a subset of the complement of L in S^3. The boundary component β_1 of δ_1 bounds a disk δ_2 in M^3, since the curve β_1 is obtained from the link component D by pushing a parallel copy via the framing. Let δ_3 equal the union of δ_1 and δ_2. Push a "parallel" copy δ_3' of δ_3 (of course, δ_3' may intersect δ_3 in isolated points). Join β, β' to δ_3, δ_3' by tubes t, t', respectively, as shown in Figure (3.F). Let δ denote the singular punctured disk in W obtained as the union of δ_0, t, t', δ_3, δ_3'. (We assume that $\delta - \alpha_1$ is normally immersed.)

The singular punctured disk δ has 16 holes whose boundary curves $\sigma_1, \sigma'_1, \ldots, \sigma_8, \sigma'_8$ are paired such that for each i, $1 \leq i \leq 8$, σ_i is a curve on δ_3, σ'_i is a curve on δ'_3, and σ_i is parallel to σ'_i. We consider $\sigma_1, \sigma'_1, \ldots, \sigma_8, \sigma'_8$ in $M^3 \times \{0\}$ as a subset of $M^3 \times I$. Let $\Delta_1, \Delta'_1, \ldots, \Delta_8, \Delta'_8$ be disjoint disks in $T^0_3 - C^0_{1-2}$, with boundaries $\sigma_1, \sigma'_1, \ldots, \sigma_8, \sigma'_8$, respectively, such that each disk is a fibre of a normal disk-bundle for the core $\bar{d}_1$ of the third stage kinky handle (see box B_1 in Figure (3.C)). For each i, $1 \leq i \leq 8$, thicken Δ_i to a 2-handle h_i attached to W such that Δ'_i is contained in h_i, Δ_i is the core of h_i, and h_i meets $\bar{d}_1$ in the cocore ξ_i of h_i; see Figures (3.G) and (3.H) for schematic drawings. We also require

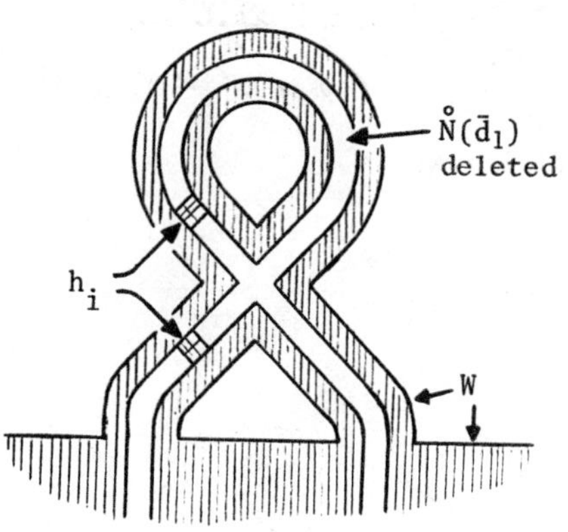

Figure (3.G)

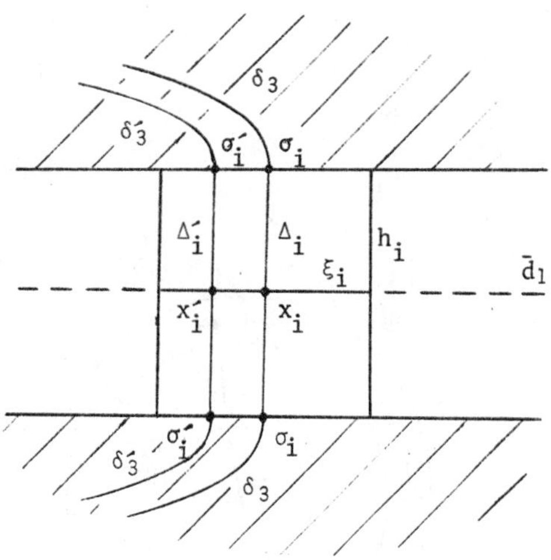

Figure (3.H)

that the handles $h_1, \ldots, h_8$ are pairwise disjoint. Define W_1 as the union of W, $h_1, \ldots,$ and h_8. Proceed in a similar manner to construct W_2 and W_3 by using α_2, box B_2, and α_3, and box B_3, respectively. Let $\hat{W}$ be the union of W_1, W_2 and W_3. Observe that $\pi_1\hat{W} \to \pi_1(T^0_3 - C^0_{1-2})$ is an isomorphism.

As above, we shall only discuss what happens inside the box B_1; a similar discussion for the boxes B_2 and B_3, which can be simultaneously carried out, will remain implicit.

Construct a singular disk $\hat{\delta}$ from δ in $\hat{W}$ by capping each σ_i, σ'_i by Δ_i, Δ'_i, respectively, where $1 \leq i \leq 8$. Observe that the points of intersection of $\hat{\delta}$ and $\bar{d}_1$ come in pairs with opposite sign. This is due to the fact that δ_3 and δ'_3 are parallel (with opposite orientations), in fact, a typical pair x_i, x'_i is obtained by intersecting Δ_i, Δ'_i with the cocore ξ_i, respectively. For each pair x_i, x'_i, $1 \leq i \leq 8$, we construct a Whitney circle λ_i and a Whitney disk ω_i. Figures (3.I) and (3.J) show λ_i and ω_i.

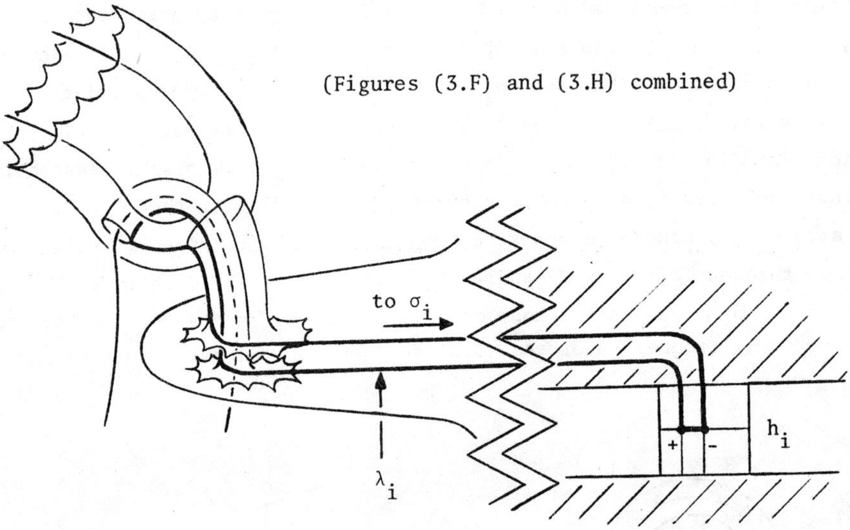

Figure (3.I)

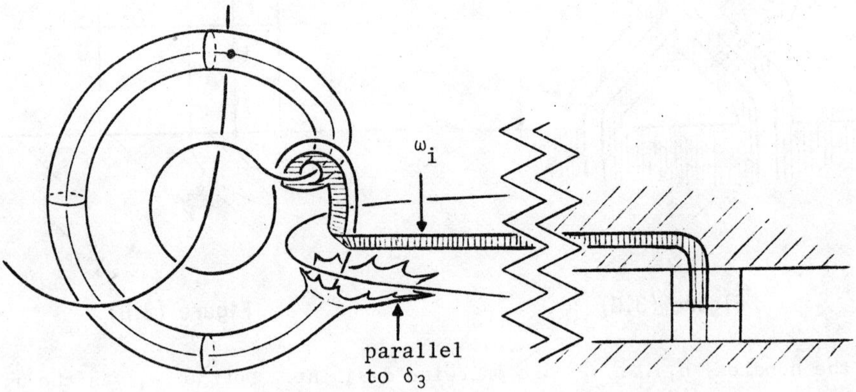

Figure (3.J)

The Whitney disk ω_i is constructed by first constructing a punctured disk whose puncture is a meridian of link component C, and then running a tube around C from this puncture to a suitable singular 2-disk in $\hat{W}$. This disk is constructed by pushing a "parallel" copy of δ_3 and filling in the punctures in $\hat{W}$ such that it is transverse to $\hat{\delta}$.

Suppose i is fixed, $1 \leq i \leq 8$. By construction, $\mathrm{Int}\omega_i$ meets ξ_j, $1 \leq j \leq 8$, in exactly one point. We make a Casson move as in Lemma (2.1.1) on each ξ_j rel boundary, $1 \leq j \leq 8$, to remove this point of intersection.

Observe that we make eight Casson moves for ω_i. Thus, we make 8×8 Casson moves in total. Note that these Casson moves can be made inside an arbitrarily small neighborhood of the union of $\omega_i, \ldots, \omega_8$, and can easily be arranged to miss the 2-complex $K_1 \cup K_2$.

Now recall $\bar{d}_1$ intersects $\hat{W}$ precisely in the union of $\xi_1, \ldots, \xi_8$. By making the above Casson moves on each ξ_i as a subset of $\bar{d}_1$, we have turned $\bar{d}_1$ into the desired $\tilde{d}_1$. By the same procedure in boxes B_2 and B_3 we obtain $\tilde{d}_2$ and $\tilde{d}_3$ from $\bar{d}_2$ and $\bar{d}_3$, respectively, where $\bar{d}_2$ and $\bar{d}_3$ denote the cores with annuli of the other third stage kinky handles.

We now verify that $\tilde{d}_1$, $\tilde{d}_2$ and $\tilde{d}_3$ have the desired properties. Consider $\tilde{d}_1$, along with the singular disks $\hat{\delta}$ and ω_i defined above. Proceed as in Lemma (2.1.1) to push the singular disk $\hat{\delta}$ across each ω_i (by the singular Whitney trick) to obtain a singular disk $\tilde{\delta}$ in $T_3^0 - \mathring{N}(C_{1-2}^0)$, bounded by α_1, which is disjoint from $\tilde{d}_1$, $\tilde{d}_2$ and $\tilde{d}_3$. Similarly, we find singular disks bounded by α_2 and α_3 in the space $X = T_3^0 - C_{1-2}^0$ minus $\tilde{d}_1$, $\tilde{d}_2$ and $\tilde{d}_3$. This proves (a) and (b) (i.e. Theorem 3.2)) as explained in the paragraph preceding (3.3.1).

Finally, we must verify that (c) holds. Note that the Whitney disk ω_i as in Figure (3.J) can be drawn entirely within the box B_1 of Figure (3.C). Since the Casson moves are made within a small neighborhood of the Whitney disks, they also lie in the box B_1 and, in particular, they cannot link the attaching curve. (They do, however, link the link component B as a determined reader may verify.) Since $\tilde{d}_1$ minus the Casson fingers lies in the top stage of T_3^0, it follows that the image $\operatorname{Im}[\pi_1\tilde{d}_1 \to \pi_1(T_3^0 - C_{1-2}^0)]$ is contained in F where $Z * F$ is the canonical decomposition of $\pi_1(T_3^0 - C_{1-2}^0)$ (see Lemma (3.1.0)). ▒

We now give an algebraic argument which may replace (3.3.1) in the above proof. (This argument was in fact constructed by carefully observing the Casson moves and null homotopy of (3.3.1).)

(3.3.2) Continuation of Proof: An Algebraic Argument. We will show that "careful Casson moves" can be made on the core d_1, after which α_1 is null homotopic. Similar arguments can be made for d_2 and d_3 to complete the proof.

We will show the existence of these Casson moves with the following fact (see [C], Lemma 1); making Casson moves corresponds to killing certain commutators. Specifically, suppose d is a disk normally immersed in a 4-manifold W, with a meridian z in $\pi_1(W-d)$. Choose an arbitrary w in $\pi_1(W-d)$. If we obtain $\tilde{d}$ by making a Casson move on d "along a loop representing w", then $\pi_1(W-\tilde{d}) \approx \pi_1(W-d)/N$, where N is the subgroup normally generated by the commutator $[z,z^w] = z(w^{-1}zw)z^{-1}(w^{-1}zw)^{-1}$.

In our case, we work with the group $\pi_1(T^0_3 - C^0_{1-3}) \approx \pi_1 M^3$. Consider Figure (3.K), an enlargement of box B_1 of Figure (3.C).

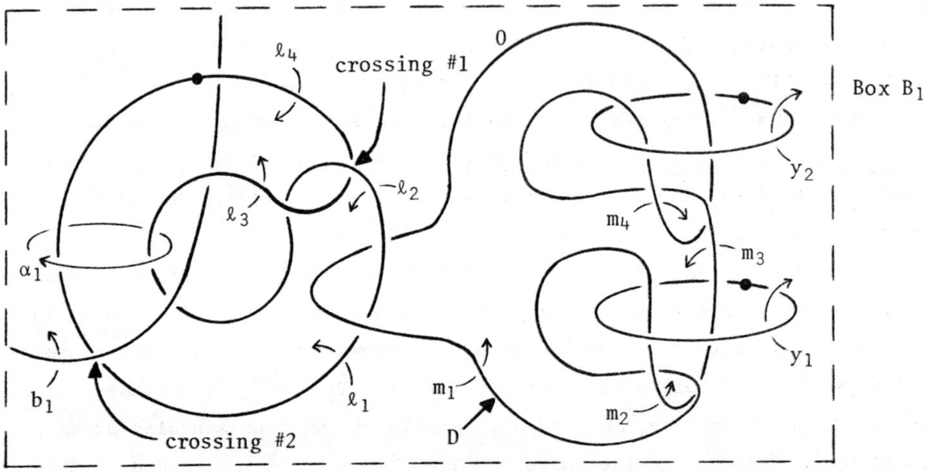

Figure (3.K)

Let $m_1, \dots, m_4$ denote the pictured meridians of $\bar{d}_1$. It is sufficient to exhibit commutators of the form $[m_i, m_j^w]$ in $\pi_1 M^3$, with w a loop lying within the box B_1, such that α_1 dies when these are all killed. For m_i and m_j are conjugate (via a loop in box B_1) for $i,j = 1,\dots,4$. Thus the above fact shows that we have killed α_1 by Casson moves which never leave the box B_1, and hence cannot link the attaching curve.

We now use some relations evident in Figure (3.K). Note that $\alpha_1 = \ell_4 \ell_3^{-1}$. Thus, it is sufficient to force the relation $\ell_3 = \ell_4$ to hold in the quotient group. But by examining crossing #1, we see that $\ell_3 = \ell_4^{\ell_2}$, so that we only need to force $\ell_4^{\ell_2} = \ell_4$. Next, notice that $\ell_1 = \ell_2$, since these meridians bound a punctured 2-sphere (which encloses link component D). From crossing #2, we now infer $\ell_4 = \ell_1^{b_1} = \ell_2^{b_1}$. Finally, we examine the surgery relation for the link component D. This relation expresses ℓ_2 as a product of conjugates of the m_i's and their inverses. Specifically, we have $\ell_2 m_1^4 m_2^{-1}(y_1^{-1} m_1^{-1} y_1) m_4^{-1}(y_2^{-1} m_3^{-1} y_2) = 1$ (where the m_1^4 factor is due to the framing). We write this relation as $\ell_2 = \beta_1 \cdot \dots \cdot \beta_8$, where each β_i is in the set $\{m_1^{-1}, m_2, m_4, m_1^{y_1}, m_3^{y_2}\}$ which we will call Σ.

We now exhibit the 25 commutators which we will kill. For each pair of elements β, γ of Σ, we kill the commutator $[\beta, \gamma^{b_1}]$. (These represent the same Casson moves which we constructed in our earlier proof, although we have eliminated some redundancy.) Note that these commutators (or conjugates of them) all have the correct form, with $w = b_1$ up to multiplication by y_i. It is immediate that in the quotient group $\pi_1(T^0_3 - C^0_{1-3})/N$ (N normally generated by the above elements), we have the relation $(\gamma^{b_1})^{\beta} = \gamma^{b_1}$ whenever β and γ

are in Σ. This tells us how we can erase the exponent β.

It is now easy to show that $\ell_4^{\ell_2} = \ell_4$, which completes the proof, as stated above. First we recall that $\ell_4 = \ell_2^{b_1}$. By the surgery relation, $\ell_2 = \beta_1 \cdot \ldots \cdot \beta_8$, hence $\ell_4 = \prod_{i=1}^{8} \beta_i^{b_1}$. (Recall the identity $(ab)^c = a^c b^c$; also $a^{bc} = (a^b)^c$.) Now $\ell_4^{\ell_2} = \left(\prod_{i=1}^{8} \beta_i^{b_1}\right)^{\ell_2} = \prod_{i=1}^{8} (\beta_i^{b_1})^{\beta_1 \cdots \beta_8} = \prod_{i=1}^{8} \beta_i^{b_1} = \ell_4$, where the third equality follows by successively "erasing the exponents" $\beta_1, \ldots, \beta_8$. ⁞⁞

4. THE "BIG" REIMBEDDING THEOREM

(4.0) Statement and Motivation. This section is devoted to the "Big" Reimbedding Theorem, which is essentially Freedman's "5-stage" reimbedding theorem except that it uses one less stage. We have used the word "big" instead of "4-stage" to avoid conflict with the existing usage in [F1], although the latter is more appropriate. This reduction of a stage is made possible by the Improved 3-Stage Reimbedding Theorem (3.3).

(4.0.0) THE BIG REIMBEDDING THEOREM. Every 4-stage tower T_4^0 contains another 4-stage tower T_4^1 satisfying:

a) (agreement) $C_{1-2}^0 = C_{1-2}^1$;

b) (π_1-neg.) $\pi_1(T_4^0 - T_4^1) \to \pi_1(T_4^0 - C_1^0)$ is an isomorphism; and

c) (nullity) $\pi_1 T_4^1 \to \pi_1 T_4^0$ is the zero map.

(4.1) Motivation of the Proof. The proof of (4.0.0) is rather lengthy (although not intrinsically difficult), so we begin with a sketch of the main ideas involved. A detailed proof will be given in subsequent sections.

The main conclusion of this theorem is nullity. The proof centers on arranging this, that is, constructing the tower T_4^1 in such a way that the nontrivial loops of its top stage "do not get caught in the top stage of T_4^0", i.e., are trivial in $\pi_1 T_4^0$. To ensure this, we will do most of our constructions within the subtower T_3^0 consisting of the first three stages of T_4^0. (Recall that $\pi_1 T_3^0 \to \pi_1 T_4^0$ is the zero map.)

(4.1.0) Key Idea: The "Singular Norman Trick." First we consider the case where the fourth stage of T_4^0 has only one kinky handle $(k, \partial^- k)$, with a single kink. Recall (2.2.6) that the core C_{1-4} of T_4^0 includes an annulus α connecting the core of k to the core C_{1-3}^0 of T_3^0. Suppose that α had a geometric dual in $T_3^0 - C_{1-3}^0$, i.e., an immersed 2-sphere s in T_3^0, called a "Norman sphere", meeting C_{1-4}^0 in exactly one point, on the annulus α. We could then obtain nullity by a singular version of the Norman trick, as indicated in Figure (4.A). Specifically, we could eliminate the self-intersection of the core of k by tubing one sheet of this down to the Norman sphere s. This gives nullity by destroying the essential loop in C^0_{1-4} (at the expense of introducing inessential loops from s). We will prove this in (4.3.9).

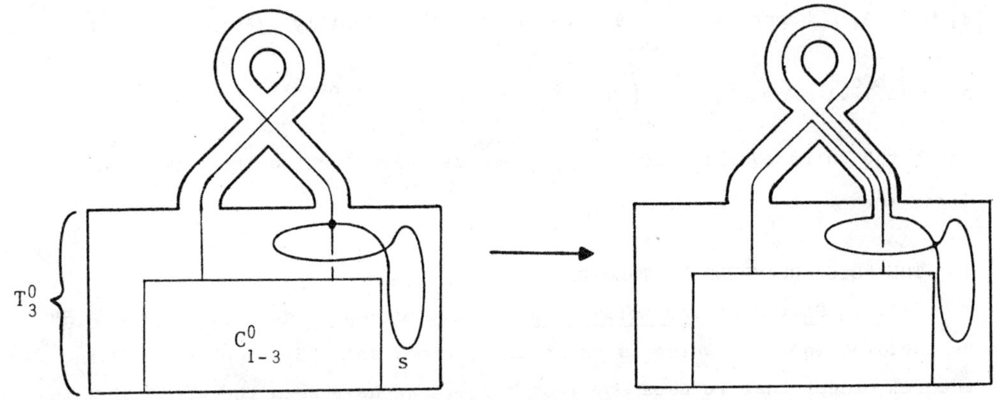

Figure (4.A)

Now consider a general T^0_4, with a total of n kinks on the fourth stage kinky handles. We would like to apply a similar "Normal trick" to obtain nullity in this case. Clearly, we need a total of n Norman spheres, paired with the annuli in correspondence to the kinks of the fourth stage. Furthermore, it is crucial that the spheres be pairwise disjoint, or else nullity will fail.

(4.1.1) Setting Up for the Norman Trick. Unfortunately, life is not this simple. Notice that the existence of a single Norman sphere is equivalent to a π_1-neg. condition on the corresponding annulus. Thus, we need to preface our discussion by making the union of the annuli π_1-neg. in $T^0_3 - C^0_{1-3}$. (This is dealt with in Lemma (4.2.2) below.) This is a Casson move argument similar to the proof of Theorem (3.2). In particular, it requires that the the third stage cores C^0_3 be π_1-neg. in $T^0_3 - C^0_{1-2}$. This condition, however, is false. Fortunately, we are rescued by the 3-stage reimbedding theorem. In particular, we are given a tower T^1_3 imbedded in T^0_3, such that the third stage is π_1-neg. in $T^0_3 - C^0_{1-2}$. Now in the "room" $R^1 = T^0_3 - \mathring{T}^1_3$, we can find a Norman sphere, as indicated above.

In general, we need n disjoint Norman spheres. We cannot accomplish this in R^1. We remedy this by beginning our proof with n successive applications of the 3-stage reimbedding theorem, obtaining imbeddings $T^n_3 \to T^{n-1}_3 \to \cdots \to T^0_3$. We now have our n spheres, one in each room $R^i = T^{i-1}_3 - \mathring{T}^i_3$, $1 \le i \le n$.

(4.1.2) Capping the Extra Kinks. The 3-stage reimbedding theorem causes more trouble. With each reimbedding, extra kinks are introduced in the third stage, adding new members to any standard family of curves. If we want T^n_3 to be the first three stages of a 4-stage tower, we must cap off these additional circles with kinky handles. We have several basic lemmas ((4.2.1) and (4.2.3) below) for such a purpose. Clearly, however, we cannot do this unless the circles in question are nullhomotopic in $T_3^0 - T_3^n$. Now $\pi_1(T_3^0 - T_3^n) = Z * F$, and by the improved 3-stage reimbedding theorem we can assume that all of the circles are in F. Unfortunately, most of the circles will represent nontrivial elements of F, so we cannot cap them off. We will instead settle for capping off a "triangular" family of curves (defined in (4.2.0)). We will do this carefully, obtaining π_1-neg. in each room R^i, so as not to lose the existence of Norman spheres.

(4.1.3) Concluding the Proof. The above constructions (Reimbeddings, cappings, and the Norman trick) give us a manifold V_4 inside T_4^0, which is almost a 4-stage tower. It is not quite a tower, as the top stage is attached via a triangular (not standard) family, and we have also been careless with framings. Nevertheless, the Norman trick has given us nullity, the key result. A simple construction now gives us an honest tower T_4^1 inside V_4, which is null inside T_4^0 because V_4 is. This is our required tower, as the other conclusions of Theorem (4.0.0) follow immediately from our construction.

(4.2) SOME LEMMAS. We now prove three similar lemmas. Two of these are for capping off circles on imbedded towers (as in (4.1.2) above), the third provides the π_1-neg. condition for annuli (as in 4.1.1)).

(4.2.0) DEFINITION. Let $\tau_1,\ldots,\tau_k$ denote the distinguished tori from the top stage of an n-stage tower T_n. A family of (unframed) circles $\{c_1,\ldots,c_k\}$ in ∂^+T_n is called triangular if

a) they are transverse to the tori,

b) for each j, $1 \le j \le k$, $c_j \cap \tau_j$ = {one point}, and

c) $c_i \cap \tau_j$ is empty when $j > i$.

Note that the definition of triangularity assumes that the curves and tori are appropriately ordered. The $k \times k$ matrix, whose entry in the ith row and j^{th} column is the geometric intersection number of c_i and τ_j (= number of points in $c_i \cap \tau_j$), is triangular with 1's along the diagonal. Clearly, every standard family is a triangular family (after forgetting the framings).

Note also that a triangular family of curves for T_n generates $\pi_1 T_n$.

Throughout the following, we suppose that T_n is an n-stage tower contained in a 4-manifold M with $\partial M \cap T_n = \partial^- T_n$.

(4.2.1) LEMMA: (THE CAPPING LEMMA). Suppose T_n and M are given as above. Suppose the meridians of the nth stage core C_n are all trivial in $\pi_1(M - \mathring{T}_n)$. Let $\{c_i\}$ be a sub-family of a triangular family of curves in T_n such that each c_i is trivial in $\pi_1(M - \mathring{T}_n)$. Then: there exists a family $\{\tilde{d}_i\}$ of pairwise disjoint normally immersed disks in $M - \mathring{T}_n$ satisfying

a) the boundary of each $\tilde{d}_i$ is c_i, and

b) the union of these disks is π_1-neg. in $M - \mathring{T}_n$.

Moreover, we can arrange the disks to avoid an arbitrary π_1-neg. 2-complex in $M - T_n$.

PROOF. By hypothesis, c_1 bounds a normally immersed disk d_1 in $M - \mathring{T}_n$. Note that d_1 intersects the corresponding distinguished torus τ_1 exactly once. The standard generators of $\pi_1 \tau_1$ are meridians of C_n, hence trivial in $\pi_1(M - \mathring{T}_n)$ by hypothesis. By Lemma (3.0), we use Casson moves (rel boundary) to convert d_1 to a π_1-neg. (normally immersed disk) $\tilde{d}_1$ in $M - \mathring{T}_n$.

Put $W = (M - T_n) - \tilde{d}_1$. The curve c_2 bounds a normally immersed disk d_2 in W. This is clear, since c_2 is trivial in $\pi_1(M - \mathring{T}_n)$ by hypothesis, and $\tilde{d}_1$ is π_1-neg. in $M - \mathring{T}_n$. Now use the argument given above to find $\tilde{d}_2$, by replacing τ_1 by τ_2 and $M - \mathring{T}_n$ by W. The general inductive step is now clear. ▥

(4.2.2) LEMMA: (THE ANNULUS LEMMA). Suppose T_n and M are given as above. Suppose that the meridians of C_n are trivial in $M - \mathring{T}_n$. Let $\{a_i\}$ be a family of pairwise disjoint normally immersed annuli in $M - \mathring{T}_n$ such that each a_i has one boundary component $\partial^+ a_i$ in ∂M, and the other boundary component $\partial^- a_i$ in $\partial^+ T_n$. Suppose $\{\partial^- a_i\}$ is a sub-family of a triangular family of curves in $\partial^+ T_n$. Then: there exists a family $\{\tilde{a}_i\}$ of pairwise disjoint normally immersed annuli in $M - \mathring{T}_n$ such that

a) each $\tilde{a}_i$ is obtained from a_i via a finite number of Casson moves (rel boundary), and

b) the union of the annuli in $\{\tilde{a}_i\}$ is π_1-neg. in $M - \mathring{T}_n$.

PROOF. This is similar to the proof of Lemma (4.2.1). Let τ_1 denote the distinguished torus corresponding to $\partial^- a_1$. As before, the map $\pi_1 \tau_1 \to \pi_1(M - \mathring{T}_n)$ is trivial, since the standard generators of $\pi_1 \tau_1$ are meridians of C_n. By Lemma (3.0), we make Casson moves on a_1, missing the union of the other annuli, to construct a π_1-neg. $\tilde{a}_1$. Our proof is finished by an easy inductive argument. ▥

(4.2.3) LEMMA: (THE TOWER EXTENSION LEMMA). Suppose T_n and M are given as above. Suppose M admits a spin structure (in particular this holds when M is a tower). Let $\partial_n T_n = \partial^+$(top stage of T_n). Suppose $\pi_1(\partial_n T_n) \to \pi_1(M - \mathring{T}_n)$ is the zero map (for all base points). Then: there is a standard family of framed curves $\{c_i\}$ bounding a family $\{\tilde{d}_i\}$ of pairwise disjoint normally immersed disks whose union is π_1-neg. in $M - \mathring{T}_n$, such that the following property holds:

(*) For each i, the kinky handle $(k_i, \partial^- k_i)$ obtained as a regular neighborhood of $\tilde{d}_i$ in $M - \mathring{T}_n$ has the property that the standard framing of $\partial^- k_i$ agrees with the framing on c_i.

(Note: This lets us extend T_n to an $(n+1)$-stage tower in M.)

REMARK. The use of spin structures can be avoided; see [F1], addendum to Theorem (3.2).

PROOF. Put a spin structure on M. By [C], see p. 6, Addendum with $W = M - \mathring{T}_n$, we choose a standard family $\{c_i\}$ of curves compatible with this spin structure. Note that each c_i is trivial in $M - \mathring{T}_n$, and $\pi_1\tau \to \pi_1(M - \mathring{T}_n)$ is trivial for any top-stage distinguished torus τ in T_n. This follows since $\pi_1(\partial_n T_n) \to \pi_1(M - \mathring{T}_n)$ is trivial. The proof is now identical to that of Lemma (4.2.1), except that we insert one additional step: each time we find a normally immersed disk d_i we immediately modify it, rel boundary, (as will be explained below), to get a new normally immersed disk $\bar{d}_i$ such that the framing is correct, i.e., $\bar{d}_i$ satisfies (*). Now return to the proof of Lemma (4.2.1), and modify $\bar{d}_i$ by Casson moves to obtain the required $\tilde{d}_i$. Note that obtaining $\bar{d}_i$ fixes up the framing while Casson moves on $\bar{d}_i$ make $\tilde{d}_i$ π_1-neg.. Recall that Casson moves preserve the framing; see (2.2.2). The resulting family $\{\tilde{d}_i\}$ is what we want. The following discussion shows how to construct $\bar{d}_i$ with correct framing.

We change d_i to $\bar{d}_i$ as follows (recall: d_i is a normally immersed disk in $M - \mathring{T}_n$ with boundary c_i). Let $(k_i, \partial^- k_i)$ be a kinky handle (reg. nbd.) in $W = M - \mathring{T}_n$ with core d_i and attaching curve c_i. The problem at hand is that the standard framing on $\partial^- k_i$ may differ from the given framing on c_i by, say, m twists. We first prove that m is even. Attach a 2-handle to W along c_i with its given framing. Let $\hat{W}$ denote the resulting manifold. Now $d_i \cup$ (core of the 2-handle) is an immersed S^2 in $\hat{W}$ representing some α in $H_2\hat{W}$. Note that $\alpha \cdot \alpha = m$ (of course, the framing is correct if $\alpha \cdot \alpha = 0$; see (2.2.2)). Since the framing of c_i is compatible with the spin structure of W, $\hat{W}$ is spin. Thus, the intersection form must be even. This proves that m is even, say, $m = 2\ell$.

We now modify d_i to $\bar{d}_i$ so that the number corresponding to m is zero. Consider the class $[\tau_i]$ in H_2W represented by the distinguished torus τ_i. Clearly, $[\tau_i] \cdot [\tau_i] = 0$ and $[\tau_i] \cdot \alpha = 1$ with α in $H_2\hat{W}$. Since $\pi_1\tau_i \to \pi_1 W$ is the trivial map, we singularly surger τ_i (and push it slightly into $\operatorname{Int} W$) to get a normally immersed 2-sphere s in W representing the class $[\tau_i]$. Now tube ℓ copies of s (with reversed orientation if ℓ is positive) to d_i to obtain the normally immersed disk $\bar{d}_i$. Let β denote the element in $H_2\hat{W}$ represented by $\bar{d}_i$ union the core of the 2-handle. Then: $\beta = \alpha - \ell[\tau_i]$, and $\beta \cdot \beta = (\alpha - \ell[\tau_i]) \cdot (\alpha - \ell[\tau_i]) = \alpha \cdot \alpha - 2\ell\, \alpha \cdot [\tau_i] + \ell^2[\tau_i] \cdot [\tau_i] = 2\ell - 2\ell + 0 = 0$. Hence, the framing of the kinky handle obtained as a reg.nbd. of $\bar{d}_i$ agrees with the given framing of c_i as desired. ⁞⁞⁞

(4.3) THE BIG REIMBEDDING THEOREM: PROOF. Since the proof is rather lengthy, we organize it into various titled sub-sections as we proceed.

(4.3.0) A Repeated Application of the Improved 3-Stage Reimbedding Theorem. Let $k^4_1, \dots, k^4_m$ denote the kinky handles attached to T^0_3 to obtain T^0_4. For each i, let n_i denote the total number of kinks (not counted by sign) in k^4_i. Put $n = n_1 + n_2 + \cdots + n_m$. Apply the Improved 3-Stage Reimbedding Theorem (3.3) n-times to construct a nest $T^n_3 \to T^{n-1}_3 \to \cdots \to T^1_3 \to T^0_3$ of 3-stage towers (i.e., the maps are the inclusions). More specifically, this nest is inductively constructed by requiring that T^{i+1}_3 be found inside T^i_3, $0 \leq i < n$, satisfying the conclusions of Theorem (3.3). For each j, $0 < j \leq n$, the "room" $T^{j-1}_3 - \mathring{T}^j_3$ is denoted by R^j.

(4.3.1) Immersed Annuli in the Room R^1. We now describe a family $\{\alpha^1_i: 1 \leq i \leq m\}$ of embedded annuli in the room R^1. For each $1 \leq i \leq m$, the upper boundary $\partial^+\alpha^1_i$ of α^1_i is the curve along which the kinky handle k^4_i is attached to T^0_3. Collapse T^0_3 to C^0_{1-3} by using the structure of the regular neighborhood on T^0_3. Recall that C^1_3 agrees with C^0_3 except near a finite number of smooth arcs in T^0_3. We use the general position of points and arcs in a 2-disk, and arcs and annuli in the room R^1, to conclude: for each i, $1 \leq i \leq m$, $\partial^+\alpha^1_i$ traces an annulus α^1_i which can be assumed properly imbedded in R^1, whose lower boundary $\partial^-\alpha^1_i$ is contained in ∂T^1_3. Observe that these annuli are pairwise disjoint. By the Annulus Lemma (4.2.2), we find a family $\{a^1_i: 1 \leq i \leq m\}$ of normally immersed pairwise disjoint annuli in R^1 such that the union of these annuli is π_1-neg. in R^1, $\partial^+\alpha^1_i = \partial^+a^1_i$, $\partial^-\alpha^1_i = \partial^-a^1_i$, and each a^1_i is obtained by applying Casson moves to α^1_i.

(4.3.2) Immersed Disks in R^1. By Theorem (3.3), each loop in the third stage of T^1_3 is in F, where $\pi_1(T^0_3 - T^1_3)$ has the canonical decomposition $Z * F$ (see Lemma (3.1.0)). Put $m_0 = m$ and $\partial^-a^1_i = c^1_i$. There is a (standard) family of curves for T^1_3 which is the union of the family $\{c^1_i: 1 \leq i \leq m_0\}$ and a family $\{\bar{c}^1_j: m_0 < j \leq m_1\}$ of curves, where the latter family comes from the Casson moves (in fact, $m_1 - m_0$ = twice the number of Casson moves). Not all the curves in $\{\bar{c}^1_j: m_0 < j \leq m_1\}$ are nullhomotopic in T^0_3. This difficulty is overcome as follows. We obtain a family of curves $\{c_j: m_0 < j \leq m_1\}$ through modification of the curves in $\{\bar{c}^1_j: m_0 < j \leq m_1\}$ by the curves in $\{c^1_i: 1 \leq i \leq m_0\}$ such that each c^1_j, $m_0 < j \leq m_1$, is nullhomotopic in T^0_3. The family of curves $\{c^1_i: 1 \leq i \leq m_1\}$ fails to be standard for T^1_3, but it is a triangular family with ordering induced by the index i. For now, assume this has been done; the details appear in (4.3.3).

By Capping Lemma (4.2.1), there exists a collection $\{d^1_j: m_0 < j \leq m_1\}$ of pairwise disjoint normally immersed disks in R^1 such that their union is π_1-neg. in R^1, for each j, $m_0 < j \leq m_1$, $\partial d^1_j = c^1_j$, and each d^1_j misses the union of the immersed annuli in the family $\{a^1_i: 1 \leq i \leq m_0\}$.

(4.3.3) A Triangular Family of Curves: Given $\{c^1_i: 1 \leq i \leq m_0\}$ and $\{\bar{c}^1_j: m_0 < j \leq m_1\}$ as in (4.3.2). Let $\varphi: \pi_1 T^1_3 \to \pi_1 T^0_3$ denote the map induced by the inclusion. Observe that $\pi_1 T_3^0$ is the free group F, where $\pi_1(T_3^0 - \mathring{T}_3^1) \approx Z * F$ as in Lemma (3.1.0). Now F is generated by the subset $\{\varphi(c_i): 1 \leq i \leq m_0\}$. Throughout these discussions we assume that the curves are carefully connected to the base point; the details are easy and we omit them. Let x equal $\bar{c}_j^1$ for some j, $m_0 < j \leq m_1$. Then $\varphi(x)$ equals a reduced word $W(\varphi(c_1^1), \ldots, (c_{m_0}^1))$ in the generators $\varphi(c_i^1)$. Put $y = x[W(c_1^1, \ldots, c_{m_0}^1)]^{-1}$. Note that $\varphi(y)$ is trivial in $\pi_1(T_3^0 - \mathring{T}_3^1)$. We use x and $c_1^1, \ldots, c_{m_0}^1$ to carefully construct a circle c_j^1 in ∂T_3^1 representing y, using its representation as $x[W(c_1^1, \ldots, c_{m_0}^1)]^{-1}$, such that $c_j^1 \cap \bar{\tau}_i = \delta_{ij}$, where $\bar{\tau}_i$ is a distinguished torus corresponding to $\bar{c}_i^1$ with $m_0 < i \leq m_1$. Thus, the family of curves $\{c_i^1: 1 \leq i \leq m_1\}$ is triangular with respect to the ordering induced by the index i.

(4.3.4) Immersed Annuli in the Room R^2. Consider the triangular family of curves $\{c_i^1: 1 \leq i \leq m_1\}$ given above. Proceed as in (4.3.1) to find a family $\{a_i^2: 1 \leq i \leq m_1\}$ of normally immersed pairwise disjoint annuli in the room R^2 such that their union is π_1-neg. in R^2 and $\partial^+ a_i^2 = c_i^1$.

(4.3.5) Immersed Disks in the Room R^2. Put $c_i^2 = \bar{\partial} a_i^2$, $1 \leq i \leq m_1$. As before, there is a (triangular) family of curves for T_3^2 which is the union of the family $\{c_i^2: 1 \leq i \leq m_1\}$ and a family $\{\bar{c}_j^2: m_1 < j \leq m_2\}$, where the latter is a subfamily of some standard family for T_3^2. We proceed as in (4.3.3) to modify $\{\bar{c}_j^2: m_1 < j \leq m_2\}$ by the family $\{c_i^2: 1 \leq i \leq m_1\}$ to obtain a family of curves $\{c_j^2: m_1 < j \leq m_2\}$ such that for each j, $m_1 < j \leq m_2$, c_j^2 is nullhomotopic in R^2, and the collective family $\{c_i^2: 1 \leq i \leq m_2\}$ in T_3^2 is triangular. (Note that $\{c_i^2: 1 \leq i \leq m_1\}$ is (up to homotopy) a triangular family for T_3^1; hence it generates $\pi_1 T_3^1$ as required.) By The Capping Lemma (4.2.1), there exists a family $\{d_j^2: m_1 < j \leq m_2\}$ of pairwise disjoint normally immersed disks in R^2 such that: their union is π_1-neg. in R^2; for each j, $m_1 < j \leq m_2$, $\partial d_j^2 = c_j^2$; and each d_j^2 misses the union of the immersed annuli in the family $\{a_i^2: 1 \leq i \leq m_1\}$.

(4.3.6) Immersed Annuli-Disks in R^k. Proceed inductively, as above, to find families of immersed annuli and immersed disks in the rooms $R^3, R^4, \ldots, R^n$. Here is what happens in a typical room R^k: there exist families $\{a_i^k: 1 \leq i \leq m_{k-1}\}$ and $\{d_j^k: m_{k-1} < j \leq m_k\}$ of pairwise disjoint normally immersed annuli and disks, respectively, such that the union of the members of these two families is π_1-neg. in R^k, and $d_j^k \cap a_i^k$ is empty for any i, $1 \leq i \leq m_{k-1}$, and j, $m_{k-1} < j \leq m_k$. Also, $\{c_i^{k-1} = \partial^+ a_i^k: 1 \leq i \leq m_{k-1}\}$ is a triangular family in T_3^{k-1}. The union of the families $\{c_i^k = \partial^- a_i^k: 1 \leq i \leq m_{k-1}\}$ and $\{c_j^k = \partial d_j^k: m_{k-1} < j \leq m_k\}$ is the triangular family $\{c_i^k: 1 \leq i \leq m_k\}$ in T_3^k. All of this can be drawn in a systematic diagram. Figure (4.B) shows this with $n = 3$.

(4.3.7) A Family of Immersed Disks in T_4^0. For each i, $1 \leq i \leq m$, let d_i^0 denote the core of the kinky handle k_i^4 (see (4.3.0)) with boundary equal to $\partial^+ a_i^1$. For each i, $1 \leq i \leq m$, let δ_i denote the normally immersed disk in $T_4^0 - \hat{T}_3^n$ obtained as the union of $d_i^0, a_i^1, a_i^2, \ldots, a_i^n$. Similarly, for each k, $1 \leq k \leq n$, and each i, $m_{k-1} < i \leq m_k$, let δ_i denote the normally immersed disk in $T_3^0 - \hat{T}_3^n$ obtained as the union of $d_i^k, a_i^{k+1}, \ldots, a_i^n$. Thus we have a family of normally immersed disks $\{\delta_i : 1 \leq i \leq m_n\}$ in $(T_4^0 - \hat{T}_3^n)$ (as indicated in Figure (4.B)).

(4.3.8) A Singular Norman Trick. The disks in the family $\{\delta_i : 1 \leq i \leq m\}$ (recall: $m = m_0$) are not acceptable, since they contain loops which are essential in T_4^0; these loops come from the disks d_i^0, corresponding to standard families of curves in k_i^4. We have all the ingredients to overcome this difficulty. The details are as follows. Recall that δ_1 is the union of $d_1^0, a_1^1, a_1^2, \ldots, a_1^n$ where the annulus a_1^i is contained in the room R^i. For each i, $1 \leq i \leq n_1$, let s_1^i denote a geometric dual of a_1^i in $\operatorname{Int} R^i$ which lies in the complement of all the remaining immersed annuli and immersed disks constructed as in (4.3.6). (Recall: s_1^i is an immersed 2-sphere which meets a_1^i in exactly one point.) As in (4.1.0), we refer to s_1^i as a Norman sphere associated with a_1^i. The existence of these Norman spheres is clear since the union of all the immersed annuli and immersed disks in a given room R^i is π_1-neg..

Consider δ_2. Proceed as above to find for each i, $n_1 < i \leq n_1 + n_2$ a Norman sphere s_2^i associated with a_2^i in the $\operatorname{Int} R^i$ missing all the remaining immersed annuli and immersed disks in R^i. Continue in this manner to handle $\delta_3, \delta_4, \ldots, \delta_m$.

Observe that there is exactly one Norman sphere in each room and, therefore, the n Norman spheres are pairwise disjoint. Also, there are n_i Norman spheres associated with δ_i so that the n_i self-intersections of d_i^0 can be removed by the singular Norman trick; see (4.1.0) and [F2; p. 189].

For each i, $1 \leq i \leq m$, let Δ_i denote the immersed disk obtained from δ_i by suitably performing the Singular Norman trick n_i times.

(4.3.9) A "Phony" Tower V_4. Define V as a reg.nbd. in T_4^0 of the union of $T_3^n, \Delta_1, \ldots, \Delta_m, \delta_{m+1}, \ldots, \delta_{m_n}$. Note that V_4 is not a tower in the usual sense, since the top stage is added along a triangular family of curves. This difficulty is overcome by finding an honest tower inside V_4. We first observe that $\pi_1 V_4 \to \pi_1 T_4^0$ is the trivial map, because of the singular Norman trick. To see this, note that the Norman spheres are all in T_3^0, and $\pi_1 T_3^0 \to \pi_1 T_4^0$ is the trivial map. Since the Norman spheres are pairwise disjoint, we can now see that any loop in V_4 can be pulled into V_4 minus the Norman spheres by a homotopy in T_4^0. We can then pull the loop into T_3^0; hence it is null homotopic in T_4^0.

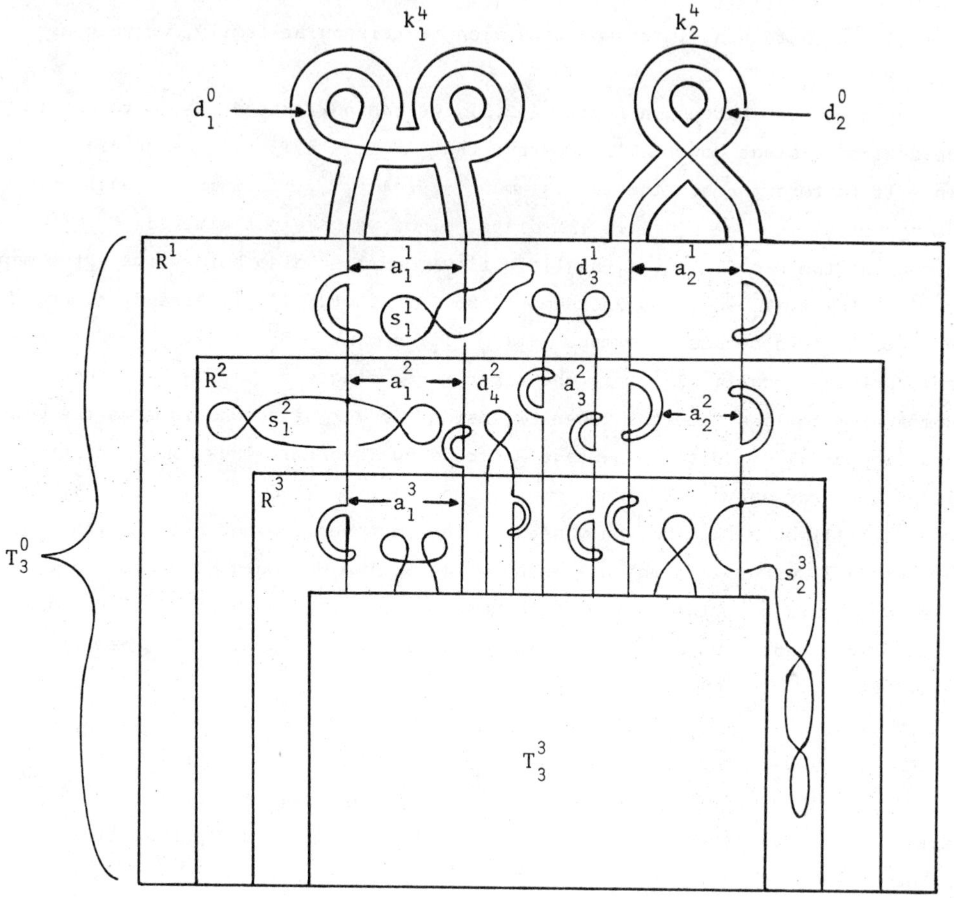
k_1^4
k_2^4
d_1^0
d_2^0
R^1
a_1^1
d_3^1
a_2^1
s_1^1
R^2
a_1^2
d_4^2
a_3^2
a_2^2
s_1^2
R^3
a_1^3
s_2^3
T_3^0
T_3^3

Figure (4.B)

(4.3.10) The Desired Tower T^1_4. Apply Theorem (3.3) to find a 3-stage tower T^{n+1}_3 inside the 3-stage sub-tower T^n_3 of the phony tower V_4. Note that any loop in the boundary of the third stage of T^{n+1}_3 maps into the part F of the canonical decomposition $\pi_1(T^n_3 - \mathring{T}^{n+1}_3) = Z * F$. But the image of F under the map $\pi_1(T^n_3 - \mathring{T}^{n+1}_3) \to \pi_1(V_4 - \mathring{T}^{n+1}_3)$ is trivial, since V_4 is obtained from T^n_3 by attaching immersed disks along a triangular family, which generates F.

By the Tower Extension Lemma (4.2.3), we can now extend T^{n+1}_3 to obtain the desired 4-stage tower T^1_4. Observe that $\pi_1 T^1_4 \to \pi_1 T^0_4$ is the trivial map since it factors through the trivial map $\pi_1 V_4 \to \pi_1 T^0_4$. This proves nullity. The agreement $C^0_{1-2} = C^1_{1-2}$ is immediate, since we have not altered C^0_{1-2} throughout the proof. The π_1-negligibility condition also holds, i.e., the map $\pi_1(T^0_4 - T^1_4) \to \pi_1(T^0_4 - C^0_1)$ is an isomorphism. In fact, T^1_4 is already π_1-neg. in $V_4 - C^0_1$. This is because the maps $\pi_1(V_4 - T^1_4) \to \pi_1(V_4 - T^{n+1}_3) \to \pi_1(V_4 - C^0_1)$ are isomorphisms (recall: T^{n+1}_3 is the first three stages of T^1_4). The first isomorphism follows from the Tower Extension Lemma; the second is from the construction of T^{n+1}_3 via the 3-Stage Reimbedding Theorem. Thus T^1_4 satisfies all of the necessary conditions, completing the proof. ⦀

(4.3.11) REMARK. Our only uses of the Improved 3-Stage Reimbedding Theorem (3.3) were as in the beginning of (4.3.2). We could have avoided this by adding a stage at the bottom and consequently obtaining a proof of Freedman's 5-Stage Reimbedding Theorem [F1]. We use a similar trick to prove Theorem (5.1).

5. APPLICATIONS

(5.0) INTRODUCTION. In this section, we prove the last of the Reimbedding Theorems, the Mitosis Theorem. We also discuss other applications of these results.

Throughout this section, we will use the following notation: when dealing with an n-stage tower T^i_n, we let T^i_m, $1 \le m \le n$, denote the union of its first m stages. For $\ell < m$, the union of stages ℓ through m (i.r., $T^i_m - \mathring{T}^i_{\ell-1}$) will be denoted by $T^i_{\ell-m}$. This is a disjoint union of $(m-\ell+1)$-stage towers.

(5.1) THEOREM. Every 5-stage tower T^0_5 contains a 6-stage tower T^1_6 satisfying:

a) (agreement) $C^0_{1-3} = C^1_{1-3}$; and

b) (π_1-neg.) $\pi_1(T^0_5 - T^1_6) \to \pi_1(T^0_5 - C^0_1)$ is an isomorphism.

PROOF. Step 1. We apply the Big Reimbedding Theorem to each component of T^0_{2-5} to attain a new 5-stage tower T^1_5. Specifically, for each component T of T^0_{2-5}, we obtain a new 4-stage tower $\bar{T}$ inside T with the same attaching curve. Note that the first stage kinky handles of T and $\bar{T}$ induce the same framing on this circle, as their cores coincide. Hence, the union of these new towers, together with T_1^0, form a 5-stage tower. We obtain T_5^1 from this tower by shrinking its first stage away from $\partial^+ T_1^0$ (i.e., we want T_1^1 to be a small regular neighborhood of the core C_1^0).

Step 2. We enlarge T_5^1 to a 6-stage tower T_6^1 via the Tower Extension Lemma (4.2.3). In order to do this, we need to check that the map $\varphi: \pi_1(\partial^+ T_5^1) \to \pi_1(T_5^0 - \mathring{T}_5^1)$, induced by the inclusion, is the zero map. The latter group has the canonical decomposition $Z * F$, since the top few stages of T_5^1 are π_1-neg.. The fifth stage of T_5^1 is contained in T_{2-5}^0, so the image of φ lies in the factor F. (This is why we need five stages. It seems impossible to avoid the Z factor in a four-stage setting.) Now the nullity conclusion of the Big Reimbedding Theorem tells us that the map φ is trivial, as required. This enables us to construct our T_6^1, which clearly satisfies the conclusions of Theorem 5.1. ∎

(5.2) THEOREM (MITOSIS). Every 5-stage tower T_5^1 contains an 11-stage tower T_{11}^* satisfying:

a) (agreement) $C_{1-3}^0 = C_{1-3}^*$ and

b) (π_1-neg.) $\pi_1(T_5^0 - T_{11}^*) \to \pi_1(T_5^0 - C_1^0)$ is an isomorphism.

PROOF. Use Theorem (5.1) to obtain a 6-stage tower T_6^1 inside T_5^0. Next apply Theorem (5.1) to the 5-stage towers composing T_{2-6}^1, obtaining a 7-stage tower T_7^2 inside T_6^1. (Use the method of Step 1 of the previous proof: T_7^2 is T_1^1 union a 6-stage tower for each component of T_{2-6}^1.) Continue in this manner with T_{3-7}^2. After several more iterations, we obtain a nest $T_{11}^6 \to T_{10}^5 \to \cdots \to T_6^1 \to T_5^0$. Let T_{11}^* equal T_{11}^6. ∎

(5.3) Other Applications. Note that the number 11 appearing in the Mitosis Theorem is arbitrary; we chose it because of its convenience for Freedman's applications [F1]. The same method of proof would give us a tower T_n^* inside T_5^0, for arbitrary $n > 5$. In fact, letting n increase without bound provides the following:

(5.3.0) THEOREM. Every 5-stage tower T_5 contains a Casson handle with the same (framed) attaching circle.

In conjunction with Freedman's theorem (every Casson handle is homeomorphic to an open 2-handle), we have:

(5.3.1) THEOREM. Every 5-stage tower T_5 contains a topological 2-handle, i.e., there is a topological imbedding $(D^2 \times D^2, \partial D^2 \times D^2) \to (T_5, \partial^- T_5)$ which maps $\partial D^2 \times \{0\}$ onto the attaching curve of T_5.

In particular, there is a flat topological imbedding $(D^2, \partial D^2) \to (T_5$, attaching circle), i.e., the attaching circle is topologically slice in T_5. Thus, to put topological 2-handles or slice disks into 4-manifolds, it is sufficient to construct 5-stage towers. As an application, we prove the following result: Let L denote the second untwisted double of the Whitehead link. (Recall that in the notation of [C, Lecture II], the Whitehead link is denoted

$D^{1/2}h$; L is then $D^{5/2}h$.)

(5.3.2) THEOREM. L is topologically slice. (That is, the two components of L in $S^3 = \partial D^4$ bound disjoint, flat, topologically imbedded 2-disks in D^4.) Furthermore, one slice disk can be taken to be smooth and unknotted in D^4.

Sketch of Proof. Let T_5 be a 5-stage tower with exactly one kink at each stage. Then T_5 is diffeomorphic to $S^1 \times D^3$. This may be explicitly seen via Kirby calculus: Draw the link picture of T_5, and cancel handles from the first stage up through fifth, leaving a single 1-handle represented by a circle c with a dot. This presents $S^1 \times D^3$ as D^4 minus a regular neighborhood of an unknotted disk d, with $\partial d = c$ in S^3. If we trace the attaching curve c' of T_5 through the above pictures, we find that c and c' form the link L in S^3. Now Theorem (5.3.1) gives a topological slice disk d' for c' in $T_5 = D^4 - \mathring{N}(d)$. The disks d and d' slice L. ▥

REMARKS. (a) Theorem (5.3.2) appears to be stronger than any previously known result.

(b) Our Big Reimbedding Theorem also supplies a missing ingredient in the proof of Theorem (8) of [F1].

6. A SUMMARY OF THE REIMBEDDING THEOREMS

For the convenience of reference, we restate in this section the main results of this paper. More specifically, Theorems (3.3), (4.0.0), (5.1), and (5.2) are restated below as Theorems (6.0), (6.1), (6.2) and (6.3), respectively.

(6.0) THEOREM: (THE IMPROVED 3-STAGE ("LITTLE") REIMBEDDING THEOREM). Every 3-stage tower T_3^0 contains another 3-stage tower T_3^1 satisfying:

a) (agreement) $C_{1-2}^0 = C_{1-2}^1$;

b) (π_1-neg.) $\pi_1(T_3^0 - T_3^1) \to \pi_1(T_3^0 - C_1^0)$ is an isomorphism; and

c) (no linking C_1^0) the image $\mathrm{Im}[\pi_1 C_3^1 \to \pi_1(T_3^0 - C_1^0)]$ lies in the image $\mathrm{Im}[\pi_1 C_3^0 \to \pi_1(T_3^0 - C_1^0)]$.

(6.1) THEOREM: (THE BIG REIMBEDDING THEOREM). Every 4-stage tower T_4^0 contains another 4-stage tower T_4^1 satisfying:

a) (agreement) $C_{1-2}^0 = C_{1-2}^1$;

b) (π_1-neg.) $\pi_1(T_4^0 - T_4^1) \to \pi_1(T_4^0 - C_1^0)$ is an isomorphism; and

c) (nullity) $\pi_1 T_4^1 \to \pi_1 T_4^0$ is the zero map.

(6.2) THEOREM. Every 5-stage tower T_5^0 contains a 6-stage tower T_6^1 satisfying:

a) (agreement) $C_{1-3}^0 = C_{1-3}^1$; and

b) (π_1-neg.) $\pi_1(T_5^0 - T_6^1) \to \pi_1(T_5^0 - C_1^0)$ is an isomorphism.

(6.3) THEOREM: (THE MITOSIS THEOREM). Every 5-stage tower T_5^0 contains an 11-stage tower T_{11}^* satisfying:

a) (agreement) $C_{1-3}^0 = C_{1-3}^*$; and

b) (π_1-neg.) $\pi_1(T_5^0 - T_{11}^*) \to \pi_1(T_5^0 - C_1^0)$ is an isomorphism.

(6.4) Concluding Remarks. Theorem (6.0) may be compared with combined Theorems (4.1) and (4.2) of [F1]. Also, Theorems (6.1), (6.2) and (6.3) may be compared with Theorems (4.3), (4.4) and (4.5), respectively.

BIBLIOGRAPHY

[C] A. Casson, Three Lectures on new constructions in 4-dimensional manifolds. Notes prepared by L. Guillou, Prepublication Orsay 81T06.

[F1] M. H. Freedman, The topology of four-dimensional manifolds. J. Differential Geometry 17 (1982), 357-453.

[F2] ————————, A fake $S^3 \times R$. Ann. of Math. 110 (1979), 177-201.

[K1] R. Kirby, A calculus for framed links in S^3. Invent. Math. 45 (1978), 35-56.

[K2] ————, Classroom lectures delivered at the University of California at Berkeley, 1981.

[R] D. Rolfsen, Knots and Links. Publish or Perish, 1976.

DEPARTMENT OF MATHEMATICS
UNIVERSITY OF CALIFORNIA
BERKELEY, CA 94720

DEPARTMENT OF MATHEMATICS
SOUTHWEST TEXAS STATE UNIVERSITY
SAN MARCOS, TX 78666

Contemporary Mathematics
Volume 35, 1984

THE HOMOLOGY OF THE MAPPING CLASS GROUP AND ITS CONNECTION TO SURFACE BUNDLES OVER SURFACES

John Harer

Let $\Gamma = \Gamma^n_{g,r}$ be the mapping class group of a connected orientable surface $F = F^n_{g,r}$ of genus g with r boundary components and n marked points. Γ is $\pi_0(\Lambda)$ where Λ is the topological group of orientation-preserving diffeomorphisms of F which fix ∂F and the n points $p_1, \ldots, p_n$. This paper is a survey of progress towards computation of $H_*(\Gamma)$.

SECTION 0: MOTIVATION

(1) Let $r = 0$ and suppose $\mathcal{M}^n_g$ is the moduli space of isometry classes of complete hyperbolic metrics on $F - \{p_1, \ldots, p_n\}$. $\mathcal{M}^n_g$ is the quotient of Teichmuller space $\mathcal{T}^n_g$ by the properly discontinuous action of Γ. $\mathcal{T}^n_g \cong B^{6g-6+2n}$ and the codimension of the fixed point set of Γ increases with g so

$$H_k(\Gamma^n_g; \mathbb{Q}) \cong H_k(\mathcal{M}^n_g; \mathbb{Q}) \quad \text{for all } k \text{ and}$$
$$H_k(\Gamma^n_g; \mathbb{Z}) \cong H_k(\mathcal{M}^n_g; \mathbb{Z}) \quad \text{when } g \gg k.$$

Furthermore, Mumford [Mu] shows that

$$H^2(\Gamma; \mathbb{Z}) \cong \text{Pic}(\mathcal{M}), \ n = 0,$$

and conjectures this group is $\mathbb{Z}$. This is proven for $g \geq 5$ in Theorem 1 below.

(2) Consider smooth fiber bundles $F \to W^4 \to X^2$ with X a closed oriented surface. Call two such bordant if they cobound a smooth F-bundle over an oriented 3-manifold; bordism classes form a group $\Omega_2(F)$ under disjoint union. The usual classifying space arguments show

$$\Omega_2(F) \cong \Omega_2(B\Lambda),$$

the latter group being the 2-dimensional bordism group of the classifying space $B\Lambda$. Homology and bordism agree in low dimensions and $\pi_i(\Lambda) = 0$ when $i > 0$, $g \geq 2$ [E-E], hence

$$\Omega_2(B\Lambda) \cong H_2(B\Lambda) \cong H_2(\Gamma).$$

0271-4132/84 $1.00 + $.25 per page

SECTION 1: RESULTS

THEOREM 1: $H_1(\Gamma) = 0$, $g \geq 3$; $r,n \geq 0$.

This was proven by Powell [P] for $r = n = 0$. We sketch a simple proof [H_1]: According to Dehn [D], Γ is generated by Dehn twists $\{\tau_C\}$ on simple closed curves in F. One first checks that for $g \geq 2$ only non-separating curves are needed. In that case, given C_1, C_2, there exists $f \in \Lambda$ with $f(C_1) = C_2$; one finds

$$\tau_{C_1} = f^{-1} \tau_{C_2} f$$

so that $H_1(\Gamma)$ is cyclic. The proof is then completed by finding on $F_{g,r}$, $g \geq 3$, a relation equating a certain product of four twists to another product of three twists.

THEOREM 2 [H_1]:

$$H_2(\Gamma;\mathbb{Z}) \cong \begin{cases} \mathbb{Z}^{n+1} & g \geq 5 , \quad r+n > 0 \\ \mathbb{Z} \oplus \mathbb{Z}/(2g-2) & g \geq 5 , \quad r = n = 0. \end{cases}$$

To interpret this result observe that a bundle $F_g^n \to W \to X$ has n canonical sections $s_1,\ldots,s_n\colon X \to W$. Define

$$H_2(\Gamma) \cong \Omega_2(F) \overset{\varphi}{\to} \mathbb{Z}^{n+1} \quad \text{by}$$

$$\varphi(\eta) = \left(\frac{\sigma(w)}{4}, [s_1(X)]^2, \ldots, [s_n(X)]^2\right)$$

where σ is the signature of W and $[s_i(X)]^2$ is the self-intersection number. φ is a surjection for $g \geq 3$ [Me], [H_1]; $\mathrm{Ker}(\varphi) = 0$, $r+n > 0$ and $\mathscr{C} = \mathbb{Z}/(2g-2)\mathbb{Z}$, $r = n = 0$. Thus if W is closed and $\sigma(W) = 0$

$$(2g-2)\begin{pmatrix} F \to W \\ \downarrow \\ X \end{pmatrix} = \partial \begin{pmatrix} F \to V^5 \\ \downarrow \\ Y^3 \end{pmatrix}.$$

<u>Sketch of the proof</u>: The proof is based on work of Hatcher and Thurston [H-T]. They construct a 1-connected 2-complex X_2 admitting a natural action of Γ, using this to give a finite presentation of Γ. Briefly, X has a vertex for each <u>cut system</u> on F, i.e. each isotopy class of collections $\mathscr{C} = \{C_1,\ldots,C_g\}$ of disjointly embedded simple closed curves in F such that $F - \mathscr{C}$ is connected. Edges are added between vertices for cut systems $\mathscr{C}$, $\mathscr{C}'$ when, say, $C_j = C_j'$ for $j > 1$ and $C_1 \cap C_1'$ is one point. Finally one adds 2-cells for three configurations of curves (see [H,T]). The proof that X_2 is 1-connected uses Cerf theory [C].

We proceed by adding 3-cells to X_2 for certain configurations of cut systems, the result is X_3 and it too admits a cellular action of Γ. Form the fiber product of X_3 with $E\Gamma$, the universal covering of the $K(\Gamma,1)$ $B\Gamma$

and consider the projections:

$$\begin{array}{ccc} & Z = X_3 \times_\Gamma E\Gamma & \\ {}^{p_1}\swarrow & & \searrow^{p_2} \\ X_3/\Gamma & & B\Gamma \, . \end{array}$$

p_2 is a fibration with fiber X_3 so $H_2(Z)$ surjects to $H_2(\Gamma)$. The cell structure of X_3/Γ allows direct computation of $H_2(Z)$ for the proof of the theorem.

For the final theorem consider two maps $\varphi: F_{g,r} \to F_{g+1,r-2}$, $\psi: F_{g,r} \to F_{g+1,r-1}$; φ is obtained by gluing two boundary components together, ψ by gluing a pair of pants (S^2 - three disks) to two boundary components of F. φ and ψ induce maps of mapping class groups.

THEOREM 3 [H_2]: Using homology with $\mathbb{Q}$-coefficients,

$$\varphi_*: H_k(\Gamma_{g,r}) \to H_k(\Gamma_{g+1,r-2}) \quad \text{and}$$

$$\varphi_*: H_k(\Gamma_{g,r}) \to H_k(\Gamma_{g+1,r-1})$$

are isomorphisms for $g \gg k$. Combining these we find $H_k(\Gamma_{g,r}) \cong H_k(\Gamma_{g+1,s})$ for all r,s, $g \gg k$.

Sketch of proof: Stability theorems require Γ-complexes whose connectivity increases with g, we construct two such complexes X and AX. X is the realization of the partially ordered set whose elements are subsets of cut systems with partial ordering given by inclusion. For AX choose a point $p \in \partial F$ and consider loops α based at p with $\alpha \cap \partial F = P$ and α $F-\partial F$ an embedded arc. Arc systems are collections $\{\alpha_1,\ldots,\alpha_{2g}\}$, defined up to isotopy, where $\alpha_i \cap \alpha_j = p$ for all i,j and $F - \{\alpha_1,\ldots,\alpha_{2g}\}$ is connected. Define AX then as the realization of the poset whose elements are subsets of arc systems, partial ordering again given by inclusion.

PROPOSITION: $X \simeq \bigvee_j S_j^{g-1}$,

$$AX \simeq \bigvee_j S_j^{2g-1} \, .$$

To prove this for X we write it as a nested union of finite subcomplexes $X_1 \subset \cdots \subset X_n \subset \cdots$. Each X_n admits an embedding φ_n into the projective lamination space $P\mathcal{L}_o \cong S^{6g-7+2r}$, φ_n simplicial with respect to a certain piecewise linear structure on $P\mathcal{L}_o$ described using a finite number of recurrent train tracks [T]. Any simplicial map $f: S^k \to X_n$ gives rise to $\varphi_n \circ f$ which extends to $\psi_n: B^{k+1} \to P\mathcal{L}_o$, $k < 6g-7+2r$. The structure of train tracks is then analyzed to obtain $\hat{f}: B^{k+1} \to X$ extending f whenever $k < g-1$.

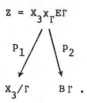

For AX a similar analysis is made, this time using the space $P\mathcal{L}_1$ whose elements are projective classes of closed subsets of $(F-\partial F) \cup \{p\}$ which are measured laminations in $F-\partial F$.

The proof now follows a standard pattern ([Q], [W], [V] and others): Γ acts on X via

$$g \cdot \{C_1,\ldots,C_k\} = \{C_1,\ldots,C_k\}.$$

The map $\varphi: F_{g,r} \to F_{g+1,r-2}$ induces a map $X(F_{g,r}) \to X(F_{g+1,r-2})$ and there is a spectral sequence converging to zero with

$$E^1_{p,q} = H_q(\Gamma^p_{g+1,r-2}; \Gamma^p_{g,r})$$

where Γ^p denotes the stabilizer of a p-cell in X of the form $\{C_o\} \subset \{C_o,C_1\} \subset \cdots \subset \{C_o,\ldots,C_p\}$. By inductively assuming φ_* is an isomorphism for all $\ell < k$ one finds $H_k(\Gamma_{g,r},\Gamma_{g-1,r+2}) \twoheadrightarrow H_k(\Gamma_{g+1,r-2},\Gamma_{g,r})$. A diagram chase then proves φ_* is an isomorphism on H_k.

A completely analogous argument works for AX.

BIBLIOGRAPHY

[C] J. Cerf, La Stratification Naturelle des Espaces de Fontions Differentiables Reeles et la Theoreme de la Pseudoisotopie, Publ. Math. I.H.E.S. 39 (1970), 5-173.

[D] M. Dehn, Die Gruppe der Abbildungsklassen, Acta Math. 69 (1938), 135-206.

[E-E] C. J. Earle, J. Eells, The Diffeomorphism Group of a Compact Riemann Surface, Bull. AMS 73 (1967), 557-559.

[H1] J. Harer, The Second Homology Group of the Mapping Class Group of an Orientable Surface, Preprint (1981).

[H2] ————, Homology Stability for the Mapping Class Group of a Surface, Preprint (1982).

[H-T] A. Hatcher, W. Thurston, A Presentation for the Mapping Class Group of a Closed Orientable Surface, Top. 19 (1980), 221-237.

[Me] W. Meyer, Die Signatur von Flachenbundeln, Math. Ann. 201 (1973), 239-264.

[Mu] D. Mumford, Abelian Quotients of the Teichmuller Modular Group, Journal d'Analyse Mathematique 18 (1967), 227-244.

[P] J. Powell, Two Theorems on the Mapping Class Group of a Surface, Proc. Amer. Math. Soc. 68 (1978), no. 3, 347-350.

[Q] D. Quillen, M.I.T. Lectures (1974-1975).

[T] W. Thurston, Princeton Lectures (1977).

[V] K. Vogtmann, Spherical Posets and Homology Stability for $O_{n,n}$, Top. 20 (1981), 119-132.

[W] J. Wagoner, Stability for Homology of the General Linear Group of a Local Ring, Top. 15 (1976), 417-423.

DEPARTMENT OF MATHEMATICS
UNIVERSITY OF MARYLAND
COLLEGE PARK, MD 20742

Contemporary Mathematics
Volume 35, 1984

ROCHLIN INVARIANT AND α-INVARIANT

Akio Kawauchi

This paper is a condensed version of the author's recent works ([10], [11], [12]) connected with the problem: how does a cyclic action on a Z_2-homology 3-sphere contribute to the Rochlin invariant of the Z_2-homology 3-sphere? This problem clearly reduces to the problem for a cyclic action on a Z_2-homology 3-sphere in the following four cases (1)-(4): (1) Free cyclic action of odd-prime order, (2) Non-free cyclic action of odd-prime order, (3) Non-free involution, (4) Free cyclic action of an order which is a power of 2. A great difference between the first three cases (1)-(3) and the case (4) is that the orbit space is also a Z_2-homology 3-sphere in the cases (1)-(3), but not in the case (4). In each case, we shall establish a congruence in Q/Z containing the Rochlin invariant and the Atiyah-Singer α-invariant of the action.

In Section 1 we introduce a notion of the slope with a value in $Q/Z \cup \{\infty\}$ of a knot in a 3-manifold. We also generalize this notion to a link there. A geometric meaning of the slope is useful for our purpose. In Section 2 we shall discuss two mutually related invariants (i.e. δ_o-invariant and δ-invariant) of a proper link in a Z_2-homology 3-sphere, generalizing an invariant of a classical knot by Robertello [16] or a knot in a Z-homology 3-sphere by Gordon [5]. The δ_o-invariant is in general an oriented link type invariant, but the δ-invariant is an unoriented link type invariant. The arguments of Sections 1-2 give a novel approach to the calculation of the Rochlin invariant. In Section 3 some results of the calculation are given. In Section 4 we shall discuss the Atiyah-Singer α-invariant of a cyclic action on a closed oriented 3-manifold. It is well-known for a free cyclic action. We shall also define it for a certain semi-free cyclic action on a closed oriented 3-manifold. In Section 5 the desired results of the cases (1) and (2) are given. Section 6 is concerned with the case (3), and Section 7 with the case (4). In Section 8 we shall give some results on the δ-invariant and δ_o-invariant of the fixed point set of a cyclic action on a Z_2-homology 3-sphere when it is a proper link and the order of the action is prime.

0271-4132/84 $1.00 + $.25 per page

Conventions: Spaces and maps are in the piecewise-linear category. Manifolds are orientable and oriented suitably. Actions on manifolds are orientation-preserving and faithful. For an oriented manifold X, $-X$ is the same manifold as X but with the opposite orientation. In case $\partial X \neq \emptyset$, ∂X is oriented by the orientation induced from X. Let $X \times [-1,1]$ have an orientation such that the natural injection $X \to X \times 1 \subset \partial(X \times [-1,1])$ is orientation-preserving (namely, $X \to X \times (-1) \subset \partial(X \times [-1,1]$ is orientation-reversing). The orbit space of a space Y with an action is denoted by $\underset{\sim}{Y}$.

All necessary definitions for our arguments of [10] and [11] are included here, but no proof is given. Full information of [12] is not given.

1. THE SLOPE OF A KNOT IN A 3-MANIFOLD

Let k be a knot in a 3-manifold M with a tubular neighborhood $T = T(k)$. An m.ℓ. pair of T (or k) is a pair (m,ℓ) of a meridiam m and a longitude ℓ of T intersecting in one point such that the intersection number is positive on ∂T. (A longitude ℓ of T is oriented so that ℓ is homotopic to k in T.) A link P in M is parallel on T (or k), if $P \subset \partial T$ and any two (oriented) components of P are isotopic on ∂T. For a link L in M, the order of the homology class $[L] \in H_1(M;Z)$ is called the order of the link L and denoted by $o(L)$. If $[L] = 0$, define $o(L) = 1$.

LEMMA 1.1 ([10]). *Given a knot $k \subset M$ of finite order with tubular neighborhood T, there exists exactly one (up to isotopy) parallel link P on T such that*

(1) $[P] = o(k)[k]$ *in* $H_1(T;Z)$,

(2) P *bounds a compact oriented surface in* $E = M - \mathrm{Int}\, T$.

The link P, any component K of P and any compact oriented surface in E, bounded by P are called the characteristic parallel link, the characteristic knot and a characteristic surface for the knot k, respectively. In case $o(k) = 1$, P is a longitude of T and we see that k bounds a surface in M, obtained by extending any characteristic surface for k, called a Seifert surface in the classical knot theory.

COROLLARY 1.2 ([10]). *The characteristic parallel link P (up to the orientation) is determined uniquely by the space $E = M - \mathrm{Int}\, T$.*

Let K be the characteristic knot of a knot k of finite order. Write $[K] = a[m] + b[\ell]$ in $H_1(\partial T;Z)$ for any m.ℓ. pair (m,ℓ) of T. Note that $b > 0$.

DEFINITION 1.3 ([10]). The fraction $a/b \pmod 1$ is called the slope of a knot k of finite order, and denoted by $s(k)$. If $s(k) = 0$, then we say the knot k is flat. When $o(k) = \infty$, we say the slope of k is infinite and denote $s(k) = \infty$

A flat knot has properties analogous to those of a classical knot. For example, any flat knot has a unique m.ℓ. pair with the longitude being the characteristic knot. For each element $s \in Q/Z$ we can have coprime positive integers a, b such that $s \equiv a/b \pmod 1$. This fraction a/b and the denominator b are called a normal presentation and the denominator of the element $s \in Q/Z$.

The following shows that the complement $M - k$ never contributes to the slope $s(k)$.

PROPOSITION 1.4 ([10, Proposition 1.5, Remark 1.6, Corollary 1.7]). *Let E be a compact oriented 3-manifold with ∂E, a torus. Then for each $s \in Q/Z \cup \{\infty\}$ there exists a knot k in M with $s(k) = s$ such that $M - \mathrm{Int}\, T(k) = E$. Moreover, if $s = \infty$, the homeomorphism type of M is uniquely determined by that of E.*

Let $\tau H_1(M)$ be the torsion part of $H_1(M;Z)$. Let $\varphi : \tau H_1(M) \times \tau H_1(M) \to Q/Z$ be the linking pairing.

LEMMA 1.5 ([10]). *For any knot $k \subset M$ of finite order we have*

$$s(k) = -\varphi([k], [k]) .$$

By this lemma, we can generalize the slope of a knot to that of a link.

DEFINITION 1.3' ([11]). The slope of a link L M, denoted by $s(L)$ is defined by the following:

$$s(L) = -\varphi([L],[L]) \quad (\text{if } o(L) < \infty) \quad \text{or} \quad \infty \quad (\text{if } o(L) = \infty).$$

If $s(L) = 0$, then we say the link L is flat. Let L be a link with components k_i, $i = 1,2\ldots,r\,(r \geq 2)$. Let $B_1, B_2, \ldots, B_{r-1}$ be mutually disjoint oriented bands in M attaching to L as 1-handles. If we obtain a knot k from L by surgery along such $B_1, B_2, \ldots, B_{r-1}$, then we say the knot k is obtained from L by a fusion. If each component k_i is a knot of finite order, the total Q-linking number $\lambda(L) \in Q$ of L is defined by the identity

$$\lambda(L) = \sum_{i>j} \mathrm{Link}_M(k_i, k_j).$$

When $r = 1$, we understand that $\lambda(L) = 0$.

LEMMA 1.6 ([11, Lemmas 1.2,1.3]). *Let k be a knot obtained from L by a fusion. Then $s(L) = s(k)$. If each k_i is of finite order, then* $s(k) = \Sigma_{i=1}^{r} s(k_i) - 2\lambda(L)$ in Q/Z.

2. A GENERALIZATION OF THE ROBERTELLO INVARIANT OF A CLASSICAL KNOT

Let M be a closed oriented 3-manifold with $H_1(M;Z_2) = 0$. Each component of M is a Z_2-homology 3-sphere. The Rochlin invariant (or μ-invariant), $\mu(M)$ of M is defined by $\mu(M) = -\mathrm{sign}\, W/16$ in Q/Z for any compact oriented spin ($w_2 = 0$) 4-manifold W with $\partial W = M$. (The well-definedness follows from the Rochlin theorem [17].) Let S be a Z_2-homology 3-sphere.

DEFINITION 2.1 ([16],[11]). A link L with components k_i, $i=1,2,\dots,r$ in S is proper, if the mod 2 linking number, $\mathrm{Link}_S(k_i, L-k_i)_2 = 0$ for all i, $1 \le i \le r$. (We understand a knot to be a proper link.)

Let W be a compact oriented 4-manifold. Let F be a locally flat, oriented (possibly disconnected) surface of (total) genus 0 in W. We say such a pair $F \subset W$ is admissible for a link $L \subset S$, if S is a component of ∂W, $\partial F = L$, $H_1(\partial W; Z_2) = 0$ and $[F_2^+] \in H_2(W;Z_2)$ is characteristic, i.e. $[F_2^+] \cdot x = x^2$ for all $x \in H_2(W;Z_2)$, where F_2^+ is a (mod 2) cycle obtained from F by attaching (mod 2) 2-chains c_i in S with $\partial c_i = -k_i$, $i = 1,2,\dots,r$.

LEMMA 2.2 ([11]). For any proper link there exists an admissible pair.

DEFINITION 2.3 ([11]). Let L be a proper link in S. Then we define

$$\delta_o(L) = ([F_Q^+]^2 - \mathrm{sign}\, W)/16 - \mu(\partial W)$$

in Q/Z for any admissible pair $F \subset W$ for $L \subset S$, where F_Q^+ is a rational 2-cycle obtained from F by attaching rational 2-chains c_i^Q in S with $\partial c_i^Q = -k_i$, $i = 1,2,\dots,r$.

This invariant was defined by Robertello [16] for a classical knot and by Gordon [5] for a knot in a Z-homology 3-sphere. In their cases, it takes the value 0 or 1/2, but in our general case, it takes more value depending on the slope of the proper link. The well-definedness of Definition 2.3 also follows from the Rochlin theorem (cf. [13]).

DEFINITION 2.4 ([11]). Two links $L_i \subset S_i$, $i = 0,1$, are said to be cobordant in the weak sense if:

(1) There exists a compact oriented 4-manifold W such that $\partial W = -S_0 \cup S_1$ and $H_*(W, S_i; Z_2) = 0$, $i = 0,1$,

(2) There exists a locally flat, compact oriented (possibly disconnected) surface F of (total) genus 0 in W such that $\partial F = -L_0 \cup L_1$.

The following is a generalization of a result of Robertello [16, Theorem 2].

THEOREM 2.5 ([11]). If proper links $L_i \subset S_i$, $i = 0,1$, are cobordant in the weak sense, then $\delta_o(L_0) = \delta_o(L_1)$.

By Proposition 1.4, we can obtain from the exterior of a knot k in S a unique closed connected oriented 3-manifold M such that $H_1(M;Z)/\text{odd torsion} \cong Z$, called a Z_2-homology handle. In [8] we defined an invariant $\varepsilon(M)$, being 0 or 1, of M, calculable from the Z_2-Alexander polynomial of M.

THEOREM 2.6 ([11, Corollary 2.8]). Let L be a proper link in S. Let M be the Z_2-homology handle of a knot k in S, obtained from L by a fusion. Let a/b be a normal presentation of the slope $s(L)$ with a odd. Then we have

$$\delta_o(L) = \delta_o(K) + (a/b - ab)/16 \quad \text{and} \quad \delta_o(K) = \varepsilon(M)/2$$

in Q/Z, *where* K *is the characteristic knot of* k *in* S.

DEFINITION 2.7 ([11]). For a proper link L in S we define

$$\delta(L) = \delta_o(L) + \lambda(L)/8$$

in Q/Z, where $\lambda(L)$ is the total Q-linking number of L.

When $r = 1$, $\delta(L) = \delta_o(L)$. The following gives an important property of the invariant $\delta(L)$.

PROPOSITION 2.8 ([11]). *The invariant* $\delta(L)$ *is an unoriented link type invariant of* L *in* S. *That is, it is independent of any particular choice of the orientations of* k_i, $i = 1,2,\ldots,r$.

3. SOME RESULTS OF THE CALCULATION OF THE ROCHLIN INVARIANT

Let T_i be oriented solid tori with m.ℓ. pairs (m_i, ℓ_i), $i = 1,2$. Let $h: \partial T_1 \to \partial T_2$ be an orientation-reversing homeomorphsim such that $h_*[m_1] = a[m_2] + b[\ell_2]$ $(b \neq 0)$. The adjunction space $T_1 \cup_h T_2$ is the lens space $-L(b,a) = L(b,-a) = L(-b,a)$. The following is obtained from Theorem 2.6 (with L being a knot).

LEMMA 3.1 (Reciprocity Law)([10]). *For coprime odd* $a,b > 0$,

$$\mu(L(a,b)) + \mu(L(b,a)) = (1 - ab)/16$$

in Q/Z.

Using that $L(b,a) \cong L(b,a')$ if and only if $a^{\pm 1} \cdot a' \equiv 1 \pmod b$, one can compute from Lemma 3.1 the μ-invariant of any lens space $L(b,a)$ with b odd.

THEOREM 3.2 ([10]). *Let* k' *be a flat knot in a* Z_2*-homology 3-sphere* S', *obtained from a knot* k *in a* Z_2*-homology 3-sphere* S *so that*

$$S' - \text{Int}\, T(k') = S - \text{Int}\, T(k)$$

and

$$[m'] = c[m] + d[\ell],$$

$$[\ell'] = [K(k)] = a[m] + b[\ell], \; ad - bc = -1,$$

in $H_1(\partial T(k);Z)$ *for m.ℓ. pairs* (m,ℓ), (m',ℓ') *of* $T(k)$, $T(k')$, *respectively. Then* b *is odd and*

$$\mu(S') = \mu(S) + \mu(L(b,a)) + d\delta(k').$$

The following is a generalization of a result of Gordon [5].

COROLLARY 3.3 ([10]). *Let* k_i *be a flat knot in a* Z_2*-homology 3-sphere* S_i *with an m.ℓ. pair* (m_i, ℓ_i) *on* $T(k_i)$ *such that* $\ell_i = K(k_i)$, $i = 1,2$. *Let* $S = S_1 - \text{Int}\, T(k_1) \cup_h S_2 - \text{Int}\, T(k_2)$ *be the adjunction space obtained by an* orientation-reversing homeomorphism $h:\partial T(k_1) \to \partial T(k_2)$ such that

$$h_*[\ell_1] = a[\ell_2] + b[m_2],$$
$$h_*[m_1] = c[\ell_2] + d[m_2],\ ad-bc = -1.$$

Suppose S is a Z_2-homology 3-sphere. Then b is odd and

$$\mu(S) = \mu(S_1) + \mu(S_2) - \mu(L(b,a)) + d\delta(k_1) + a\delta(k_2).$$

4. THE ATIYAH-SINGER α-INVARIANT OF A CYCLIC ACTION ON A 3-MANIFOLD

First we consider a free Z_n-action on a closed connected oriented 3-manifold M. It is well-known by Casson-Gordon [2] that M is the equivariant boundary of a compact connected oriented 4-manifold W with a semi-free Z_n-action such that the fixed point set $F = F(Z_n,W)$ is $\emptyset$ or a locally flat closed orientable surface.

DEFINITION 4.1. $\alpha(Z_n,M) = -\operatorname{sign} W + n \operatorname{sign} \underset{\sim}{W} - [F]^2(n^2-1)/3$, where $[F]^2 = 0$ if $F = \emptyset$.

It follows from the Novikov addition theorem of signatures and the Atiyah-Singer G-signature theorem (cf. [6]) that $\alpha(Z_n,M)$ is an invariant of the equivariant orientation-preserving homeomorphism type of (Z_n,M). Two kinds of finer invariants but depending on each $t \in Z_n$ are widely known. One is the Atiyah-Singer α-invariant $\alpha(t,M)$ (cf. [6, p. 72]) and the other, the Casson-Gordon invariant, say $\sigma(t,M)$ (cf. [2, p. 42]). From these definitions, we see the following:

4.2. $\alpha(Z_n,M) = \sum_{t(\neq 1)\in Z_n} \alpha(t,M) = \sum_{t(\neq 1)\in Z_n} \sigma(t,M)$.

When $n = 2$, $\alpha(Z_2,M) = \alpha(t,M) = \sigma(t,M)$, $t \neq 1$, is an integer and called the Browder-Livesay invariant (cf. [6]).

Next we consider a closed connected oriented 3-manifold M with semi-free Z_n-action such that $F(Z_n,M) = L \neq \emptyset$. Note that L is a link in M. We shall define an analogous invariant $\alpha(Z_n,M)$ of the equivariant, orientation-preserving homeomorphism type invariant of (Z_n,M) only when each component of L is a knot of finite order. A difficult point is that $\alpha(Z_n,M)$ should not depend on any particular choice of the orientations of the components of L.

LEMMA 4.3 ([10]). *Let M be a closed oriented 3-manifold with a semi-free or free Z_n-action. Let k be a Z_n-invariant knot in M such that k is a component of $L = F(Z_n,M)$ or $L \cap k = \emptyset$. Then k is of finite order in M if and only if $\underset{\sim}{k}$ is so in $\underset{\sim}{M}$. Further, in this case, we have* $ns(k) = s(\underset{\sim}{k})$ *(if* $k \subset L$) *or* $s(k) = ns(\underset{\sim}{k})$ *(if* $L \cap k = \emptyset$).

Thus, for example, if $H_1(\underset{\sim}{M};Q) = 0$, each component of L is a knot of finite order in M. We assume the components, k_i, $i = 1,2,\ldots,r$, of $L = F(Z_n,M)$ are knots of finite order in M. Let W be a compact connected oriented 4-manifold with semi-free Z_n-action such that $\partial(Z_n,W) = (Z_n,M)$ and $F = F(Z_n,W)$ is a locally-flat, compact proper orientable surface. Such a 4-manifold always exists (cf. [10]). Orient F and then $k_1,k_2,\ldots,k_r$ so that $\partial F = L$. Let F_Q^+ be a rational cycle in W obtained from F by attaching rational 2-chains $c_1^Q,c_2^Q,\ldots,c_r^Q$ in M with $\partial c_i^Q = -k_1$, $i = 1,2,\ldots,r$. Let $\lambda(L)$ be the Q-total linking number of L in M.

DEFINITION 4.4)[10]).

$$\alpha(Z_n,M) = -\operatorname{sign} W + n \operatorname{sign} \underset{\sim}{W} - ([F_Q^+]^2 + 2\lambda(L))(n^2-1)/3.$$

The well-definedness for L with a fixed orientation follows also from the Novikov addition theorem of signatures and the Atiyah-Singer G-signature theorem. Then one can check that $\alpha(Z_n,M)$ is not altered by any change of the orientations of k_i, $i = 1,2,\ldots,r$.

REMARK 4.5. Let $S(L)_2$ be the double branched covering space of a Z-homology 3-sphere S, branched over a link L. Then we have $\alpha(Z_2,S(L)_2) = -\sigma(L) - \lambda(L)$, where $\sigma(L)$ is the Murasugi signature of L, i.e., $\sigma(L) = \operatorname{sign}(A+A')$ for a link matrix A associated with a Seifert surface for L. It follows that $\sigma(L) + \lambda(L)$ is an invariant of the unoriented link type of L, since $\alpha(Z_2,S(L)_2)$ is such (cf. [14],[7]).

5. THE CASE OF A CYCLIC ACTION OF ODD-PRIME ORDER

THEOREM 5.1 ([10]). *Let S be a Z_2-homology 3-sphere with free Z_p-action for an odd prime p. Then $\underset{\sim}{S}$ is a Z_2-homology 3-sphere and we have*

$$\mu(S) = \begin{cases} 9\alpha(Z_3,S)/16 + 3\mu(\underset{\sim}{S}) & (p=3) \\ \alpha(Z_p,S)/16 + p\mu(\underset{\sim}{S}) & (p>3) \end{cases}$$

in Q/Z.

THEOREM 5.2 ([10]). *Let S be a Z_2-homology 3-sphere with non-free Z_p-action for an odd prime p. Let $k_1,k_2,\ldots,k_r$ be the components of $L = F(Z_p,S)$. Then $\underset{\sim}{S}$ is a Z_2-homology 3-sphere, and for a normal presentation $2a_i/b_i$ of the slope $s(k_i)$, $i = 1,2,\ldots,r$, we have*

$$\mu(S) = \begin{cases} 9\alpha(Z_3,S)/16 + 3\mu(\underset{\sim}{S}) + 3\,\Sigma_{i=1}^r a_i/b_i & (p=3) \\ \alpha(Z_p,S)/16 + p\mu(\underset{\sim}{S}) + \{(p^2-1)/24\}\,\Sigma_{i=1}^r a_i/b_i & (p>3) \end{cases}$$

in Q/Z. *[Note that $(p^2-1)/24$ is an integer for $p > 3$.]*

The key to the proofs of these theorems is the following lemma:

LEMMA 5.3 ([10]). *Let W be a compact oriented 4-manifold with free or semi-free Z_n-action such that n is odd and $F = F(Z_n,W)$ is $\emptyset$ or a locally flat surface. Then W is spin if and only if $\underset{\sim}{W}$ is spin. [Note that $\underset{\sim}{W}$ is a 4-manifold.]*

REMARK 5.4. When n is even, this lemma is not true even for the case of a free Z_n-action (cf. [10]).

6. THE CASE OF A NON-FREE INVOLUTION

Let S be a Z_2-homology 3 sphere with non-free Z_2-action. Since the action is assumed to be orientation-preserving, it follows from Smith theory that $k = F(Z_2, S)$ such that $ab \equiv 1 \pmod 4$ we have

$$\mu(S) = \alpha(Z_2, S)/16 + 2\mu(\underset{\sim}{S}) + (ab + a/b + a^2 - 3)/16$$

in Q/Z. In particular, if k is flat, then

$$\mu(S) = \alpha(Z_2, S)/16 + 2\mu(\underset{\sim}{S}) \ .$$

COROLLARY 6.2 ([10]). *Let S be a Z_2-homology 3-sphere with semi-free Z_2n-action such that $k = F(Z_2n, S)$ is a knot. Let b be the denominator of the slope $s(k)$. Then*

$$b\mu(S) = b\alpha(Z_2n, S)/16 + b2^n\mu(\underset{\sim}{S}) \ .$$

The following generalizes a result of Contreras-Caballero [1] (cf. [9], [18]).

COROLLARY 6.3 ([10]). *Let S be a Z_2-homology 3-sphere with Z_2-action such that $k = F(Z_2, S)$ is a knot. If $\underset{\sim}{k} \subset \underset{\sim}{S}$ is amphicheiral (i.e., $\exists$ an orientation-reversing homeomorphism of S onto itself sending $\underset{\sim}{k}$ to $\pm\underset{\sim}{k}$), then* $\mu(S) = 0$.

Let a/b be a normal presentation of the slope $s(k)$ of $k = F(Z_2, S)$ with a odd. Let (m, ℓ) be an m.ℓ. pair of a Z_2-invariant $T(k)$ such that m is Z_2-invariant and $t\ell \cap \ell = \emptyset$, $t(\neq 1) \in Z_2$, and $[K(k)] = a[m] + b[\ell]$ in $H_1(\partial T(k); Z)$, where $K(k)$ is the characteristic knot. Construct a 4-manifold $W = S \times [-1,1] \ \ D^2 \times D^2$ identifying $T(k) \times 1$ with $\partial D^2 \times D^2$ so that $m \times 1 \equiv p \times \partial D^2$ and $\ell \times 1 \equiv \partial D^2 \times q$ $(p, q \in \partial D^2)$. Then the Z_2-action on S extends to a Z_2-action on W with $F(Z_2, W) = k \times [-1,1] \ \ D^2 \times 0$, a disk. This action induces a free Z_2-action on $S^* = \partial W - S \times (-1)$. Note that S^* is a Z_2-homology 3-sphere since a is odd.

THEOREM 6.4 ([12]). *Let a/b be a normal presentation of the slope $s(k)$ such that $ab \equiv 1 \pmod 4$ and $\delta(K(k)) = (a^2-1)/16$ in Q/Z. Then there exists a compact connected oriented 4-manifold W^* with Z_2-action fixing only one point, say x, such that $\delta(Z_2, W^*) = (Z_2, S^*)$ and $\underset{\sim}{W}^* - \{\underset{\sim}{x}\}$ is spin. For any such W^* we have*

$$\mu(\underset{\sim}{S}) = -\alpha(Z_2, S)/32 - (ab + a/b + a^2 - 3)/32 + (ab + a^2 - 2 - \operatorname{sign} W^*)/32$$

in Q/Z. [Note that $(ab + a/b + a^2 - 3)/32 \pmod 1$ does not depend on any choice of a normal presentation a/b of $s(k)$ such that $ab \equiv 1 \pmod 4$.]

7. THE CASE OF A FREE CYCLIC ACTION OF AN ORDER WHICH IS A POWER OF 2

Let S be a Z_2-homology 3-sphere with free Z_{2^n}-action. Let k be a Z_{2^n}-invariant knot in S. Let b be the denominator of the slope $s(k)$. In [10] it is shown that $\delta(K(k)) + (b^2-1)/16 \in \{0,1/2\} \subset Q/Z$ does not depend on any choice of a Z_{2^n}-invariant knot k, where $K(k)$ is the characteristic knot of k.

DEFINITION 7.1 ([10]). For any Z_{2^n}-invariant knot k in S

$$\delta(Z_{2^n},S) = \delta(K(k)) + (b^2-1)/16$$

in Q/Z.

$\delta(Z_{2^n},S)$ is an invariant of the equivariant homemorphism type of (Z_{2^n},S). Next note that the slope $s(k)$ of a knot k in $\tilde{S}$ with $[k] \neq 0$ in $H_1(\tilde{S};Z_2) = Z_2$ has a normal presentation of type $a/2^n b$ with odd a,b. Let (m,ℓ) be an m.ℓ. pair of $T(k)$ such that $[K(k)] = a[m] + 2^n b[\ell]$ in $H_1(\partial T(k);Z)$. Construct $W = \tilde{S} \times [-1,1] \cup D^2 \times D^2$ identifying $T(k) \times 1$ with $\partial D^2 \times D^2$ such that $m \times 1 \equiv p \times \partial D^2$, $\ell \times 1 \equiv D^2 \times q (p,q \in \partial D^2)$. $\partial W - \tilde{S} \times (-1)$ is a Z_2-homology 3-sphere. Denote it by $\tilde{S}(k,-2^n b/a)$. The knot $\bar{k} = 0 \times \partial D^2 \subset \tilde{S}(k,-2^n b/a)$ with orientation specified by $K(\bar{k}) = K(k)$ is called the <u>dual knot</u> of k in $\tilde{S}$ with respect to the normal presentation $a/2^n b$ of $s(k)$. Note that $s(\bar{k}) = -2^n b/a$. Let k' be another knot in $\tilde{S}$ with $[k'] \neq 0$ in $H_1(\tilde{S};Z_2) = Z_2$, and $a'/2^n b'$ be a normal presentation of $s(k')$.

LEMMA 7.2 ([10]). <u>For</u> $n = 1$ $\mu(\tilde{S}(k';-2b'/a')) = \mu(\tilde{S}(k;-2b/a))$ <u>if</u> $a'b' \equiv ab \pmod 4$. <u>For</u> $n = 2$ $\mu(\tilde{S}(k';-4b'/a')) = \mu(\tilde{S}(k;-4b/a))$ <u>if</u> $a'b'r'^2 \equiv abr^2 \pmod 8$, <u>where</u> r, r' <u>are the numbers of the components of the characteristic parallel links</u> P,P' <u>of</u> k,k', <u>respectively</u>.

It follows from Wall [19] that there is an element $e \in H_1(S;Z)$ of order 2^n such that $\varphi(e,e) = -u/2^n$, where $u = 1$ (if $n=1$), or 3 (if $n=2$), or ± 1 or ± 3 (if $n \geq 3$), and that the integer u is uniquely determined by the linking pairing φ on $\tilde{S}$. We shall use this integer u.

DEFINITION 7.3 ([10]. Let $n = 1$. Then we define $\mu(Z_2,S) = \mu(\tilde{S}(k;-2b/a))$ for any knot k in $\tilde{S}$ with $[k] \neq 0$ in $H_1(\tilde{S};Z_2)$ and any normal presentation $a/2b$ of $s(k)$ with $ab \equiv 1 \pmod 4$. Let $n = 2$. Then we define $\mu(Z_4,S) = \mu(\tilde{S}(k;-4b/a))$ for any knot k in $\tilde{S}$ with $[k] \neq 0$ in $H_1(\tilde{S};Z_2)$ and any normal presentation $a/4b$ of $s(k)$ such that $abr^2 \equiv u \pmod 8$, where r is the number of the components of the characteristic parallel link P of k.

THEOREM 7.4 ([10]). <u>We have the following congruences in</u> Q/Z:

(1) $\mu(S) = \alpha(Z_2,S)/16 + 2\mu(Z_2,S) + \sigma(Z_2,S)$ $(n = 1)$

(2) $\mu(S) = \alpha(Z_4,S)/16 + u/8 + 4\mu(Z_4,S) + \delta(Z_4,S)$ $(n = 2)$,

(3) $\mu(S) = \alpha(Z_{2^n},S)/16 + \nu(n,u)/18 + \delta(Z_{2^n},S)$ $(n \geq 3)$,

where

$$\nu(n,u) = \begin{cases} -u & (n=3,\ u=\pm 1) \\ u/|u| & (n=3,\ u=\pm 3) \\ 3u & (n>3,\ u=\pm 1) \\ -u & (n>3,\ u=\pm 3). \end{cases}$$

REMARK 7.5. In general, $\alpha(Z_2,S)/16$, $2\mu(Z_2,S)$ and $\delta(Z_2,S)$ have the forms $m_1/8$, $m_2/4$ and $m_3/2$ for integers m_1, m_2 and m_3, respectively. We can show that for any integers m_1, m_2 and m_3 there exists a Z_2-homology 3-sphere S with free Z_2-action such that $\alpha(Z_2,S)/16 = m_1/8$, $2\mu(Z_2,S) = m_2/4$ and $\delta(Z_2,S) = m_3/2$ in Q/Z. Therefore, the invariants $\alpha(Z_2,S)/16$ ($\in Q/Z$), $2\mu(Z_2,S)$ and $\delta(Z_2,S)$ appearing in Theorem 7.4(1) are mutually independent (cf. [10]).

Special cases of (Z_2,S) produce variations of the congruence of Theorem 7.4(1). For example, the congruence $\mu(S) = \alpha(Z_2,S)/16$ in Q/Z for any Z-homology 3-sphere S with free Z_2-action (cf. [20]) and the congruence $\mu(L(b,a)) = -\alpha(Z_2,\tilde{L}(b,a))/16$ in Q/Z (cf. [15]), where $\tilde{L}(b,a) = L(2b,a)$, are consequences of our congruence (cf. [10]).

THEOREM 7.6 ([10]). *For any two* Z_2-*homology 3-spheres* S,S' *with free* Z_2-*action,* $\delta(Z_2,S) = \delta(Z_2,S')$ *if and only if there exists a compact connected oriented 4-manifold* W *with free* Z_2-*action such that* $\partial(Z_2,W) = (Z_2,S'\cup -S)$ *and* $\tilde{W}$ *is spin. In this case, we have*

$$\mu(Z_2,S') - \mu(Z_2,S) + (\alpha(Z_2,S') - \alpha(Z_2,S))/32 = -\operatorname{sign} W/32$$

in Q/Z *for any such 4-manifold* W. *Further we can take* W *so that* $H_1(W;Z_2) = 0$.

Analogous arguments have been made by Cappell-Shaneson [3] and Fintushel-Stern [4] to find a fake P^4 and an exotic free involution on S^4.

8. THE δ-INVARIANT OF THE FIXED POINT SET.

Let S be a Z_2-homology 3-sphere with semi-free Z_n-action such that $L = F(Z_n,S)$ is a link. First we consider the case $n = 2$. Then by Smith theory L is a knot. Let $L = k$.

THEOREM 8.1 ([11]). *Let* $2a/b$ *be a normal presentation of the slope* $s(k)$. *Then*

$$\delta(k) = \delta(\tilde{k}) - (a/b - ab)/8$$

in Q/Z. *In particular, if* k *is flat, then* $\delta(k) = \delta(\tilde{k})$.

Next, to consider the case that n is an odd prime p, we remark the following:

LEMMA 8.2 [11]). L *is proper if and only if* $\tilde{L}$ *is proper.*

THEOREM 8.3 ([11]). *Let* L *be a proper link in* S *with components* $k_i, i = 1,2,\ldots,r$. *Let* n *be an odd prime* p. *Let* $2a_i/b_i$ *be a normal presentation of the slope* $s(k_i)$, $i = 1,2,\ldots,r$. *Then*

$$\delta(L) = p\delta(\underset{\sim}{L}) - \{(p^2-1)/8\}\,\Sigma_{i=1}^{r} a_i/b_i$$

in Q/Z.

Now we orient L suitably. Let $\underset{\sim}{L}$ have the induced orientation. For a normal presentation 2a/b of the slope s(L), we define $s^*(L) \equiv a/b \pmod 1$ and call it the *half-slope* of the link L. [This is well-defined, since b is odd.]

THEOREM 8.4 ([11]). *Assume* L *is proper*. *Let* n *be an odd prime* p. *Then we have*

$$\delta_o(L) = p\delta_o(\underset{\sim}{L}) - \{p^2-1)/8\}\, s^*(L)$$

in Q/Z. *In particular, if* L *is flat, then* $\delta_o(L) = \delta_o(\underset{\sim}{L})$.

BIBLIOGRAPHY

[1] L. Contreras-Caballero, Periodic transformations in homology 3-spheres and the Rohlin invariant, Low-Dimensional Topology, London Math. Soc. Lecture Note Series 48, Cambridge Univ. Press, 1982, 39-47.

[2] A. J. Casson and C.McA. Gordon, On slice knots in dimension three, Proc. Sympos. Pure Math. 32, Part 2, Amer. Math. Soc., Providence, R. I., 1978, 39-53.

[3] S. E. Cappell and J. L. Shaneson, Some new four-manifolds, Ann. of Math. 104 (1976), 61-72.

[4] R. Fintushel and R. J. Stern, An exotic free involution on S^4, Ann. of Math. 113 (1982), 357-365.

[5] C.McA. Gordon, Knots, homology spheres, and contractible 4-manifolds, Topology 14 (1975), 151-172.

[6] F. Hirzebruch and D. Zagier, The Atiyah-Singer Theorem and Elementary Number Theory, Math. Lec. Series 3, Publish or Perish, 1974.

[7] L. H. Kauffman and L. R. Taylor, Signature of Links, Trans. Amer. Math. Soc. 216 (1976), 351-365.

[8] A. Kawauchi, On 3-manifolds admitting orientation-reversing involutions, J. Math. Soc. Japan 33 (1981), 579-589.

[9] ————, Vanishing of the Rochlin invariants of some Z_2-homology 3-spheres, Proc. Amer. Math. Soc. 79 (1980), 303-308.

[10] ————, On the Rochlin invariants of Z_2-homology 3-spheres with cyclic actions, Jap. J. Math. (to appear).

[11] ————, On the Robertello invariants of proper links (preprint).

[12] ————, Dihedral coverings and the Rochlin invariant (in preparation).

[13] M. A. Kervaire and J. W. Milnor, On 2-spheres in a 4-manifold, Proc. Nat. Acad. Sci. U. S. A. 49 (1961), 1651-1657.

[14] K. Murasugi, On the signature of links, Topology 9 (1070), 283-298.

[15] W. D. Neumann and F. Raymond, Seifert manifolds, plumbing, μ-invariant and orientation-reversing maps, Algebraic and Geometric Topology, Lecture Notes in Math. 664, Springer-Verlag, 163-196.

[16] R. Robertello, An invariant of knot cobordism, Comm. Pure Appl. Math. 18 (1965), 543-555.

[17] V. A. Rochlin, New results in the theory of 4-dimensional manifolds, Dokl. Akad. Nauk. SSSR 84 (1952), 221-224 (Russian).

[18] L. C. Siebenmann, On vanishing of the Rohlin invariant and non-finitely amphicheiral homology 3-spheres, Topology Symposium, Siegen, 1979, Lecture Notes in Math. 788, Springer-Verlag, 172-222.

[19] C. T. C. Wall, Quadratic forms on finite groups, and related topics, Topology 2 (1964), 281-298.

[20] T. Yoshida, On the Browder-Livesay invariant of free involutions on homology 3-spheres, Math. J. Okayama Univ. 22 (1980), 91-106.

OSAKA CITY UNIVERSITY

Contemporary Mathematics
Volume 35, 1984

COBORDISM OF SATELLITE KNOTS

R. A. Litherland

0. INTRODUCTION

In this paper we study the Casson-Gordon invariants of satellite knots. Other cobordism invariants of such knots have been studied by various authors: the (ordinary) signature [23], the Tristram-Levine signatures [15] and the Milnor signatures [10]. In fact, in the last reference the Blanchfield pairing, and so (implicitly) the algebraic cobordism class, of a satellite is determined. See also [17]. The most striking feature to emerge is that the algebraic cobordism class of a satellite depends only on the constituent knots and the winding number. It is intuitively clear that this is not true of the geometric cobordism class, and one motivation for computing the Casson-Gordon invariants is to verify this intuition, which we do in Theorem 3.

We also apply our results to Kawauchi's group of $\tilde{H}$-cobordism classes of homology $S^1 \times S^2$'s [9]. The homomorphism from knot cobordism to algebraic cobordism factors through this group, and we show that the first factor has kernel containing a $\subset \mathbb{Z}^\infty$.

1. TERMINOLOGY, AND AN EXAMPLE

All manifolds will be oriented. Our statements may be interpreted in the PL or the smooth category, according to taste.

Let K be a knot in S^3. By an <u>axis</u> for K of <u>winding number</u> w we mean an unknotted simple closed curve A in $S^3 - K$ having linking number w with K. Let V be a solid torus complementary to a tubular neighborhood of A, with K contained in the interior of V. There is a preferred generator x for $H_1(V)$, specified by the condition $Lk(x,A) = +1$. For any knot C in S^3 there is an untwisted, orientation-preserving embedding $h:V \to S^3$ taking V onto a tubular neighborhood of C so that C represents $h_*(x)$ in $H_1(hV)$. We say that the knot $S = h(K)$ is a <u>satellite</u> of C with <u>orbit</u> K, <u>axis</u> A and <u>winding number</u> w. (In [17], the term "embellishment" is used where we use "orbit".)

The knot S is determined (up to isotopy) by C and the link $K \cup A$. We write $S = \mathscr{S}(K,C;A)$. We also denote the set of all satellites of C with orbit

K and winding number w by $\mathcal{S}_w(K,C)$. Thus we can rephrase the qualitative result on the algebraic cobordism class mentioned in the introduction by saying that (for given K, C and w) the image of $\mathcal{S}_w(K,C)$ in the algebraic cobordism group consists of a single element. We remark that the original examples of non-slice, algebraically slice knots [1] show that there are some $\mathcal{S}_0(K,C)$ containing knots from more than one cobordism class. Take K to be the $n(n+1)$-twist knot, for $n>1$, and C to be the torus knot of type $(n,n+1)$. Then $\mathcal{S}_0(K,C)$ contains the $n(n+1)$-twist double of C, which is slice (Casson, unpublished; see [16] for a proof). But $\mathcal{S}_0(K,C)$ also contains K itself, by taking a trivial axis A, i.e. one such that $K \cup A$ is a split link. It was proved by Casson and Gordon in [1] that K is not slice. If one disallows trivial satellites, it is still easy to construct an element of $\mathcal{S}_0(K,C)$ cobordant to K, by using for instance the axis shown in Fig. 1.

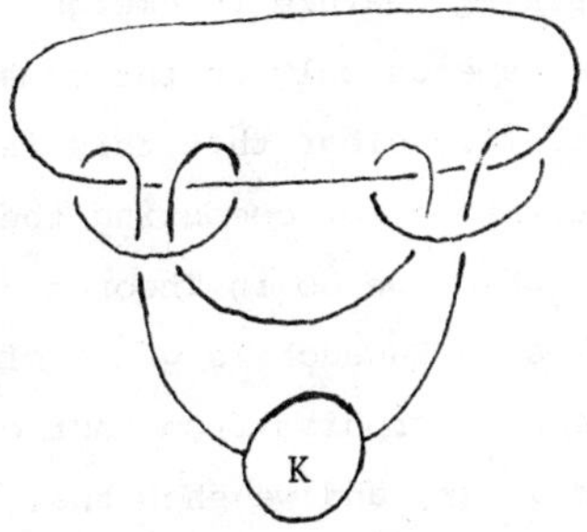

Figure 1

2. ALGEBRAIC COBORDISM

In some cases, our formula for the Casson-Gordon invariants of a satellite involves the algebraic cobordism class (Corollary 2) and for this it is necessary to put the two kinds of invariant on a similar footing. This we shall do in this section by giving a "Casson-Gordon type" definition of the algebraic cobordism class. That this can be done is probably well-known to the experts.

We denote the ring $\mathbb{Q}[t,t^{-1}]$ of Laurent polynomials with rational coefficients by Γ, and its field of fractions $\mathbb{Q}(t)$ by $Q\Gamma$. The involution $f(t) \to f(t^{-1})$ of Γ or $Q\Gamma$ will be denoted by J. The (multiplicative) infinite cyclic group is written as C_∞, and we assume that a generator t is fixed once for all. Let (M,φ) be a closed 3-manifold over C_∞; that is, M is a closed 3-manifold and φ is a homomorphism $H_1(M) \to C_\infty$. Suppose that M has the rational homology of $S^1 \times S^2$. Since $\Omega_3(K(C_\infty,1)) = 0$, we have $(M,\varphi) = \partial(W,\psi)$ for some compact 4-manifold (W,ψ) over C_∞.

REMARK. We do not assume that φ is onto, and it may be that $\varphi = 0$. In that case we always take W so that $H_1(M;\mathbb{Q}) \to H_1(W;\mathbb{Q})$ is injective and $\psi = 0$. Note that the injectivity is automatic if $\varphi \neq 0$. Here $\varphi = 0$ means that $\varphi(x) = 1$ for all x; in general we write $\mathrm{Hom}(A,B)$ additively even when B is multiplicative.

We define twisted homology and a twisted intersection pairing just as in [1]: if $\tilde{W}$ is the infinite cyclic covering of W determined by ψ, then $C_*(\tilde{W};\mathbb{Q})$ is a complex of Γ-modules, and we set

$$C_*^t(W;Q\Gamma) = C_*(\tilde{W};\mathbb{Q}) \otimes_\Gamma Q\Gamma .$$

The homology of this complex is written $H_*^t(W;Q\Gamma)$. There is a pairing $H_2^t(W;Q\Gamma) \times H_2^t(W;Q\Gamma) \to Q\Gamma$, Hermitian with respect to J, given at the chain level by

$$\langle x \otimes f, y \otimes g\rangle = fg^J \sum_{i=-\infty}^{\infty} (x \cdot t^i y)t^i , \; x,y \in C_2(\tilde{W};\mathbb{Q}), \; f,g \in Q\Gamma .$$

Here $x \cdot t^i y$ is the ordinary intersection number. The pairing is non-singular and so represents an element $t(W) = t_\psi(W)$ of the Witt group $W(Q\Gamma;J)$. For $\varphi \neq 0$ this is because the Milnor exact sequence for the infinite cyclic covering of (M,φ) [19] shows that $H_1^t(M;Q\Gamma) = 0$ (even if φ is not onto). If $\varphi = 0$, we have $H_1^t(M;Q\Gamma) = H_1(M;\mathbb{Q}) \otimes_{\mathbb{Q}} Q\Gamma$ and $H_1^t(W;Q\Gamma) = H_1(W;\mathbb{Q}) \otimes_{\mathbb{Q}} Q\Gamma$, so that $H_1^t(M;Q\Gamma) \to H_1^t(W;Q\Gamma)$ is injective. (See the remark above.) The ordinary intersection form on $H_2(W;\mathbb{Q})$ is also non-singular; let $t_0(W)$ be its image in $W(Q\Gamma;J)$. Define

$$\alpha(M,\varphi) = t(W) - t_0(W) \in W(Q\Gamma;J) .$$

The proof that this is well-defined is just like that for the Casson-Gordon invariants (for which see [1]).

REMARK. If $\varphi = 0$ then $\psi = 0$ so $t_\psi(W) = t_0(W)$. Hence $\alpha(M,0) = 0$. Our reason for being careful about the "trivial" case is that we have to deal with 4-manifolds over C_∞ of the form $(W_1,\psi_1) \cup (W_2,\psi_2)$ where one of ψ_1,ψ_2 may be zero.

Now let K be a knot in S^3. The manifold M_K obtained by 0-framed surgery along K comes with a preferred isomorphism $\varphi_K\colon H_1(M_K) \to C_\infty$ determined by the orientations of S^3 and K. We write $\alpha(K)$ or α_K for $\alpha(M_K,\varphi_K)$. It is not hard to see that $\alpha_K = 0$ if K is slice; together with Theorem 1 below this shows that α induces a homorphism

$$\alpha\colon \mathscr{C}^{3,1} \longrightarrow W(Q\Gamma;J)$$

where $\mathscr{C}^{3,1}$ is the knot cobordism group.

We now indicate why this invariant is equivalent to the algebraic cobordism class. Let $(M_K,\varphi_K) = \partial(W,\psi)$. Under the boundary homomorphism

$$\partial : W(Q\Gamma;J) \longrightarrow W(Q\Gamma/\Gamma;J)$$

of the localization exact sequence for $W(Q\Gamma;J)$, $t_0(W)$ dies and $t_\psi(W)$ is sent to minus the Witt class of the Blanchfield pairing on $H_1(\tilde{M}_K;\mathbb{Q})$, where $\tilde{M}_K$ is the infinite cyclic covering of M_K. According to Trotter [25] the isomorphism class of the Blanchfield pairing determines the rational S-equivalence class of a Seifert matrix V for K. It follows that $\partial\alpha_K$ determines the (rational) Witt class of V, which is to say, the algebraic cobordism class of K. (Recall that the homorphism from the integral algebraic cobordism group $W_S(\mathbb{Z})$ to the corresponding rational group $W_S(\mathbb{Q})$ is injective [12]). In the other direction we have:

PROPOSITION 1. If V is a Seifert matrix for K then $\alpha(K)$ is represented by the matrix $(1-t)V+(1-t^{-1})V^T$.

Here V^T is the transpose of V. Actually it can be shown that there is an isomorphism $W(Q\Gamma;J) \cong W_S(\mathbb{Q}) \oplus W(\mathbb{Q})$ under which $\alpha(K)$ corresponds to $([V],0)$; an account of this will be found in Appendix A.

Before giving the proof we describe an additivity property that we shall need frequently. Recall that if W_1 and W_2 are 4-manifolds with $\partial W_1 \cong -\partial W_2$ and W is the closed 4-manifold $W_1 \cup_\partial W_2$ then the signature of W is given by

$$\text{sign}(W) = \text{sign}(W_1) + \text{sign}(W_2)$$

(Novikov additivity). However, if W_1 and W_2 are glued along only part of their boundaries, this may fail. This situation was studied by Wall [28]. Suppose that $\partial W_1 \cong M_1 \cup M_0$ and $\partial W_2 \cong M_2 \cup -M_0$, where for $i=1,2$, M_i and M_0 are 3-manifolds meeting only in their common boundary, and let $W = W_1 \cup_{M_0} W_2$. Let $F = \partial M_0 = \partial M_1 = \partial M_2$, and let $A_i = \ker(H_1(F;\mathbb{Q}) \to H_1(M_i;\mathbb{Q}))$ for $i=0,1,2$. Wall showed that the failure of additivity is measured by the signature of a bilinear form on

$$\frac{A_i \cap (A_j+A_k)}{(A_i \cap A_j)+(A_i \cap A_k)}, \quad \{i,j,k\} = \{1,2,3\}.$$

In fact, this holds on the level of the Witt classes of the intersection forms, and for twisted homology as well. We shall need only the special case in which at least two of A_0, A_1, A_2 are equal, when additivity does hold. We shall refer to this as Wall additivity.

We remark that Wall's result can be derived from Novikov additivity (or rather, the easy generalization to the case of gluing along some whole boundary components) by decomposing $W_1 \cup W_2$ into three pieces as indicated in Fig. 2. The "correction term" is the intersection form of the Θ-shaped piece.

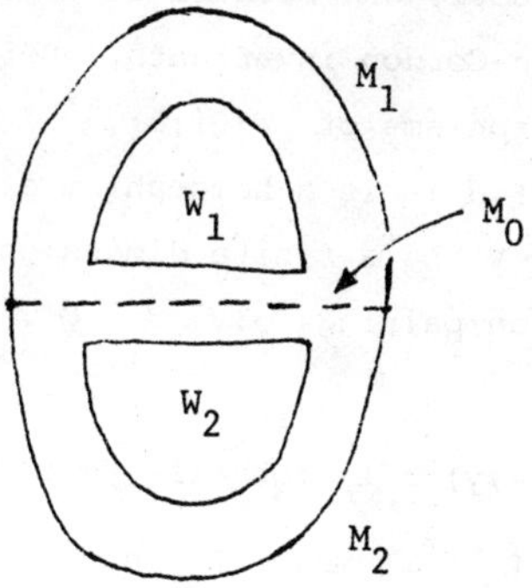

Figure 2

PROOF OF PROPOSITION 1. Let F be a spanning surface for K giving the Seifert matrix V. Let $\hat{F} \subset D^4$ be obtained by pushing $\operatorname{int} F$ into $\operatorname{int} D^4$. Set $W_1 = D^4 - (\hat{F} \times \operatorname{int} D^2)$ and let $\psi_1 : H_1(W_1) \longrightarrow C_\infty$ be given by linking number with $\hat{F}$. We shall show by a cut-and-paste construction of the infinite cyclic cover $\tilde{W}_1$ that $t(W_1)$ is represented by $(1-t)V + (1-t^{-1})V^T$; c.f. [8] Section 5, [27] Section 5. Let $F \times [-1,1]$ be a bicollar of F in S^3. Cutting W_1 open along the trace of the push yields D^4, and the faces exposed by the cut are $F \times [-1,-1/2]$ and $F \times [1/2,1]$. Thus $\tilde{W}_1$ is obtained by taking copies $t^i D^4$ of D^4 for $i \in \mathbb{Z}$ and identifying $t^i(F \times [-1,-1/2])$ with $t^{i+1}(F \times [1/2,1])$. It follows that $H_2(\tilde{W}_1;\mathbb{Q}) \cong H_1(F;\mathbb{Q}) \otimes_{\mathbb{Q}} \Gamma$. If x is a cycle on F, let $C_\pm x$ be the cone on $x \times \pm 1$ in D^4, and let

$$Sx = (C_- x) - t(C_+ x) ,$$

a 2-cycle in $\tilde{W}_1$. This represents the element of $H_2(\tilde{W}_1;\mathbb{Q})$ corresponding to $[x] \otimes 1$. If Θ is the Seifert form on $H_1(F;\mathbb{Q})$ it follows easily that

$$\langle Sx, Sy \rangle = (1-t)\Theta([x],[y]) + (1-t^{-1})\Theta([y],[x]) .$$

This gives the result claimed. Note also that the ordinary intersection form on $H_2(W_1;\mathbb{Q})$ is identically zero.

Now let H be a solid handlebody with $\partial H = F \cup E^2$, where E^2 is a disc, and let $W = W_1 \cup_{F \times \partial D^2} H \times \partial D^2$. Then ψ_1 extends to $\psi : H_1(W) \longrightarrow C_\infty$, and $\partial(W,\psi) = (M_K, \varphi_K)$. We have $H_2^t(H \times \partial D^2; \mathbb{Q}\Gamma) = 0$, and the intersection form on $H_2(H \times \partial D^2;\mathbb{Q})$ is identically zero. Finally, Wall additivity applies to $W = W_1 \quad H \times \partial D^2$ in both ordinary and twisted homology (in this case, all three kernels are the same) to complete the proof. ▯

Our next aim is to determine the algebraic cobordism class of a satellite knot. Although this follows from the results on the Blanchfield pairing in [10] and [17], we include a proof because it seems particularly simple from the point of view introduced above, and because it serves as a model for the proof of our theorem on the Casson-Gordon invariants. Before stating the result, we must discuss induced homomorphisms of $W(Q\Gamma;J)$.

Suppose $f:(\Gamma,J) \longrightarrow (F,J')$ is a homorphism of rings-with-involution, where F is a field. Let V be a finite dimensional vector space over $Q\Gamma$ with a non-singular Hermitian pairing $\varphi: V\times V \to Q\Gamma$. For any Γ-lattice L in V let

$$L^{\#} = \{x \in V \mid \varphi(x,y) \in \Gamma \text{ for all } y \in L\} .$$

We can choose L so that $L \leq L^{\#}$. Make F into a Γ-module via f. Then we have an induced Hermitian pairing φ_f on the F-vector space $L \otimes_\Gamma F$, namely

$$\varphi_f(x\otimes\alpha,\ y\otimes\beta) = \alpha\beta^{J'} f\varphi(x,y),\ x,y \in L,\ \alpha,\beta \in F .$$

In general, φ_f may be singular. However, one can show that the set of elements of $W(Q\Gamma;J)$ represented by (V,φ) for which L can be so chosen as to make φ_f non-singular forms a subgroup $\mathrm{Def}(f_*)$, say, and that the assignment

$$[V,\varphi] \longrightarrow [L\otimes_\Gamma F,\varphi_f]$$

is a well-defined homomorphism $f_*:\mathrm{Def}(f_*) \longrightarrow W(F;J')$. If $\alpha \in \mathrm{Def}(f_*)$ is represented by a matrix A over Γ then $f_*(\alpha)$ is represented by $f(A)$, provided this is non-singular. Clearly $\mathrm{Def}(f_*) = W(Q\Gamma;J)$ if f is injective; the same is true if $f(t) = 1$. (See Appendix A, where $\mathrm{Def}(f_*)$ is determined.)

If $f(t) = x$ and $\alpha \in \mathrm{Def}(f_*)$ we shall also write $\alpha[x]$ instead of $f_*(\alpha)$. In particular, we shall sometimes write an element α of $W(Q\Gamma;J)$ as $\alpha[t]$. For $\alpha \in W(Q\Gamma;J)$ and $\zeta \in S^1$, $\alpha[\zeta] \in W(\mathbb{C},\text{ conjugation})$ is defined for all but finitely many ζ. We define $\sigma_\zeta(\alpha)$ to be the signature of $\alpha[\zeta]$ whenever possible, and elsewhere to be the average of the one-sided limits. (These signatures were introduced in a slightly different context by Casson and Gordon [1].) This gives a step function $\sigma_\bullet(\alpha): S^1 \to \mathbb{Z}$, all of whose discontinuities occur at points where $\alpha[\zeta]$ is not defined. In view of Proposition 1, for a knot K, $\sigma_\zeta(\alpha_K)$ is equal to the Tristram-Levine signature of K at ζ, except perhaps at finitely many points of S^1. We abbreviate $\sigma_\zeta(\alpha_K)$ to $\sigma_K(\zeta)$.

THEOREM 1. *Let S be a satellite of C with orbit K and winding number w. Then*

$$\alpha_S[t] = \alpha_K[t] + \alpha_C[t^w] .$$

We shall need the following lemma.

LEMMA 1. *Let (M,φ) be a closed 3-manifold over C_∞, and suppose that M has the rational homology of $S^1 \times S^2$. Then*

$$\alpha(M,w\varphi)[t] = \alpha(M,\varphi)[t^w]$$

for any integer w. In particular,

$$\alpha(M,\varphi)[1] = 0 .$$

PROOF OF THEOREM 1, ASSUMING LEMMA 1. Let (W_K,ψ_K) and (W_C,ψ_C) be compact 4-manifolds over C_∞ such that

$$\partial(W_K,\psi_K) = (M_K,\varphi_K)$$

$$\text{and} \quad \partial(W_C,\psi_C) = (M_C,\varphi_C) .$$

Let $U_C \subset M_C$ be the surgery solid torus, and let $U_K \subset M_K$ be a small tubular neighborhood of the axis of K used to form S. We can construct

$$(1) \qquad (W_S,\psi_S) = (W_K,\psi_K) \bigcup_{U_K \equiv U_C} (W_C,w\psi_C)$$

so that $\partial(W_S,\psi_S) = (M_S,\varphi_S)$. Wall additivity applies to (1) in both ordinary and twisted homology, since the kernels corresponding to the two pieces of ∂W_C are the same. Therefore

$$\begin{aligned}\alpha(M_S,\varphi_S) &= \alpha(M_K,\varphi_K) + \alpha(M_C,w\varphi_C)\\ &= \alpha(M_K,\varphi_K) + \alpha(M_C,\varphi_C)[t^w]\end{aligned}$$

by Lemma 1. This is the assertion of the theorem. ||||

PROOF OF LEMMA 1. Let $(M,\varphi) = \partial(W,\psi)$. First suppose that $w \neq 0$. Let $\tilde{W}_\psi$, $\tilde{W}_{w\psi}$ be the infinite cyclic coverings of W determined by ψ, $w\psi$ respectively. Since $t_o(W)[t^w] = t_o(W)$, we need to show that

$$t_{w\psi}(W)[t] = t_\psi(W)[t^w] .$$

But this is easy, since $\tilde{W}_{w\psi}$ consists of $|w|$ copies of $\tilde{W}_\psi$; t permutes these copies cyclically, with t^w acting on each copy like t on $\tilde{W}_\psi$.

It remains to prove that $\alpha(M,\varphi)[1] = 0$. We may assume that φ is onto, since if $\varphi = 0$ there is nothing to prove, and otherwise we can use the previous case to replace φ by an epimorphism. Since $Q\Gamma$ is torsion-free over the PID Γ,

$$H_*^t(W;Q\Gamma) = H_*^t(W;\Gamma) \otimes_\Gamma Q\Gamma .$$

The intersection pairing on $H_*^t(W;Q\Gamma)$ comes from a pairing $H_2^t(W;\Gamma) \times H_2^t(W;\Gamma) \longrightarrow \Gamma$ by tensoring with $Q\Gamma$. This pairing induces one on

$$L = H_2^t(W;\Gamma)/\Gamma\text{-torsion},$$

and L is a Γ-lattice in $H_2^t(W;Q\Gamma)$. Now $H_*^t(W;\Gamma)$ is just the ordinary rational homology of the infinite cyclic covering $\tilde{W}$, regarded as a Γ-module. By doing surgery on W we may assume that $\pi_1(W) \cong C_\infty$. Then $\tilde{W}$ is simply connected. Also $H_1(M) \longrightarrow H_1(W)$ is onto, so $H_3(W;\mathbb{Q}) = 0$. From the exact sequence for the covering $\tilde{W} \to W$ [19] we therefore have

$$0 \to H_2^t(W;\Gamma) \xrightarrow{1-t} H_2^t(W;\Gamma) \longrightarrow H_2(W;\mathbb{Q}) \longrightarrow 0$$

exact. Thus $H_2(W;\mathbb{Q}) \cong H_2^t(W;\Gamma) \otimes_\Gamma \mathbb{Q}$; note that the intersection form on $H_2(W;\mathbb{Q})$ comes from that on $H_2^t(W;\Gamma)$ by tensoring with $\mathbb{Q}$. Also $H_2^t(W;\Gamma)$ has no (1-t)-torsion, so $H_2^t(W;\Gamma) \otimes_\Gamma \mathbb{Q} \cong L \otimes_\Gamma \mathbb{Q}$. Therefore $t_\psi(W)[1] = t_o(W)$, proving that $\alpha(M,\varphi)[1] = 0$. ⁞⁞

<u>REMARK</u>. We could have defined $\alpha(M,\varphi)$ without the assumption that M has the rational homology of $S^1 \times S^2$, since this was only used to ensure non-singularity of the intersection forms and any Hermitian form over a field gives rise to a non-singular form on the quotient by its radical. However, the case $w = 0$ of Lemma 1 would no longer hold. For instance, if M is the manifold obtained by 0-surgery on both components of the Whitehead link, and if φ sends the meridians of the components to t and 1 respectively, then $\alpha(M,\varphi)$ is the rank 1 form $\langle 1\rangle$.

There is a related result that we shall need. In [6], Section 13 an invariant $\sigma(M,\varphi) \in \mathbb{Q}$ is associated to any closed 3-manifold (M,φ) over C_m, the finite cyclic group of m'th roots of unity.

<u>LEMMA 2</u>. <u>Let</u> K <u>be a knot</u>, <u>let</u> m <u>be a power of a prime</u>, <u>and let</u> $g: C_\infty \to C_m$ <u>be a homomorphism</u>. <u>Let</u> $\zeta = g(t)$. <u>Then</u> $\alpha_K[\zeta] \in W(\mathbb{C};$ conjugation) <u>is defined and</u>

$$\sigma_K(\zeta) = \sigma(M_K, g\varphi_K) .$$

<u>PROOF</u>. That $\alpha_K[\zeta]$ is defined follows from Proposition 1 and the fact that, if Δ is the Alexander polynomial of K, $\Delta(\zeta) \neq 0$ since ζ is a prime-power root of unity. (See [24], Lemma 2.5.) The second assertion follows from the identification of $\sigma(M,\varphi)$ with an eigenspace signature ([1], pp. 5-6), Lemma 3.1 of [2] and Proposition 1. ⁞⁞

We conclude this section with a remark on surgery presentations of a knot K. In [22], Rolfsen shows how such a description gives rise to a presentation matrix $A(t)$ for the Alexander invariant of K. This matrix satisfies $A(t)^T = A(t^{-1})$, and $A(1)$ is a diagonal matrix with diagonal entries ± 1. It is evident from the definition that $A(t)$ represents the intersection form on $H_2^t(W;Q\Gamma)$ for a certain 4-manifold (W,ψ) over C_∞ with $\partial(W,\psi) = (M_K,\varphi_K)$; W is obtained by attaching 2-handles to B^4 as specified by the surgery description, and removing a neighborhood of an unknotted 2-disc spanning K. The intersection form on $H_2(W;\mathbb{Q})$ is represented by $A(1)$, and so

$$\alpha_K = [A(t)] - [A(1)]$$

where [..] denotes Witt class. (That the Tristram-Levine signatures of K can be computed from $A(t)$ was observed in [14], Section 12.)

3. THE CASSON-GORDON INVARIANTS...

In this section we set out our notation for these invariants and prove a technical lemma. If (M,φ) is a closed 3-manifold over $C_m \times C_\infty$, there is an invariant $\tau(M,\varphi) \in W(\mathbb{C}(t),J) \otimes_{\mathbb{Z}} \mathbb{Q}$ defined as in [6], Section 13. (The involution J of $\mathbb{C}(t)$ is given by $f(t)^J = \bar{f}(t^{-1})$.) Let K be a knot in S^3. Let $L = L_{K,n}$ be the n-fold branched cyclic covering of K, and let $M = M_{K,n}$ be obtained from L by 0-surgery along the lift $\tilde{K}$ of K. Thus $M_{K,1}$ is the manifold M_K of the last section, and $M_{K,n}$ is an n-fold cyclic covering of $M_{K,1}$. We identify $H_1(M)$ with $H_1(L) \oplus C_\infty$, where the generator t of the C_∞ summand is represented by a meridian of $\tilde{K}$. Let $Ch_n(K) = \mathrm{Hom}(H_1(L), \mathbb{C}^*)$ be the group of characters of $H_1(L)$. We shall always assume that n is a power of a prime, so that L is a rational homology sphere, and any $\chi \in Ch_n(K)$ takes values in C_m for some m. Define $\chi^+ : H_1(M) \longrightarrow C_m \times C_\infty$ by

$$\chi^+(x,y) = (\chi(x),y) \ , \ x \in H_1(L),\ y \in C_\infty \ ,$$

and set

$$\tau(K,\chi) = \tau(M,\chi^+) \ .$$

(In [6] it is assumed that m is the order of χ, but it is easy to see that the choice of m is immaterial.)

Linking number gives a non-singular symmetric pairing $Lk: H_1(L) \times H_1(L) \longrightarrow \mathbb{Q}/\mathbb{Z}$, which yields another such pairing on $Ch_n(K)$, also denoted by Lk. We shall always think of $Ch_n(K)$ as carrying this form, and $-Ch_n(K)$ will denote $Ch_n(K)$ with the form $-Lk$. The theorem of Casson and Gordon ([1],[6]) is that if K is slice then (for any prime power n) $Ch_n(K)$ has a metaboliser $\mathcal{M}$ such that $\tau(K,\chi) = 0$ for all $\chi \in \mathcal{M}$ of prime-power order. (A metaboliser is a subgroup which is equal to its orthogonal complement.) The case $n = 2$ has received most attention; in Section 5 we shall have need of odd primes. We remark that the above makes sense for $n = 1$; in this case there is only one character, 0, and $\tau(K,0)$ is the image $\alpha_K^{\mathbb{C}}$ of α_K in $W(\mathbb{C}(t),J) \otimes \mathbb{Q}$. (If 0_n is the zero of $Ch_n(K)$, $\tau(K,0_n)$ may be non-zero for some $n > 1$ as well; it is determined by the algebraic cobordism class of K. See Appendix B.)

If K is a composite knot $K_1 \# K_2$, $Ch_n(K)$ may be identified with the orthogonal direct sum $Ch_n(K_1) \oplus Ch_n(K_2)$, and then

$$\tau(K,\chi_1 \oplus \chi_2) = \tau(K_1,\chi_1) + \tau(K_2,\chi_2) \ .$$

This is proved in [5], Proposition 3.2 for the case $n = 2$; it is a special case of Corollary 1 below.

Induced homomorphisms on $W(\mathbb{C}(t),J)$ are defined just as for $\mathbb{Q}(t)$ in Section 2.

LEMMA 3. Let $K \subset S^3$ be a knot. Let $\chi \in Ch_n(K)$ take values in C_m, and suppose that m and n are both powers of primes. Let $x \in C_m \times C_\infty \subset \mathbb{C}(t)$. If x has finite order suppose further that $n = 1$. Then $\tau(K,\chi)[x]$ is defined and

$$\tau(K,\chi)[x] = \tau(M_{K,n}, f\chi^+)$$

where $f: C_m \times C_\infty \longrightarrow C_m \times C_\infty$ is defined by $f(y) = y$ for $y \in C_m$ and $f(t) = x$.

PROOF. By $\tau(K,\chi)[x]$ we mean the image of $\tau(K,\chi)$ under the homomorphism $W(\mathbb{C}(t),J) \otimes \mathbb{Q} \longrightarrow W(\mathbb{C}(t),J) \otimes \mathbb{Q}$ induced by

$$\hat{f}: \mathbb{C}[t,t^{-1}] \longrightarrow \mathbb{C}(t);$$

$$\hat{f}(\alpha) = \alpha \ , \ \alpha \in \mathbb{C},$$

$$\hat{f}(t) = x \ .$$

If x has infinite order, $\hat{f}$ is injective, and the proof is similar to the case $w \neq 0$ of Lemma 1. We leave this case to the reader. Suppose then that x has finite order, i.e. $x \in C_m$. By assumption $n = 1$, and so $\chi^+: H_1(M_K) \longrightarrow C_m \times C_\infty$ is given by $\chi^+(z) = (1, \varphi(z))$ where $\varphi = \varphi_K: H_1(M_K) \rightarrow C_\infty$ is the canonical isomorphism of Section 2. Define $g: C_\infty \rightarrow C_m$ by $g(t) = x$. Choose a compact 4-manifold (W,ψ) over C_m such that $\partial(W,\psi) = r(M_K, g\varphi)$ for some $r > 0$. Let $j: C_m \rightarrow C_m \times C_\infty$ be the inclusion. Then $\partial(W, j\psi) = r(M, f\chi^+)$ over $C_m \times C_\infty$. Now the $C_m \times C_\infty$ covering of W determined by $j\psi$ is a trivial infinite cyclic covering of the C_m covering determined by ψ. It follows that $\tau(M, f\chi^+)$ lies in the image of $W(\mathbb{C}, \text{conjugation}) \otimes \mathbb{Q}$ and has signature $\sigma(M_K, g\varphi)$. On the other hand, $\tau(K,\chi) = \alpha_K^{\mathbb{C}}$. By Lemma 2, $\alpha_K[x] \in W(\mathbb{C}, \text{conjugation})$ is defined and has signature $\sigma(M_K, g\varphi)$. The result follows (remembering that signature gives an isomorphism $W(\mathbb{C}, \text{conjugation}) \cong \mathbb{Z}$). ▥

4. ....OF SATELLITE KNOTS

First we identify the character groups of a satellite.

LEMMA 4. Let S be a satellite of C with orbit K and winding number w. Let n be a power of a prime, and set $h = \text{h.c.f.}(n,w)$ and $k = n/h$. Then

$$Ch_n(S) \cong Ch_n(K) \oplus h(Ch_k(C)) \ ,$$

with the linking form on $Ch_n(S)$ being the orthogonal sum of the forms on the summands.

PROOF. We prove the corresponding statement for the dual groups $H_1(L_{-,n})$.

Let A be the axis of K used to construct S. In $L_{K,n}$, A is covered by h curves $\tilde{A}_1,\ldots,\tilde{A}_h$. Let $U_1,\ldots,U_h$ be disjoint tubular neighborhoods of the $\tilde{A}_i$. Let $L^u_{C,k}$ be the unbranched k-fold cyclic covering of S^3 less an open tubular neighborhood of C. Let $X = L_{K,n} - \mathrm{int}(U_1 \cup\cdots\cup U_h)$. We can construct $L_{S,n}$ by gluing a copy of $L^u_{C,k}$ to X along each ∂U_i via an appropriate gluing map. A Mayer-Vietoris argument gives

$$H_1(L_{S,n}) \cong H_1(L_{K,n}) \oplus h(H_1(L_{C,k})) .$$

It remains to determine the linking form. First let x belong to the i'th copy of $H_1(L_{C,k})$. Then x can be represented by a cycle ξ which lies in $L^u_{C,k}$ and represents a torsion element of $H_1(L^u_{C,k}) \cong H_1(L_{C,k}) \oplus \mathbb{Z}$. Let D be a 2-chain in $L^u_{C,k}$ with $\partial D = r\xi$, $r > 0$. For any $y \in H_1(L_{S,n})$ represented by a cycle η,

$$Lk(x,y) \equiv \frac{1}{r}(D\cdot\eta) \mod \mathbb{Z} .$$

It follows that each $H_1(L_{C,k})$ is an orthogonal summand and inherits the correct form from $H_1(L_{S,n})$.

Finally, if $x \in H_1(L_{K,n})$, represent x by a cycle ξ missing $U_1 \cup\cdots\cup U_n$, and let D be a 2-chain in $L_{K,n}$ with $\partial D = r\xi$, $r > 0$, and transverse to the $\tilde{A}_i$. We get a 2-chain D' in $L_{S,n}$ with $\partial D' = r\xi$ by replacing each component of $D \cap U_i$ with a 2-chain in a copy of $L_{C,k}$. It follows that for $y \in H_1(L_{K,n})$ we get the same value of $Lk(x,y)$ by working in $L_{K,n}$ or $L_{S,n}$.

THEOREM 2. *Let S be a satellite of C with orbit K, axis A and winding number w. Let n be a power of a prime,* $h = \mathrm{h.c.f.}(n,w)$, $k = n/h$. *Let $x_i \in H_1(L_{K,n})$ be represented by the* i'th *lift of* A, $i = 1,\ldots,h$. *Identify* $Ch_n(S)$ *with* $Ch_n(K) \oplus h(Ch_k(C))$ *as in Lemma 4. Let* $\chi_S = (\chi_K,\chi_1,\ldots,\chi_h) \in Ch_n(S)$ *be of prime-power order. Then*

$$\tau(S,\chi_S)[t] = \tau(K,\chi_K)[t] + \sum_{i=1}^{h} \tau(C,\chi_i)[\chi_K(x_i)t^{w/h}] .$$

NOTES. (1) The terms under the summation sign are defined by Lemma 3, since either $w \neq 0$ or $\chi_i \in Ch_1(C)$.

(2) It is understood that the i'th lift of A corresponds to the i'th copy of $Ch_k(C)$.

The two extreme cases of this theorem embodied in the following corollaries are probably of most interest.

COROLLARY 1. *In the situation of Theorem 2, suppose that n is coprime to* w. Then $Ch_n(S) \cong Ch_n(K) \oplus Ch_n(C)$ *and for* $\chi_S = (\chi_K, \chi_C) \in Ch_n(S)$ *of prime-power order we have*

$$\tau(S,\chi_S) = \tau(K,\chi_K) + \tau(C,\chi_C)[t^w] .$$

If $w=1$ this is always the case, and the Casson-Gordon invariants are the same as those of $K \# C$. In general, if n is coprime to w, the invariants associated to Ch_n cannot distinguish between elements of $\mathcal{S}_w(K,C)$.

PROOF. In the situation of Theorem 2 we always have $x_1 + \cdots + x_h = 0$. In the present case, $h=1$ and so $x_1 = 0$. ▒

COROLLARY 2. *In the situation of Theorem 2, suppose that n divides w. Then $Ch_n(S) \equiv Ch_n(K)$ and for $\chi \in Ch_n(S)$ of prime-power order we have*

$$\tau(S,\chi)[t] = \tau(K,\chi)[t] + \sum_{i=1}^{n} \alpha_C[\chi(x_i)t^{w/n}] .$$

PROOF. Here $Ch_k(C) = Ch_1(C) = \{0\}$ and $\tau(C,0)$ is the image of α_C in $W(\mathbb{C}(t),J) \otimes \mathbb{Q}$. ▒

PROOF OF THEOREM 2. Let m be the order of χ_S, and regard $\chi_K, \chi_1, \ldots, \chi_h$ as taking values in C_m. Take compact 4-manifolds (W_K,ψ_K), $(W_1,\psi_1), \ldots, (W_h,\psi_h)$ over $C_m \times C_\infty$ such that

$$\partial(W_K,\psi_K) = r(M_{K,n}, \chi_K^+)$$

$$\partial(W_i,\psi_i) = r(M_{C,k}, \chi_i^+) \quad , \quad i=1,\ldots,h ,$$

for some $r>0$. Note that $\chi_K^+(x_i) = (\chi_K(x_i), t^{w/h})$ for $i=1,\ldots,h$. Recall that $M_{C,k}$ is obtained from $L_{C,k}$ by 0-surgery on the lift of C. Let $U \subset M_{C,k}$ be the surgery solid torus, and let $V_i \subset M_{K,n}$ be a tubular neighborhood of the i'th lift of A, with $V_1,\ldots,V_h$ disjoint. For $i=1,\ldots,h$ and $j=1,\ldots,r$, let U_{ij} be the copy of U in the j'th boundary component of W_i, and let V_{ij} be the copy of V_i in the j'th boundary component of W_K. We can construct

$$W_S = W_K \cup \bigcup_{i=1}^{h} W_i \tag{2}$$

where each U_{ij} is glued to V_{ij}, so that $\partial W_S = rM_{S,n}$. Define $f_i : C_m \times C_\infty \longrightarrow C_m \times C_\infty$ by $f_i(y) = y$ for $y \in C_m$ and $f_i(t) = (\chi_K(x_i), t^{w/h})$. Then $\psi_K | H_1(V_{ij})$ and $f_i\psi_i | H_1(U_{ij})$ agree under the identification, so we can combine ψ_K and the $f_i\psi_i$ to give $\psi_S : H_1(W_S) \longrightarrow C_m \times C_\infty$, and then

$$\partial(W_S,\psi_S) = r(M_S,\chi_S^+) .$$

Wall additivity applies to (2) in both ordinary and twisted homology, since the kernels corresponding to the pieces of the ∂W_i are the same. Therefore

$$\tau(M_{S,n}, \chi_S^+) = \tau(M_{K,n}, \chi_K^+) + \sum_{i=1}^{h} \tau(M_{C,k}, f_i\chi_i^+) .$$

But $\tau(M_{C,k}, f_i\chi_i^+) = \tau(C,\chi_i)[\chi_K(x_i)t^{w/h}]$ by Lemma 3, completing the proof. ▒

5. SATELLITES WHICH ARE INDEPENDENT IN $\mathscr{C}^{3,1}$.

In this section we shall prove:

THEOREM 3. *Let* $w \in \mathbb{Z}$ *be given. There exist knots* K *and* C *such that* $\mathscr{S}_w(K,C)$ *contains* $r \geq 2$ *knots representing linearly independent elements of* $\mathscr{C}^{3,1}$ *provided that at least* $2^{r-1}-1$ *distinct primes divide* w.

Note in particular that for $w \neq \pm 1$ the condition holds with $r = 2$. Further, if $w = 0$ then r can be any integer. On the other hand, we know that the algebraic cobordism class and all the Casson-Gordon invariants are consistent with a positive answer to the following:

QUESTION. Is every member of $\mathscr{S}_1(K,C)$ cobordant to $K \# C$?

Combining Theorems 1 and 3 we have the following result of Jiang [7].

COROLLARY 3. *The cobordism group of algebraically slice knots contains a free abelian group of infinite rank.*

We are going to use Corollary 2. To get any mileage from this we need axes for a knot K whose lifts represent different elements of $H_1(L_{K,n})$ for some factor n of the winding number. This motivates the following definitions. Let A be an axis for K of winding number w, and let n divide w. Let $L = L_{K,n}$, and let $x_1,\dots,x_n \in H_1(L)$ be represented by the lifts of A. We say that A is *n-trivial* if $x_i = 0$ for $i = 1,\dots,n$, and that A is *n-generating* if $x_1,\dots,x_n$ generate $H_1(L)$. Note that, for any factor n' of n, if A is n-trivial then it is n'-trivial and also (since $H_1(L_{K,n}) \longrightarrow H_1(L_{K,n'})$ is onto) if A is n-generating then it is n'-generating. Given K, w and n it is easy to find an n-trivial axis for K of winding number w. In order for K to have an n-generating axis it is necessary for $H_1(L)$ to be cyclic as a $\mathbb{Z}[t,t^{-1}]$-module. It is not hard to see that this is also sufficient, but we shall not make use of this, as we now give specific examples of n-generating axes. If K is a torus knot of type (p,q) then K has two obvious "standard" axes A_p and A_q of winding numbers p and q respectively. (The satellite $\mathscr{S}(K,C;A_q)$ is the (p,q)-cable of C.)

LEMMA 5. *Let* K *be a torus knot of type* (p,q), *where* $p,q > 1$, *and let* n *be a factor of* q.

(i) $H_1(L_{K,n}) \cong (n-1)(\mathbb{Z}/p)$.

(ii) *The standard axis* A_q *is n-generating.*

PROOF. Let $L = L_{K,n}$, and let $\tilde{A}_1,\dots,\tilde{A}_n$ be the lifts of A_q to L, representing $x_1,\dots,x_n \in H_1(L)$. Let $\tilde{A}'$ be the single lift of A_p. Let U be a small tubular neighborhood of K, and let $L^u \subset L$ be the n-fold cyclic covering of $S^3 - \text{int}(U)$. Let $y_1,\dots,y_n, y' \in H_1(L^u)$ be represented by $\tilde{A}_1,\dots,\tilde{A}_n, \tilde{A}'$, respectively. The decomposition of $S^3 - \text{int}(U)$ into two solid tori with cores A_p and A_q lifts to a decomposition of L^u into $(n+1)$ solid tori. The Mayer-Vietoris sequence yields

$$H_1(L^u) = \langle y_1,\dots,y_n,y' \mid py_i = (q/n)y',\ i=1,\dots,n\rangle$$

$$\cong \mathbb{Z} \oplus (n-1)(\mathbb{Z}/p)$$

since p and q/n are coprime. Hence $H_1(L) \cong (n-1)(\mathbb{Z}/p)$. Since y_i maps to x_i and y' to 0 in $H_1(L)$, $x_1,\dots,x_n$ generate $H_1(L)$. ▦

PROPOSITION 2. *Let* K *be a torus knot of type* (p,q), *where* $p,q>1$. *Let* $q=q'q''$ *be a factorization of* q *into coprime integers, and let* w *be a multiple of* q. *Then there is an axis for* K *of winding number* w *which is* q'*-generating and* q''*-trivial.*

PROOF. Let $L' = L_{K,q'}$ and $L'' = L_{K,q''}$. Let A''' be the standard axis for K of winding number q; it is both q'-generating and q''-generating by Lemma 5(ii). Modify A''' by winding it locally around K as in Fig. 3 to give an axis A'' of winding number q'. Because the modification

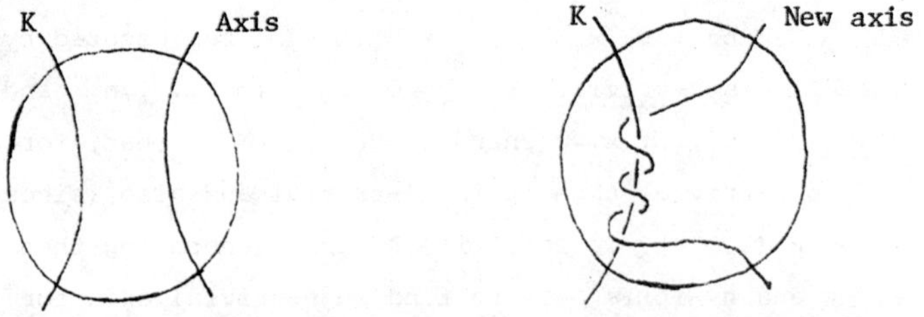

Figure 3

lifts to L' to give an isotopy of each lift of A''', A'' is also q'-generating. On the other hand, A'' is covered by a single, null-homologous curve in L''. Now let A' be a $(1,q'')$-cable about A''; A' is an axis of winding number q. Since each lift of A' to L' is homologous to q'' times the corresponding lift of A'', A' is still q'-generating (by Lemma 5(i)). Each lift of A' to L'' is homologous to the single curve over A'', and hence to zero; i.e. A' is q''-trivial. Finally use the modification of Fig. 3 again to give an axis A of winding number w. This time the modification lifts to both L' and L'', so A has the desired properties. ▦

The whole process is illustrated in Fig. 4 for the case $p=5$, $q=6$, $q'=3$, $q''=2$ and $w=12$.

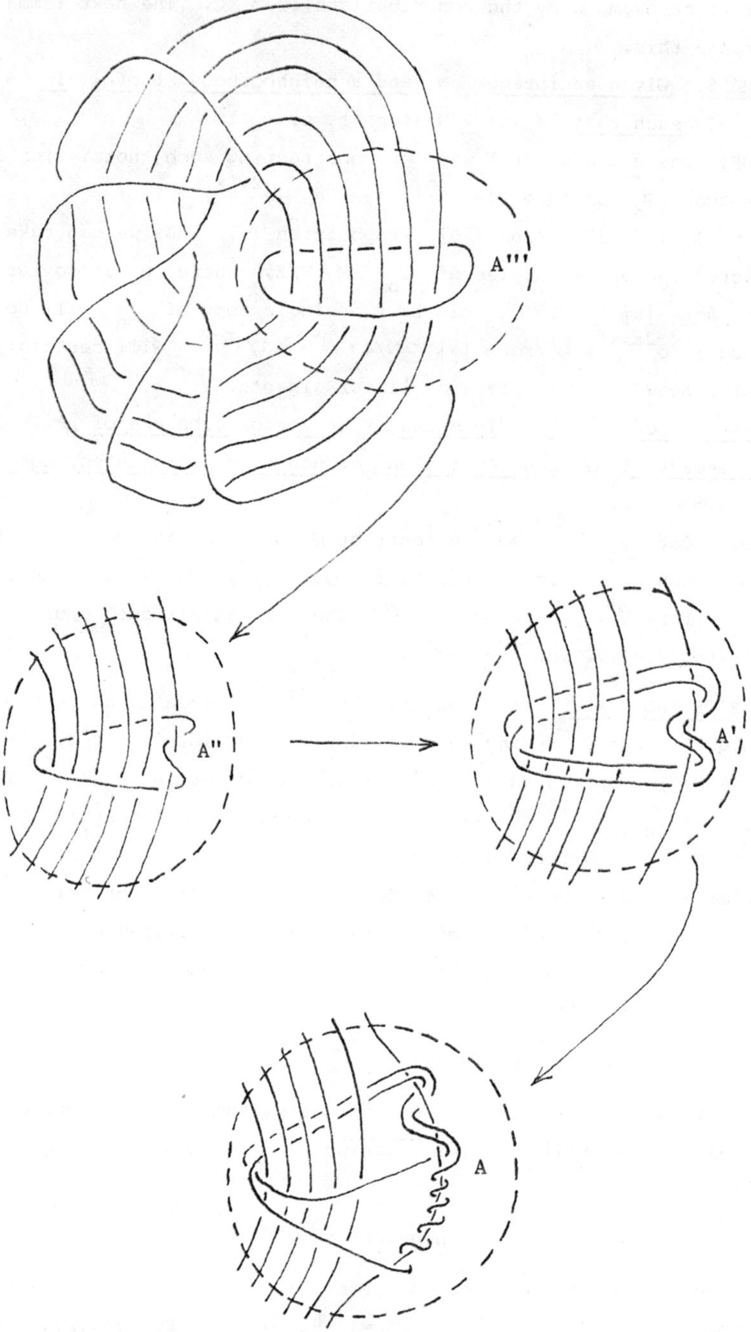

Figure 4

In order to avoid calculating $\tau(K,\chi)$ when applying Corollary 2, we want that term to be swamped by the contributions from C. The next lemma enables us to arrange this.

LEMMA 6. Given an integer N and a neighborhood U of 1 in S^1, there is a knot C such that $\sigma_C(\zeta) \geq N$ for $\zeta \notin U$.

PROOF. There are several ways of constructing such knots. For instance, the torus knot T_n of type $(-2,3^n)$ has $\sigma_{T_n}(e^{2\pi i x}) \geq 2$ for $1/2 \cdot 3^n < x < 1 - 1/2 \cdot 3^n$. (See [15], Proposition 1.) Thus we can take C to be the connected sum of M copies of T_n, $M \geq N/2$, where n is so large that $e^{2\pi i x} \in U$ for $|x| \leq 1/2 \cdot 3^n$. In fact a single copy of T_n will do if n is large enough; $e^{2\pi i x} \in U$ for $|x| \leq N/4 \cdot 3^{n-1} + 1/3^{n-1}$ will certainly suffice. ▯

We also need a simple piece of linear algebra.

LEMMA 7. Let F be a field and V a vector subspace of F^ν. Suppose that any element of V has fewer than μ non-zero coordinates, for some fixed μ. Then $\dim V < \mu$.

PROOF. Let $\pi_i : V \to F$ be the restriction of the i'th coordinate function, $i=1,\ldots,\nu$. Then $\pi_1,\ldots,\pi_\nu$ generate the dual V^*, so we can pick out a basis $\pi_{i_1},\ldots,\pi_{i_d}$ for V^*, where $d = \dim V$. There is an element v of V such that $\pi_{i_j}(v) = 1$ for $j=1,\ldots,d$, so $d < \mu$. ▯

PROOF OF THEOREM 3. First observe that if K is a torus knot of type (p,q) $(p,q > 1)$ then, for any knot C and any multiple w of q, each satellite $S \in \mathscr{S}_w(K,C)$ maps to an element of infinite order in $W_S(\mathbb{Z})$. This is because $\sigma_S(\zeta) = \sigma_K(\zeta)$ if $\zeta^w = 1$ (by [15] Theorem 2, or Theorem 1 above), while $\sigma_K(e^{2\pi i/q}) \neq 0$ (by [15], Proposition 1).

Now let w and r be as in the theorem. Choose distinct primes q_I dividing w, one for each non-empty subset I of $\{1,\ldots,r-1\}$, and let p be any prime distinct from all the q_I. Let $q = \prod_I q_I$, and let K be the torus knot of type (p,q). Set

$$N = \max|\sigma_1 \tau(K,\chi)|$$

where χ ranges over $\bigcup_I Ch_{q_I}(K)$. For the definition of $\sigma_\zeta \tau$, where $\tau \in W(\mathbb{C}(t),J) \otimes \mathbb{Q}$, see [1] or [6], Section 13.) By Lemma 6, we can take a knot C so that

$$\sigma_C(\zeta) > 4N \quad \text{whenever} \quad \zeta^p = 1 \ , \ \zeta \neq 1 \ .$$

It remains to choose r axes for K. For $i=1,\ldots,r$ we have a factorization $q = q_i' q_i''$ where $q_i' = \prod_{I: i \in I} q_I$ and $q_i'' = \prod_{I: i \notin I} q_I$. By Proposition 2 there is an axis A_i for K of winding number w which is q_i'-generating and q_i''-trivial. Let $S_i = \mathscr{S}(K,C;A_i)$. We claim that $S_1,\ldots,S_r$ represent linearly independent elements of $\mathscr{C}^{3,1}$.

Suppose they do not. Any non-trivial relation can be written in the form

(3) $$\sum_{k=1}^{\ell} S_{i_k} \sim \sum_{k=1}^{\ell'} S_{j_k}$$

where "~" means "is cobordant to", $\ell > 0$, $1 \leq i_k, j_k \leq r$ and $i_k \neq j_{k'}$ for any k, k'. Since all the S_i map to the same element of infinite order in $W_S(\mathbb{Z})$, we have $\ell' = \ell$. Let $I = \{i_k | k=1,\ldots \ell\}$; I is non-empty and we may assume (by switching the sides of (3) if necessary) that $r \notin I$. Set $n = q_I$. Note that A_i is n-generating if $i \in I$, and n-trivial if not.

We shall use the τ-invariants associated to Ch_n. Let

$$\mathscr{W} = \sum_{k=1}^{\ell} Ch_n(S_{i_k}) \oplus \sum_{k=1}^{\ell} (-Ch_n(S_{j_k})) .$$

Since $n|w$, Lemma 4 gives

$$Ch_n(S_{i_k}) = Ch_n(S_{j_k}) = Ch_n(K) ,$$

which is isomorphic to $(n-1)(\mathbb{Z}/p)$ by Lemma 5(i). In particular, $\mathscr{W}$ has prime-power order, so the relation (3) implies that $\mathscr{W}$ has a metaboliser $\mathscr{M}$ such that

(4) $$\sum_{k=1}^{\ell} \tau(S_{i_k}, \chi_{i_k}) = \sum_{k=1}^{\ell} \tau(S_{j_k}, \chi_{j_k})$$

whenever $(\chi_{i_k}) \oplus (\chi_{j_k}) \in \mathscr{M}$. For any S_i and $\chi \in Ch_n(K)$, Corollary 2 gives

$$\sigma_1 \tau(S_i, \chi) = \sigma_1 \tau(K, \chi) + \sum_{s=1}^{n} \sigma_C(\chi x_s^{(i)})$$

where $x_1^{(i)}, \ldots, x_n^{(i)} \in H_1(L_{K,n})$ are represented by the lifts of A_i. If $i \notin I$ this simplifies to

$$\sigma_1 \tau(S_i, \chi) = \sigma_1 \tau(K, \chi)$$

(A_i being n-trivial). Therefore (4) gives

(5) $$\sum_{k=1}^{\ell} \sum_{s=1}^{n} \sigma_C(\chi_{i_k} x_s^{(i_k)}) = \sum_{k=1}^{\ell} (\sigma_1 \tau(K, \chi_{j_k}) - \sigma_1 \tau(K, \chi_{i_k})) \leq 2\ell N$$

for $(\chi_{i_k}) \oplus (\chi_{j_k}) \in \mathscr{M}$. Now for any $\chi \in Ch_n(K)$ and $x \in H_1(L_{K,n})$ we have $\chi(x)^p = 1$, so either $\sigma_C(\chi(x)) > 4N$ or $\chi(x) = 1$. Thus (5) implies that there are fewer than $\frac{1}{2}\ell$ values of k for which some $\chi_{i_k} x_s^{(i_k)} \neq 1$. Since A_{i_k} is n-generating

$$\chi_{i_k} x_s^{(i_k)} = 1 \text{ for } s=1,\ldots,n \implies \chi_{i_k} = 0 .$$

Thus we have shown that if $(\chi_{i_k}) \oplus (\chi_{j_k}) \in \mathscr{M}$ then χ_{i_k} is non-zero for fewer than $\frac{1}{2}\ell$ values of k.

Write $\mathcal{W}$ as $\mathcal{W}_1 \oplus \mathcal{W}_2$, where $\mathcal{W}_1 = \Sigma\, Ch_n(S_{i_k})$ and $\mathcal{W}_2 = \Sigma(-Ch_n(S_{j_k}))$. Let $\mathcal{M}_1$ be the projection of $\mathcal{M}$ into $\mathcal{W}_1$ and let $\mathcal{M}_2 = \mathcal{M} \cap \mathcal{W}_2$, so that $\mathcal{M}_1 \cong \mathcal{M}/\mathcal{M}_2$. Identify $Ch_n(K)$ with $(n-1)(\mathbb{Z}/p)$, and hence identify $\mathcal{W}_1$ with $\ell(n-1)(\mathbb{Z}/p)$. Under this identification, each element of $\mathcal{M}_1$ has fewer than $\frac{1}{2}\ell(n-1)$ non-zero coordinates, so by Lemma 7

$$\dim_{\mathbb{Z}/p} \mathcal{M}_1 < \tfrac{1}{2}\ell(n-1) = \tfrac{1}{2} \dim_{\mathbb{Z}/p} \mathcal{W}_1$$

or

$$|\mathcal{M}_1| < |\mathcal{W}_1|^{\frac{1}{2}}$$

Hence

$$|\mathcal{M}_2| > |\mathcal{W}_2|^{\frac{1}{2}} .$$

But $\mathcal{M}_2$ is a self-annihilating subgroup of $\mathcal{W}_2$, so this is impossible. This contradiction establishes the independence of $S_1, \ldots, S_r$ in $\mathcal{C}^{3,1}$. ▦

6. HOMOLOGY HANDLES.

In [9], Kawauchi defined a group $\Omega(S^1 \times S^2)$ which fits into a commutative diagram

$$\begin{array}{ccc} \mathcal{C}^{3,1} & \xrightarrow{e} & \Omega(S^1 \times S^2) \\ & \searrow \quad \swarrow & \\ & W_S(\mathbb{Z}) & \end{array}$$

and which may be described as follows. A <u>homology handle</u> is a 3-manifold with the integral homology of $S^1 \times S^2$, and a <u>special homology handle</u> is a homology handle M together with an isomorphism $\varphi : H_1(M) \to C_\infty$. (In comparing this with Kawauchi's definition, remember that all our manifolds are oriented.) Two such objects (M_1, φ_1), (M_2, φ_2) are $\tilde{H}$-<u>cobordant</u> if there is a compact 4-manifold (W, ψ) over C_∞ such that

$$\partial(W, \psi) = (M_1, \varphi_1) \cup (-M_2, \varphi_2)$$

and

$$H_*^t(W; Q\Gamma) = 0 .$$

The elements of $\Omega(S^1 \times S^2)$ are the $\tilde{H}$-cobordism classes of special homology handles. The group operation will be described later; the zero is represented by $S^1 \times S^2$ (with either φ), and the inverse of $[M, \varphi]$ is $[-M, \varphi]$. The homomorphism $e: \mathcal{C}^{3,1} \longrightarrow \Omega(S^1 \times S^2)$ is given by $[K] \longrightarrow [M_K, \varphi_K]$.

Now if (M_1, φ_1) and (M_2, φ_2) are $\tilde{H}$-cobordant, it is clear that

$$\alpha(M_1, \varphi_1) - \alpha(M_2, \varphi_2) = \alpha(M_1, \varphi_1)[1] - \alpha(M_2, \varphi_2)[1] ,$$

which is zero by Lemma 1. Therefore we have a commutative diagram

$$\begin{array}{ccc} \mathscr{C}^{3,1} & \xrightarrow{e} & \Omega(S^1 \times S^2) \\ & \searrow^{\alpha} \quad \swarrow_{\alpha} & \\ & W(Q\Gamma, J) & \end{array}$$

It can be shown (see below) that $\alpha:\Omega(S^1 \times S^2) \longrightarrow W(Q\Gamma,J)$ is a homomorphism. It can also be shown, exactly as in Section 2, that it is equivalent to the homomorphism $\Omega(S^1 \times S^2) \longrightarrow W_S(\mathbb{Z})$ defined by Kawauchi.

We now define operations on $\Omega(S^1 \times S^2)$ analogous to forming satellite knots. Let (M_1,φ_1) and (M_2,φ_2) be special homology handles, and let w be an integer. Choose embeddings

$$j_i: S^1 \times D^2 \longrightarrow M_i \quad , \quad i=1,2,$$

such that j_1 is orientation preserving, j_2 is orientation reversing and

$$\varphi_1 j_{1*}[S^1 \times 0] = t^w \ ,$$

$$\varphi_2 j_{2*}[S^1 \times 0] = t \ .$$

Define a special homology handle (M,φ) by

$$M = (M_1 - j_1(S^1 \times \operatorname{int} D^2)) \underset{\substack{j_1(x) \equiv j_2(x) \\ x \in S^1 \times \partial D^2}}{\cup} (M_2 - j_2(S^1 \times \operatorname{int} D^2)) \ ,$$

$$\varphi | H_1(M_1) = \varphi_1 \ , \ \varphi | H_1(M_2) = w\varphi_2 \ .$$

We write

$$(M,\varphi) = (M_1,\varphi_1) \circ_w (M_2,\varphi_2) \ .$$

The case $w=1$ was considered by Kawauchi, who called it <u>circle union</u>. This construction is not well-defined, since in general (M,φ) depends on the choice of j_1 and j_2. However, we have:

<u>PROPOSITION 3</u>. (Kawauchi [9] in case $w=1$). <u>Fix</u> $w \neq 0$. <u>Then the</u> $\tilde{H}$-<u>cobordism class of</u> $(M_1,\varphi_1) \circ_w (M_2,\varphi_2)$ <u>depends only on those of</u> (M_1,φ_1) <u>and</u> (M_2,φ_2).

<u>PROOF</u>. (cf [9], Lemma 1.6) For $i = 1,2$ let (W_i,ψ_i) be an $\tilde{H}$-cobordism from (M_i,φ_i) to (M_i',φ_i'), and let

$$j_i: S^1 \times D^2 \longrightarrow M_i \quad , \quad j_i': S^1 \times D^2 \longrightarrow M_i'$$

be embeddings as in the definition of $\circ_w$. Let

$$(M,\varphi) = (M_1,\varphi_1) \circ_w (M_2,\varphi_2) \ ,$$

$$(M',\varphi') = (M_1',\varphi_1') \circ_w (M_2',\varphi_2') \ ,$$

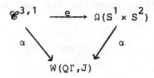

constructed using j_i, j_i' respectively. We must show that (M,φ) is $\tilde{H}$-cobordant to (M',φ'). Let W be obtained from the disjoint union $W_1 \sqcup W_2$ by making the identifications

$$j_1(x) \equiv j_2(x), \quad j_1'(x) \equiv j_2'(x), \quad x \in S^1 \times D^2$$

Define $\psi: H_1(W) \longrightarrow C_\infty$ by

$$\psi | H_1(W_1) = \psi_1, \quad \psi | H_1(W_2) = w\psi_2 .$$

Then

$$\partial(W,\psi) = (M,\varphi) \cup (-M',\varphi') .$$

That $H_*^t(W,Q\Gamma) = 0$ follows from the Mayer-Vietoris sequence. (Note that, expanding the notation for twisted homology to indicate the twisting homorphism, $H_*^t((W_2,w\psi_2);Q\Gamma) = 0$ because the infinite cyclic covering of W_2 determined by $w\psi_2$ consists of $|w|$ copies of the one determined by ψ_2. This fails for $w=0$; $H_*^t((W_2,0);Q\Gamma) = H_*(W_2;\mathbb{Q}) \otimes_{\mathbb{Q}} Q\Gamma$.) ⁞⁞⁞

Thus we have, for each $w \neq 0$, a well-defined binary operation O_w in $\Omega(S^1 \times S^2)$. The addition in $\Omega(S^1 \times S^2)$ is O_1.

The next result if immediate from the definitions.

PROPOSITION 4. *Let* S *be a satellite of the knot* C *with orbit* K *and winding number* w. *Then*

$$(M_S,\varphi_S) = (M_K,\varphi_K)\ O_w\ (M_C,\varphi_C) . \quad ⁞⁞⁞$$

Observe that the proof of Theorem 1 actually shows that if $(M,\varphi) = (M_1,\varphi_1)\ O_w\ (M_2,\varphi_2)$ then

$$\alpha(M,\varphi)[t] = \alpha(M_1,\varphi_1)[t] + \alpha(M_2,\varphi_2)[t^w]$$

(even for $w=0$). The case $w=1$ justifies our earlier claim that α induces a homomorphism $\Omega(S^1 \times S^2) \longrightarrow W(Q\Gamma;J)$.

From Propositions 3 and 4 we see that for any knots K and C and any non-zero integer w, $\mathscr{S}_w(K,C)$ maps to a single element of $\Omega(S^1 \times S^2)$. Together with Theorem 3 this gives:

THEOREM 4. *The kernel of the homorphism* $e:\mathscr{C}^{3,1} \longrightarrow \Omega(S^1 \times S^2)$ *contains a free abelian group of infinite rank.* ⁞⁞⁞

Of course, whether or not this is really stronger than Corollary 3 depends on whether $\Omega(S^1 \times S^2) \longrightarrow W_S(\mathbb{Z})$ has kernel or not, which is an open question.

APPENDIX A. The Witt group of Hermitian forms over a function field.

A0. PREAMBLE. Let k be a field of characteristic different from 2, provided with an involution $x \mapsto \bar{x}$ (which may be trivial). Let the involution J of the rational function field $k(t)$ be given by $f(t)^J = \bar{f}(t^{-1})$. We study the Witt group $W(k(t),J)$ of Hermitian forms over $k(t)$, and prove a version of Milnor's exact sequence for the Witt group of symmetric forms [20]. If the involution of k is trivial then Milnor's proof can be carried over virtually unchanged, because the fixed field of $k(t)$ is $k(t+t^{-1})$. In the general case this does not work; our approach was suggested by Trotter's proof that the Blanchfield pairing of a knot determines its Seifert form. As special cases we obtain the isomorphism $W(\mathbb{Q}(t),J) \cong W_S(\mathbb{Q}) \oplus W(\mathbb{Q})$ mentioned in Section 2, and a computation of the group $W(\mathbb{C}(t),J)$ used by Casson and Gordon.

At the heart of Trotter's proof is his "trace" function $\mathbb{Q}(t)/Q[t,t^{-1}] \longrightarrow \mathbb{Q}$ ([25],[26]). We use a slightly different function which has a nice geometric interpretation, given in Section A4. We also give a new proof of a result of Matumoto [18].

A1. GENERALITIES

We recall the definitions and elementary results on Witt groups that we need. General references for this are [21] and [3]. In what follows, Γ is a PID with involution J, $\operatorname{char}\Gamma \neq 2$, $Q\Gamma$ is the field of fractions and $\Gamma^{\bullet}$ is the group of units. Also $(k,-)$ is a field-with-involution, $\operatorname{char} k \neq 2$, and $k[t,t^{-1}]$ is given the involution $f(t)^J = \bar{f}(t^{-1})$. The case of trivial involution is allowed.

Let N be a Γ-module with an involution, also called J, such that $(\gamma n)^J = \gamma^J n^J$ for $\gamma \in \Gamma$, $n \in N$. If M is a Γ-module and $\varphi: M \times M \to N$ is a sesquilinear pairing (linear in the first variable, conjugate linear in the second) we denote by φ^* the pairing $\varphi^*(x,y) = \varphi(y,x)^J$. Let $u \in \Gamma$ have form $uu^J = 1$. If $\varphi = u\varphi^*$ then φ is said to be u-<u>Hermitian</u>. We have the adjoint homomorphism

$$\operatorname{ad}\varphi: M \longrightarrow \overline{\operatorname{Hom}}(M,N) \quad ;$$
$$\operatorname{ad}\varphi(x)(y) = \varphi(x,y) \quad ,$$

where $\overline{\operatorname{Hom}}$ denotes the module of conjugate-linear maps. We say that φ is <u>non-singular</u> if $\operatorname{ad}\varphi$ and $\operatorname{ad}\varphi^*$ are isomorphisms.

We need four kinds of Witt groups; we now list the objects from which they are formed.

(a) $W_u(\Gamma,J)$ for $u \in \Gamma$ of norm 1. The objects are u-<u>Hermitian</u> <u>spaces</u> (V,φ); i.e. V is a finitely-generated free Γ-module and $\varphi: V \times V \to \Gamma$ is u-Hermitian and non-singular.

(b) $W(Q\Gamma/\Gamma,J)$. <u>Torsion forms</u> (M,φ); i.e. M is a finitely-generated torsion Γ-module and $\varphi: M \times M \longrightarrow Q\Gamma/\Gamma$ is Hermitian and non-singular.

(c) $W_-(C_\infty;k,-)$. <u>Skew-isometric structures</u> (V,φ,t); i.e. (V,φ) is a skew-Hermitian space over k and t is an isometry of (V,φ). Here a metaboliser is required to be t-invariant.

(d) $W_S(k,-)$. <u>Seifert forms</u> $(V,\mathcal{S})$, i.e. V is a finite dimensional k-vector space and $\mathcal{S}: V \times V \longrightarrow k$ is sesquilinear and non-singular, with $\mathcal{S} - \mathcal{S}^*$ also non-singular.

REMARK. The last group was defined by Levine [12] for the case of trivial involution, with two differences. Namely, he did not require that $\mathscr{S}$ be non-singular, but did require $\mathscr{S}+\mathscr{S}^*$ to be so. As to the first, it is shown in [13] that every Witt class has a non-singular representative. It will follow from Section A5 below that the second change does not affect the Witt group either.

If $v \in \Gamma$ and $\varphi: V \times V \longrightarrow \Gamma$ is u-Hermitian then $v\varphi$ is (uv/v^J)-Hermitian, and we get an isomorphism $v: W_u(\Gamma, J) \longrightarrow W_{uv/v^J}(\Gamma, J)$. By Hilbert's Theorem 90 this gives:

PROPOSITION A1.

$$W_u(k,-) \cong \begin{cases} 0 & \text{if } - \text{ is trivial and } u = -1 ; \\ W(k,-) & \text{otherwise} . \end{cases}$$

Since $W(k,-)$ is generated by rank 1 forms, the same is true of $W_u(k,-)$. For $\gamma \in \Gamma^{\bullet}$ with $\gamma = u\gamma^J$ we denote the corresponding rank 1 form, or its class in $W_u(\Gamma, J)$, by $\langle\gamma\rangle$.

The following remarks apply to both (b) and (c). In case (b), let (M, φ) be a torsion form over Γ. In case (c), let (V, φ, t) be a skew-isometric structure over k. Set $\Gamma = k[t, t^{-1}]$, and think of (V, t) as a finitely generated torsion Γ-module M. If we restrict M to be -torsion, where is a symmetric ($= {}^J$) prime ideal of Γ, we obtain Witt groups $W(Q\Gamma/\Gamma, J)$ and $W_-(C_\infty; k, -)$. For any prime ideal of Γ let M denote the -torsion part of M. Then M is the orthogonal sum of M for $= {}^J$ and $M \oplus M_J$ for $\neq {}^J$, and the latter summands are metabolic. This gives canonical isomorphisms

(A1) $\quad W(Q\Gamma/\Gamma, J) \cong \bigoplus_{= J} W(Q\Gamma/\Gamma, J) \quad ,$

(A2) $\quad W_-(C_\infty; k, -) = \bigoplus_{= J} W_-(C_\infty; k, -) \quad .$

We denote by $W^0(Q\Gamma/\Gamma, J)$, $W^0_-(C_\infty; k, -)$ the sum of those terms on the right-hand side of (A1), (A2) (respectively) for which $\neq (t-1)$.

One can further show that any element of $W(Q\Gamma/\Gamma, J)$ or $W_-(C_\infty; k, -)$ can be represented by a form for which $M = 0$. In case (c) it follows that $W_-(C_\infty; k, -)_{(t-1)}$ can be identified with $W_-(k, -)$. In particular, if the involution of k is trivial, $W_-(C_\infty; k) = W^0_-(C_\infty; k)$. In case (b) it follows that the summands of (A1) are (non-canonically) isomorphic to groups of type (a). For let π be a generator of . Then $\pi^J = u\pi$ for some $u \in \Gamma^{\bullet}$. Since φ takes values in $(\frac{1}{\pi}\Gamma)/\Gamma$, we have a pairing

$$\pi^J\varphi: M \times M \longrightarrow \Gamma/ \quad .$$

Regarding M as a $\Gamma/$ - vector space, this is a non-singular, $\hat{u}$-Hermitian form, where $\hat{}$ denotes reduction modulo . It can be shown that $(M,\varphi) \longrightarrow (M,\pi^J\varphi)$ (for $M = O$) induces an isomorphism.

$$\pi_*^J: W(Q\Gamma/\Gamma,J) \longrightarrow W_{\hat{u}}(\Gamma/\ ,J) .$$

Finally, we recall the "localization" exact sequence and deal more fully with the subject of induced homorphisms treated in Section 2. The sequence is

$$O \longrightarrow W(\Gamma,J) \xrightarrow{i_*} W(Q\Gamma,J) \xrightarrow{\partial} W(Q\Gamma/\Gamma,J) .$$

The first homomorphism is induced by the inclusion $\Gamma \longrightarrow Q\Gamma$. (Homomorphisms induced by injections cause no problems, of course.) The definition of the second runs as follows. Let (V,φ) be a Hermitian space over $Q\Gamma$. If L is a Γ-lattice in V with $L \leq L^\#$ (definition as in Section 2) then $\partial[V,\varphi]$ is represented by the torsion form

$$\varphi': L^\#/L \times L^\#/L \longrightarrow Q\Gamma/\Gamma ;$$

$$\varphi'([x],[y]) \equiv \varphi(x,y) \mod \Gamma .$$

We denote the composition of ∂ with the projection to $W(Q\Gamma/\Gamma,J)$ by ∂ , and if π is a generator of we write ∂_π for $\pi_*^J\partial : W(Q\Gamma,J) \longrightarrow W_{\hat{u}}(\Gamma/\ ,J)$. Now $W(Q\Gamma,J)$ is generated by $<\gamma>$ for $\gamma \in Q\Gamma^{\bullet}$, $\gamma = \gamma^J$. We may assume that $\gamma \in \Gamma$ and that either γ is coprime to π or $\gamma = \pi\delta$ with δ coprime to π. Computation shows that

$$\partial_\pi <\gamma> = O \qquad \text{for } \gamma \text{ coprime to } \pi ;$$

$$\partial_\pi <\pi\delta> = <\hat{\delta}> \qquad \text{for } \delta \text{ coprime to } \pi .$$

(cf. [21], Chapter IV, (1.2).)

If $L \leq L^\#$ we obtain a Hermitian pairing φ on the $\Gamma/$ vector space $L = L \otimes_\Gamma \Gamma/$ by setting

$$\varphi\ (x \otimes \alpha, y \otimes \beta) = \alpha\beta^J\varphi(x,y)^{\wedge} , \quad x,y \in L,\ \alpha,\ \beta \in \Gamma/ \ .$$

PROPOSITION A2. <u>We can choose</u> L <u>so that</u> φ <u>is non-singular if and only if</u> $\partial\ [V,\varphi] = O$.

PROOF. Identifying $L^\#$ with $\overline{\mathrm{Hom}}_\Gamma(L,\Gamma)$ we have an exact sequence

$$O \rightarrow L \xrightarrow{\mathrm{ad}\,\varphi} \overline{\mathrm{Hom}}_\Gamma(L,\Gamma) \longrightarrow L^\#/L \longrightarrow O .$$

Tensoring with $\Gamma/$ gives an exact sequence

$$\mathrm{Tor}_\Gamma(L^\#/L,\Gamma/\ ) \longrightarrow L \xrightarrow{\mathrm{ad}\,\varphi} \overline{\mathrm{Hom}}_{\Gamma/}(L\ ,\Gamma/\ ) \longrightarrow (L^\#/L) \otimes_\Gamma \Gamma/ \longrightarrow O.$$

Thus φ is non-singular iff $L^\#/L$ has no -torsion. If this is the case, certainly $\partial\ [V,\varphi] = O$. For the converse, let L_1 be any lattice with $L_1 \leq L_1^\#$. Set $M = L_1/L_1^\#$, and write $M = M \oplus M$ where M is the part of M with torsion coprime to .

This is an orthogonal sum. If $\partial\,[V,\varphi] = 0$, M has a metabolizer N. Let $p: L_1^{\#} \to M$ be the quotient map. Then $L = p^{-1}(N)$ is a lattice with $L_1 \leq L \leq L_1^{\#}$.

$$L^{\#} = p^{-1}(N^{\perp}) = p^{-1}(N \oplus M \quad) ,$$

so $L \leq L^{\#}$ and $L^{\#}/L \cong M$ has no -torsion. ||||

It is now not hard to show that one can define a homomorphism $\ker \partial \longrightarrow W(\Gamma/\ ,J)$ by $[V,\varphi] \longrightarrow [L\ ,\ \varphi\ ]$, where L is chosen so that φ is non-singular. If $f:(\Gamma,J) \longrightarrow (\Gamma',J')$ is a homomorphism of PID's-with-involution and $\ = \ker f$, one gets an induced homomorphism

$$f_*: \ker \partial \longrightarrow W(\Gamma/\ ,J) \longrightarrow W(\Gamma',J') .$$

Thus $\ker \partial$ is the subgroup called $\mathrm{Def}(f_*)$ in Section 2. If $\Gamma = k[t,t^{-1}]$ we use the notation $\tau[x] = f_*(\tau)$ introduced in Section 2.

A2. THE LOCALIZATION SEQUENCE FOR A FUNCTION FIELD

For the rest of this appendix, $(k,-)$ is a field-with-involution, $\mathrm{char}(k) \neq 2$, $\Gamma = k[t,t^{-1}]$, $Q\Gamma$ is the quotient field $k(t)$ and J is the involution $f(t)^J = \bar{f}(t^{-1})$ of Γ or $Q\Gamma$.

LEMMA A1. The sequence

$$0 \longrightarrow W(k,-) \xrightarrow{i_*} W(Q\Gamma,J) \xrightarrow{\partial} W(Q\Gamma/\Gamma,J)$$

is exact, where i_* is induced by the inclusion $k \longrightarrow Q\Gamma$.

PROOF. This amounts to showing that the map $W(k,-) \longrightarrow W(\Gamma,J)$ induced by inclusion is an isomorphism. It has a left inverse π given by $\pi(\tau) = \tau[1]$, so it is enough to show that π is injective. First we show that $\ker(\pi)$ is generated by forms of rank 2. Let (L,φ) be a Hermitian space over Γ, and suppose that $\pi[L,\varphi] = 0$. This means that (L,φ) becomes metabolic upon tensoring with k, so there exists a non-zero x in L such that $\varphi(x,x) = f$ and $f(1) = 0$. Without loss of generality we may assume that x is primitive, so there exists y in L with $\varphi(x,y) = 1$. Let W be the submodule of L spanned by x and y. Suppose that $z \in W \cap W^{\perp}$, and let $z = ax + by$, $a,b \in \Gamma$. Then

$$0 = \varphi(z,x) = a\varphi(x,x) + b$$

and

$$0 = \varphi(z,y) = a + b\varphi(y,y) .$$

It follows from the first equation that $b(1) = 0$, and then from the second that $a(1) = 0$. Hence $z = (1-t)z'$ with $z' \in W \cap W^{\perp}$. Since this process can be repeated indefinitely, z must be zero, and we have $W \cap W^{\perp} = 0$. Therefore $L = W^{\perp} \oplus W^{\perp\perp}$, and $\varphi|W^{\perp\perp}$ is a rank 2 form representing an element of $\ker(\pi)$. By induction, φ is a sum of such forms.

Now consider a rank 2 form (L,φ) representing an element of $\ker(\pi)$. Let A be a matrix for φ. Then $\det A(1) = -\alpha\bar{\alpha}$ for some $\alpha \in \dot{k}$. Since φ is non-singular over Γ, $\det A \in \dot{k}$, so $\det A = -\alpha\bar{\alpha}$. Changing basis, we may assume that $\det A = -1$. Let the corresponding basis of L be x, y. Set $f = \varphi(x,x)$, $g = \varphi(x,y)$ and $h = \varphi(y,y)$. If $h = 0$, φ is metabolic. If not, let $z = hx + (1-g)y \neq 0$. We have

$$\begin{aligned}\varphi(z,z) &= h^2 f + h(1-g)^J g + h(1-g)g^J + (1-g)(1-g)^J h \\ &= h^2 f + h(1-gg^J) \\ &= 0 \quad \text{since} \quad fh - gg^J = \det A = -1 .\end{aligned}$$

Thus φ is metabolic in any case, and so $\ker(\pi) = 0$ as claimed. ▯

The formula (A1) and the discussion following it show that the computation of $W(Q\Gamma/\Gamma, J)$ reduces to the computation of the Hermitian Witt groups of finite extensions of k. These are known if k is a finite extension of the rationals (Landherr [11]), or $\mathbb{R}$ or $\mathbb{C}$. Below we determine the image of ∂ and show that the sequence splits, which determines $W(Q\Gamma, J) = W(k(t), J)$ in these cases.

A3. A TRACE FUNCTION, TORSION FORMS AND SKEW ISOMETRIC STRUCTURES

We are going to define a k-linear function $\chi: Q\Gamma/\Gamma \to k$. Let Γ^* be the Γ-module of all Laurent power series $\sum_{i=-\infty}^{\infty} a_i t^i$, $a_i \in k$, and extend J to Γ^*. There are two fields

$$\Gamma_+ = \{\sum_{i=m}^{\infty} a_i t^i \mid m \in \mathbb{Z},\ a_i \in k\} ,$$

and

$$\Gamma_- = \{\sum_{i=-\infty}^{n} b_i t^i \mid n \in \mathbb{Z},\ b_i \in k\}$$

inside Γ^*. These give rise to two Γ-linear embeddings $i_+, i_-: Q\Gamma \to \Gamma^*$. Since $\Gamma_+ \cap \Gamma_- = \Gamma$, $i_+ - i_-$ induces a Γ-linear embedding $j: Q\Gamma/\Gamma \to \Gamma^*$. Let $\text{const}: \Gamma^* \to k$ be given by $\text{const}(\Sigma a_i t^i) = a_0$, and let χ be the k-linear map $(\text{const})\, j$. The properties of χ that we need are:

PROPOSITION A3.

(i) $\chi(x^J) = -\overline{\chi(x)}$ _for_ $x \in Q\Gamma/\Gamma$;

(ii) $\chi_*: \text{Hom}_\Gamma(M, Q\Gamma/\Gamma) \longrightarrow \text{Hom}_k(M,k)$ _is an isomorphism for every torsion_ _Γ-module_ M.

Part (ii) says that χ is a universal element for the functor $\text{Hom}_k(-,k)$ of torsion Γ-modules.

PROOF (i) This follows from the observation that

$$i_+(f^J) = i_-(f)^J \quad \text{for} \quad f \in Q\Gamma .$$

(ii) First we show that χ_* is injective. Note that if N is a Γ-submodule of Γ^* and $\text{const}(N) = 0$ then $N = 0$. Let $\varphi \in \ker \chi_*$. Then $\text{const}(j\varphi M) = 0$, so $j\varphi M = 0$ and since j is injective $\varphi M = 0$, i.e. $\varphi = 0$. If M is finitely generated then $\text{Hom}_\Gamma(M, Q\Gamma/\Gamma)$ and $\text{Hom}_k(M,k)$ have the same finite dimension over k, so χ_* is an isomorphism. The general case follows because M is the union of its finitely generated submodules. ∎

Let M be a finitely generated torsion Γ-module. We claim that there is a bijection between Hermitian forms $\varphi: M \times M \longrightarrow Q\Gamma/\Gamma$ and skew-Hermitian forms $\psi: M \times M \to k$ with the property that $\psi(tx,y) = \psi(x,t^{-1}y)$ for $x,y \in M$, given by $\varphi \to \chi\varphi$. To see this, let $\bar{M}$ denote M with the conjugate action of Γ. Regard forms of the first type as elements of $\mathrm{Hom}_\Gamma(M \otimes_\Gamma \bar{M}, Q\Gamma/\Gamma)$ such that

$$\begin{array}{ccc} M \otimes_\Gamma \bar{M} & \xrightarrow{\ \varphi\ } & Q\Gamma/\Gamma \\ \big\downarrow{\scriptstyle \sigma} & & \big\uparrow{\scriptstyle J} \\ M \otimes_\Gamma \bar{M} & \xrightarrow{\ \varphi\ } & Q\Gamma/\Gamma \end{array}$$

commutes, where σ switches the factors. Similarly, forms of the second type are elements of $\mathrm{Hom}_k(M \otimes_\Gamma \bar{M}, k)$ such that

$$\begin{array}{ccc} M \otimes_\Gamma \bar{M} & \xrightarrow{\ \psi\ } & k \\ \big\downarrow{\scriptstyle \sigma} & & \big\uparrow{\scriptstyle \alpha} \\ M \otimes_\Gamma \bar{M} & \xrightarrow{\ \psi\ } & k \end{array}$$

commutes, where $\alpha(x) = -\bar{x}$. The claim follows on using Proposition A3. Moreover, φ is non-singular if and only if $\chi\varphi$ is, as one sees by regarding them as elements of $\mathrm{Hom}_\Gamma(M, \mathrm{Hom}_\Gamma(\bar{M}, Q\Gamma/\Gamma))$ and $\mathrm{Hom}_\Gamma(M, \mathrm{Hom}_k(\bar{M}, k))$ respectively and using the universal property of χ. Therefore if (M,φ) is a torsion form over Γ then $(M,\chi\varphi)$ is a skew-isometric structure over k. Further, (M,φ) is metabolic if and only if $(M,\chi\varphi)$ is. (Use the universal property again for the "if" part.) Thus we have:

<u>LEMMA A2</u>. <u>The trace function</u> χ <u>induces on isomorphism</u>

$$\chi_*: W(Q\Gamma/\Gamma, J) \longrightarrow W_-(C_\infty; k,-) \ . \ ∎$$

This isomorphism respects the splittings (A1) and (A2); in particular it takes $W^0(Q\Gamma/\Gamma, J)$ onto $W_-^0(C_\infty; k,-)$. Recall the homomorphism $\partial: W(Q\Gamma, J) \longrightarrow W(Q\Gamma/\Gamma, J)$. Let $\mathrm{forget}: W_-(C_\infty; k,-) \longrightarrow W_-(k,-)$ be the homomorphism which forgets the action of t.

<u>LEMMA A3</u>. $\chi_*(\mathrm{Im}\,\partial) \leq \mathrm{Ker}(\mathrm{forget})$.

<u>PROOF</u>. Consider an arbitrary generator $\langle\gamma\rangle$ of $W(Q\Gamma, J)$, where $\gamma \in Q\Gamma^{\bullet}$ and $\gamma = \gamma^J$. We may assume that $\gamma \in \Gamma$. Then $\partial\langle\gamma\rangle$ is represented by a form φ on a cyclic Γ-module M of order γ, where for a generator x we have

$$\varphi(x,x) \equiv 1/\gamma \mod \Gamma \ .$$

Let $\gamma = \sum_{i=-n}^{n} a_i t^i$ where $a_{-i} = \bar{a}_i$ and $a_n \neq 0$. Then as a k-vector space M

has a basis $t^i x$, $-n \leq i < n$, and

$$\chi\varphi(t^i x, t^j x) = \chi(\frac{t^{i-j}}{\gamma}) .$$

Now $i_+(1/\gamma) = \sum_{i=n}^{\infty} b_i^+ t^i$, $i_-(1/\gamma) = \sum_{i=-\infty}^{-n} b_i^- t^i$ for some $b_i^+, b_i^- \in k$, and so $j(1/\gamma) = \sum_{i=-\infty}^{\infty} b_i t^i$ with $b_i = 0$ for $-n < i < n$. It follows that

$$\chi\varphi(t^i x, t^j x) = 0 \quad \text{for} \quad |i-j| < n .$$

In particular, $x, tx, \ldots, t^{n-1}x$ span a metaboliser for $\chi\varphi$ (considered just as a skew-Hermitian form over k). ||||

Consider the splitting

(A3) $$W(Q\Gamma/\Gamma, J) \cong W^0(Q\Gamma/\Gamma, J) \oplus W(Q\Gamma/\Gamma, J)_{(t-1)} .$$

Since the restriction of forget to $W_-(C_\infty; k, -)_{(t-1)}$ is an isomorphism it follows that $(\mathrm{Im}\,\partial) \cap W(Q\Gamma/\Gamma, J)_{(t-1)} = 0$, so if we let $\partial^0 : W(Q\Gamma, J) \longrightarrow W^0(Q\Gamma/\Gamma, J)$ be the composite of ∂ and projection on the first factor in (A3), we have proved:

<u>LEMMA A4</u>. <u>The sequence</u>

$$0 \longrightarrow W(k,-) \longrightarrow W(Q\Gamma, J) \xrightarrow{\partial^0} W^0(Q\Gamma/\Gamma, J)$$

<u>is exact</u>. ||||

A4. A KNOT-THEORETIC INTERLUDE

In this section the base field k will be the rationals. Let $K \subset S^3$ be a knot. The rational Blanchfield pairing β of K represents an element of $W(Q\Gamma/\Gamma, J)$, while the skew-symmetric Milnor pairing μ defined in [19], represents an element of $W_-(C_\infty; \mathbb{Q})$. It can be seen by computing matrix representatives in terms of a Seifert matrix for K that $\chi\beta = -\mu$ (cf. Section A6; this is also true for Trotter's trace function). We give here a direct geometric proof of this; in fact this suggested our definition of χ in the first place.

We first recall the definitions of β and μ. Let $M = M_K$, the result of 0-surgery along K, and let $\tilde{M}$ be the infinite cyclic covering of M. Let $H = H_1(\tilde{M}; \mathbb{Q})$, a finitely generated torsion Γ-module. Let x be a (rational) 1-cycle in $\tilde{M}$. There exist $f \in \Gamma$, $f \neq 0$, and a 2-chain C such that $\partial C = fx$. Then one defines

$$\beta([x],[y]) = \frac{1}{f} \sum_{i=-\infty}^{\infty} (t^i C \cdot y) t^{-i}$$

where $\cdot$ denotes ordinary intersection number. The Milnor pairing arises from an isomorphism $\partial : H_2^\infty(\tilde{M}; \mathbb{Q}) \longrightarrow H_1(\tilde{M}; \mathbb{Q})$ where H_*^∞ is homology based on infinite chains. The desired pairing μ is the composite

$$H \times H \xrightarrow{\partial^{-1} \times \mathrm{id}} H_2^\infty(\tilde{M}; \mathbb{Q}) \times H \longrightarrow \mathbb{Q}$$

where the final arrow is the ordinary intersection pairing. To see μ geometrically we need the definition of ∂. Let ε_+ (respectively ε_-) be the end of $\tilde{M}$ such that for $x \in \tilde{M}$, $t^i x \to \varepsilon_+$ (respectively ε_-) as $i \to +\infty$ (respectively $-\infty$). The chain complex $C_*^\infty(\tilde{M};\mathbb{Q})$ has subcomplexes $C_*(\tilde{M},\varepsilon_+;\mathbb{Q})$, $C_*(\tilde{M},\varepsilon_-;\mathbb{Q})$ consisting of those chains whose support lies outside some neighborhood of ε_-, ε_+ respectively. There is an exact sequence

$$0 \to C_*(\tilde{M};\mathbb{Q}) \longrightarrow C_*(\tilde{M},\varepsilon_+;\mathbb{Q}) \oplus C_*(\tilde{M},\varepsilon_-;\mathbb{Q}) \longrightarrow C_*^\infty(\tilde{M};\mathbb{Q}) \to 0 .$$

One shows that $H_*(\tilde{M},\varepsilon_\pm;\mathbb{Q}) = 0$, and defines ∂ to be the connecting homomorphism in the long exact homology sequence. Thus given (finite) 1-cycles x and y there are 2-chains $C_+ \in C_2(\tilde{M},\varepsilon_+; \ )$, $C_- \in C_2(\tilde{M},\varepsilon_-; \ )$ with $\partial C_+ = x = \partial C_-$ and

$$\mu([x],[y]) = (C_- - C_+) \cdot y .$$

<u>REMARKS</u>. (1) There is lots of scope in this area for conflicting sign conventions. The one we use means that for a fibered knot μ is the same as the intersection pairing on the fiber.

(2) Milnor [19] used the dual cohomology pairing.

<u>THEOREM A1</u>. <u>If β,μ are the Blanchfield and Milnor pairings of the knot</u> K, <u>then</u> $\mu = -\chi\beta$. <u>In particular</u>, $\chi_*\partial\alpha_K = [\mu]$ <u>in</u> $W_-(C_\infty;\mathbb{Q})$.

<u>PROOF</u>. Note that if $\Delta \in C_i(\tilde{M};\mathbb{Q})$ and $\gamma \in \Gamma_+$ there is a chain $\gamma\Delta \in C_i(\tilde{M},\varepsilon_+;\mathbb{Q})$ and $\partial(\gamma\Delta) = \gamma\partial\Delta$. Similarly if $\gamma \in \Gamma_-$.

Let x,y be (finite) 1-cycles; and choose f and C as in the definition of β. Then we have

$$i_\pm(1/f)C \in C_2(\tilde{M},\varepsilon_\pm;\mathbb{Q}) ,$$

$$\partial(i_\pm(1/f)C) = i_\pm(1/f)fx = x .$$

Hence

$$\mu([x],[y]) = \{(i_-(1/f) - i_+(1/f))C\} \cdot y$$

$$= - \sum_{i=-\infty}^{\infty} a_i(t^iC\cdot y)$$

where $i_+(1/f) - i_-(1/f) = \sum_{i=-\infty}^{\infty} a_i t^i$. On the other hand,

$$\chi\beta([x],[y]) = \chi\{\frac{1}{f}\sum_{i=-\infty}^{\infty}(t^iC\cdot y)t^{-i}\}$$

$$= \text{const}\{(\sum_{i=-\infty}^{\infty} a_i t^i)(\sum_{i=-\infty}^{\infty}(t^iC\cdot y)t^{-i})\}$$

$$= \sum_{i=-\infty}^{\infty} a_i(t^iC\cdot y) .$$

That is, $\chi\beta([x],[y] = -\mu([x],[y])$. ▒

A5. SKEW-ISOMETRIC STRUCTURES AND SEIFERT FORMS

Let V be a finite dimensional vector space over k. There is a 1-1 correspondence between skew-isometric structures (V,φ,t) such that $1-t$ is an automorphism of V, and Seifert forms $(V,\mathscr{S})$, given by the formulae

$$\varphi = \mathscr{S}-\mathscr{S}^*: V\times V\to k$$

$$t = (\mathrm{ad}\mathscr{S})^{-1}\mathrm{ad}\mathscr{S}^*: V\to V$$

and

(A4) $$\mathscr{S}(x,y) = \varphi((1-t)^{-1}x,y) \quad , \; x,y\in V .$$

Moreover a subspace W of V is a metaboliser for $\mathscr{S}$ iff it is a t-invariant metaboliser for φ. (We leave it to the reader to supply the easy proofs of these assertions, which are straightforward generalisations from the case of a trivial involution.) Thus we have:

<u>LEMMA A5</u>. <u>There is an isomorphism</u>

$$\lambda: W_-^0(C_\infty;k,-) \longrightarrow W_S(k,-)$$

<u>given by the formula</u> (A4). ▒

Note that if (V,φ,t) and $(V,\mathscr{S})$ correspond as above then

$$(1+t) = (\mathrm{ad}\mathscr{S})^{-1}\mathrm{ad}(\mathscr{S}+\mathscr{S}^*) \quad .$$

Thus $\mathscr{S}+\mathscr{S}^*$ is non-singular if and only if $(1+t)$ is an automorphism of V. If the involution on k is trivial then $W_-(C_\infty;k)_{(1+t)} \cong W_-(k) = 0$, so we can always assume that $1+t$ is an automorphism. This justifies the assertion made in Section A1 that it does not affect $W_S(k)$ if we insist that $\mathscr{S}+\mathscr{S}^*$ is non-singular.

A6. DETERMINATION OF $W(Q\Gamma,J)$.

<u>THEOREM A2</u>. (cf. [20] Theorem 5.3). <u>The sequence</u>

$$0\to W(k,-) \longrightarrow W(Q\Gamma,J) \xrightarrow{\partial^0} W^0(Q\Gamma/\Gamma,J) \to 0$$

<u>is split exact</u>.

<u>PROOF</u>. In view of Lemmas A4, A2 and A5, it is enough to produce a homomorphism $\nu: W_S(k,-) \longrightarrow W(Q\Gamma,J)$ such that $\lambda\chi_*\partial^0\nu$ is the identity of $W_S(k,-)$, where λ is as in Lemma A5. Let $\sigma\in W_S(k,-)$ be represented by a matrix S, and define $\nu(\sigma)$ to be the element of $W(Q\Gamma,J)$ represented by

$$S_t = (1-t)S + (1-t^{-1})S^* .$$

Note that $\det(S_t) = (1-t)^n\det(S-t^{-1}S^*)$ if S is $n\times n$, and since $\det(S-S^*)\neq 0$, S_t is non-singular. It is straightforward to check that ν is a well-defined homomorphism. The proof that $\lambda\chi_*\partial^0\nu = \mathrm{id}$ is a simple matrix calculation. First, $\partial\nu(\sigma) = \partial[S_t]$ is represented by a form φ on the

Γ-module M presented by S_t, and φ is given by

$$\varphi(\tilde{x},\tilde{y}) \equiv xS_t^{-1}y^* \mod \Gamma$$

where $\tilde{x},\tilde{y}$ are the images in M of the row vectors x,y. Since $S_t = (1-t)(S-t^{-1}S^*)$ and $\det(S-t^{-1}S^*)$ is coprime to $1-t$, $\partial^o[S_t]$ is represented by the restriction φ^o of φ to $M^o = (1-t)M$. A presentation matrix for M^o is $S-t^{-1}S^*$, and relative to this presentation φ^o is given by

$$\begin{aligned}\varphi^o(\tilde{x},\tilde{y}) &\equiv (1-t)(1-t^{-1})xS_t^{-1}y^* \\ &\equiv (1-t^{-1})x(S-t^{-1}S^*)^{-1}y^* \mod \Gamma .\end{aligned}$$

Making a change of basis we see that M^o also has a presentation matrix $tI-S^*S^{-1}$, and the corresponding representation of φ^o is

$$\begin{aligned}\varphi^o(\tilde{x},\tilde{y}) &\equiv (1-t^{-1})(t^{-1}xS)(S-t^{-1}S^*)^{-1}(t^{-1}yS)^* \\ &\equiv (1-t)x(S^*S^{-1}-tI)^{-1}\ S^*y^* \mod \Gamma\end{aligned}$$

Thus as a vector space over k, M^o has dimension equal to the size of S, and the automorphism t has matrix S^*S^{-1}. In other words, if S represents the Seifert form $\mathscr{S}$, $t = (\mathrm{ad}\mathscr{S})^{-1}(\mathrm{ad}\mathscr{S}^*)$. (The order is reversed since matrices act on the right of row vectors.) If ξ and η are row vectors over k we have

$$\begin{aligned}\chi\varphi^o(\xi,\eta) &= \chi\{(1-t)\xi(S^*S^{-1}-tI)^{-1}S^*\eta^*\} \\ &= \mathrm{const}\{(1-t)\xi(\sum_{i=-\infty}^{\infty}(S^*S^{-1})^{-(i+1)}\ t^i)S^*\eta^*\} \\ &= \xi(S-S^*)\eta^* \ .\end{aligned}$$

Comparing this with Section A5 we see that $\chi_*\partial^o[S_t] = \lambda^{-1}[S]$, as claimed. ▒

We have isomorphisms

$$\begin{aligned}W(Q\Gamma,J) &\cong W(k,-) \oplus W^o(Q\Gamma/\Gamma,J) \\ &\cong W(k,-) \oplus W^o_-(C_\infty;k,-) \\ &\cong W(k,-) \oplus W_S(k,-)\end{aligned}$$

<u>ADDENDUM TO THEOREM.</u> <u>Let</u> $K \subset S^3$ <u>be a knot with Seifert form</u> Θ, <u>Blanchfield pairing</u> β <u>and (skew-symmetric) Milnor pairing</u> μ. <u>Under the above isomorphisms with</u> $k = \mathbb{Q}$, α_K <u>corresponds to</u> $(0,[-\beta])$, $(0,[\mu])$ <u>and</u> $(0,[\Theta])$ <u>respectively.</u>

<u>PROOF.</u> By Theorem A1, it is enough to show that $\nu[\Theta] = \alpha_K$. But this is Proposition 1. ▒

A7. THE GROUP $W(\mathbb{C}(t),J)$.

In this section we take $(k,-) = (\mathbb{C}$, conjugation), so that $\Gamma = \mathbb{C}[t,t^{-1}]$ and $Q\Gamma = \mathbb{C}(t)$. By a <u>balanced function</u> we mean a function $f:S^1 \to \mathbb{Z}$ with a finite number of discontinuities such that (in an obvious notation) $f(\xi) = 1/2(f(\xi+) + f(\xi-))$ for all $\xi \in S^1$. Recall that for $\tau \in W(Q\Gamma, J)$ and $\xi \in S^1$, $\sigma_\xi \tau$ is defined to be $\text{sign}(\tau[\xi])$ whenever $\tau[\xi]$ exists, and for the remaining ξ it is defined to make $\sigma_\bullet \tau : S^1 \to \mathbb{Z}$ a balanced function (see [1]). Thus we have a homomorphism $\sigma_\bullet$ from $W(Q\Gamma, J)$ to the group of balanced functions.

We can also associate signatures to an element υ of $W(Q\Gamma/\Gamma, J)$, in two equivalent ways. The symmetric prime ideals of Γ are $(t-\xi)$ for $\xi \in S^1$, so we have

$$W(Q\Gamma/\Gamma, J) \cong \bigcup_{\xi \in S^1} W(Q\Gamma/\Gamma, J)_{(t-\xi)}$$

and isomorphisms

$$W(Q\Gamma/\Gamma, J)_{(t-\xi)} \xrightarrow{(t-\xi)^J_*} W_{(-t^{-1}\bar{\xi})}\hat{}(\Gamma/(t-\xi), J)$$

$$\xrightarrow{=} W_{-\bar{\xi}^2}(\mathbb{C}, \text{conjugation})$$

$$\xrightarrow{i\xi} W(\mathbb{C}, \text{conjugation}) \quad .$$

Denote the image of the ξ'th component of υ in $W(\mathbb{C}, \text{conjugation})$ by υ_ξ. Then we have the signatures $\text{sign}(\upsilon_\xi)$.

Secondly, we have

$$W_-(C_\infty; \mathbb{C}, \text{conjugation}) \cong \bigoplus_{\xi \in S^1} W_-(C_\infty; \mathbb{C}, \text{conjugation})_{(t-\xi)}$$

and each summand is isomorphic to $W_-(\mathbb{C}, \text{conjugation})$ by forgetting the action of t. Denote the ξ'th component of $\chi_* \upsilon$ by $(\chi_* \upsilon)_\xi$. We claim that $(\chi_* \upsilon)_\xi = -i\upsilon_\xi$. It is enough to check this when υ is represented by the form φ on a cyclic Γ-module M of order $t-\xi$, $\xi \in S^1$, given by

$$\varphi(x,x) \equiv it/(t-\xi) \mod \Gamma$$

where x generates M. For $\zeta \neq \xi$ we have $\upsilon_\zeta = 0 = (\chi_* \upsilon)_\zeta$, while

$$\upsilon_\xi = \langle 1 \rangle$$

$$(\chi_* \upsilon)_\xi = \langle \chi(\tfrac{it}{t-\xi}) \rangle = \langle -i \rangle \ .$$

Thus we can define

$$\sigma_\zeta \upsilon = \text{sign}(\upsilon_\zeta) = \text{sign}(i(\chi_* \upsilon)_\zeta) \quad .$$

REMARK. In [18] Matumoto studies two families of signatures associated to a Seifert matrix S over $\mathbb{C}$. These are essentially the same as the signatures of $\nu[S]$ and $\partial\nu[S]$ defined above. Thus our next result, which says that the signatures of $\partial\tau$ are the jumps in $\sigma_\bullet \tau$, $\tau \in W(Q\Gamma, J)$, is just that part of

Matumoto's theorem which does not consider the value at a discontinuity. Our proof is a trivial computation.

THEOREM A3. (Matumoto [18])
For $\tau \in W(\mathbb{C}(t),J)$ and $\zeta \in S^1$ we have

$$\sigma_{\zeta+}\tau - \sigma_{\zeta-}\tau = 2\sigma_{\zeta}(\partial\tau) .$$

PROOF. It suffices to consider the cases

$$\tau = \langle\gamma\rangle , \ \gamma \in \Gamma \text{ coprime to } (t-\zeta)$$

and
$$\tau = \langle(t-\zeta)\delta\rangle , \ \delta \in \Gamma \text{ coprime to } (t-\zeta) .$$

In the first, $(\partial\tau)_{\zeta} = 0$ so $\tau[\zeta]$ is defined and both sides of the asserted equality are zero. In the second, for $\xi \in S^1$ close to ζ we have

$$\tau[\xi] = \langle(\xi-\zeta)\delta(\xi)\rangle = \langle(\xi-\zeta)\delta(\xi)/|\xi-\zeta|\rangle ,$$

whence

$$\sigma_{\zeta\pm}\tau = \text{sign}(\pm i\zeta\delta(\zeta)) .$$

On the other hand

$$(\partial\tau)_{\zeta} = \langle i\zeta\hat{\delta}\rangle = \langle i\zeta\delta(\zeta)\rangle . \quad ▥$$

Combining this with Theorem A2 we have:

COROLLARY A1. The map $\sigma_{\bullet}$ is an isomorphism from $W(\mathbb{C}(t),J)$ to the group of balanced functions. ▥

One can deal similarly with the cases $k = \mathbb{R}$ or k an algebraic number field. In the first case, $W(\mathbb{R}(t),J)$ is isomorphic to the group of balanced functions f for which $f(\zeta) = f(\bar{\zeta})$. In the second, $W(k(t),J)$/torsion is determined by the functions $\sigma_{\bullet}$ associated to the involution-preserving embeddings of k in $\mathbb{C}$.

APPENDIX B. RELATIONS BETWEEN CASSON-GORDON INVARIANTS.

Let K be a knot, and let n and N be powers of the same prime with $n < N$. Let $p: L_{K,N} \to L_{K,n}$ be the covering projection. Each $\chi \in Ch_n(K)$ gives rise to $\chi p_* \in Ch_N(K)$. We show how $\tau(K,\chi)$ determines $\tau(K,\chi p_*)$. The main purpose of this is to shed some light on the multiplicative behavior of $\sigma(K,\chi)$ for certain K noted in [2]. In fact, we show that $\tau(K,\chi)$ has the same behavior, and identify the properties of the knots responsible. Throughout, $\Gamma = \mathbb{C}[t,t^{-1}]$, and for $\gamma \in \Gamma$ we write $\gamma|x$ instead of $\gamma(x)$.

THEOREM B1. In the above situation, let $\nu = N/n$. Then we have

$$\sigma_{\zeta}\tau(K,\chi p_*) = \sum_{\xi:\xi^{\nu}=\zeta} \sigma_{\xi}\tau(K,\chi) - \sum_{\omega:\omega^N=1} \sigma_K(\omega) + \nu \sum_{\eta:\eta^n=1} \sigma_K(\eta) .$$

REMARK. This determines $\tau(K,\chi p_*)$ by Corollary A1.

COROLLARY B1. In the situation of the theorem, suppose further that K is algebraically slice and that $\tau(K,\chi)$ is in the image of $W(\mathbb{C}$, conjugation$) \otimes \mathbb{Q}$. Then

$$\tau(K,\chi p_*) = \nu\tau(K,\chi) \quad .$$

REMARK. By Theorem (3.5) of [5], the last hypothesis is satisfied when $n=2$ and K has genus 1. This is the case in [2].

COROLLARY B2. Let K be a knot and ν a prime power. Let O_ν denote the zero of $Ch_\nu(K)$. Then

$$\sigma_\zeta\tau(K,O_\nu) = \sum_{\xi:\xi^\nu=\zeta}\sigma_K(\xi) - \sum_{\omega:\omega^\nu=1}\sigma_K(\omega) \quad .$$

PROOF. In the theorem take $n=1$, $N=\nu$, and recall that $\tau(K,O_1)=\alpha_K^{\mathbb{C}}$.

REMARK. Of course, one is only interested in $\tau(K,\chi)$ for K algebraically slice. However, if K is a sum of two non-algebraically-slice knots then this result shows that some care must be taken. Note however that $\sigma_1\tau(K,O_\nu)$ is always zero.

PROOF OF THEOREM B1. Let χ take values in C_m. Let p denote also the projection $M_{K,N}\to M_{K,n}$; we have

$$\chi^+p_* = (\chi p_*)^+ : H_1(M_{K,N}) \to C_m \times C_\infty \quad .$$

Choose (W^4,ψ) such that

$$\partial(W,\psi) = r(M_{K,n},\chi^+) \quad , \quad r>0 \quad .$$

There is a ν-fold cyclic covering $q:W_\nu\to W$ such that

$$\partial(W_\nu, \psi q_*) = r(M_{K,N},\chi^+p_*) \quad .$$

From the approach to knot signatures via branched covering spaces (see [4] or [15]) it follows that

$$\frac{1}{r}(\mathrm{sign}(W_\nu) - \mathrm{sign}(W)) = \sum_{\omega:\omega^N=1}\sigma_K(\omega) - \nu\sum_{\eta:\eta^n=1}\sigma_K(\eta)$$

and hence that the desired result is equivalent to

$$\text{(B1)} \qquad \sigma_\zeta t_{\psi q_*}(W_\nu) = \sum_{\xi:\xi^\nu=\zeta}\sigma_\xi t_\psi(W) \quad .$$

In what follows, ω and η will be variables ranging over the ν'th roots of unity. Let $e, e_\omega:\Gamma\to Q\Gamma$ be the injections given by

$$e(\gamma) = \gamma|t^\nu \quad , \quad e_\omega(\gamma) = \gamma|\omega t \quad .$$

We show that

$$e_*t_{\psi q_*}(W_\nu) = \sum_\omega e_{\omega *}t_\psi(W) \quad ,$$

from which (B1) follows upon taking σ_ξ for some νth root ξ of ζ.

Let $L = H_2^t(W;\Gamma)/\text{torsion}$ and $L_* = H_2^t(W_\nu;\Gamma)/\text{torsion}$, which are Γ-lattices in $H_2^t(W;Q\Gamma)$ and $H_2^t(W_\nu;Q\Gamma)$ respectively, with $L \leq L^\#$, $L_* \leq L_*^\#$. Now the $C_m \times C_\infty$ covering $\tilde{W}$ of W determined by ψ is also the covering of W_ν determined by ψq_*, so as $\mathbb{C}$-vector spaces

$$H_2^t(W;\Gamma) = H_2^t(W_\nu;\Gamma) = H_2(\tilde{W};\mathbb{C}) .$$

For $x \in H_2^t(W;\Gamma)$, let x_* be the corresponding element of $H_2^t(W_\nu;\Gamma)$. The Γ-module structures are related by

$$t \cdot x_* = (t^\nu x)_* .$$

Then L and L^* are related in the same way. Further, the intersection forms φ, φ_* on L, L_* are connected by the formulae

$$\text{(B2)} \qquad \varphi(x,y) = \sum_{i=-\infty}^{\infty} a_i t^i , \quad \varphi_*(x_*,y_*) = \sum_{i=-\infty}^{\infty} a_{\nu i} t^i$$

for $x,y \in L$. Let $Q\Gamma$ with the Γ-module structures induced by e, e_ω be denoted by $Q\Gamma_e$ $Q\Gamma_\omega$. Then $e_* t_{\psi q^*}(W_\nu)$ is represented by $(L_* \otimes Q\Gamma_e, \varphi_e)$ where

$$\varphi_e(x_* \otimes \gamma, y_* \otimes \delta) = \gamma\delta^J(\varphi_*(x_*,y_*)|t^\nu) ,$$

and $e_{\omega *} t_\psi(W)$ by $(L \otimes Q\Gamma_\omega, \varphi_\omega)$ where

$$\varphi_\omega(x \otimes \gamma, y \otimes \delta) = \gamma\delta^J(\varphi(x,y)|\omega t) .$$

There is an isometry T of $(L_* \otimes Q\Gamma_e, \varphi_e)$ defined by

$$T(x_* \otimes \gamma) = (tx)_* \otimes \gamma$$

and T^ν is multiplication by t^ν. Therefore $L_* \otimes Q\Gamma_e$ splits as an orthogonal direct sum $\oplus_\omega E_\omega$ where E_ω is the ωt-eigenspace of T. The proof is completed by showing that

$$(E_\omega, \varphi_e|E_\omega) \cong (L \otimes Q\Gamma_\omega, \varphi_\omega) .$$

Define homomorphisms $\alpha_\omega : L_* \otimes Q\Gamma_e \to L \otimes Q\Gamma_\omega$ and $\beta_\omega : L \otimes Q\Gamma_\omega \to L_* \otimes Q\Gamma_e$ by

$$\alpha_\omega(x_* \otimes \gamma) = \frac{1}{\sqrt{\nu}}(x \otimes \gamma) ,$$

$$\beta_\omega(x \otimes \gamma) = \frac{1}{\sqrt{\nu}} \sum_{i \bmod \nu} (\omega t)^{-i} T^i(x_* \otimes \gamma) .$$

We leave it to the reader to verify that these are well-defined and satisfy

$$\beta_\omega(L \otimes Q\Gamma_\omega) \leq E_\omega ,$$

$$\alpha_\eta \beta_\omega = \begin{cases} \text{id} & \text{if } \eta = \omega , \\ 0 & \text{if } \eta \neq \omega , \end{cases}$$

$$\sum_\omega \beta_\omega \alpha_\omega = \text{id} .$$

Thus β_ω maps $L \otimes Q\Gamma_\omega$ isomorphically onto E_ω. Finally, we have

$$\varphi_e(\beta_\omega(x \otimes \gamma)\ ,\ \beta_\omega(y \otimes \delta))$$

$$= \frac{\gamma\delta^J}{\nu} \sum_{i,j \bmod \nu} (\omega t)^{j-i}\ \varphi_e(x_* \otimes 1\ ,\ T^{j-i}(y_* \otimes 1))$$

$$= \gamma\delta^J \sum_{k \bmod \nu} (\omega t)^k (\varphi_*(x_*,(t^k y)_*)\,|\,t^\nu)\ .$$

From (B2),

$$\varphi(x,y) = \sum_{k \bmod \nu} t^k(\varphi_*(x_*,(t^k y)_*)\,|\,t^\nu)\ ,$$

so

$$\varphi_e(\beta_\omega(x \otimes \gamma)\ ,\ \beta_\omega(y \otimes \delta)) = \gamma\delta^J(\varphi(x,y)\,|\,\omega t)$$

$$= \varphi_\omega(x \otimes \gamma\ ,\ y \otimes \delta)\ .$$

BIBLIOGRAPHY

[1] A. J. Casson and C. McA. Gordon, Cobordism of classical knots, mimeographed notes, Orsay, 1975.

[2] A. J. Casson and C. McA. Gordon, On slice knots in dimension three, Proc. Symp. in Pure Math. XXX, 1978, part two, 39-53.

[3] P. E. Conner, Notes on the Witt classification of Hermitian inner-product spaces over a ring of algebraic integers, University of Texas Press, Austin and London, 1979.

[4] P. M. Gilmer, Configurations of surfaces in 4-manifolds, Trans. Amer. Math. Soc. 264 (1981), 353-380.

[5] P. M. Gilmer, Slice knots in S^3, to appear.

[6] C. McA. Gordon, Some aspects of classical knot theory, Knot Theory, Proc. Plans-sur-bex, Switzerland, 1977, Lecture Notes in Mathematics 685, Springer-Verlag, Berlin-Heidelberg-New York, 1978, 1-60.

[7] B. J. Jiang, A simple proof that the concordance group of algebraically slice knots is infinitely generated, Proc. Amer. Math. Soc. 83 (1981), 189-192.

[8] L. H. Kauffman and L. R. Taylor, Signature of links, Trans. Amer. Math. Soc. 216 (1976), 351-365.

[9] A. Kawauchi, $\tilde{H}$-cobordism I; the groups among three-dimensional homology handles, Osaka J. Math. 13 (1976), 567-590.

[10] C. Kearton, The Milnor signatures of compound knots, Proc. Amer. Math. Soc. 76 (1979), 157-160.

[11] W. Landherr, Äquivalenz Hermetischer Formen über einen beliebigen algebraischen Zahlkörper, Abh. Math. Sem. Univ. Hamburg 11 (1935), 245-248.

[12] J. Levine, Knot cobordism groups in codimension 2, Comment. Math. Helv. 45 (1970), 185-198.

[13] J. Levine, Invariants of knot cobordism, Inventiones Math. 8 (1969), 98-110.

[14] R. A. Litherland, Topics in knot theory, thesis, Cambridge University, 1978.

[15] R. A. Litherland, Signatures of iterated torus knots, Topology of low-dimensional manifolds, Proceedings, Sussex 1977, Lecture Notes in Mathematics 722, Springer-Verlag, Berlin-Heidelberg-New York, 1979, 71-84.

[16] R. A. Litherland, Slicing doubles of knots in homology 3-spheres, Inventiones Math. 54 (1979), 69-74.

[17] C. Livingston and P. Melvin, Algebraic knots are algebraically dependent, Proc. Amer. Math. Soc. 87 (1983), 179-180.

[18] T. Matumoto, On the signature invariants of a non-singular complex sesqui-linear form, J. Math. Soc. Japan 29 (1977), 67-71.

[19] J. Milnor, Infinite cyclic coverings, Conf. on the topology of manifolds, Michigan State University, 1967, Prindle, Weber and Schmidt, Boston, Mass., 1968, 115-133.

[20] J. Milnor, Algebraic K-theory and quadratic forms, Inventiones Math. 9 (1970), 318-344.

[21] J. Milnor and D. Husemoller, Symmetric Bilinear Forms, Ergebnisse der Mathematik und ihrer Grenzgebiete Band 73, Springer-Verlag, Berlin-Heidelberg-New York, 1973.

[22] D. Rolfsen, A surgical view of Alexander's polynomial, Geometric Topology, Proc. of the Utah Conf. 1974, Lecture Notes in Mathematics 438, Springer-Verlag, Berlin-Heidelberg-New York, 1975, 415-423.

[23] Y. Shinohara, On the signature of knots and links, Trans. Amer. Math. Soc. 156 (1971), 273-285.

[24] A. G. Tristram, Some cobordism invariants for links, Proc. Camb. Phil. Soc. 66 (1969), 251-264.

[25] H. Trotter, On S-equivalence of Seifert matrices, Inventiones Math. 20 (1973), 173-207.

[26] H. Trotter, Knot modules and Seifert matrices, Knot Theory, Proc. Plans-sur-bex, Switzerland, 1977, Lecture Notes in Mathematics 685, Springer-Verlag, Berlin-Heidelberg-New York, 1978, 291-299.

[27] O. Ja. Viro, Branched coverings of manifolds with boundary and link invariants I, Math. USSR Izvestija 7 (1973), 1239-1256.

[28] C. T. C. Wall, Non-additivity of the signature, Inventiones Math. 7 (1969), 269-274.

DEPARTMENT OF MATHEMATICS
LOUISIANA STATE UNIVERSITY
BATON ROUGE, LA 70803

Contemporary Mathematics
Volume 35, 1984

COMPLEX STRUCTURES ON 4-MANIFOLDS

Richard Mandelbaum

Suppose V^n is a smooth compact manifold. When does V admit a complex structure? If it does admit a complex structure how many different structures does it admit and what can be said about the space of these structures? These two basic questions have been the source of much of the most beautiful work in mathematics in the past hundred years. Yet, even in the simplest cases, $n=2$ and $n=4$ we are still very much in the dark! Recently the work of Thurston [T] exploring the relationship between the space of complex structures on 2-manifolds and 3-dimensional topology has broadened the interest in understanding this space. Similarly the new work of Taubes [Ta] and Donaldson [D] exploring the closely related question of different differential-geometric structures on 4-manifolds promises to stimulate interest in this area also. In this paper we will briefly and broadly survey some of what is known about the existence of comlex structures on 4-manifolds and its relationship to the smooth topology of 4-manifolds.

Returning to our first question of when does V admit a complex structure it is easy to see that first of all n must be even and secondly V must be orientable! We can thus set $n=2m$ and begin by examining the case of $m=1$. We have

THEOREM 1 <u>Let</u> R <u>be a smooth compact orientable connected</u> 2-<u>manifold</u>. <u>Then</u> R <u>admits a complex structure</u>.

This is a classic theorem essentially due to Riemann whose proof can be found in any text on Riemann Surfaces. See, for example [S]. Riemann also studied the question of different complex structures on the underlying manifold and claimed that there were "3g-3 complex parameters which controlled the complex structure", where g is the genus of R and $g>1$. This question, the modulii problem, was not returned to until the later work of Teichmüller, Ahlfors and Bers which clarified precisely what is meant by saying that $3g-3$ complex parameters determine the complex structure. For further details on Modulii problems see [A].

We now turn to $m=2$. Here the existence problem is much, much more complicated. We can begin to approach it by noting that if V^{2n} has a complex structure, then its tangent bundle TV also has the natural induced structure

0271-4132/84 $1.00 + $.25 per page

of a complex vector bundle. That is we can reduce the structure group $GL(2m,\mathbb{R})$ of TV to the group $GL(m,\mathbb{C})$.

Conversely if V^{2m} is a smooth manifold such that TV can be given the structure of a complex vector bundle, we say that V has an almost complex structure. Clearly a necessary condition for V to admit a complex structure is that it admit an almost complex one!

Looking at S^4 it can be shown that it doesn't even admit an almost complex structure! More precisely we have

THEOREM 2 (See Ehresmann [E], Wu [W]). S^{2m} admits an almost complex structure iff $m=1$ or 3. (Actually S^2 admits a complex structure while the case of S^6 is as yet unsettled!) Thus S^4 cannot be made into a complex manifold and life at $m=2$ promises to be much harder than at $m=1$. However, we can pursue the question of almost-complex structure rather fully for $m=2$ to obtain

THEOREM 3 (Ehresmann-Wu, see [E]). Let V be a smooth compact orientable 4-manifold and suppose $h \in H^2(V,\mathbb{Z})$. Then there exists an almost complex structure on V with first chern class $c_1(V) = h$ if and only if

1) $w_2(V)$ is congruent to h mod 2 (i.e. The mod 2 reduction of h, $[h]_2 \in H^2(V,\mathbb{Z}_2)$ equals the second Steifel-Whitney class $w_2(V)$.)

2) $h \cdot h = 3\sigma(V) + 2e(V)$ where $h \cdot h = h \cup h[Y]$ is the cup-product of h with itself evaluated on V, $\sigma(V)$ is the signature of V and $e(V)$ is its Euler-Poincare number.

(Note that if V is almost-complex $c_2(V) = e(V)$). Thus we have a complete solution to the question of almost-complex structures on 4-manifolds. We can now reformulate and extend our basic existence question as follows:

QUESTION A Suppose $(p,q) \in \mathbb{Z} \times \mathbb{Z}$. When does there exist a complex (almost-complex) 4-manifold V with $c_1^2(V) = p$, $c_2(V) = q$? (where c_i is the ith Chern-class of V).

In the almost-complex context one can see as a consequence of work of Milnor on cobordism theory (see [M]) that $p+q \equiv 0(12)$ is a necessary and sufficient condition for the existence of a not-necessarily connected almost complex manifold V with $c_1^2(V) = p$ and $c_2(V) = q$. A more explicit answer is however provided by the work of Van-de-Ven.

THEOREM 4 (Van-de-Ven). see [VV1]. For every pair of integers $(p,q) \in \mathbb{Z} \times \mathbb{Z}$ with $p+q = 12d$ for some $d \in \mathbb{Z}$ there exists a connected almost complex surface V with $c_1^2(V) = p$, $c_2(V) = q$.

Furthermore if $P = \mathbb{CP}^2$, $Q = \overline{\mathbb{CP}^2}$, $R = F_2 \times S^2$ (where F_2 is a surface of genus 2) and $w_{\ell,m,n} = \ell P \# mQ \# nR$ then every realizable pair (p,q) can be realized by some $w_{\ell,m,n}$.

We note that if $q \geq 2d-1 \geq 1$ then (p,q) can be realized by the simply-connected almost complex 4-manifold $w_{2x+1,y}$ where $x = d-1$ and $y = q-1-2d$. (A straightforward calculation shows that $c_1^2(w_{2b+1,c,2a}) = 9-24a+10b-c$, $c_1^2(w_{2b,c,2a+1}) = -8-24a+10b-c$, $c_2(w_{2b+1,c,2a}) = 3-12a+2b-c$, $c_2(w_{2b,c,2a+1}) = -4-12a+2b+c$, and we can realize any admissible (p,q) by some choice of non-negative (a,b,c).)

Thus the almost-complex problem is solved. The complex situation is a bit more subtle. It was already noticed by Van-de Ven in [VV] that there exist almost-complex manifold not admitting complex structures. He showed that in fact $p \leq 8q$ was a necessary condition for an almost-complex manifold realizing (p,q) to admit a complex structure. Thus for example $P \# 2(S^1 \times S^3)$ is an almost-complex manifold which doesn't admit a complex structure.

More precisely let $D = \{(p,q) \in \mathbb{Z} \times \mathbb{Z} \mid p+q \equiv 0 \pmod{12}$. For any $n \in \mathbb{Z}_+$ let $D_n = D \cap \{p \leq nq\}$ Then we have:

THEOREM 5 [Van-de-Ven]

1) *Suppose* $(p,q) \in D_2$. *Then there exists a complex manifold realizing* (p,q).

2) *Suppose* M *is a complex manifold realizing* (p,q). *Then* $(p,q) \in D_8$.

3) *There exist almost-complex manifolds with* $(c_1^2, C_2) \in D_2$ *which do not admit complex structures.*

In particular

a) $S^2 \times S^2 \# 2(S^1 \times S^3)$

b) $(2k+1)V_4$ *(with* $k > o$ *and* V_4, *the Kummer Surface,) are almost-complex manifolds which do not admit complex structures.*

Van-de-Ven's necessary condition was improved by Bogomolov to $c_1^2 \leq 4c_2$ and by Miyaoka and Yau to $c_1^2 \leq 3c_2$. This last condition is best possible as by [Mi, Y] if G is a proper discontinuous group of hyperbolic isometries of the unit Ball B $\mathbb{C}^2$ then B/G is a compact complex manifold with $c_1^2 = 3c_2$ and there are an infinite number of such groups.

Summarizing we state:

THEOREM 6 (Miyaoka, Yau)

Let M *be a complex manifold. Then* $c_1^2(M) \leq 3c_2(M)$ *with equality if and only if either* $M = CP^2$ *or* M *is the quotient of the unit ball in* $\mathbb{C}^2$ *by a group of properly discontinuous transformations.*

We pause to draw some implications from the above.

1) X realizes $(p,q) \in D_2$ if and only if $\sigma(X) \leq 0$.

PROOF:

By Theorem 3

$$c_1^2(X) = 3\sigma(X) + 2c_2(X)$$

Thus $c_1^2 - 2c_2 = 3\sigma(X)$ so $\sigma(X) \leq 0 \leftrightarrow c_1^2 \leq 2c_2$.

2) Suppose X is a simply-connected complex manifold with $w_2(X) = 0$. Then X is homeomorphic to $\pm a(X)V_4 \oplus b(X)(S^2 \times S^2)$ where $a(X) = |\sigma(X)/16|$ and $b(X) = \frac{1}{2}c_2(X) - 11a(X) - 1$.

PROOF:

As a consequence of Freedman's results [F] it suffices to show that

$$\beta_2(X)/|\sigma(X)| \geq 11/8$$

where $\beta_2(X)$ is the second Betti Number of X. To do this we only need the fact [ ] that if X is a simply connected complex manifold with $w_2(X) = 0$ then $c_1^2(X) \geq 0$ except if $X \simeq S^2 \times S^2$. Since the statement is trivially true for $X \simeq S^2 \times S^2$ we assume without loss of generality that $c_1^2(X) \geq 0$.

Now we have 2 cases.

Case I: $\sigma(X) < 0$. Then $3\sigma(X)+2c_2(X) = c_1^2(X) \geq 0$. So $2c_2(X) \geq -3\sigma(X) = 3|\sigma(X)|$. So $8\beta_2+16 \geq 11|\sigma(X)| + |\sigma(X)|$. Since $w_2(X) = 0$ we have by Rokhlin $16|\sigma(X)$. So $8\beta_2 \geq 11|\sigma(X)|$ as desired.

Case II: $\sigma(X) \geq 0$. Then $c_1^2(X)-3c_2(X) = 3\sigma(X)-c_2(X) \leq 0$. Thus $3|\sigma(X)| = 3\sigma(X) \leq c_2(X)$ or $24\,\sigma(X) \leq 8\beta_2(X) + 16$. Clearly then $8\beta_2(X) \geq 11|\sigma(X)|$.

3) We note that the realization of $(p,q) \in D_2$ is quite coarse. In fact if $X_g = F_2 \times F_g$, where F_j is a 2-manifold of genus j. Then $c_1^2(X_g) = 8(g-1)$ $c_2(V_g) = 4(g-1)$. If $X_{g,n} = X_g \# nQ$ then

$$c_1^2(X_{g,n}) = 8(g-1)-n \qquad c_2(X_{g,n}) = 4(g-1)+n\ .$$

Setting $g = \frac{p+q}{12} - 1$ and $n = \frac{2q-p}{3} \geq 0$ we see that if $p+q \geq 12$ then

$$c_1^2(X_{g,n}) = p \qquad c_2(X_{g,n}) = q\ .$$

By using other combinations $F_{p1} \times F_{p2}$ it is easily seen that all $(p,q) \in D_2$ can be realized.

4) The question of which connected sums admit complex structures is an intriguing one. It is a standard fact in algebraic geometry that if M is a complex surface, then so is $M \# Q$ ($Q = \overline{CP^2}$ as before). Conversely the theorem of Castelnuevo-Kodaira [K1] says that 'embedded Q's' can be 'excised' from complex surfaces still leaving complex manifolds. More precisely $Q - D^2$ can be thought of as a tubular neighborhood of an embedded sphere with self-intersection -1 (i.e. a (-1)-Hopf-disc-bundle over S^2). The Castelnuevo-Kodaira theorem then states that if N is a complex surface and L is a holomorphically embedded 2-sphere with self-intersections -1 in N then there exists a complex surface M and a holomorphic surjection $\varphi: N \to M$ with $\varphi|N-L$ a biholomorphic isomorphism and $\varphi(L)$ a point in M. In this case $N = M \# Q$. This process is called 'blowing down' while the reverse process of going from M to $M \# Q$ is called 'blowing up'.

We note that $c_1^2(N) = c_1^2(M) - 1$

$$c_2(N) = c_2(M) + 1$$

A complex surface M which has no holomorphic embedded (-1)-spheres is called a minimal surface.

There are no known examples of simply-connected minimal surfaces which are diffeomorphic to connected sums, and an attractive conjecture is:

CONJECTURES (1) _Let N be a simply-connected compact smooth 4-manifold which is diffeomorphic to a connected sum $A \# B$ of smooth manifolds. Then either N does not admit a complex structure or $N = M \# rQ$ for some complex manifold M. In addition,_

(2) _Simply-connected minimal complex surfaces do not admit smooth connected sum decompositions._

By the work of Freedman, this conjecture is of course false in the topological category where such decompositions do of course exist. The 2nd conjecture is false if we do not restrict ourselves to simply-connected surfaces. In [Ka] Kato shows that the Inoue Surfaces of [In], which have $\pi_1 \simeq \mathbb{Z}$, though minimal, are in fact diffeomorphic to blow-ups of Hopf-Surfaces.

If we restrict ourselves to minimal surfaces we can ask more precise questions about the realizability of points in D_3. We shall further restrict ourselves to surfaces V of general type. For our purpose this is equivalent to demanding that $V \neq CP^2$ and $c_1^2(V) > 0$. Under these added hypothesis we state:

THEOREM 7 (Bombieri) [B]

Let V be a minimal surface of general type. Then

$$c_1^2(V) \geq \frac{c_2(V)-36}{5}$$

It turns out that it is generally more convenient to work with $\frac{c_1^2+c_2}{12}$ rather than with c_2. We set $\chi(V) = \frac{c_1^2(V)+c_2(V)}{12}$. $\chi(V)$ is called the complex Euler Characteristic. Then restating Theorems 6 and 7 in terms of χ we obtain:

THEOREM 8 _Let V be a minimal surface of general type. Then_

$$2\chi(V)-6 \leq c_1^2(V) \leq 9\chi(V) .$$

More exactly, if $c_1^2(V)$ is even, then $c_1^2(V) \geq 2\chi(V)-6$, while if $c_1^2(V)$ is odd,

$$c_1^2(V) \geq 2\chi-5 .$$

If equality holds on the lower end then by work of Horikawa [H] V must be simply-connected. Thus in terms of the (χ, c_1^2) invariants the admissible region is sketched below. We shall henceforth denote it by D^* in Fig. 1. We now ask again

QUESTION <u>Which points in the admissible region are in fact realizable by minimal surfaces of general type? What can be said about</u> π_1 <u>of those surfaces?</u>

There is a tremendous difference in the quality as well as quantity of our results depending on whether $\sigma \leq 0$ or $\sigma > 0$. We note that $\sigma(V) \leq 0 \Leftrightarrow c_1^2(V) \leq 8\chi(V)$. Restricting ourselves to these surfaces we obtain:

THEOREM 9 (Persson) [P]. <u>Suppose</u> $(x,y) \in D^*$ <u>and</u> $y \leq 8x$. <u>Then provided</u> $y \neq 8x-k$ <u>for</u> $k \in \{1,2,3,5,6,7,9,11,13,15,19\}$, (x,y) <u>is realizable by a surface</u> V <u>with</u> $\chi(V) = x$, $c_1^2(V) = y$. <u>Furthermore if</u> $y \leq 8(x - \sqrt{30x})$ <u>then</u> (x,y) <u>is realizable by a simply-connected minimal surface of general type.</u>

Looking at Fig. 1 we remark that

1) Double coverings of CP^2 and hypersurfaces of CP^3 have invariants which tend to the line $y=4\chi$

2) Complete intersection surfaces tend to the region $6\chi \leq y \leq 8\chi$.

3) Persson has shown that if V admits a genus 2 pencil the $c_1^2(V) \leq 7\chi(V)$ and he conjectures that

CONJECTURE (Persson)

(A) <u>If</u> X <u>admits a hyperelliptic pencil then</u> $c_1^2 \leq 8\chi$.

(B) <u>If</u> X <u>is simply-connected then</u> $c_1^2 \leq 8\chi$.

We now turn to Complex Surfaces with Positive Signature. Here much less is known!

1. In [Z] Zappa conjectured that there does not exist a family of algebraic surfaces with arbitrarily large positive signature. This was disproved by Borel in [Bo] where he constructed infinite families of groups Γ of analytic isomorphisms acting on the unit ball $B_2 \subset \mathbb{C}^2$. These give rise to infinite families of complex surfaces $V = B_2/\Gamma$ with $c_1^2(V) = 9\chi(V)$ and $\sigma(V) \to \infty$. Recently, partially as a result of Yau's work, interest in these types of surfaces has been reviewed and new examples have been found by Mumford, Hirzebruch and others (see [Mum, Hz2).

2. In the region $8\chi(V) < c_1^2(V) < 9\chi(V)$ or equivalently $2c_2 < c_1^2 \leq 3c_2$ until 1979 the only known examples were Fg_1-bundles over Fg_2. These were originally constructed by Kodaira [K2] with extensions by Atiyah and Hirzebruch [At] [Hz1]).

3. In 1980 Mostow and Siu [MS] gave new examples of a completely different class of positive signature surfaces. They constructed infinite families of groups Γ acting freely on bounded domains $B \subset \mathbb{C}^2$ which were not biholomorphic to the unit ball in $\mathbb{C}^2$. These then give rise to surfaces $V = B/\Gamma$ with invariants satisfying $8\frac{8}{11}\chi(V) \leq c_1^2(V) \leq 8.727\chi(V)$ or equivalently $2\frac{2}{3}c_2(V) \leq c_1^2(V) \leq 2.942c_2(V)$. Their surfaces admit Kähler metrics with negative sectional curvature and are thus also counterexamples to the conjecture that all such surfaces have Universal Covering Surface biholomorphic to the

unit ball.

All of the previously mentioned examples are obtained either as bundles of 2-surfaces over 2-surfaces or as quotients of groups acting freely on a 4-cell. In particular, they are all $K(\pi,1)$'s. In [VV2] Van-de-Ven asked whether all minimal surfaces of general type with $c_1^2 > 8\chi$ were $K(\pi,1)$'s? Recently, we have shown that there exist infinitely many families of surfaces with $c_1^2 > 8\chi$ which are not $K(\pi,2)$'s. In order to explain our construction we first review the Kodaira-Hirzebrvch examples.

The basic idea of the construction is to construct surfaces of positive signatures as branched covers of simpler surfaces. The G-signature theorem says that if $X \to V$ is an n-fold branched covering of surfaces with ramification divisor $E \subset V$ then $\sigma(X) = n\sigma(V) - (n^2-1)/3n\ (E \cdot E)$. The key step in constructing branched coverings with positive signature then turns out to be finding divisors $E \subset V$ with $E \cdot E \ll 0$ and such that E is "divisible" in the sense that as a homology class $[E] = n[D]$ for some integer $n > 1$ and homology class $[D] \in H_2(V)$.

Given such an E we can in a straightforward fashion construct the n-fold cyclic branched cover of V over E and if $[D]^2 = -d$ then

$$\sigma(X) = \frac{d}{3}\, n(n^2-1) + n\sigma(V)$$

which will be positive provided n is sufficiently large.

The problem with this approach is that divisors with negative self-intersection are 'rigid' in complex surfaces and usually are not divisible.

In the Kodaira-Hirzebrvch examples we begin with a compact Riemann-surface (compact smooth 2-manifold with complex structure) R_o of genus $g(R_o) > 1$ and construct a 2-fold unramified covering $R \to R_o$ with involution τ $(R/\tau = R_o)$.

We then let $T \xrightarrow{f} R$ be the universal $H_1(R,\mathbb{Z}_n)$ covering of R. This is a regular $n^{2g(R)}$-fold covering of R. We now let Γ_f be the graph of f in $T \times R$ and $\Gamma_{\tau f}$ be the graph of τf.

It can then be checked that

1) $\Gamma_f \cdot \Gamma_f = \Gamma_{\tau f} \cdot \Gamma_{\tau f} = n^{2g(R)} \cdot (2-2g(R)) < 0$.

2) If $[E] = \Gamma_f - \Gamma_{\tau f}$ then looking at $[E]$ as a class in $H_1(T\times R, \mathbb{Z}_n)$ we see $[E] \sim 0$.

Thus $n \mid [E]$ as a class in $H_1(T\times R)$. Furthermore $E\cdot E = 2n^{2g(R)} \cdot (2-2g(R))$.

3) Since $R \to R_o$ is an unbranched cover, τ has no fixed points and so $[E]$ has a non-singular representative E.

We can now construct the n-fold branched cover $X \to (T\times R)$ ramified over E, which will be the desired surface of positive signature. It is clear that S is an F_{g_1}-bundle over F_{g_2} where $F_{g_2} = T$ and F_{g_1} is an n-fold cyclic cover

over R ramified at 2 points.

Thus setting $g = \text{genus}(R_o)$, so $g(R) = 2g-1$ and $X = X(n,g)$, we find

$$g_1 = \text{genus}(F_{g_1}) = n(2g-1)$$

$$g_2 = \text{genus}(F_{g_2}) = 2n^{4g-2}(g-1)+1$$

$$\sigma(X(n,g)) = 8/3\ (n^2-1)(g-1)n^{4g-3}$$

$$c_1^2(X(n,g)) = 8n^{4g-3}(g-1)[n^2(4g-1)-(2n+1)]$$

$$c_2(X(n,g)) = 8[n^2(2g-1)-n](g-1)\cdot n^{4g-2}$$

$$\chi(X(n,g)) = \frac{2}{3}(g-1)n^{4g-3}[2n^2(3g-1)-(3n+1)]$$

We note that

$$c_1^2/\chi = 8 + \frac{1}{3}\left[\frac{n^2-1}{(2n^2(3g-1)-(3n+1))}\right], \quad c_1^2/c_2 = 2 + \left[\frac{n^2-1}{n^2(2g-1)-n}\right]$$

so that

$$\lim_{n\to\infty} c_1^2/\chi = 8 + \frac{2}{3g-1} \quad ; \quad \lim_{g\to\infty} c_1^2/\chi = 8$$

while

$$\lim_{n\to\infty} c_1^2/c_2 = 2 + \frac{1}{2g-1} \quad ; \quad \lim_{g\to\infty} c_1^2/c_2 = 2 .$$

The 'smallest' such example is $X(2,2)$ which is a bundle of genus 6 curves over a genus 129 base with

$$\sigma(2,2) = 4.64 \qquad c_1^2 = 92.64 \qquad c_2 = 40.64 \qquad \chi = 11.64 .$$

In order to modify this construction and produce surfaces which are not-simply fiber bundles we recall the classic construction of the Kummer Surface. The Kummer Surface is constructed by letting $X = T^2 \times T^2$ be the product of two tori and letting $\sigma = i \times i$ be the product of the canonical involutions on them. We then let $Y = X/\sigma$. Since σ has 16 fixed points, Y has 16 points where it is not a manifold but rather a cone C on RP^3. We then 'resolve' these singularities by replacing the C's with disc-bundles which have RP^3 as boundary. This can be done in a complex analytic fashion and we wind up with the Kummer Surface V, a simply-connected complex analytic manifold which is certainly not a $K(\pi,1)$.

The basic idea is therefore to use quotienting by a non-free group action to introduce singularities. Resolving these singularities thus produces a manifold which is not a $K(\pi,q)$. This construction is carried out in [Man]. Whole new families of minimal complex surfaces, as with positive signatures, are constructed, none of which are $K(\pi,q)$'s. The smallest such example

$X(3,1,1)$ satisfies $\sigma(X) = 90$ $c_2(X) = 2106$ $c_1^2(X) = 4482$ $\chi(X) = 549$. All of these surfaces while not being F_{g_1}-bundles over F_{g_2} mimic the Kummer Surface in being singular F_{g_1} fibrations over on F_{g_2}. The singularities can be constructed to be sufficiently complex to kill all of the π_1 coming from the fiber so that $\pi_1(V) = \pi_1(F_{g_2})$, while the resolution of the singularities introduces non-trivial elements into π_2.

These new surfaces though not $K(\pi,q)$'s still have infinite fundamental group and all lie in the sector $8 < \frac{c_1^2}{\chi} \leq 8\frac{2}{5}$ or equivalently $2 < \frac{c_1^2}{c_2} \leq 2\,1/3$.

We thus still do not have a single example of a minimal complex surface with invariants within the region $2\frac{1}{3}c_2 \leq c_1^2 \leq 2\frac{2}{3}c_2$ (equivalently $8\frac{2}{5}\chi \leq c_1^2 \leq 8\frac{8}{11}\chi$). Which is not a $K(\pi,1)$. In addition every example of a minimal surface of general type with positive signature has infinite fundamental group. We can thus ask whether any such surfaces with finite π_1 exist? Lastly, it is still of immense interest to determine just how closely the positive signature region is packed and what are its 'smallest' elements.

BIBLIOGRAPHY

[A] L. Ahlfors, Quasiconformal Mappings, Teichmuller Spaces and Kleinian Groups, Proc. Int. Conf. of Mathematicians, Helsinki (]978), 71-84.

[At] M. F. Atiyah, The signature of fibre-bundles, Global Analysis, Princeton Univ. Press, Princeton, N.J., 1969, pp. 73-84.

[B] E. Bombieri, Canonical models of surfaces of general type, Publ. Math. I.H.E.S., 42 (1973), 171-219.

[Bo] A. Borel, Compact Clifford-Klein forms of symmetric spaces, Topology 2 (1963), 111-122.

[D] S. K. Donaldson, An application of gauge theory to the topology of 4-manifolds, J. Diff. Geo. (1983),

[E] C. Ehresmann, Sur Les Varietes Presque Complexes

[F] M. H. Freedman, The topology of four-dimensional manifolds, J. Diff. Geo. 17 (1982), 357-454.

[Hzl] Hirzebruch, The signature of ramified coverings, Global Analysis, Princeton Univ. Press, Princeton, N.J., 1969, pp. 253-265.

[Hz2] ——————, Preprint

[H] E. Horikawa, Algebraic surfaces of general type with small c_1^2, I and II, Ann. Math., 104 (1976), 357-387 and Inv. Math., 31 (1976), 43-85.

[I] M. Inoue, New surfaces with no neromorphic functions, Proc. Int. Congress of Math. Vancouver, 1974.

[Ka] Ma. Kato, Compact complex manifolds containing "global" spherical I, Intl. Symp. on Algebraic Geometry, Kyoto, 1977, 45-89.

[Kl] K. Kodaira, On the structure of compact complex analytic surfaces, I, Amer. J. of Math., 86 (1964), 751-798.

[K2] K. Kodaira, A certain type of irregular algebraic surface, J. Analyse Math. 19 *1967), 207-215.

[Man] R. Mandelbaum, Some new durfaces of General Type, AMS Proceedings of Symposia in Pure Math., vol. 40 (1980) 882, 193-197.

[M] J. Milnor, On the cobordism ring Ω^+ and a complex analogue, I, Am. J. Math., 82, (1960), 505-521.

[Mi] Y. Miyaoka, On the Chern numbers of surfaces of general type, Invent. Math. 42 (1977), 225-237.

[Ms] G. D. Mostow and Y. T. Siu, A compact Köhler surface of negative curvature not covered by the ball, Annals of Math. 112 (1980), 321-360.

[Mum] D. Mumford, An algebraic surface with K ample, $(K^2) = 9$, $p_g = q = 0$, pp. 232-242.

[P] Ulf Persson, On Chern invariants of surfaces of general type, preprint.

[Sml] A. Sommesse, On the density of ratios of Chern numbers of algebraic surfaces (preprint).

[Ta] C. H. Taubes, Self-dual Yang-Mills connections on non-self-dual n-manifolds, J. Diff. Geo 17 (1982).

[T] Thurston, Lectures on hyperbolic geometry and 3-manifolds (mimeo'd notes).

[VV1] A Van de Ven, On the Chern numbers of certain complex and almost complex manifolds, Proc. Nat. Acad. Sci. U.S.A., 55 (1966), 1624-1627.

[VV2] A Van de Ven, On the Chern numbers of surfaces of general type, Inv. Math., 36 (1976), 285-293.

[VV3] A. Van de Ven, Some recent results on surfaces of general type, Seminar Bourbaki 1976-77, No. 500, February 1977.

[W] Wen-Tsui Wu, Sur la structure presque complexe d'ane variete differentiable re elle de dimension 4, C. R. Acad. Sci. Paris, 227 (1948).

[Y] S. T. Yau, Calabi's conjecture and some new results in algebraic geometry, Proc. Nat. Acad. Sci, U.S.A. 74 (1977), 1798-1799.

[Z] G. Zappa, Sopra une probabile diseguaglianza tra i cautteri invariantivi di una superficie algebrica, Rend. Mat. 14 (1955), 455-464.

DEPARTMENT OF MATHEMATICS
UNIVERSITY OF ROCHESTER
ROCHESTER, NY 14627

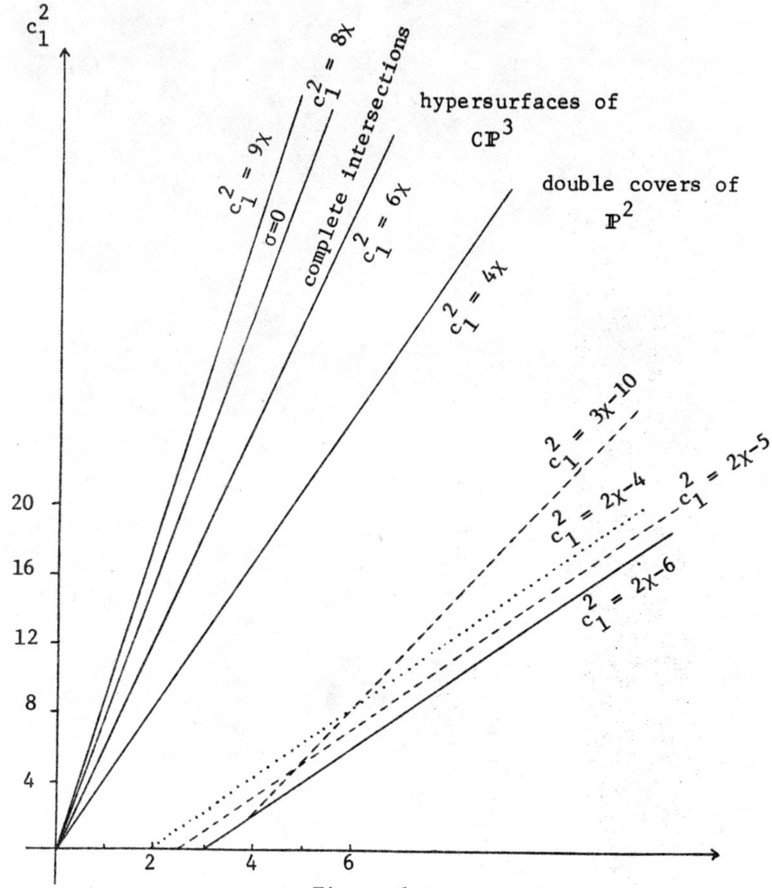
c_1^2
$c_1^2 = 9\chi$
$\sigma=0$
$c_1^2 = 8\chi$
complete intersections
$c_1^2 = 6\chi$
hypersurfaces of $\mathbb{CP}^3$
double covers of $\mathbb{P}^2$
$c_1^2 = 4\chi$
$c_1^2 = 3\chi-10$
$c_1^2 = 2\chi-4$
$c_1^2 = 2\chi-5$
$c_1^2 = 2\chi-6$
20
16
12
8
4
2
4
6

Figure 1

Contemporary Mathematics
Volume 35, 1984

GOOD TORUS FIBRATIONS

Yukio Matsumoto

1. INTRODUCTION

Fiberings over surfaces with fiber the 2-torus, which admit certain singular fibers, have been studied by several authors. Thornton [Th] defined such singular fibrations (with fiber any dimensional torus) and Zieschang [Z] studied their homeomorphism types. The singular fibers they considered were essentially the multiple tori (in the terminology of Section 3). Moishezon [Msh] studied Lefschetz and Kodaira fibrations of 2-tori, and Harer [H] generalized them to fibrations with fiber an arbitrary surface. Sakamoto and Fukuhara [SF] dealt with T^2-bundles over T^2, one of the exceptional cases which Zieschang [Z] did not treat.

In a previous note [Mt1], the author introduced torus fibrations with more general singular fibers. The only restriction on singular fibers, there, was that they should locally look like a 'cone' over a 'multiple fibered link'. The following is proved: Let M be a closed smooth 4-manifold which has a special handle decomposition without 1 and 3-handles, then there exists a torus fibration over the 2-sphere $f:M \to S^2$. (For the proof, see [Mt2].)

Unfortunately, the singular fibers of our torus fibrations are too complicated in general to attack directly. Thus a natural program would be: First to set up a reasonable subclass of torus fibrations with 'easy' singular fibers, and then to develop certain deformation theory through which general singular fibers are simplified.

In this paper we are concerned with the former step.

Following Neumann [N], we say a torus fibration is good if the only singularities of the singular fibers are normal crossings. The purpose of this paper is to study some basic properties of good torus fibrations, having the following problem in mind: Classify all closed, oriented smooth 4-manifolds that admit good torus fibrations, up to orientation preserving (but not necessarily fiber preserving) diffeomorphism.

In Sections 2 and 3, we classify all possible types of singular fibers of good torus fibrations. In Section 4, we describe their monodromy matrices. In Section 5, it is proved that a homology 4-sphere that admits a good torus

0271-4132/84 $1.00 + $.25 per page

fibration without multiple fibers is diffeomorphic to S^4. In Section 6, we discuss some replacement techniques of singular fibers. A theorem proved in Section 7 would be worthwhile to state here:

THEOREM 7.1 Let M be a closed, oriented, smooth 4-manifold which admits a good torus fibration without multiple fibers. Suppose that $H_1(M;\mathbb{Z}) = \{0\}$ and that the intersection form $H_2(M) \otimes H_2(M) \to \mathbb{Z}$ is positive definite. Then M is degree (+1)-diffeomorphic to $\mathbb{C}P_2 \# \cdots \# \mathbb{C}P_2$.

This result illustrates the Donaldson theorem [D] in the class of good torus fibrations. In Section 8, we give a theorem which generalizes Kas' classification of regular elliptic surfaces.

2. GENERAL PROPERTIES OF SINGULAR FIBERS

First we give a precise definition of good torus fibrations.

DEFINITION. Let M and B be oriented 4 and 2-dimensional smooth manifolds, $f:M \to B$ a proper, surjective, smooth map. We call $f:M \to B$ a good torus fibration (GTF) if it satisfies the following: (i) at each point $p \in \mathrm{Int}(M)$, the germ (f,p) is smoothly (+)-equivalent (in other words, is conjugate via orientation preserving diffeomorphisms) to the germ at $\mathbb{O} = (0,0)$ of the complex valued functions $z_1^m z_2^n$ or $\bar{z}_1^m z_2^n : \mathbb{C}^2 \to \mathbb{C}$, where m and n are non-negative integers, not necessarily coprime, satisfying $m+n \geq 1$; (ii) there exists a set of isolated points, σ in $\mathrm{Int}(B)$ so that $f|f^{-1}(B-\sigma):f^{-1}(B-\sigma) \to B-\sigma$ is a smooth T^2-bundle over $B-\sigma$, where $T^2 = S^1 \times S^1$.

We call f, M and B the projection, the total space and the base space of the GTF, respectively.

Those points $p \in \mathrm{Int}(M)$, at which the exponents of the germ of f satisfy the inequality $m+n \geq 2$, make a nowhere dense subset $\Sigma \subset M$. We assume $f(\Sigma) = \sigma$ and call σ the set of singular values.

Let $f:M \to B$ be a GTF.

DEFINITION. The fiber $F_x = f^{-1}(x)$ is called a general or singular fiber according as $x \in B-\sigma$ or $x \in \sigma$.

Let $D_\delta(x)$ be a 2-disk of radius $\delta > 0$ in $\mathrm{Int}(B)$ centered at $x \in \mathrm{Int}(B)$. If $\delta > 0$ is small enough, $N_{\delta,x} = f^{-1}(D_\delta(x))$ is a regular neighborhood of the fiber F_x.

Let F_x be a singular fiber. It is easy to see that F_x is connected. F_x is written as a union $F_x = \Theta_1 \cup \cdots \cup \Theta_s$ of its irreducible components, each of which is a smoothly immersed surface, (cf. [N, p.337]). The mutual and self intersections between the components are transverse. Thus $N_{\delta,x}$ is a plumbed manifold, (cf. [N]).

LEMMA 2.1. The fundamental group $\pi_1(F_x)$ is isomorphic to either $\mathbb{Z} \oplus \mathbb{Z}$, $\mathbb{Z}$ or $\{1\}$.

PROOF. Let F be a general fiber in $N_{\delta,x}$. We have $\pi_1(F) = \mathbb{Z}\oplus\mathbb{Z}$. It is shown that the image of $\pi_1(F)$ under the homomorphism induced by the inclusion $\pi_1(F) \to \pi_1(N_{\delta,x}) \cong \pi_1(F_x)$ is a subgroup of finite index (see [Mt2]). This together with the fact that $\pi_1(F_x)$ is isomorphic to a free product of $\mathbb{Z}$'s and surface groups prove the lemma. ⦀

COROLLARY 2.1.1. All the irreducible components of F_x except one at most are smoothly embedded 2-spheres. The exceptional one, if any, is either an immersed 2-sphere with a single transverse self-intersection or a smoothly embedded 2-torus.

Since the total space M and the base space B are oriented, the general fibers and each irreducible component of a singular fiber are naturally oriented. In particular, the components $\Theta_1,\dots,\Theta_s$ of F_x determines a basis $[\Theta_1],\dots,[\Theta_s]$ of $H_2(N_{\delta,c};\mathbb{Z})$. If F is a general fiber in $N_{\delta,x}$, we have

$$[F] = m_1[\Theta_1] + \cdots + m_s[\Theta_s] \in H_2(N_{\delta,x};\mathbb{Z}) ,$$

where $m_i \geq 1$. We call m_i the multiplicity of Θ_i, see [N,p.337].

In what follows, singular fibers will be classified by the weighted graphs. They are essentially plumbing diagrams, but instead of the self-intersection number, the multiplicity is attached to each vertex. (The vertices are in one to one correspondence to the irreducible components of the singular fiber F_x in question.) Each edge is assigned the sign ε (+1 or -1) of the corresponding intersection. If a vertex corresponds to a component which is T^2, we attach [1] (the genus) to the vertex.

LEMMA 2.2. If a vertex v with multiplicity m is the common end point of the k-edges as shown in Fig. 2.1, then the component Θ corresponding to v has the self-intersection number $[\Theta]\cdot[\Theta] = -(\sum_{i=1}^{k} \varepsilon_i m_i)/m$. (We disregard a loop if v has any.)

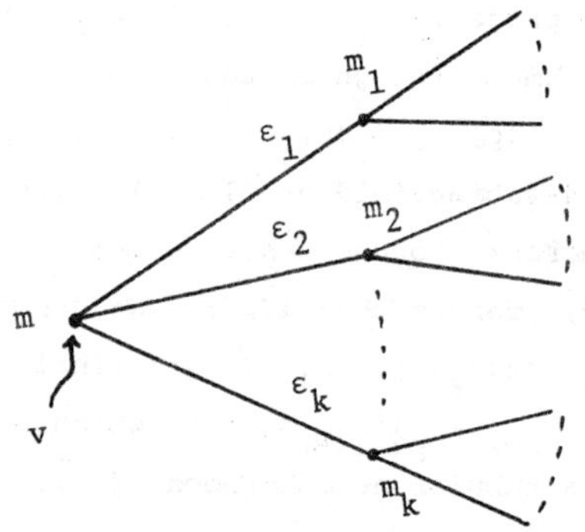

Fig. 2.1

The proof is well-known: Let F be a general fiber in $N_{\delta,x}$. Since F doesn't intersect any irreducible component of F_x, we have

$$0 = [F]\cdot[\Theta] = (m[\Theta] + m_1[\Theta_1] + \cdots + m_k[\Theta_k] + \cdots)\ [\Theta]$$
$$= m[\Theta]\ [\Theta] + \sum_{i=1}^{k} \varepsilon_i m_i .$$

Thus the lemma follows.

LEMMA 2.3. *Let* Γ *be a chain as in Fig.* 2.2(i). *Then we have* $\gcd(m_0,m_1) = \gcd(m_1,m_2) = \cdots = \gcd(m_{\nu-1},m_\nu)$. *If* Γ *is a linear branch as in Fig.* 2.2(ii), *then* $\gcd(m_0,m_1) = \cdots = \gcd(m_{\nu-1},m_\nu) = m_\nu$.

(i)

(ii)

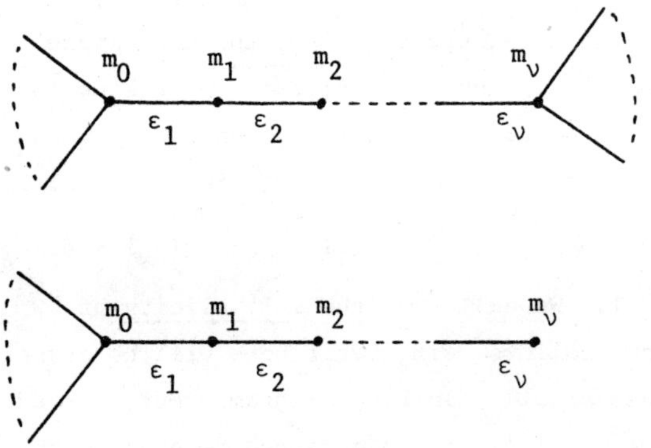

Fig. 2.2

REMARK. In Lemma 2.3 we assume that the vertex v_i with multiplicity m_i is labelled no index of genus, [1], $1 \leqq i \leqq \nu$. Therefore, the corresponding irreducible component Θ_i is a 2-sphere for $1 \leqq i \leqq \nu$.

DEFINITION. For a linear branch Γ, define a positive integer $p(\Gamma)$ by $p(\Gamma) = m_0/m_\nu$.

Before proving 2.3, we introduce some notation. Let $I(F_x)$ denote the set of the intersection points between different irreducible components of a singular fiber F_x. Let $p \in I(F_x)$. Choose 'local complex coordinates' z_1, z_2 in which the projection $f: M \to B$ is written locally as $f = z_1^m z_2^n$ or $f = \bar{z}_1^m z_2^n$. Choose δ, ε small enough so that $0 < \delta < \varepsilon$.

Define $\Delta_\delta(p)$ by $\Delta_\delta(p) = \{(z_1,z_2) \mid |f| \leqq \delta,\ |z_i| \leqq \varepsilon,\ i = 1,2\}$. Then $\Delta_\delta(p)$ is a smooth 4-submanifold of $\mathrm{Int}(M)$ with corners along the boundary. $\Delta_\delta(p)$ is PL homeomorphic to the 4-disk. Let Θ be an irreducible component of F_x and let $I(\Theta)$ denote $\Theta \cap I(F_x)$. We denote the punctured surface $\Theta - \bigcup_{p \in I(\Theta)} \mathrm{Int}(\Delta_\delta(p))$ by $\check{\Theta}$ and finally we denote the connected component of closure $(N_{\delta,x} - \bigcup_{p \in I(\Theta)} \Delta_\delta(p))$ which contains $\check{\Theta}$, by $N_{\delta,x}(\check{\Theta})$. Note that $N_{\delta,x}(\check{\Theta})$ is a tubular neighborhood of $\check{\Theta}$. Now $N_{\delta,x} = f^{-1}(D_\delta(x))$ is decomposed as follows:

$$N_{\delta,x} = \Big(\bigcup_{i=1}^{s} N_{\delta,x}(\check{\Theta}_i)\Big) \cup \Big(\bigcup_{p \in I(F_x)} \Delta_\delta(p)\Big),$$

see Figure 2.3.

Fig. 2.3

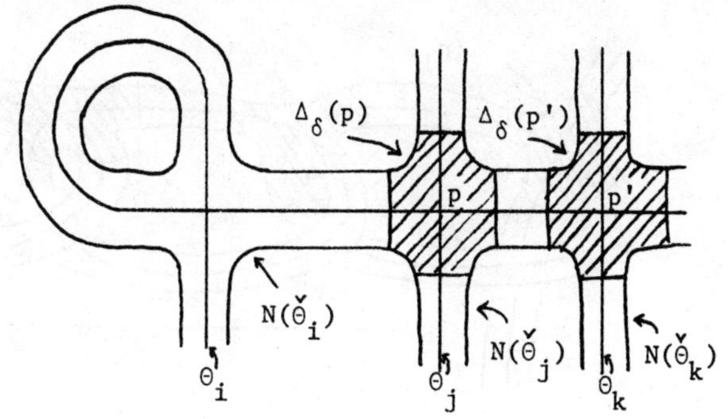

PROOF OF 2.3. Let Θ_i be the irreducible component corresponding to the vertex v_i of multiplicity m_i, $0 \leqq i \leqq \nu$. For $1 \leqq i \leqq \nu$, we assume that Θ_i is a 2-sphere. Let $\{p_i\} = \Theta_{i-1} \cap \Theta_i$.

For a nearby general fiber F, $F^{(i)} = F \cap N_\delta(\check{\Theta}_i)$ is an m_i-fold covering space over $\check{\Theta}_i$ which is an annulus for $1 \leqq i \leqq \nu-1$. Thus $F^{(i)}$ is a disjoint union of annuli. Let r_i be the number of the components. Since $F^{(i)} \cong (F^{(i)} \cap \partial\Delta_\delta(p_i)) \times [0,1]$, r_i is equal to the number of the connected components of $F^{(i)} \cap \partial\Delta_\delta(p_i)$. This intersection is a torus link defined by $z_1^{m_{i-1}} z_2^{m_i} = \text{const.}$ (or $\bar{z}_1^{m_{i-1}} z_2^{m_i} = \text{const.}$). Thus r_i is equal to $\gcd(m_{i-1}, m_i)$. Similarly, from the fact $F^{(i)} \cong (F^{(i)} \cap \partial\Delta_\delta(p_{i+1})) \times [0,1]$, it follows that $r_i = \gcd(m_i, m_{i+1})$. This proves $\gcd(m_{i-1}, m_i) = r_i = \gcd(m_i, m_{i+1})$ for $1 \leqq i \leqq \nu-1$.

In the case of linear branch, $F^{(\nu)}$ consists of m_ν 2-disks, because $\check{\Theta}_\nu$ is a 2-disk. Thus m_ν is equal to the number of the components of $\partial F^{(\nu)} = F^{(\nu)} \cap \partial\Delta_\delta(p_\nu)$, that is $\gcd(m_{\nu-1}, m_\nu)$. ⦙⦙⦙

DEFINITION. A <u>removable linear branch</u> (RLB) is a linear branch Γ for which $p(\Gamma) = m_0/m_\nu = 1$.

For a linear branch Γ as in Fig. 2.2(ii), define the neighborhood $N_\delta(\Gamma)$ by

$$N_\delta(\Gamma) = \left(\bigcup_{i=1}^{\nu} N_{\delta,x}(\check{\Theta}_i)\right) \cup \left(\bigcup_{i=1}^{\nu} \Delta_\delta(p_i)\right),$$

see Fig. 2.4.

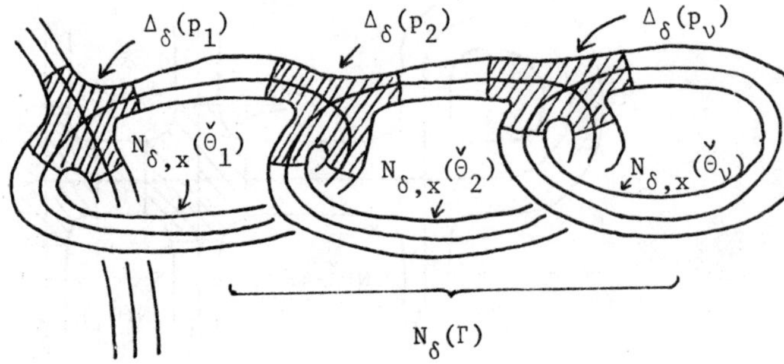

Fig. 2.4

Let $f: M \to B$ be a GTF, F_x a singular fiber.

THEOREM 2.4. *Suppose that the weighted graph for* F_x *has an RLB,* Γ. *Then the boundary* $\partial N_\delta(\Gamma)$ *is diffeomorphic to* $\partial(D^2 \times D^2)$. *By removing* $\operatorname{Int} N_\delta(\Gamma)$ *from* M *and filling in the hole with* $D^2 \times D^2$, *we have a new GTF* $f': M' \to B$. *The singular fibers do not change except for* F_x. F_x *changes to* F'_x *whose weighted graph lacks just the RLB,* Γ.

The diffeomorphism types of M and M' are related as follows:

ADDENDUM. $M \cong M' \# \pm\mathbb{C}P_2 \# \cdots \# \pm\mathbb{C}P_2$ *or* $M \cong M' \# S^2 \times S^2 \# \cdots \# S^2 \times S^2$ *according as* $N_\delta(\Gamma)$ *is a spin manifold or not. Moreover,* $e(M) = e(M') + \nu$ *and* $\sigma(M) = \sigma(M') - \sum_{i=1}^{\nu} \varepsilon_i$, *where* e *and* σ *denote the euler number and the signature, respectively.*

The first assertion of the addendum follows from [NW]. The numerical relations are proved by standard calculus on plumbing diagrams, see [N].

PROOF OF 2.4. Let Γ be the RLB, F a general fiber in $N_{\delta,x}$.

ASSERTION. $F \cap N_\delta(\Gamma)$ *is diffeomorphic to a disjoint union of* m_ν *2-disks.*

PROOF. Clearly we have

$$F \cap N_\delta(\Gamma) = \Big(\bigcup_{i=1}^{\nu} F \cap N_{\delta,x}(\check{\Theta}_i)\Big) \cup \Big(\bigcup_{i=1}^{\nu} F \cap \Delta_\delta(p_i)\Big).$$

$F \cap N_{\delta,x}(\check{\Theta}_i)$ consists of m_ν-annuli, for $1 \leqq i \leqq \nu-1$, (see the proof of 2.3). $F \cap \Delta_\delta(p_i)$ is essentially the Milnor fiber for the function $z_1^{m_{i-1}} z_2^{m_i}$ (or $\bar{z}_1^{m_{i-1}} z_2^{m_i}$), (see [Mi]). Since $\gcd(m_{i-1}, m_i) = m_\nu$, it consists of m_ν annuli. Finally, $F \cap N_{\delta,x}(\check{\Theta}_\nu)$ consists of m_ν 2-disks. Therefore, $F \cap N_\delta(\Gamma)$ is obtained by pasting these annuli successively and then capping off the boundaries by m_ν disks. This proves the assertion. ⁞⁞⁞

Let $D_F = F \cap N_\delta(\Gamma)$ be the disjoint union of m_ν disks. To see ∂D_F, let us take local complex coordinates z_1, z_2 around p_1 in which $\Delta_\delta(p_1)$ is defined as before. The boundary ∂D_F appears in the solid torus $T_\delta = \Delta_\delta(p_1) \cap N_{\delta,x}(\check{\Theta}_0)$. T_δ is given by $|z_1|^{m_0}|z_2|^{m_1} \leqq \delta, |z_2| = \varepsilon$, and in

T_δ, ∂D_F is a torus link defined by

$$z_1^{m_0} z_2^{m_1} = \text{const.}, \ |z_2| = \varepsilon, \quad \text{or}$$

$$\bar{z}_1^{m_0} z_2^{m_1} = \text{const.}, \ |z_2| = \varepsilon,$$

according as $\varepsilon_1 = +1$ or -1.

Here we make use of the assumption that Γ is an RLB, thus $m_0 = m_\nu$. Then $m_0 | m_1$, and let $k = m_1/m_0$. The above equation becomes $(z_1 z_2^k)^{m_0} = \text{const.}$ or $(\bar{z}_1 z_2^k)^{m_0} = \text{const.}$, and $|z_2| = \varepsilon$.

Take $D^2 \times D^2 = \{(z_1, z_2) \in \mathbb{C}^2 \mid |z_1| \leqq 1, |z_2| \leqq 1\}$ and define a diffeomorphism $h: D^2 \times (\partial D^2) \to T_\delta$ by

$$h(z_1, e^{i\theta}) = (\sqrt[m_0]{\delta}\, \sqrt[k]{\varepsilon}^{-1} z_1 e^{-ik\theta}, \varepsilon e^{i}) \qquad \text{or}$$

$$= (\sqrt[m_0]{\delta}\, \sqrt[k]{\varepsilon}^{-1} z_1 e^{ik\theta}, \varepsilon e^{i})$$

according as $\varepsilon_1 = +1$ or -1, where $i = \sqrt{-1}$.

Let $\tilde{N}_{\delta,x} = \text{closure } (N_{\delta,x} - N_\delta(\Gamma)) \cup_h D^2 \times D^2$. Then a GTF $\tilde{f}: \tilde{N}_{\delta,x} \to D_\delta(x)$ is defined by setting

$$\tilde{f} | \text{closure } (N_{\delta,x} - N_\delta(\Gamma)) = f | \text{closure } (N_{\delta,x} - N_\delta(\Gamma)), \quad \text{and}$$
$$\tilde{f} | D^2 \times D^2 = \delta z_1^{m_0}.$$

It is not difficult to see that h extends to a fiber preserving diffeomorphism $\tilde{h}: \partial(D^2 \times D^2) \to \partial N_\delta(\Gamma)$. Also $\tilde{f} | \partial N_{\delta,x}: \partial\tilde{N}_{\delta,x} \to \partial D_\delta(x)$ and $f | \partial N_{\delta,x}: \partial N_{\delta,x} \to \partial D_\delta(x)$ are isomorphic T^2-bundles. Therefore, $M' = (M - \text{Int}N_{\delta,x}) \cup \tilde{N}_{\delta,x}$ is a GTF over B. Now Theorem 2.4 is clear by the above construction. ⦀

3. CLASSIFICATION OF SINGULAR FIBERS

For $i = 1,2$, let $F_i = f_i^{-1}(x_i)$ be a singular fiber of a GTF $f_i: M_i \to B_i$. F_1 and F_2 are said to be equivalent if there exist neighborhoods D_1, D_2 of x_1, x_2 and orientation preserving diffeomorphisms $H: f_1^{-1}(D_1) \to f_2^{-1}(D_2)$ and $K: D_1 \to D_2$, so that $K(x_1) = x_2$ and $K\ f_1 = f_2\ H$.

THEOREM 3.1. Singular fibers of GTF's which are free from RLB are classified up to equivalence into the following six classes:

(i) class mI_0 (multiple tori, that is Seifert's fibered neighborhood $\times S^1$, [Th]). Their graphs are $\overset{m}{\underset{[1]}{\bullet}}$, $m \geq 1$,

(ii) class $\tilde{A}$ in which the graphs are cyclic

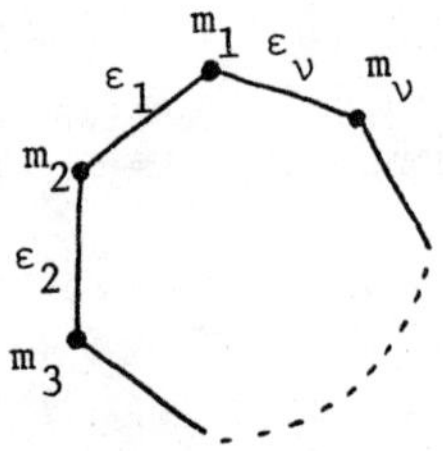

(iii) <u>class</u> $\tilde{D}$ <u>in which the graphs are of the form</u>

$\tilde{D}_4$ =

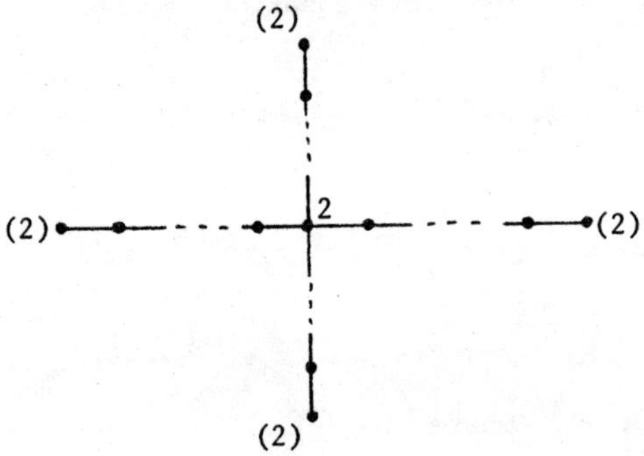

<u>or</u>

$\tilde{D}_{5+\nu}$ =

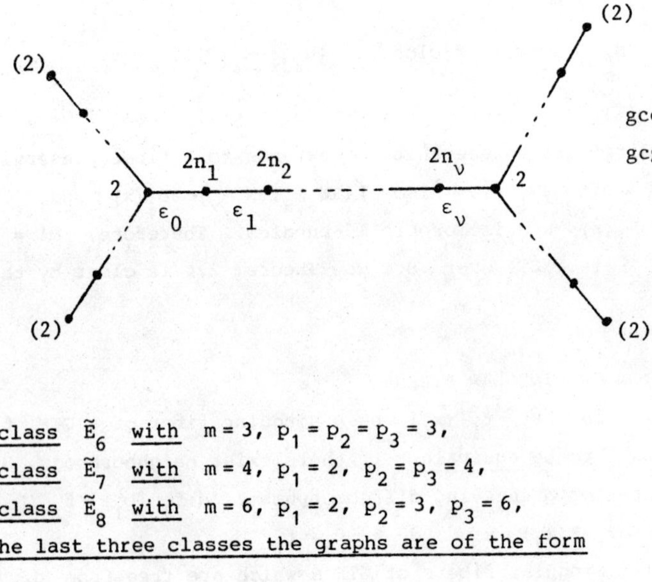

$\gcd(n_1,n_2) = \ldots$
$\mathrm{gcg}(n_{\nu-1},n_\nu) = 1$

$\nu \geq 0$

(iv) <u>class</u> $\tilde{E}_6$ <u>with</u> $m = 3,\ p_1 = p_2 = p_3 = 3,$

(v) <u>class</u> $\tilde{E}_7$ <u>with</u> $m = 4,\ p_1 = 2,\ p_2 = p_3 = 4,$

(vi) <u>class</u> $\tilde{E}_8$ <u>with</u> $m = 6,\ p_1 = 2,\ p_2 = 3,\ p_3 = 6,$

<u>where in the last three classes the graphs are of the form</u>

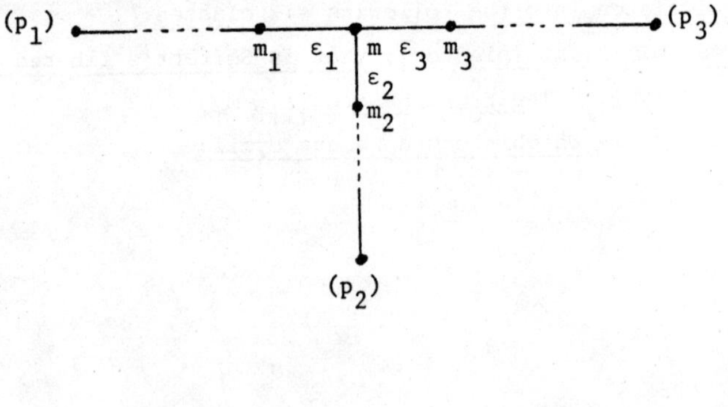

and the number in the parentheses at the end of a linear branch is the number $p(\Gamma)$ defined before. The weights must satisfy (a) the properties stated in Lemma 2.3 and (b) the divisibility $m \mid \Sigma \varepsilon_i m_i$ required by Lemma 2.2 at each vertex.

Moreover, each graph in (i) ~ (vi) can be realized as the graph of a singular fiber if its weights satisfy the above conditions (a), (b).

REMARK. Closely related classifications of diagrams are done in Scharf [Sch] and Neumann [N]. However, their classifications differ from ours in some points. They are mainly studying boundary 3-manifolds, while we are interested in GTF structures on 4-manifolds. Scharf considers only those diagrams for which, in our notation, each ε_i is +1.

PROOF OF 3.1. We will proceed along the following scheme:

- Case (I) Θ_i = embedded S^2 $\forall i$
  - Case (I-1) $\exists$ branch point
    - Case (I-1-1) $\exists$ type (iii) or (iv) $\longrightarrow$ $\tilde{E}_6, \tilde{E}_7, \tilde{E}_8, \tilde{D}_4$
    - Case (I-1-2) all branch points are of type (ii) $\longrightarrow$ $\tilde{D}_{5+\nu}$ $\nu \geq 0)$
  - Case (I-2) No branch point $\longrightarrow$ $\tilde{A}$
- Case (II) $\exists \Theta_i \neq$ embedded S^2 $\longrightarrow$ $mI_0, \tilde{A}$.

Let F_x be a singular fiber.

Case (I). All the irreducible components of F_x are smoothly embedded 2-spheres.

In what follows, the notation is as introduced before the proof of 2.3. Let Θ_0 be an irreducible component of F_x. Suppose that $I(\Theta_0) = \{q_1, \ldots, q_k\}$. The number k is called the valency of Θ_0 or of the vertex v_0 corresponding to Θ_0. Let Θ_i be the irriducible component which intersects Θ_0 at q_i, with sign $\varepsilon_i (= \pm 1)$. Let m_i be the multiplicity of Θ_i, $i = 0, 1, \ldots, k$. Take a nearby general fiber F. The intersection $\check{F} = F \cap N(\check{\Theta}_0)$ is a punctured surface and is an m_0-fold covering space over $\check{\Theta}_0$. We call $\check{F}$ the punctured surface at the vertex v_0. $N(\check{\Theta}_0)$ is a tubular neighborhood of $\check{\Theta}_0$. Let D_u denote the 2-disk fiber of $N(\check{\Theta}_0) \longrightarrow \check{\Theta}_0$ over $u \in \check{\Theta}_0$. D_u intersects $\check{F}$ in m_0 points sitting on a circle $\subset D_u$ centered at u. As u moves on $\check{\Theta}_0$ along a loop γ, the positions of the m_0 points on D_u change continuously and when u comes back to the original point, they are affected by the 'monodromy' σ_γ.

LEMMA 3.2. *Let $\gamma(i)$ be a small loop on $\check{\Theta}_0$ which goes around q_i once in the positive direction. Then the monodromy $\sigma_{\gamma(i)}$ is the rotation on D_u through the angle $-2\pi\varepsilon_i m_i / m_0$.*

This follows from the fact that the intersection $\partial\check{F} \cap \partial\Delta(q_i)$ is a torus link of type $(m_0, -\varepsilon_i m_i)$, (see the proof of 2.3).

Since $\check{\Theta}_0$ is a punctured sphere, a product of the loops $\gamma(1),\ldots,\gamma(k)$ in a certain order must be null homotopic on $\check{\Theta}_0$. Thus by Lemma 3.2, $\sum_{i=1}^{k} (-2\pi\varepsilon_i m_i / m_0)$ is an integral multiple of 2π.

COROLLARY 3.2.1. m_0 <u>divides</u> $\Sigma\varepsilon_i m_i$.

This divisibility has appeared in Lemma 2.2. This is a necessary and sufficient condition for the neighborhood $N(\Theta_0) = (\bigcup_{i=1}^{k} \Delta(q_i)) \cup N(\check{\Theta}_0)$ to be 'singularly fibered' over $D^2 \subset \mathrm{Int}(B)$ with general fiber a punctured surface $\cong F$ and with singular fiber a 2-sphere Θ_0 pierced by k 2-disks $\Theta_i \cap \Delta(q_i)$, $i=1,\ldots,k$.

The number of the components of $\partial\check{F} \cap \partial\Delta(q_i)$ is equal to $\gcd(m_0, m_i)$. Thus the number of the components of ∂F is $\sum_{i=1}^{k} \gcd(m_0, m_i)$. Let $\bar{F}$ be the closed surface obtained from $\check{F}$ by capping off the boundaries with 2-disks. The euler number $e(\bar{F})$ is given by

$$e(\bar{F}) = m_0(2-k) + \sum_{i=1}^{k} \gcd(m_0, m_i) .$$

$e(\bar{F})$ must be non-negative, for F is a codimension 0 submanifold of a torus F.

Putting $p_i = m_0 / \gcd(m_0, m_i)$, we have

$$(*) \qquad (2-k) + \sum_{i=1}^{k} (1/p_i) \geqq 0 .$$

We call $(p_1,\ldots,p_k)$ the <u>type</u> of the vertex v_0 corresponding to Θ_0.

LEMMA 3.3. <u>All the possible types of vertices are as follows</u>:

(i) (p_1), $k=1$;

(ii) $(p_1,\ldots,p_k) = (p,p,1,\ldots,1)$, $k \geqq 2$, $p \geqq 1$;

(iii) $(p_1,\ldots,p_k) = (3,3,3,1,\ldots,1)$, $(2,4,4,1,\ldots,1)$, $(2,3,6,1,\ldots,1)$, $k \geqq 3$;

(iv) $(p_1,\ldots,p_k) = (2,2,2,2,1,\ldots,1)$, $k \geqq 4$.

In fact, the inequality (*) produces these combinations plus five extra ones: $(p_1,p_2,1,\ldots,1)$, $p_1 \neq p_2$; $(2,2,p_3,1,\ldots,1)$, $p_3 \geqq 2$; $(2,3,3,1,\ldots,1)$; $(2,3,4,1,\ldots,1)$; $(2,3,5,1,\ldots,1)$. However, it is shown that these extra cases contradict Cor. 3.2.1.

Observe that the punctured surface $\check{F}$ at a vertex of type (i) is a disjoint union of disks. At a vertex of type (ii), each component of F is a punctured sphere, for $e(\bar{F}) > 0$. At a vertex of type (iii) or (iv) we have $e(\bar{F}) = 0$, and $\check{F}$ is proved to be connected, thus it is a punctured torus.

Since a torus F cannot contain two or more punctured tori, the weighted graph has at most one vertex of type (iii) or (iv), and if it contains one, the graph has no cycle.

A vertex v_0 is said to be a branch point if the valency $k \geqq 3$.

<u>Case</u> (I-1). <u>The graph contains a branch point.</u>

Case (I-1-1). The branch point, v_0 say, is of type (iii) or (iv).

In this case the graph is a tree. First assume that v_0 is of type (iii). We will show that the valency $k=3$. Suppose, on the contrary, $k \geqq 4$. Then the corresponding irreducible component Θ_0 would have an intersection point q_i for which $p_i = 1$. The branch Γ_i (of the graph) attached to Θ_0 at q_i cannot be linear, because a linear branch with $p_i = 1$ is an RLB. Thus Γ_i contains a branch point v_1. v_1 is no longer of type (iii) or (iv). Thus it is of type (ii), and a branch Γ with $p=1$ is attached to v_1. Γ is not linear (not being an RLB), thus Γ contains another branch point v_2 of type (ii), and so on. This argument continues endlessly. This absurdity shows that $k=3$.

Likewise we can show that the three branches attached to v_0 are linear. Thus the graph is in one of the classes $\tilde{E}_6$, $\tilde{E}_7$, $\tilde{E}_8$.

In case v_0 is of type (iv), a similar argument shows that the graph is in $\tilde{D}_4$.

Case (I-1-2). All the branch points are of type (ii).

Let $\{v_1,\dots,v_n\}$ be the totality of the branch points.

Let $k_i (\geqq 3)$, m_i be the valency and the multiplicity of v_i, respectively. The euler number of the punctured surface at v_i is equal to $m_i(2-k_i) < 0$.

Let Γ be a linear branch, if any, attached to v_i. Γ is not an RLB, thus $p(\Gamma) \geqq 2$. By Assertion in the proof of 2.4, $F \cap N(\Gamma)$ is a disjoint union of $m_i / p(\Gamma)$ 2-disks. We have $e(F \cap N(\Gamma)) = m_i / p(\Gamma)$. The branch point v_i has the type $(p_i, p_i, 1, \dots, 1)$. Thus there are at most two linear branches attached to v_i, since an RLB $(p(\Gamma)=1)$ is prohibited. Therefore, considering the sum of the euler numbers of the punctured surface at v_i and of $F\ N(\Gamma)$ for all the linear branches Γ attached to v_i, we know that the sum does not exceed $m_i(2-k_i) + 2(m_i / p_i)$. Since $k_i \geqq 3$, $p_i \geqq 2$, the latter sum is non-positive.

Note that the punctured surface at a non-branch, non-terminal vertex is a union of annuli, thus the euler number is equal to 0.

Now we obtain

$$0 = e(F) = \sum_{i=1}^{n} \{m_i(2-k_i) + 2(m_i / p_i)\} \leqq 0 ,$$

which implies $m_i(2-k_i) + 2(m_i / p_i) = 0$ for $\forall i = 1,\dots,n$. Thus $k_i = 3$, $p_i = 2$ for $\forall i = 1,\dots,n$, and v_i has exactly 2 linear branches. Each branch point has the form:

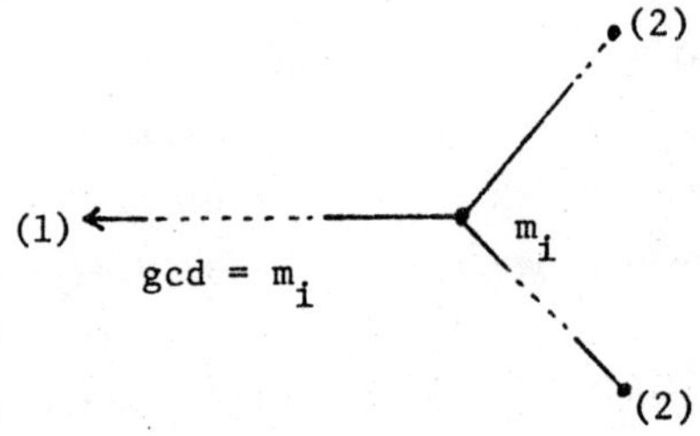

Since the graph is connected, $n = 2$. The graph must be of the form:

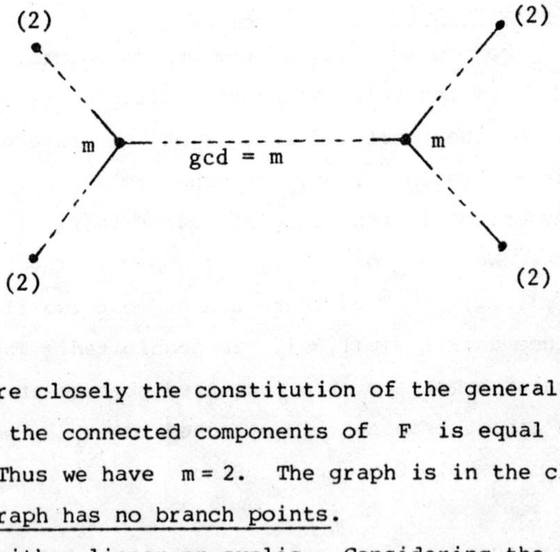

Inspecting more closely the constitution of the general fiber F, we see that the number of the connected components of F is equal to $m/2$. However, F is connected. Thus we have $m = 2$. The graph is in the class $\tilde{D}_{5+\nu}$ $(\nu \geqq 0)$.

Case (I-2). The graph has no branch points.

The graph is either linear or cyclic. Considering the euler number of F, a linear graph is impossible. Thus we obtain a graph in the class $\tilde{A}$.

Case (II). There exists an irreducible component which is an embedded torus or an immersed sphere with one self-intersection point.

By the euler number argument, there is no branch point. Also we can show linear branch attached to the 'exceptional' component is an RLB. This contradicts our assumption. Thus the graph has only one vertex. The graph is now in the class mI_0 or is a special case of $\tilde{A}$.

We have classified all the possible graphs. The existence and the uniqueness (up to equivalence) of the singular fibers that realize these graphs are clear from our constructive argument. ▥

REMARKS (1) The singular fiber consisting of an immersed S^2 is a special case of $\tilde{A}$. If its multiplicity $m=1$, we denote it by I_1^+ or I_1^- according as the sign ε of the self-intersection is $+1$ or -1. It is well-known that the monodromy around I_1^+ or I_1^- is $\begin{bmatrix} 1 & 1 \\ 0 & 1 \end{bmatrix}$ or $\begin{bmatrix} 1 & -1 \\ 0 & 1 \end{bmatrix}$, (see Section 4).

(2) In case all the intersections have sign $+1$, the above singular fibers reduce to Kodaira's ones or their blown ups [Ko1] according to the rule: $mI_0 \to {}_mI_0$, $\tilde{A} \to I_b$, $\tilde{D} \to I_b^*$, $\tilde{E}_6 \to$ {IV blown up, IV^*}, $\tilde{E}_7 \to$ {III blown up, III^*}, $\tilde{E}_8 \to$ {II blown up, II^*}.

4. MONODROMY

By the <u>monodromy around a singular fiber</u> F_x is meant the monodromy of the T^2-bundle over S^1, $\partial N_{\delta,x}(F_x) \to \partial D_\delta(x)$. Choose a positive basis (μ,λ) in a fiber with $\mu \cdot \lambda = +1$. Then the monodromy is represented by a 2×2 matrix A: $(h(\mu), h(\lambda)) = (\mu,\lambda)A$, where $\partial N_{\delta,x}(F_x) \cong T^2 \times [0,2\pi] / (t,2\pi) \sim (h(t),0)$, $h: T^2 \to T^2$ being an orientation preserving diffeomorphism. A is unique up to conjugation in $SL(2,\mathbb{Z})$.

THEOREM 4.1. <u>The monodromy matrices are given as follows (with the same m_i, n_i, ε_i as in Thm. 3.1):</u>

Class		monodromy matrix
mI_0		$\begin{bmatrix} 1 & 0 \\ 0 & 1 \end{bmatrix}$
$\tilde{A}$		$\begin{bmatrix} 1 & b \\ 0 & 1 \end{bmatrix}$ where $b = (\varepsilon_1/n_1n_2) + (\varepsilon_2/n_2n_3) + \cdots + (\varepsilon_\nu/n_\nu n_1)$ and $n_i = m_i/\gcd(m_1,\ldots,m_\nu)$.
$\tilde{D}$	$\tilde{D}_4$	$\begin{bmatrix} -1 & 0 \\ 0 & -1 \end{bmatrix}$
	$\tilde{D}_{\nu+5}$ $(\nu \geq 0)$	$\begin{bmatrix} -1 & -b \\ 0 & -1 \end{bmatrix}$ where $b = (\varepsilon_0/n_1) + (\varepsilon_1/n_1n_2) + \cdots + (\varepsilon_{\nu-1}/n_{\nu-1}n_\nu) + (\varepsilon_\nu/n_\nu)$.

$\tilde{E}_6$	$\begin{bmatrix} -1 & -1 \\ 1 & 0 \end{bmatrix}$ or $\begin{bmatrix} 0 & 1 \\ -1 & -1 \end{bmatrix}$ according as $\varepsilon_3 m_3 \equiv -1 \pmod 3$ or $\varepsilon_3 m_3 \equiv 1 \pmod 3$.
$\tilde{E}_7$	$\begin{bmatrix} 0 & -1 \\ 1 & 0 \end{bmatrix}$ or $\begin{bmatrix} 0 & 1 \\ -1 & 0 \end{bmatrix}$ according as $\varepsilon_3 m_3 \equiv -1 \pmod 4$ or $\varepsilon_3 m_3 \equiv 1 \pmod 4$.
$\tilde{E}_8$	$\begin{bmatrix} 0 & -1 \\ 1 & 1 \end{bmatrix}$ or $\begin{bmatrix} 1 & 1 \\ -1 & 0 \end{bmatrix}$ according as $\varepsilon_3 m_3 \equiv -1 \pmod 6$ or $\varepsilon_3 m_3 \equiv 1 \pmod 6$.

This theorem is proved by looking at the GTF structure of $N_{\delta,x}(F) \to D_\delta(x)$, closely. The details are omitted.

REMARK. Note that the trace $\mathrm{Tr}(A)$ characterizes the class of the singular fibers:

Class	mI_0 or $\tilde{A}$	$\tilde{D}$	$\tilde{E}_6$	$\tilde{E}_7$	$\tilde{E}_8$
Tr(A)	2	-2	-1	0	1

5. HOMOLOGY 4-SPHERES.

DEFINITION. A singular fiber $F_x = \sum_{i=1}^{s} m_i \Theta_i$ is called a <u>multiple fiber</u> if $\gcd(m_1,\ldots,m_s) > 1$.

F_x can be a multiple fiber only if F_x is in the class mI_0 or $\tilde{A}$ (see [Ko1, Lemma 6.1], Thm 3.1). The study of torus fibrations with multiple fibers seem to require a deeper theory, (cf. [Ko2]). In this paper, we will confine ourselves to GTF's without multiple fibers.

THEOREM 5.1. <u>Let</u> $f: M \to B$ <u>be a GTF without multiple fibers. If</u> $H_*(M;\mathbb{Z}) = H_*(S^4;\mathbb{Z})$,, <u>then</u> M <u>is diffeomorphic to the 4-sphere</u> S^4.

PROOF. From the fact $H_1(M;\mathbb{Z}) = \{0\}$ it follows $H_1(B;\mathbb{Z}) = \{0\}$, and B is a 2-sphere. A GTF over S^2 has the abelian fundamental group (which is generated by π_1(general fiber)), thus $\pi_1(M) = H_1(M) = \{0\}$, and M is a homotopy 4-sphere.

If $M = M' \# \pm \mathbb{C}P_2 \# \cdots$ (Addendum to Thm. 2.4). This is impossible. Therefore no RLB appears.

The euler number $e(M)\,(=2)$ is the sum of the euler numbers of the singular fibers $\Sigma e(F_i)$. There are two possibilities:

<u>Case</u> (1). M <u>has only one singular fiber</u> F_0 <u>with</u> $e(F_0) = 2$.

<u>Case</u> (2). M <u>has two singular fibers</u> F_1, F_2 <u>with</u> $e(F_1) = e(F_2) = 1$.

In Case (1), F_0 has the following graph (Thm. 3.1),

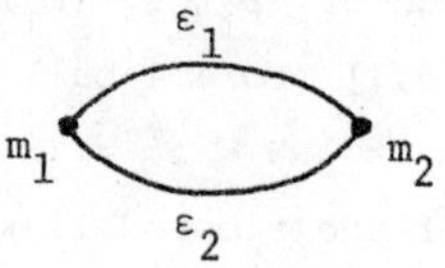

By Lemma 2.2, the self-intersection number of the 2-sphere corresponding to the m_1-vertex is equal to $-(\varepsilon_1 m_2 + \varepsilon_2 m_2)/m_1$. Since $H_2(M;\mathbb{Z}) = \{0\}$, this must be 0. Thus $\varepsilon_1 + \varepsilon_2 = 0$. F_0 is not a multiple fiber, thus $\gcd(m_1, m_2) = 1$. We call such a singular fiber (with $\varepsilon_1 + \varepsilon_2 = 0$, $\gcd(m_1, m_2) = 1$) a <u>twin singular fiber</u>, because its regular neighborhood $N_{\delta,0}(F_0)$ is a <u>twin</u> in the sense of Montesinos [Mo].

By Thm. 4.1, the monodromy of F_0 is trivial, and we have the decomposition: $M = N_{\delta,0}(F_0) \cup T^2 \times D^2$. Now $M \cong S^4$ follows from [Mo, Cor. 5.6].

For an explicit construction of a GTF: $S^4 \to S^2$, see Section 3 of [Mt1].

In Case (2), the singular fibers F_1, F_2 are of the types I_1^+, I_1^-. The total monodromy around F_1, F_2 must be trivial, thus one of the two singular fibers, say F_1, is of type I_1^+ and the other F_2 is of type I_1^-. Let D be a 2-disk on B which contains $x_1 = f(F_1)$ and $x_2 = f(F_2)$ in $\mathrm{Int}(D)$.

LEMMA 5.2. <u>One can deform the GTF structure of</u> $f|f^{-1}(D)\colon f^{-1}(D) \to D$ <u>without altering it in a neighborhood of the boundary</u> $\partial(f^{-1}(D))$, <u>so that the resulting GTF</u> $f'\colon f^{-1}(D) \to D$ <u>has a single singular fiber of the type</u>

1  1, i.e. <u>a twin</u>.

By Lemma 5.2, Case (2) is reduced to Case (1). Theorem 5.1 is proved. ⁞⁞

PROOF OF 5.2. Divide D into the two disks D_1, D_2 as in the figure:

Fig. 5.1

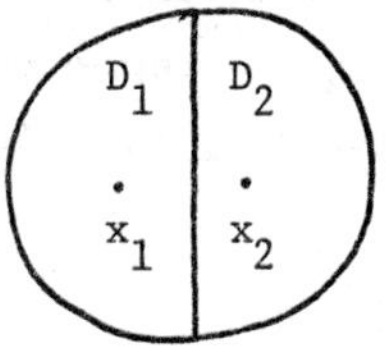

= D

Let $I = D_1 \cap D_2$. Then $f^{-1}(I) \cong T^2 \times I$. Choose a basis (μ_1, λ_1) (or (μ_2, λ_2)) of $H_1(f^{-1}(I);\mathbb{Z})$ with which the monodromy around x_1 (or x_2) is represented by $\begin{bmatrix} 1 & 1 \\ 0 & 1 \end{bmatrix}$ (or $\begin{bmatrix} 1 & -1 \\ 0 & 1 \end{bmatrix}$). Let A be a 2×2 matrix defined by $(\mu_1, \lambda_1) = (\mu_2, \lambda_2)A$. Then, since the monodromy around ∂D is trivial, we have $A^{-1}\begin{bmatrix} 1 & -1 \\ 0 & 1 \end{bmatrix} A \begin{bmatrix} 1 & 1 \\ 0 & 1 \end{bmatrix} = \begin{bmatrix} 1 & 0 \\ 0 & 1 \end{bmatrix}$ From this, $A = \pm\begin{bmatrix} 1 & b \\ 0 & 1 \end{bmatrix}$, and $\mu_1 = \pm\mu_2$, $\lambda_1 = b\mu_2 \pm \lambda_2$.

Being a regular neighborhood of F_1, $f^{-1}(D_1)$ is obtained by attaching a round 1-handle $RH = (S^1 \times D^2) \times [0,1]$ to a 4-ball Δ^4 along a (+)-Hopf link L_+ in $\partial\Delta^4$. The attaching place is $RH \cap \Delta^4 = (S^1 \times D^2) \times \{0,1\}$. In order to extend the fibering structure of the fibered link L_+ to the T^2-bundle $f|f^{-1}(\partial D_1): f^{-1}(\partial D_1) \to \partial D_1$, we have to choose the framing -1. See Fig. 5.2.

Let $RH_+ = (S^1 \times D^2) \times [0, \frac{1}{3}] \cup (S^1 \times D^2) \times [\frac{2}{3}, 1]$, and $RH_- = (S^1 \times D^2) \times [\frac{1}{3}, \frac{2}{3}]$.

We call $\Delta^4 \cup RH_+$ the <u>upper half</u> (UH) and RH_- the <u>lower half</u> (LH) of $f^{-1}(D_1)$, respectively. (Fig. 5.2). Also we call $RH_+ \cap RH_- = (S^1 \times D^2) \times \{\frac{1}{3}, \frac{2}{3}\}$ the <u>mid-level</u>. We decompose $f^{-1}(D_2)$ similarly.

Fig. 5.2

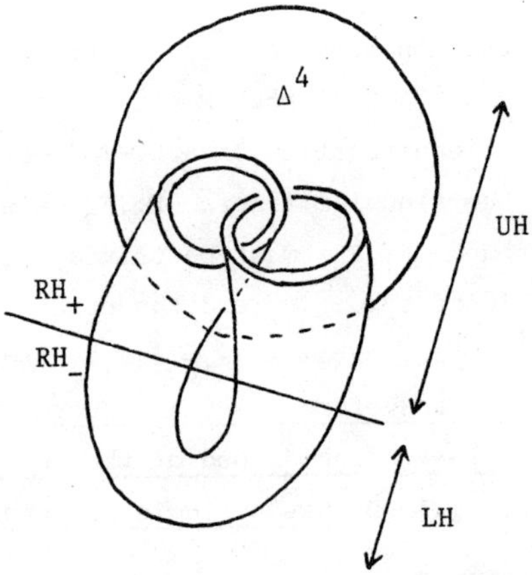

Each (singular or general) fiber of $f: f^{-1}(D_1) \to D_1$ intersects the mid-level in a longitudinal circle. These sections make a foliation of $RH_+ \cap RH_-$ with every leaf a circle. By inspecting the monodromy of I_1^+, we see that we can take one of such sectional circles as a loop representing $\mu_1 \in H_1(f^{-1}(I); \mathbb{Z})$. Also we may assume that λ_1 is represented by a loop which transverses RH once and intersects μ_1 transversely in a point. The same remark applies to (μ_2, λ_2).

Recall that $\mu_1 = \pm\mu_2$, $\lambda_1 = \pm\lambda_2 + b\mu_2$. Thus the pasting of $f^{-1}(D_1)$ and $f^{-1}(D_2)$ along $f^{-1}(I)$ can be arranged so that it preserves the mid-level.

Turning $f^{-1}(D_2)$ upside down, if necessary, we may assume that the UH (resp. LH) of $f^{-1}(D_1)$ matches LH (resp. UH) of $f^{-1}(D_2)$ through the pasting above. Note that the union of the UH of $f^{-1}(D_1)$ (resp. of $f^{-1}(D_2)$) and the LH of $f^{-1}(D_2)$ (resp. of $f^{-1}(D_1)$) is diffeomorphic to the UH of $f^{-1}(D_1)$ (resp. of $f^{-1}(D_2)$). Therefore, $f^{-1}(D)$ is obtained by gluing the two UH's of $f^{-1}(D_1)$ and $f^{-1}(D_2)$ along the mid-level under a certain orientation reversing diffeomorphism g which preserves the foliations by sectional circles.

One can deform g through diffeomorphisms which preserve the sectional foliations to obtain a new diffeomorphism g' which sends the section of the singular fiber F_1 to that of F_2. This is possible because all the leaves of the foliation are parallel longitudinal circles in the mid-level.

It is easily seen that the new GTF of $f^{-1}(D)$ obtained by sewing up the halves of fibers via g' has a single singular fiber of the type $1 \overset{+}{\underset{-}{\Leftrightarrow}} 1$. This completes the proof of Lemma 5.2. ▥

Schematically, the deformation of Lemma 5.2 is described as

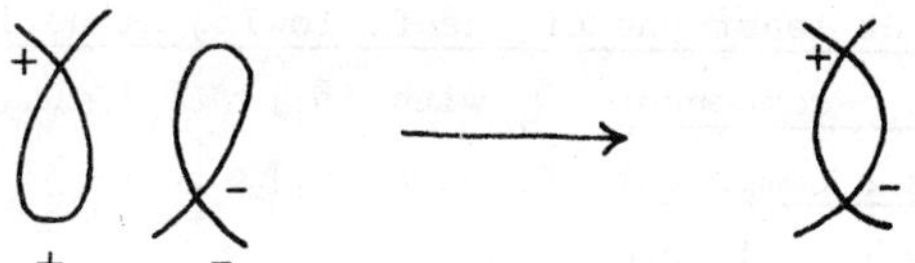

$I_1^+ \cup I_1^-$

6. REPLACEMENT OF SINGULAR FIBERS

Let p,q be positive coprime integers. Let $f:N(p,q) \to D_\delta$ be a regular neighborhood of a twin singular fiber with multiplicities p,q. If $f':N(m,n) \to D_\delta$ is another such neighborhood, we can replace $N(p,q)$ in a GTF $M \to B$ by $N(m,n)$. This is because $\partial N(p,q) \to \partial D_\delta$ and $\partial N(m,n) \to \partial D_\delta$ are both trivial T^2-bundles over a circle.

However, this replacement might change the diffeomorphism type of M.

LEMMA 6.1. *There exists an orientation preserving diffeomorphism* $\varphi:N(p,q) \to N(m,n)$ *with* $\varphi|\partial: \partial N(p,q) \to \partial N(m,n)$ *preserving fibers, if and only if* $p+q \equiv m+n \pmod 2$.

PROOF. This follows from the three assertions:

(1) *If* $p+q \equiv 0 \pmod 2$, *then there exists such a diffeomorphism* $N(1,1) \to N(p,q)$.

(2) *If* $p+q \equiv 1 \pmod 2$, *then there exists such a diffeomorphism* $N(1,2) \to N(p,q)$.

(3) *There is no such diffeomorphism* $N(1,1) \to N(1,2)$.

The proofs of these assertions are based on Montesinos' extension theorem of a diffeomorphism of the boundary of a twin, [Mo, Theorem 5.3]. ▥

By Lemma 6.1, we can replace the multiplicities p,q of a twin singular fiber by $1,1$ or $1,2$ according as $p+q \equiv 0$ or $1 \pmod 2$, without affecting the diffeomorphism type of M.

Next we will study general singular fibers of type $\tilde{A}$:

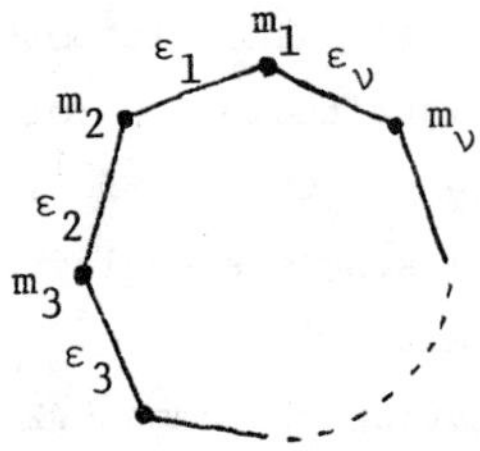

<u>LEMMA</u> 6.2. <u>Assume that a singular fiber of type</u> $\tilde{A}\,(\sum_{i=1}^{\nu} m_i\,\Theta_i)$ <u>is not a multiple fiber</u>. <u>Then at least one of the following situations occurs</u>:

(1) <u>there exists a component</u> Θ_i with $[\Theta_i]\cdot[\Theta_i]=0$,

(2) <u>there exists a component</u> Θ_i with $[\Theta_i]\cdot[\Theta_i]=\pm 1$,

(3) $m_1=m_2=\cdots=m_\nu=1$ <u>and</u> $\varepsilon_1=\varepsilon_2=\cdots=\varepsilon_\nu$.

<u>PROOF</u>. First assume $\nu \geqq 3$. By Lemma 2.2, we have $[\Theta_i]\cdot[\Theta_i]=-(\varepsilon_{i-1}m_{i-1}+\varepsilon_i m_{i+1})/m_i$, where the indices i are understood by modulo ν. Assume that $|[\Theta_i]\cdot[\Theta_i]|\geqq 2$ for all i, then $m_{i-1}+m_{i+1}\geqq 2m_i$. If $m_1<m_2$, then $m_2<m_3$, since $m_1+m_3\geqq 2m_2$. Similarly, $m_2<m_3$ implies $m_3<m_4$, and so on. This is absurd, thus $m_1\geqq m_2$. However, $m_1>m_2$ is also impossible. Thus we have $m_1=\cdots=m_\nu=1$, because the singular fiber is not a multiple fiber. The assumption $|[\Theta_i]\cdot[\Theta_i]|\geqq 2$ implies that $\varepsilon_1=\cdots=\varepsilon_\nu$.

The case $\nu\leqq 2$ is treated similarly. ⫴

If the situation (2) in Lemma 6.2 occurs, we can blow down the component Θ_i with $[\Theta_i]\cdot[\Theta_i]=\pm 1$. If (3) occurs then through the deformation indicated by the scheme below (inverse to the deformation of Lemma 5.2), the singular fiber splits into ν number of I_1^+'s (or ν number of I_1^-'s).

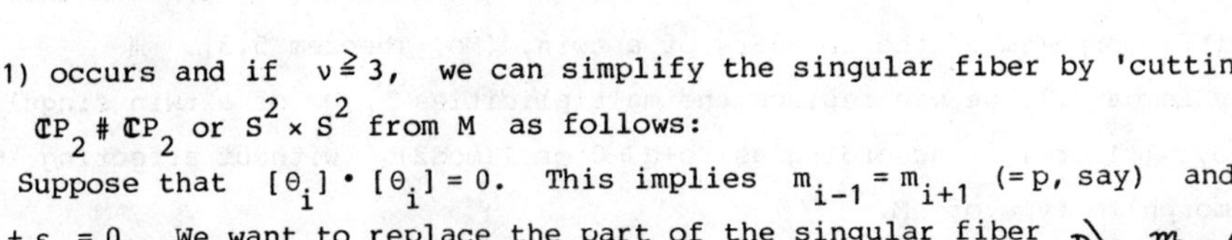

If (1) occurs and if $\nu\geqq 3$, we can simplify the singular fiber by 'cutting off' $\mathbb{C}P_2 \# \mathbb{C}P_2$ or $S^2\times S^2$ from M as follows:

Suppose that $[\Theta_i]\cdot[\Theta_i]=0$. This implies $m_{i-1}=m_{i+1}$ (= p, say) and $\varepsilon_{i-1}+\varepsilon_i=0$. We want to replace the part of the singular fiber

by

.

This replacement is explained by the picture:

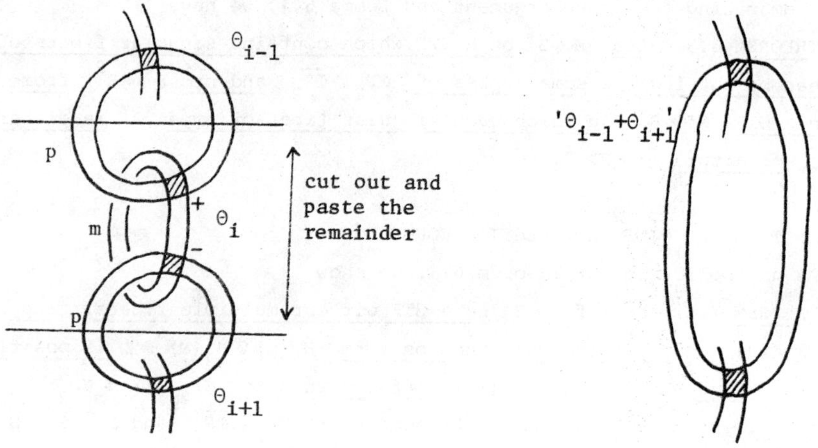

Fig. 6.1

The 'upper half' of Θ_{i-1} and the 'lower half' of Θ_{i+1} are pasted together to form a new irreducible component '$\Theta_{i-1} + \Theta_{i+1}$'. In the section of a regular neighborhood of Θ_{i-1}, the sections of fibers make a foliation whose 'general' leaves are torus knots of type $(p,-m)$. Similar sectional foliation for the section of a regular neighborhood of Θ_{i+1} has torus knots of type (p,m) as 'general' leaves. These are sewn together via an orientation reversing diffeomorphism to give a GTF structure on the regular neighborhood of '$\Theta_{i-1} + \Theta_{i+1}$'.

Let N and N' be the regular neighborhoods of the old and new singular fibers, respectively. Then in terms of framed links [Ki], we have

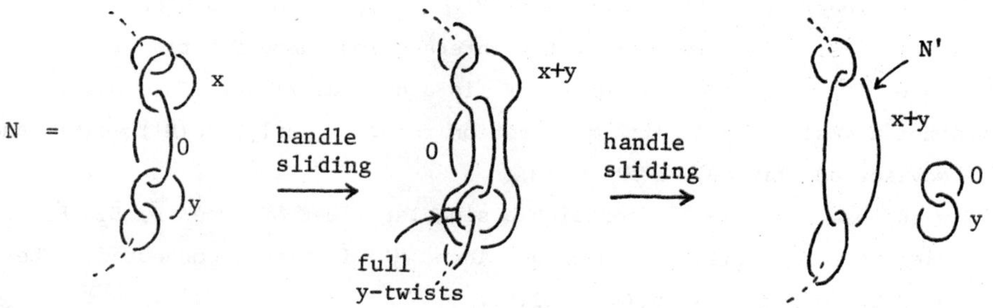

where $x = [\Theta_{i-1}] \cdot [\Theta_{i-1}]$, $y = [\Theta_{i+1}] \cdot [\Theta_{i+1}]$.

Therefore, $N \cong N' \# S^2 \times S^2$ or $N \cong N' \# \mathbb{C}P_2 \# \overline{\mathbb{C}P}_2$ according as $y \equiv 0$ or 1 (mod2).

Summarizing the above argument and Lemma 6.1, we have

THEOREM 6.3. *Let* $M \to B$ *be a GTF which contains singular fibers of type* $\tilde{A}$. *Then by cutting off some copies of* $\mathbb{C}P_2$, $\overline{\mathbb{C}P}_2$ *and/or* $S^2 \times S^2$ *from* M, *we obtain a GTF* $M' \to B$, *in which each singular fiber of type* $\tilde{A}$ *is either* I_1^+, I_1^- *or a twin* $1 \overset{+}{\underset{-}{\rightleftarrows}} 2$.

7. POSITIVE DEFINITE INTERSECTION FORMS

As an application of Theorem 6.3, we show

THEOREM 7.1. *Let* $f: M \to B$ *be a GTF without multiple fibers. Suppose that* $H_1(M; \mathbb{Z}) = \{0\}$ *and that the intersection form* $H_2(M) \otimes H_2(M) \to \mathbb{Z}$ *is positive definite. Then* M *is degree* $(+1)$-*diffeomorphic to* $\mathbb{C}P_2 \# \cdots \# \mathbb{C}P_2$.

PROOF. By cutting off a finite number of $\mathbb{C}P_2$, $\overline{\mathbb{C}P}_2$ and/or $S^2 \times S^2$ from M, we may assume that each singular fiber does not contain RLB. (Since M has the positive definite intersection form, the manifolds cut off are in fact $\mathbb{C}P_2$'s.) We need a lemma.

LEMMA 7.2. *Let* $F: M \to B$ *be as in* 7.1. *Then it has no singular fibers of types* $\tilde{D}$, $\tilde{E}_6$, $\tilde{E}_7$, $\tilde{E}_8$.

PROOF. First, note that a non-multiple singular fiber admits a local cross-section. In fact, a singular fiber of type $\tilde{D}$, $\tilde{E}_6$, $\tilde{E}_7$ or $\tilde{E}_8$ has an irreducible component with multiplicity 1 (corresponding to a terminal vertex). A local cross-section is found so that its image is a 2-disk intersecting this component transversely in a point. For a singular fiber of type $\tilde{A}$, one can find a (continuous) cross-section whose image is a cone over a torus knot with cone-vertex an intersection point of irreducible components.

Now $H_1(M; \mathbb{Z}) = \{0\}$ implies that $B = S^2$. Remove all the singular fibers from $M \to S^2$ to get a T^2-bundle over $S^2 - \sigma$. $S^2 - \sigma$ has the homotopy type of a 1-complex, and T^2 is connected. So there is a cross-section $S^2 - \sigma \to M - f^{-1}(\sigma)$.

By rechoosing the diffeomorphisms which sew back the regular neighborhoods of singular fibers, we obtain a new GTF $M' \to S^2$ that admits a global cross-section $s: S^2 \to M'$. Note that the intersection numbers satisfy: $[F] \cdot [F] = 0$, $[F] \cdot [s(S^2)] = \pm 1$, where F is a general fiber. This means that M' has not a positive definite intersection form. Clearly, $e(M') = e(M)$ and by the Novikov additivity, $\sigma(M') = \sigma(M)$.

Now suppose that $M \to S^2$ contains a singular fiber of type $\tilde{D}$, $\tilde{E}_6$, $\tilde{E}_7$ or $\tilde{E}_8$, which is necessarily 1-connected. Then M' is also 1-connected. Therefore, $e(M') = e(M)$ implies $b_2(M') = b_2(M)$.

We have $b_2(M) = b_2(M') > |\sigma(M')| = |\sigma(M)|$. This contradicts the positive definiteness of M. The lemma is proved. ▥

By Lemma 7.2, the only singular fibers of $M \to S^2$ are in class $\tilde{A}$. By Thm. 6.3, we can cut off a finite number of $\mathbb{C}P_2$, $\overline{\mathbb{C}P}_2$ and/or $S^2 \times S^2$ from M to obtain a GTF $\tilde{M} \to S^2$ whose singular fibers are I_1^+, I_1^- and/or twins. (By the positive-definiteness of M, we have only to cut off $\mathbb{C}P_2$'s.) The following signature theorem is essentially due to Harer [H]:

THEOREM 7.3 (Harer). *Let* a,b,c *be the numbers of singular fibers of types* I_1^+, I_1^-, *twin, in the GTF* $\tilde{M} \to S^2$, *respectively. Then* $a-b \equiv 0 \pmod{12}$ *and we have* $\sigma(\tilde{M}) = -(2/3)(a-b)$.

Now $e(\tilde{M}) = a+b+2c$. Since $H_1(\tilde{M}; \mathbb{Z}) = \{0\}$, we have $b_2(\tilde{M}) = a+b+2c-2$. From the positive-definiteness of $\tilde{M}$, we have $a+b+2c-2 = -(2/3)(a-b)$. This together with $a-b \equiv 0 \pmod{12}$ implies $(a,b,c) = (0,0,1)$ or $(1,1,0)$. In both cases, $\tilde{M}$ is a homology 4-sphere, thus a natural S^4 (Thm. 5.1). We obtain S^4 by cutting off $\mathbb{C}P_2$'s from M, so M must be diffeomorphic to $\mathbb{C}P_2 \# \cdots \# \mathbb{C}P_2$. ▥

8. A THEOREM OF KAS

Kas' classification of regular elliptic surfaces [Ka] can be extended in the class of GTF as follows:

THEOREM 8.1. *Let* $f_i : M_i \to S^2$, $i = 1,2$, *be GTF's over* S^2 *with at least one singular fiber. Suppose that each singular fiber is of type* I_1^+, I_1^- *and that* $\sigma(M_1) \neq 0$. *Then* M_1 *is diffeomorphic to* M_2 *if and only if* $\sigma(M_1) = \sigma(M_2)$ *and* $e(M_1) = e(M_2)$.

REMARK. If $\sigma(M_1) \neq 0$, then M_1 is 1-connected. (cf. [Msh, Part II, Section 2]

Our proof simply follows Moishezon's approach [Msh] to the Kas theorem. For this, we slightly generalize Livne's theorem on modular groups and Moishezon's complement to it.

Let $G = \langle a,b \mid a^3 = b^2 = 1 \rangle$. Put $s_0 = a^2b$, $s_1 = aba$, $s_2 = ba^2$, $s_0^{-1} = ba$, $s_1^{-1} = a^2ba^2$, $s_2^{-1} = ab$.

THEOREM 8.2. *Let* $g_1, \ldots, g_n \in G$ *be conjugates of* s_1 *or* s_1^{-1} *such that* $g_1 \cdots g_n = 1$. *Then, by successive application of elementary transformations in the sense of* [Msh, p.223], *the* n-*tuple* $(g_1, \ldots, g_n)$ *can be transformed into an* n-*tuple* $(h_1, \ldots, h_n)$ *with each* $h_i \in \{s_0, s_1, s_2, s_0^{-1}, s_1^{-1}, s_2^{-1}\}$.

THEOREM 8.3. *Let* $y_1, \ldots, y_n \in G$ *be such that each of* y_i *is equal to one of the elements* $s_0, s_1, s_2, s_0^{-1}, s_1^{-1}, s_2^{-1}$ *and* $y_1 \cdots y_n = 1$. *Then by successive application of elementary transformations,* $(y_1, \ldots, y_n)$ *can be transformed into* $(z_1, \ldots, z_n)$ *such that at least one of the following holds:*

(1) n <u>is even and</u> $(z_1,\ldots,z_n) = (s_1, s_2,\ldots,s_1, s_2)$;

(2) n <u>is even and</u> $(z_1,\ldots,z_n) = (s_2^{-1}, s_1^{-1},\ldots,s_2^{-1}, s_1^{-1})$;

(3) <u>there exists</u> j <u>such that</u> $z_j z_{j+1} = 1$.

As a corollary of Theorems 8.2 and 8.3, we have

<u>COROLLARY</u> 8.4. <u>Let</u> $g_1,\ldots,g_n \varepsilon G$ <u>be conjugates of</u> s_1 <u>or</u> s_1^{-1} <u>such that</u> $g_1 g_2 \cdots g_n = 1$. <u>Then by successive application of elementary transformations</u> $(g_1,\ldots,g_n)$ <u>can be transformed into</u> $(h_1,\ldots,h_n)$ <u>such that one of the following holds</u>:

(1) $(h_1,\ldots,h_n) = (h_1, h_1^{-1}, h_3, h_3^{-1},\ldots,h_k, h_k^{-1}, s_1, s_2,\ldots,s_1, s_2)$

(2) $(h_1,\ldots,h_n) = (h_1, h_1^{-1}, h_3, h_3^{-1},\ldots,h_k, h_k^{-1}, s_2^{-1}, s_1^{-1},\ldots,s_2^{-1}, s_1^{-1})$

To prove these results, one has only to follow Moishezon's arguments in [Msh] almost word for word. Details are omitted.

BIBLIOGRAPHY

[D] S. Donaldson; An application of the Yang-Mills equations to the geometry of 4-manifolds (manuscript, 1982).

[H] J. Harer; Pencils of curves on 4-manifolds, Thesis, UCB (1979).

[Ka] A. Kas; On the deformation types of regular elliptic surfaces, Complex Analysis and Algebraic Geometry, Camb. Univ. Press (1977), 107-112.

[Ki] R. Kirby; A calculus for framed links in S^3, Inventiones math. 45(1978), 35-56.

[Ko] K. Kodaira; On Compact analytic surfaces, II, Ann. of Math., 77(1963), 563-626.

[Ko] K. Kodaira; On homotopy K3-surfaces, Essays on Topology and Related Topics, Memoires dedies a G. de Rham, Springer, (1970), 58-69.

[Mi] J. Milnor; <u>Singular points of complex hypersurfaces</u>, Ann. Math. Studies, 61, Princeton Univ. Press and the Univ. of Tokyo Press (1968).

[Mo] J. M. Montesinos; On twins in the four-sphere, preprint, Univ. de Zaragoza, September (1980).

[Msh] B. Moishezon; <u>Complex Surfaces and Connected Sums of Complex Projective Planes</u>, Lecture Notes in Math. Vol. 603, Springer (1977).

[Mt] Y. Matsumoto; On 4-manifolds fibered by tori, Proc. Japan Acad., 58, 1982.

[Mt] Y. Matsumoto; A topological study of torus fibrations, (in preparation).

[N] W. D. Neumann; A calculus for plumbing applied to the topology of complex surface singularities and degenerating complex curves, Trans. AMS. 268(1981), 299-344.

[NW] W. D. Neumann and S. H. Weintraub; Four-manifolds constructed via plumbing, Math. Ann. 238(1978), 71-78.

[Sch] A. Scharf; Zur Faserung von Graphenmannigfaltigkeiten, Math. Ann. 215(1975), 35-45.

[SF] K. Sakamoto and S. Fukuhara; Classification of T^2-bundles over T^2, preprint, Tsuda College, October (1982).

[Th] M. C. Thornton; Singularly fibered manifolds, Illinois J. Math., 11(1967), 189-201.

[Z] H. Zieschang; On toric fiberings over surfaces, Math. Zametki 5(1969) = Math. Notes 5(1969), 341-345.

DEPARTMENT OF MATHEMATICS
FACULTY OF SCIENCE
UNIVERSITY OF TOKYO
HONGO, TOKYO (113)
JAPAN

Contemporary Mathematics
Volume 35, 1984

4-DIMENSIONAL ORIENTED BORDISM

Paul Melvin*

In 1952 Rohlin [4] (see appendix) outlined a proof of the following result:

THEOREM. Every closed oriented smooth 4-manifold M of signature zero is the boundary of a compact oriented smooth 5-manifold.

Two years later Thom [6] gave a proof using stable homotopy theory as part of his general program for computing the oriented bordism groups. Although his methods are of fundamental importance, the proof is unnecessarily complicated in this particular case.

In a lecture at IHES in 1976, John Morgan proposed a more geometric proof of the theorem. (A sketch is given in Remark 1.) Morgan's proof followed Rohlin's outline, but used a fact not known to Rohlin: a simply connected cobordism of dimension ≥ 6 has a handlebody structure which reflects its homology structure [5].

We present a new proof of the theorem, also following Rohlin's outline. Our proof partially incorporates Morgan's (step 2 below) but avoids the handlebody theorem by using the Whitney immersion theorem and a transversality argument (steps 1 and 3). This is perhaps closer to what Rohlin had in mind.

The author wishes to thank John Hughes for his valuable suggestions during the preparation of this paper.

PROOF OF THE THEOREM. We shall work in the smooth category.

Observe that M is bordant to a simply connected manifold, obtained for example by surgery on a set of normal generators of the fundamental group of M [2]. So we may assume that M is simply connected.

STEP 1. Find a submanifold M_1 of S^7 which is bordant to M.

By a theorem of Whitney [7] M immerses in S^7 with singular set consisting of double circles at which the sheets of M meet transversely. Each double circle C may be eliminated at the cost of a surgery on M, as follows. Since M is orientable, C is the image of two circles C_1 and C_2 in M (rather than one circle by a double cover). As M is simply connected, C_1 bounds a disc D missing the rest of the singular set, so in fact D is

*Supported in part by NSF Grant MCS82-05450

embedded in S^7. D has a tubular neighborhood $D \times B^5$ in S^7 which intersects M in tubular neighborhoods $D \times (B^2 \times 0)$ of D and $\partial D \times (0 \times B^3)$ of C_2. Now remove $\partial D \times (0 \times B^3)$ from M and replace it with $D \times (0 \times \partial B^3)$. See Figure 1.

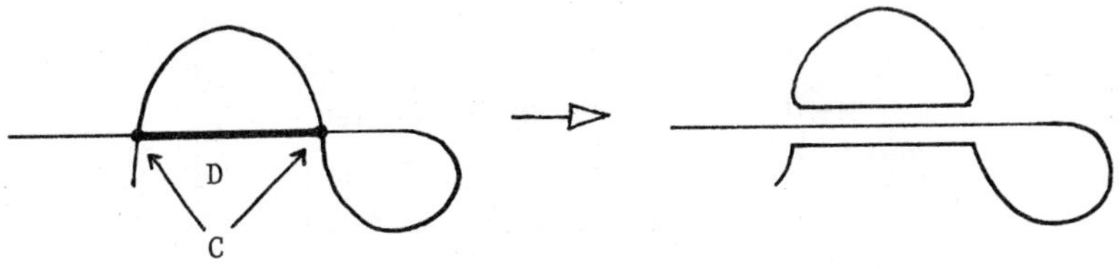

Figure 1

This leaves a simply connected 4-manifold bordant to M and immersed in S^7 with fewer double curves. The bordism is across the 2-handle $D \times (0 \times B^3)$. Continuing in this way we obtain M_1 bordant to M and embedded in S^7.

STEP 2. *Let N be a tubular neighborhood of M_1 in S^7, and $W = S^7 - \mathrm{int}(N)$. Find a submanifold M_2 of ∂W which is null homologous in W and is diffeomorphic to $M_1 \# n\mathbb{C}P^2$ (for some integer n).**

We shall denote by $[Q]$ the class in $H_4(W)$ represented by a closed 4-manifold Q embedded in ∂W. The null-homologous condition above means that $[M_2] = 0$.

The geometric key to this step is the following observation of Morgan.

LEMMA. *Let $p: E \to B$ be the trivial S^2-bundle over a 4-ball B. Then the image of any partial section $s: \partial B \to E$ bounds a submanifold K of E which is diffeomorphic to $k\mathbb{C}P^2 -$ (open 4-ball) (for some integer k).*

PROOF. Using a trivialization, identify $p: E \to B$ with the projection map $p_1: B^4 \times S^2 \to B^4$.

Let $h: S^3 \to S^2$ be the Hopf map. Observe that the image of the partial section $S^3 = \partial B^4 \to B^4 \times S^2$ given by $x \to (x, h(x))$ bounds a Hopf disc bundle

$$H = \{(tx, h(x)) : t \in [0,1] , x \in S^3\}$$

in $B^4 \times S^2$.

Now let k be the Hopf degree of the map $p_2 s: S^3 \to S^2$, where $p_2: B^4 \times S^2 \to S^2$ is the projection map. We may assume $k > 0$. (If $k < 0$ then the argument is analogous, and if $k = 0$ then s extends to a global section $t: B^4 \to B^4 \times S^2$ and so we may take $K = t(B^4)$.) Let B_i $(i=1,\ldots,k)$ be disjoint 4-balls in B^4. Choose trivializations $T_i: B^4 \times S^2 \to p_1^{-1}(B_i)$ covering diffeomorphisms $t_i: B^4 \to B_i$. Then s extends to a partial section

---

* If $n > 0$ then nM denotes the connected sum of n copies of M. If $n < 0$ then $nM \equiv (-n)(-M)$, where $-M$ is M with the opposite orientation. Finally $0M \equiv S^4$.

$$t: B^4 - (\bigcup_i \operatorname{int}(B_i)) \longrightarrow B^4 \times S^2$$

with $p_2 T_i^{-1} t(t_i | S^3) = h$ for all i. Each 3-sphere $t(\partial B_i)$ bounds a Hopf bundle $H_i = T_i(H)$ in $B^4 \times S^2$. Set

$$K = \operatorname{im}(t) \cup (\bigcup_i H_i) .$$

PROOF OF STEP 2. A straightforward computation shows that the Euler class of the bundle

$$p: N \to M_1$$

is zero [3,§11.4]. It follows that there is a partial section

$$s: M_1 - \operatorname{int}(B) \to \partial N$$

where B is a 4-ball in M_1 [3,§12.5]. The lemma provides a submanifold $K = k\mathbb{C}P^2$ - (open 4-ball) of $p^{-1}(B) \cap \partial N$ with boundary $s(\partial B)$. Thus

$$L = \operatorname{im}(s) \cup K$$

is a submanifild of $\partial N = \partial W$ diffeomorphic to $M_1 \# k\mathbb{C}P^2$. See Figure 2.

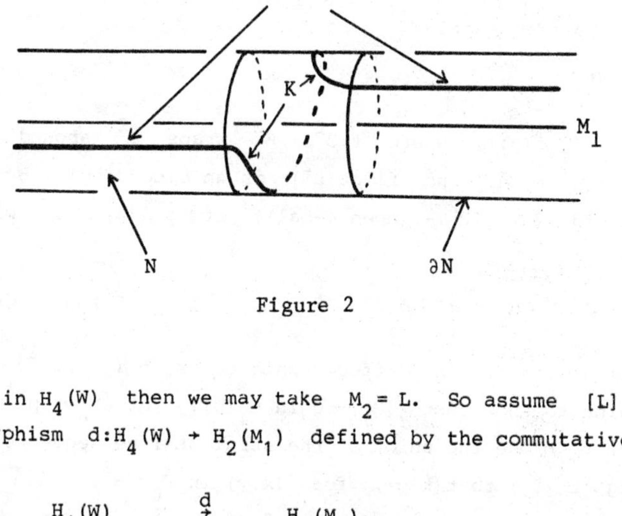

Figure 2

If $[L] = 0$ in $H_4(W)$ then we may take $M_2 = L$. So assume $[L] \neq 0$. Consider the isomorphism $d: H_4(W) \to H_2(M_1)$ defined by the commutative diagram

$$\begin{array}{ccc} H_4(W) & \overset{d}{\to} & H_2(M_1) \\ \partial \uparrow & & \uparrow \text{Thom isomorphism} \\ H_5(S^7, W) & \underset{\text{excision}}{\leftarrow} & H_5(N, \partial N) \end{array}$$

Represent $d([L])$ by an embedded surface F in $M_1 - B$. (One may think of F as the intersection of M_1 with a 5-cycle bounded by L in S^7.) To get M_2, we shall modify L near $s(F)$.

Let D be a 4-ball in $M_1 - B$ which intersects F in a trivial 2-disc. Set $M_0 = M_1 - \mathrm{int}(D)$. Then $F_0 = F \cap M_0$ is a surface with boundary, properly embedded in M_0 (Figure 3).

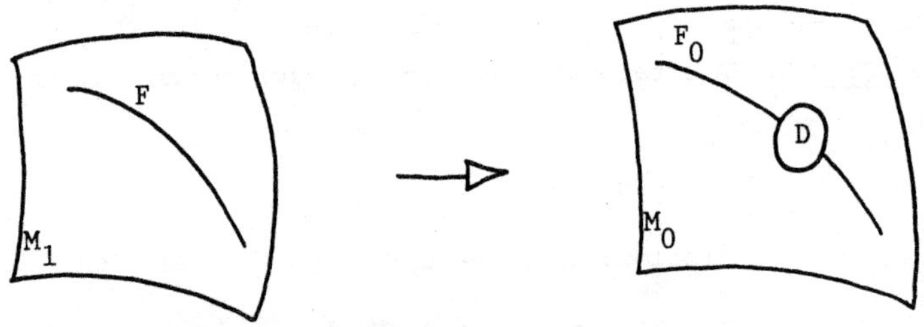

Figure 3

As F_0 is homotopy equivalent to a 1-complex, it has a tubular neighborhood $F_0 \times B^2$ in M_0, and N restricts to a trivial bundle over $F_0 \times B^2$. Pick a trivialization $(F_0 \times B^2) \times B^3$ (we suppress the map) so that $s(x,y) = (x,y,n)$ (n = the north pole of B^3) for all (x,y) in $F_0 \times B^2$.

Define a partial section

$$t: F_0 \times B^2 \to \partial N$$

by $t(x,y) = (x,y,f(y))$, where $f: B^2 \to S^2$ wraps B^2 around S^2 (i.e. $f(\partial B^2) = n$, $f(0) = -n$, and $f|\mathrm{int}(B^2)$ is an embedding). By the lemma, there is a submanifold $J = j\mathbb{C}P^2$ - (open 4-ball) of $p^{-1}(D) \cap \partial N$ with boundary $t(\partial D)$. Set

$$M_2 = (L - s(F_0 \times B^2 \cup D)) \cup t(F_0 \times B^2) \cup J .$$

M_2 is a submanifold of ∂W diffeomorphic to $M_1 \# (j+k)\mathbb{C}P^2$.

It remains to show that $[M_2] = 0$ in $H_4(W)$. Put $N_0 = p^{-1}(M_0)$ and $C = p^{-1}(D)$. Let X be the union of the straight line segments in each B^3 fiber joining $s(x,y)$ to $t(x,y)$, for (x,y) in $F_0 \times B^2$.

X can be extended across C to a 5-cycle $\bar{X}$ in N whose boundary represents $[L]-[M_2]$ in $H_4(W)$. Furthermore $\bar{X}$ intersects M_1 in F, and so $d([M_2]) = d([L]) - d([L]-[M_2]) = [F]-[F] = 0$ in $H_2(M_1)$. Thus $[M_2] = 0$ in $H_4(W)$.

As $\bar{X}$ is perhaps hard to visualize, we provide an alternative algebraic argument that $[M_2] = 0$. Consider the isomorphism $c: H_4(W \cup C) \to H_2(M_0, \partial M_0)$

defined by the commutative diagram

$$\begin{array}{ccc} H_4(W \cup C) & \xrightarrow{c} & H_2(M_0, \partial M_0) \\ \partial \uparrow & & \uparrow \text{ Thom isomorphism} \\ H_5(S^7, W \cup C) & \xrightarrow[\text{excision}]{} & H_5(N_0, \partial N_0) \,. \end{array}$$

One readily verifies that X represents the element of $H_5(N_0, \partial N_0)$ corresponding to $[L] - [M_2]$ in $H_4(W \cup E)$. Since X and M_0 intersect in F_0, $c([L] - [M_2]) = [F_0]$. It follows that $d([L] - [M_2]) = [F]$, by the commutativity of the following diagram

$$\begin{array}{ccccc} H_4(W) & & \xrightarrow{d} & & H_2(M_1) \\ \cong \downarrow & & & & \downarrow \cong \\ H_4(W \cup C) & \xrightarrow[c]{} & H_2(M_0, \partial M_0) & \xrightarrow[\text{excision}]{} & H_2(M_1, D) \,. \end{array}$$

Thus $[M_2] = 0$ by the same argument as above.

STEP 3. Show that M_2 bounds a 5-manifold V.

This will prove the theorem. For then $\sigma M_2 = 0$. But $\sigma M_2 = \sigma M_1 + n$ (by step 2) and $\sigma M_1 = \sigma M = 0$ (by step 1), so $n = 0$. Thus M_2 is bordant to M, and so M bounds.

To prove that M_2 bounds, first construct a map

$$f: W \to \mathbb{C}P^n$$

(for large n) with $f \pitchfork \mathbb{C}P^{n-1}$ and $f^{-1}(\mathbb{C}P^{n-1}) = M_2$. For example, define f on an open tubular neighborhood U of M_2 in ∂W to be the classifying map $U \to \mathbb{C}P^n - x$ of the normal bundle of M_2 in ∂W. (Here x is a point in $\mathbb{C}P^n$, and so $\mathbb{C}P^n - x$ is the canonical complex line bundle over $\mathbb{C}P^{n-1}$.) Extend f to ∂W by mapping $\partial W - U$ to x.

The only obstruction to extending f to a map

$$F: W \to \mathbb{C}P^n$$

lies in $H^3(W, \partial W; \pi_2(\mathbb{C}P^n))$. Since $\pi_2(\mathbb{C}P^n) = \mathbb{Z}$ is generated by a $\mathbb{C}P^1$ intersecting $\mathbb{C}P^{n-1}$ transversely in one point, this obstruction is Poincaré dual to $[M_2] \in H_4(W)$. Since $[M_2] = 0$, F exists.

Now homotop $F(\text{rel } \partial W)$ transverse to $\mathbb{C}P^{n-1}$ and set

$$V = F^{-1}(\mathbb{C}P^{n-1}) \,.$$

The proof is complete.

REMARK 1. Morgan's proof follows the same three step outline, but the proofs of steps 1 and 3 are different. Here is a sketch.

To achieve step 1, first embed M in S^8 by the Whitney embedding theorem. Using a normal vector field, push M out to the boundary of a tubular neighborhood N. Set $W = S^8 - \mathrm{int}(N)$. Since M may be taken simply connected (as in step 1 above), $H_*(W, \partial W)$ vanishes except in dimensions 3, 5 and 8. Build W as a handlebody on ∂W with handles of index 3,5 and 8. M misses the attaching 2-spheres of the 3-handles by general position, but may meet the attaching 4-spheres of the 5-handles in circles. Surgery on M (as in the proof of step 1 above) produces M_1 missing these as well. Thus M_1 lies in the boundary S^7 of an 8-handle.

Step 3 is similar to step 1. The only difficulty is in pushing M_2 off of the attaching 2-spheres of the 3-handles. But there is no algebraic obstruction to doing this since M_2 is null homologous in W, and so the Whitney trick applies. Finally we have M_3 (bordant to M_2 after pushing past the 5-handles) lying in S^6. A standard transversality argument shows that M_3 bounds.

REMARK 2. There is also an immersion theoretic proof of the theorem, worked out by Kirby and Freedman [1].

We conclude with a problem.

PROBLEM. (D. Ruberman) Modify some variant of Rohlin's proof to give a topological computation of the 4-dimensional oriented spin bordism group.

APPENDIX

For the convenience of the reader, here is an English translation of the French translation by L. Guillou and V. Sergiercu of Section 2 of Rohlin's article [4]:

THEOREM. M^4 <u>bounds if and only if</u> $\sigma(M^4) = 0$ . . . [M^4 is an oriented, closed smooth 4-manifold of signature $\sigma(M^4)$.] This follows from:

LEMMA A. <u>For every</u> M^4 <u>there exists an integer</u> s <u>such that</u> $M^4 \sim s\mathbb{C}P^2$. [$\sim$ denotes "is bordant to"]

LEMMA B. <u>If</u> $M^4 \sim N^4$, <u>then</u> $\sigma(M^4) = \sigma(N^4)$.

LEMMA C. $\sigma(s\mathbb{C}P^2) = s$.

<u>PROOF OF A.</u> One shows easily that $M^4 \sim M_1^4 \subset \mathbb{R}^7$. On M_1^4 one can find a normal vector field with isolated singularities of index ± 1. We seek $M_2^4 \sim M_1^4 + n\mathbb{C}P^2$, $M_2^4 \subset \mathbb{R}^7$, having a nonzero normal vector field. To achieve this, form the connected sum about each singularity with a $\mathbb{C}P^2$. Let L^7 be the complement of a tubular neighborhood of M_2^4 in S^7, and U^5 be the generator of $H_5(L^7, \partial L^7) \cong H_5(S^7, M_2^4) \cong H_4(M_2^4)$ determined by the orientation of M_2^4. Among the cycles representing U^5 one can find a manifold whose boundary

is bordant to $M_2^4 + m\mathbb{C}P^2$ for some integer m. Thus $M^4 \sim M_1^4 \sim M_2^4 + m\mathbb{C}P^2 - (m+n)\mathbb{C}P^2 \sim -(m+n)\mathbb{C}P^2$.

BIBLIOGRAPHY

1. Kirby, R., handwritten notes.

2. Milnor, J., "A procedure for killing homotopy groups of differentiable manifolds", Proc. Symp. Pure Math. Vol. III, Amer. Math. Soc., Providence, R.I. (1961), 39-55.

3. Milnor, J., and J. Stasheff, Characteristic classes, Annals of Math. Studies no. 76, Princeton (1974).

4. Rohlin, V. A., "New results in the theory of 4-dimensional manifolds", Dokl. Akad. Nauk SSSR 84 (1952), 221-224 (Russian). Translation by L. Guillou and V. Sergiercu (1976).

5. Smale, S., "On the structure of manifolds", Amer. Jour. Math. 84 (1962), 387-399.

6. Thom, R., "Quelques propriétés globales des variétés différentiables", Comm. Math. Helv. 28 (1954), 17-86.

7. Whitney, H., "The singularities of a smooth n-manifold in (2n-1)-space", Ann. of Math. 45, no. 2 (1944), 247-293.

DEPARTMENT OF MATHEMATICS
BRYN MAWR COLLEGE
BRYN MAWR, PENNSYLVANIA 19010

Contemporary Mathematics
Volume 35, 1984

A NEW PROOF OF THE HOMOTOPY TORUS AND ANNULUS THEOREM

Richard T. Miller

0. INTRODUCTION

It is a remarkable fact about Haken 3-manifolds that each one contains a Seifert fibered submanifold into which any sufficiently non-trivial map of a torus deforms. This is the content of the celebrated Homotopy Torus Theorem proved originally by Jaco and Shalen [6] and independently by Johannson [7]. The analogous theorem for proper maps of annuli is the Homotopy Annulus Theorem of the same authors. I consider the original proofs of these theorems nothing short of stupendous. They are long and hard. In this paper I offer proofs that are considerably simpler owing principally to a focusing of effort obtained by imitating the general form of Stallings proof of the Loop Theorem. Unknown to me, Cannon and Feustel [1] used similar techniques and obtained partial results.

Scott [10] has also, previously, reworked this material getting both theorems in their entirety.

I benefited greatly from conversations with many people. Discussions with Bus Jaco played what was obviously a unique role. In historical order, Marko Kranjc, Bob Edwards, Ulrich Oertel, Ray Lickorish, and Michael Handel all made very significant contributions to the final substance and form of this work. I am extremely grateful to them. In addition, I would like to thank the members of the topology seminars at the University of California at Los Angeles, Santa Barbara, and Berkeley, and at Michigan State University for the enthusiastic and critical hearings I received from them.

1. THE TOOLS AND THE THEOREMS

We begin this section with some definitions. After that we state our main theorem and discuss its proof a bit. Finally we list the tools we shall need to use.

Throughout this paper, I, D^2, A^2, T^2 denote the interval, the disc, the annulus, and the torus respectively. If Y is a subset of X, $\bar{Y}$ and $\mathrm{Fr}(Y)$ denote its <u>closure</u> and <u>frontier</u>. If M is a manifold, ∂M is its <u>boundary</u> and $\mathring{M}$ its <u>interior</u>. A <u>submanifold</u> means a locally-flat submanifold. We assume that immersions of one manifold in another are likewise locally-flat.

0271-4132/84 $1.00 + $.25 per page

At a crucial point in our argument we have to work with manifold triads. This being unavoidable, we find it expedient to work entirely in that category. A triad $(M;P,W)$ is a *manifold triad* if M is a manifold and P and W are top-dimensional submanifolds of ∂M which intersect precisely in a top-dimensional submanifold of each of their boundaries. If $(X;Y,Z)$ is another such triad, we call a map $f:(X;Y,Z) \to (M;P,W)$ *proper* if $f(\mathring{X};\mathring{Y},\mathring{Z}) \subset (\mathring{M};\mathring{P},\mathring{W})$. We abbreviate $(M;P,\emptyset)$ or $(M;\emptyset,P)$ by (M,P) and call this object a *manifold pair*.

A *strip* is a surface pair homeomorphic to $(I^2, I \times \partial I)$.

Homotopies of maps of pairs or triads shall always be through maps of pairs or triads.

Let M be a 3-manifold and let F be a surface such that either $(F,\partial F)$ is properly embedded in $(M,\partial M)$ or F lies in ∂M. We say F is *compressible* in M if either $F = S^2$ and F bounds a 3-cell in M or if there is a disc embedded in M whose interior lies in $M - F$ and whose boundary lies in F and does not bound a disc in F.

Note that a properly embedded disc in $(M,\partial M)$ is always incompressible.

Let (M,P) be a 3-manifold pair, and $(F,\partial F)$ be a surface properly embedded in $(M,\partial M)$, or such that $(M;P,F)$ forms a manifold triad. We say F is *boundary compressible* in (M,P) if F is a disc in the boundary of a 3-cell in M the rest of whose boundary lies in P, or if there is a disc embedded in M whose interior lies in $M - F$ and whose boundary is a two-point union of two arcs, one lying in F and the other in P, and if the arc in F is not the frontier of a disc in F the rest of whose boundary lies in P.

It is convenient to extend the above definitions slightly by allowing the pair (M,P) to be a surface pair and F to be embedded in M with $F \cap \partial M$ $(= \partial F \cap \partial M)$ a 1-submanifold of ∂M and ∂F. In this case we must of course relax the condition that the interior of the compressing disc misses F.

A 3-manifold M is *irreducible* if every sphere in M is compressible.

Let M be a compact, oriented, irreducible 3-manifold. If each component of M contains a non-empty two-sided incompressible surface, M is a *Haken manifold*. If M is connected and $\partial M \neq \emptyset$, M is Haken since it contains a properly embedded disc; such discs are always two-sided and incompressible. In particular, the 3-cell is a Haken manifold; the 3-sphere is not Haken. If M is Haken, it is not hard to show that M is a $K(\pi,1)$ and that $\pi_1(M)$ has no elements of finite order.

A *manifold triad* $(M;P,W)$ is Haken if M is a Haken manifold, if P is incompressible in M, and if the components of $(W, W \cap P)$ are strips that are essential in (M,P).

A 3-manifold is *Seifert fibered* if it is foliated by circles each of which has a saturated neighborhood foliated as the mapping torus of a periodic rotation of the disc. The reader can find out about Seifert fibered manifolds in the books by Orlik [9] and Hempel [4].

We say a Haken triad $(X;Y,Z)$ is <u>Seifert</u> if for each component $(X_0;\ X_0 \cap Y,\ X_0 \cap Z)$, which we abbreviate $(X_0;\ Y_0,\ Z_0)$, either X_0 is a Seifert fibered manifold, Y_0 is a saturated submanifold of ∂X_0, and $Z_0 = \emptyset$, or (X_0, Y_0) is an $(I,\partial I)$-bundle pair and Z_0 is saturated by I-fibers.

Observe that the components of $(\overline{\partial X - Y \cup Z},\ \overline{\partial Y - Z},\ \overline{\partial Z - Y})$ are of the form $(I^2;\ I \times \partial I,\ \partial I \times I)$, $(A^2;\ \partial A^2, \emptyset)$, or $(T^2;\ \emptyset, \emptyset)$. We call these forms the <u>fundamental surface triads</u>.

If $(C;E,F)$ is a fundamental surface triad, we say a map $f:(C;E,F) \to (M;P,W)$ is <u>essential</u> if the induced homomorphisms $f_*:\pi_1(C) \to \pi_1(M)$, $f_*:\pi_1(C,E) \to \pi_1(M,P)$; and $f_*:\pi_1(C,F) \to \pi_1(M,W)$ are all monomorphisms.

In our work, embeddings of Seifert triads in Haken triads will be very important. We say that a Seifert triad $(X;Y,Z)$ embedded in a Haken triad $(M;P,W)$ is <u>essential</u> if

(i) Each component of $(\overline{\partial X - Y \cup Z};\ \overline{\partial Y - Z},\ \overline{\partial Z - Y})$ lies properly in $(M;P,W)$ or equals a component of $(\overline{\partial M - P \cup W};\ \overline{\partial P - W},\ \overline{\partial W - P})$.

(ii) For each component $(X_0;\ Y_0,\ Z_0)$ of $(X;Y,Z)$ there is a map of a fundamental surface triad into $(X_0;\ Y_0,\ Z_0)$ that is essential in $(M;P,W)$.

The first condition is not very important. We include it to improve the statements of the subsequent theorems, where it has the effect of allowing us to make desired moves of Seifert triads by ambient isotopies of $(M;P,W)$ rather than just by isotopies. The second condition is more meaningful. It insures that the Seifert triads with which we deal are homotopically more complicated than essential embedded circles or essential properly embedded intervals in (M,P). These latter objects are too common in 3-manifold triads; they exist anywhere there is non-trivial first homotopy or relative first homotopy. Essential Seifert triads, as we shall see in subsequent theorems, can occupy only very restricted positions in Haken triads, and it is this exclusiveness that makes them so important to us.

We can put a partial ordering on the set of Seifert triads $(X;Y,Z)$ in $(M;P,W)$ by defining $(X;Y,Z) \leq (X';Y',Z')$ if $(X;Y,Z)$ is contained in a regular neighborhood of $(X';Y',Z')$ in $(M;P,W)$. A maximal essential Seifert triad in $(M;P,W)$ that is strictly larger than any proper subcollection of its components is said to be <u>characteristic</u> for $(M;P,W)$.

We can now state the main theorem of this paper.

THEOREM 1. <u>Let</u> $(\Sigma;\Phi,\Omega)$ <u>be a characteristic triad for the Haken triad</u> $(M;P,W)$. <u>Let</u> $(C;E,F)$ <u>be a fundamental surface triad and let</u> $f:(C;E,F) \to (M;P,W)$ <u>be an essential map</u>. <u>Then</u> f <u>homotops into</u> $(\Sigma;\Phi,\Omega)$.

If $(C;E,F) = (T^2,\emptyset,\emptyset)$ this is the Homotopy Torus Theorem; if $(C;E,F) = (A^2,\partial A^2,\emptyset)$ it is the Homotopy Annulus Theorem; and if $(C;E,F) = (I^2;I\times\partial I, \partial I\times I)$ we get what we call the Homotopy Disc Theorem.

We shall prove Theorem 1 by proving these last named theorems in reverse order, each proof using the previously proved result. We do this in Sections 4, 5, and 6.

In the rest of this section we lay out the tools we shall use in proving Theorem 1. Foremost among these are the characteristic triads themselves. To see that they exist at all, we first establish the fact that each essential Seifert triad in $(M;P,W)$ can be enlarged until its frontier components are all essential. This is done by checking the handful of cases that arise if one of the frontier components is not essential. We leave this to the reader; it is a good warm-up exercise. From there the existence of maximal Seifert triads in $(M;P,W)$ follows from the famous theorem of Haken (see Jaco [5, III.20] easily modified to encompass triads) that if $(M;P,W)$ is a compact manifold triad, there is a upper bound on the number of disjoint, non-parallel surface triads incompressible in (M,P) and (M,W) that can be properly and simultaneously embedded in $(M;P,W)$. We obtain a characteristic triad by eliminating any extraneous components from a maximal Seifert triad.

The next theorem implies, among other things, the uniqueness, up to ambient isotopy and possible change of structure, of the characteristic triad $(\Sigma;\Phi,\Omega)$. It is our principal tool. It is also due to Jaco-Shalen [6] and Johannson [7]. Hatcher [3] gives a nice proof in a special case.

THEOREM 2. Let $(\Sigma;\Phi,\Omega)$ be a characteristic triad for the Haken triad $(M;P,W)$. Let $(X;Y,Z)$ be an essential Seifert triad in $(M;P,W)$. Then $(X;Y,Z)$ ambient isotops into $(\Sigma;\Phi,\Omega)$.

We briefly outline a proof, primarily to show the sorts of things that go into it, and incidentally to give the reader interested in filling in the details a moderately difficult exercise to work over. Subsequent arguments in this paper will be much less sketchy. By the discussion three paragraphs back, we might as well assume that $(X;Y,Z)$ is also maximal Seifert in $(M;P,W)$. In particular, this makes its frontier essential. Move $(X;Y,Z)$ to eliminate as many components of $\mathrm{Fr}(X;Y,Z)\cap \mathrm{Fr}(\Sigma;\Phi,\Omega)$ as possible; do this by the usual innermost circle and edgemost arc arguments, and then make further reductions using the fact that a fundamental surface triad properly and essentially embedded in a Seifert triad either ambient isotops so that it is saturated or is parallel into the boundary of the Seifert triad. A little imagination is necessary in setting up these last moves. At this point $\mathrm{Fr}(X;Y,Z)\cap \mathrm{Fr}(\Sigma;\Phi,\Omega) = \emptyset$, for otherwise, after a possible change of Seifert structure, either $(X;Y,Z)$ or $(\Sigma;\Phi,\Omega)$ could be extended to a strictly larger Seifert triad. By a further application of the fact stated above we can arrange that $\mathrm{Fr}(X;Y,Z)$ is saturated in $(\Sigma;\Phi,\Omega)$ and that $\mathrm{Fr}(\Sigma;\Phi,\Omega)$ is saturated in $(X;Y,Z)$, so in particular $(X;Y,Z) \cap (\Sigma;\Phi,\Omega)$ inherits a Seifert triad structure from both X and Σ. Finally, without loss, we can suppose that among all Seifert triads in their ambient isotopy classes in $(M;P,W)$ having this last property, $(X;Y,Z) \cap (\Sigma;\Phi,\Omega)$ has the fewest components. In that case we discover that the structures on $(\Sigma;\Phi,\Omega)$ and $(X;Y,Z)$ can be chosen to agree on $(X;Y,Z) \cap (\Sigma;\Phi,\Omega)$. That makes $(X;Y,Z) \cup (\Sigma;\Phi,\Omega)$ an extension of $(\Sigma;\Phi,\Omega)$, so the union, and therefore $(X;Y,Z)$ must lie in a regular neighborhood of $(\Sigma;\Phi,\Omega)$ in $(M;P,W)$. This clearly implies the theorem.

Theorem 2 yields a nice characterization of maximal Seifert triads which we give in Appendix A.

We shall use the next theorem in an important way in our proof of Theorem 1.

THEOREM 3. *If $(M;P,W)$ is a Haken triad with characteristic triad $(\Sigma;\Phi,\Omega)$ and if $p:M' \to M$ is a double covering, then $(M';p^{-1}(P),p^{-1}(W))$ is a Haken triad, and a subcollection of the components of $p^{-1}(\Sigma;\Phi,\Omega)$ is characteristic for it.*

In Appendix B we offer a proof that uses the characterization of maximal essential Seifert triads established in Appendix A. Theorem 3 was known to Jaco-Shalen [6]. Scott [10] has given a more recent and more general proof.

The next theorem is due to Nielsen [8] except in one special case where his argument was flawed. The general result is part of Thurston's classification of surface homeomorphisms [2]. The theorem in the case of a torus was apparently known to Poincare.

THEOREM 4. *Let (S,U) be a compact orientable surface pair, where S is either a torus or a hyperbolic surface. Let τ be a self-homeomorphism of (S,U). Let $\mathscr{C}$ be a finite collection of non-trivial homotopy classes of maps of circles and arcs into S with the ends of the arcs going into U. Suppose τ preserves $\mathscr{C}$. Then there is an incompressible, boundary incompressible surface (S',U') in (S,U) and a homeomorphism τ' isotopic to τ such that (S',U') contains a representative of $\mathscr{C}$ and the restriction of τ' to (S',U') is periodic.*

In our proof of the Homotopy Annulus and Torus Theorems, Theorem 4 will be used to set up an application of the next result.

THEOREM 5. *Let (S,U) be a compact orientable surface pair and let $h:(S,U) \to (S,U)$ be a periodic homeomorphism. Then the 1-dimensional foliation induced on the mapping torus of h by the product foliation on $(S,U) \times [0,1]$ is a Seifert fibering.*

We outline a proof. In this case, filling in the details is easy. To begin, by taking a suitable finite covering of the mapping torus, establish that the leaves of the mapping torus are all circles which have saturated solid torus neighborhoods. Then show that periodic homeomorphisms of the closed disc are either the identity or have just one fixed point, and all other points have exactly the same period. This can be done by showing first that all points on the boundary of the disc have the same period as a boundary point having minimal period. And then, unless the result sought is true, taking an appropriate power of the homeomorphism on the disc to reduce to one of the cases illustrated in Figure 1.

<u>Figure 1</u>. The period of h is k.

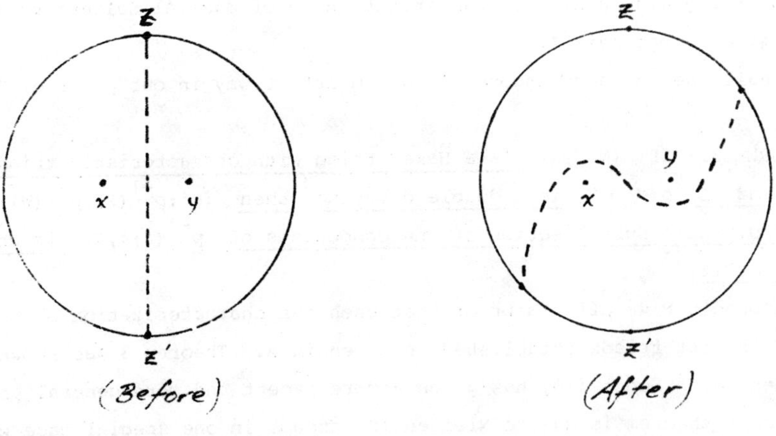

<u>Case 1</u>. Both x and y fixed by the homeomorphism h while ∂D undergoes a non-trivial periodic rotation.

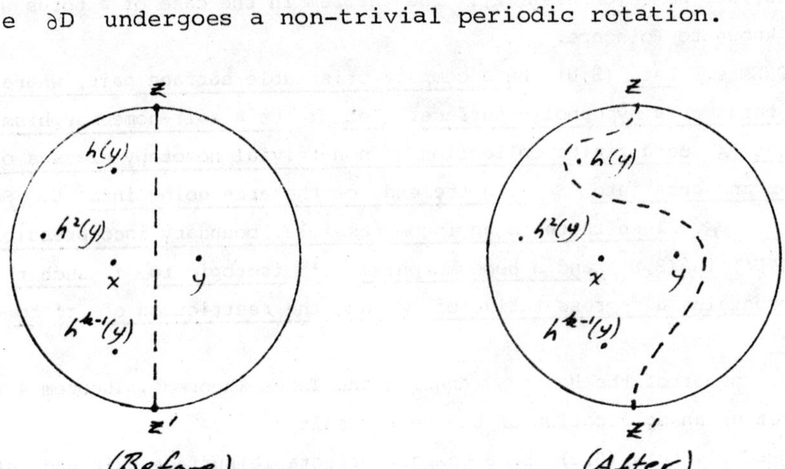

<u>Case 2</u>. The point x is fixed, as is the entire boundary. There is a non-fixed point y.

In both cases the homotopy class in $(D-x \cup y, z \cup z')$ represented by the restriction of the dashed diameter mod its endpoints under the restriction of h^k is not the class represented by the restriction of the identity. This contradicts the assumption that h is periodic of period k.

Finally, our work is made easier by a little knowledge of the geometry of surfaces. What we need is listed in the next theorem. Its proof is classical.

THEOREM 6. Let (M,P) be a compact connected oriented surface pair. Let B be a compact 1-manifold. If $\chi(M) < 0$, there is a hyperbolic metric on M in which ∂M is totally geodesic. If $f:(B,\partial B) \to (M,P)$ is a map, there is a unique minimal length geodesic map g homotopic to f. If f is a proper embedding, then f ambient isotops so its image consists of ε-parallels of the components of the image of f. (Note that in a hyperbolic metric, an ε-parallel of a geodesic is a geodesic only if $\varepsilon = 0$.) If the image $g(B)$ is connected and is not just a single embedded circle or arc, then the image $f(B)$ is connected.

If $\chi(M) = 0$, there is a euclidean metric on M for which all the above properties hold except the uniqueness of the geodesic representative. In this case there is a whole family of geodesics representing a given map, but they are all parallel.

2. AUGMENTED REGULAR NEIGHBORHOODS

We shall need to jazz up regular neighborhoods in compact orientable surface pairs and in Haken triads. In both cases we use essentially the same procedure to obtain a particular sort of manifold neighborhood. However, because of the very different properties of the ambient manifolds in the two cases, the close similarity in construction is not paralleled by as striking a resemblance in the properties of the neighborhoods. We define the neighborhoods immediately below for surface pairs followed by propositions displaying the properties we use later on. After that we do the same for the neighborhoods in Haken triads.

Let (M,P) be a compact orientable triangulated surface pair and let (J,K) be a polyhedral subpair in (M,P). We iteratively define a (n, as we shall see, finite) sequence of surface subpairs $(\hat{M}_0,\hat{P}_0) \subset \cdots \subset (\hat{M}_j, \hat{P}_j) \subset \cdots$ where $\hat{P}_j = \hat{M}_j \cap P$. We begin with a regular neighborhood $(\hat{M}_0,\hat{P}_0)$ of (J,K) in (M,P). Suppose $\hat{M}_{j-1}$ has been defined. If there is a disc D in M whose frontier in M is a properly embedded 1-manifold pair in (M,P) consisting either of the entire boundary of D or of a single arc in ∂D whose complement in ∂D lies in P, if $Fr(D,M)$ is contained in $\hat{M}_{j-1}$, and if D is not contained in $\hat{M}_{j-1}$, then define $\hat{M}_j \equiv \hat{M}_{j-1} \cup D$. If no such D exists, terminate the sequence with $(\hat{M}_{j-1}, \hat{P}_{j-1})$.

Since the disc D, if it exists, contains each component of $M-\hat{M}_{j-1}$ that it hits, and since it must hit at least one component, we can conclude that $M-\hat{M}_j$ has fewer components than $M-\hat{M}_{j-1}$. In particular, this means that the sequence we obtain is finite. We define the last pair in such a sequence to be an augmented regular neighborhood of (J,K) in (M,P).

PROPOSITION 1. *Let* (M,P) *be a compact, orientable triangulated surface pair and let* (J,K) *be a polyhedral subpair of* (M,P). *Let* $(\hat{M},\hat{P})$ *be an augmented regular neighborhood of* (J,K) *in* (M,P). *Then*

(i) *The components of* $M-\hat{M}$ *are a subcollection of the components of* $M-\hat{M}_0$.

(ii) *Each component of* $\hat{P}$ *intersects* J; *each component of* $\hat{M}\cap\partial M$ *intersects* J.

(iii) $(\hat{M},\hat{P})$ *is incompressible and boundary incompressible in* (M,P).

(iv) *If each component of* J *is homotopically non-trivial in* M, *then each component of* $\hat{M}$ *contains precisely one component of* J. *In that case* $\mathrm{incl}_*:\pi_1(J)\to\pi_1(\hat{M})$ *is surjective.*

PROOF. Property (i) follows from the fact that the components of $M-\hat{M}_j$ are a subcollection of those of $M-\hat{M}_{j-1}$. Property (ii) holds for components of $\hat{P}_j$ since it is true for $\hat{P}_0$, and is clearly not lost at any stage of the construction since $\hat{P}_j-\hat{P}_{j-1}$ is contained in an arc in P_j whose endpoints at least are in $\hat{P}_{j-1}$ and so is connected through them to J. The argument for $\hat{M}\cap\partial M$ is analogous. If (iii) were to fail, we could extend the sequence beyond $(\hat{M},\hat{P})$. As for Property (iv), the first part holds since it does for $\hat{M}_0$, since at no stage does the construction add new components, and since if two components of $\hat{M}_{j-1}$ were in the same component of $\hat{M}_j$, one of them would lie in D, making a component of J trivial in M. The second part of (iv) holds since for each j, $\pi_1(D,\mathrm{Fr}(D,M))=0$ and $\mathrm{Fr}(D,M)\subset\hat{M}_{j-1}$.

PROPOSITION 2. *Notation as in the hypothesis of Proposition* 1. *Suppose in addition that each component of* J *is homotopically non-trivial in* M. *If* $(N,N\cap P)$ *is a surface pair neighborhood of* (J,K) *in* $(\hat{M},\hat{P})$ *that is incompressible and boundary incompressible in* $(\hat{M},\hat{P})$, *and such that each component of* N *contains a component of* J, *then* $(N,N\cap P)$ *ambient isotops to* $(\hat{M},\hat{P})$ *in* (M,P) *fixing* (J,K). *In particular this means* $(N,N\cap P)$ *is an augmented regular neighborhood of* (J,K) *in* (M,P).

PROOF: By the hypotheses of this proposition and by (iv) of Proposition 1, we need deal only with the case that J, N, and $\hat{M}$ are connected. By (iv) we have that $\mathrm{inclusion}_*:\pi_1(N)\to\pi_1(\hat{M})$ is onto. From this it follows by Van Kampen's Theorem and the classification of surface groups that each component of $\overline{\hat{M}-N}$ is either a disc D that hits N in the whole boundary of D or in a single arc in ∂D, or is an annulus that hits N precisely along one of its boundary circles. We can eliminate the first possibility since by hypothesis N is incompressible in $\hat{M}$. Thus $\overline{\hat{M}-N}$ is a collar neighborhood of $Fr(N,\hat{M})$ in $\hat{M}$.

From the hypothesis that $(N, N\cap P)$ is boundary incompressible in $(\hat{M},\hat{P})$, it follows that no disc component of $\overline{\hat{M}-N}$ can hit P in the entire arc B in its boundary that is complementary to its arc of intersection with N.

By Property (ii) of Proposition 1, we can conclude, since N contains J, that in fact P intersects B, if at all, in a regular neighborhood of one or both points of ∂B in B. Similarly for $\partial\hat{M}\cap B$. Together with the conclusion of the previous paragraph, these facts imply that $(N, N\cap P)$ ambient isotops across the collar to $(\hat{M},\hat{P})$ in (M,P).

We now turn to augmented regular neighborhoods in Haken manifolds. Let $(M;P,W)$ be a triangulated Haken triad and let $(J;K,L)$ be a subpolyhedron of $(M;P,W)$ such that $J\cap W = L$ and $(L;L\cap K)$ is a disjoint union of proper essential arcs in $(W, W\cap P)$. We construct an augmented regular neighborhood $(\hat{M};\hat{P},\hat{W})$ $(\equiv \hat{M};\hat{M}\cap P,\hat{M}\cap W)$ just as we did for surfaces, except in this case D is a 3-cell whose frontier in M is either the sphere ∂D or a single disc whose complement in ∂D lies in P.

PROPOSITION 3. $(M;P,W)$, $(J;K,L)$ and $(\hat{M};\hat{P},\hat{W})$ as above. Then

(i) The components of $M-\hat{M}$ are a subcollection of the components of $M-\hat{M}_0$.

(ii) Each component of $\hat{P}$ intersects J; each component of $\hat{M}\cap\partial\hat{M}$ intersects J; the components of $(\hat{W},\hat{W}\cap P)$ are strips that are essential in the strips $(W, W\cap P)$.

(iii) $\hat{M}$ is irreducible and $\hat{P}$ is incompressible in $\hat{M}$. Together with the last assertion in (ii) this makes each component of $(\hat{M};\hat{P},\hat{W})$ a Haken triad.

(iv) If each component of J is homotopically non-trivial in M, then each component of $\hat{M}$ contains precisely one component of J. In that case, $incl_*: \pi_1(J) \to \pi_1(\hat{M})$ is surjective.

PROOF. Except for the last assertion of (ii) the proof is precisely the same as that of Proposition 1. The proof of the exceptional assertion is trivial from the hypotheses.

3. DOUBLE COVERINGS AND AUGMENTED REGULAR NEIGHBORHOODS

Let J be a connected complex and M be a simplicial Haken manifold. Let $f:J\to M$ be a simplicial map. We define the complexity of f, $c(f)$, to be the number of pairs of distinct simplexes in J, the f images of whose interiors intersect in M. If $\hat{M}$ is an augmented regular neighborhood of $f(J)$ in M it is easy to find a simplicial structure on $\hat{M}$ which extends that on $f(J)$. The map $f:J\to\hat{M}$ is therefore simplicial. Of course the complexity of

f is the same whether its range is $\hat{M}$ or M.

THEOREM 1. Let $f: J \to M$ be a simplicial map. Then unless $f_{\#}: H_1(J;\mathbb{Z}_2) \to (H_1(\hat{M};\mathbb{Z}_2)$ is surjective, there is a connected double covering $p: \hat{M}' \to \hat{M}$ and a lift $f': J \to \hat{M}'$ that is simplicial into the lifted structure on $\hat{M}$ and that has lower complexity than f.

The proof of Theorem 1 is just part of Stallings proof of the Loop Theorem.

PROOF: Consider the commutative diagram

$$\begin{array}{ccc}
\pi_1(J) & \xrightarrow{H} & H_1(J;\mathbb{Z}_2) \\
{\scriptstyle f_*}\downarrow & & \downarrow{\scriptstyle f_{\#}} \\
\pi_1(\hat{M}) & \xrightarrow{H} & H_1(\hat{M};\mathbb{Z}_2) \\
 & & /\!/ \\
 & & f_{\#}(H_1(J;\mathbb{Z}_2)) \oplus G \overset{q}{\dashrightarrow} \mathbb{Z}_2
\end{array}$$

where the horizontal maps H are Hurewicz homomorphisms and the vertical maps are induced by f. The homology group $H_1(\hat{M};\mathbb{Z}_2)$ is a finite dimensional $\mathbb{Z}_2$- vector space, so it decomposes into $f_{\#}(H_1(J;\mathbb{Z}_2))$ and a complement G. Suppose $f_{\#}$ is not surjective. Then $G \neq 0$. This allows us to define q to be a non-trivial projection whose kernel contains $f_{\#}(H_1(J;\mathbb{Z}_2))$. Thus $q\ H: \pi_1(\hat{M}) \to \mathbb{Z}_2$ is a non-trivial homomorphism whose kernel is of index 2 in $\pi_1(\hat{M})$ and contains $f_*(\pi_1(J))$. It follows that $\hat{M}$ has a connected double covering $p: \hat{M}' \to \hat{M}$, and f lifts to a map $f': J \to \hat{M}'$. Clearly f' satisfies the hypotheses of the theorem. It remains to show that $c(f') < c(f)$. Clearly $c(f') \leq c(f)$ since f' is a lift of f to the covering space $\hat{M}'$ of $\hat{M}$. We now argue that $c(f) \neq c(f')$. Otherwise the map $p \circ \mathrm{incl}|f'(J): f'(J) \to f(J)$ is a homeomorphism: In that case we would have the following commutative diagram:

$$\begin{array}{ccc}
\pi_1(f'(J)) & \xrightarrow{\mathrm{incl}_*} & \pi_1(\hat{M}') \\
\cong\ \downarrow{\scriptstyle (p \circ \mathrm{incl}_*)} & & \downarrow{\scriptstyle p_*} \\
\pi_1(f(J)) & \underset{(\text{onto})}{\xrightarrow{\mathrm{incl}_*}} & \pi_1(\hat{M})
\end{array}$$

The lower horizontal map is onto since that is a characteristic of augmented regular neighborhoods, and the vertical map on the left is an isomorphism since it is induced by a homeomorphism. That forces p_* to be surjective. This is only possible if p is the trivial covering projection. This concludes the argument that $c(f) > c(f')$, and the proof of Theorem 1.

We need to recall the following result.

LEMMA 2. *If* M *is a compact 3-manifold, and if* $\mathrm{incl}_{\#}: H_1(\partial M;\mathbb{Z}) \to H_1(M;\mathbb{Z})$ *is the induced homomorphism, then*

$$\mathrm{Rank}(\mathrm{Ker}(\mathrm{incl}_{\#})) = \frac{1}{2}\,\mathrm{Rank}(H_1(\partial M;\mathbb{Z}))\,.$$

With $\mathbb{Z}_2$ *coefficients,* $\mathrm{Ker}(\mathrm{incl}_{\#})$ *splits off as a direct summand having half the dimension of* $H_1(\partial M;\mathbb{Z}_2)$.

The proof of Lemma 2 is a standard exercise in the use of duality.

4. THE PROOF OF THE HOMOTOPY DISC THEOREM

PROOF. In broad outline, the proof of this theorem and those of the Homotopy Annulus are Torus Theorems are all the same. Without loss, the map f is assumed to be proper and simplicial. The proofs are by induction on the complexity of f. They all end by showing that f homotops through proper maps into a connected Seifert triad which can be trivially modified to miss $\overline{\partial M - P \cup W}$ and therefore satisfy condition (i) in the definition of essential Seifert triad in (M;P,W). Since the Seifert triad contains f, condition (ii) is also satisfied, so it is essential in (M;P,W). Then by Theorem 1.2 the essential Seifert triad ambient isotops into $(\Sigma;\Phi,\Omega)$ carrying f with it.

If the complexity $c(f) = 0$, f is an embedding, so its regular neighborhood in (M;P,W) is already an essential Seifert triad, and the theorem follows immediately.

We specialize to the Homotopy Disc Theorem. We suppose that $c(f) > 0$. By the discussion in Section 2 and at the beginning of Section 3, there is an augmented regular neighborhood $(\hat{M};\hat{P},\hat{W})$ of $f(I^2;\ I\times\partial I,\ \partial I\times I)$ in (M;P,W) triangulated so that f is simplicial into $(\hat{M};\hat{P},\hat{W})$. There are two cases, depending on whether $f_{\#}: H_1(I^2;\mathbb{Z}_2) \to H_1(\hat{M};\mathbb{Z}_2)$ is surjective. Suppose $f_{\#}$ is not surjective. Then by Theorem 3.1, f lists to a map
$f': (I^2;\ I\times\partial I,\ \partial I\times I) \to (\hat{M}';\hat{P}',\hat{W}')$ where $\hat{M}' \xrightarrow{\ p\ } \hat{M}$ is a double covering; $\hat{P}' \equiv p^{-1}(\hat{P})$ and $\hat{W}' \equiv p^{-1}(\hat{W})$. By Theorem 1.3, $(\hat{M}';\hat{P}',\hat{W}')$ is a Haken triad with characteristic triad contained in $p^{-1}(\hat{\Sigma};\hat{\Phi},\hat{\Omega})$ where $(\hat{\Sigma};\hat{\Phi},\hat{\Omega})$ is characteristic for $(\hat{M};\hat{P},\hat{W})$. The map f' satisfies the hypothesis of this theorem and $c(f') < c(f)$. By induction on the complexity, we can homotop f' into $p^{-1}(\hat{\Sigma};\hat{\Phi},\hat{\Omega})$. This homotopy covers one taking f into $(\hat{\Sigma};\hat{\Phi},\hat{\Omega})$. We finish the proof as indicated above.

Now suppose $f_{\#}$ is surjective. Since $\partial I^2 \neq \emptyset$, we know $\partial\hat{M} \neq \emptyset$. From Lemma 3.2, $H_1(\partial\hat{M};\mathbb{Z}_2) = 0$, so $\partial\hat{M} = S^2$. Since $\hat{M}$ is Haken, $\hat{M}$ is a 3-cell. The conditions placed on $\hat{P}$ and $\hat{W}$ by the fact that $(\hat{M};\hat{P},\hat{W})$ is a Haken triad imply that the components of $\hat{P}$ and $\hat{W}$ are all discs, and that each disc in $\hat{P}$ hits each disc in $\hat{W}$ if at all, in precisely one interval. The hypothesis of the theorem that f is essential implies that the components intersecting $f(\partial I^2)$ are precisely two discs in $\hat{P}$ and two discs in $\hat{W}$ that alternate to form an annulus in $\partial\hat{M}$ (see Figure 1). Consequently, there is no difficulty in homotoping f to an embedding and then finishing the proof as in the case $c(f) = 0$.

Figure 1. Case $f_\#$ is surjective:

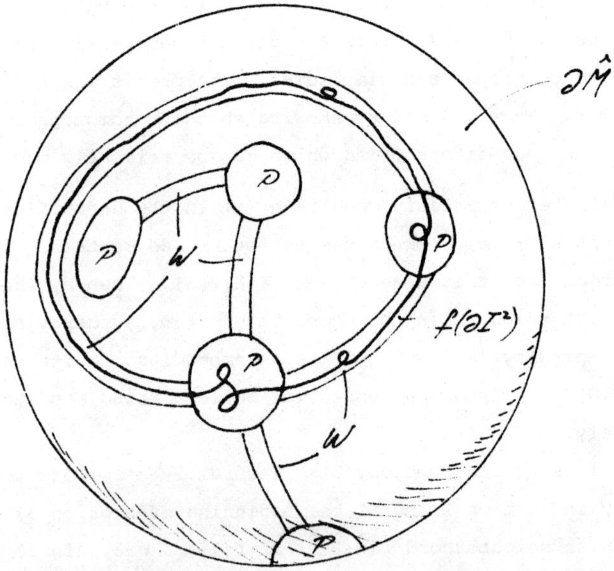

5. THE PROOF OF THE HOMOTOPY ANNULUS THEOREM

PROOF. We set up the proof and complete it as described in the first paragraph of Section 4. The proof here is identical to that proof up to the case that $f_\#$ is surjective.

We suppose $f_\#: H_1(A^2;\mathbb{Z}_2) \to H_1(\hat{M};\mathbb{Z}_2)$ is surjective. $H_1(\hat{M};\mathbb{Z}_2)$ cannot be zero since that would make $\hat{M}$ a 3-cell which in turn would force f to be inessential. Thus $f_\#$ is an isomorphism. Together with Lemma 3.2, this implies that $\partial\hat{M} = T^2$.

Since $\hat{P}$ is incompressible in $\hat{M}$ and since each component of $\hat{P}$ must contain a component of $f(\partial A^2)$ (this last since $(\hat{M},\hat{P})$ is an augmented regular neighborhood of $f(A^2,\partial A^2)$ in (M,P)), $\hat{P}$ is either all of $\partial\hat{M}$ or it consists of one or two parallel annuli representing a non-trivial homotopy class in the torus $\partial\hat{M}$.

With no loss of generality we can suppose that $\partial\hat{M}$ is equipped with a euclidean metric and that the frontier of $\hat{P}$ in $\partial\hat{M}$ is totally geodesic.

By Lemma 3.2, we know that the kernel of the map on $\mathbb{Z}$-homology by the inclusion of $\partial\hat{M}$ into $\hat{M}$ has a single generator which, however, may not be indivisible in the homology of $\partial\hat{M}$.

By pulling that generator back into $H_2(\hat{M},\partial\hat{M})$ by the homology boundary homomorphism, then taking it into $H^1(\hat{M})$ by duality and finally interpreting $H^1(\hat{M})$ as a class in $[\hat{M},S^1]$ we get a map of $\hat{M}$ onto S^1, transverse to a point $*$ in S^1 for which the total preimage of $*$ is an oriented surface properly embedded in $\hat{M}$ whose boundary represents the original generator of the kernel in $H_1(\partial\hat{M})$. Since $\partial\hat{M}=T^2$, by a homotopy of the map of M to S^1, we can arrange that the intersection of the surface with $\partial\hat{M}$ consists of parallel geodesic circles that all inherit the same orientation from the surface when viewed in $\partial\hat{M}$. Compress the surface in $\hat{M}$, and eliminate a maximal union of components whose boundary represents the zero class in $H_1(\partial\hat{M})$. A connected incompressible surface V remains. We call it the <u>kernel surface</u>.

Since the frontier of $\hat{P}$ in $\partial\hat{M}$ is also a collection of parallel oriented geodesics, it is easy to check that regardless of the form of $\hat{P}$, any arc in $\hat{P}$ with endpoints in ∂V and interior missing ∂V that begins and ends on the same side of V is inessential in $(\hat{P},\hat{P}\cap\partial V)$. This means that V is boundary incompressible in $(\hat{M},\hat{P})$, since the intersection of any boundary compressing disc with $\hat{P}$ would be an arc of the sort just described except that it would be essential in $(\hat{P},\hat{P}\cap\partial V)$.

We begin to homotop f to improve its situation in $(\hat{M},\hat{P})$. Without loss we can suppose f is transverse to V and that among all maps homotopic to it, $f^{-1}(V)$ has the fewest components. We claim that $f^{-1}(V)$ is non-empty, and in fact, that $f^{-1}(V)\cap\partial_+A^2\neq\emptyset$, where ∂_+A^2 is a component of ∂A^2. Otherwise the $\mathbb{Z}_2$-algebraic intersection of $f(\partial_+A^2)$ with V would be zero. But since $f(\partial_+A^2)$ generates $f_\#(H_1(A^2;\mathbb{Z}_2))$ and $(V,\partial V)$ generates $H_2(\hat{M},\partial\hat{M};\mathbb{Z}_2)$ and since the dual pairing between $H_1(\hat{M};\mathbb{Z}_2)$ with $H_2(\hat{M},\partial\hat{M};\mathbb{Z}_2)$ is given by algebraic intersection, this is impossible.

By the usual innermost circle argument, using the fact that V is incompressible, we find that $f^{-1}(V)$ contains no inessential circles; by the edgemost arc analog, using V boundary incompressible in $(\hat{M},\hat{P})$, there are no inessential arcs. There must be arcs, since we found at least one endpoint above; these arcs are essential. Finally, since there are essential arcs, there can be no essential circles.

We now eliminate the case that V is a disc. If it were, $\partial\hat{M}$ would be compressible in $\hat{M}$ making $\hat{M}$ a solid torus; (M,P) would then be Seifert fibered and we could skip immediately to the usual ending. We can also assume that V is not an annulus, since because $\hat{M}$ is orientable, the boundary of such an annulus would be homologically trivial in $\partial\hat{M}$. Thus, we can assume that V comes equipped with a hyperbolic structure of the sort described in Theorem 1.6.

Finally, we can suppose that f has been homotoped leaving $f^{-1}(V)$ unchanged so that $f|f^{-1}(V)$ is geodesic in the hyperbolic structure on V.

We cut the Haken pair $(\hat{M},\hat{P})$ along V, obtaining the manifold triad $(M^*;V^*,P^*)$. See Figure 1. Since V is two-sided, V^* consists of two disjoint copies V_-^* and V_+^* of V. The manifold M^* is clearly irreducible and $\partial M^* \neq \emptyset$, so it is Haken. The surface V^* is incompressible in M^* since V is incompressible in M.

Figure 1.

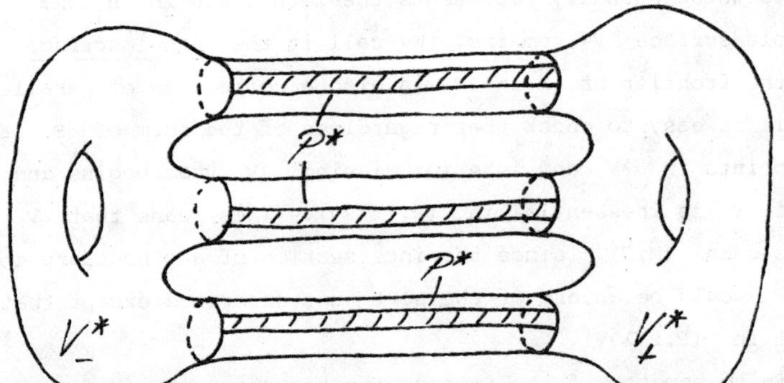

The surface represents ∂M^*.

If $\hat{P} \neq \partial\hat{M}$, then each component of $(P^*,P^* \cap V^*)$ is a strip. It is essential in (M^*,V^*) since its ends are in different components of V^*. If $\hat{P} = \partial\hat{M}$, clearly the components of $(P^*,P^* \cap V^*)$ are annulus pairs. They, too, are essential since their ends are in different components of V^*.

Let $(W,W\cap\hat{P})$ be an augmented regular neighborhood of $\partial V \cup f(f^{-1}(V))$ in $(V,V\cap\hat{P})$. Of course $W\cap\hat{P} = V\cap\hat{P}$. By Theorem 1.6, we can arrange that the frontier of $(W,W\cap\hat{P})$ consists of geodesics and ε-parallels in $(V,V\cap\hat{P})$. Let $W^* \subset V^*$ be the surface obtained from W in splitting M along V.

Cut $(A^2,\partial A^2)$ along $f^{-1}(V)$ obtaining the disc triads $(H_j;K_j,H_j\cap\partial A^2)$, $j=1,\ldots,n$. The H_j are ordered cyclically around the annulus A^2; $K_j = K_j^- \cup K_j^+$ where K_j^- and K_{j+1}^+ are the result of splitting the component of $f^{-1}(V)$ that lies between H_{j-1} and H_j in the cyclic ordering. Let $(H;K,H \cap \partial A^2) = \bigcup_{j=1}^{n} (H_j;K_j,H_j \cap \partial A^2)$. Let $g = f| : (H;K,H \cap \partial A^2) \to (M^*;W^*,P^*)$. The homomorphism $g_*: \pi_1(H,H \cap \partial A^2) \to \pi_1(M^*,P^*)$ is monic on each component since $f_*: \pi_1(A^2,\partial A^2) \to \pi_1(\hat{M},\hat{P})$ is, and $g_*: \pi_1(H,K) \to \pi_1(M^*,W^*)$ is monic since otherwise the number of components in $f^{-1}(V)$ could be reduced by a homotopy; this makes g essential.

Here we come to the heart of the proof. We shall produce a usually disconnected and not necessarily Seifert (the strip condition may fail) product I-bundle triad in $(M^*;V^*,P^*)$ that contains $\overline{\partial M^* - V^*}$ and whose corresponding ∂I-bundle pair coincides with $(W^*,W^* \cap P^*)$. We shall also produce a homotopy of $g: (H;K,H \cap \partial A^2) \to (M^*;W^*,P^*)$ fixed on K that takes $g(H)$ into this product I-bundle triad. During the construction we make use of auxiliary homotopies that move $g|K$.

Using the produce structure on $(P^*,P^* \cap V^*)$, homotop $g|(H,K) \cap \partial A^2$ to an embedding in $(P^*,P^* \cap V^*)$. By the construction of W, this homotopy can be made to extend to a homotopy of $g: (H;K,H \cap \partial A^2) \to (M^*;W^*,P^*)$. Let Q^* be the regular neighborhood of the homotoped $g(H \cap \partial A^2)$ in P^*. Note that $(M^*;W^*,Q^*)$ is Haken. By the Homotopy Disc Theorem, g further homotops to h mapping into the characteristic triad $(\Gamma;\chi,\Lambda)$ of $(M^*;W^*,Q^*)$. Let $(\Gamma^{**};\chi^{**},\Lambda^{**})$ be the union of the components of $(\Gamma;\chi,\Lambda)$ that actually intersect the image of h. These components are necessarily product I-bundle triads since they each have components of χ^{**} in different components of W^*.

Notice that Γ^{**} must contain $\overline{\partial M^* - V^*}$ and χ^{**} must contain ∂V^*, for otherwise we could enlarge $(\Gamma^{**};\chi^{**},\Lambda^{**})$ by adjoining a collar of $\overline{\partial M^* - V^*}$ fibered so that Q^* is saturated in it, thereby violating the maximality of $(\Gamma^{**};\chi^{**},\Lambda^{**})$.

Since each component of $g(K) \cup \partial V^*$ contains at least one component of ∂V^*, since the components of ∂V^* are homotopically non-trivial in V^* (otherwise V would be a disc), and since W^* is the augmented regular neighborhood of $g(K) \cup \partial V^*$ in V^*, we can conclude from Proposition 2.1 (iv) that $g(K) \cup \partial V^*$ hits each component of W^* in a connected set.

We now show that each component of W^* contains precisely one component of χ^{**}. Clearly, it contains at least one component.

First of all, if the images of the restrictions of g to two components of $(K,\partial K)$ have an endpoint of each in a common component of ∂V^* then the h images have the same property. This of couse means that the union of the two h images with the component(s) of ∂V^* they intersect is connected. If the two g image restrictions do not have an endpoint of each in a common component of ∂V^*, but if the two g images intersect, then the h images must also intersect. From these two observations and the definition of W^*, it follows that the intersection of $h(K) \cup \partial V^*$ with each component of W^* is a connected set. Since the union of $h(K)$ with those components of ∂V^* that $h(K)$ intersects lies in χ^{**}, we conclude that each component of W^* contains just one component of χ^{**}.

Now we move $(\chi^{**}, \chi^{**} \cap P^*)$ to $(W^*, W^* \cap P^*)$ in $(V^*, V^* \cap P^*)$ fixing ∂V^*. We do this in two steps. First, working inside $(W^*, W^* \cap P^*)$, we ambient isotop $(\chi^{**}, \chi^{**} \cap P^*)$ so that its frontier consists of geodesics and ε-parallels of geodesics. The important feature of this move is that if ε is chosen sufficiently small and with the correct sign, after the isotopy, $g(K) \cup \partial V^*$ is contained in χ^{**}. We note that since $\partial V^* \subset \chi^{**}$, this move can be chosen to fix ∂V^*. The first move sets up the second move, which is effected by an application of Proposition 2.2. It is only necessary to check that the hypotheses of that proposition are satisfied, which we do now. We saw two paragraphs back that each component of $g(K) \cup \partial V^*$ is homotopically non-trivial in V^*. In the previous paragraph we saw that each component of W^* contains precisely one component of χ^{**}. Since each component of W^* contains a component of $g(K) \cup \partial V^*$, and this latter set is now in χ^{**}, we have that each component of χ^{**} contains a component of $g(K) \cup \partial V^*$. Finally, by the discussion preceding Theorem 1.2, $Fr(\Gamma;\chi,\Lambda)$ is essential in $(M^*;W^*,Q^*)$. In fact, in the present situation the maximality of $(\Gamma;\chi,\Lambda)$ in $(M^*;W^*,Q^*)$ implies a bit more, namely that $Fr(\Gamma;\chi,\Lambda)$ hence $Fr(\Gamma^{**};\chi^{**},\Lambda^{**})$ is essential in $(M^*;W^*,P^*)$. We conclude from this that $(\chi^{**}, \chi^{**} \cap P^*)$ is incompressible and boundary incompressible in $(W^*, W^* \cap P^*)$. Thus the hypotheses of Proposition 2.2 are satisfied and we get our second move. This move fixes $g(K) \cup \partial V^*$.

Let φ denote the composition of the two ambient moves. Let $(\Gamma^*;\chi^*,\Lambda^*) = \varphi(\Gamma^{**};\chi^{**},\Lambda^{**})$. Note that the homotopy taking $g|(K, K \cap \partial A^2)$ to $\varphi\, h|(K, K \cap \partial A^2)$ lives in $(W^*, W^* \cap P^*)$, and so by the usual collaring argument we can modify the homotopy taking g to $\varphi\, h$ so that it is actually fixed on K, and draw the weaker conclusion that $\varphi\, h(H;K,H \cap \partial A^2) \subset (\Gamma^*;W^*,P^*)$. We suppose this has been done.

We now reglue $(M^*;V^*,P^*)$ along V^* regaining the Haken pair $(\hat{M},\hat{P})$. As things are now arranged, the components of $(\Gamma^*;\chi^*,P^*)$ glue up to a 3-manifold pair $(\hat{X},\hat{Y})$ where $\hat{Y} = \hat{P}$. The collection of disc triads $(H;K,H \cap \partial A^2)$ reglues to the pair $(A^2, \partial A^2)$, and the homotopy from g to $\varphi\, h$ becomes a homotopy in $(\hat{M},\hat{P})$ of f to a map $\hat{f}:(A^2,\partial A^2) \to (\hat{X},\hat{Y})$.

Next we identify $(\hat{X},\hat{Y})$ as a Seifert fibered pair. We begin with the fact that each component of $(\Gamma^*;\chi^*,P^*)$ is a product I-bundle triad.

Since $h(H)$ intersects each component of Γ^*, $(\hat{X},\hat{Y})$ is connected. Since $(\hat{X},\hat{Y})$ is obtained by gluing product $(I,\partial I)$-bundles end to end with no ends left over, all the components of $(\chi^*,\chi^* \cap P^*)$ are homeomorphic, and $(\hat{X},\hat{Y})$

is the mapping torus of some self-homeomorphism τ of $(W_0, W_0 \cap \hat{P})$ where W_0 is a component of W in V.

We can discover the effect of τ on the homotopy classes $\hat{f}|\hat{f}^{-1}(W_0, W_0 \cap P)$ by sliding each such relative class along $f|(A^2, \partial A^2)$, in the direction of the cyclic ordering until it again lies in $\hat{f}^{-1}(W_0, W_0 \cap \hat{P})$. It is clear that this procedure preserves the set of classes $\hat{f}|\hat{f}^{-1}(W_0, W_0 \cap P)$. Of course τ preserves the classes represented by $\partial V \cap W_0$. Since $(W_0, W_0 \cap \hat{P})$ is an augmented regular neighborhood of $\hat{f}(\hat{f}^{-1}(W_0, W_0 \cap \hat{P})) \cup (\partial V \cap W_0)$ in $(V, V \cap \hat{P})$, we can apply Theorem 1.4 to conclude that τ isotops to a homeomorphism $\hat{\tau}$ that is of finite order on $(W_0, W_0 \cap P)$. The mapping torus of $\hat{\tau}$ is Seifert fibered by Theorem 1.5.

We now complete the proof of this theorem as outlined at the beginning of Section 4.

6. THE PROOF OF THE HOMOTOPY TORUS THEOREM

The proof is much the same as that given in Section 5. Here we follow through that proof, commenting on differences as they appear and making additions when necessary. If anything, the proof in this case is easier since the image $f(T^2)$ and all the characteristic submanifolds stay away from ∂M. We first suppose $\partial M \neq \emptyset$. The proof to the case $f_\#$ surjective is identical to that of the Homotopy Annulus Theorem. If $f_\#$ is surjective, $\partial \hat{M}$ may be more complicated than a single torus: it could be two tori or one two-holed torus; but as long as it is not empty, the construction of the kernel surface V goes through as before. After homotoping f to minimize the number of components in $f^{-1}(V)$ we can still conclude that $f^{-1}(V) \neq \emptyset$; here we argue that at least one of two embedded circles in T^2 that generate $H_1(T^2; \mathbb{Z}_2)$ must intersect $f^{-1}(V)$. There can be no inessential circles in $f^{-1}(V)$.

The kernel surface V cannot be a disc because f is essential. It may be a annulus; if so, put a euclidean metric of it. Otherwise equip it with a hyperbolic metric. In any case arrange that the boundary is totally geodesic. Homotop f so $f|f^{-1}(V)$ is geodesic. Let W be the augmented regular neighborhood of $f(f^{-1}(V))$ in V. Cut $\hat{M}$ along V obtaining the Haken pair (M^*, V^*); W^* is W split along V; (M^*, W^*) is a Haken pair. Cutting T^2 along $f^{-1}(V)$ yields a collection of annulus pairs (H_j, K_j) where $K_j = \partial H_j$. Define g as before; g homotops to h, but here we quote the Homotopy Annulus Theorem instead of the Homotopy Disc Theorem. The characteristic pair for (M^*, W^*) is (Γ, χ). The components of (Γ^{**}, χ^{**}) may at this stage be $(I, \partial I)$-bundle pairs or Seifert fibered pairs.

The proof bifurcates here depending on whether any component of W^* is an annulus. We first work under the assumption that there are no annulus components. Then the construction of φ works just as before. It follows that each component of (Γ^*, χ^*) is an $(I, \partial I)$-bundle pair. If none of these is twisted, we proceed just as in Section 5 and conclude that $\hat{X}$ is Seifert fibered.

If some component, say (Γ_0^*, χ_0^*), of (Γ^*, χ^*) is twisted, then in particular, χ_0^* is connected. It follows that there is precisely one component (Γ_1^*, χ_1^*) of (Γ^*, χ^*) adjacent to (Γ_0^*, χ_0^*) in the sense that after regluing one of its χ^* components is identified with χ_0^*.

If (Γ_1^*, χ_1^*) is also a non-trivial $(I\text{-}\partial I)$-bundle, then Γ_0^* and Γ_1^* glue along χ_0^* and χ_1^* to form $\hat{X}$. Let $(\Gamma_0^{*\prime}, \chi_0^{*\prime}) \to (\Gamma_0^*, \chi_0^*)$ and $(\Gamma_1^{*\prime}, \chi_1^{*\prime}) \to (\Gamma_1^*, \chi_1^*)$ be double coverings associated with the orientable coverings of the base spaces of Γ_0^* and Γ_1^* respectively. It follows that $(\Gamma_0^{*\prime}, \chi_0^{*\prime})$ and $(\Gamma_1^{*\prime}, \chi_1^{*\prime})$ are product $(I, \partial I)$-bundles. The gluing of χ_0^* to χ_1^* induces a gluing of $\chi_0^{*\prime}$ to $\chi_1^{*\prime}$ that produces a double covering $\hat{X}' \to \hat{X}$. The map $f: T^2 \to \hat{X}$ lifts to $\hat{X}'$ since its image in χ_0^* (and in χ_1^*) is orientation preserving. In fact it has precisely two lifts, $\hat{f}', \hat{f}'': T^2 \to \hat{X}'$. We can now proceed as in Section 5, using both lifts to when setting up the application of Theorem 1.4, and conclude that $\hat{X}'$ is Seifert fibered. Once we know this, Theorem 1.3 tells us that $\hat{X}$ is Seifert fibered.

If (Γ_1^*, χ_1^*) is a product $(I, \partial I)$-bundle, then the object obtained by gluing it to (Γ_0^*, χ_0^*) is bundle equivalent to (Γ_0^*, χ_0^*). It has a single component of χ_1^* for its associated ∂I-bundle. Choose (Γ_2^*, χ_2^*) to have a component of χ_2^* glued to the free component of χ_1^*. This process eventually stops when we encounter a non-trivial $(I, \partial I)$-bundle component of (Γ^*, χ^*), and that returns us to the case settled above.

It remains to deal with the case that some component of W^* is an annulus. Note that if W_0^* is a component of W^* that merely contains an annulus C with frontier properly embedded in W_0^*, and if $h(K) \cap \mathrm{Fr}(C) = \emptyset$ but $h(K) \cap C \neq \emptyset$, then since the geodesic representative of $h|h^{-1}(W_0^*)$ is connected, it must be a single embedded circle. This makes W_0^* an annulus.

We now show that if W^* has an annulus component, then all its components are annuli. Let (Γ_j^*, χ_j^*) be the component of (Γ^*, χ^*) that contains $h(H_j, K_j)$. Of course, it is possible that $\Gamma_k = \Gamma_j$, $k \neq j$. Let W_j^{*-} and W_j^{*+} be the components of W^* containing $h(K_j^-)$ and $h(K_j^+)$ respectively. We can suppose that W_1^{*-} is an annulus. The component of χ_1^* containing $h(K_1^-)$ is contained in W_1^{*-} and is therefore also an annulus. This implies that (Γ_1^*, χ_1^*) admits a Seifert fibering. So all the components of χ_1^* are annuli; there are no tori since V^* has boundary. By the argument of the previous paragraph, we conclude that W_1^{*+} is also an annulus. Now W_2^{*-} is homeomorphic to W_1^{*+}, so it too is an annulus. Continuing in this fashion, we find that all components of W^* are annuli.

Since each component of W^* is an annulus, the components of χ in each of these must be parallel essential annuli. This means that every component of (Γ,χ) admits a Seifert fibering. Since (Γ,χ) is characteristic and therefore maximal, we conclude that each component of W^* contains precisely one component of χ. Thus W^* is a regular neighborhood of χ. As in the previous cases, there is an ambient isotopy of (M^*,V^*) that makes $\chi = W^*$. Upon regluing (M^*,V^*) along V^*, the components of (Γ,χ) glue up to the manifold $\hat{X}$. The Seifert fiberings on the components of (Γ,χ) induce a Seifert fibering on $\hat{X}$. Again, the reglued homotopies carry f into $\hat{X}$.

From here, the proof in the case $\partial M \neq \emptyset$ ends in the usual way.

Now suppose $\partial M = \emptyset$. Since M is Haken, there is a connected, two-sided incompressible surface V in M. If f homotops off V, cut M along V, returning to the case with boundary just done. If f does not homotop off V, there are two cases depending on whether V is a torus. Suppose not. Then we again return to the above proof, using this V rather than constructing one as a kernel surface.

If V is a torus, and if W is an annulus in V, the appropriate case of the above proof again does the trick. The only other possibility is that $W = V$. If this happens we shall show that (Γ^*,χ^*) consists either of a single product $(I,\partial I)$-bundle over the torus or of two twisted $(I,\partial I)$-bundles over the Klein bottle. We first note that these are the only $(I,\partial I)$-bundles that could appear as components of (Γ^*,χ^*), since they are the only ones with torus boundary. Since both these bundles also admit Seifert fiberings, we know that all components of (Γ^*,χ^*) admit Seifert fiberings. Now, in order for the components of W^* to be tori, it is necessary that no matter what Seifert fibering we have on a component (Γ_0^*,χ_0^*) of (Γ^*,χ^*), for some j, at least one map $h|K_j^-$ or $h|K_j^+$ must homotop transverse to that fibering.

We shall show that (Γ_0^*,χ_0^*) is either the product S^1-bundle over $(A^2,\partial A^2)$ or the twisted S^1-bundle over $(\text{Möb}, \partial\,\text{Möb})$. First note that (Γ_0^*,χ_0^*) cannot be a product S^1-bundle over $(D^2,\partial D^2)$ since the restriction $h|(H_j,K_j)$ could not be essential in that case. This means that (Γ_0^*,χ_0^*) is sufficiently complex that it contains a properly embedded saturated essential annulus pair $(A,\partial A)$. Without loss we can suppose that $h|K_j^-$ homotops transverse to the fibers of (Γ_0^*,χ_0^*) and choose A so that one of its boundary components, ∂A^-, lies in the same component of χ_0^* as $h(K_j^-)$ and so necessarily hits $h(K_j^-)$. Homotop $h|(H_j,K_j)$ to a map h^* that is transverse to $(A,\partial A)$. Then $(h^*)^{-1}(A) \neq \emptyset$ and by the usual methods we find that it is, in fact, a finite disjoint union of essential arcs in $(A,\partial A)$. These arcs cut $(A,\partial A)$ into rectangles. Cut (Γ_0^*,χ_0^*) along $(A,\partial A)$ obtaining another Seifert fibered pair $(\Gamma_0^\vee,\chi_0^\vee)$. Each of the rectangles cut from A maps properly into $(\Gamma_0^\vee,\partial\Gamma_0^\vee)$ under the restriction of h^* and implies the compressibility of $\partial\Gamma_0^\vee$ in $\Gamma_0^\vee$, making $\Gamma_0^\vee$ a solid torus. Since the four sides of such a rectangle map alternately into X_0^v and $\overline{\partial\Gamma_0^v - X_0^v}$, we can conclude further that $\overline{\partial\Gamma_0^v - X_0^v}$ consists of either two annuli whose winding numbers in Γ_0^v are both one, or of a single annulus whose winding number is exactly two. Regluing (Γ_0^v, X_0^v) along $\overline{\partial\Gamma_0^v - X_0^v}$, we find that (Γ_0^*, X_0^*) has one of the forms claimed at the beginning of this paragraph. Notice that these Seifert fibered pairs can also be fibered as $(I, \partial I)$-bundles with torus boundary.

There are only enough components in $V^* = W^* = \chi^*$ for one product $(I, \partial I)$- or two twisted $(I, \partial I)$-bundles. Upon regluing (M^*, V^*) along V^* we obtain $\hat{X}$ containing a homotoped image of f. $\hat{X}$ is Seifert fibered just as in the case $\partial M \neq \emptyset$ and W having no annulus components. The usual ending completes the proof of the present case and with it the proof of the Homotopy Torus Theorem.

APPENDIX A - A CHARACTERIZATION OF MAXIMAL ESSENTIAL SEIFERT TRIADS

Given the Existence and Uniqueness Theorem for characteristic triads, it is not hard to get a useful characterization of maximal Seifert triads in terms of a sort of engulfing of spines of some very simple Seifert triads. As will become apparent in the proof, the spines we use are the least complicated objects that insure the sufficiency of our characterization. We call the spines "spectacle triads".

A *spectacle triad* is either a pair formed by identifying the boundary components of a *core annulus* fibered by circles with interior fibers of one or two fundamental surface pairs fibered by circles, or a triad formed by identifying the boundary fibers of a *core strip* fibered by essential intervals with integer fibers of one or two fundamental surface triads fibered by intervals.

A spectacle triad embedded in a Haken triad (M;P,W) is *essential* if its corresponding fundamental triads are essential in (M;P,W), if they are either properly embedded in (M;P,W) or equal to a component of $(\overline{\partial M - P \cup W}; \overline{\partial P - W}, \overline{\partial M - P})$, if the interior of the core lies in the interior of M, and in the case of a core strip, if its ends lie in P.

THEOREM 1. *Let* (M;P,W) *be a connected Haken triad, and let* (X;Y,Z) *be a non-empty essential Seifert triad in* (M;P,W). *Then* (X;Y,Z) *is maximal if and only if these conditions are satisfied:*

(i) *Each component of* Fr(X;Y,Z) *in* (M;P,W) *is essential in* (M;P,W)

(ii) *If* (C;E,F) *is an essential spectacle triad in* $\overline{(M;P,W) - (X;Y,Z)}$ *having the property that each fundamental triad of* (C;E,F) *that is also a component of* Fr(X;Y,Z) *inherits the same fibering from both* C *and* X, *then* (C;E,F) *lies in a regular neighborhood of* Fr(X;Y,Z) *in* $\overline{(M;P,W) - (X;Y,Z)}$.

Note that condition (ii) implies (but is not implied by) the same condition where $(C;E,F)$ is taken to be a single fundamental triad.

PROOF. The necessity of the conditions is obvious, for if any were to fail, $(X;Y,Z)$ could be extended to a strictly larger Seifert triad.

As for the sufficiency. Let $(\Sigma;\Phi,\Omega)$ be a characteristic triad for $(M;P,W)$. We shall show that after an ambient isotopy, $(\Sigma;\Phi,\Omega)$ lies in a regular neighborhood of $(X;Y,Z)$ in $(M;P,W)$. That will be enough. We begin by using the Uniqueness Theorem to find an ambient isotopy of $(M;P,W)$ that moves $(X;Y,Z)$ into $(\Sigma;\Phi,\Omega)$; in fact, with almost no more effort we can move $(X;Y,Z)$ into the set-theoretic interior of $(\Sigma;\Phi,\Omega)$ in $(M;P,W)$. At that point $(X;Y,Z)$ is essentially embedded in $(\Sigma;\Phi,\Omega)$ in the sense of Section 1. Because of the following lemma (whose proof we leave as an exercise), we can suppose that the Seifert triad structure on $(\Sigma;\Phi,\Omega)$ extends that on $(X;Y,Z)$.

LEMMA 2. If $(\Sigma;\Phi,\Omega)$ is a Seifert triad, and if $(X;Y,Z)$ is a Seifert triad embedded essentially in $(\Sigma;\Phi,\Omega)$, then there are (generally new) structures on these triads with respect to which $(X;Y,Z)$ is a Seifert subtriad of $(\Sigma;\Phi,\Omega)$.

We shall complete the proof of Theorem 1 by showing that each component $(Q;R,S)$ of $\overline{(\Sigma;\Phi,\Omega) - (X;Y,Z)}$ is contained in a regular neighborhood of $(X;Y,Z)$ in $(M;P,W)$.

First observe that no component of $\partial Q - R \cup S$ intersects ∂M. If one did, by condition (i) in the definition of essential Seifert triad and by the definition of $(Q;R,S)$, it would coincide with a component of $(\partial M - X) - P \cup W$. On intersecting the closure of that component with the triad $(M;P,W)$ we would obtain a fundamental surface triad. If it were inessential in $(M;P,W)$, since it lies in ∂M, $(\Sigma;\Phi,\Omega)$ itself would be inessential in $(M;P,W)$ (compare the exercise in the third paragraph after the statement of Theorem 1.1). By the note just after condition (ii) in the definition of spectacle triad (this appendix), our fundamental triad would lie in a regular neighborhood of $\mathrm{Fr}(X;Y,Z)$ in $\overline{(M;P,W) - (X;Y,Z)}$ as well as in ∂M, which is impossible. Thus, $(\overline{\partial Q - R \cup S}; \overline{\partial R - S}, \overline{\partial S - R}$ equals the frontier of $(Q;R,S)$ in $(M;P,W)$.

Next notice that each component of $\mathrm{Fr}(Q;R,S)$ in $(M;P,W)$ is a component of either $\mathrm{Fr}(X;Y,Z)$ or $\mathrm{Fr}(\Sigma;\Phi,\Omega)$ in $(M;P,W)$ and hence is essentially embedded in $(M;P,W)$. Further, each of these components of $\mathrm{Fr}(Q;R,S)$ is either properly embedded in $\overline{(M;P,W) - (X;Y,Z)}$ or coincides with a component of $\mathrm{Fr}(X;Y,Z)$. Finally, since the Seifert triad structure on Σ extends that on X, for those components in $\mathrm{Fr}(Q;R,S)$ that are also components of $\mathrm{Fr}(X;Y,Z)$, the Seifert triad structures induced by Q and X agree.

We must now deal with three possibilities depending on the number of components of $\mathrm{Fr}(Q;R,S)$. First suppose there are two or more. If that happens we can connect two distinct components by a saturated properly embedded annulus or strip obtaining a spectacle triad $(C;E,F)$ in $\overline{(M;P,W)-(X;Y,Z)}$. In light of the discussion in the previous paragraph, it satisfies the hypotheses of condition (ii) and so it lies in a collar neighborhood of $\mathrm{Fr}(X;Y,Z)_0$ in $\overline{(M;P,W)-(X;Y,Z)}$, where $\mathrm{Fr}(X;Y,Z)_0$ is a component of $\mathrm{Fr}(X;Y,Z)$. Notice in particular that this means no two components of $\mathrm{Fr}(Q;R,S)$ can lie in $\mathrm{Fr}(X;Y,Z)$. Since the fundamental triads in $(C;E,F)$ are essential in $\overline{(M;P,W)-(X;Y,Z)}$ and lie in the collar neighborhood of $\mathrm{Fr}(X;Y,Z)_0$, they are in fact parallel to $\mathrm{Fr}(X;Y,Z)_0$. At least one of them separates the other from the part of $\overline{(M;P,W)-(X;Y,Z)}$ outside the collar of $\mathrm{Fr}(X;Y,Z)_0$, which means since Q is connected that Q lies in the collar of $\mathrm{Fr}(X;Y,Z)_0$.

Second we consider the case that $\mathrm{Fr}(Q;R,S)$ has just one component. Again we construct a spectacle triad $(C;E,F)$ this time by adjoining to the single component $\mathrm{Fr}(Q;R,S)$ a saturated annulus or strip. By condition (ii), $(C;E,F)$ lies in a collar neighborhood of $\mathrm{Fr}(X;Y,Z)_0$, and as before $\mathrm{Fr}(Q;R,S)$ is parallel to $\mathrm{Fr}(X;Y,Z)_0$ in that collar. If Q lies between these two surface triads we are done. If not, we can at least conclude that $\overline{\overline{(M;P,W)-(X;Y,Z)}-(Q;R,S)}$ is precisely the part of the collar between the two triads. Now unless Q is a solid torus and R is a single annulus whose winding number in Q is precisely one, or (Q,R) is an $(I,\partial I)$-bundle over the disc and S is either a single disc or is empty, we can choose the core of $(C;E,F)$ so that $(C;E,F)$ is not contained in a collar neighborhood of $\mathrm{Fr}(Q;R,S)$ in $(Q;R,S)$. But then since $\overline{\overline{(M;P,W)-(X;Y,Z)}-(Q;R,S)}$ is a collar of $\mathrm{Fr}(X;Y,Z)_0$, $(C;E,F)$ is not contained in a collar neighborhood of $\mathrm{Fr}(X;Y,Z)_0$ in $\overline{(M;P,W)-(X;Y,Z)}$ either, which contradicts the conclusion of condition (ii).

The exceptional cases listed in the previous paragraph do not arise since they would cause $\mathrm{Fr}(Q;R,S)$ to be inessential in $(M;P,W)$.

Third and finally, $\mathrm{Fr}(Q;R,S)$ can be empty only if $Q=\Sigma=M$ and $X=\emptyset$. We have ruled out this case by hypothesis.

APPENDIX B – DOUBLE COVERINGS

We start with an easy exercise.

LEMMA 1. *Let* $(M;P,W)$ *be a manifold triad and let* $(C;E,F)$ *be a surface triad properly embedded in* $(M;P,W)$. *Let* $M' \xrightarrow{p} M$ *be a finite covering. Then* $(M';P',W') \equiv p^{-1}(M;P,W)$ *is a manifold triad and* $(C';E'F') \equiv p^{-1}(C;E,F)$ *is a surface triad properly embedded in* $(M';P',W')$. *If* M *is compact and orientable, then* M' *is compact and orientable. If, in addition,* (C,E) *is two-sided, incompressible and boundary incompressible in* (M,P) *then* (C',E') *has these properties in* (M',P').

THEOREM 2. *If* $(M;P,W)$ *is a Haken triad, and if* $M' \xrightarrow{p} M$ *is a double covering, then* $(M';P',W')$ *is a Haken triad.*

PROOF. In light of Lemma 1 and the obvious fact that the lift of an essential strip in (M,P) is an essential strip in (M',P') it is only necessary to check that M' is irreducible.

Accordingly, let S^2 be a sphere embedded in M'. If $p|S^2$ is also an embedding, then $p(S^2)$ bounds a 3-cell in M. This 3-cell necessarily lifts to two disjoint 3-cells in M', and one of them is bounded by S^2. Thus to prove the theorem, it is sufficient to show that S^2 ambient isotops so that its projection into M is an embedding. We begin by moving S^2 so that $p|S^2$ is in general position. Then, since p is a double cover, $p|S^2$ is an immersion and its singularities are all exactly double points which occur along a finite disjoint collection of circles. The image of these circles is a disjoint collection of embedded circles in M. A priori, it is possible that a singular circle γ might cover its image in M, but that would imply first that the image circle would not be null homotopic in M, and then, since γ bounds a disc in S^2, we would know that the image circle would be precisely of order two in $\pi_1(M)$. But this is impossible since Haken manifolds have no elements of finite order in their fundamental groups.

Thus the singular circles come in disjoint pairs, each pair projecting to a single circle in M. We work by induction on the number of singular circles, supposing that any sphere that projects with fewer singular circles bounds a 3-cell in M'. Figure 1 illustrates the following discussion. Let γ be a singular circle that is innermost on S^2 (that is, one which bounds a disc D in S^2 whose interior contains no singular points). Let γ' be the companion singular circle to γ (that is, $p(\gamma) = p(\gamma')$), and let D' be a disc in S^2 bounded by γ' that misses D. Notice that $p|D$ is an embedding, so in particular $p^{-1}p(D)$ consists of two disjoint discs. Let D'' be the one that is not D. Then $D' \cup D''$ forms an embedded sphere in M' since D' and D'' are each embedded in M', since their boundary circles coincide, and since their interiors are disjoint. (The last assertion follows from the fact that $p(\mathring{D}') \cap p(\mathring{D}'') = p(\mathring{D}') \cap p(\mathring{D}) = \emptyset$.) Since $D \cap D' = \emptyset$, we know that $p|D' \cup D''$ is in general position and has fewer singular circles than $p|S^2$.

By induction, $D' \cup D''$ bounds a 3-ball B in M'. We claim that $B \cap S^2 = D'$. Since $p(S^2 - D' - D) \cap p(D) = \emptyset$, we have $(S^2 - D' - D) \cap D'' = \emptyset$, which implies since $D \cap D'' = \emptyset$ that $(S^2 - D') \cap D'' = \emptyset$. Since S^2 is embedded, $(S^2 - D') \cap D' = \emptyset$. Hence $(S^2 - D') \cap \partial B = \emptyset$. Now we can suppose that $(S^2 - D')$ is not contained in the 3-ball B since otherwise the entire 2-sphere S^2 would lie in B, making it compressible and cutting short our task. It follows immediately from $(S^2 - D') \cap \partial B = \emptyset$, $(S^2 - D') \not\subset B$ and the fact that $S^2 - D'$ is connected that $(S^2 - D') \cap B = \emptyset$.

We can therefore ambient isotop S^2 along B to just beyond D''. Doing so removes at least two singular circles and adds no new singularities. This completes the induction step and the proof of the theorem.

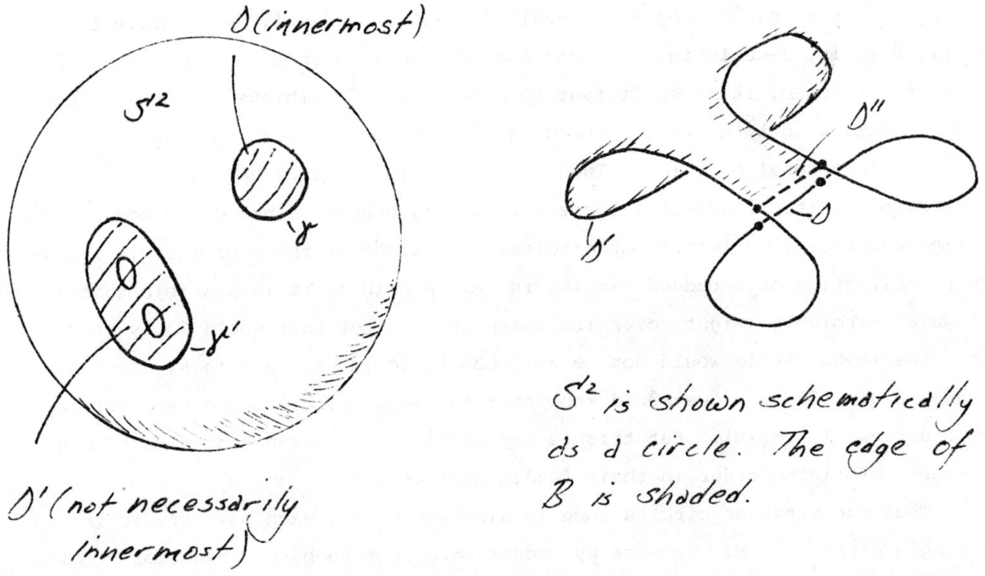

Figure 1

COROLLARY 3. *If* $(X;Y,Z)$ *is a Seifert triad, and if* $X' \xrightarrow{p} X$ *is a double covering, then* $(X';Y',Z')$ *is a Seifert triad.*

PROOF. The only non-trivial question is whether $(X';Y',Z')$ is a Haken triad, and that is settled by Theorem 2.

To this point we have worked on lifting information up a double covering. Now we start reversing that process. The reader will see that our technique was previewed in the proof of Theorem 2.

LEMMA 4. *Let* $(M';P',W') \xrightarrow{p} (M;P,W)$ *be a double covering of Haken triads. Let* $(C';E',F')$ *be a compact surface triad essentially and properly embedded in* $(M';P',W')$ *whose components are fundamental surface triads. Then* $(C';E',F')$ *ambient isotops in* (M',P',W') *so that its projected image in* $(M;P,W)$ *is a saturated subset of an essential Seifert triad.*

PROOF. By a small move in $(M';P',W')$, we can suppose that $p|(C';E',F')$ is in general position and consequently that its singularities, $\mathrm{Sing}(p|C')$ is a disjoint union of properly embedded arcs and circles in $(C';E',F')$. Since the components of $(W',W'\cap P')$ are strips that embed under covering projection,

$\text{Sing}(p|C') \cap F' = \emptyset$. If there is a circle γ in $\text{Sing}(p|C)$ that is inessential in C', bounding, say, the disc D, it cannot double cover its image under p just as in the proof of Theorem 2; hence it must be paired under projection to another circle γ' in $\text{Sing}(p|C')$. This circle is also inessential in C', for otherwise the lift $D'' \neq D$ of $p(D)$ would violate the incompressibility of C' in M'. Starting with an innermost such circle, the procedure in Theorem 2 produces an ambient isotopy of (M',P',W') guided by an embedded 3-ball B that removes at least one pair of singular circles. In the present case, we can not in general choose D' such that $D \cap D' = \emptyset$. The other possibility is that $D \cap D' = D$. It is illustrated in Figure 2. If that happens, we must parallel displace D'' a small amount so that $D' \cup D''$ projects in general position. Displacing to the correct side insures that $p|D' \cup D''$ has fewer singular circles than $p|S^2$. We can then continue the argument in Theorem 2.

The case $D' \cap D = D$.

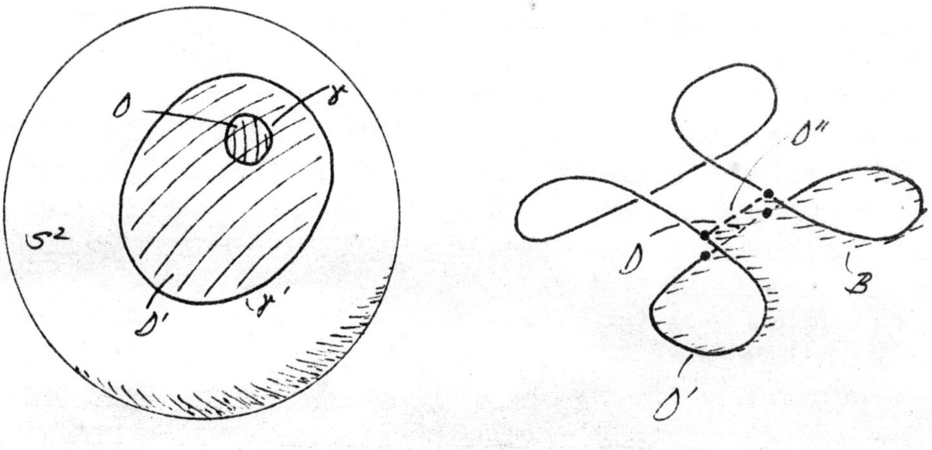

Figure 2

No induction is needed in the current proof since we now know that M' is Haken. We also know that no component of C' is contained in B since C' is incompressible in M'. Iteration of this procedure removes all the inessential circles from $\text{Sing}(p|C')$.

Similarly we can remove the inessential arcs of $\text{Sing}(p|C')$. Inessential arcs must be paired with one another or the boundary incompressibility of (C',E') in (M',P') would be violated. Pairs of inessential arcs can be removed, starting with one with an edgemost member, by just "half" of the procedure used for inessential circles, half in the sense that the objects involved are cut in half by P'. One must keep in mind that P' is incompressible in M'.

Thus we can suppose that Sing(p|C') consists entirely of essential circles and arcs in (C',E'). Consequently, all those in a given component of (C',E') are of the same sort and, in fact, parallel. If we foliate each component of (C',E') by arcs or circles parallel to these, it is clear that by projection we obtain an incipient Seifert triad structure on p(C';E',F'). This extends to a real Seifert triad structure on the regular neighborhood of p(C';E',F') in (M';P',W'). The incompressibility and boundary incompressibility of (C';E',F') in (M';P',W') together with the properness of the embedding implies that the Seifert triad is essential.

We now call attention to a result that is easy to prove, but which is central to understanding the structure of Seifert fibered manifolds.

PROPOSITION 5. *If* M *is a Seifert fibered manifold whose orbit space is a disc with fewer than two exceptional points, then* M *is a solid torus. If the orbit space is a disc with precisely two exceptional fibers, then* ∂M *does not compress in* M.

An (admittedly involved, but standard) corollary is

PROPOSITION 6. *Let* (M,P) *be a Seifert fibered Haken pair. Then, unless the orbit space of* M *is a disc and the number of components of* P *union the exceptional fibers is less than two, or the orbit space of* M *is a sphere with exactly three exceptional points, there is a saturated properly embedded essential torus or annulus in* (M,P).

The analogous proposition for I-bundle triads is trivial. We state it below.

PROPOSITION 7. *Let* (M;P,W) *be an I-bundle Seifert triad. Then unless the base space of the bundle is a disc and* P *has fewer than two components, there is a saturated, properly embedded, essential disc triad or annulus pair in* (M;P,W).

The next theorem deals with an annoying special case.

THEOREM 8. *If* $p:M' \to M$ *is a double covering of Haken manifolds, and if* M' *is Seifert fibered over* S^2 *with three exceptional fibers, then* M *is Seifert fibered.*

We would like to have an easy geometric proof of this fact, but do not. Jaco-Shalen [6, Section 6] give an algebraic proof of the generalization of Theorem 8 where p is allowed to be an arbitrary finite covering and M' is an arbitrary Seifert manifold.

The next proposition completes the proof of Theorem 1.3.

PROPOSITION 9. *If* (M;P,W) *is a Haken triad with characteristic triad* $(\Sigma;\Phi,\Omega)$, *and if* $M' \xrightarrow{p} M$ *is a connected double covering, then* $p^{-1}(\Sigma;\Phi,\Omega)$ *is maximal for* (M';P',W').

PROOF. We prove the proposition by showing that the hypotheses of Theorem A1 are fulfilled by $p^{-1}(\Sigma;\Phi,\Omega)$. Let $(\Sigma';\Phi',\Omega')$ be characteristic for $(M';P',W')$. If $\Sigma' = \emptyset$ there is nothing to do. If not, and if $M' = \Sigma'$ is Seifert fibered over S^2 with three exceptional fibers, then by Theorem 8, M is Seifert fibered. Since $\Sigma' \neq \emptyset$, there is an essential map of a torus into M'. The composition of that map with p is essential in M. Since M is obviously maximal Seifert fibered in itself, $\Sigma = M$. So Σ and $p^{-1}(\Sigma)$ are non-empty. In any other case (Σ' is still non-empty), by Propositions 6 and 7, $(M';P',W')$ contains an essential, properly embedded torus or annulus whose projected image, by Lemma 4, is contained in an essential Seifert triad in $(M;P,W)$. So, again, Σ and $p^{-1}(\Sigma)$ are non-empty.

Property (i) is immediate from Lemma 1.

As for property (ii), let $(C';E',F')$ be a spectacle triad embedded in $\overline{(M';P',W') - p^{-1}(\Sigma;\Phi,\Omega)}$ so that it satisfies the hypothesis of (ii). Then we can apply Lemma 4 to the fundamental triads of $(C';E',F')$ to move them so their projections lie in an essential Seifert triad in $(M;P,W)$, and consequently, by Theorem 1.2 in a collar neighborhood of $(\Sigma;\Phi,\Omega)$ in $(M;P,W)$. It follows that the fundamental triads of $(C';E',F')$ lie in a collar of $p^{-1}(\Sigma;\Phi,\Omega)$ in $(M';P',W')$. Since they are essential, each component is parallel to a component of $\text{Fr}(p^{-1}(\Sigma;\Phi,\Omega))$. Using this fact, we can move them in $\overline{(M';P',W') - p^{-1}(\Sigma;\Phi,\Omega)}$ so their projected images precisely parallel the corresponding components of $\text{Fr}(\Sigma;\Phi,\Omega)$.

Now there are two cases depending on whether a bounding fiber of the core of $(C';E',F')$ ambient isotops in its fundamental triad so that its projection is an embedding. Suppose it does. (This always happens for I-fibers.) Then by the construction of Lemma 4 applied over $\overline{(M;P,W) - ((\Sigma;\Phi,\Omega) \cup (N;Q,R))}$, where $(N;Q,R)$ is a small regular neighborhood of the projection of the fundamental triads we can move $(C';E',F')$ in $\overline{(M';P',W') - p^{-1}(\Sigma;\Phi,\Omega)}$, leaving the fundamental triads fixed, so that the singularities of the projection of the core are all fibers, which means that $p(C';E',F')$ has a Seifert triad neighborhood in $\overline{(M;P,W) - (\Sigma;\Phi,\Omega)}$. This Seifert triad is essential since the fundamental triads of $(C';E',F')$ are. By Theorem 1.2, it lies in an outside collar of $(\Sigma;\Phi,\Omega)$ in $(M;P,W)$, so $(C';E',F')$ lies in one of $p^{-1}(\Sigma;\Phi,\Omega)$ in $(M';P',W')$.

If a bounding fiber cannot be moved so its projection embeds, it can be moved so its projection precisely double covers its image. The same can be done to the other bounding fiber. Consequently, it is easy to arrange that in a small collar neighborhood of the projected fundamental triads, the projected image of $(C;E,F)$ is as illustrated in Figure 3a. Consequently, after the moves prescribed in the proof of Lemma 4, the image of the connecting annulus is precisely the set drawn in Figure 3b.

Figure 3a.

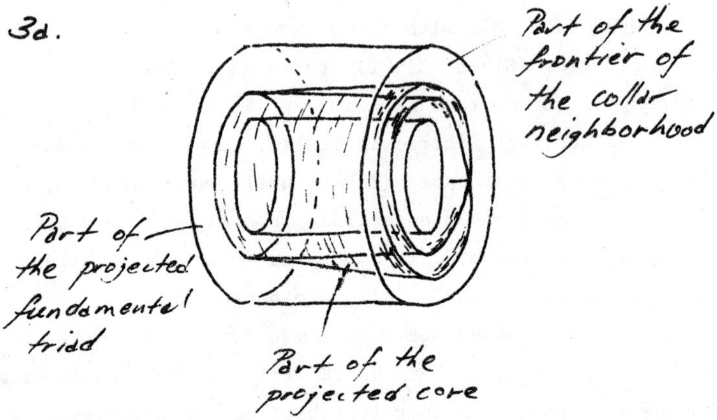

Figure 3b.

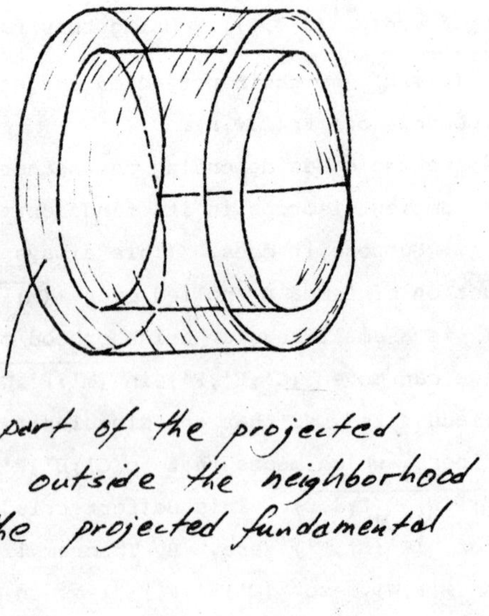

The regular neighborhood of the projected annulus in $(M;P,W)$ is therefore of the form $A^2 \times I$ with a 3-cell missing from its interior. The boundary of the 3-cell lies in $\overline{(M;P,W) - ((\Sigma;\Phi,\Omega) \cup (N;Q,R))}$. Since M is irreducible, so is the last triad, so the 3-cell can be filled back in, yielding an $A^2 \times I$ neighborhood. The union of this neighborhood with $(N;Q,R)$ is clearly Seifert fiberable, so the result follows just as in the previous case.

This completes the proof of Proposition 9 and of Theorem 1.3.

BIBLIOGRAPHY

[1] J. Cannon and C. Feustel, Essential annuli and Möbius bands in M^3, Trans. Amer. Math. Soc. 215 (1976), 219-239.

[2] A. Fathi, F. Laudenbach, and V. Poenaru Travaux de Thurston sur les Surfaces, Asterisque 66-67 (1979).

[3] A. Hatcher, Chapter III, preprint.

[4] J. Hempel, 3-Manifolds, Annals of Math. Studies 86 (1976), Princeton University Press.

[5] W. Jaco, Lectures on 3-manifold topology, Regional Conference Series 43 (1980), Amer. Math. Soc.

[6] W. Jaco and P. Shalen, Seifert fibered spaces in 3-manifolds, Memoirs 220 (1979), Amer. Math. Soc.

[7] K. Johannson, Homotopy equivalences of 3-manifolds with boundary, Lecture Notes in Mathematics 761 (1979), Springer-Verlag.

[8] J. Nielsen, Untersuchung zur Topologie der geschlossenen zweiseitigen Flachen, Acta Math. 58 (1931).

[9] P. Orlik, Seifert manifolds, Lecture Notes in Mathematics 291 (1972), Springer-Verlag.

[10] P. Scott, A new proof of the Annulus and Torus Theorems, Amer. J. Math. 102 (1980), 241-277.

DEPARTMENT OF MATHEMATICS
MICHIGAN STATE UNIVERSITY
EAST LANSING, MI 48824

Contemporary Mathematics
Volume 35, 1984

FIBERED KNOTS IN S^4 - TWISTING, SPINNING, ROLLING, SURGERY, AND BRANCHING

Steven P. Plotnick*

0. INTRODUCTION

One of the best known and most geometrically appealing ways to construct knots in the 4-sphere is to twist-spin classical knots. This process was discovered in 1965 by Zeeman [12] who proved the beautiful theorem that the resulting knots are fibered. Several years ago, Litherland [6] gave a reformulation of twist-spinning which allowed him to extend the motion to a more general class of deformations. Those which he called "untwisted" again yield fibered knots, provided one twists as one deforms. In this paper we reformulate Litherland's construction in order to extend it to a much larger class of knots - again, as long as one twists a non-trivial amount, the resulting knots are fibered.

Let K be a smooth knot in S^3. Removing an unknotted ball pair (B_-,K_-) leaves a knotted ball pair (B_+,K_+).

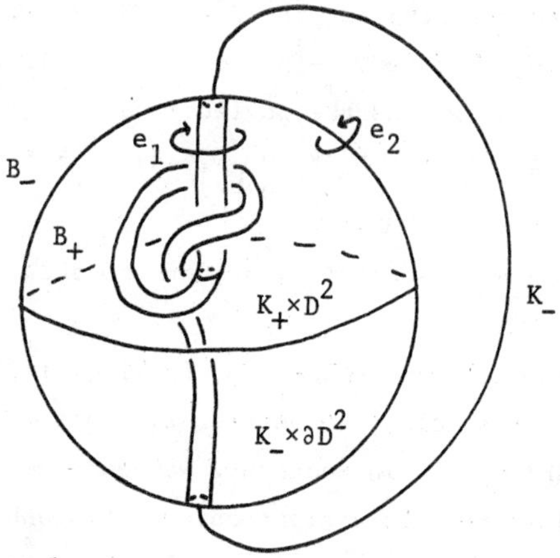

Rotating B_+ through S^1, keeping ∂B_+ fixed, yields S^4. If, during this process, K_+ does not move, we have the classical spin of K. Zeeman twists K_+ about itself k times as B_+ is rotated through S^1 to define the k-twist-spin of K. Litherland absorbs this twist into the knot exterior X: Let τ be a diffeomorphism of X, fixed on ∂X and with support in a collar

*This work was partially supported by NSF grant MCS-82-01045.

of ∂X, that induces conjugation by a meridian. Let $X \times_\tau S^1$ be $X \times I/(x,0) \sim (\tau(x),1)$. Litherland constructs the exterior of the k-twist-spin of K by gluing $K_- \times \partial D^2 \times B^2$ to $X \times_\tau S^1$. To roll, he replaces τ by a diffeomorphism ρ, fixed on ∂X, that induces conjugation by a longitude. The α-roll, k-twist-spin of K is constructed by gluing $K_- \times \partial D^2 \times B^2$ to $X \times_{\rho^\alpha \tau^k} S^1$.

Now, there are two natural 2-spheres above - the one produced by spinning K_+ through S^1, keeping ∂K_+ fixed, and the one pictured, ∂B_+. They intersect in two points, ∂K_+. Hence, a regular neighborhood of these spheres, which we call P, can be described as the result of plumbing two copies of $S^2 \times D^2$ at two points. Montesinos calls P a twin. The meridians to these spheres are pictured above: $e_1 = \partial D^2$, $e_2 = \partial B^2$.

Litherland's construction, then, involves attaching P to $X \times_{\rho^\alpha \tau^k} S^1$ to produce S^4. However, since ρ and τ are isotopic to the identity, we can un-roll, un-twist the bundle to a product, absorbing the rolls and twists into the attaching map. Specifically, we attach P to $X \times S^1$ by sending $e_1 \to m$, $e_2 \to m^k h \ell^\alpha$, where h generates the S^1 factor. The third generator of $H_1(\partial P = T^3)$ is glued to ℓ, so we can describe the construction by the matrix $\begin{pmatrix} 1 & k & 0 \\ 0 & 1 & 0 \\ 0 & \alpha & 1 \end{pmatrix}$.

It so happens that these same ingredients are present in the context of S^1-actions on homotopy 4-spheres. Fintushel [1] showed that all S^1-actions on homotopy 4-spheres are constructed as $(S^1, \Sigma^4) = (S^1, P \cup X \times S^1)$, where X is the exterior of a knot in a homotopy 3-sphere, and the gluing is equivariant for an appropriate S^1-action on P and the obvious one on $X \times S^1$. In the above language, the gluing is given by $\begin{pmatrix} p & k & 0 \\ -\gamma & \beta & 0 \\ 0 & 0 & 1 \end{pmatrix}$, $k\gamma + p\beta = 1$.

Pao used this description and geometric arguments to show that, if $\Sigma^4/S^1 = S^3$ or B^3 then $\Sigma^4 = S^4$ [8].

Our starting point is the determination of all gluings $P \cup X \times S^1$, equivariant or not, that yield homology 4-spheres Σ. This we do in Section 1, deriving an element $A \in GL(3,\mathbb{Z})$. In Section 2 we describe several simplifications of A, originally used in the equivariant case by Fintushel and Pao, which allow a simple description of Σ (Theorem 2.5). The construction produces infinite families of homotopy 4-spheres; we are able to show that roughly two-thirds of them are smoothly S^4 (Corollary 2.7). Section 3 extends the construction to include deformations g of X, $P \cup (X \times_g S^1)$, as in [6]. Interestingly, it turns out that to carry out the simplifications of Section 2 in this context, we need to assume the deformation is untwisted, the same condition Litherland needs to show his knots are fibered.

In Sections 4-7 we turn our attention to the knots determined by the cores of P. Section 4 describes fundamental groups. In Section 5 we give the generalization of Litherland's theorem - namely, if the construction involves an untwisted deformation and also twists a non-trivial amount, the resulting knot is fibered (Theorem 5.1). We then specialize to the case of a trivial deformation, $P \underset{A}{\cup} X \times S^1$, and explicitly describe the fibering in terms of branched covers (Theorem 5.6). The general matrix we consider is

$$A = \begin{pmatrix} p & k & 0 \\ -\gamma & \beta & 0 \\ -\alpha\gamma+pb & \alpha\beta+bk & 1 \end{pmatrix}, \quad k\gamma+p\beta = 1, \qquad \alpha, b \text{ arbitrary}.$$

Roughly speaking, k determines twisting, α gives rolling, b corresponds to $1/b$ surgery on $K \subset S^3$, and p determines branched covers - hence the title of this paper. Section 6 considers those knots of Section 5 that we can show live in S^4. Many new examples arise here, and we include several remarks concerning knots in S^4 with the same fundamental group. Finally, Section 7 describes cyclic branched covers. Surprisingly, it turns out that most of these knots are themselves cyclic branched covers, so that most of them provide counterexamples to the higher-dimensional Smith Conjecture.

Many of these ideas originally appeared in my thesis. At about the same time, Montesinos independently discovered some of these results [7]. His emphasis is somewhat different. He considers the general problem of surgery on twins in S^4, while I restrict to those twins whose exterior is $X \times S^1$.

NOTATION. We work in the smooth category. K will be a smooth, oriented knot in S^3, with exterior $X = S^3 - K \times \mathring{D}^2$. We let m and ℓ denote a meridian and preferred longitude to K, lying on ∂X. X_b denotes the homology 3-sphere obtained by $1/b$ surgery on K - attach a solid torus to ∂X so that $m\ell^b$ bounds a disk.

We follow [6] for bundle notation. Given a diffeomorphism $g: X \to X$, write $X \underset{g}{\times} S^1$ for the space

$$X \times \mathbb{R}/(x,\theta) \sim (g(x), \theta+1), \qquad x \in X, \ \theta \in \mathbb{R}.$$

Equivalently, $X \underset{g}{\times} S^1$ is $X \times I/(x,0) \sim (g(x),1)$, a fiber bundle over S^1 with fiber X and characteristic map g. The equivalence class of (x,θ) in $X \underset{g}{\times} S^1$ is written $x \tilde{\times} \theta$. When the bundle is trivial, we simply write (x,θ). Finally, the covering $\mathbb{R} \to S^1 = \mathbb{R}/\theta \sim \theta+1$ is written $\theta \to \bar{\theta}$.

1. A CONSTRUCTION OF SOME HOMOLOGY 4-SPHERES

We describe a construction from [9]. Let K be a smooth, oriented knot in S^3, with exterior X. Let P be the manifold obtained by plumbing together two copies of $S^2 \times D^2$ at two points. Both $X \times S^1$ and P have a

3-dimensional torus as boundary, and we seek all gluings of P to $X \times S^1$ that produce homology spheres. We first give an explicit parameterization of P so that we can describe a gluing by an element of $GL(3,\mathbb{Z})$.

We can write

$$P = S^2 \times D^2 \underset{S^0 \times D^2 \times S^1}{\cup} S^1 \times I \times D^2 .$$

We use coordinates $((\psi,\theta),(r,\varphi))$ for $S^2 \times D^2$, where (ψ,θ) are spherical coordinates, (r,φ) are polar coordinates. For $S^1 \times I \times D^2$ we use $(\varphi,t,(s,\theta))$, $-1 \leq t \leq 1$, $0 \leq \varphi \leq 2\pi$, (s,θ) polar coordinates. To build P, we identify $((\psi,\theta),(1,\varphi)) \equiv (\varphi,\pm 1,(\sqrt{2}\sin\psi,\theta))$, with $\psi \in [0,\pi/4] \cup [3\pi/4,\pi]$, $0 \leq \theta$, $\varphi \leq 2\pi$, and we use $+1$ if $\psi \in [0,\pi/4]$, -1 if $\psi \in [3\pi/4,\pi]$. The sphere $S^2 \times \{0\} \subset S^2 \times D^2$ is a <u>core</u> of P. The other core is $((\psi,0),(r,\varphi)) \cup S^1 \times I \times \{0\}$, $0 \leq r \leq 1$, $0 \leq \varphi \leq 2\pi$, $\psi = 0,\pi$.

Pick $(0,0,(1,0)) \in \partial(S^1 \times I \times D^2)$ as basepoint. Define

$e_1 = \{(0,0,(1,\theta)):\ 0 \leq \theta \leq 2\pi\}$, a meridian of $S^1 \times \{0\} \times D^2$

$e_2 = \ (\varphi,0,(1,0)):\ 0 \leq \varphi \leq 2\pi\}$, a longitude of $S^1 \times \{0\} \times D^2$

$e_3 = \{((\psi,0),(1,0)):\ \frac{\pi}{4} \leq \psi \leq 3\pi/4\} \cup \{(0,t,(1,0)):\ -1 \leq t \leq 1\}$,
a generator of $\pi_1 P \cong \mathbb{Z}$.

These curves form a basis for $H_1(\partial P)$. The curve e_1 is a meridian of the solid torus lying over the equator of one of the cores of P. The curve e_2 is the unique longitude of the solid torus which, when isotoped past a plumbing point, becomes a meridian of the solid torus lying over the equator of the other core.

Pick a meridian m and preferred longitude ℓ for K (oriented so that $\ell k(m,K) = 1$). Let $h = * \times S^1 \subset X \times S^1$, where $*$ is a basepoint on ∂X. Then m and h generate $H_1(X \times S^1) \cong \mathbb{Z}^2$.

Suppose we glue P to $X \times S^1$ by a diffeomorphism g to obtain Σ^4. The Mayer-Vietoris sequence yields

$$\begin{array}{ccccccc} H_1(T^3) & \xrightarrow{i_*} & H_1(X \times S^1) \oplus H_1(P) & \longrightarrow & H_1(\Sigma) & \longrightarrow & 0 . \\ \cong & & \cong \qquad\quad \cong & & & & \\ \mathbb{Z}^3 & & \mathbb{Z}(m) \oplus \mathbb{Z}(h) \oplus \mathbb{Z}(e_3) & & & & \end{array}$$

To insure $H_1(\Sigma) = 0$, we must be able to pick a basis $\{x,y,z\}$ for $H_1(\partial P)$ so that, in the above sequence, $i_*(x) = m$, $i_*(y) = h$, $i_*(z) = e_3$. This means x and y are null homotopic in P, so we have

$$y = e_1^k e_2^{-p}$$
$$x = e_1^{-\beta} e_2^{-\gamma}, \quad k\gamma + p\beta = \pm 1$$
$$z = e_1^r e_2^s e_3^{\pm 1}, \quad r,s \text{ arbitrary} .$$

To describe g, we must include ℓ: $g(x) = m\ell^b$, $g(y) = h\ell^\alpha$, $g(z) = \ell^{\pm 1}$, b and α arbitrary integers. This allows us to specify g by a matrix

$$A = \begin{matrix} \\ m \\ h \\ \ell \end{matrix} \begin{matrix} \begin{matrix} e_1 & e_2 & e_3 \end{matrix} \\ \begin{pmatrix} p & k & -rp-sk \\ -\gamma & \beta & r\gamma - s\beta \\ -\alpha\gamma + bp & \alpha\beta + bk & 1-r(bp-\alpha\gamma) \end{pmatrix} \end{matrix} .$$

In finding A we assumed the signs in $z = e_1^r e_2^s e_3^{\pm 1}$ and $g(z) = \ell^{\pm 1}$ were the same. There is no harm in doing this, since P admits a self-diffeomorphism taking e_3 to $-e_3$.

It is straightforward to show that these matrices in fact yield homology spheres Σ_A [9]. The fundamental group is given by

$$\pi_1(\Sigma_A) \cong \langle \pi_1 X \mid 1 = m\ell^b = [\ell^\alpha, x], \ \forall x \in \pi_1 X \rangle ,$$

which we will write as $\langle \pi_1 X \mid 1 = m\ell^b, \ \ell^\alpha \text{ central} \rangle$.

2. SIMPLIFICATION OF A

We want to simplify A, without changing Σ_A, by self-diffeomorphisms of either P or $X \times S^1$. There are three such.

(2.1) PROPOSITION: The manifold Σ_A is unchanged if we replace the third column of A by $\begin{pmatrix} 0 \\ 0 \\ 1 \end{pmatrix}$.

PROOF: In [1, Lemma 3.3], Fintushel shows that there is a diffeomorphism of P which takes $e_1 \to e_1$, $e_2 \to e_2$, $e_3 \to e_1^{-r} e_2^{-s} e_3$, for any r,s. Essentially, we cut P open along the solid torus lying over an equator, define an isotopy in a collar of one resulting copy of $S^1 \times D^2$ that starts at the identity and flows $S^1 \times D^2$ around $e_1^{-r} e_2^{-s}$, and reglue. Using f, we have

$$\begin{array}{ccc} P \supset \partial P & \xrightarrow{A} & X \times S^1 \\ f \downarrow & & \| \\ P \supset \partial P & \xrightarrow{B} & X \times S^1 \end{array} , \quad \text{so that}$$

$$B = \begin{pmatrix} p & k & 0 \\ -\gamma & \beta & 0 \\ -\alpha\gamma+bp & \alpha\beta+bk & 1 \end{pmatrix},$$ proving the proposition. ⁞⁞

This is analogous to the situation in dimension 3, where different framings of $S^1 \times D^2$ allow us to ignore where the longitude goes in a surgery. The diffeomorphism f is analogous to a Dehn twist about a meridian. Here we are doing surgery on P, and the various framings provided by f allow us to ignore where the "longitude" e_3 goes. Notice also that f preserves both cores of P.

The second simplification, due to Pao[8] in the equivariant case, exploits the kernel of $\pi_1(SO(2)) \to \pi_1(SO(3))$. Given an integer i, define a diffeomorphism f of P by the following:

$$\begin{array}{ccccc} P & = & S^2 \times D^2 & \cup & S^1 \times I \times D^2 \qquad (\varphi,t,(s,\Theta)) \\ f\downarrow & & \bar{f}\downarrow & & \downarrow \qquad\qquad \downarrow \\ P & = & S^2 \times D^2 & \cup & S^1 \times I \times D^2 \qquad (\varphi,t,(s,\Theta+2i\varphi)) \end{array}$$

where $\bar{f}$ is as follows. On $\partial(S^2 \times D^2)$, $\bar{f}(((\psi,\Theta),(1,\varphi))) = ((\psi,\Theta+2i\varphi),(1,\varphi))$. This represents $2i \in \mathbb{Z} \cong \pi_1(SO(2))$, so is in the kernel of $\pi_1(SO(2)) \to \pi_1(SO(3))$ and is isotopic to the identity. Do this isotopy in a collar of the boundary, and extend to the rest of $S^2 \times D^2$ by the identity.

Notice that f preserves one core of P, namely $S^2 \times \{0\} \subset S^2 \times D^2$, but alters the other one. (The two disks of the other core, which are {north pole, south pole}$\times D^2$ in $S^2 \times D^2$, are altered by the belt trick.)

It is easy to see that the effect of f on $\{e_1,e_2,e_3\}$ is $\begin{pmatrix} 1 & 2i & 0 \\ 0 & 1 & 0 \\ 0 & 0 & 1 \end{pmatrix}$.

Altering A by f gives

(2.2) <u>PROPOSITION:</u> <u>The homology spheres constructed using</u>

$$A = \begin{pmatrix} p & k & 0 \\ -\gamma & \beta & 0 \\ -\alpha\gamma+bp & \alpha\beta+bk & 1 \end{pmatrix} \text{ and } B = Af = \begin{pmatrix} p & k+2ip & 0 \\ -\gamma & \beta-2i\gamma & 0 \\ -\alpha\gamma+bp & \alpha(\beta-2i\gamma)+b(k+2ip) & 1 \end{pmatrix} \text{ are}$$

<u>diffeomorphic.</u> ⁞⁞

We can vary the above f. Reverse the orientation of $S^2 \times D^2$ by flipping D^2, and also flip the S^1 in $S^1 \times I \times D^2$. This reverses e_2, so that

$$f = \begin{pmatrix} 1 & 2i & 0 \\ 0 & -1 & 0 \\ 0 & 0 & 1 \end{pmatrix},$$ changing B accordingly. Notice that this changes the sign of $k\gamma+p\beta$.

Of course, we could equally well have used the other core of P to define f. This would change f to $\begin{pmatrix} \pm 1 & 0 & 0 \\ 2i & 1 & 0 \\ 0 & 0 & 1 \end{pmatrix}$, and change p,γ accordingly.

The third and final simplification takes place in $X \times S^1$.

(2.3) PROPOSITION: Given $j \in \mathbb{Z}$, the homology sphere Σ_A is diffeomorphic to Σ_B, where

$$B = \begin{pmatrix} p & k & 0 \\ -\gamma+pj & \beta+kj & 0 \\ -\alpha(\gamma-pj)+(b-\alpha j)p & \alpha(\beta+kj)+(b-\alpha j)k & 1 \end{pmatrix},$$ and the diffeomorphism preserves both cores of P.

PROOF: Let $\pi: X \to S^1$ be a projection for K, and let $j: S^1 \to S^1$ be the j-fold cover: $\bar{\theta} \to \overline{j\theta}$. Define a diffeomorphism $\tilde{j}: X \times S^1 \to X \times S^1$ by $\tilde{j}(x,\bar{\theta}) = (x, \overline{j\pi(x)+\theta})$. On π_1, $\tilde{j}$ induces $\sigma \to \sigma h^{j \cdot \ell k(\sigma,K)}$, $\sigma \in \pi_1 X$, $h \to h$. Thus, we have

$$\begin{array}{ccc} P \supset \partial P & \xrightarrow{A} & X \times S^1 \\ \| \quad \| & & \downarrow \tilde{j} \\ P \supset \partial P & \xrightarrow{B} & X \times S^1 \end{array},$$

with $B = \tilde{j}A = \begin{pmatrix} 1 & 0 & 0 \\ j & 1 & 0 \\ 0 & 0 & 1 \end{pmatrix} A$ as above. Since we use the identity on P, both cores are preserved. This proposition is essentially [1, Lemma 3.4]. ∎

REMARK: These various simplifications were originally used by Fintushel and Pao in studying S^1 actions on homotopy spheres. As described in [9], when $\alpha = 0$ we can always put an S^1 action on P so that the gluing is equivariant, where we use the natural S^1 action on $X \times S^1$. Orbits are matched via $e_1^k e_2^{-p} \to h$. The curve $e_1^{-\beta} e_2^{-\gamma}$ is a section to the action over a meridian. The quotient of this action is the homology 3-sphere X_b obtained by $1/b$ surgery on K, and by a meridian we mean a meridian in $X_b (= m\ell^b$ in $X)$. We can always alter a section by adding multiples of an orbit, which amounts to changing β,γ by multiples of k,p. The above proposition, then, asserts that the choice of section is irrelevant, if $\alpha = 0$. When $\alpha \neq 0$ the gluing is "non-equivariant", and from (2.3) we see that b changes by multiples of α. This implies that $\pi_1(\Sigma_A)$ only depends on $|\alpha|$ and the equivalence class of $b(\mathrm{mod}\,\alpha)$. In fact, we will almost prove (Theorem 2.5) that the manifold itself only depends on $|\alpha|$ and $b(\mathrm{mod}\,\alpha)$, up to a framing question.

(2.4) We now use the above results to simplify A. First of all, if we modify A by the map $X \times S^1 \to X \times S^1$ that flips S^1, we change the sign of α, so we can assume $\alpha \geq 0$. (The map also changes the signs of β,γ, but we will see that this is irrelevant.)

Let F be the subgroup of $GL(2,\mathbb{Z})$ generated by the matrices $\begin{pmatrix} 1 & 2i \\ 0 & \pm 1 \end{pmatrix}$, $\begin{pmatrix} \pm 1 & 0 \\ 2i & 1 \end{pmatrix}$, as in (2.2), plus $\begin{pmatrix} 0 & 1 \\ 1 & 0 \end{pmatrix}$. We identify F with its image in $GL(3,\mathbb{Z})$ via the natural inclusion. The matrix $\begin{pmatrix} 0 & 1 & 0 \\ 1 & 0 & 0 \\ 0 & 0 & 1 \end{pmatrix}$ comes from the self-diffeomorphism of P that interchanges e_1 and e_2 (interchange the cores). Let J be the subgroup of $GL(3,\mathbb{Z})$, isomorphic to $\mathbb{Z}$, given by the matrices $\begin{pmatrix} 1 & 0 & 0 \\ j & 1 & 0 \\ 0 & 0 & 1 \end{pmatrix}$ of (2.3). We seek a "simplest" representative of the double coset JAF.

Given A, examine the left coset AF. Since elements of F leave α, b unchanged, we just indicate the $\begin{pmatrix} p & k \\ -\gamma & \beta \end{pmatrix}$ part of A. Observe that multiplying A on the right by a generator of F leaves the parity of both $k+p$ and $\beta+\gamma$ unchanged. F has index 3 in $GL(2,\mathbb{Z})$, the matrices $\begin{pmatrix} 1 & 0 \\ 0 & 1 \end{pmatrix}$, $\begin{pmatrix} 1 & 1 \\ 0 & 1 \end{pmatrix}$, and $\begin{pmatrix} 1 & 0 \\ 1 & 1 \end{pmatrix}$ are coset representatives, and the coset to which $\begin{pmatrix} p & k \\ -\gamma & \beta \end{pmatrix}$ belongs is determined by the parity of $k+p$ and $\beta+\gamma$. A procedure for reducing A to one of these three, derived from [8], is the following: We can assume $p > 0$. Given $(k,p) = 1$, we can find $0 < k' < p$ with either $k' \equiv k \pmod p$ and $(k-k')/p \equiv 0 \pmod 2$, or $k' \equiv -k \pmod p$ and $(k+k')/p \equiv 0 \pmod 2$, so that we can use (2.2) to modify A, arriving at a new A satisfying $0 < k < p$. Now reverse the roles of k,p and reduce p. Continuing in this fashion, we eventually arrive at $\{k,p\} = \{0,1\}$ or $\{1,1\}$. Further use of (2.2) will now reduce β and γ, and we eventually arrive at one of the three possibilities listed above. These facts are also proved in [7, Corollary 2.5, Proposition 2.6] via continued fractions. In fact, Montesinos shows [7, Theorem 5.3] that the elements of F are <u>precisely</u> those automorphisms of ∂P that extend to P, so no further simplifications of A can be achieved using self-diffeomorphisms of P.

Multiplying A on the left by an element of J, as in (2.3), allows us to assume $0 \le b < \alpha$ (unless $\alpha=0$). This may change the coset AF, however. There are several cases. Let A be as usual, and let j be the unique integer with $0 \le \bar{b} \equiv b+j\alpha < \alpha$. (We assume $\alpha > 0$.)

<u>CASE I:</u> $p+k$ even

Using (2.3), we change b to $\bar{b}$. This changes $\beta+\gamma$ to $\beta+\gamma+j(k-p)$. Since the parity is unchanged, the coset AF is unchanged. Multiplying by an appropriate element of F, we simplify A to $\begin{pmatrix} 1 & 0 & 0 \\ 0 & 1 & 0 \\ \bar{b} & \alpha+\bar{b} & 1 \end{pmatrix}$.

CASE II: $p+k$ odd

1. j even

(i) $\beta+\gamma$ odd: Changing $\beta+\gamma$ to $\beta+\gamma+j(k-p)$ does not change parity. As in Case I, we simplify A to $\begin{pmatrix} 1 & 0 & 0 \\ 0 & 1 & 0 \\ \bar{b} & \alpha & 1 \end{pmatrix}$.

(ii) $\beta+\gamma$ even: As above, we reduce A to $\begin{pmatrix} 1 & 0 & 0 \\ 1 & 1 & 0 \\ \alpha+\bar{b} & \alpha & 1 \end{pmatrix}$. We now use (2.3) to change this to $\begin{pmatrix} 1 & 0 & 0 \\ 0 & 1 & 0 \\ \alpha+\bar{b} & \alpha & 1 \end{pmatrix}$.

2. j odd

(i) $\beta+\gamma$ odd: Changing $\beta+\gamma$ to $\beta+\gamma+j(k-p)$ changes parity. Thus, A reduces via F to $\begin{pmatrix} 1 & 0 & 0 \\ 1 & 1 & 0 \\ \alpha+\bar{b} & \alpha & 1 \end{pmatrix}$, which simplifies to

$$\begin{pmatrix} 1 & 0 & 0 \\ 0 & 1 & 0 \\ \alpha+\bar{b} & \alpha & 1 \end{pmatrix}.$$

(ii) $\beta+\gamma$ even: As above, we reduce to $\begin{pmatrix} 1 & 0 & 0 \\ 0 & 1 & 0 \\ \bar{b} & \alpha & 1 \end{pmatrix}$.

It is not hard to see that this reduction process is unique. The point of this procedure is explained by the following.

(2.5) THEOREM: Let A be as in (2.1), $\alpha > 0$, and let j be such that $0 \leq \bar{b} = b+j\alpha < \alpha$. Then $\Sigma_A = P \underset{A}{\cup} X \times S^1$ is diffeomorphic to one of Σ_{A_i}, $i = 1,2,3$, with

$$A_1 = \begin{pmatrix} 1 & 0 & 0 \\ 0 & 1 & 0 \\ \bar{b} & \alpha & 1 \end{pmatrix} \qquad A_2 = \begin{pmatrix} 1 & 1 & 0 \\ 0 & 1 & 0 \\ \bar{b} & \alpha+\bar{b} & 1 \end{pmatrix} \qquad A_3 = \begin{pmatrix} 1 & 0 & 0 \\ 0 & 1 & 0 \\ \alpha+\bar{b} & \alpha & 1 \end{pmatrix},$$

according to the procedure in (2.4). Σ_{A_1} is the result of doing surgery on the curve $h\ell^{\alpha}$ in $X_{\bar{b}} \times S^1$, with the natural framing induced from the product structure in $X_{\bar{b}} \times S^1$. Σ_{A_2} is the result of surgery on $h\ell^{\alpha}$ in $X_{\bar{b}} \times S^1$ with the other framing. Σ_{A_3} is the result of surgery on $h\ell^{\alpha}$ in $X_{(\bar{b}+\alpha)} \times S^1$, with the natural framing. If $\alpha = 0$, we have A_1 or A_2, with b instead of $\bar{b}$.

PROOF: It only remains to show Σ_{A_i} is as stated. Consider A_1. We construct Σ_{A_1} by first gluing $S^1 \times I \times D^2$ along $S^1 \times I \times S^1$ according to $e_1 \to m\ell^{\bar{b}}$ $e_2 \to h\ell^{\alpha}$, $e_3 \to \ell$, and then adding $S^2 \times D^2$. With coordinates

$(\varphi,t,(1,\theta))$, the map is $(\varphi,t,(1,\theta)) \to (\theta,\varphi,\bar{b}\theta+\alpha\varphi)$. For fixed φ_0, we glue $\{\varphi_0\}\times I\times S^1$ to $m\ell^{\bar{b}}$, which amounts to doing $1/\bar{b}$ surgery on K and deleting a ball. We do this for each φ_0, and note that the balls we remove form a neighborhood of $h\ell^\alpha$. Gluing in $S^2\times D^2$ then accomplishes surgery on $h\ell^\alpha$ in $X_{\bar{b}}\times S^1$, and it is easy to see that we use the natural framing.

Adding the $S^2\times D^2$ with the other framing means simply that e_2 goes where e_1+e_2 used to go, i.e. we use A_2. The description of Σ_{A_3} is similar to Σ_{A_1}.

REMARK: This theorem corrects the overly optimistic footnote in [9, page 399]. Of course, it may be true that the Σ_{A_i}, $i=1,2,3$, are diffeomorphic, but it cannot be proved by the methods here.

(2.6) COROLLARY: *Let $\Sigma_{\alpha,b}$ be the homology sphere obtained by surgery on $h\ell^\alpha$ in $X_b\times S^1$ with the natural framing, and let $\Sigma'_{\alpha,b}$ be obtained by using the other framing. Then*

$$\Sigma'_{\alpha,b_1} \cong \Sigma'_{\alpha,b_2} \quad \text{if} \quad b_1 \equiv b_2 \pmod{\alpha}$$

$$\Sigma_{\alpha,b_1} \cong \Sigma_{\alpha,b_2} \quad \text{if} \quad b_1 \equiv b_2 \pmod{2\alpha}\ .$$

(2.7) COROLLARY: $\Sigma'_{\alpha,b} \cong S^4$ if $b\equiv 0 \pmod{\alpha}$

$$\Sigma_{\alpha,b} \cong S^4 \quad \text{if} \quad b\equiv 0 \pmod{2\alpha}\ .$$

PROOF: By (2.6), these manifolds are the result of surgery on $h\ell^\alpha$ in $X_0\times S^1 = S^3\times S^1$, hence S^4. ∎

Corollary 2.7 leads to new classes of fibered knots, which we describe in Sections 5 and 6.

REMARK: If $b\not\equiv 0 \pmod{\alpha}$, one does not expect, in general, Σ_A to be simply-connected. Using calculations of [9, Section 3], we can always find torus knots so that $\pi_1\Sigma_A \neq \{1\}$. Of course, for certain knots and certain α,b, one might end up with a homotopy sphere. Examples where this occurs, again using torus knots, are in [9, Section 3]. The examples are all smoothly S^4.

3. DEFORMATIONS

We now allow the additional complication of a deformation (see [6]). Let $g:(S^3,K)\to(S^3,K)$ be a diffeomorphism which is the identity on a neighborhood of K. Such a g is called a deformation of K. Litherland considers two deformations equivalent if they are pseudo-isotopic, relative to some neighborhood of K. Homologically, $X\times_g S^1$ is identical to $X\times S^1$, so we can consider $P\cup_A X\times_g S^1$ just as in Sections 1,2.

Both (2.1) and (2.2) apply unchanged. Proposition (2.3) requires an additional hypothesis. Following Litherland, call a deformation untwisted if there is a projection $\pi: X \to S^1$ so that $\pi \circ (g|_X) = \pi$.

(3.1) PROPOSITION: Let g be an untwisted deformation of K, and let $j \in \mathbb{Z}$. Then there is a diffeomorphism $\tilde{j}: X \times_g S^1 \to X \times_g S^1$ inducing $\sigma \to \sigma h^{j \cdot \ell k(\sigma, K)}$, $\sigma \in \pi_1 X$, $h \to h$.

PROOF: As in (2.3) define $\tilde{j}(x \tilde{\times} \bar{\theta}) = x \tilde{\times} (\overline{j\pi(x)+\theta})$. This is well defined since g is untwisted, and induces the required map on π_1. ⦀

Thus, when g is untwisted, all simplifications of Section 2 apply (except that we cannot assume $\alpha \geq 0$. The analogue of Theorem 2.5 is

(3.2) THEOREM: The homology sphere $\Sigma_A = P \bigcup_A X \times_g S^1$ is diffeomorphic to one of Σ_{A_i}, $i=1,2,3$ as in (2.3), where $0 \leq b + j\alpha < |\alpha|$. Σ_{A_1} is the result of surgery on $h\ell^\alpha$ in $X_b \times_g S^1$ with the natural framing, and similarly for Σ_{A_2}, Σ_{A_3}. ⦀

When $b \equiv 0 \pmod{\alpha}$, we have homotopy spheres. The analogue of (2.7) is true, since $(S^3\text{-ball}) \times_g S^1 \cong (S^3\text{-ball}) \times S^1$, by the Alexander trick (see [6, Lemma 1.2]).

4. THE KNOTS - FUNDAMENTAL GROUPS

In $P \bigcup_A (X \times_g S^1)$, the two cores of P determine knots in homology 4-spheres. Consider the core given by $\{(\psi,0),(r,\varphi)\} \cup S^1 \times I \times \{0\}$, $0 \leq r \leq 1$, $0 \leq \varphi \leq 2\pi$, $\psi = 0,\pi$. We write this knot as $(A,g)K$, or simply as $A(K)$ when g = identity.

The exterior of $(A,g)K$ is obtained by gluing $I \times \partial D^2 \times B^2$, along $I \times \partial D^2 \times \partial B^2$, to $X \times_g S^1$, where e_1 generates $\pi_1(I \times \partial D^2 \times B^2)$ and e_2 bounds B^2. Note that e_1 is a meridian to $(A,g)K$. (To build the exterior corresponding to the other core, reverse e_1 and e_2.)

By using (2.2), and possibly reversing orientation of the knot, we can assume $k\gamma + p\beta = +1$, $0 \leq k$. This assumption will be made throughout the rest of the paper.

By Van Kampen's theorem,

$$\pi_1((A,g)K) \cong \langle \pi_1 X,\ h \mid h x h^{-1} = g_*(x),\ 1 = m^k h^\beta \ell^{\alpha\beta+bk},\ x \in \pi_1 X \rangle .$$

There is another useful presentation. One verifies that $e_1^{-k} = h\ell^\alpha$ and $e_1^\beta = m\ell^b$, which leads to the following isomorphism:

$$\begin{array}{ccc}
\langle \pi_1 X, h \mid h x h^{-1} = g_*(x),\ 1 = m^k h^\beta \ell^{\alpha\beta+bk},\ x \in \pi_1 X \rangle \\
x \downarrow \mapsto x, \quad h \downarrow \mapsto e_1^{-k}\ell^{-\alpha} \qquad \updownarrow \cong \qquad x \uparrow \mapsfrom x, \quad m^p h^{-\gamma}\ell^{-\alpha\gamma+bp} \uparrow \mapsfrom e_1 \\
\langle \pi_1 X, e_1 \mid e_1^{-k}\ell^{-\alpha} x \ell^\alpha e_1^k = g_*(x),\ e_1^\beta = m\ell^b,\ x \in \pi_1 X \rangle .
\end{array}$$

If $g = \text{identity}$, and $\alpha = b = 0$, then

$$\pi_1 \begin{pmatrix} p & k & 0 \\ -\gamma & \beta & 0 \\ 0 & 0 & 1 \end{pmatrix} K \cong \langle \pi_1 X, h \mid h \text{ central}, \; 1 = m^k h^\beta \rangle$$

$$\cong \langle \pi_1 X, e_1 \mid e_1^k \text{ central}, \; e_1^\beta = m \rangle .$$

When $\beta = 1$, we get $\langle \pi_1 X \mid m^k \text{ central} \rangle$, the group of the k-twist spin of K. When $\beta \neq 1$, we have a branched cover of the k-twist spin of K. The kth power of the meridian is central, but does not correspond to the meridian of K. Non-zero values of α and b correspond to rolling and other variations, which we will explain in Sections 5,6.

5. FIBERING THE KNOTS

We generalize Litherland's theorem [6, Theorem 2.4] as follows:

(5.1) THEOREM: Consider $\Sigma = P \cup_A (X \times_g S^1)$, where g is an untwisted deformation of K, $k > 0$. Then the knot $(A,g)K$ is fibered.

We need the following result of Litherland:

(5.2) LEMMA [6, Lemma 3.2]: Let k be non-zero. If $q: Y \to S^1$ is a map onto S^1, and if $\mathbb{R}$ acts on Y so that $q(t \cdot y) = q(y) + \overline{kt}$, $t \in \mathbb{R}$, $y \in Y$, then $q: Y \to S^1$ is a fiber bundle with characteristic map given by the action of $-(1/k) \in \mathbb{R}$.

PROOF OF THEOREM 5.1: Define $q: X \times_g S^1 \to S^1$ by $q(x \tilde{\times} \varphi) = \beta\pi(x) - \overline{k\varphi}$ where π is a projection for X, with $(\pi \circ g|_X) = \pi$. This is well-defined since g is untwisted. Let $\mathbb{R}$ act on $X \times_g S^1$ by $t \cdot (x \tilde{\times} \varphi) = x \tilde{\times} (\varphi - t)$. Then

$$q(t \cdot (x \tilde{\times} \varphi)) = \beta\pi(x) - \overline{k\varphi} + \overline{kt} = q(x \tilde{\times} \varphi) + \overline{kt} ,$$

so (5.2) shows that $q: X \times_g S^1 \to S^1$ is a fiber bundle.

Now, we construct the exterior of $(A,g)K$ by gluing $I \times \partial D^2 \times B^2$, along $I \times \partial D^2 \times \partial B^2$, to $\partial(X \times_g S^1)$ via the map

$$A(y,\theta,\varphi) = ((y + (-\alpha\gamma + bp)\theta + (\alpha\beta + bk)\varphi, \; p\theta + k\varphi), \; -\gamma\theta + \beta\varphi) .$$

Here our coordinates in $\partial(X \times_g S^1)$ are $((\ell, m), h)$. The composite qA sends (y,θ,φ) to $\beta\overline{(p\theta + k\varphi)} - k\overline{(-\gamma\theta + \beta\varphi)} = \overline{\theta}$, i.e. projection on the ∂D^2 factor. Thus, q extends over $I \times \partial D^2 \times B^2$ to give a fibering of the knot exterior. ▥

REMARK: (1) Our proof essentially starts in the middle of Litherland's proof [6, Theorem 2.4]. For him, twisting is part of the deformation, and he first "untwists", i.e. writes $X \times_{t_g^k} S^1 \approx X \times_g S^1$, thereby twisting the gluing. For us, the twist has already been accounted for by A, so we do not need to do this.

(2) The proof shows that the fiber is the interior of

$$q^{-1}(\overline{0}) = I \times \{0\} \times B^2 \cup \{x \tilde{\times} \varphi \mid \beta\pi(x) = \overline{k\varphi}\} .$$

Equivalently, the fiber is $\mathrm{punc}(M) = M - \{pt\}$, where

$$M = S^1 \times B^2 \cup_{A'} \{x \tilde{\times} \varphi \mid \beta\pi(x) = \overline{k\varphi}\},$$

for

$$A': \begin{cases} S^1 \times \partial B^2 \xrightarrow{\cong} \{x \tilde{\times} \varphi \in \partial X \times_g S^1 \mid \beta\pi(x) = \overline{k\varphi}\} \\ y \times \bar{\varphi} \longrightarrow (y + (\alpha\beta + bk)\varphi, \overline{k\varphi}) \tilde{\times} \overline{\beta\varphi} . \end{cases}$$

(5.3) Off of $A(I \times \partial D^2 \times B^2) \cup \partial(X \times_g S^1) \times I$, the characteristic map is given by the action of $(-1/k)$. Extending over the rest of the fiber is somewhat complicated - the action of $-1/k$ may not extend over the attached ball, and the form of A forces us to build an inner automorphism into the characteristic map. Rather than do this in complete generality, we will assume $g = \text{identity}$, in which case the fiber is more understandable.

For the rest of this section, then, assume $g = \text{identity}$. Then $q^{-1}(\bar{0}) \cap (X \times S^1)$ is just $\{(x,\varphi) \mid \beta\pi(x) = \overline{k\varphi}\}$. This is the pullback of

$$\begin{array}{c} S^1 \\ \downarrow k \\ X \xrightarrow{\pi} S^1 \xrightarrow{\beta} S^1 \end{array},$$

recognizable as the k-fold cyclic unbranched cover of X. The action of $-(1/k) \in \mathbb{R}$ corresponds to a certain power of the canonical covering transformation, which we compute as follows (see also [7, Section 10]):

Let M^k denote the k-fold cyclic unbranched cover of X, and let N^k denote the corresponding branched cover. Let σ be the canonical generator of the group $\mathbb{Z}_k$ of covering transformations, i.e. σ induces rotation by $2\pi/k$ about the branch set. Suppose $k\gamma + p\beta = +1$, $k > 0$.

Since σ^p is a covering transformation of order k, there is a natural free circle action on $M^k \times_{\sigma^p} S^1$ with quotient space X, so that $M^k \times_{\sigma^p} S^1$ is a principal S^1-buncle over X. The action is given by $\bar{\theta} \cdot (x \tilde{\times} \bar{\varphi}) = x \tilde{\times} \overline{k\theta + \varphi}$. Since $H^2(X) = 0$, the bundle is trivial, so that $M^k \times_{\sigma^p} S^1 \approx X \times S^1$. We must compute the correspondence on the boundary.

We choose generators for $H_1(\partial(M^k \times_{\sigma^p} S^1))$ as follows. Let R be a meridian (so R maps to m^k in X). Let L be a longitude, i.e. a lift of the longitude in X. Let S be the unique curve so that an orbit of the above action corresponds to $S^k R^{-p}$. Note that S maps to the oriented generator of $H_1(S^1)$ via the bundle projection.

Let Q be a section to the S^1-action over a meridian in X. We can select Q so that $Q = S^\beta R^\gamma$. These selections will look familiar to anyone acquainted with Seifert manifolds. The picture below is drawn for $k = 5$, $p = -7$, $\beta = 2$, $\gamma = 3$:

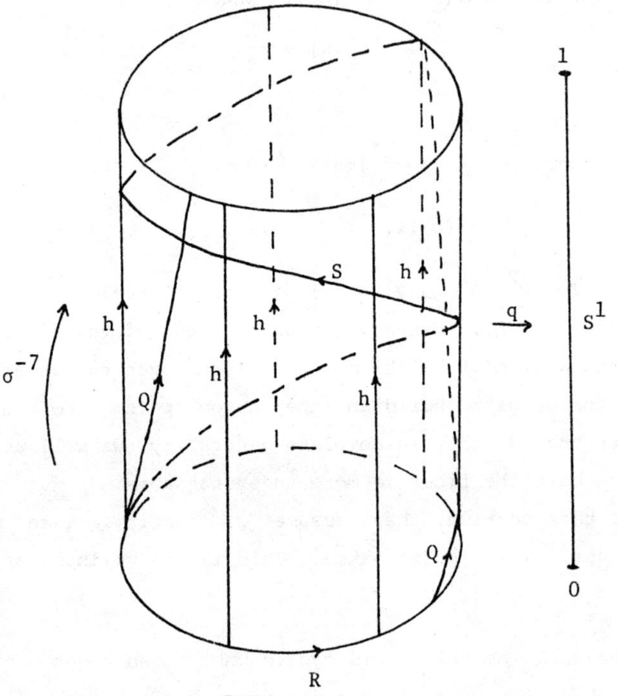

$h = S^5R^7$

$Q = S^2R^3$

(orientation conventions: $\uparrow y$ $\rightarrow x$ $x \cdot y = 1$)

The choice of S may look somewhat strange. One might expect it to instead look like  This would be the case if $0 < p < k$. For other values of p, S changes by multiples of R.

With this selection, the following is easy:

(5.4) <u>LEMMA</u>: <u>Let</u> $k\gamma + p\beta = 1$, $0 < k$. <u>There is an S^1-bundle equivalence</u> $\Delta: X \times S^1 \rightarrow M^k \underset{\sigma^p}{\times} S^1$ <u>taking</u> $h \rightarrow S^kR^{-p}$, $m \rightarrow S^\beta R^\gamma$, $\ell \rightarrow L$.

<u>PROOF</u>: The bundle equivalence matches orbits, hence $h \rightarrow S^kR^{-p}$. Since $S^\beta R^\gamma$ is a section over a meridian, we can arrange for the equivalence to match $m \rightarrow S^\beta R^\gamma$. (Changing the section by multiples of an orbit will change β, γ by multiples of k, p, not surprising in light of (2.3).) Finally, $\ell \rightarrow L$, since these generate the kernels of $H_1(\partial X \times S^1) \rightarrow H_1(X \times S^1)$ and $H_1(\partial(M^k \underset{\sigma^p}{\times} S^1)) \rightarrow H_1(M^k \underset{\sigma^p}{\times} S^1)$. ▥

(5.5) The projection q used in the proof of (5.1) takes a meridian β times around S^1. The map $X \times S^1 \to M^k \times_{\sigma^p} S^1$ above, followed by the natural projection to S^1, does the same. The only difference is that, in the proof of (5.1) we let $\mathbb{R}$ act via $t(x,\varphi) = (x,\varphi - t)$. This gives the opposite circle action to the one in (5.4). In other words, the following diagram commutes:

$$\begin{array}{ccccc} X \times S^1 & \xrightarrow{\text{id}\times\text{flip}} & X \times S^1 & \xrightarrow{\Delta} & M^k \times_{\sigma^p} S^1 . \\ & \searrow^{q} & \downarrow & \swarrow_{q} & \\ & & S^1 & & \end{array}$$

We have shown, then, that we can replace $X \times S^1 \overset{q}{\to} S^1$ in (5.1) by $M^k \times_{\sigma^p} S^1 \to S^1$ under the correspondence:

$$h \to S^{-k}R^p$$
$$m \to S^\beta R^\gamma$$
$$\ell \to L .$$

It is now fairly straightforward to explicitly describe the fibering. Let $\partial M^k \times I = S^1 \times S^1 \times I$ be a collar of ∂M^k in M^k, with $L = S^1 \times \{\bar{0}\} \times \{0\}$, $R = \{\bar{0}\} \times S^1 \times \{0\}$.

(5.6) <u>THEOREM</u>: <u>Let</u> $\Sigma = P \underset{A}{\cup} X \times S^1$, $k\gamma + p\beta = 1$, $k > 0$. <u>The fiber of</u> $A(K)$ <u>is</u> punc $((N^k)_{\alpha\beta+bk})$, <u>where</u> $(N^k)_{\alpha\beta+bk}$ <u>is obtained by</u> $1/(\alpha\beta+bk)$ <u>surgery on the branch set. The characteristic map is given by</u>

$$\begin{cases} \sigma^p(x) & ; \quad x \in M^k - \partial M^K \times I \\ \overline{((\alpha\gamma - bp)(1-s)+\Theta,\ (ps/k)+\varphi, s)} & ; \quad x = (\bar{\Theta},\bar{\varphi},s) \in \partial M^k \times I \\ x & ; \quad x \in I \times \{\bar{0}\} \times B^2 . \end{cases}$$

<u>PROOF:</u> We construct the exterior by attaching $I \times \partial D^2 \times B^2$ along $I \times \partial D^2 \times \partial B^2$ to $\partial(X \times S^1) \approx \partial(M^k \times_{\sigma^p} S^1)$ via $e_1 \to m^p h^{-\gamma} \ell^{-\alpha\gamma+bp} \to SL^{-\alpha\gamma+bp}$, $e_2 \to RL^{\alpha\beta+bk}$, using (5.5). Thus we see that the fiber is $I \times \{\bar{0}\} \times B^2 \underset{e_2}{\cup} M^k = $ punc$((N^k)_{\alpha\beta+bk})$.

We now modify σ^p in $\partial M^k \times I$. Define an isotopy $G_t : M^k \to M^k$, $0 \leq t \leq 1$ by

$$G_t(x) = \begin{cases} (\overline{(\alpha\gamma-bp)t(1-s)+\Theta},\ \overline{-p(1-s)t/k)+\varphi}, s) & ; \quad x = (\bar{\Theta},\bar{\varphi},s)\ 0 \leq s \leq 1 \\ x & ; \quad x \in M^k - \partial M^k \times I . \end{cases}$$

Then $G_0 = $ identity, $G_1 \circ \sigma^p \big|_{\partial M^k} = $ identity. The isotopy provides a diffeomorphism $G: M^k \times_{\sigma^p} S^1 \to M^k \times_{G_1 \circ \sigma^p} S^1$ that "straightens" S, i.e. takes it to the S^1 factor in $\partial(M^k \times_{G_1 \circ \sigma^p} S^1) = \partial M^k \times S^1$. The attaching map followed by G_1 is given by

$$I\times \partial D^2\times \partial B^2 \longrightarrow \partial(M^k \times_{G_1\circ\sigma^p} S^1) = \partial M^k\times S^1$$

$$(y,\bar\theta,\bar\varphi) \longrightarrow \overline{(y+(\alpha\beta+bk)\varphi,\bar\varphi,\bar\theta)} \quad .$$

The characteristic map $G_1\circ\sigma^p$ on M^k now matches nicely with the identity on $I\times\{\bar 0\}\times B^2$, and the result follows. ⁞⁞⁞

(5.7) REMARKS: (1) Suppose for the moment that $\alpha=0$. Then, as described in the remark following (2.3), Σ_A admits an S^1 action with $\Sigma_A/S^1 = X_b$. This suggests that we should regard M^k as the unbranched cover of a knot in X_b. Letting K' denote the core of the solid torus added in $1/b$ surgery on $K\subset S^3$, $X_b - K'\times \overset{\circ}{D}{}^2 = X$, but a meridian to K' is $m\ell^b$ when seen from X.

Let $(M_b)^k$ be the k-fold cyclic unbranched cover of (X_b, K'), and let $\underline{S},\underline{R},\underline{L}$ be the analogues, in $(M_b)^k\times_{\sigma^p}S^1$, of S,R,L in $M^k\times_{\sigma^p}S^1$. Tracing around the diagram

$$\begin{array}{ccc} X\times S^1 & \approx & (X_b - K'\times \overset{\circ}{B}{}^2)\times S^1 \\ \wr\wr & & \wr\wr \\ M^k\times_{\sigma^p} S^1 & \approx & (M_b)^k \times_{\sigma^p} S^1 , \end{array}$$

we find the bottom correspondence to be $S\leftrightarrow \underline{SL}^{-pb}$, $R\leftrightarrow \underline{RL}^{-bk}$, $L\leftrightarrow \underline{L}$. The knot exterior, then, is constructed by gluing $I\times\partial D^2\times B^2$ to $(M_b)^k\times_{\sigma^p}S^1$ via $e_1\leftrightarrow \underline{S}$, $e_2\leftrightarrow\underline{R}$. It is not hard to see that we can completely dispense with the isotopy G_t in (5.6). The rotation induced by σ^p extends over the attached ball. The fiber is the punctured k-fold cyclic branched cover of (X_b, K'), and the characteristic map is σ^p.

When $b=0$, these are the examples discovered by Pao [8] - the p-fold cyclic branched cover of the k-twist spin of K. His result, that these branched covers are S^4, follows from Section 2 (see (6.1)). Indeed, these examples motivated much of this work. When $b\neq 0$, we simply have the analogue of twist-spinning to knots in homology 3-spheres.

In $\alpha\neq 0$, we can still make the above identifications. This simplifies the gluing somewhat, but we are still forced to modify σ^p by an isotopy. Since little seems to be gained, we will not do this.

(2) The characteristic map involves conjugation by powers of M that depend on P/k. Using (2.2), we can first "normalize" the gluing so that $-k<p<k$. That is, we can use the matrix

$$\begin{pmatrix} p+2ik & k & 0 \\ -\gamma+2i\beta & \beta & 0 \\ -\alpha(\gamma-2i\beta)+b(p+2ik) & \alpha\beta+bk & 1 \end{pmatrix},$$

for appropriate i. This normalizes the section S of (5.3) to either

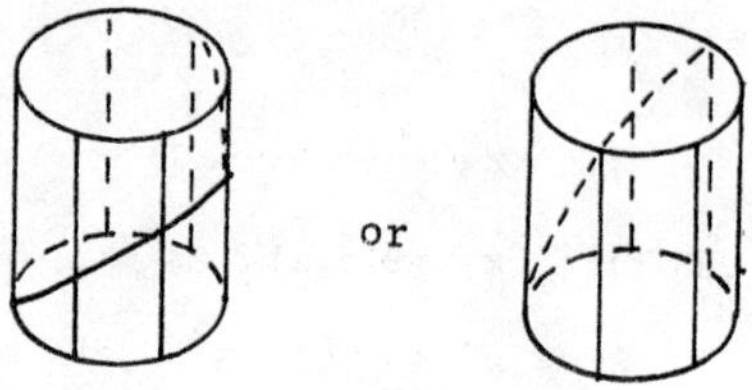

but does not change the knot type. In the description of the fibering, notice that the characteristic map "loses" conjugation by $2i$ meridians, but "gains" conjugation by $2i(\alpha\beta+bk)$ longitudes. Since $M^{-1} = L^{\alpha\beta+bk}$, there is no change. If we are just interested in the knot exterior, as opposed to the knot, we are free to alter p by odd multiples of k. This change is induced by a framing change in $I \times \partial D^2 \times B^2$ taking $e_1 \leftrightarrow e_1 + (2i+1)e_2$, $e_2 \leftrightarrow e_2$, and has the effect of replacing $A(K)$ by its "Gluck companion". See Section 6.

Using "unnormalized" values of p makes the computation of certain cyclic branched covers of $A(K)$ particularly easy. See Section 7.

6. NEW FIBERED KNOTS IN S^4

The results of (2.4), (2.5), (2.7), (5.6) give many fibered knots in S^4. We collect these results in the following

(6.1) COROLLARY: Consider the knot $A(K)$, $k > 0$, $k\gamma+p\beta = +1$, where either

(i) $\alpha=b=0$,

(ii) $p+k \equiv 0 \pmod 2$, $b\equiv 0 \pmod \alpha$,

(iii) $p+k\equiv 1 \pmod 2$, $\beta+\gamma\equiv 1 \pmod 2$, $b\equiv 0 \pmod{2\alpha}$, or

(iv) $p+k\equiv 1 \pmod 2$, $\beta+\gamma\equiv 0 \pmod 2$, $b\equiv 0 \pmod\alpha$, $b/\alpha\equiv 1 \pmod 2$, $\alpha\neq 0$

Then $A(K)$ is a fibered knot in S^4. ||||

EXAMPLES: (1) $A = \begin{pmatrix} 1 & k & 0 \\ 0 & 1 & 0 \\ 0 & 0 & 1 \end{pmatrix}$, so $b=\alpha=0$. From (5.5), the fiber is $\text{punc}(N^k)$. By Remark (1) of (5.7), the characteristic map is σ. This is the k-twist spin of K.

(2) $A = \begin{pmatrix} p & k & 0 \\ -\gamma & \beta & 0 \\ 0 & 0 & 1 \end{pmatrix}$. The fiber is $\text{punc}(N^k)$, the characteristic map is σ^p. This is the p-fold cyclic branched cover of the k-twist spin of K.

(3) $A = \begin{pmatrix} 1 & k & 0 \\ 0 & 1 & 0 \\ 0 & \alpha & 1 \end{pmatrix}$. This is the matrix arising from Litherland's

α-rolled, k-twist spun K [6,p.323]. See Section 0. The fiber is $\mathrm{punc}((N^k)_\alpha)$. The characteristic map is

$$\begin{cases} \sigma(x) & ; \quad x \in M^k - (\partial M^k \times I) \\ (\bar\theta, \overline{s/k + \varphi}, s) & ; \quad x = (\bar\theta, \bar\varphi, s) \in \partial M^k \times I \\ x & ; \quad x \in I \times \{\bar 0\} \times B^2 \quad . \end{cases}$$

Notice that, for $k=1$, the fiber is $\mathrm{punc}(X_\alpha)$, and the characteristic map is just conjugation by $m = \ell^{-\alpha}$ [6, Corollary 5.3]. In general, the kth power of the characteristic map will be conjugation by $L^{-\alpha}$ in $\mathrm{punc}((N^k)_\alpha)$. The fundamental group is given by $\langle\pi_1 X | m^k \ell^\alpha \text{ central}\rangle$.

$$(4)\quad A = \begin{pmatrix} 1 & k & 0 \\ 0 & 1 & 0 \\ j\alpha & \alpha+j\alpha k & 1 \end{pmatrix}, \text{ where } j \text{ is even if } k \text{ is even.}$$

The fiber is $\mathrm{punc}((N^k)_{\alpha(1+jk)})$. The group is $\langle\pi_1 X | m^k \ell^{\alpha(1+jk)} \text{central}\rangle$, isomorphic to that of the $\alpha(1+jk)$-rolled, k-twist spin. The characteristic map of this knot differs from that of the $\alpha(1+jk)$-roll, k-twist spin, however, by an additional conjugation along $L^{j\alpha}$.

Let $k=1$, and let $a(j+1) = \alpha$. Then the knot determined by

$$A = \begin{pmatrix} 1 & 1 & 0 \\ 0 & 1 & 0 \\ ja & a(1+j) & 1 \end{pmatrix}$$ has the same group as the α-roll, 1-twist spin. The characteristic map is conjugation by $\ell^{ja-a(1+j)} = \ell^{-a}$, so its $(j+1)$st power gives the characteristic map for the α-roll, 1-twist spin. The fundamental group is isomorphic to $\pi_1(X_\alpha) \times \mathbb{Z}(h)$, with a meridian corresponding to $\ell^{-a}h$. This generalizes [6,Cor. 5.3]. For each divisor of α, then, we can construct a knot as above. All fundamental groups are isomorphic, but all meridians are different. In general, we expect these knots to have distinct $\mathbb{Z}\pi_1$-module structures on π_2. (The fibers are all $\mathrm{punc}(X_\alpha)$, so all π_2's are identical as abelian groups.) In [10], I used branched covers of twist-spin torus knots to produce arbitrarily many examples of knots in S^4 with the same π_1 but distinct π_2. The examples here show that the phenomena is quite widespread. Unfortunately, these only yield finite collections. In [11], an infinite family is given. The construction is similar to the above, but the elements along which the conjugations are performed are general "weight elements", not powers of ℓ as above, so we are unable to conclude the resulting homotopy spheres are S^4 without resorting to Freedman's proof of the Poincaré conjecture.

(5) The examples of (4) work more generally. Fix k,p,β,γ, and now vary α and $b = j\alpha$ so that $\alpha\beta+bk$ remains constant and j satisfies the requirements of (6.1). This will give a (finite) collection of knots in S^4. The groups and fibers are the same, but the characteristic maps differ by peripheral conjugations.

Given a knot in S^4, a "Gluck construction" [3] means that we remove a tubular neighborhood of the knot and replace it by a twist coming from the generator of $\pi_1(SO(3))$. It is easy to see, for our examples, that this amounts to replacing A by

$$A' = \begin{pmatrix} \pm p+k & k & 0 \\ \mp\gamma+\beta & \beta & 0 \\ \mp\alpha\gamma+\alpha\beta\pm bp+bk & \alpha\beta+bk & 1 \end{pmatrix}.$$ Using (6.1), a case-by-case analysis gives

(6.2) COROLLARY: *A Gluck construction on* $A(K)$, $k\gamma+p\beta=1$, *yields* S^4 *if either*

(i) $\alpha=b=0$

(ii) $p+k$ even: γ odd, β even, $b\equiv 0 \pmod{2\alpha}$,

γ even, β odd, $b\equiv 0 \pmod{\alpha}$, $b/\alpha \equiv 1 \pmod 2$, $\alpha\neq 0$,

(iii) p even, k odd, $b\equiv 0 \pmod{\alpha}$,

(iv) p odd, k even: γ even, $b\equiv 0 \pmod{\alpha}$, $b/\alpha \equiv 1 \pmod 2$, $\alpha\neq 0$,

γ odd, $b\equiv 0 \pmod{2\alpha}$. ⁞⁞⁞

EXAMPLES: (1) $\alpha=b=0$. A Gluck construction on the p-fold cyclic branched cover of the k-twist-spin of K yields S^4, as proved by Pao [8], and, when $p=1$, by Gordon [5].

(2) This says nothing for the α-roll, k-twist-spin.

Notice that a Gluck construction on the k-twist spin yields $\begin{pmatrix} \pm 1+(2i+1)k & k \\ 2i+1 & 1 \end{pmatrix}$, the $(\pm 1 + (2i+1)k)$-fold cyclic branched cover of the k-twist-spin, while an even framing charge, i.e. one that preserves the knot, gives $\pm 1 + 2ik$. See [2],[4]. For the special case $k=2$, $-1 + 1\cdot 2 = 1$, so these covers are all the same knot. Hence,

(6.2) COROLLARY: *The 2-twist-spin of any knot is determined by its complement.* ⁞⁞⁞

This has also been observed by Litherland (see [5, footnote p.595]) and Montesinos [7, Corollary 9.2].

7. CYCLIC BRANCHED COVERS

Suppose we take a cyclic branched cover of the knot $A(K)$ in (5.6). In order to insure that the result be a homology sphere, the order of branching must be prime to k.

Let $q > 0$, $(q,k) = 1$. It is easy to see that the q-fold cyclic branched cover of $A(K)$ can be written as $P \underset{A_q}{\cup} X \times S^1$, for some A_q. An easy way to find A_q is the following:

Let $$A_q = \begin{pmatrix} p' & k' & 0 \\ -\gamma' & \beta' & 0 \\ -\alpha'\gamma'+b'p' & \alpha'\beta'+b'k' & 1 \end{pmatrix}, \quad k'\gamma' + p'\beta' = 1.$$ The characteristic map of the q-fold cover of $A(K)$ is the qth power of the characteristic map for $A(K)$. Using (5.6), this gives:

$$\begin{aligned} k' &= k \\ p' &= qp \\ \alpha'\beta'+b'k' &= \alpha\beta+bk \\ -\alpha'\gamma'+b'p' &= q(-\alpha\gamma+bp) \\ k'\gamma'+p'\beta' &= k\gamma+p\beta = 1 \quad . \end{aligned}$$

Solving these yields

(7.1) PROPOSITION: *The q-fold branched cover of* $A(K)$ *is given by* A_q *above, with* $k'=k$, $p'=qp$, $\alpha'=q\alpha$, $b'=b+\alpha(\beta\gamma'-\beta'q\gamma)$. ▯

EXAMPLES: (1) $\alpha=b=0$. These are just the k-twist spin of K, and n-fold covers as mentioned earlier. Here is an example: Let k be an odd prime, let p be even. When is this the q-fold cover, $(q,k) = 1$, of a knot in S^4, say the one given by k and p'? By (7.1) and (2.2), we must be able to write $qp' = p-2ik$ for some i, i.e. find i so that $q \mid (p-2ik)$. Write $kx + qy = 1$. If q is even, $q = p-2i$, so that $p-2ixk = p-2i(1-qy) = q(1+2iy)$. If q is odd, $q = p-2i-1$, we can assume x is even, and then $p-x(2i+1)k = q(1+y+y2i)$. Summing up, we have shown

(7.2) THEOREM: *Let* k *be an odd prime. Let* p *be even,* $(p,k) = 1$. *Then the* p*-fold cover of the* k*-twist-spin of any knot* $K \subset S^3$ *is the fixed point set of a semi-free* $\mathbb{Z}_q$ *- action (which embeds in an* S^1*-action) on* S^4, *for any* q *such that* $k \nmid q$. ▯

For $k=2$, $p=1$, the analogous statement is due to Giffen [2,Theorem 3.5]. Gordon produced knots in S^n, $n \geq 5$, which are the fixed point sets of $\mathbb{Z}_q$-actions (that embed in S^1-actions) on S^n, for all q such that $k \nmid q$, k a given prime [4,Theorem 3]. This example settles the question, raised in [4,Section 4, Remark (2)] as to whether n can be lowered to 4.

(2) Let $A = \begin{pmatrix} 1 & 1 & 0 \\ 0 & 1 & 0 \\ ja & a(j+1) & 1 \end{pmatrix}$, as in example 4 of Section 6. The $(j+1)$-fold cyclic branched cover is given by

$$A_{j+1} = \begin{pmatrix} j+1 & 1 & 0 \\ j & 1 & 0 \\ (j+1)ja & a(j+1) & 1 \end{pmatrix}, \text{ so that } \alpha' = (j+1)a,\ b' = 0.$$ If j is even,

we have S^4; if j is odd, a Gluck construction on $A_{j+1}(K)$ gives S^4. If j is even, this is the same knot as the one determined by $\begin{pmatrix} 1 & 1 & 0 \\ 0 & 1 & 0 \\ 0 & a(j+1) & 1 \end{pmatrix}$, the $a(j+1)$-roll, 1-twist spin of K. This shows that there is a $\mathbb{Z}_{j+1}$-action on S^4 with fixed point set the $a(j+1)$-roll, 1-twist spin of K, as long as j is even, so we have more counterexamples to the higher-dimensional Smith conjecture.

Consider the general problem of deciding whether a knot $A(K)$ is the q-fold cyclic branched cover of another knot $A'(K)$. Assume A is as usual, and let $A' = \begin{pmatrix} p' & k' & 0 \\ -\gamma' & \beta' & 0 \\ -\alpha'\gamma'+b'p' & \alpha'\beta'+b'k' & 1 \end{pmatrix}$. In view of (7.1), we let $k'=k$, $q\alpha'=\alpha$, $(q,k) = 1$. Then the q-fold cover of $A'(K)$ is determined by

$$A'' = \begin{pmatrix} qp' & k & 0 \\ -\gamma'' & \beta'' & 0 \\ -\alpha\gamma''+b''qp' & \alpha\beta''+b''k & 1 \end{pmatrix}, \text{ where } k\gamma'' + qp'\beta'' = 1,$$

$b'' = b' + \frac{\alpha}{q}(\beta'\gamma'' - \beta''q\gamma')$. In order that A'' gives $A(K)$, we must be able to find i,j so that

(1) $\beta'' + jk = \beta$

(2) $qp' + 2ik = p$,

(3) $-\gamma'' + jqp + 2i\beta = -\gamma$,

(4) $\alpha\beta'' + b''k = \alpha\beta + bk$,

(5) $-\alpha\gamma'' + b''qp' + 2i(\alpha\beta + bk) = -\alpha\gamma + bp$.

From (2), $p' = (p-2ik)/q$. Equations (1),(3),(4) give

$$\beta'' = \beta - jk$$

$$\gamma'' = \gamma + 2i\beta + jp - 2ijk$$

$$b'' = b + j\alpha.$$

Substituting these values above and simplifying, we find

(6) $b' = b - \alpha'\beta'(\gamma+2i\beta) - \beta\gamma'\alpha$.

In other words, we first let $k' = k$, $p' = (p-2ik)/q$, $k'\gamma' + p'\beta' = \alpha' = \alpha/q$, and then let (6) define b'. Now take a q-fold cover of $A'(K)$. Our selections insure that (1)-(4) are satisfied; (5) then follows automatically. We have proved:

(7.3) THEOREM: *Let* $A(K)$ *be the knot determined by*

$$A = \begin{pmatrix} p & k & 0 \\ -\gamma & \beta & 0 \\ -\alpha\gamma+bp & \alpha\beta+bk & 1 \end{pmatrix}, \quad k\gamma + p\beta = 1,\ k > 0.$$

Let $q > 0$, $q|\alpha$, $(q,k) = 1$. *If there is an integer* i *so that* $q|(p-2ik)$, *then* $A(K)$ *is the* q-*fold cyclic branched cover of* $A'(K)$, *where* A' *is determined as above. In particular,* $A(K)$ *is the fixed point set of a* $\mathbb{Z}_q$*-action on* Σ_A. ▒

REMARK: A slight extension of Example (1) above shows that we can always find an i so that $q|(p-2ik)$, *except* when p is odd, q is even. In general, if α is divisible by many primes that don't divide k, then $A(K)$ is the fixed point set of many different $\mathbb{Z}_q$-actions. If $\alpha=0$, α is divisible by all primes so $A(K)$ is the fixed point set of infinitely many $\mathbb{Z}_q$-actions. Examples (1) and (2) above follow from the proof (and procedure) of this theorem. Almost all of the knots in S^4 provided by (6.1) are counterexamples to the higher-dimensional Smith conjecture.

BIBLIOGRAPHY

1. R. Fintushel, Locally smooth circle actions on homotopy 4-spheres. *Duke Math. J.*, 43 (1976), 63-70.

2. C. H. Giffen, The generalized Smith Conjecture. *Amer. J. Math.* 88 (1966), 187-198.

3. H. Gluck, The embedding of two-spheres in the four-sphere. *Trans. AMS*, 104 (1962), 308-333.

4. C. McA. Gordon, On the higher-dimensional Smith Conjecture. *Proc. London Math. Soc.* (3), 29 (1974), 98-110.

5. ————————, Knots in the 4-sphere. *Comm. Math. Helvetici*, 39 (1977), 585-596.

6. R. A. Litherland, Deforming twist-spin knots. *Trans. AMS*, 250 (1979), 311-331.

7. J. M. Montesinos, On twins in the four-sphere. Preprint.

8. P. S. Pao, Nonlinear circle actions on the 4-sphere and twisting spun knots. *Topology* 17 (1978), 291-296.

9. S. P. Plotnick, Circle actions and fundamental groups for homology 4-spheres. *Trans. AMS*, 273 (1982), 393-404.

10. ———————, The homotopy type of four-dimensional knot complements, Math. Z., 183 (1983), 447-471.

11. ———————, Infinitely many disk knots with the same exterior. Proc. Camb. Phil. Soc., 93 (1983), 67-72.

12. E. C. Zeeman, Twisting spun knots. Trans. AMS, 115 (1965), 471-495.

DEPARTMENT OF MATHEMATICS
COLUMBIA UNIVERSITY
NEW YORK, NY 10027

Contemporary Mathematics
Volume 35, 1984

THE EMBEDDING THEOREM FOR TOWERS

Frank Quinn†

The purpose of this note is to give a proof of M. Freedman's [1] embedding theorem for towers of immersed disks in a 4-manifold: a 7-stage tower can be embedded in a neighborhood of a 6-stage tower. This is a principal ingredient of his proof of the 4-dimensional topological Poincare conjecture.

The proof given here does not use the calculus of links. Instead it depends on a more complete development of transverse spheres. The idea here, as in [1] is to use connected sum with a transverse sphere to pull sixth stage self-intersections down into a lower stage. This gives homotopy classes of discs for the seventh stage. Then transverse spheres are used to make them disjoint from everthing else. We work in the smooth category, since a neighborhood of a PL or topological tower can be smoothed.

Circles and discs (from [2]). Suppose A is a surface immersed in a 4-manifold with transverse self-intersections. A Whitney circle is a circle in the image which passes through exactly two intersections points, changing sheets at each one. We also require that the intersections have opposite signs. A Whitney disc is an immersed disc with boundary a Whitney circle, and such that the framing of the disc when restricted to the boundary agrees with the framing determined by the surfaces. An accessory circle is a circle which passes through exactly one self-intersection point (changing sheets). An accessory disc is an immersed disc with boundary on accessory circle. There are no framing restrictions on accessory discs.

DEFINITION. (Towers). A 1-stage tower is a collection of disjoint discs immersed in a 4-manifold with boundary in the boundary. An n-stage tower is an n-1 stage tower union n^{th} stage immersed discs. The intersections in the n-1 stage discs are arranged in pairs, and the nth stage discs consist of a Whitney disc and an accessory disc for each pair. The nth stage discs are disjoint except for the necessary points of intersection (where Whitney and accessory discs

†Partially supported by an NSF Grant, VPI&SU, Blacksburg, VA 24061-4097, January 1982.

pass through the same intersection points in the n-1 stage).

THEOREM. *Suppose* $C_{1,6} \subset M^4$ *is a* 6-*stage tower. Then there is a* 7-*stage tower* $C'_{1,7}$ *contained in a regular neighborhood of* $C_{1,6}$, *with the same first four stages*.

The first step is to generalize towers slightly, by allowing an inferior type of accessory disc. These appear temporarily during the proof.

DEFINITION (errant accessory circles). Suppose $C_{1,j} \subset M$ is a j-stage tower, and suppose Whitney circles are given for some of the intersections in the jth stage. Then an *errant collection of accessory circles* for these Whitney discs is:

1) an ordering $S_1,\dots,S_n$ of the Whitney circles.
2) maps $\alpha_i : S^1 \to C_{1,j} \cup \partial M$, $i = 1,\dots,n$ such that $\alpha_i(S^1)$ is disjoint from S_k if $i < k$, it passes through exactly one of the intersection points of S_i, and does not pass through a j-stage intersection point not on one of the S_k.

Thus errant accessory circles can intersect earlier Whitney circles, and can go down into lower stages and out into ∂M. Henceforth "tower" means a tower with errant accessory discs. We say a tower has *pure* k^{th} *stage* if the accessory circles in the first k-stages are not errant, and accessory circles in higher stages do not intersect the kth stage. A pure tower is therefore a tower in the sense of the first definition.

The first two lemmas give some homotopy information about neighborhoods of towers.

LEMMA 1. *Suppose* $A \subset M$ *is an immersed disc with* $\partial A \subset \partial M$, *and suppose immersed discs* D_* *are attached to* $\partial M \cup A$. *Let* $D_1,\dots,D_j$ *be the ones attached on curves passing through a self-intersection point* p *of* A *(discs listed once for each passage through* p). *Let* N *be a neighborhood of* A. *Then a product of the linking circles of* $D_1,\dots,D_j$ *is a commutator in* $\pi_1(N - A \cup D_*)$.

PROOF. A small ball about p has boundary a 3-sphere, which intersects A in two linking circles. It intersects the D_* in arcs joining the circles. The homotopy to a commutator in the complement of $A \cup D_*$ is seen in the picture.

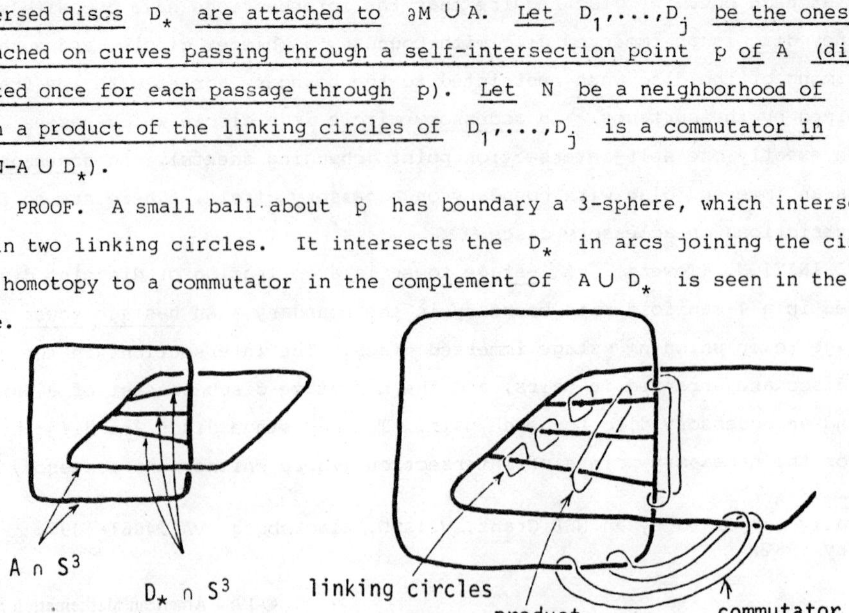

LEMMA 2. Suppose $C_{1,j} \subset M$ is a j-stage tower with only one first stage disc, $j \geq 2$, and whose accessory discs do not intersect ∂M. Let N be a regular neighborhood of the first j-1 stages, $C_{1,j-1}$. Then the image of $\pi_1(N - C_{1,j}) \to \pi_1(M - C_{1,j-1})$ is cyclic, and is generated by the linking circle of the first stage.

PROOF. Let $\hat{M}$ denote the manifold obtained from N by attaching 2-handles to the attaching regions for the top stage of $C_{1,j}$. This is the same as a neighborhood of $C_{1,j}$ except without intersections in the top stage. There is therefore a map $\hat{M} \to M$ which is an isomorphism onto a neighborhood of $C_{1,j}$, except in the top stage. This gives a factorization of the homomorphism of the lemma through $\pi_1(\hat{M} - C_{1,j})$. We will show that this is cyclic, generated by the first stage linking circle.

In $\hat{M}$ the topmost Whitney discs are embedded. Use these for Whitney moves to remove intersections in the j-1 stage, and denote the result by $C'_{1,j-1}$. The Whitney move shows that $\hat{M} - C_{1,j}$ is isomorphic to the complement of $C'_{1,j-1}$ union some arcs (where the Whitney discs used to be) union the accessory discs.

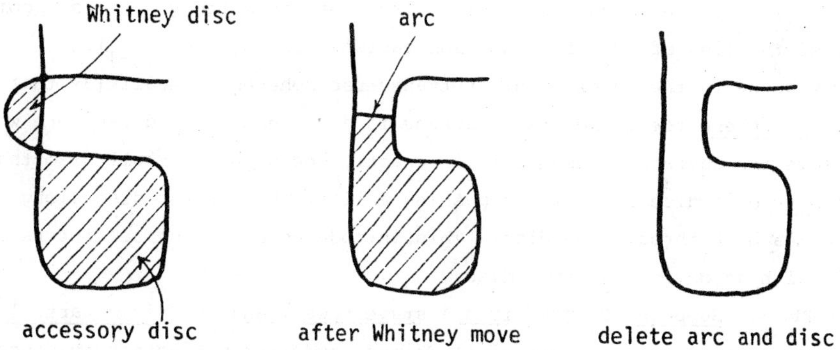

The accessory discs which used to go through the intersection points now go across the arcs. Since the accessory discs are errant (at worst), the arc left by the last Whitney disc has only one accessory disc passing through, and only once. The union therefore collapses to the object obtained by deleting that arc and disc. The collapse can be done ambiently, so the complements are the same. Now the next to the last arc intersects only one disc, so it can be collapsed. Continuing we see that the complement is the same as $\hat{M} - C'_{1,j-1}$, and that $\hat{M}$ is a regular neighborhood of $C'_{1,j-1}$.

The tower $C'_{1,j-1}$ has no intersections in the top stage, so the argument above applies to show it has the same complement as a smaller tower $C'_{1,j-2}$. Continuing we eventually get down to the first stage, and the statement: $\hat{M}$ is a regular neighborhood of an embedded disc, and $\hat{M} - (\text{disc}) \simeq \hat{M} - C'_{1,j}$. Therefore $\pi_1(\hat{M} - C'_{1,j})$ is cyclic, generated by the linking circle of C_1.

This completes the proof of Lemma 2.

The first application is to combine Lemmas 1 and 2 to get a vanishing result (commutator in a cyclic group).

COROLLARY. *Suppose* $C_{1,j} \subset M$ *is a* j *stage tower,* $j \geq 3$, *whose first stage is pure. Let* $D_{2,j} \subset M - C_1$ *be the* $j-1$ *stage tower of discs attached to one of the second stage discs of* $C_{1,j}$, *and let* W *be a regular neighborhood of* $D_{2,j-1}$ *in* $M - C_1$. *Then* $\pi_1(W-D_{2,j}) \to \pi_1(M-C_{1,j-1})$ *is trivial.*

PROOF. Let P be a neighborhood of $D_{2,j}$ in $M - C_1$. Then the homomorphism of the lemma factors through $\pi_1(W-D_{2,j}) \to \pi_1(P-D_{2,j-1})$. According to Lemma 2 the image of this is generated by the linking circle of the bottom disc D_2. It therefore suffices to see that this element is trivial.

If D_2 is a Whitney disc, then since the first stage is pure there is no other second stage disc passing through one of the intersection points. Lemma 1 applied at this point shows that the linking circle is a commutator in $\pi_1(N - C_{1,j})$, where N is a neighborhood of $C_{1,j-1}$. But by Lemma 2 again the image of this group is cyclic, so a commutator has trivial image.

If D_2 is an accessory disc, then it shares its intersection point with a Whitney disc. By Lemma 1 a product of the linking circles is a commutator in $\pi_1(N - C_{1,j})$. By the above the linking circle of the Whitney disc is a commutator, so the link of D_2 is also, and vanishes in $\pi_1(M - C_{1,j-1})$.

Now we begin the development of transverse spheres. Recall [3] that a *transverse sphere* for an immersed surface is a framed immersed 2-sphere which intersects the surface transversely in exactly one point. Recall also that a *Casson move* on a disc consists of pushing a little bit of the disc along an arc which comes back through the disc. This introduces a pair of selfintersection points, with an obvious Whitney disc.

LEMMA 3. *Suppose* $C_1 \subset M$ *is a 1-stage tower, and* C_2 *is a partial collection of (not necessarily disjoint) Whitney and (errant) accessory discs for intersections in* C_1. *Suppose* $N \subset M$ *is a submanifold which intersects each disc in a connected set. Suppose there are transverse spheres* $C_1^{\perp}$ *in* N *which intersect* C_1 *in only the canonical points. Then the discs* C_2 *can be changed to* C_2' *by Casson moves in* N *so that all the discs have transverse spheres in* N, *which intersect* $C_1 \cup C_2'$ *in only the canonical points.*

ADDENDUM. *The* C_2' *transverse spheres have Whitney discs in* $N - C_1 \cup C_2'$ *for all their intersections and self-intersections.*

PROOF. Let D be a disc in C_2, and choose a point in ∂D lying in C_1. Then the linking 2-sphere of ∂D at that point intersects D in exactly one point, intersects a C_1 disc in 2 points, and is disjoint from everything else. We can make it disjoint from C_1 by connected sum with copies of $C_1^{\perp}$.

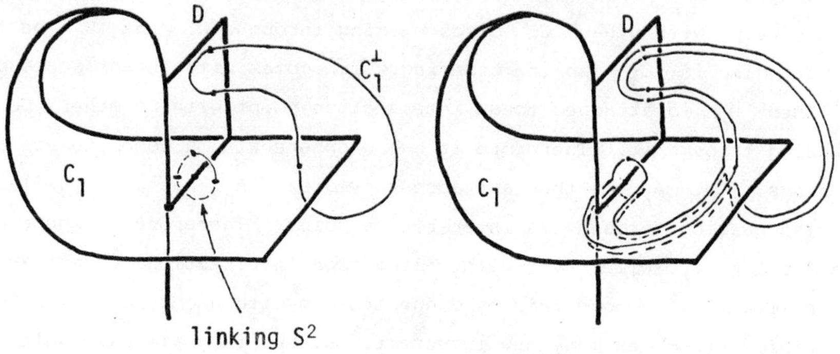

This may introduce many intersections with C_2, but we have some control over them.

A potential transverse sphere for D has a <u>complete set of Whitney discs</u> if it is disjoint from C_1, there is one distinguished point of intersection with D, and all other intersections (with everything, including itself) are arranged in pairs with Whitney discs. At this point we are not concerned with what the Whitney discs might intersect.

D is a Whitney or accessory disc for intersection points in C_1. Begin with a linking sphere near one of these intersection points, say p. As in the picture above, the intersections with C_1 appear on opposite sides of ∂D. Choose arcs along ∂D past p, then extend by parallel arcs to parallel copies of $C_1^\perp$. Do the connected sums along these arcs. Since these are parallel, there are Whitney discs for all intersections occuring past p.

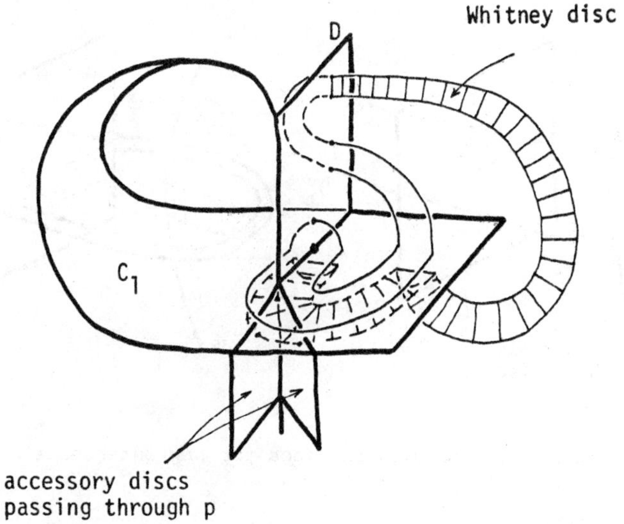

The only intersection points which do not have such Whitney discs are intersections near p with other C_2 discs passing through p. Let D_n be the last Whitney disc in C_2 (in the ordering which comes with errant accessory discs). Then D_n is attached to an intersection point with no other C_2 discs passing through it. Therefore it has a sphere with a complete set of Whitney discs. Let A_n be the last accessory disc. A_n has only the last Whitney disc passing through its intersection point. Therefore a sphere constructed for the A_n will have Whitney discs for intersections except an intersection with D_n. Remove this by connected sum with a copy of the sphere for D_n. This introduces many new intersections, but they all have Whitney discs (parallels of the ones for the sphere for D_n). Therefore A_n has a sphere with a complete set of Whitney discs.

In a similar manner we can work our way down through C_2, using spheres for the later ones to remove intersections without Whitney discs. Therefore we can conclude that C_2 has a set of spheres (in $N - C_1$) with complete sets of Whitney discs. Use the transverse spheres $C_1^{\perp}$ to make the Whitney discs disjoint from C_1. Then these are also contained in $N - C_1$.

The next step is to pick one of these discs, call it D_1, and use the Whitney discs to remove excess intersections with its associated spheres, call it S. Let W be one of the Whitney discs. If D_1 intersects W then choose an arc on W to the edge which lies on D_1. Push a bit of D_1 along this arc. This is a Casson move, removes an intersection of D_1 with W, and introduces two selfintersections of D_1. Repeat to obtain D_1 disjoint from W except for the boundary. Then push S across W (a Whitney move on S). Since $D_1 \cap \text{int}(W) = \emptyset$ this reduces $S \cap D_1$ by 2 points. It introduces intersections of S with everything which intersected W, but these occur in pairs with Whitney discs. We therefore maintain our hypotheses.

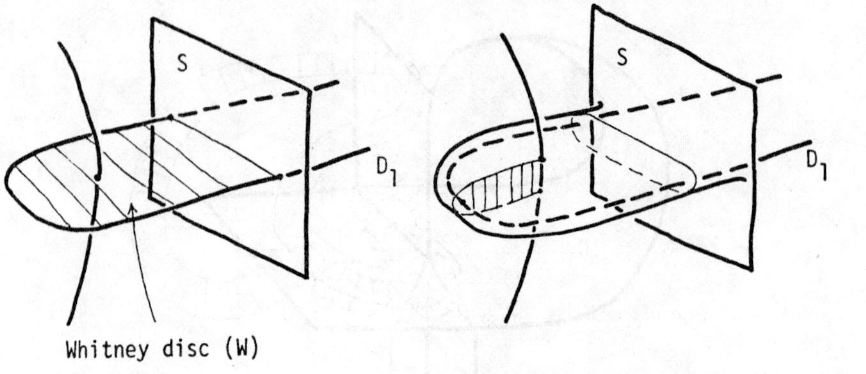

the Whitney move, with discs for new intersection points.

Continuing in this way we can obtain a transverse sphere $D_1^\perp$ in $N - C_1$, still with a complete set of Whitney discs. Use $D_1^\perp$ to make all the candidate transverse spheres, all the $C_1^\perp$, and all the Whitney discs, disjoint from D_1. As above sums of spheres with complete sets of Whitney discs again have such complete sets. The hypothesis $D_1 \cap N$ connected ensures that these operations can be carried out inside N.

Now we proceed by induction. Suppose that discs $D_1, \ldots, D_k$ have transverse spheres $D_i \subset N$, disjoint from $C_1 \cup D_1 \cup \cdots \cup D_k$ except for the canonical points. The other discs in C_2 have "associated" spheres with complete sets of Whitney discs, in $N - C_1 \cup D_1 \cup \cdots \cup D_k$. Let D_{k+1} be one of the other discs. Go through the process above changing D_{k+1} by Casson moves to get a transverse sphere $D_{k+1}^\perp$. $D_{k+1}^\perp$ is still disjoint from $D_1, \ldots, D_k$ so it can be used to make everything disjoint from the new D_{k+1}. This replicates the induction hypothesis. When all of the discs have been improved we have the conclusion of Lemma 3 (and the addendum).

As an application of Lemma 3 we show how transverse spheres and some homotopy information can be used to add discs to a tower.

COROLLARY. *Suppose* $C_{1,j} \subset M$ *is a* j *stage tower, and the* j *stage discs have transverse spheres in* $M - C_{1,j}$. *Suppose* $S_1, \ldots, S_k$ *are Whitney and accessory circles for some of the* j *stage intersections, and are nullhomotopic in* $M - C_{1,j-1}$. *Then there are immersed Whitney and accessory discs spanning these circles, with interiors disjoint from each other and* $C_{1,j}$.

ADDENDUM. *If some Whitney and accessory discs are already given, and the* $C_j^\perp$ *are disjoint from them, then the new discs will also be disjoint from them. Also, the new discs have transverse spheres in* $M - C_{1,j}$, *and algebraic self-intersection* 0.

PROOF. By hypothesis we can span the discs by discs immersed in $M - C_{1,j-1}$. Discs on Whitney circles can be spun [3] to correct the framing. Then copies of $C_j^\perp$ can be added to make them disjoint from $C_{1,j}$. Call these discs $E_1, \ldots, E_k$.

We apply Lemma 3. The discs E_* are changed by Casson moves, and acquire transverse spheres disjoint from $C_{1,j} \cup E_*$ except for the canonical points. Further these spheres have a complete set of Whitney discs for their intersections and self-intersections.

If the algebraic self-intersections of $E_*^\perp$ are not zero, this can be corrected by sums with the $E_*^\perp$ (note algebraic intersections of the E_i are 0). This ruins the spheres, so repeat the last step. This does not change algebraic intersection since the discs are changed by Casson moves.

Add copies of $E_i^\perp$, $i > 1$, to E_1 to obtain E_1' disjoint from E_i, $i > 1$. E_1' has excess intersections with $E_1^\perp$, but there are Whitney discs for these intersections. We can change E_1' further by Casson moves to be disjoint from the Whitney discs. Then push $E_1^\perp$ across the discs to remove the extra intersections. This gives a transverse sphere $(E_1')^\perp$ disjoint from E_i, $i > 1$. Use $(E_1')^\perp$ to make all the $E_i^\perp$ and Whitney discs disjoint from E_1'.

As in the proof of Lemma 3 we can repeat this process, improving the E_i, $E_i^\perp$ one at a time. The end result satisfies the conclusion of the lemma.

The next lemma is the main step in the proof.

LEMMA 4. *Suppose* $C_{1,5} \subset M$ *is a 5-stage tower with first stage pure. Then there is a 6-stage tower in* M *with first stage pure, and with the same first three stages as* $C_{1,5}$.

ADDENDUM. (a) *If* $C_{1,j} \subset M$ *is a* j *stage tower with* $j \geq 5$ *and first stage pure, then we can get a* $j+1$ *stage tower.* (b) *The attaching circles for the fourth stage are also the same as in* $C_{1,5}$. (c) *There are transverse spheres for the top three stages, intersecting* $C_{1,5}'$ *in only the canonical points.*

Notice that (a) can be used to get towers of arbitrary height. These are not very useful to us because the accessory discs are errant.

PROOF. Let N be a neighborhood of $C_{2,4}$ in $M - C_1$. We will improve the 5th stage intersections by induction. The induction hypothesis is that there is a tower $C_{1,5}'$ satisfying:

1) The first three stages and the circles for the fourth stage are the same as $C_{1,5}$, $C_{1,4}' \subset N$, and the first stage is pure.
2) The 4th and 5th stage discs in $C_{1,5}'$ have transverse spheres contained in N, and intersect $C_{1,5}'$ in only the canonical points. The spheres $(C_5')^\perp$ have Whitney discs for their self-intersections.
3) The 5th stage intersections are in two groups: one with Whitney and (standard) accessory circles contained in N, and the others.

First note that if there are no "other" intersections, then we can obtain the conclusion of Lemma 4 and the addendum. The corollary to Lemmas 1 and 2 shows that $\pi_1(N) \to \pi_1(M - C_{1,3})$ is trivial. Therefore all the Whitney and accessory circles given in (3) are nullhomotopic in $M - C_{1,3}'$. Span the circles by immersed discs in $M - C_{1,3}'$. The transverse spheres to C_4' can be used to get discs in $M - C_{1,4}'$. Now the corollary to Lemma 3 applies, to give a 6th stage.

The next thing to note is that we can remove one "other" intersection point by connected sum with a transverse sphere $(C_5')^\perp$.

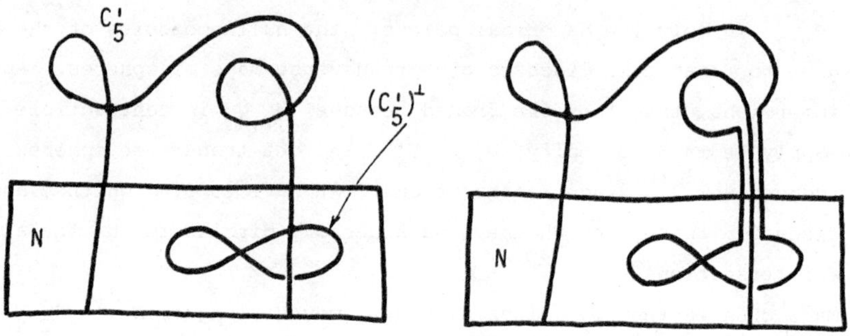

This introduces new 5th stage intersections, but by the hypothesis on $(C_5')^{\perp}$ these have Whitney and accessory circles in N. Therefore the "other" intersections are reduced. Unfortunately this operation ruins the transverse sphere hypothesis. If this hypothesis can be restored then the induction step will be complete.

What we have to prove, then, is that if $C_{1,5}$ satisfies the induction hypothesis except for the transverse sphere hypothesis (2), then there is tower $C'_{1,5}$ satisfying all the hypotheses and which has the same number of "other" intersections.

Because of the purity hypothesis, the first stage will not be involved. Therefore let $D_{2,5}$ be one of the second stage discs, together with the upper stages built on it. Let $N_{2,i}$ be a regular neighborhood of $D_{2,i}$ in M-(open regular neighborhood of C_1). Then $D_{2,5}$ is a 4-stage tower in $N_{2,5}$, with upper stages disjoint from the boundary. Also $N_{2,4} \subset N$.

The first step is to obtain transverse spheres for the D_3 discs. By Lemma 2, the image of $\pi_1(N_{2,3} - D_{2,3}) \to \pi_1(N_{2,4} - D_{2,3})$ is cyclic. Therefore, if we show that the linking circles of the D_3 are commutators in the first group, they will vanish in the second group. There is an ordering on the Whitney discs in D_3 from the errant condition on the accessory discs. Shuffle in the accessory discs by letting the kth one immediately precede the kth Whitney disc. Then for each disc there is a D_2 intersection point, through which it passes exactly once, and such that only later discs also pass through it. Assume as induction hypothesis that linking circles of later discs are commutators. Then Lemma 1 shows that the product of the current linking circle and a bunch of commutators is a commutator. The linking circle itself is therefore a commutator. Therefore by induction they are all commutators.

We conclude that the D_3 discs have transverse spheres in $N_{2,4}$ intersecting $D_{2,3}$ in only the canonical points: the nullhomotopies of the linking circles glue to transverse discs to give transverse maps of spheres. Approximate by immersions, then they are framed because $N_{2,4}$ is contractible in M.

Now apply Lemma 3 to modify D_4 to D_4', and get transverse spheres $(D_4')^{\perp}$. This introduces new 4th stage intersections. We have to find Whitney and accessory discs for these, and Whitney and accessory circles in N for all new 5th stage intersections.

Lemma 3 changes the D_4 discs by Casson moves along arcs in $N_{2,4}$ from a D_4 disc to itself. Therefore the new intersections occur in pairs, with Whitney discs disjoint from everything except $D_4^{\perp}$. Further these arcs lie in $N_{2,4} - C_{2,4}$, except for their endpoints. Since $N_{2,4}$ is a regular neighborhood of $C_{2,4}$, we can use the mapping cylinder structure to extend the arcs to embedded discs, with the rest of the boundary on $C_{2,4}$. This gives errant accessory discs for the new Whitney discs (when they are put after the old ones in the ordering).

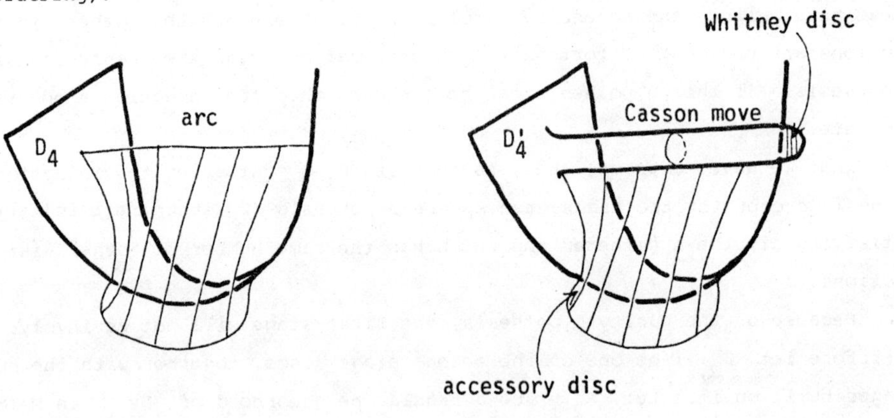

We add these to the old discs to get D_5', a complete set of 5th stage Whitney and accessory discs. This is not quite a tower, since the new 5th stage discs may not be disjoint from the old ones.

Next we apply Lemma 3 again, to get transverse spheres for the 5th stage. First note that the discs in D_5' either intersect $N_{2,4}$ in an annulus (the old ones) or are contained in $N_{2,4}$ (the new ones). Therefore the connectivity hypotheses are satisfied to get transverse spheres contained in $N_{2,4}$.

These transverse spheres also have complete sets of Whitney discs for their intersections and self-intersections. We can therefore use them as in the proof of Lemma 3 and its corollary to make the new 5th stage discs disjoint from the old ones, and each other.

We observe that all the new intersections in D_5' are good ones. The new intersection in the old D_5 discs come from Casson moves inside $N_{2,4}$. The new

D_5' discs lie entirely in $N_{2,4}$, and they have Whitney and accessory circles because they were constructed from embedded discs by adding transverse spheres with complete sets of Whitney discs, and then some Casson moves. In particular we have the same number of "other" intersection points that we started with, and the induction step is complete.

This completes the proof of Lemma 4.

PROOF OF THE THEOREM. Suppose $C_{1,6}$ is a pure 6-stage tower. Consider it as a bunch of 5-stage towers attached to C_1. Apply Lemma 4 to each of these (inside disjoint neighborhoods) to get 6-stage towers. All together we get a 7-stage tower $C_{1,7}'$ with first two stages pure, and first four and attaching circles of the fifth equal to those of $C_{1,6}$.

Let $N_{i,j}$ denote a regular neighborhood of $C_{i,j}'$ in $M - C_{1,i-1}'$. The first errant discs occur in C_6'. Choose standard accessory circles for the Whitney circles in C_5'. By the corollary to Lemmas 1 and 2, $\pi_1(N_{3,5} - C_{3,5}') \to \pi_1(N_{2,6} - C_{2,5}')$ is zero. This means first of all that we can get transverse spheres for C_5' (by gluing nullhomotopies in $N_{2,6} - C_{2,5}'$ on boundaries of transverse discs). Second, it means that all the Whitney and accessory circles are nullhomotopic in $N_{2,6} - C_{2,5}'$. Therefore the corollary to Lemma 3 applies to span these by disjoint immersed discs. This gives a pure 6-stage tower, $C_{1,6}''$, with 6th stage discs with transverse spheres inside $N_{2,6}$.

Now we repeat this argument. The corollary to Lemmas 1 and 2 implies that $\pi_1(N_{2,6} - C_{2,5}') \to \pi_1(N_{1,7} - C_{1,5}')$ is trivial. Therefore we can get nullhomotopies of the Whitney, accessory, and linking circles. At first these may intersect C_6'', but they can be made disjoint by adding transverse spheres $(C_6'')^{\perp}$. Then we have the data required to apply the corollary to Lemma 3. Again this yields disjoint immersed discs, which give a pure 7-stage tower.

BIBLIOGRAPHY

[1] M. Freedman, The Topology of Four-Dimensional Manifolds, Journal of Differential Geometry 17(1982) 357-453.

[2] M. Freedman and F. Quinn, Slightly Singular 4-Manifolds, Topology 20 (1981) 161-173.

[3] F. Quinn, The Stable Topology of 4-Manifolds, Topology and its Applications 15(1983) 71-77.

DEPARTMENT OF MATHEMATICS
VIRGINIA POLYTECHNIC INSTITUTE AND STATE UNIVERSITY
BLACKSBURG, VA 24061

Contemporary Mathematics
Volume 35, 1984

SMOOTH STRUCTURES ON 4-MANIFOLDS

Frank Quinn

ABSTRACT. The purpose of this note is to record the main points of the current (end 1982) understanding of structures on topological 4-manifolds. The results are contrasted with higher dimensional analogs to emphasize the peculiarities of dimension 4. As background we recall that every 3-manifold has a smooth structure, unique up to isotopy. Therefore the existence and uniqueness questions considered here are all relative to the boundary. Also recall [5] that smooth and PL structures are the same (up to isotopy) up through dimension 6, so we do not distinguish between the two types of structures. It is customary to refer to smooth structures, but in a number of ways the PL ones are more basic and might be more appropriate.

1. EXISTENCE

Suppose M is a topological 4-manifold. The Kirby-Siebenmann obstruction to smoothing M is a class $k \in H^4(M, \partial M; \mathbb{Z}/2)$. $M \times \mathbb{R}$ is smoothable if and only if $k = 0$, but unlike higher dimensions [5,p.3] this does not imply that M itself is smoothable.

If M is not compact the obstruction group is trivial. The high dimensional theory therefore predicts M should be smoothable. This much, at least, does hold.

1.1 THEOREM. [8,2.2.3] _Let_ M_0 _be obtained from_ M _by deleting an interior point from each compact component. Then_ M_0 _has a smooth structure._

Unfortunately, extending the structure across these remaining points involves more than the $\mathbb{Z}/2$ obstruction. The following was announced by S. Donaldson (August 1982):

1.2 THEOREM. [2] _Suppose_ M _is a closed 1-connected smooth 4-manifold whose quadratic form on_ $H_2(M;\mathbb{Z})$ _is positive definite. Then the form is equivalent over_ $\mathbb{Z}$ _to the form given by the identity matrix._

The 8×8 matrix E_8 is unimodular, symmetric and positive definite, but not equivalent over $\mathbb{Z}$ to I_8. Freedman [4] has shown that there is a closed 1-connected manifold with this form. This manifold was already known to be nonsmoothable since the high dimensional obstruction $k(E_8)$ is $\neq 0$. However Donaldson's theorem also applies to $2E_8$, even though $k(2E_8) = 0$. To explore the gap between 1.1 and 1.2 we introduce some notation. An _almost_ smoothing of M is a smooth structure in the complement of a discrete set. The exceptional points are referred to as the singular points. Note that by collapsing topologically flat arcs joining singular points we can arrange that there is only one in each compact component (the M_0 of 1.1). This is often required in the definition of an almost smoothing, but we will find the slight additional generality useful.

A singular point has a neighborhood homeomorphic to $\mathbb{R}^4$, with the singular point corresponding to 0. The almost smooth structure induces a smooth structure on $S^3 \times (0,\infty)$, which we refer to as the end of the singularity. Two ends are equivalent if there is a diffeomorphism of the smooth structures "near $S^3 \times \{0\}$"; diffeomorphism of open sets containing some $S^3 \times (0,\varepsilon)$, $\varepsilon > 0$.

Some useful examples of ends and singular points can be obtained this way: the complement of the standard $S^2 \subset CP^2$ is $\mathbb{R}^4$ (with the standard smooth structure). Let S^2_d denote the image of S^2 under a homeomorphism $CP^2 \to CP^2$ which is topologically isotopic to the identity. (We call S^2_d a displacement of S^2). The complement is still homeomorphic to $\mathbb{R}^4$, and has a smooth structure since it is open in CP^2. However the smooth structure may no longer be standard. We say that a singular point is resolvable if its end is equivalent the end of one of these structures on $\mathbb{R}^4$. The terminology is inspired by the resolution of singularities of algebraic geometry: if $p \in M$ is a resolvable singular point then there is $r: N \to M$ defined by using the equivalence of ends to glue together $M - p$ and a neighborhood of $S^2_d \subset CP^2$. N is smooth, r is a diffeomorphism away from $r^{-1}(p)$, $r^{-1}(p) = S^2$, and a neighborhood of $r^{-1}(p)$ is diffeomorphic to a neighborhood of a displacement $S^2_d \subset CP^2$. Topologically $N \approx M \# CP^2$.

1.3 THEOREM. *A compact connected 4-manifold has an almost smoothing such that*

(a) *if* $k(M) = 0$, *all the singular points are resolvable.*

(b) *if* $k(M) \neq 0$, *all but one are resolvable, and the exceptional one has end isomorphic to Freedman's fake* $S^3 \times \mathbb{R}$ [3].

This will be proved in Section 3. Combined with Donaldson's theorem it implies an observation of Freedman.

1.4 COROLLARY. *There is a smooth structure on* $\mathbb{R}^4$ *not diffeomorphic to the standard one: There is a compact set* $K \subset \mathbb{R}^4$ *which is not contained in a compact contractible submanifold smooth in the strange structure.*

The contractible manifold property shows that the structure is strange, since a compact set is contained in a ball in the standard structure. By contrast, in every other dimension a smooth structure on $\mathbb{R}^4$ is not only diffeomorphic to the standard structure but isotopic to it.

PROOF OF 1.4: As noted above, [4] implies that there is a closed 1-connected 4-manifold with quadratic form $2E_8$. $k(2E_8) = 0$ so by 1.2 it has a resolvable almost smoothing. Let $\{p_i\}$ be the singular points, with ends equivalent to complements of displacements $S_i^2 \subset CP^2$. We claim that one of the complements $CP^2 - S_i^2 (\approx \mathbb{R}^4)$ satisfies 1.3.

Let U_i be a neighborhood of S_i^2 in CP^2 so that $U_i - S_i^2$ is diffeomorphic to a neighborhood of the end at p_i. $CP^2 - U_i$ is a compact subset of $CP^2 - S_i^2$. Suppose that for all i there is a compact contractible smooth submanifold W_i, $CP^2 - U_i \subset W_i \subset CP^2 - S_i^2$. Then $\partial W_i \subset U - S_i^2$, and its image under the diffeomorphism is a smooth codimension 1 submanifold of the almost smoothing of $2E_8$. The image of ∂W_i bounds an acyclic submanifold of $2E_8$ containing the singular point p_i. Replacing this acyclic manifold with W_i gives a smooth 1-connected closed 4-manifold with form $2E_8$. Since $2E_8$ is positive definite this contradicts Donaldson's theorem. Therefore for at least one i such a contractible submanifold does not exist.

We remark that 1.3 shows that smoothability of M cannot be determined from the ends near the singular points: the same ends occur in almost smoothings of $S^4 (= CP^2/S_d^2)$. Further, by taking connected sums with these almost smoothings we can introduce "arbitrarily bad" singular points into a manifold. Finally, call an end "Euclidean" if it is equivalent to the end of a smooth structure on $\mathbb{R}^4$. By joining the singular points of an almost smoothing (dividing out flat arcs between them) we get from 1.3: <u>If $k(M) = 0$ then M has an almost smoothing with one singular point, which has a Euclidean end.</u>

The proof of 1.4 from 1.2 was given to emphasize the commonness of Euclidean ends. A proof somewhat closer to basic facts is obtained this way: The Kummer surface K_3 is a smooth manifold which topologically is a connected sum $M \# 3S^2 \times S^2$ [4]. Use 2.1 (b) to identify smooth structures near the skeleta $3(S^2 \vee S^2)$ as neighborhoods of displacements of the standard $S^2 \vee S^2 \subset S^2 \times S^2$. Since $S^2 \times S^2 - S^2 \vee S^2 \approx \mathbb{R}^4$, the proof given above shows that at least one of these displacements has complement which does not have the smooth contractible manifold property.

2. UNIQUENESS

Suppose $f: M \to N$ is a homeomorphism of smooth manifolds, and suppose $W_0 \subset W \subset M$ are locally flat polyhedra with f smooth on a neighborhood of W_0. Then we say <u>f can be smoothed near $W \text{ rel } W_0$</u> if there is an isotopy of $f \text{ rel } W_0$ to a homeomorphism which is smooth on a neighborhood of W.

We recall that if the dimension is 3, f can be smoothed near any W. If the dimension of M is ≥ 5 and $\dim W \leq 5$ then f can be smoothed if and only if the Kirby-Siebenmann obstruction $k(f) \in H^3(W, W_0; \mathbb{Z}/2)$ is zero [5].

(k(f) measures the obstruction to making f PL, without any restriction on the dimension of W).

Unfortunately this is false in dimension 4. To explain what is true, some weaker notions are needed. A displacement of $W(\mathrm{rel}\, W_0)$ is the image of W under homeomorphism of M ambient isotopic (rel W_0) to the identity. Therefore we can speak of smoothing f near a displacement of W. This is not completely unnatural. For example a smoothing of f near a displacement of W trivially gives a smoothing of f^{-1} near a displacement of f(W). The corresponding statement without "displacement" is false in dimension 4, and in higher dimensions requires the use of a deep theorem.

Weaker yet is the idea of sliced concordance. A sliced concordance is a smooth structure on $M \times I$ such that the projection $P: M \times I \to I$ is a smooth submersion. This can be thought of as a continuous family of smooth structures, since each $P^{-1}(t)$ has a smooth structure. A sliced concordance of f to a map smooth near W is an isotopy $F: M \times I \to N$ together with a sliced concordance (a structure on $M \times I$) so that $F_0 = f$, the structure on $M \times \{0\}$ is the original one, and F_1 is smooth near W with respect to the structure on $M \times \{1\}$. Notice that a topological isotopy defines a sliced concordance simply by pulling back the smooth structure; $(F,1): M \times I \to N \times I$.

2.1 THEOREM. Suppose $f: M \to N$ is a homeomorphism of smooth 4-manifolds, $M \supset W \supset W_0$ locally flat polyhedra, and f is smooth near W_0.

a) If $\dim W \leq 1$ then f can be smoothed near W, rel W_0.

b) If $\dim W = 2$ then f can be smoothed near a displacement W_d of $W(\mathrm{rel}\, W_0)$. Generally f cannot be smoothed near W itself.

c) If $\dim W = 3$ and the high dimensional obstruction in $H^3(W, W_0; \mathbb{Z}/2)$ vanishes, then f is sliced concordant to a map smooth near $W(\mathrm{rel}\, W_0)$. Generally f cannot be smoothed near a displacement of W.

Statements a and b are [8,2.2.2]. Statement c is a result of Lashof and Taylor [7]. We give a proof of this result in Section 3.

The negative results come from contradictions of Donaldson's theorem. If we could smooth near $D^2 \subset D^2 \times \mathbb{R}^2 \to M$ then Freedman's 2-handles could be smoothed, and all of [4] would work smoothly. In particular we would obtain a smooth manifold with form $2E_8$. Similarly it is shown in [4] that the Kummer surface is a topological connected sum $K_3 \simeq N\#(3S^2 \times S^2)$ reflecting the decomposition of the quadratic form as $2E_8 \oplus 3\begin{bmatrix} 0 & 1 \\ 1 & 0 \end{bmatrix}$. There is an embedding $S^3 \times \mathbb{R} \to K_3$ separating the components of the connected sum. If this could be smoothed on a displacement of $S^3 \times \{0\} \subset S^3 \times \mathbb{R}$, then this could be used to glue the $2E_8$ side of the displacement to one side of $S^3_d \subset S^4$ and obtain a smooth manifold with form $2E_8$.

The high dimensional obstruction arises in the following way. First there is the theorem that concordance implies isotopy (which fails in dimension 4). Then there is a very general formal result from immersion theory which states that on an open manifold, sliced concordance classes of structures correspond bijectively with homotopy classes of reductions of the tangent microbundle to a vector bundle [6]. The topological tangent microbundle of an n-manifold determines a classifying map $M \to B_{top(n)}$. A reduction corresponds to a map to $B_{0(n)}$ so that the diagram

$$\begin{array}{ccc} & & B_{0(n)} \\ & \nearrow & \downarrow \\ M & \longrightarrow & B_{top(n)} \end{array}$$

homotopy commutes. Obstructions to existence and uniqueness of such maps involve the fiber $top(n)/0(n)$. Specifically uniqueness obstructions are given by $H^j(M;\pi_j(top(n)/0(n)))$. This theory applies in dimension 4.

The final ingredient is the stability result that $top(n)/0(n) \to top/0$ is n-connected $(n \geq 5)$, and that up to dimension 5 $top/0$ is a $K(\mathbb{Z}/2,3)$. These ingredients together give the obstruction to isotopy in $H^3(M;\mathbb{Z}/2)$. The concordance statements of 2.1 together with the immersion theory result extends the stability theorem to dimension 4:

2.2 THEOREM. $top(4)/0(4) \to top/0$ <u>is</u> 4-<u>connected</u>.

That this is 3-connected is [8,2.2.3], the π_2, π_3 results are in [7]. This result implies 1.1, via the existence aspect of the immersion theory (for <u>open</u> manifolds).

We remark that 2.2 shows that the nonsmoothability of $2E_8$ cannot be detected by the bundle constructions which work in higher dimensions. The new phenomenon is therefore considerably more subtle than what we are used to, and presumably will require further developments in differential geometry to be understood in more detail.

3. PROOFS OF 1.3 and 2.1c:

The first step in 1.3 is to reduce to the $k = 0$ case by introducing a singularity if $k = 1$. If M is a compact connected 4-manifold then $k(M)$ can be measured this way: take an almost smoothing of M, with singular points p_i. The end at p_i is a smoothing of $S^3 \times \mathbb{R}$. Perturb the projection to $\mathbb{R}$ to be smoothly transverse to 0, and let N_i be the inverse image. N_i is an orientable 3-manifold, so bounds a framed smooth 4-manifold, say W_i. Then $k(M) = \sum_i \frac{1}{8}$ index W_i, mod 2.

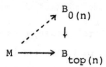

Now suppose $k(M) \neq 0$. Use the smoothing of $S^3 \times \mathbb{R}$ given in [3] to define an almost smoothing of a neighborhood of $p \in M$, with p corresponding to the $+\infty$ end of $S^3 \times \mathbb{R}$. As above make the projection to $\mathbb{R}$ transverse to 0. Let N_p denote the inverse image of 0, and M' the complement of the inverse image of $(0,\infty)$. M' is a compact 4-manifold. An almost smoothing of M' fits together with the almost smoothing of the neighborhood of p to give an almost smoothing of M. By the description of $k(M)$ given above, we have $k(M') = k(M) - \frac{1}{8}$ index W_p, where W_p bounds N_p as above. But for the example of [3], index $W_p = 8$. Therefore $k(M') = 0$. A resolvable almost smoothing of M' will extend to an almost smoothing of M satisfying the statement of 1.3b.

Suppose $k(M) = 0$, and use 1.1 to obtain a smooth structure on $M - p$. Let D^4 denote a disc with center p. The product structure on $(M - \frac{1}{2} D^4) \times \mathbb{R} \subset (M-p) \times \mathbb{R}$ extends to a smooth structure on all of $M \times \mathbb{R}$ (because $k(M) = 0$). The projection $M \times \mathbb{R} \to \mathbb{R}$ is smoothly transverse to 0 on $(M - \frac{1}{2} D^4) \times \mathbb{R}$. Approximate it rel this set to be transverse to 0 on all of

The inverse image N is a smooth manifold which is topologically $M \# P$ since it still contains $D^4 - \frac{1}{2} D^4 \subset M - \frac{1}{2} D^4$. By doing 0 and 1 surgeries (smoothly) we may assume P is 1-connected. Next note index $P = 0$ (this is because $N \times CP^2$ is bordant to (a smooth structure on $M) \times CP^2$, so index N = index M). By [4], $P \simeq \#^k S^2 \times S^2$ for some k. We therefore have a smooth structure on $M \# k S^2 \times S^2$, for some k.

Since $S^2 \times S^2 \# CP^2$ is diffeomorphic to $2CP^2 \# (-CP^2)$, $N \# CP^2 \simeq M \# i CP^2 \# j(-CP^2)$. This almost defines a "resolution" since $N \# CP^2$ is a smooth manifold mapping to $M = M \# i CP^2 \# j(-CP^2)/(i+j) S^2$. The only ingredient missing is the identification of the smooth structure near the copies of S^2 as neighborhoods of displacements of the standard $S^2 \subset CP^2$. This however can be obtained by application of 2.1(b).

This completes the proof of 1.3. Note we could also use displacements of $S^2 \vee S^2 \subset S^2 \times S^2$ as models for the singular points.

PROOF OF 2.1c. This is essentially the argument given in [7]. Given 2.1 a,b, the immersion theory formulation shows 2.1c is equivalent to: the kernel of $\pi_3 \mathrm{top}(4)/O(4) \to \pi_3 \mathrm{top}/O$ is trivial. Again by immersion theory an element of $\pi_3 \mathrm{top}(4)/O(4)$ is represented by a smooth structure on $D^3 \times (0,1)$ which is standard on the boundary. Take the union over the boundary with the standard structure to get a structure on $S^3 \times (0,1) = (\mathrm{int}\, D^4) - \{0\}$. Since the homotopy class is trivial in top/O, the stabilization $S^3 \times (0,1), S^3_- \times (0,1) \to \mathrm{top}(5)/O(5)$ extends to a map of $\mathrm{int}\, D^4$. As above this implies that the product structure on $(S^3 \times (\frac{1}{2}, 1)) \times \mathbb{R}$ extends to a smooth structure on $(\mathrm{int}\, D^4) \times \mathbb{R}$. As above make the projection to $\mathbb{R}$ transverse to 0, make the inverse image 1-connected, and recognize it as homeomorphic to $(\#^k S^2 \times S^2) - p$. This gives a third smooth structure on the end. The first structure is the possibly exotic one, which by construction extends to a smooth structure on $(\#^k S^2 \times S^2) - p$. The classifying map comparing the first and third structures therefore factors through the skeleton $\vee^{2k} S^2$.

Since $\mathrm{top}(4)/O(4)$ is already known to be 2-connected, this implies that these two structures are concordant. The proof is completed by showing that the second and third structures on the end are also concordant. The second structure was a standard structure on $S^3 \times (0,1)$. Both it and the third structure extend across the singular point, so again the classifying map comparing the two is nullhomotopic.

BIBLIOGRAPHY

1. S. Cappell, R. Lashof, and J. Shaneson, A splitting theorem and the structure of 5-manifolds, Inst. Maz. Mat. Symp. Vol. X (1972) 47-58.

2. S. Donaldson, in preparation.

3. M. Freedman, A fake $S^3 \times R$. Ann. Math. 110 (1979) 177-201.

4. M. Freedman, The topology of four-dimensional manifolds, J. Diff. Geometry.

5. R. Kirby and L. Siebenmann, Foundational essays on topological manifolds, smoothings, and triangulations, Ann. Math. Studies 88 (1977).

6. R. Lashof, The immersion approach to triangulation and smoothing, Proc. AMS Symp. Pure Math. 22 (1971) 131-164.

7. R. Lashof and L. Taylor, Smoothing theory and Freedman's work on four-manifolds, preprint 1982.

8. F. Quinn, Ends of maps II: dimensions 4 and 5, J. Diff. Geometry.

DEPARTMENT OF MATHEMATICS
VIRGINIA POLYTECHNIC INSTITUTE AND STATE UNIVERSITY
BLACKSBURG, VA 24061

Contemporary Mathematics
Volume 35, 1984

CONCORDANCE OF LINKS IN S^4

Daniel Ruberman

Concordance of links is known to involve more than simply concordance of the individual components; it is not known, for instance, if every even-dimensional link is slice. Null-concordance of even-dimensional links would follow if one knew that every link were concordant to a boundary link. (See (1) or apply Kervaire's argument (3) that $C_{2n} = 0$ to the disjoint Seifert surfaces.) N. Sato (4) has defined an invariant $\beta(L)$ such that L concordant to a boundary link implies $\beta(L)$ trivial, however, no example of a link with $\beta(L) \neq 0$ is known in other than the classical dimension. He observes that $\beta(L)$ can be defined for some links for which the components aren't spheres; the condition is given below. $\beta(L)$ lies in $\pi_{n+2}(S^2)$; Sato constructs non-spherical links to realize any element of $\pi_{n+2}(S^2)$. In S^4, his construction yields a link with both components genus two surfaces. In this note we construct an example of a link of a 2-sphere and a torus with non-trivial β; this is either best possible or next-best.

β is defined for oriented links $L = \{L_1, L_2\}$ in S^{n+2} for which $\mathrm{lk}(H_1 L_i, H_n L_j) = 0\ (i \neq j)$, where lk denotes linking number between cycles in S^{n+2}. In this case, $L_i = \partial M_i^{n+1}$, $M_i \cap L_j = \emptyset\ (i \neq j)$, and we set $\beta(L) = (M_1 \cap M_2, F)$ in $\pi_{n+2}(S^2)$ = the framed cobordism group. Here F is the framing of $\nu(M_1 \cap M_2)$ given by the normals to M_1 and M_2 restricted to $M_1 \cap M_2$. Sato shows this is well-defined and indeed an invariant of concordance; since it clearly vanishes for boundary links, $\beta(L) = 0$ for L concordant to a boundary link.

Construction of the example.

Start with the arc α and closed curve $\gamma \subset B^3$ as pictured in Figure 1. Define an isotopy γ_ϑ of γ by rotating γ by ϑ in the direction given by the arrow in Figure 1b. Then $(S^4, K, T) = (S^4, S^2, S^1 \times S^1)$ is defined to be $(S^1 \times B^3,\ S^1 \times \alpha,\ \bigcup_\vartheta e^{2\pi i\vartheta} \times \gamma_\vartheta) \cup (D^2 \times S^2,\ D^2 \times \partial\alpha, \emptyset)$. γ_ϑ has the obvious genus one Seifert surface F_ϑ; then $M = \bigcup_\vartheta e^{2\pi i\vartheta} \times F_\vartheta$ is a Seifert surface for T. Let G_ϑ be the surface in Figure 1c), where the tube goes along γ_ϑ, and δ is a fixed arc in ∂B^3. Then $N = \bigcup_\vartheta e^{2\pi i\vartheta} \times G_\vartheta \cup D^2 \times \delta$ is a Seifert surface for K; note that $N \cap T = M \cap K = \emptyset$ so that $\beta(L)$ is defined.

0271-4132/84 $1.00 + $.25 per page

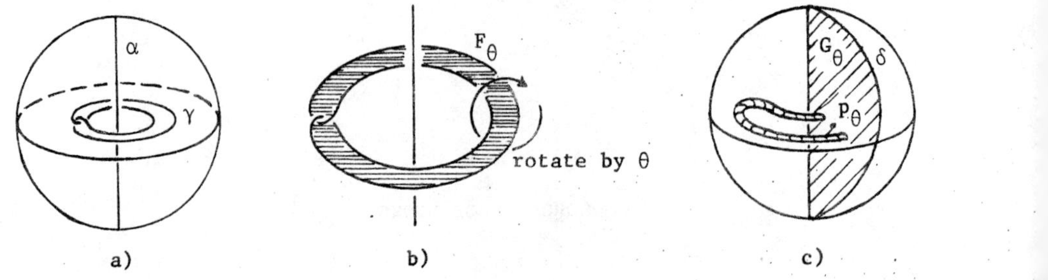

Figure 1

CLAIM: $\beta(L) \neq 0$

PROOF: In $\pi_4(S^2) = \mathbb{Z}_2$, the non-trivial element is represented by a torus with the Arf-invariant one framing; i.e. the induced framing is non-trivial on both members of a symplectic basis for H_1. To show $\beta(L) \neq 0$, it suffices to identify $\beta(L)$ with this element. Now $M \cap N = \bigcup_\vartheta e^{2\pi i\vartheta} \times (F_\vartheta \cap G_\vartheta)$ is certainly $S^1 \times S^1$. A symplectic basis of $H_1(M \cap N)$ is given by $a = \bigcup_\vartheta e^{2\pi i\vartheta} \times p_\vartheta$, $b = 1 \times F_0 \cap G_0$, where p_ϑ is a point on $F_\vartheta \cap G_\vartheta$ that doesn't move during the isotopy.

The framing on a is the suspension of the framing of $F_0 \cap G_0$ in $1 \times B^3$; this latter is the framing $+1$ in $\pi_3(S^2)$ (this is shown by Sato and demonstrates that β(Whitehead link) = 1) and hence its suspension is non-trivial. The non-trivial framing on a circle in S^4 is the one which differs from the framing extending over a disc by a single rotation as you go around the circle, so to calculate the framing $\psi_{|b}$ we first see the trivial framing on b. But this is given by a fixed frame (relative to B^3) at p_ϑ, for we can certainly extend that over a disc. The framing ψ is pictured in Figure 2; ψ_ϑ is just ψ_0 rotated by $2\pi\vartheta$. We conclude that $\beta \neq 0$.

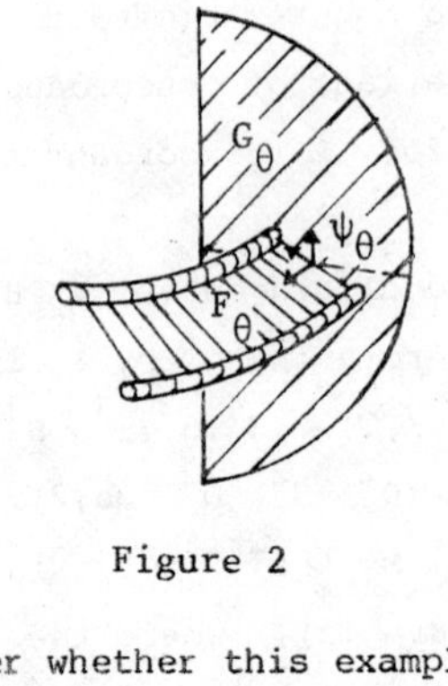

Figure 2

REMARK. One might wonder whether this example could be improved to give a spherical link in S^4 with $\beta \neq 0$, perhaps by surgering the torus T in the complement of M and N. A result of T. Cochran (2) precludes this; he shows that $\beta(L) = 0$ for L a spherical link with an unknotted component. Since K

in our link is the unknot (being simply the spin of the unknot), but $\beta \neq 0$, such a surgery is not possible.

BIBLIOGRAPHY

1. Cappell, S. and Shaneson, J. Link Cobordism, Comm. Math. Helv. 55 (1980) 20-49.

2. Cochran, T. On an Invariant of Link Cobordism in Dimension 4, preprint (1981).

3. Kervaire, M. Les Noeuds de Dimensions Superieures, Bull. Soc. Math. de France 93 (1965) 225-272.

4. Sato, N. Concordance of Manifold Links, preprint (1981).

DEPARTMENT OF MATHEMATICS
BRANDEIS UNIVERSITY
WALTHAM, MA 02154

Contemporary Mathematics
Volume 35, 1984

CONSTRUCTIONS OF QUASIPOSITIVE KNOTS AND LINKS, II

Lee Rudolph[1]

1. INTRODUCTION

A positive band in Artin's braid group B_n is any conjugate of the standard generator σ_1. (On the interpretation of B_n as the "knot group" of the discriminant locus Δ in the space E_n of unordered n-tuples of complex numbers, a positive band is simply a positively-oriented meridian of Δ.) A quasipositive braid in B_n is any product of positive bands. A quasipositive closed braid in (an unknotted solid torus in) S^3 or $\mathbb{R}^3$ is the closure of a quasipositive braid; these are precisely the closed braids naturally associated [Ru 1,2,3] to n-valued (complex) algebraic functions without poles. Finally, a quasipositive link is an oriented link which has some representation as a quasipositive closed braid.

Many such links can be constructed as boundaries of "quasipositive braided surfaces" in S^3. (These are the models in S^3 of those surfaces $S(\vec{b})$ of [Ru 2] for which $\vec{b}$ is an "embedded quasipositive band representation". They are also the quasipositive O-braided surfaces of Part I of this paper [Ru 3], with O denoting the braid axis, an unknot; since here we won't consider more general K-braided surfaces, K a fibred knot other than O, we drop O from the notation.) By such a construction, we see in Section 2 that many doubled knots are quasipositive. Then, returning to the braid-theoretical description, we see in Section 3 and Section 4 that the link used to describe a closed oriented 3-manifold as an irregular 3-sheeted branched cover of S^3 (respectively, as the boundary of a 4-dimensional (0,2)-handlebody) may always be taken to be quasipositive.

These constructions were conceived of as further evidence for the ubiquity of quasipositive links (see Part I). It would be interesting now to see whether the process can be turned around, and something proved -- say, about 3-manifolds -- from the knowledge that an auxiliary link in some construction can be assumed to be quasipositive and consequently tinged, however lightly, with complex algebraic geometry.

---

[1]Research partially supported by NSF MCS 76-08230.

2. MANY DOUBLED KNOTS ARE QUASIPOSITIVE

Given a knot K and an integer t, there are four oriented knot types (*a priori* distinct in non-trivial cases) which can be called "a double of K with t twists". They are constructed as follows: let A_t be an annulus in S^3 with core circle of the type of K, and so that the linking number of the two boundaries, identically oriented, is t; let $B_\pm$ be an annulus with unknotted core and linking number ± 1 (a Hopf band so-called); and let $S_\pm$ be the once-punctured torus in S^3, unique up to isotopy, obtained by "plumbing" A_t and $B_\pm$ together. Then the boundaries of $S_\pm$, with their two orientations, are *the doubles* of K *with* t *twists*.

PROPOSITION 1. *For any knot* K, *there exists* $t(K) \in \mathbb{Z} \cup \{-\infty\}$ *such that one of the orientations of the positive double* ∂S_+ *of* K *with* t *twists is a quasipositive knot whenever* $t > t(K)$.

PROOF: This follows from an inspection of Figure 1, and an application of the next, more general, proposition. ⁞⁞⁞

Let S be an oriented surface in (oriented) S^3, given with a handlebody decomposition $S = h_1^0 \cup \cdots \cup h_n^0 \cup h_1^1 \cup \cdots \cup h_k^1$ without 2-handles. Orient the core arc of each 1-handle. Then, given integers $t_1,\ldots,t_k$, it is clear how to define a reimbedding of S which is the inclusion on the 0-handles and which "inserts t_j twists in h_j^1" for $j = 1,\ldots,k$. This reimbedding is well-defined up to isotopy relative to the 0-handles. In particular, the various positive (resp., negative) doubles of K can be obtained by inserting twists into the 1-handle $A_0 - \operatorname{Int} B_\pm$ with core arc of knot type K, of the surface $S_\pm = h_1^0 \cup h_1^1 \cup h_2^1$, $h_1^0 = A_0 \cap B_\pm$, $h_1^1 = B_\pm - \operatorname{Int} A_0$, $h_2^1 = A_0 - \operatorname{Int} B_\pm$.

PROPOSITION 2. *Given* S *as above, there are* $t_1^*,\ldots,t_k^* \in \mathbb{Z} \cup \{-\infty\}$ *so that, if* $t_j > t_j^*$ *for* $j = 1,\ldots,k$, *then the surface obtained from* S *by inserting* t_j *twists in* h_j^1, $j = 1,\ldots,k$, *is bounded by a quasipositive link.*

PROOF (sketch): Use the method of Part I [Ru 3, Sections 2-3] first to "braid" S and then to do the requisite "twist insertion" in a braided manner. (An upper bound for the "modulus of quasipositivity" t_j is the number of negative bands involved in h_j^1 once S has been braided; for changing each such to a positive band, in each 1-handle, makes S a quasipositive braided surface, while inserting as many twists as there are changes of sign. Additional positive twisting is, of course, possible without losing quasipositivity of the surface once it is attained.) ⁞⁞⁞

REMARKS. (1) The given method of proof can never produce $-\infty$ as a "modulus of quasipositivity", and one may well doubt that *all* the positive doubles of *any* knot can be quasipositive; but it is not clear how this could be proved.

(2) Likewise, the given method of proof seems ill-adapted to obtaining either negative double (with any number of twists), or the oppositely-oriented double to that constructed. This raises the interesting questions, as yet

untouched, of invertibility and amphicheirality of quasipositive knots and links (but see Part I, Section 5, for a remark on amphicheirality).

(3) In a sense, the more "positive" the knot K is, the more negative $t(K)$ can be. The following is true.

SCHOLIUM. Let $\beta \in B_n$ be a braid on n strings, with exponent sum e, and closure $\hat{\beta} = K$. Then in Proposition 1 we may take $t(K) = n - e - 1$. ▯

Figure 2 illustrates the case of $\beta = \sigma_1^5 \in B_2$. (If β includes negative letters, the procedure becomes slightly more complicated at the negative crossings.)

3. QUASIPOSITIVE BRANCH LOCI.

The theorem of Alexander [A], that every (closed, connected, oriented) 3-manifold is a branched covering space of the 3-sphere, branched over a link, has in recent years been reproved and improved; it is now known that the covering may be taken to be three-sheeted, and the branch locus in S^3 taken to be a knot, [H], [M].

PROPOSITION 3. If M is a closed, connected oriented 3-manifold then M admits a representation $f: M \to S^3$ as a 3-sheeted covering branched over a quasipositive knot K.

PROOF: We make a straightforward application of the basic move used in [M] to reduce the number of components of the branch locus until it becomes a knot; we will phrase the move in braid-theoretical terms.

Let $\beta \in B_n$ be a braid, $L = \hat{\beta}$ its closure in $S^1 \times \mathbb{C} \subset S^3$, $p: S^1 \times \mathbb{C} \to [0,2\pi] \times \mathbb{R}: (e^{i\vartheta}, x+iy) \to (\vartheta, x)$ the standard projection, $p(L)$ a braid diagram for β (assumed to be in general position). An admissible 3-coloring of $p(L)$ is an assignment of one of 3 colors to each of the over-arcs of $p(L)$, in such a way that at each crossing either one or three colors are present, and of course for $j = 1,\dots,n$ the same color is assigned to the jth string at the top of the diagram as to the jth string at the bottom. Then it is well-known that admissible 3-colorings of $p(L)$ correspond to dihedral 3-sheeted coverings of S^3 branched over L. Suppose given an admissible 3-coloring of $p(L)$, so that for some j, the colors of the jth and $(j+1)$st strings at the bottom of $p(L)$ are different. Then if L' is the closure of $\beta\sigma_j^{\pm 3}$, and $p(L')$ is the obvious braid diagram with three new crossings at the bottom and no other changes, evidently there is a unique admissible 3-coloring of $p(L')$ which "extends" the given coloring of $p(L)$; and it is not hard to prove that the corresponding dihedral cover of S^3 branched over L' is homeomorphic to the original cover branched over L. Note that the number of components of L and of L' differ by exactly one.

Now take any representation of M as a 3-sheeted dihedral cover of S^3 branched over a link L. We will first perform basic moves until L becomes quasipositive, then if necessary perform more until it becomes a knot without losing quasipositivity.

Orient L arbitrarily, then represent it as the closure of some braid β in some B_n. Let $\vec{b} = (b(1),\ldots,b(s))$ be a band representation of β in B_n (cf. [Ru 2]); that is, for $j = 1,\ldots,s, b(j) = w(j)\sigma_{i(j)}^{\varepsilon(j)}w(j)^{-1}$ for some $w(j) \in B_n$, $1 \leq i(j) \leq n-1$, $\varepsilon(j) = \pm 1$, and $\beta = b(1)\cdots b(s)$. Let $N(\vec{b})$ be the number of negative bands in $\vec{b}$, that is, the number of indices j with $\varepsilon(j) = -1$. Then if $N(\vec{b}) = 0$, $\vec{b}$ and so L are quasipositive. Suppose $N(\vec{b}) \neq 0$. Then by conjugating $\vec{b}$ and β, if necessary, we may assume $b(s) = \sigma_1^{-1}$ without changing the link type of L. (The number n of strings is at least 2 since M is connected.) There are now two cases. The easier case is that, in an admissible 3-coloring of $p(L)$, the first and second strings at the bottom of the braid diagram are of different colors. In this case, the direct application of the basic move replaces $\vec{b}$ with $\vec{b}' = (b(1),\ldots,b(s-1),\sigma_1,\sigma_1)$, with $N(\vec{b}') = N(\vec{b}) - 1$.

Suppose, on the contrary, that the first and second strings at the bottom of the braid diagram are of the same color, say, red. Since M is connected, there must be at least one more string at the bottom, of a different color, say, blue. Again conjugating if necessary, we may assume this to be the third string. A sketch will aid the reader to confirm that two basic moves (with an intervening application of the braid relation $\sigma_1^{-1}\sigma_2 = \sigma_2\sigma_1\sigma_2^{-1}\sigma_1^{-1}$) replace $\vec{b}$ with $\vec{b}'' = (b(1),\ldots,b(s-1), \sigma_2,\sigma_1,\sigma_2^{-1}\sigma_1\sigma_2, \sigma_2^{-1}\sigma_1\sigma_2,\sigma_2)$, $N(\vec{b}'') = N(\vec{b}) - 1$.

Thus in either case L may be made quasipositive. Clearly, more basic moves will convert it to a quasipositive knot. ⁞⁞⁞

4. QUASIPOSITIVE SURGERY INSTRUCTIONS.

Again, we use the braid-theoretical description of quasipositivity to show that by applying basic moves of the "calculus of framed links" [K], [Cr], one may convert the link in a(n integral) surgery description of a 3-manifold to a quasipositive link; even the framings may be taken to be "natural" in a sense. --A more complete account of the translation of the link calculus into a "braid calculus" will be postponed to a later date.

PROPOSITION 4. *If M is a closed, connected, oriented 3-manifold, then there is a quasipositive braid β in some B_n so that M can be represented as the result of performing surgery on the link $\hat{\beta}$, framed "naturally".*

(The "natural" framing on a closed braid in the solid torus $S^1 \times \mathbb{C} \subset S^3$ is that induced by a constant vectorfield in the $\mathbb{C}$ factor.)

PROOF (sketch): Here, the two basic moves are the introduction (or suppression) of an unknot, split from the rest of the framed link, framed by ± 1; and the "band moves", corresponding to handle-sliding, which proceed by first taking the twisted longitude of one component of the framed link (twisted according to its framing), then joining this by a band to a second component, and finally adjusting the framings of the resulting link. It is well-known these moves, in combination, allow one to "change crossings", at the expense of adding a new component, linked but unknotted. The reader should have no trouble verifying that, if $\beta \in B_n$ has band representation $\vec{b} = (b(1),\ldots,b(s))$, $b(s) = \sigma_{n-1}^{-1}$, and β (with suitable framings) gives surgery instructions for M, then (again with suitable framing) the braid with band representation $\vec{b}' = (b(1),\ldots,b(s-1), \sigma_n,\sigma_{n-1},\sigma_n,\sigma_{n-1},\sigma_n)$ in B_{n+1} also gives surgery instructions for M. Thus, framing aside, M has quasipositive surgery instructions. As to the framing, it is readily verified that the "natural" framing of a component of a closed braid is the self-winding of that component (in the language of Laufer [L]), that is, its exponent sum after all other components are erased. Therefore it is easy to increase the natural framing of a component, by increasing the number of strings and joining them to that component by Markov moves which do not change the link type. But it is not much harder to decrease the natural framing, at the expense of adding extra components and performing band moves. ⁞⁞⁞

BIBLIOGRAPHY

[A] Alexander, J. W., "Note on Riemann spaces", Bull. Amer. Math. Soc. 26 (1920), 370-372.

[Cr] Craggs, R., "Stable equivalence in Heegaard and surgery presentations for 3-manifolds", to appear.

[H] Hilden, H., "Every closed orientable 3-manifold is a 3-fold branched covering space of S^3", Bull. Amer. Math. Soc. 80 (1974), 1243-1244.

[K] Kirby, R., "A calculus for framed links in S^3", Inv. Math. 45 (1978), 35-56.

[L] Laufer, Henry B., "On the number of singularities of an analytic curve", Trans. Amer. Math. Soc. 136 (1969), 527-535.

[M] Montesinos, J., "A representation of closed, orientable 3-manifolds as 3-fold branched coverings of S^3", Bull. Amer. Math. Soc. 80 (1974), 845-846.

[Ru 1] Rudolph, Lee, "Algebraic functions and closed braids", to appear in Topology (1983).

[Ru 2] ———————, "Braided surfaces and Seifert ribbons for closed braids", to appear in Comm. Math. Helv. (1983).

[Ru 3] ———————, "Constructions of quasipositive knots and links, I", to appear in L'enseignement mathématique (1983), volume of Proceedings of a conference on knot theory and algebraic geometry, Les Plans sur Bex, March, 1982.

DEPARTMENT OF MATHEMATICS
BRANDEIS UNIVERSITY
WALTHAM, MASSACHUSETTS

Current address:
P. O. Box 251
Adamsville, R. I. 02801

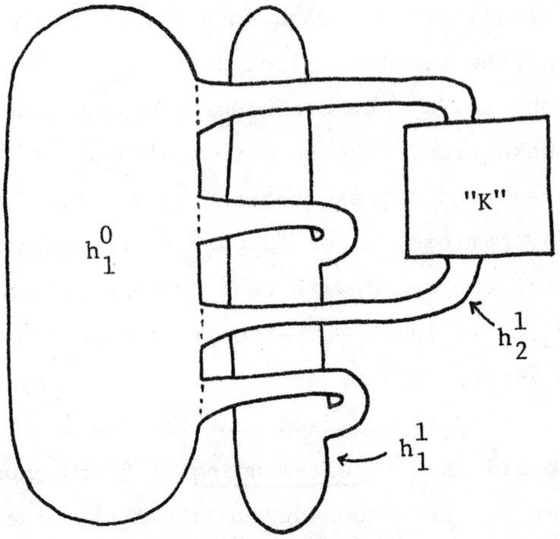

1A - generic

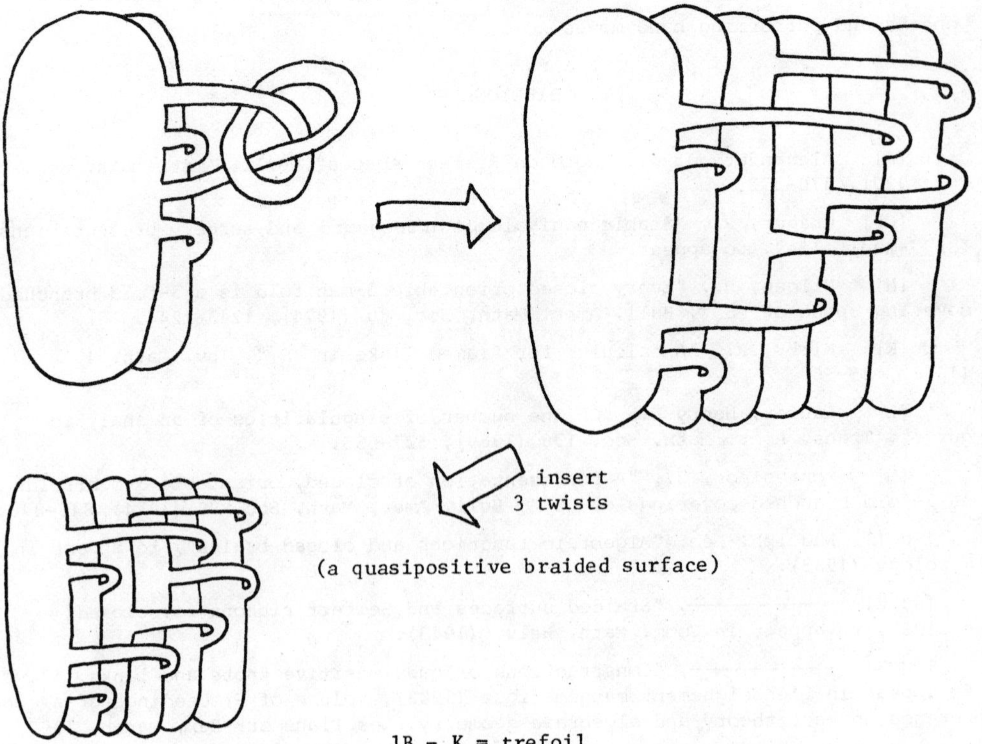

1B - K = trefoil

FIGURE 1

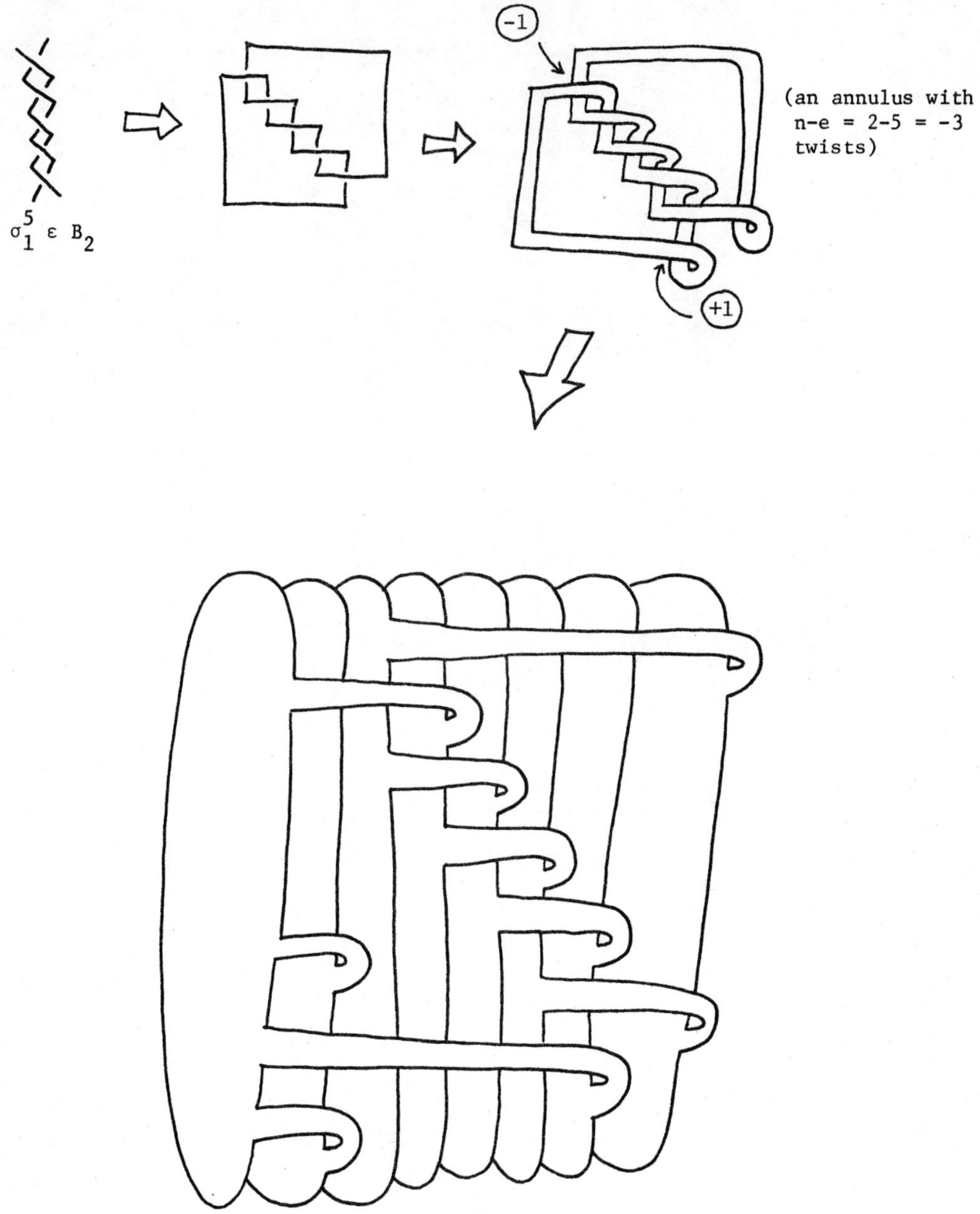

$\sigma_1^5 \in B_2$
-1
+1
(an annulus with
n-e = 2-5 = -3
twists)

FIGURE 2

Contemporary Mathematics
Volume 35, 1984

AN INTRODUCTION TO SELF-DUAL CONNECTIONS*

Clifford Henry Taubes†

I. GAUGE THEORIES

The Yang-Mills equations are differential equations for connections on principal Lie group bundles over 4-dimensional manifolds. The equations have been a subject for study for less than ten years, principally by physicists [1], but in the last five years or so, by mathematicians interested in their geometric [2] and analytic aspects [3]. There has now appeared an intimate relationship between the topology of 4-dimensional Riemannian manifolds and the Yang-Mills equations, see S. Donaldson's lectures. In order to help you to follow Donaldson's lectures, my lectures will introduce you to the concepts and terminology that are minimally necessary to communicate with a "gauge theorist". The lecture naturally splits into two parts. The first part contains a brief introduction to the mechanics of principal bundles, connections and curvature. This material is, hopefully, a review of your graduate course on differential geometry. Good references are Steenrod [4], Husemoller [5], and the recent article by Atiyah & Bott [6]. You may also find Kobayashi & Nomizu [7] useful. The second part of the lecture is an introduction to the recent results of Atiyah, Hitchin & Singer [8] and myself [9] on the self-duality equations - a special case of the Yang-Mills equations. An excellent new reference is [10].

In these lectures, M will denote a smooth, oriented, compact 4-manifold without boundary. I will assume that M has a fixed, Riemannian metric, m. Because M is smooth, I am doing <u>differential geometry</u>. The smoothness of M allows me to utilize many additional tools which are naturally applicable upon the choice of a C^∞ structure. This smooth structure is to be fixed once, and for all.

The first consequence of having a smooth structure on M is that we can define the notion of a smooth fibre bundle, or prinicpal Lie group bundle

$$\pi : P \to M .$$

†Junior Fellow of the Harvard University Society of Fellows.

*Supported in part by the National Science Foundation under Grant PHY79-16812.

There are some elegant difinitions around, but let me follow Steenrod since he was a topologist.

Let G be a simple, compact Lie group. (Much more general Lie groups are also allowed.) Then a principal G-bundle is defined by specifying the following data:

<u>PRINCIPAL G-BUNDLES</u>: (1) An open cover $\{U_\alpha\}$ of M. (2) Clutching functions (transition functions)

$$\{g_{\alpha\beta}: U_\alpha \cap U_\beta \to G\}$$

such that (3) in $U_\alpha \cap U_\beta \cap U_\alpha$ the cocycle condition

$$g_{\alpha\beta}\, g_{\beta\gamma}\, g_{\gamma\alpha} = 1 \ .$$

(4) By necessity $g_{\alpha\beta}^{-1} = g_{\beta\alpha}$ and $g_{\alpha\alpha} = 1$.

So we are to think of a point $p \in P$ as a pair $p = (x, h_\alpha)$ where $x \in U_\alpha$ and $h_\alpha \in G$. If $x \in U_\alpha \cap U_\beta$ then h_α is related to h_β by

$$h_\alpha = g_{\alpha\beta}\, h_\beta \ .$$

It is important to note that for $x \in M$, $\pi^{-1}(x) = P|_x$ is diffeomorphic to G, but not canonically so. However, P admits a smooth G action by right multiplication: $(p,g) \to pg^{-1}$. Thus if $p = (x, h_\alpha)$ then $pg = (x, h_\alpha g)$. This is consistently defined, as for $x \in U_\alpha \cap U_\beta$,

$$pg = (x, h_\alpha g) = (x, (g_{\alpha\beta} h_\beta) g) = (x, g_{\alpha\beta}(h_\beta g)) \ .$$

Along with the notion of a principal bundle, there is also the notion of an <u>associated bundle</u>. Let V be a vector space on which G acts via a representation ρ. Then the associated vector bundle,

$$\pi: \hat{V} (\equiv P \times_\rho V) \to M \ ,$$

is defined by

$$\hat{V} = P \times V/\sim \quad \text{where}$$

$$(p,v) \sim (pg, \rho(g)v) \ .$$

The cohomological data is as follows: A point $\hat{v} \in \hat{V}$ is given by $\hat{v} = (x, v_\alpha)$ when $\pi(\hat{v}) = x \in U_\alpha$. In $U_\alpha \cap U_\beta$, $v_\alpha = \rho(g_{\alpha\beta}) v_\beta$. Schematically, a principal G-bundle is

P: 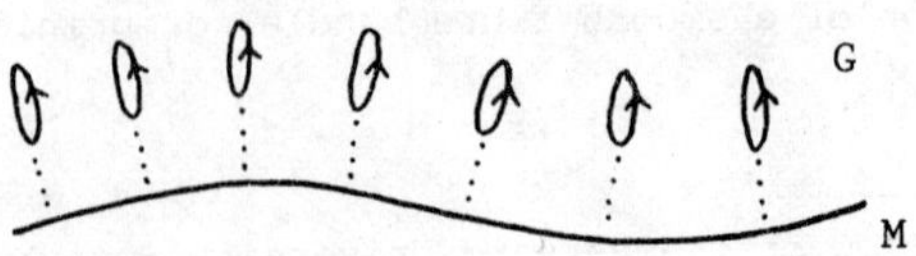

while an associated vector bundle is

$\hat{V}$:

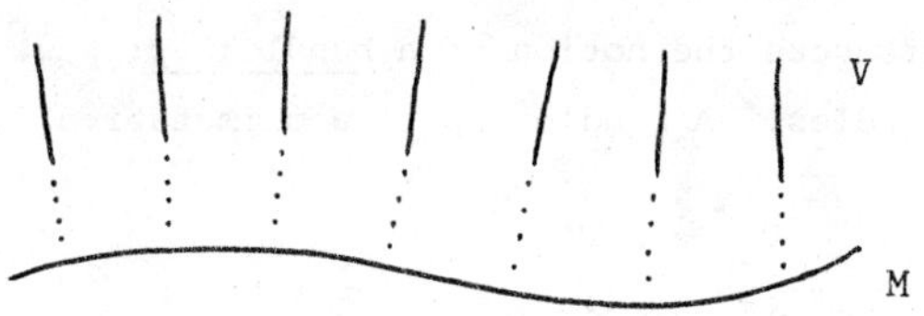

For example, M being a differentiable manifold, its <u>tangent bundle</u>,

$$\pi : T_M \to M .$$

a vector bundle with fibre $\pi^{-1}(x) \simeq \mathbb{R}^4$. The Riemannian metric is a smooth choice of metric $m_x(\cdot,\cdot)$ on $\pi^{-1}(x)$.

Let $F_M|_x$ denote the set of orthonormal frames in $T_M|_x$ with positive orientation as defined by the metric m_x and the given orientation of M. It is a fact that $F_M|_x$ is diffeomorphic to the Lie group $SO(4)$. Indeed, the set

$$F_M = \bigcup_{x \in M} F_M|_x$$

can be readily given the structure of a principal $SO(4)$-bundle in a way that makes T_M its associated vector bundle. The bundle $F_M \to M$ is called the <u>frame bundle</u> of M.

F_M 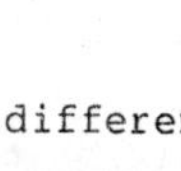   frame

$\downarrow$

M $\bullet$ x point

A concrete, hands on example of a principal bundle is given as follows: Take $M = S^4$ and $G = SU(2) =$ the group of unit quaternions. As a manifold $SU(2) \simeq S^3$. Cover S^4 by $U_+ = S^4 - s$ and $U_- = S^4 - n$ where $s =$ south pole and $n =$ north pole. Take $g_{+-} : U_+ \cap U_- \to S^3$ to be the projection map $U_+ \cap U_- = S^3 \times (0,\pi) \to S^3$. (This is the identity map on S^3's of constant latitude.) We call this bundle P_1. As a manifold, $P_1 = S^7$. Then $S^7 \to S^4$ is the Hopf fibration.

Another useful notion is that of the <u>pull-back bundle</u>. Let $\varphi : M \to N$ be a smooth map, and let $\pi : P \to N$ be a principal G-bundle. The pull-back bundle $\varphi^* P \to M$ is the principal G-bundle over M that is defined by the cohomological data $\{\varphi^{-1}(U_\alpha) \,;\, \varphi^* g_{\alpha\beta} : \varphi^{-1}(U_\alpha) \cap \varphi^{-1}(U_\beta) \to G\}$.

$$\begin{array}{ccc} \varphi^* P & \overset{\hat{\varphi}}{\cdots\cdots>} & P \\ \downarrow & & \downarrow \\ M & \overset{\varphi}{\to} & N \end{array}$$

Note, there is a natural map $\hat{\varphi} : \varphi^* P \to P$ which covers φ:

$$\hat{\varphi}(x,h_\alpha) = (\varphi(x),h_\alpha) \quad \text{for} \quad x \in \varphi^{-1}(U_\alpha) .$$

Note also that $\hat{\varphi}(pg) = \hat{\varphi}(p)g$ for $g \in G$.

This introduces the notion of a <u>bundle map</u>: Let $P \to M$, and $P' \to N$ be principal G-bundles. A bundle map is a commutative diagram

$$\begin{array}{ccc} P & \overset{\hat{\varphi}}{\to} & P' \\ \downarrow\pi & & \downarrow\pi \\ M & \overset{\varphi}{\to} & N \end{array} ,$$

defined by a smooth pair of maps $(\hat{\varphi},\varphi): (P,M) \to (P',N')$ such that $\hat{\varphi}$ commutes with the G actions on P and P'.

It is logical to define two bundles $P \to M$ and $P' \to M$ to be isomorphic, $P \sim P'$, when there exists a bundle map $\hat{\varphi}: P \to P'$ which covers the identity map from M to itself. Suppose that P is given by data $\{U_\alpha, g_{\alpha\beta}\}$ and P is given by $\{U'_\alpha, g'_{\alpha\beta}\}$. By taking a refinement of the covers $\{U_\alpha\}$ and $\{U'_\alpha\}$ of M, we may assume that $U_\alpha = U'_\alpha$. A <u>bundle isomorphism</u> $\hat{\varphi}: P \to P'$ is given by data $\{U_\alpha; \varphi_\alpha: U_\alpha \to G\}$ which satisfies in each $U_\alpha \cap U_\beta$,

$$\varphi_\alpha\, g_{\alpha\beta} = g'_{\alpha\beta}\, \varphi_\beta \; .$$

Principal G-bundles over M are usually classified up to isomorphism. A theorem on the subject is

THEOREM 1.1: <u>Let M be a compact, oriented, smooth 4-manifold without boundary. Every principal SU(2)-bundle over M is isomorphic to the pull-back of $P_1 \to S^4$ by a degree k map $\varphi_k: M \to S^4$ for some integer $k \in \mathbb{Z}$. Any two degree k maps from M to S^4 pull-back isomorphic bundles.</u>

A <u>bundle automorphism</u> is a bundle isomorphism $\hat{\varphi}: P \to P$; i.e. it is given by data $\{U_\alpha, \varphi_\alpha: U_\alpha \to G\}$, where in $U_\alpha \cap U_\beta$, $\varphi_\alpha = g_{\alpha\beta}\, \varphi_\beta\, g_{\alpha\beta}^{-1}$. The set of automorphisms of P is a group, in fact, it can readily be given the structure of a smooth, infinite dimensional Lie group. [11]

$$\text{Aut}\, P \equiv \mathcal{G}(P) = \text{"group of gauge transformations"},$$

Note that $\mathcal{G}$ is the set of sections of the associated bundle of groups $P \times_{AdG} G$ (Prove this.)

What is Lie alg. $\mathcal{G}$? This is the space of sections of the vector bundle

$$\hat{\mathfrak{g}} = P \times_{AdG} \mathfrak{g} \; .$$

Here $\mathfrak{g}$ denotes Lie alg G. Thus $\sigma \in \Gamma(\hat{\mathfrak{g}})$ is given by data $\{U_\alpha, \sigma_\alpha: U_\alpha \to \mathfrak{g}\}$ where in $U_\alpha \cap U_\beta$,

$$\sigma_\alpha = g_{\alpha\beta}\, \sigma_\beta\, g_{\alpha\beta}^{-1} \; .$$

Notice that $\mathcal{G}$ acts on $\hat{\mathfrak{g}}$ and hence on $\Gamma(\hat{\mathfrak{g}})$. For $(\varphi,\sigma) \in \mathcal{G} \times \Gamma(\hat{\mathfrak{g}})$ we have $\varphi\sigma = \{U_\alpha, \varphi_\alpha\, \sigma_\alpha\, \varphi_\alpha^{-1}: U_\alpha \to \mathfrak{g}\} \in \Gamma(\hat{\mathfrak{g}})$.

Principal bundles, associated bundles, bundle maps etc. can be defined in the topological category. However, the notions of connection and curvature are essentially C^2 phenomena. There are many equivalent ways to define a connection, c.f. [6],[7].

CONNECTIONS: If $v \in \mathcal{g}$, then v defines a vector field $\hat{v}$ on P as follows: For a C^1 function $f: P \to \mathbb{R}$,

$$(\tilde{v}f)(p) \equiv \frac{d}{dt} f(p \cdot \exp tv) \Big|_{t=0}$$

where $\exp: \mathcal{g} \to G$ is the exponential map. Note that $\pi_* \tilde{v} = 0$, hence $\tilde{v}$ is called a vertical vector. For $g \in G$, we observe that the right action of G on P induces

$$(R_{g*})\tilde{v}\Big|_p = \widetilde{(Adg^{-1}v)}\Big|_p .$$

Let $\tilde{V} \subset T_P$ denote the sub-bundle of vertical vectors. There exists the exact sequence of vector bundles

$$0 \to \tilde{V} \to T_P \to \pi^* T_M \to 0.$$

A connection, A, on P is a G-equivariant splitting of this sequence.

From the cohomological point of view, a connection A is given by specifying the following data:

$$A = \{U_\alpha, a_\alpha \in \Gamma(T^*\big|_{U_\alpha}) \times \mathcal{g}\},$$

and in $U_\alpha \cap U_\beta$, we require the cocycle relation

$$a_\alpha = g_{\alpha\beta}\, a_\beta\, g_{\alpha\beta}^{-1} + g_{\alpha\beta}\, d(g_{\alpha\beta}^{-1}).$$

Since $g_{\alpha\beta}^{-1}: U_\alpha \cap U_\beta \to G$, we can think of $g_{\alpha\beta}^{-1}$ as a G-valued function and the exterior derivative $d(g_{\alpha\beta}^{-1})$ makes since in this context.

Let $\mathcal{C} = \mathcal{C}(P)$ be the set of connections on P. The set $\mathcal{C}$ is naturally an affine space; one can see from the above cocycle relation that for $A, A' \in \mathcal{C}$,

$$a = A - A' = \{U_\alpha, a_\alpha = a'_\alpha\} \in \Gamma(\hat{\mathcal{g}} \otimes T^*) \equiv \Omega^1(\hat{\mathcal{g}}) .$$

A connection A defines a horizontal sub-bundle of T_P called H_A, and H_A is isomorphic to $\pi^* T_M$. Let $p \in P$ and $X \in T_M\big|_{\pi(x)}$. The isomorphism $\pi^* T_M \sim H_A$ defines the horizontal lift of X at p, $X_A \in T_P\big|_p$ which is the unique vector in $T_P\big|_p$ satisfying both $X_A \in H_A$ and $\pi_* X_A = X$.

When H_A is an integrable sub-bundle, the connection A is called flat. The product bundle $M \times G$ with the connection $A = 0$ (all $a_\alpha = 0$) gives an example of a flat connection. By Frobenious' theorem, H_A is integrable iff for all vector fields $X_A, Y_A \in H_A$,

$$[X_A, Y_A] \in H_A .$$

We see that the obstruction to the integrability of H_A at $p \in P$ is measured by

$$\pi^* F_A(X,Y)\Big|_p = \mathrm{Vert}([X_A, Y_A])\Big|_p ,$$

where $X,Y \in T_{\pi(p)}$. The above notation implies the fact that $\mathrm{Vert}([\cdot,\cdot])$ actually defines a two form, $F_A \in \Gamma(\wedge^2 T^* \otimes \hat{\mathcal{G}}) \equiv \Omega^2(\hat{\mathcal{G}})$, which is called the <u>curvature</u> of A.

When A is represented by data $\{U_\alpha, a_\alpha\}$, then

$$F_A = \{U_\alpha, (F_A)_\alpha = da_\alpha + a_\alpha \wedge a_\beta\}$$

where $\wedge$ is the exterior product on 1-forms. This is often written

$$F_A = dA + \frac{1}{2}[A,A] .$$

In $U_\alpha \cap U_\beta$, $(F_A)_\alpha = g_{\alpha\beta}(F_A)_\beta\, g_{\alpha\beta}^{-1}$. By definition, A is flat iff $F_A \equiv 0$. The following schematic might help.

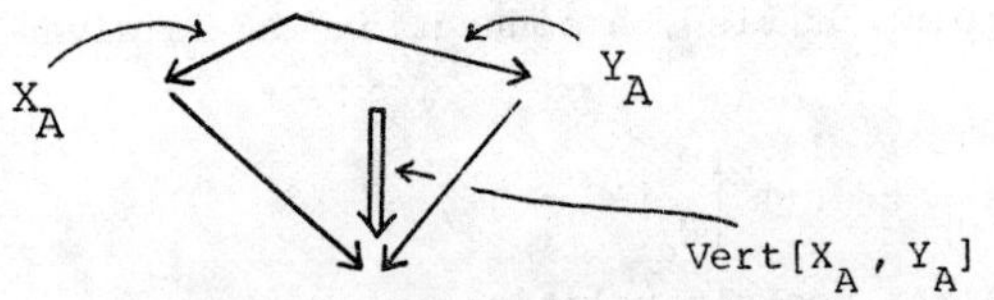

Since F_A measures how much H_A fails to be integrable, it is an interesting object for study.

As $\mathcal{G}$ acts on P, $\mathcal{G}$ acts on $\mathcal{C}$ as well. Indeed, if $g = \{U_\alpha, g_\alpha\} \in \mathcal{G}$ and $A = \{U_\alpha, a_\alpha\} \in \mathcal{C}$, then

$$gA = \{U_\alpha, g_\alpha a_\alpha g_\alpha^{-1} + g_\alpha dg_\alpha^{-1}\} ,$$

and

$$F_{gA} = \{U_\alpha, g_\alpha (F_A)_\alpha g_\alpha^{-1}\} .$$

Now suppose that $P \sim M \times G$, so P admits a flat connection, A_{flat}. Then for each $g \in \mathcal{G}$, gA_{flat} is still flat so there exists an infinite number of flat connections on P. It is silly to consider these flat connections as distinct. In fact, if M is simply connected, or if $\mathrm{Hom}(\pi_1(M),G) = (0)$, then

$$\{[A] \in \mathcal{C}/\mathcal{G} : A \text{ is flat}\}$$

contains one element. We see that when $\pi_1(M) = (0)$, there exists, up to equivalence, a unique, natural connection on the product bundle.

Suppose that $P \not\sim M \times G$, so P has no flat connections. Is there still a natural choice of orbit in the infinite dimensional space $\mathcal{C}/\mathcal{G}$? If we mimic the preceding example, we will look for connections which in some sense minimize F_A.

But in what sense? First of all, we want to minimize F_A in a $\mathcal{G}$ invariant way, that is, we want a $\mathcal{G}$ invariant norm on $\Omega^2(\hat{\mathfrak{g}}) = \Gamma(\wedge^2 T^* \otimes \hat{\mathfrak{g}})$. Notice that for each $p \in (0,\ldots,4)$, $\wedge^p T^* \otimes \hat{\mathfrak{g}}$ has a natural inner product. The Cartan form on $\mathfrak{g}$ gives a $\mathcal{G}$-invariant metric on $\hat{\mathfrak{g}}$, while the Riemannian metric on T_M defines metrics on $\wedge^p T^*$. The product metric is well defined, and $\mathcal{G}$-invariant. We see that $\Omega^2(\hat{\mathfrak{g}})$ has a natural, $\mathcal{G}$-invariant norm given by

$$\|\omega\|_2^2 \equiv \int_M d\mathrm{vol}\, |\omega|^2(x)$$

where $|\cdot|(x)$ is the aforementioned norm on $(\wedge^2 T^* \otimes \hat{\mathfrak{g}})\big|_x$. The above norm has the distinct advantage that A is flat iff $\|F_A\|_2 = 0$.

For $A \in \mathcal{C}(P)$, define the Yang-Mills functional by

$$\mathcal{YM}(A) = \frac{1}{32\pi^2} \|F_A\|_2^2 . \tag{1.1}$$

As per the previous discussion, $\mathcal{YM}(\cdot)$ is a reasonable way to measure the non-integrability of the horizontal subspaces. Note also that if P is not isomorphic to $M \times G$, then

$$\mathcal{YM}(A) > 0 \quad \text{for all} \quad A \in \mathcal{C}(P) .$$

We shall see that there is an estimate for the lower bound of $\mathcal{YM}(\cdot)$ on $\mathcal{C}(P)$ which is obtained from the Pontrjagin class of the bundle $\hat{\mathfrak{g}}$.

The Pontrjagin class, $p_1(\hat{\mathfrak{g}})$ is a characteristic class in $H^4(M;\mathbb{Z}) \simeq \mathbb{Z}$ which can be computed using any connection on P by the Chern-Weil prescription [7], [12]

$$p_1(\hat{\mathfrak{g}}) = c(\mathfrak{g}) \int_M \mathrm{tr}\,(F_A \wedge F_A) , \tag{1.2}$$

where $c(\mathfrak{g})$ is a group theoretic constant (c.f. [8]) and tr is the trace that the Cartan form on $\mathfrak{g}$ defines. We see using the triangle inequality that for every $A \in \mathcal{C}(P)$,

$$\mathcal{YM}(A) \geq (32\pi^2 c(\mathfrak{g}))^{-1} |p_1(\hat{\mathfrak{g}})| . \tag{1.3}$$

When $G = SU(2)$, $p_1(\hat{\mathfrak{g}}) = 8k$, and $k \in \mathbb{Z}$ is the degree of the map $f: M \to S^4$ such that $P \sim f^* P_1$ (c.f. Theorem 1.1.) The integer k is also minus the Chern class of the complex vector bundle $P \times_{SU(2)} \mathbb{C}^2$. Thus, when $G = SU(2)$,

$$\mathcal{YM}(A) \geq |k| .$$

Equation 1.3 is described by the following diagram:

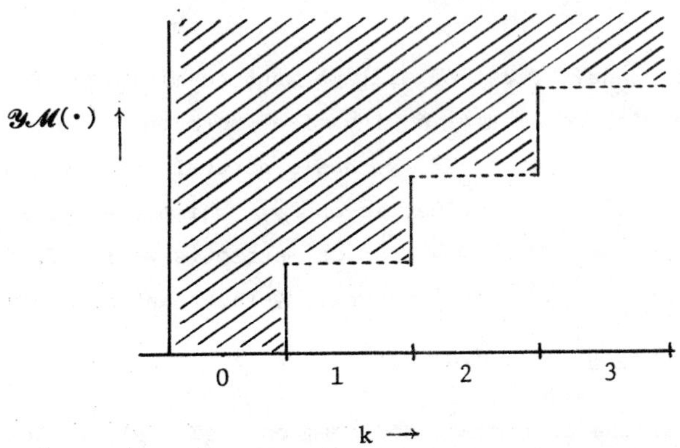

Motivated by the case of the flat connection on the trivial bundle, we hope that the set $[A] \in \mathscr{C}/\mathscr{G}$ which minimize $\mathscr{YM}(\cdot)$ forms a nice, finite dimensional space. The first question to ask is: What conditions on $A \in \mathscr{C}(P)$ are necessary and sufficient for Eq. (1.3) to be an equality? In order to answer this question, we must digress to define self-duality. Recall that for each $p \in (0,\dots,4)$ $\hat{\wedge}_p T^*$ has a Riemannian metric, m, that is induced from the given metric on T_M. Define the Hodge duality operator

$$*: \hat{\wedge}_p T^* \to \hat{\wedge}_{4-p} T^*$$

as follows:

(1) $*\mathrm{dvol} = 1$ and $*1 = \mathrm{dvol}$.

(2) For $\omega, \eta \in \hat{\wedge}_p T^*$, $*\eta$ is defined uniquely by the requirement that $\omega \wedge *\eta = m(\omega,\eta) \cdot \mathrm{dvol}$.

This is a differential form version of duality. Extend $*$ to $\hat{\wedge}_p T^* \otimes \hat{\mathscr{g}}$ by linearity.

In 4-dimensions,

$$*: \hat{\wedge}_2 T^* \otimes \hat{\mathscr{g}} \to \hat{\wedge}_2 T^* \otimes \hat{\mathscr{g}}$$

and $*^2 = 1$. If $e^1,\dots,e^4$ are a local, orthonormal basis for T^*, then

$$*(e^1 \wedge e^2 \pm e^3 \wedge e^4) = \pm(e^1 \wedge e^2 \pm e^3 \wedge e^4) ,$$

etc. Thus $*$ splits $\hat{\wedge}_2 T^* \otimes \hat{\mathscr{g}} \sim (\hat{\wedge}_2 T^* \otimes \hat{\mathscr{g}})_+ \oplus (\hat{\wedge}_2 T^* \otimes \hat{\mathscr{g}})_-$ and hence $\Omega^2(\hat{\mathscr{g}}) = \Omega^2_+(\hat{\mathscr{g}}) \oplus \Omega^2_-(\hat{\mathscr{g}})$. For example, $\omega \in \Omega^2_+(\hat{\mathscr{g}})$ iff at each $x \in M$,

$$*\omega\Big|_x = +\omega\Big|_x .$$

We observe the following equalities:

$$\mathscr{YM}(A) = \frac{1}{32\pi^2} (\|P_+F_A\|_2^2 + \|P_-F_A\|_2^2)$$

and

$$p_1(\hat{\mathscr{g}}) = c(\mathscr{g}) \cdot (\|P_+F_A\|_2^2 - \|P_-F_A\|_2^2) .$$

where $P_\pm = \frac{1}{2}(1 \pm *)$ are the projections onto the $\pm$ eigenspaces of $*$. It is evident that $\mathscr{YM}(\cdot)$ achieves the lower bound of Eq. (1.3) at $A \in \mathscr{C}(P)$ iff

$$F_A = \pm * F_A \tag{1.4}$$

with the + occurring only if $p_1(\hat{\mathscr{g}}) > 0$, and the - only if $p_1(\hat{\mathscr{g}}) < 0$. A connection whose curvature satisfies Eq. (1.4) with the +(-) sign is said to be (anti) self-dual (an "instanton" in the physics literature.) The study of self-dual connections leads us to Part II of the lecture.

II. SELF-DUALITY

I argued in the first lecture that the self-duality condition arises naturally in the study of principal bundles on Riemannian 4-manifolds. One hopes that the set of self-dual connections on a bundle $P \to M$ is somehow nice. In this section, I will describe some of the properties of this set.

Because self-dual connections pull back under bundle isomorphisms, it is sufficient to restrict attention to one representative bundle from each isomorphism class of principal G-bundles over M. Given one self-dual $A \in \mathscr{C}(P)$, one can generate an infinite number through the action of $\mathscr{G}$. For this reason, it is natural to investigate the moduli space of self-dual connections on P,

$$\mathscr{M} = \mathscr{M}(P) \equiv \{[A] \in \mathscr{C}/\mathscr{G} : F_A = *F_A\} . \tag{2.1}$$

I only discuss self-dual connections, as opposed to anti-self dual connections - the two cases are interchanged by reversing the orientation of M. The questions that arise are

(1) What are necessary and sufficient conditions on M, the metric m(,) and $P \to M$ in order that $\mathscr{M}(P) \neq \emptyset$?

(2) If $\mathscr{M} \neq \emptyset$, what is its structure?

Immediately, we observe

PROPOSITION 2.1: A necessary condition for $\mathscr{C}(P)$ to admit a self-dual connection is that $p_1(\hat{\mathscr{g}}) \geq 0$.

PROOF: Indeed, if $A \in \mathscr{C}(P)$ is self-dual, then

$$0 < \mathscr{YM}(A) = (32\pi^2 c(\mathscr{g}))^{-1}\, p_1(\hat{\mathscr{g}}) .$$

Some sufficient conditions for $\mathscr{M} \neq \emptyset$ are given by

THEOREM 2.2 ([9]): Suppose that the intersection form, $\omega: H_2(M;\mathbb{Z}) \otimes H_2(M;\mathbb{Z}) \to \mathbb{Z}$ is positive definite. Let $P \to M$ be a principal G-bundle where G is a simple and simply connected compact Lie group. If $p_1(\hat{\mathscr{g}}) \geq 0$, then $\mathscr{M} \neq \emptyset$.

Since the conference, the author has proved additionally [13]

THEOREM 3.3 (C. H. Taubes): Let ω be the intersection form of M and let $n = \frac{1}{2}(\text{rank } \omega - \text{signature } \omega)$. Let m be a generic metric on T_M. Let $P \to M$ be a principal G-bundle where G is a simple and simply connected Lie group. Let $k = 1/8\ P_1(\hat{\mathscr{G}})$. If $n \in (0,1,3)$ and if $k > n$, then $\mathscr{M}(P) \neq \emptyset$. If $n \notin (0,1,3)$ and if $k \geq n$, then $\mathscr{M}(P) \neq \emptyset$.

S. Donaldson has proved that principal G-bundles over elliptic surfaces admit self-dual connections, and G. 't Hooft has considered self-dual connections on T^4 [14]. Self-dual connections also exist on $S^2 \times \mathbb{R}^2$ [15],

The condition in Theorem 2.2 that the intersection form ω be definite appears in [9] as the condition that $P_- H^2_{\text{DeRham}}(M) = (0)$. Let me explain. The DeRham cohomology of M is the cohomology of the complex

$$0 \to C^\infty(M) \overset{d}{\to} \Omega^1(M) \overset{d}{\to} \cdots \Omega^4(M) \to 0 ,$$

where d = exterior derivative. The DeRham theorem states that $H^*_{DR} = H^*(M;\mathbb{Q})$. The signature matrix, $\omega: H^2(M;\mathbb{Z}) \oplus H^2(M;\mathbb{Z}) \to \mathbb{Z}$ is a symmetric, unimodular matrix which is defined by

$$\omega(X_1 X_2) = X_1 \cup X_2(M)$$

where $X_1, X_2 \in H^2(M;\mathbb{Z})$ are generators. In the DeRham complex, $\cup \to \wedge$: For $X_1, X_2 \in H^2_{DR}(M)$;

$$(X_1, X_2) = \int_M X_1 \wedge X_2 .$$

Furthermore, every symmetric matrix over $\mathbb{Q}$ is diagonalizable. This is accomplished for the signature matrix by choosing an L_2-orthonormal basis, $\{X_i\}$ for H^2_{DR}. Thus

$$\int_M X_i \wedge *X_j = \delta_{ij}$$

By diagonalizing ω, we observe that

$$\omega(X_i, X_j) = \varepsilon^i \delta^{ij} \quad \text{where } \varepsilon^i = +1 \text{ iff } \omega_i = *\omega_i ,$$
$$\text{and } \quad \varepsilon^i = -1 \text{ iff } \omega_i = -*\omega_i .$$

Thus $P_- H^2_{DR} = 0$ iff ω is definite.

In order to discuss the structure of the moduli space, I need to introduce the notion of a reducible bundle and connection. Let $H \subset G$ be a proper subgroup. The principal G-bundle $P \to M$ is reducible to a principal H-bundle $P' \to M$ iff there exists a G-bundle isomorphism

$$P \sim P' \times_H G .$$

Here H acts on G by left multiplication. The reducibility of a bundle P is a homotopy condition, and it is equivalent to the fibre bundle $P\times_G(G/H)$ admitting a global section. The idea is that $P=\{U_\alpha, g_{\alpha\beta}\}$ is reducible iff there exists $\varphi=\{U_\alpha, \varphi_\alpha : U_\alpha \to G\}$ such that in $U_\alpha \cap U_\beta$,

$$\varphi_\alpha g_{\alpha\beta}\varphi_\beta^{-1} = h_{\alpha\beta} : U_\alpha \cap U_\beta \to H \subset G .$$

A connection A on a G-bundle P which is reducible to an H bundle P' is a reducible connection iff there exists $A' \in \mathscr{C}(P')$ and $A=\varphi^* A'$ where $\varphi : P \to P' \times_H G$ is the reduction.

For example, an $SU(2)$ bundle $P\to M$ could be reducible to a $U(1)$ bundle. The simplest case is over S^2. Take $U_+ = S^2 - s$, $U_- = S^2 - n$ and

$$g_{+-} = \begin{pmatrix} e^{i\theta} & 0 \\ 0 & e^{-i\theta} \end{pmatrix}$$

where θ is longitude on S^2.

It is a fact that irreducible connections are dense in $\mathscr{C}(P)$. Whether or not the moduli space contains irreducible connections can be determined too:

THEOREM 2.4 [9]: Under the same conditions that Theorem 2.2 requires, there exists irreducible self-dual $SU(2)$ connections on $P\to M$ when $p_1(\hat{\mathscr{G}}) > 0$. These connections are constructable.

I remark that the self-dual connections of Theorem 2.3 are irreducible too.

A theorem similar to Theorem 2.4 holds with $SU(2)$ replaced by G, a simple and simply connected compact Lie group [9].

The local structure of $\mathscr{M}$ is described in [8],[10],[16] whose exposition I follow. There are certain preliminary facts to establish [11].

FACT: $\mathscr{G}$ and $\mathscr{C}$ can be completed as infinite dimensional, paracompact Banach manifolds such that

(1) $\mathscr{G}$ is a smooth Lie group.

(2) The action of $\mathscr{G}$ on $\mathscr{C}$ is

(a) smooth

(b) free away from $\mathscr{R}$ = {reducible connections}

(c) $\bar{\mathscr{C}} = (\mathscr{C}-\mathscr{R})/\mathscr{G}$ is a smooth, paracompact Banach manifold, and the projection: $0\to\mathscr{G}\to\mathscr{C}-\mathscr{R}\to\bar{\mathscr{C}}$ is a principal bundle.

Schematically,

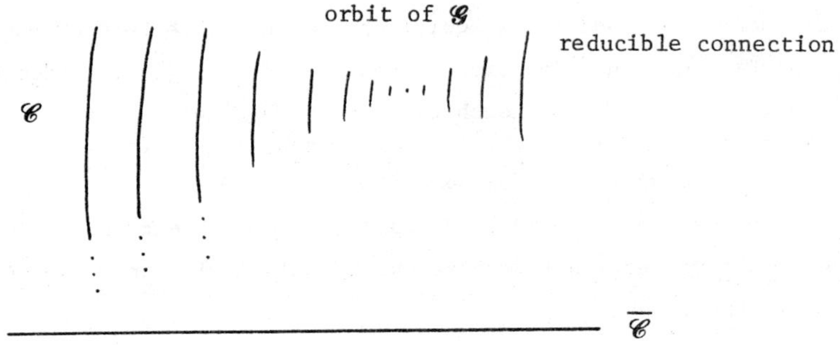

As an exercise, I compute the vertical vector fields for this principal bundle. Let $A \in \mathcal{C}$ and $g_t = \exp(tu)$ with $u \in \Omega^0(\hat{\mathcal{g}})$ and $t \in \mathbb{R}$. Thus $g \in \mathcal{G}$. Next, I represent $A = \{U_\alpha, a_\alpha\}$ and $u = \{U_\alpha, u_\alpha\}$. Then

$$g_t A = \{U_\alpha, \exp(tu_\alpha) a_\alpha \exp(-tu_\alpha) + \exp(tu_\alpha) d \exp(-tu_\alpha)\},$$

$$\frac{d}{dt} g_t A \Big|_{t=0} = \{U_\alpha, -(du_\alpha + [a_\alpha, u_\alpha])\} \equiv \{U_\alpha, (\nabla_A u)_\alpha\}.$$

Here we have defined the covariant derivative

$$\nabla_A : \Omega(\hat{\mathcal{g}}) \to \Omega^1(\hat{\mathcal{g}}).$$

Recall that $T_{\mathcal{C}}\Big|_A \simeq \Omega^1(\hat{\mathcal{g}})$. The preceding calculation tells us the vertical vector fields are

$$\tilde{\nabla}\Big|_{\mathcal{C}\,A} \simeq \nabla_A(\Omega(\hat{\mathcal{g}})) \subset \Omega^1(\hat{\mathcal{g}}).$$

EXERCISE. $A \in \mathcal{C}$ is a reducible connection if $\exists u \neq 0$ in $\Omega^0(\hat{\mathcal{g}})$ such that $\nabla_A u = 0$. Alternatively, iff the stabilizer of A in $\mathcal{G}$ is a subgroup of dimension > 0.

For $G = SU(2)$, the space $\mathcal{C}/\mathcal{G}$ is singular at reducible connections; as the vertical vector space "drops dimension" at $A \in \mathcal{R}$.

THEOREM 2.5 ([9]): In a neighborhood of the connections constructed in Theorem 2.4, $\mathcal{M} \cap \overline{\mathcal{C}}$ is a C^∞ Hausdorff manifold with the induced C^∞ structure from $\mathcal{C}$. When $G = SU(2)$,

$$\begin{aligned} \dim \mathcal{M} &= 8k - 3/2(\chi(M) - \tau(M)), \\ &= 8k - 3(1 - h^1 + h^2_-). \end{aligned}$$

Here $\chi(M)$ = Euler characteristic of M; $\tau(M)$ = signature of M; $h^1 = \dim H^1_{DR}$ and $h^2_- = \dim P_- H^2_{DR}$; $k = \frac{1}{8} p_1(\hat{\mathcal{g}})$.

For the remainder of my lecture, I set $k = 1$, $h^2_- = 0$ and I assume that $\pi_1(M) = 0$ so $h^1 = 0$. Under these circumstances, where $\mathcal{M}$ is smooth, dim $\mathcal{M} \cap \overline{\mathcal{C}} = 5$. The preceding theorem states that there exists an open set $\mathcal{K} \subseteq \mathcal{M}$ which is a smooth 5-manifold:

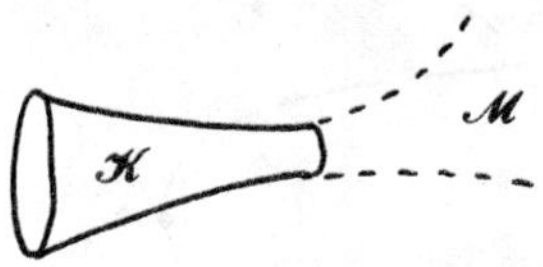

What else do we know about $\mathcal{K}$?

THEOREM 2.6: ([16],[10]) There is a diffeomorphism

$$\Phi: M \times (0,1) \simeq \mathcal{K}$$

THEOREM 2.7: ([16],[10]) Every sequence $\{[A_i]\} \in \mathcal{M}$ either

(1) Has a limit point in $\mathcal{M} \cap \mathcal{C}/\mathcal{G}$, or

(2) $[A_i] \in \mathcal{K}$ for all i sufficiently large.

Thus, $\mathcal{M}$ has a natural, collared boundary, $\partial\mathcal{M} \simeq M$.

The proofs of these theorems are rather detailed. The map Φ of Theorem 2.6 is actually constructed during the existence proof of Theorem 2.2. Theorem 2.5 is proved using information acquired in proving Theorem 2.2 and the Atiyah-Singer Index theorem. The arguments to prove Theorems 2.4, 2.5 are slight generalizations of arguments in [8]. The fact that $\partial\mathcal{M} \simeq M$ requires two crucial theorems of K. Uhlenbeck [17],[18].

I will outline the formal strategy for the proof of Theorem 2.2. Consider the map $T:\mathcal{C} \to \Omega^2_-(\hat{\mathcal{g}})$ defined by

$$T(A) = P_-F_A .$$

The space $\mathcal{C}$ is a smooth Banach manifold and the tangent space to $\mathcal{C}$ at A, $T_{\mathcal{C}}\big|_A \simeq \Omega^1(\hat{\mathcal{g}})$. (I must complete $\mathcal{C}, \Omega^1(\hat{\mathcal{g}})$ so they have compatible Banach space structures, but allow me to skip these technicalities.) The differential $T_*\big|_A : \Omega^1(\hat{\mathcal{g}}) \to \Omega^2_-(\hat{\mathcal{g}})$ of the map T at $A \in \mathcal{C}$, is the first order differential operator $T_*\big|_A v = \mathcal{D}_{A_1} v = \{U_\alpha, P_-(dv_\alpha + a_\alpha \wedge v_\alpha + v_\alpha \wedge a_\alpha)\}$, where $A = (U_\alpha, a_\alpha)$ and $v = (U_\alpha, v_\alpha) \in \Omega^1(\hat{\mathcal{g}})$. If $T_*\big|_A$ is a surjection, then I can use the implicit function theorem to conclude that there exists a neighborhood $\mathcal{Y} \subseteq \Omega^2_-(\hat{\mathcal{g}})$, of P_-F_A, in the image of T:

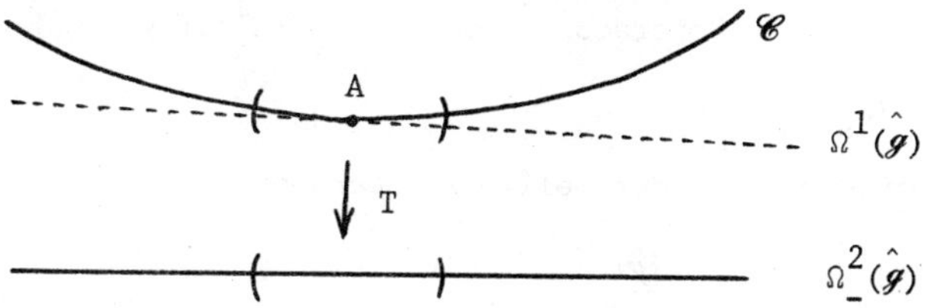

If $T_*\big|_A$ is not surjective, then I have to work much harder to conclude anything:

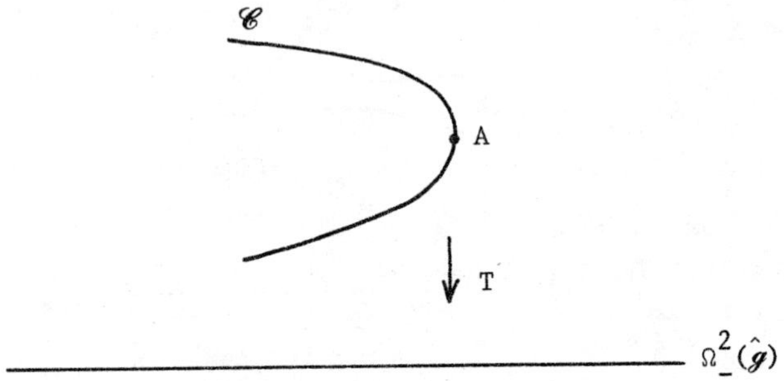

When $T_*\big|_A$ is surjective, there will be some radius $\rho(A)$ such that for all

$$\beta \in \{\beta \in \Omega^2_-(\hat{\mathscr{g}}) : |||P_-F_A - \beta||| < \rho(A)\} ,$$

there exists $A' \in \mathscr{C}$ with $T(A') = \beta$ and $|||A - A'|||$ small. The number $\rho(A)$ determines the size of the neighborhood of A in $\mathscr{C}$ that the tangent plane to $\mathscr{C}$ at A approximates. The following picture gives the idea:

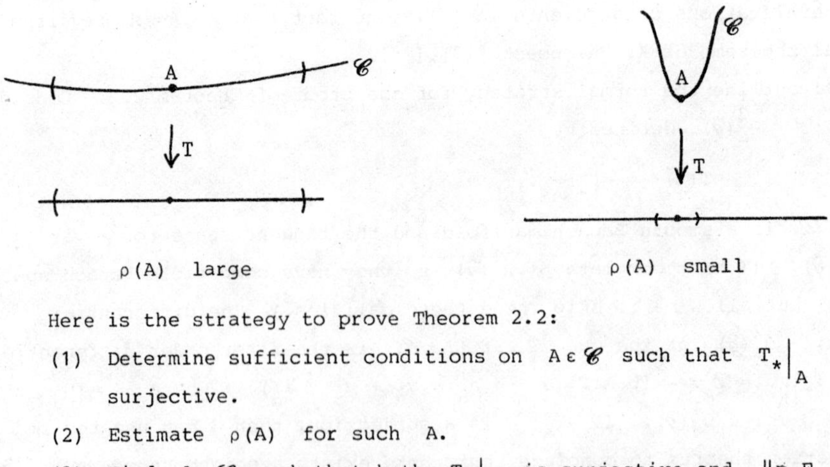

$\rho(A)$ large $\rho(A)$ small

Here is the strategy to prove Theorem 2.2:

(1) Determine sufficient conditions on $A \in \mathscr{C}$ such that $T_*\big|_A$ is surjective.

(2) Estimate $\rho(A)$ for such A.

(3) <u>Find</u> $A \in \mathscr{C}$ such that both $T_*\big|_A$ is surjective <u>and</u> $\|P_-F_A\| < \rho(A)$. Then the implicit function theorem implies that $T^{-1}(0) \neq \emptyset$.

Parts (1) and (2) require a reasonable amount of analysis. To give you the results, I define two new operators: Define $D_A : \Omega^2_-(\hat{\mathscr{g}}) \to \Omega^3(\hat{\mathscr{g}})$ by

$$D_A\beta = \{U_\alpha , (D_A\beta)_\alpha = d\beta_\alpha + a_\alpha \wedge \beta_\alpha - \beta_\alpha \wedge a_\alpha\} .$$

I can now form the second order, elliptic operator

$$T_*T_*^+ = \mathscr{D}_A^* D_A : \Omega^2_-(\hat{\mathscr{g}}) \to \Omega^2_-(\hat{\mathscr{g}}) .$$

The answer to (1) is

THEOREM 2.7: [9] The map $T_*:\Omega^1(\hat{\mathscr{g}}) \to \Omega^2_-(\hat{\mathscr{g}})$ is surjective at A whenever the lowest eigenvalue, $\mu(A)$ of $T_*T_*^+$ is strictly positive.

The answer to (2) is

THEOREM 2.8: [9] For $A \in \mathscr{C}$, define

$$\delta(A) = \|P_-F_A\|_2 + \mu(A)^{-\frac{1}{2}} \|P_-F_A\|_{4/3}(1+\|F_A\|_4) .$$

There exists constants $\varkappa < \infty$, $\varepsilon > 0$ depending only on (M,m) such that when $\delta(A_0) < \varepsilon$, there exists a self-dual connection $A \in \mathscr{C}$ with

$$\|A - A_0\| < \varkappa\delta(A_0) .$$

In fact, this last theorem gives somewhat more, if you study the proof carefully you observe that

COROLLARY 2.9: Let $\varepsilon, \varkappa$ be the constants of Theorem 2.8. Define $\mathscr{C}_\varepsilon = \{A \in \mathscr{C} : \delta(A) < \varepsilon\}$. For each $A_0 \in \mathscr{C}_\varepsilon$, there exists a self-dual connection, $\hat{A}(A_0) \in \mathscr{C}$ with

(1) $\|\hat{A}(A_0) - A_0\| < \varkappa\delta(A_0)$

(2) The assignment $A_0 \to \hat{A}(A_0)$ is a C^∞ map from $\mathscr{C}_\varepsilon$ to $\mathscr{C}_\varepsilon$.

(3) The map $\hat{A}(A_0)$ is $\mathscr{G}$-equivariant.

(4) If A_0 is sufficiently far from being reducible, then $\hat{A}(A_0)$ is irreducible.

The Corollary paints the following picture:

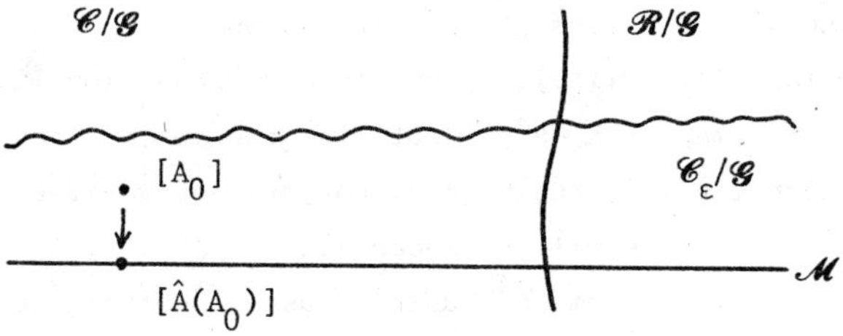

In a very standard way, the implicit function theorem for T implies that if $\mathscr{C}_\varepsilon \neq \emptyset$, then $\mathscr{M} \neq \emptyset$ and that $\mathscr{C}_\varepsilon/\mathscr{G}$ is a tubular neighborhood of $\mathscr{M} \subset \mathscr{C}/\mathscr{G}$.

In the proof that $\mathscr{C}_\varepsilon \neq \emptyset$, the map $\Phi : M \times (0,1) \to \mathscr{M}$ is constructed. Let me outline this construction. Recall that P is isomorphic to the pull back

$$\begin{array}{ccc} f^{-1}P_1 & \to & P_1 \\ \downarrow & & \downarrow \\ M & \overset{f}{\to} & S^4 \end{array}$$

by a degree 1 map, f, of the bundle $P_1 \to S^4$. Since connections pull back also, we can do the following. Find a self-dual connection W on $P_1 \to S^4$ and attempt to find $f : M \to S^4$ such that f^*W is close to self-dual. This requires us to first understand self-dual connections on $P_1 \to S^4$.

THEOREM 2.10: [8] The moduli space $\mathcal{M}$ for $P_1 \to S^4$ is a smooth manifold, diffeomorphic to the 5-ball B^5 (Note that $S^4 = \partial B^5$.) Let x^1, x^2, x^3, x^4 denote stereographic coordinates on $U_+ = S^4 - s$ (Cartesian coordinates on $\mathbb{R}^4$.) Look at $\mathbb{R}^4$ as the vector space of quaternions, $\mathbb{H}$. Let $\lambda \in (0,\infty)$. Then up to isomorphism, and rotations of S^4, the self-dual connections on $P_1 \to S^4$ are given by $W = (U_+, W_\pm(\lambda))$, where

$$W_+ = \frac{|x|^2}{\lambda^2+|x|^2} \frac{\bar{x}}{|x|} d\left(\frac{x}{|x|}\right) , \qquad x \in U_+ ;$$

$$W_- = \frac{\lambda^2}{\lambda^2+|x|^2} \frac{x}{|x|} d\left(\frac{\bar{x}}{|x|}\right) , \qquad x \in U_- .$$

The facts to notice are that $|F_{W(\lambda)}|(x) \sim \lambda^2(|x|^2+\lambda^2)^{-2}$, and it has the following extreme behavior:

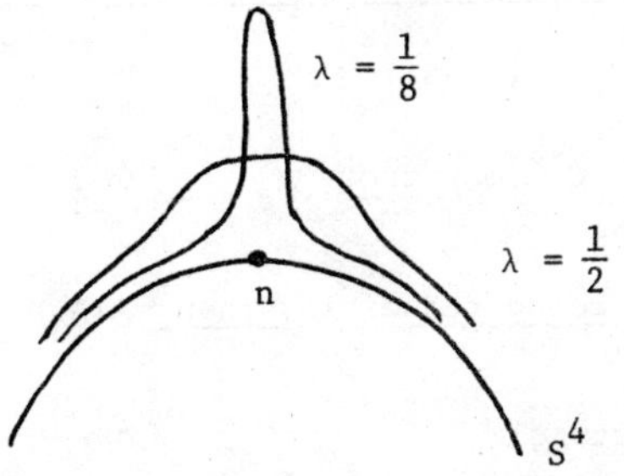

: Behavior of $|F_{W(\lambda)}|$ as $\lambda \to 0$.

Note also that SO(4) rotations of S^4 (rotations which fix n and s) leave the isomorphism class $[(P,W(\lambda))]$ of bundle and connection invariant.

To construct the map $f: M \to S^4$, I utilize the observation that self-duality is a condition which involves the metric, not its derivatives. This suggests that the following procedure will be successful:

(1) Pull-back $W(\lambda)$ from S^4 using Gaussman Normal Coordinates, in the neighborhood of a point $p \in M$.

(2) Choose λ very small, so that most of the pulled-back curvature is situated in a small neighborhood about p where the metric is close (in the C^0 norm) to the flat metric.

Schematically the strategy is

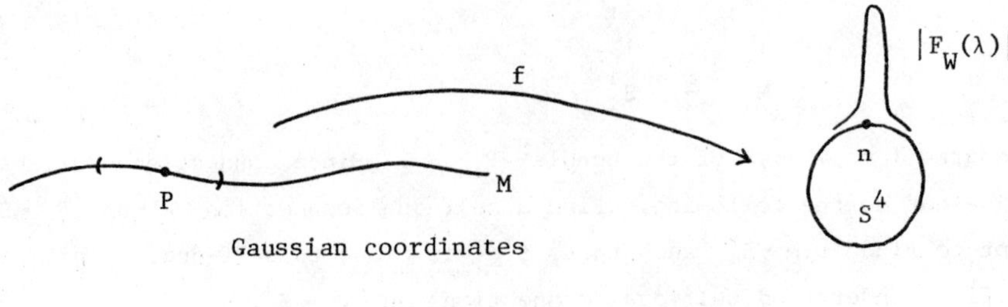

It is not surprising that the diffeomorphism, $\Phi : M \times (0,1) \to \mathcal{M}$ is given by specifying the point $p \in M$ for the Gaussian coordinate system, and a scale size λ for $W(\lambda)$.

Let me remind you about Gaussian Normal Coordinates [7]. Given a point $p \in M$, and an ortho-normal frame, $e \in F_M\big|_p$ (the frame bundle at p), there exists a coordinate chart,

$$\varphi_{(p,e)} : B_\rho(p) \to B_\rho(n) \subset \mathbb{R}^4 \simeq S^4/s$$

where $B_\rho(p)\,(B_\rho(n))$ is the geodesic ball of radius ρ about p $(n \in \mathbb{R}^4)$. The radius ρ depends on M and the metric m. The coordinates chart $\varphi_{(p,e)}$ is uniquely specified by requiring that

(1) $\varphi(p) = 0 \in \mathbb{R}^4$

(2) $\varphi_*\big|_p e = \{\frac{\partial}{\partial x^1}, \frac{\partial}{\partial x^2}, \frac{\partial}{\partial x^3}, \frac{\partial}{\partial x^4}\}$

(3) $m(\varphi^*(dx^i), \varphi^*(dx^j)) = \delta^{ij} + \mathcal{O}(|x|^2)$ for $i,j = (1,\ldots,4)$

(4) $dm(\varphi^*(dx^i), \varphi^*(dx^j)) = \mathcal{O}(|x|)$ for $i,j = (1,\ldots,4)$

(5) $\varphi_*^{-1}(\frac{x^i}{|x|}\frac{\partial}{\partial x^i})$ is a unit speed, geodesic vector field.

Schematically, we have

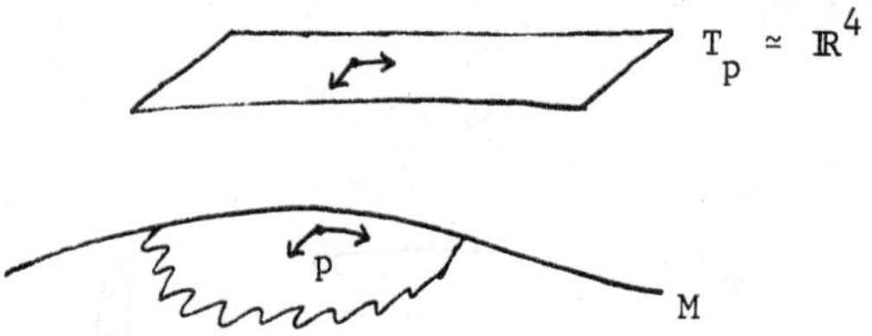

To avoid singularities, let $\beta(x)$ be the bump function

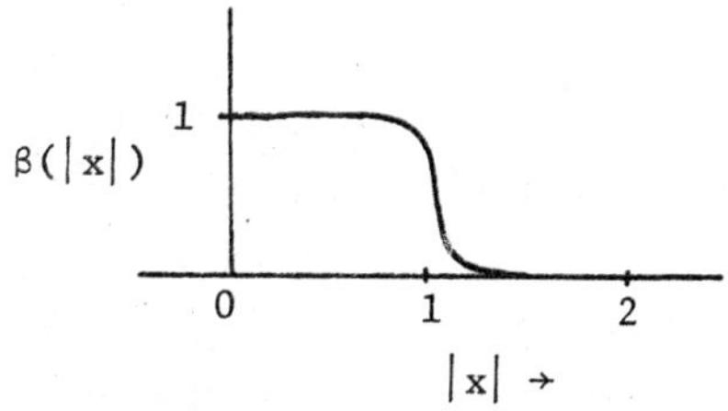

Define, for $\lambda_0 \ll \rho$ a map $\Phi' : F_M \times (0,\lambda_0) \to \mathcal{C} - \mathcal{R}$ by

$$\Phi'(p,e,\lambda) = \begin{cases} \varphi^*_{(p,e)} W_+(\lambda) & \text{in } \varphi^{-1}(B_{\sqrt{\lambda}}(0)), \\ \varphi^*_{(p,e)} B(x/\sqrt{\lambda})W_-(\lambda) & \text{in } M - \varphi^{-1}(B_{1/2\sqrt{\lambda}}(0)). \end{cases}$$

It is rather easy to check that $\Phi'(p,e,\lambda)$ is a connection on a bundle $P_{(p,e)}$ which has Pontrjagin index $k = 1$.

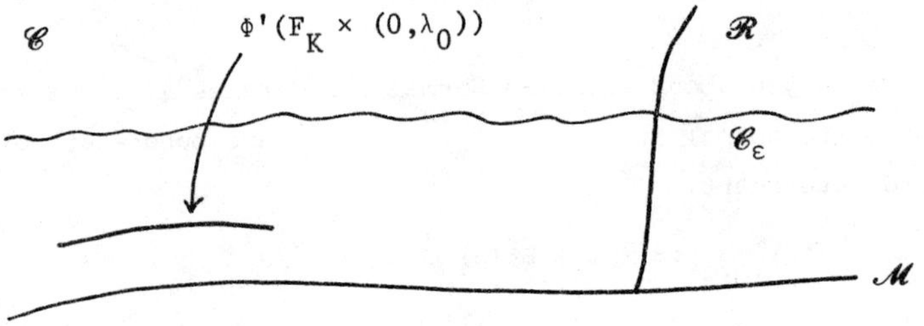

Now, when one project Φ' into $\mathscr{C}/\mathscr{G}$ one observes that because $W(\lambda)$ is SO(4) equivariant on S^4, one obtains the following commutative diagram:

$$\begin{array}{ccc} F_M \times (0,\lambda_0) & \xrightarrow{\Phi'} & \mathscr{C}-\mathscr{R} \\ \downarrow SO(4) & & \downarrow \mathscr{G} \\ M \times (0,\lambda_0) & \xrightarrow{\Phi''} & \overline{\mathscr{C}} \end{array} \tag{2.2}$$

The above diagram defines the smooth map Φ'':

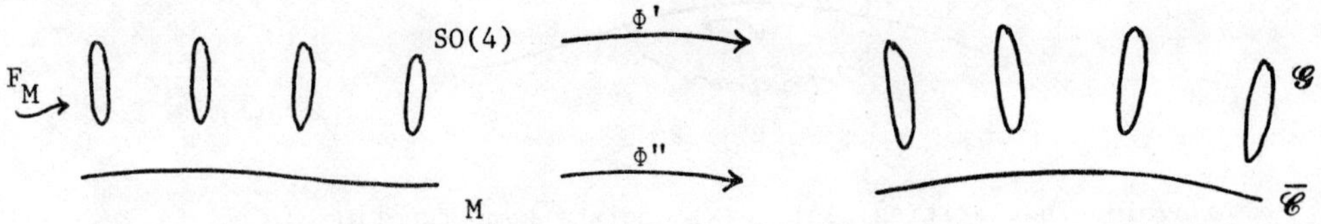

One can establish the following facts:

FACTS:

(1) $\operatorname{Im}\Phi' \cap \mathscr{R} = \emptyset$

(2) Φ' is an embedding

(3) If M has definite intersection form, then for λ_0 sufficiently small, $\operatorname{Im}\Phi' \subset \mathscr{C}_\varepsilon$.

From the above facts, one concludes that $\mathscr{M} \neq \emptyset$. Because of Fact (3) above, $\operatorname{Im}\Phi' \subset$ domain of $\hat{A}:\mathscr{C}_\varepsilon \to \mathscr{M}$. Define the map

$$\Phi: M \times (0,\lambda_0) \to \mathscr{M}$$

by $\Phi(p,\lambda) = [\hat{A}(\Phi'(p,e,\lambda))]$. Because of the commutative diagram in Eq. (2.2) and the fact that $\hat{A}$ is $\mathcal{G}$-equivariant, the map Φ makes good sense:

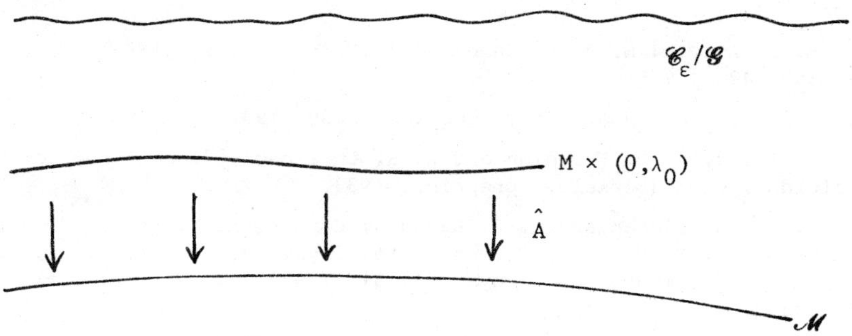

The map Φ, as the composition of smooth maps, is itself smooth. The map Φ'' is an embedding; in fact, the quantity $|F_{\Phi''(p,\lambda)}|$ has its supremum on M at p with value λ^{-2} which demonstrates that Φ'' is 1-1. The map Φ is an embedding also. The proof uses the fact that Φ'' is an embedding, and Statement (1) of Corollary 2.9.

Therefore, Φ establishes a diffeomorphism between an open subset $\mathcal{K} \subset \mathcal{M}$ and $M \times (0,\lambda_0)$.

At the end, one must prove that every sequence $\{[A_i]\} \subset \mathcal{M}$ with no convenient subsequences enters $\mathcal{K}$. The proof of this fact uses the removable singularities theorem and the weak compactness theorem of K. Uhlenbeck [17], [18]. An application of K. Uhlenbeck's theorems in a similar context can be found in a short paper by S. Sedlacek [19]. I understand that S. Donaldson gives some details in his lectures.

Thank you.

BIBLIOGRAPHY

[1] A. A. Belavin, A. M. Polyakov, A. Schwartz and Y. Tyupkin, Phys. Lett. 59B, 85 (1975); E. Witten, Phys. Rev. Lett. 38, 121 (1977); G. 't Hooft, Nucl. Phys. B79, 276 (1976), and unpublished. See also C. N. Yang and R. L. Mills, Phys. Rev. 96, 191 (1954).

[2] R. Ward, Phys. Lett. 61A, 81-82 (1977); M. F. Atiyah and R. Ward, Commun. Math. Phys. 55, 117 (1977); M. F. Atiyah, V. A. Drinfeld, N. J. Hitchin and Yu. I. Munin, Phys. Lett. 65A, 185 (1978); N. Christ, E. Weinberg and N. Stanton, Phys. Rev. D18, 2013 (1978).

[3] J. P. Bourgurgnon, H. B. Lawson and J. Simons, Proc. Nat. Acad. Sci. USA 76, 1550 (1979); J. P. Bourguignon and H. B. Lawson, Commun. Math. Phys. 79, 189 (1980); I. M. Singer, Commun. Math. Phys. 60, 7 (1978); M. F. Atiyah and J. D. S. Jones, Commun. Math. Phys. 61, 97 (1978).

[4] N. Steenrod, Topology of Fibre Bundles, Princeton (1951).

[5] D. Husemöller, Fibre Bundles, McGraw-Hill, (1966).

[6] M. F. Atiyah and R. Bott, "The Yang-Mills Equations over Riemann Surfaces", to appear in Phil. Trans. Roy. Soc. London.

[7] S. Kobayashi and K. Nomizu, Foundations of Differential Geometry, Interscience (1963).

[8] M. F. Atiyah, N. J. Hitchin and I. M. Singer, Proc. Roy. Soc. (London) A362, 425 (1978).

[9] C. H. Taubes, Jour. Diff. Geom. 17, 139 (1982).

[10] D. Freed, M. H. Freedman and K. K. Uhlenbeck, "Gauge Theories and Four Manifolds", MSRI (Berkeley) preprint, 1983.

[11] M. S. Narasimhan and T. R. Ramadas, Commun. Math. Phys. 67, 21 (1979); P. K. Mitter and C. M. Viallet, Commun. Math. Phys. 79, 457 (1981); O. Babelon and C. M. Viallet, Commun. Math. Phys. 81, 515 (1981); T. Parker, Commun. Math. Phys. 85, 563 (1982).

[12] J. Milnor and J. Stasheff, Characteristic Classes, Princeton University Press (1974).

[13] C. H. Taubes, "Self-dual Connections on 4-manifolds with Indefinite Intersection Matrix" preprint, 1983.

[14] G 't Hooft, Commun. Math. Phys. 81, 267 (1981).

[15] A. Jaffe and C. H. Taubes, Vortices and Monopoles, Birkhauser (1980).

[16] S. K. Donaldson, "An Application of Gauge Theory to the Topology of 4-manifolds" Jour. Diff. Geom., to appear.

[17] K. K. Uhlenbeck, Commun. Math. Phys. 83, 11 (1982).

[18] ————————, Commun. Math. Phys. 83, 31 (1982).

[19] S. Sedlacek, "A Direct Method for Minimizing the Yang-Mills functional over 4-manifolds". Commun. Math. Phys., to appear.

DEPARTMENT OF MATHEMATICS
HARVARD UNIVERSITY
CAMBRIDGE, MA 02138

Contemporary Mathematics
Volume 35, 1984

4-MANIFOLD PROBLEMS

edited by Rob Kirby

We begin by updating the 1976 list of 4-dimensional problems in [R. Kirby, Proc. Sym. Pure Math., 32(1978), 273-312], as well as a few slice knot theory problems. This update starts with a brief review of the theorems of Freedman, Donaldson and Quinn which have affected the subject so greatly. Next follows a list of new problems, N1.52 - N1.57 on knot theory and N4.41-N4.68 on 4-manifolds, which is based on problem sessions in July 1982 in Durham N. H. and in August 1983 in Santa Barbara, California

Much of the progress on the old problems depends on the work of Freedman, Donaldson and Quinn. Much of their work has appeared remarkably quickly in journals: [Freedman, J. Diff. Geom., 17(1982), 357-453], [Donaldson, Bull. AMS 8(1983), 81-83] and [Quinn, J. Diff. Geom. 17(1982), 503-521] (Quinn's handwritten manuscript was ready 10 days after the proof of the annulus conjecture on 7 July, 1983 — is that a record for such a paper?). We summarize some of the more useful (to us) theorems here:

[Freedman, op.cit.] THEOREM. To each even (odd), symmetric, unimodular, integral bilinear form, there corresponds exactly one (two), closed, simply connected, 4-manifold. The two 4-manifolds in the odd case are distinguished by their Kirby-Siebenmann invariants.

THEOREM. A proper 5-dimensional h-cobordism is a topological product.

THEOREM. "Surgery works" in dimension 4 for simply connected topological manifolds.

[Donaldson, op.cit.] Let M^4 be a smooth, closed, simply connected 4-manifold with positive definite intersection form of rank n. Then the form is isomorphic to $\overset{n}{\oplus} \langle 1 \rangle$. Using [Freedman], M^4 is homeomorphic to $\overset{n}{\#} CP^2$.

[Quinn, op.cit.] A thin h-cobordism theorem gives the following applications:

(1) Zero and 1-handles can be smoothed (this includes the annulus conjecture), and any 4-manifold is smooth off a point,

(2) map transversality holds in the missing 4-dimensional cases,

(3) any 5-manifold is a handlebody.

0271-4132/84 $1.00 + $.25 per page

A remarkable corollary of [Freedman] and [Donaldson] is the existence of exotic structures on R^4 [R. Gompf, J. Diff. Geom. 18(1983), 317-328], [notes by D. Freed of lectures by K. Uhlenbeck, et al., Berkeley MSRI, 1983], [R. Kirby, An exotic structure on R^4 and the work of Freedman and Donaldson, 1983].

THEOREM. R^4 has an exotic differential structure Θ (in fact, at least two [Gompf, op.cit.]) satisfying

(1) R^4_Θ does not smoothly imbed in any smooth structure on S^4, but it does smoothly imbed in $S^2 \times S^2$.

(2) R^4_Θ contains a compact set K such that K is not in the bounded complement of any smoothly imbedded 3-sphere in R^4_Θ, i.e. K is not surrounded by a smooth S^3.

(3) There does not exist a homeomorphism $h: R^4 \longrightarrow R^4_\Theta$ for which h or h^{-1} is C^∞.

Since the above works, progress has been made by Freedman in the non-simply connected case. As of June 1984, he can prove the topological s-cobordism theorem and do non-simply connected surgery for fundamental groups in the collection G; G contains all finite groups, the integers Z, and is closed under the operations of forming short exact sequences (if $A, B \in G$ and $0 \to A \to C \to B \to 0$ then $C \in G$), taking quotients, subgroups, and ascending unions. Thus G contains all virtually solvable groups (groups with a solvable group of finite index). Also, a group belongs to G if all its finitely generated subgroups do.

In the remainder of the text, a reference to Freedman, Donaldson, or Quinn will refer to the above work, and all other references will mention a journal or preprint title.

OLD KNOT THEORY PROBLEMS

Problem 1.25. A proof that $v_{io} \equiv 2(4)$ for branching curves in the p-fold irregular cover of a knot is given in these proceedings by Cappell and Shaneson. A framed link description of the irregular branched cover of the 4-ball along a complex (extending the above cover of S^3) is given in D. Schorow's thesis, University of California, Berkeley, 1983.

Problem 1.30. The classical PL (=DIFF) and TOP knot concordance groups are different since (see Problem 1.36 update) there exist Alexander polynomial one knots which are not smoothly slice, but are topologically slice.

Problem 1.36. Not all Alexander polynomial one knots are slice in the smooth case. Casson gives an example as follows: choose an Arf invariant zero knot K for which ± 1 surgery gives a homology 3-sphere bounding a smooth definite 4-manifold; thus the homology 3-sphere cannot bound a smooth contractible 4-manifold so K cannot be smoothly ± 1 shake slice (see Problem 1.41). But Casson shows that K is ± 1 shake concordant to an Alexander polynomial one knot (see Problem 1.46A) which then cannot be smoothly slice.

However, in the topological category, Freedman has proven that all such knots are slice (by a topological 2-ball in B^4 having a trivial normal bundle).

Problem 1.37. First note that the figure is drawn incorrectly — the top three right half twists should be left half twists. The knots are topologically slice as remarked above in the update of Problem 1.36.

Problem 1.38. The untwisted double of a knot K is always topologically slice (Freedman) but it is still possible that it is smoothly slice iff K is.

Problem 1.39. The Whitehead link W_1 =  is not topologically slice. If the link W_2 is obtained by doubling one strand (in an untwisted fashion), then W_2 is not topologically slice for it has non-zero signature. Doubling one strand again gives W_3 which is unknown to be slice. W_4 is topologically slice (Freedman) by ad hoc methods. W_k, $k \geq 5$, is topologically slice as are ramified versions of W_k (i.e. components of W_k may be repeated).

Problem 1.40 (B). The statement attributed to L. Rudolph is incorrect. He showed [Rudolph, Topology 22(1983), 191-202] that the "links" of algebraic functions without poles are precisely the quasipositive closed braids (a composition of conjugates of positive braids). So the natural question to ask now is: which knots or links are quasipositive closed braids? All of them???

Problem 1.42. It is easy to represent $(m,n) \in H_2(S^2 \times S^2; z)$ by a smoothly imbedded S^2 if $|m| \leq 1$ or $|n| \leq 1$. K. Kuga [Not. AMS 4(1983), 401; 83T-57-347] has shown that these are the only cases when (m,n) is so represented. For if Σ^2 represents (m,n) when $|m|$ or $|n| > 1$, then $\Sigma^2 \#^{2mn} CP^1$ has zero self-intersection in $S^2 \times S^2 \#^{2mn} (-CP^2)$, and surgery on it gives a smooth 4-manifold with non-trivial definite intersection form, contradicting Donaldson.

In the topological case we can represent (m,n) if m and n are relatively prime.

Problem 1.43. This is true for torus knots and others [A. J. Casson and C. McA. Gordon, Inv. Math. 74(1983), 119-137].

Problem 1.46. Let K be a knot in S^3 with Arf invariant zero. Then there is a $(-1,-1)$ twist changing K to an algebraically slice knot [S. J. Kaplan, Pac. J. Math., 102(1982), 55-60]. K is also concordant to a knot which can be $(-1,-1)$-twisted to an Alexander polynomial one knot (A. J. Casson, 1978), which is then topologically slice (Freedman). K cannot always be $(-1,-1)$-twisted to a ribbon knot, for $+1$ surgery on the $(2,7)$-torus knot, N^3, bounds a definite, index 16, 4-manifold V; if some $(-1,-1)$ twist of this knot was ribbon, then N^3 would bound an acyclic 4-manifold W with $\pi_1(N) \twoheadrightarrow \pi_1(W)$ so that $V \cup W$ would contradict Donaldson's Theorem.

OLD 4-MANIFOLD PROBLEMS

Problem 4.1. All such forms are realized by TOP 4-manifolds (in the case of an odd form, by two 4-manifolds), (Freedman). In the smooth case, definite forms other than $\pm \oplus \langle 1 \rangle$ are ruled out by Donaldson. Perhaps the next most interesting case is to find a smooth, simply connected, closed 4-manifold with $\chi/\sigma < 3/2$ where χ = Euler class and σ = index.

Problem 4.2. All homology 3-spheres bound contractible TOP 4-manifolds (Freedman). A homology 3-sphere does not bound a smooth contractible 4-manifold if it also bounds a smooth 4-manifold with definite intersection form other than $\pm \oplus \langle 1 \rangle$ (Donaldson).

Note that the figure should have four left half twists, not right.

Casson and Harer [Pac. J. Math. 96(1981), 23-36] show that $\Sigma(p,ps-1,ps+1)$ for p even, s odd and $\Sigma(p,ps\pm1,ps\pm2)$ for p odd, s arbitrary, bound contractible manifolds. Stern [Not. AMS, 25(1978), p. A448] shows that the following classes bound contractible manifolds: $\Sigma(2,2s\pm1, 4(2s\pm1) + 2s\mp1)$ for s odd, and for any s, $\Sigma(3,3s\pm1, 6(3s\pm1) + 3s\pm2)$ and $\Sigma(3,3s\pm2, 6(3s\pm2) + 3s\pm1)$.

Problem 4.3. Those homology 3-spheres which bound smooth contractible 4-manifolds do not bound smooth definite 4-manifolds (Donaldson). Note that all homology 3-spheres bound TOP contractible, hence even, definite 4-manifolds.

Problem 4.6. Surgery works in dimension 4 in the topological case (i.e. the answer to the (A) part of this problem is yes) when $\pi_1 = 0$ (by Freedman's published work) and for the class of fundamental groups described in the introduction under Freedman's later work.

Problem 4.7. There is such an index 8 4-manifold, e.g. the Poincare homology 3-sphere bounds both a TOP contractible 4-manifold and plumbing according to the E_8 diagram. Incidentally, TOP map transversality now holds in all cases; the heretofore missing case when the expected preimage is dimension 4 is due to Freedman, and the case when the domain is dimension 4 is due to Quinn. Still unknown however is whether two submanifolds of a 4-manifold can be made transverse if the expected intersection would have dimension ≥ 1.

Problem 4.8. Yes, there exist exotic smooth structures on $S^3 \times R$. The first example [Freedman, Ann. Math., 110(1979), 177-201] is fake because it has a "transverse" smooth imbedding of the Poincare homology 3-sphere, but not S^3. There is a growing list of other examples (check with R. Gompf) which have in various combinations: ends like the fake R^4's and/or imbedded homology 3-spheres with Arf invariants 0 or 1.

Problem 4.11. Freedman gives a complete answer to the homeomorphism question (yes, if their triangulation obstructions are equal). The smooth case is still wide open; as yet there are no known exotic smooth structures on simply connected, compact 4-manifolds. Candidates for exotic smooth manifolds abound, e.g. the Gluck construction on a knotted S^2 in S^4 (see Problem 4.24), the boundary (homotopy 4-sphere) of the 5-manifold built according to a "nontrivial" presentation of the trivial group (see Problem 5.2), logarithmic transforms of elliptic surfaces (see [J. Harer, A. Kas and R. Kirby, Handlebody structures for complex surfaces, Memoirs A.M.S., 1984]).

Problem 4.12. From Freedman's classification, we have that M_1 is homeomorphic to M_2 iff they have the same triangulation obstruction, and this answer is not affected by connected summing with copies of $\pm CP^2$.

Problem 4.13. Freedman shows that any homotopy RP^4 is homeomorphic to RP^4.

Akbulut and Kirby [Topology, 18(1979), 1-15] only prove that the double cover of one of the Cappell-Shaneson fake RP^4's is homeomorphic to S^4 (the error was found by and is explained in [I. Aitchison and J. H. Rubenstein, these Proceedings]); they also show that the double cover is the Gluck construction on a certain knotted 2-sphere, is homeomorphic to S^4, and has a particularly simple handle decomposition [A-K, A potential smooth counterexample in dimension 4 to the Poincare conjecture, the Schoenflies conjecture and the Andrews-Curtis conjecture, Topology, 1984].

R. Fintushel and R. Stern [Ann. Math. 113(1981), 357-366] give a different description of an exotic smooth involution on S^4.

Problem 4.14. The answer to part (A) is yes; this is essentially a surgery problem for $\pi_1 = Z/2$ and this case falls into the collection of groups for which Freedman's methods work (see Introduction).

For part (B), Akbulut shows (these Proceedings) that the homotopy equivalence is exotic for the $(T^3 - B^3)$-bundle over S^1 of Cappell and Shaneson (he shows the manifold is diffeomorphic to $S^2 \times RP^2$).

Problem 4.16. Every diffeomorphism (or homeomorphism) of the boundary extends to a homeomorphism of the contractible manifold (use Freedman's h-cobordism theorem); whether or not it extends to a diffeomorphism is wide open.

Problem 4.18. Trace has two relevant papers concerning 3-handles in simply connected, closed 4-manifolds: [B. Trace, Proc. AMS 79(1980), 155-156] and [B. Trace, Pac. J. Math. 99(1982), 175-181]. An interesting example to study is a logarithm transform of the "half Kummer" surface which is simply connected and appears to need one 1-handle and one 3-handle [J. Harer, A. Kas, and R. Kirby, Handlebody structures for complex structures, Memoirs AMS 1984].

Problem 4.19 (B). Yes, for simply connected manifolds. For, by the two old remarks, it suffices to consider the case of 4-manifolds with definite intersection form ; but Donaldson has shown all such manifolds have intersection form $\pm \overset{k}{\oplus} \langle 1 \rangle$ and these have a characteristic element α (the sum of the generators) such that $\alpha \cdot \alpha$ = index. The topological case remains open. A reference for the Remark (iii), that M^4 smoothly imbeds in $R^6 \Leftrightarrow M$ is spin and index $M=0$, is [D. Ruberman, Math. Proc. Cam. Phil. Soc. 91(1982), 107-110]. For codimension one imbeddings, see Problem N4.63.

Problem 4.20. No in the orientable case. R. Herbert [Memoirs AMS Vol. 34, number 250] and J. White [Proc. Sym. Pure Math. XXVII (1975), 429-437] proved that the number of triple points of a generic immersion (counted algebraically in the preimage) is $-p_1(M^4) = -3\ \text{index}(M^4)$. Also $p_1(\tau_M \oplus \nu_M) = p_1(\tau_M) + \chi^2(\nu_M) = 0$, so $-p_1(M^4) = \chi^2(\nu_M)$. But the double point set is an integral dual ξ to $w_2(M)$, so $\xi \cdot \xi = \text{index}\ M^4 (\text{mod}\ 8)$, and $\chi^2(\nu_M) = \xi \cdot \xi$. Thus -3 index M = index M(8), so p_1 and the number of triple points is even.

Yes in the non-orientable case [J. Hughes, Quart. J. Math., 1983]. Immerse $RP^2 \times RP^2$ as Boy's surface cross itself, which has an odd number of triple points (see last paragraph). By ambient surgeries one can remove pairs of triple points to get $RP^2 \times RP^2 \# n(S^1 \times S^3)$ immersed with a single triple point.

Problem 4.22. Either (B) is false or there exists a curve in the boundary of the contractible 4-manifold which does not bound a PL disk [A. J. Casson and C. McA. Gordon, Inv. Math. 74(1983), 119-137].

Problem 4.23. A locally flat S^2 in CP^2 which represents the generator of $H_2(CP^2;Z)$ is unknotted, i.e. (CP^2,S^2) is pairwise homeomorphic to $(CP^2;CP^1)$. The 4-dimensional topological Poincare conjecture (Freedman) implies this.

Problem 4.24. The "Gluck construction" on a knotted 2-sphere in S^4 gives a homotopy S^4 which is then homeomorphic to S^4 (Freedman). This problem is open in the smooth case; also see the remarks about Problem 4.13.

Problem 4.29. Examples of surfaces $F \to S^4$ with $H_2(\pi_1(S^4-F); Z) \neq 0$ have been given by [T. Maeda, Math. Sem. Notes, Kwansei Gakuin Univ., 1977], [A. Brunner, E. Mayland Jr., and J. Simon, Pac. J. Math., 103(1982), 315-324], [C. McA. Gordon, Math. Proc. Camb. Phil. Soc. 81(1979), 113-117], and [R. Litherland, Quart. J. Math. 32(1981), 425-434]. In particular, Litherland shows that if A is an abelian group with $2g$ generators, then there is a closed surface of genus g, F_g, and a smooth imbedding $F_g \to S^4$ such that $H_2(\pi_1(S^4-F_g); \mathbb{Z}) = A$.

Problem 4.31. A locally flat surface in a 4-manifold has a normal bundle; the proof uses Quinn's work and Freedman's s-cobordism theorem for $\pi_1 = Z$.

Problem 4.32. The smooth Schoenflies conjecture is known if there exists a smooth function $f:S^4 \to R$ whose restriction f/S^3 is Morse with k 0-handles and $\leq k+1$ 1-handles (then the middle level has genus ≤ 2). [M. Scharlemann, Topology, 1984].

Problem 4.33. An S^3 in $S^2 \times S^2$ bounds a topological 4-ball because of the topological h-cobordism theorem (Freedman).

Problem 4.40. (A): The conjecture is true; an algebraic proof is given in [W. Fulton, Ann. Math. 111(1980), 407-409], and a geometric version in [P. Deligne, Sem. Bour. 1979/80, Lect. Notes Math. v. 842, Springer, 1-10].

Problem 5.3. Every 2-complex imbeds topologically in R^4. Any abstract 4-dimensional regular neighborhood N of the 2-complex has boundary a homology 3-sphere S which then bounds a contractible 4-manifold W^4 (Freedman). Since $\pi_1(S) \to \pi_1(N)$ is onto, it follows that $N \cup W$ is a homotopy 4-sphere, hence S^4.

PROBLEMS IN KNOT THEORY

These new problems are numbered N1.52 - N1.57 following the numbered problems 1.1 - 1.51 in the earlier problem list.

Problem N1.52 (L. Kauffman). Conjecture: If K is a slice knot in S^3 and F^2 is an orientable Seifert surface for K, then there exists a simple closed curve α in F such that

1) α is null (meaning that the linking number of α is zero and $0 \neq \alpha \in H_1(F;\mathbf{Z})$),

2) the Arf invariant of α is zero.

If true, can one then find a null α which is slice?

Problem N1.53. Does mutation preserve the concordance type of a knot in S^3? (Mutation is the operation on a knot which removes a tangle, twists it 180°, and glues it back in).

Problem N1.54 (Hillman). When is the result of surgery on a knot in S^4 aspherical?

Remark: The knot group must be an orientable Poincare duality group of formal dimension four (Hillman, Houston J. of Math., 6(1980), 67-76), but is this condition sufficient?

Problem N1.55. (A) If a smooth 2-sphere K in S^4 has group $\pi_1(S^4 - K) = Z$ (this implies that $S^4 - K \simeq S^1$), is it smoothly unknotted?

Remark: K is unknotted in the topological category (Freedman). Also, see Problem N4.41.

(B) Let L be a link in S^4 with unknotted components and let $\pi_1(S^4 - L)$ be free on a set of meridians. Is L trivial (topologically or smoothly)?

Remark: A. Swarup [J. Pure App. Alg., 11(1977), 75-82] has shown that the exterior $S^4 - L$ has the right homotopy type rel boundary. However, the homotopy type rel boundary of a knot exterior does not in general determine the homeomorphism type of the knot exterior [S. Plotnick, Homotopy type of 4-dimensional knot complements, Math. Z. 183(1983), 447-471].

(C) When is a 2-link splittable? In particular, is it sufficient that the group be a free product with each factor normally generated by a meridian?

Problem N1.56. Are all 2-links slice?

Remarks: All 2-knots are slice [Kervaire, Bull. Soc. Math., France 93(1965), 225-271]. All boundary links are slice ([Kervaire, op.cit.], [Gutierrez, Bull. AMS, 79(1973), 1299-1302], [Cappell and Shaneson, Comm. Math. Helv. 55(1980), 30-49), so the problem is to show that every 2-link is concordant to a boundary link. Note that L is a boundary link iff there exists a homomorphism $\varphi:\pi_1(S^4-L) \to F_\mu$ (= free group on number of components) taking meridians to generators [Gutierrez, op.cit.]. More generally, it is sufficient to find $\varphi:\pi_1(S^4-L) \to P$ where the normal closure of image φ is P, P is a higher dimensional μ-component link group, and $H_3(P;Z/2) \cong H_4(P;Z) = 0$, [T. Cochran, Slice links in the 4-sphere, 1981].

An easier problem is: does the Z/2 invariant of Sato-Levine [N. Sato, Concordances of manifold links, 1981] vanish for all 2-links? It vanishes for certain classes of 2-component links, e.g. when one of the components is unknotted [T. Cochran, On an invariant of link cobordism in dimension 4, 1981].

Problem N1.57. (A) Is the center of a 2-knot group finitely generated?

Remark: The only known centers are $\mathbb{Z}$, $\mathbb{Z} \oplus \mathbb{Z}/2$, $\mathbb{Z} \oplus \mathbb{Z}$ and are realized by twist spun trefoil knots, [J. Hillman, Comm. Math. Helv. 56(1981), 465-473].

(B) Is the center of the group of a 2-link with more than one component trivial?

Remark: The argument of Hausmann and Kervaire may be readily modified to show that any finitely generated abelian group is the center of the group of some μ-component n-link for each $\mu \geq 1$, $n \geq e$. In the classical case, $n=1$, the center must be $0, \mathbb{Z}$, or $\mathbb{Z} \oplus \mathbb{Z}$.

PROBLEMS CONCERNING 4-MANIFOLDS

These new problems are numbered N4.41 - N4.68 following the numbered problems 4.1 to 4.40 in the earlier problem list.

Problem N4.41. There exists a smooth, proper imbedding of the Poincare homology sphere P minus a point in R^4 with a possibly exotic smooth structure [Freedman]. Exhibit this smooth imbedding, or (easier) ignore the differentiability and construct a locally flat imbedding into R^4. Is there a smooth proper imbedding of P-pt. into R^4?

Remark: If yes, that would give an example of a smooth S^2 in S^4 which is topologically unknotted, but smoothly knotted since it would have the punctured Poincare homology sphere as "Seifert surface".

Problem N4.42. Let $r\dot{B} = \{x \in R^4 \,|\, |x| < r\}$ and give $r\dot{B}$ the smooth structure inherited from R^4_θ, one of the fake R^4's.

(A) What is the largest value of r, say ρ, for which $r\dot{B}$ is diffeomorphic to R^4, and what happens at ρS^3? (This depends on fixing an atlas representing θ.)

(B) Is $r\dot{B}$ diffeomorphic to R^4_θ or to $s\dot{B}$ for any $\rho < r < s$?

Remark: If so, then a furling argument gives an exotic structure on $S^3 \times S^1$. If not, then the reals inject into the moduli space of sm-oth structures on R^4.

(C) Does every smoothly imbedded S^3 in R^4_θ bound a smooth B^4? Or, avoiding the smooth 4-dimensional Schoenflies conjecture, can it be engulfed in a standard R^4 in R^4_θ.

Problem N4.43. (A) Can any exotic R^4 be covered by a finite number of coordinate charts? In particular, can an exotic R^4 be the union of two copies of R^4?

(B) Find a handlebody decomposition of an exotic R^4.

(C) Describe in some usable way a complete Reimannian metric on an exotic R^4. What can be said about the topology of the cut locus for this metric?

(D) Does there exist an exotic R^4 which cannot be split by a smooth proper R^3 into two exotic pieces?

Problem N4.44. (A) Can every homeomorphism of R^4 be approximated by a Lipschitz homeomorphism?

(B) Does Donaldson's theorem hold in the Lipschitz category?

Remark (Sullivan): If the answer to (A) is yes, then every topological 4-manifold has a Lipschitz structure, (see [D. Sullivan, Proc. 1977 Georgia Conf., ed. J. Cantrell, 543-555]), negating (B). Recall that in higher dimensions the answer to (A) is yes; in fact Lipschitz can be replaced by PL or DIFF [E. H. Connell, Ann. Math. 78(1963), 326-338] in (A), and furthermore, TOP = Lipschitz in dimensions $\neq 4$ [Sullivan, op.cit].

Problem 4.45. Does there exist an exotic differentiable structure on S^4? On $S^3 \times S^1$? On any other closed orientable, smooth 4-manifold?

Remarks: There are plenty of candidates. E.g. the Gluck construction on any knotted S^2 in S^4 gives a homotopy 4-sphere (for a specific example without 3-handles, see [Akbulut and Kirby, A potential smooth counterexample in dimension 4 to the Poincare conjecture, the Schoenflies conjecture and the Andrews-Curtis conjecture]), or any presentation of the trivial group which cannot be trivialized by Andrews-Curtis moves gives a smooth homotopy 5-ball whose boundary may be fake. For possible fake $S^3 \times S^1$'s, see Problem N4.42.

The existence of many fake smooth structures on non-compact 4-manifolds makes an affirmative answer seem likely.

Exotic smooth structures on non-orientable 4-manifolds abound; several topologists have found isolated examples, and M. Kreck [Some closed 4-manifolds with exotic differential structure] has found large classes.

Problem N4.46 (Freedman): Is a positive untwisted double of the Borromean rings topologically slice?

Remark: This is a simple case of the kind of slicing problem one runs into with some approaches to the topological s-cobordism conjecture. The answer is yes if either non-simply connected surgery or the proper s-cobordism theorem holds.

Problem N4.47 (Freedman): Let X be the cone on the unlink of n components in S^3. Suppose X is imbedded properly in B^4 and is locally flat except at the cone point $*$. Suppose the local homotopy at $*$ is free. Does this imply that the imbedding is "flat", i.e. has a neighborhood homeomorphic to a neighborhood of the standard imbedding of X in B^4?

Remarks: If yes, then topological non-simply connected surgery works and we "almost" get the s-cobordism theorem; conversely, the s-cobordism theorem for all π_1 would imply "yes". Note that each disk in X is flat by itself.

Problem N4.48 (Freedman): Find a homotopy theoretic criterion for when $M^3/Y \subset R^4$ has a one-sided mapping cylinder neighborhood, where Y is an acyclic set in the 3-manifold M.

Remarks: [Quinn] has such a criterion when Y is a CE set. A "reasonable" criterion would give the topological s-cobordism theorem. An interesting acyclic set is obtained by starting with a genus two handlebody Y_0; get Y_1 by reimbedding Y_0 in itself according to any two distinct words in the commutators of the two generators of $\pi_1(Y_0)$; get Y_2 by reimbedding Y_0 in Y_1 according to the same two words, or any other such pair. Continue, and let $Y = \bigcap_{k=0}^{\infty} Y_k$.

Problem N4.49. If M^3 is a homology 3-sphere, does $M \# (-M)$ bound a smooth contractible 4-manifold?

Remarks: It bounds a topological, contractible 4-manifold [Freedman] and it smoothly bounds $M^3 \times I$.

Problem N4.50. Is each simply connected, smooth, closed 4-manifold (other than S^4) realized as a connected sum of complex surfaces (with or without their preferred orientations)?

Remark: Probably the answer is no, but Donaldson's work makes "yes" a bit more likely. Furthermore, yes is indicated by analogy with dimension 2 where every orientable closed 2-manifold is a complex curve.

Problem N4.51 (Akbulut): (A) If M_1^4 and M_2^4 are simple homotopy equivalent, closed, smooth 4-manifolds, can we pass from M_1 to M_2 by a series of Gluck twists on imbedded 2-spheres?

Remarks: No for certain lens spaces cross S^1 (S. Weinberger). Yes in a few examples, e.g. $M_1 = S^4$, M_2 = double cover of a Cappell-Shaneson fake RP^4 (Akubulut-Kirby, see update to Problem 4.13).

(B) Same question for a generalized Gluck twist, which is defined as follows: split M^4 along a smooth submanifold N^3 with closed complements W_1 and W_2. In W_1, find a properly imbedded, smooth 2-ball D_1. Twist D_1 by removing $D_1 \times B^2$ and sewing back by spinning D_1 k-times while traversing ∂B^2. Then find a ∂D_2 with $\partial D_2 = \partial D_1$ and twist back by $-k$. Thus N remains unchanged and we can reglue along N.

In this way, a Cappell-Shaneson fake RP^4 can be changed to RP^4 by splitting $RP^4 = S^1 \tilde{\times} B^3 \cup RP^2 \tilde{\times} B^2$ and twisting $RP^2 \tilde{\times} B^2$ along $* \times B^2$ to $RP^2 \times B^2$ (k=1) and then twisting back by a strange B^2 in $RP^2 \times B^2$ [S. Akbulut, these Proceedings].

Problem N4.52. Given M^m and N^n imbedded in Q^4, is there an isotopy making M^m topologically transverse to N^n when $m=3$, $n=2$ or $m=3$, $n=3$?

Remark: The answer is yes for other m and n [F. Quinn, J. Diff. Geom. 17(1982), 503-521]. When Q is higher dimensional, see [A. Marin, Ann. Math., 106(1977), 269-294].

Problem N4.53 (Mandlebaum). What (minimal) knowledge of homotopy groups, intersection pairings, etc. determines the homotopy type of a closed, compact 4-manifold?

Remarks: For $\pi_1(M^4) = 0$, the intersection form determines. For $\pi_1(M^4) = Z/p$, p prime, then π_1 and the intersection pairing $\pi_2(M) \otimes \pi_2(M) \longrightarrow Z[\pi_1]$ determine, [C. T. C. Wall]. Is this theorem true for a larger class of fundamental groups? Give an example where π_1 and the intersection form do not suffice.

Let a generalized Lefschetz torus fibration $M^4 \xrightarrow{f} F_g$ be a map which is a torus bundle off a finite number of points in F_g (= surface of genus g) and over those points $f^{-1}(p)$ is an immersed 2-sphere with one transverse double point. Examples of these are complex elliptic surfaces with no multiple fibers, and (Y. Matsumoto) simply connected, smooth 4-manifolds without one and 3-handles. R. Mandlebaum and J. Harper have shown that the homotopy type of a generalized Lefschetz torus fibration is determined by the genus g and the intersection pairing $H_2(M;Z) \otimes H_2(M;Z) \longrightarrow Z$.

Problem N4.54. Find a geometric proof that $\Omega^4_{spin} = Z$.

Remark: There exists such a proof that $\Omega^4 = Z$ ([P. Melvin, 4-dimensional oriented bordism, these Proceedings and his references)], but it is not clear how to modify it to get the spin case.

Problem N4.55. Describe the Fintushel-Stern involution on S^4 in "equations". (See their paper in [Ann. Math., 113(1981), 357-366]).

Problem N4.56 (Melvin). Let M^4 be a smooth closed orientable 4-manifold which supports an effective action of a compact connected Lie group G.

(A) Suppose that $\pi_1 M$ is a free group. Is M diffeomorphic to a connected sum of copies of $S^1 \times S^3$, $S^2 \times S^2$ and $S^2 \tilde{\times} S^2$?

Remark: The answer to both questions is yes for $G \neq S^1$ or T^2; also for $G = T^2$ provided the orbit space of the action (a compact orientable surface) is not a disc with ≥ 2 holes [Melvin, Math. Ann. 256(1981) 255-276].

Problem N4.57. Classify closed 4-manifolds which fiber

(A) over a circle with fiber an S^1-manifold,

(B) over a surface.

Remark: If (in (A)) the monodromy is periodic and equivariant, then M supports a nonsingular T^2-action and is generally classified by $\pi_1 M$ [Orlik-Raymond, Topology 13(1974) 89-112]. Exceptions arise when $\pi_1 M/\text{center}$ is finite, e.g. for $S^1 \times L$, L a lens space.

Problem N4.58 (Melvin). Let $P \subset S^4$ be the standardly embedded $\mathbb{RP}^2$ (e.g. $P = q(\mathbb{RP}^2)$, where $q: \mathbb{CP}^2 \to S^4$ is the quotient map by complex conjugation) and $K \subset S^4$ be an odd twist spun knot. Denote by $(S^4, P\#K)$ the pairwise connected sum $(S^4, P) \# (S^4, K)$. Is $(S^4, P\#K)$ pairwise diffeomorphic to (S^4, P)?

Remarks: (i) They have the same 2-fold branched covers, namely $(\mathbb{CP}^2, \mathbb{RP}^2)$ (Melvin), so a negative answer yields an exotic involution on $\mathbb{CP}^2$ with fixed point set $\mathbb{RP}^2$ and quotient S^4.

(ii) $\pi_1(S^4 - P\#K) = \mathbb{Z}_2$, so $S^4 - N(P\#K)$ is s-cobordant rel boundary to $S^4 - N(P)$ (T. Lawson), where $N(\ )$ denotes an open tubular neighborhood.

Problem N4.59 (Hillman). Minimize the Euler characteristic over all closed 4-manifolds M with $\pi_1(M^4) = G$ given.

Remarks: Hopf's Theorem gives $H_2(\tilde{M}) \to H_2(M) \to H_2(\pi_1(M)) \to 0$ which puts a lower bound on the rank of $H_2(M)$, given $\pi_1(M)$. But this minimum is not always achieved (Hillman), e.g. let $G = Z \oplus Z$ so that $H_2(Z \oplus Z) = Z$, but $\chi(M) = -1$ is not possible by an Euler characteristic argument on the equivariant homology of the universal covering space. Note that this problem generalizes the problem of which which groups are the fundamental group of a homology 4-sphere.

Problem N4.60 (Hass). Let M^4 be closed, smooth and satisfy $\pi_2(M) = 0$ but $\pi_3(M) \neq 0$, i.e. $L(p,q) \times S^1$. Is there a smooth imbedded 3-manifold L^3, with finite cover S^3, representing a non-zero element of $\pi_3(M)$?

Problem N4.61 (Hughes). (A) Find representatives for each regular homotopy class of immersions of S^n in R^{n+k}.

Remarks: This is trivial for $k > n$ and solved by Whitney-Graustein for S^1 in R^2. For $n = k$, Smale's solution is to add double points to get Z for n even or one, and $Z/2$ otherwise. For S^2 in R^3, Smale's famous theorem (that $\mathrm{Imm}(S^n, R^{n+k}) = \pi_n(V_{n+k,n})$) shows there is just one class. For S^3 in R^4, Hughes [thesis, Berkeley 1982] gives two generators gotten by capping off the track of an eversion of S^2 in R^3, and capping off twice an eversion. The inclusion of the first of these solves the case S^3 in R^5. The next interesting case is S^4 in R^5.

(B) Find representatives for all bordism classes of immersions of n-manifolds in R^{n+k}.

Remarks: This group is $\pi^s_{n+k}(MSO(k))$ (assuming orientability) [R. Wells, Topology 5(1966), 281-294]. This has been solved for $n = 1$, all k, and 2-manifolds in R^3 [J. Hass and J. Hughes, Immersions of surfaces in 3-manifolds, 1982]. Several bordism invariants have been developed ([J. S. Carter, thesis, Yale 1982] gives a good summary of n-tuple point invariants).

(C) For a surface in R^3, a neighborhood of a double curve is an immersed $B^1 \times B^1$ - bundle over S^1. In general a k-tuple set will have an immersed $B^n \vee \overset{k}{\dots} \vee B^n$-bundle. Does the multiple point set with this structure determine the bordism class of the immersion?

Remarks: Yes for 2-manifolds in R^3. The codimension one case is investigated in [Eccles, Math. Proc. Camb. Phil. Soc. 87(1980), 312-220] and [J. S. Carter, op.cit.].

(D) Can one find explicit coordinates for Boy's surface, i.e. find a smooth function from S^2 to R^3 taking S^2 onto Boy's surface as a 2-1 cover.

Remark: Morin and Francis [Not. AMS 25 (1978), A-718] have a complicated function whose image is not the standard Boy's surface.

(E) The number of quadruple points of an immersed S^3 in R^4 is a Z/2 invariant under regular homotopy. Is it a Z/24 (= π_3^s) or even a Z invariant?

Problem N4.62. (A) Do the cyclic branched covers of 2-spheres in S^4 imbed in S^5?

(B) Does every mapping torus at a 3-manifold imbed in S^5?

Remark: If a knot K is doubly null concordant (the slice of an unknotted S^3 in S^5) then all of its cyclic branched covers imbed in S^5, so (A) concerns an obstruction to K being doubly null-concordant.

Problem N4.63. (A) Find a smooth, closed, spin, index zero 4-manifold X^4 which does not imbed punctured in S^5.

(B) Find an X^4 such that $X \# kS^2 \times S^2$ imbeds smoothly in S^5 but X does not.

Remarks: X smoothly imbeds in S^5 if its fundamental group is simple enough, e.g. $H_1(X;Z)$ is the direct sum of no more than two cyclic groups [T. Cochran, Imbedding 4-manifolds in S^5, Topology]. However there are examples with $\pi_1 = Z/p \oplus Z/p \oplus Z/p$, p odd where X does not smoothly imbed in S^5, does not imbed stably ($\# kS^2 \times S^2$), and sometimes is known to imbed punctured [T. Cochran, 4-manifolds which imbed in R^6 but not R^5, and Seifert manifolds for fibered knots].

Problem N4.64. (A) What 4-manifolds have a symplectic structure?

Remarks: A symplectic structure is given by a 2-form Ω with $d\Omega = 0$ for which $\Omega \wedge \Omega$ is a volume form. Thus it is necessary that $H^2(M^4;R)$ contain an element Ω with $\Omega \wedge \Omega \neq 0$.

(B) Does every contact structure on a 3-manifold M^3 extend to symplectic structure on a bounding 4-manifold?

Remark: A contact structure is a 1-form α such that $\alpha \wedge d\alpha$ is nowhere zero. We would require that $\alpha(v) = \Omega(v,n)$ for n an outward pointing normal to $M^3 = \partial W^4$ and $v \in T_M$.

Problem N4.65. (A) Find a differential geometric invariant which distinguishes the ends of smooth non-compact 4-manifolds. For example, if X^4 is a simply connected topological manifold with a definite intersection form, then X^4-point is smooth but its end is not standard; can this be detected in a direct differential geometric way?

(B) Find a differential geometric proof of Rohlins theorem.

Remark: There is such a proof using the $\tilde{A}$ genus, and Taubes has found a nice proof by getting the quaternions to act on the sequence $\Omega^0(\mathfrak{g}) \xrightarrow{d_A} \Omega^1(\mathfrak{g}) \xrightarrow{P-d_A} \Omega^2_-(\mathfrak{g})$ from Donaldson's work, but maybe there is a proof more in the spirit of (A).

Problem N4.66. How do metrics (e.g. Riemannian, Lorentz, constant curvature) behave under standard topological constructions such as connected sum, plumbing, handle addition? Same question for η-invariants, moduli spaces, etc.

Problem N4.67 (Hopf). Does there exist a metric of strictly positive sectional curvature on $S^2 \times S^2$?

Problem N4.68. (A) There exists a self-dual Einstein metric on the Kummer surface. Describe it explicitly.

Remark: Its existence follows from Yau's proof of the Calabi conjecture [P. N. A. C. 74(1977), 1798-1799].

(B) If M^4 is compact, closed and has an Einstein metric, then $\chi(M) \geq 3/2|\text{index } M|$ (N. Hitchin, J. Diff. Geom. 9(1974), 435-441). Are there any other topological restrictions?

(C) Does $\overset{p}{\#}\, CP^2 \overset{q}{\#}(\overline{CP}^2)$ have an Einstein metric?

Remarks: If $p > 3$ and $q = 0$, then no. If $p = 1$ then yes because the manifold is complex.

An Einstein metric has the property that sectional curvatures are equal on orthogonal 2-planes. A good reference is [J. P. Bourguignon, Inv. Math. 63(1981), 263-286].

DEPARTMENT OF MATHEMATICS
UNIVERSITY OF CALIFORNIA
BERKELEY, CA 94720

ABCDEFGHIJ–AMS–8987654